7th International Conference on Pumps and Fans (ICPF 2015)

AA006529

IOP Conference Series: Materials Science and Engineering
Volume 129

Hangzhou, China
18 - 21 October 2015

ISBN: 978-1-5108-2449-2
ISSN: 1757-8981

Printed from e-media with permission by:

Curran Associates, Inc.
57 Morehouse Lane
Red Hook, NY 12571

Some format issues inherent in the e-media version may also appear in this print version.

Printed by Curran Associates, Inc. (2016)

For permission requests, please contact the Institute of Physics
at the address below.

Institute of Physics
Dirac House, Temple Back
Bristol BS1 6BE UK

Phone: 44 1 17 929 7481
Fax: 44 1 17 920 0979

techtracking@iop.org

Additional copies of this publication are available from:

Curran Associates, Inc.
57 Morehouse Lane
Red Hook, NY 12571 USA
Phone: 845-758-0400
Fax: 845-758-2633
Email: curran@proceedings.com
Web: www.proceedings.com

Table of contents

Volume 129

7th International Conference on Pumps and Fans (ICPF2015)
18–21 October 2015, Hangzhou, China

Accepted papers received: 12 April 2016
Published online: 19 May 2016

Preface

Papers

Foreword

The 2015 International Conference on Pumps and Fans (IPCF) builds on the experiences gained from, and successes of, the previous International Conferences on Pumps and Fans from 1992 to 2013. This 7th conference convened from October 18–21 in Hangzhou, China.

This conference was organized by Zhejiang University, Tsinghua University, Jiangsu University, State Key Laboratory of Fluid Power Transmission and Control, Xi'an University of Technology and Zhejiang Sci-Tech University. The objective of the ICPF conferences is to provide an opportunity for experts, researchers, engineers and students in the field to report on the latest developments in pumps, fans, compressors and other fluid machinery. The conference encourages the presentation of work in progress and promotes discussion of the problems encountered. Distinguished scholars from all over the world are invited to give reports on the new research areas of fluid machinery, and to honour Professor Zuyan Mei for his contribution to these areas and his organization of the previous conferences.

The conference comprised 134 highly informative technical papers, of which 14 plenary lectures were delivered by 14 distinguished scholars, and 120 technical papers were presented in 14 oral sessions and one poster session. They comprehensively cover the latest practice and findings in the field of pumps, fans and compressors, including pump design and simulation, multi-phase flows and cavitation, experimental study, transient behavior and unsteady flow, optimal design and theory analysis, numerical simulation and CFD application, power machinery, bearings and seals and so on.

Sixty-eight papers of ICPF2015, which were published in this volume of *IOP Conference Series: Materials Science and Engineering*, have been peer reviewed through processes administered by the editors of the ICPF2015 (Yulin WU, Shouqi YUAN, Leqin WANG, Weirong HONG, Zhengwei WANG, Shuhong LIU, Xingqi LUO, Zuchao ZHU and Dazhuan WU).

We sincerely hope that the 7th International Conference on Pumps and Fans with Compressors and Wind Turbines is a significant step forward in the worldwide efforts to address the present challenges facing modern fluid mechanics.

Yulin Wu, Professor

Shouqi Yuan, Professor

Chairmen of the Organizing Committee
7th International Conference on Pumps and Fans with Compressors and Wind Turbines (ICPF2015)
October 2015

IOP Conf. Series: Materials Science and Engineering **129** (2016) 011001 doi:10.1088/1757-899X/129/1/011001

COMMITTEE MEMBERS

Honorary Chairman: Jianzhong XU
Chairmen: Yulin WU (Tsinghua University)
Shouqi YUAN (Jiangsu University)
Executive Chairman: Jinyang ZHENG (Zhejiang University)

Co-Chairmen: Weirong HONG (Zhejiang University)
Zhengwei WANG (Tsinghua University)
Houlin LIU (Jiangsu University)
Weidong XING (National Key Laboratory of Diesel Engine Turbo-charging Technique)
Xingqi LUO (Xi'an University of Technology)
Zuchao ZHU (Zhejiang Sci-Tech University)

Co-organizers
Chao LIU (Yangzhou University)
Dazhuan WU (Zhejiang University)
Dezhong WANG (Shanghai Jiaotong University)
Fengchen LI (Harbin Institute of Tech.)
Fujun WANG (China Agriculture University)
Guang XI (Xi'an Jiaotong University)
Guoyu WANG (Beijing Institute of Tech.)
Huashu DOU (Zhejiang Sci-Tech University)
Jianping YUAN (Jiangsu University)
Jianjun FENG (Xi'an University of Tech.)
Jiegang Mu (Zhejiang University of Technology)
Kun LI (Hefei General Machinery Research Institute)
Mingming ZHANG (Institute of Engineering Thermophysics, CAS)
Ren DAI (University of Shanghai for Sci and Tech)
Rennian LI (Lanzhou University of Tech.)
Shuliang CAO (Tsinghua University)
Shuhong LIU (Tsinghua University)
Xinping LONG (Wuhan University)
Yuning ZHANG (North China Electric Power University)
Zhongdong QIAN (Wuhan University)

INTERNATIONAL TECHNICAL COMMITTEE

Principals: Yangjun ZHANG (Tsinghua University)
Hyung Hee CHO (Yonsei University, Korea)
Shuiying ZHENG (Zhejiang University)

Members: D. BOHN (Aachen University of Technology, Germany)
G. WEBB (University of Nevada, USA)
Guoyi PENG (Nihou University, Japan)
Kazuyoshi MIYAGAWA (WASEDA University, Japan)
Kwang-Yong KIM (Inha University, Korea)
Paul U. THAMSEN (TU of Berlin, Germany)
Satoshi WATANABE (Kyushu University, Japan)

IOP Conf. Series: Materials Science and Engineering **129** (2016) 011001 doi:10.1088/1757-899X/129/1/011001

Sönke BRODERSEN (KSB AG, Germany)
Tom I-P. SHIH (Purdue University USA)
Toshiaki KANEMOTO (Kyushu Institute of Technology, Japan)
Weidong SHI (Jiangsu University)
Yoshinobu TSUJIMOTO (Osaka University, Japan)
Young-Ho LEE (Korea Maritime University, Korea)
Young-Seok CHOI (Korea Institute of Industrial Technology, Korea)

INTERNATIONAL ADVISOR COMMITTEE

Principals: Guang XI (Xi'an Jiaotong University)
Leqing WANG (Zhejiang University)

Members: Andy C. C. TAN (Queensland University of Technology, Australia)
Byung Sun HWANG (Korea Institute of Material Science, Korea)
Christoph SINGRTÜN (VDMA, Germany)
Clements BLEYT (Sulzer Pumps Ltd., Switzerland)
D.-H. HELLMANN (KSB AG, Germany)
E. TURKEL (Tel Aviv University, Israel)
Fengqing HAN (Hunan Univ. of Sci. & Tech., China)
Gerard BOIS (ENSAM, Lille, France)
Martin BÖHLE (Kaiserslautern Univ., Germany)
Michihiro NISHI (Kyushu Institute of Technology, Japan)
Minguan YANG (Jiangsu University)
Luca d'AGOSTINO (University of Pisa, Italy)
Peter PELZ (Darmstadt University, Germany)
Rene Van den BRAEMBUSSCHE (VKI, Belgium)
Ricardo F. MARTINEZ-BOTAS (Imperial College, UK)
Smaine KOUIDRI (Université Pierre et Marie Curie, Paris 6, France)
Shengcai LI (Warwick University, UK)
S.H. WINOTO (National University of Singapore, Singapore)
Vigor YANG (Georgia Institute of Technology, USA)

SECRETARY GENERAL:
Dazhuan WU (Zhejiang Univ., wudazhuan@zju.edu.cn)

SECRETARIAT MEMBERS:
Haijun XUAN (Zhejiang Univ., marine@zju.edu.cn)
Peng WU (Zhejiang Univ., roc@zju.edu.cn)
Linlin CAO (Zhejiang Univ., caolinlin@zju.edu.cn)
Dmitriy LIKHACHEV(Zhejiang Univ., dima@zju.edu.cn)

INSTITUTE OF PROCESS EQUIPMENT
College of Chemical and Biological Engineering
Zhejiang University
No.38 Zheda Rd., Hangzhou, Zhejiang, 310027, China
E-mail: icpf2015@zju.edu.cn
Phone: +86-571-87953186
Fax: +86-571-87953186

IOP Conf. Series: Materials Science and Engineering **129** (2016) 011002 doi:10.1088/1757-899X/129/1/011002

Peer review statement

All papers published in this volume of *IOP Conference Series: Materials Science and Engineering* have been peer reviewed through processes administered by the proceedings Editors. Reviews were conducted by expert referees to the professional and scientific standards expected of a proceedings journal published by IOP Publishing.

Comparison of rotordynamic fluid forces in axial inducers and centrifugal turbopump impellers

Luca d'Agostino[1]

Dipartimento di Ingegneria Civile ed Industriale
Università di Pisa, Pisa, Italy

E-mail: luca.dagostino@ing.unipi.it

Abstract. The paper illustrates and compares the results of the experimental campaigns carried out in the Cavitating Pump Rotordynamic Test Facility (CPRTF) at Alta, Italy, under ESA funding for the characterization of the lateral rotordynamic fluid forces acting on high-head axial inducers and centrifugal turbopump impellers for space propulsion applications. The configurations presented here refer to a three-bladed tapered-hub, variable-pitch, inducer (DAPROT3) and a single-stage centrifugal pump (VAMPIRE) with vaneless diffuser and single spiral volute. Both the centrifugal pump and the inducer have been designed by means of reduced order models specifically developed by the author and his collaborators for the geometric definition and performance prediction of this kind of hydraulic turbomachinery. Continuous spectra of the rotordynamic forces acting on the impellers as functions of the whirl frequency have been obtained by means of the novel technique recently developed and demonstrated at Alta. The influence of the rotor whirl motion, flow rate, cavitating conditions, and liquid temperature (thermal cavitation effects) on the rotordynamic fluid forces is illustrated and the observed differences in their behavior in axial inducers and centrifugal turbpumps are discussed and interpreted in the light of the outcome of recent cavitation visualization experiments carried out by the Chemical Propulsion Team at Alta.

Introduction

Liquid-fed rocket engines still play a crucial role in primary space propulsion systems. In these applications propellant feed turbopumps are one of the most important components, whose design and operation are critical for the success of the mission. In order to minimize the total weight of the propulsion system, propellant feed turbopumps must be designed for high head and suction performance. This objective is invariably achieved by reducing the dimensions of the machine, increasing its rotational speed and by operating it under "controlled cavitation" conditions. The drawbacks of this choice consists in the performance degradation effects of cavitation, and more so in the possible development of fluid dynamic and rotordynamic instabilities, which can easily responsible for catastrophic failures of the machine.

Current rocket propellant feed turbopumps often employ an inducer in order to improve the suction performance and avoid unacceptable cavitation levels in the main centrifugal stage(s), thus allowing for the reduction of the propellant tank pressure and weight. This is typically achieved by

[1] To whom any correspondence should be addressed.

proper hydrodynamic design of the inducer, which is characterized by fewer blades than centrifugal pump impellers, large stagger angles with respect to the axial direction, and significantly high blade solidities. Long blades and small incidence angles allow for the reduction of the blade loading, and consequently for significant improvements of the suction performance. Radial and mixed-flow pumps, conversely, exploit the work developed by centrifugal forces on the fluid for generating most of the required pressure rise. Hence, their blades must be sturdier (i.e. thicker) than those of inducers in order to sustain the higher loads exerted by the liquid. With respect to centrifugal compressors, radial turbopump impellers usually have a smaller number of blades (typically six or eight) in order to avoid excessive flow blockage.

According to Brennen ([1]), flow instabilities in turbopumps can be divided in system, global and local oscillations (respectively involving the entire pumping line, the whole pump and a more limited region of the pump flow), as well as instabilities caused by radial or rotordynamic forces. Rotordynamic forces, together with flow instabilities whether or not generated by cavitation, are one of the most recognized and dangerous sources of vibrations in turbomachines. These forces can affect all of the components of the machine, including its bearings, seals and, obviously, the impeller itself ([2]). Even if the rotordynamic fluid forces acting on noncavitating centrifugal pump impellers and inducers have been extensively investigated (see for example [3], [4], [5], [6], [7], [8]), the experimental information on the influence of cavitation on these forces is still very limited. Available data mainly come from the work carried out in the 1980's at the California Institute of Technology, Pasadena, California, ([9] [10], [11], [11] and [12]). More recently, Yoshida and his coworkers have investigated the relationship between the size of the bubble in the uneven rotating cavitation and the associated vibration ([13] and [14]). On the other hand, not so many numerical approaches have been proposed so far (see, for example, [15], [16] and [17]), probably due to the complexity of the flowfield. In general, the lack of data is particularly evident for unshrouded inducers under cavitating condition, which is a common operating condition in space rockets applications. Finally, in the last five years more extensive experimental investigations of rotordynamic impeller forces on unshrouded inducers and centrifugal turbopump impellers under cavitating conditions have been presented by the Chemical Propulsion Team at Alta, Pisa, Italy (see, for instance, [18], [19] and [20]). In their experiments, just like as the pioneering ones carried out at Caltech in the 1980's, the impeller is subject to a whirl motion of given constant eccentricity and angular velocity. A special procedure ([21]) has been used to measure the continuous spectrum (behavior) of rotordynamic forces as functions of the whirl-to-rotational speed ratio.

Experimental Apparatus

The experimental activity reported in the present paper has been carried out in Alta's CPRTF, which is illustrated in Figure 1 and specifically designed for characterizing the performance of cavitating and/or noncavitating turbopumps in a wide variety of alternative configurations (axial, radial, or mixed flow, with or without an inducer ([22], [23]). The facility uses water as working fluid at temperatures up to 90 °C and is intended as a flexible apparatus readily adaptable to conduct experimental investigations on virtually any kind of fluid dynamic phenomena relevant to the design of high performance turbopumps under fluid dynamic and cavitation similarity conditions. The pump housing and the inlet section can easily be configured to host machines and inducers with actual geometries of different sizes. Unsteady pressure sensors can be flush-mounted on the casing walls of the pump to detect, identify, and monitor the occurrence of flow instabilities developing in the machine.

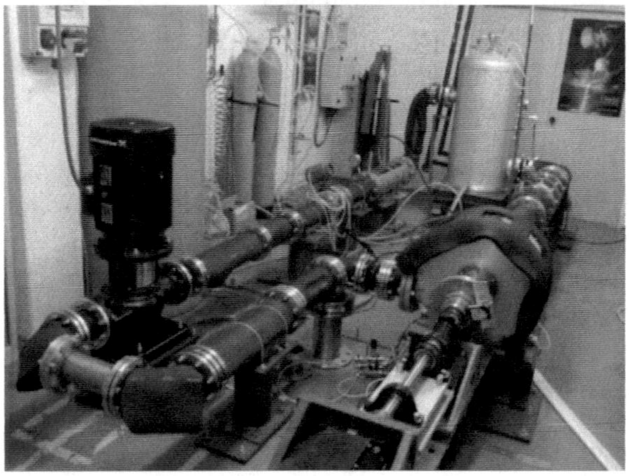

Figure 1. Alta's Cavitating Pump Rotordynamic Test Facility (CPRTF).

For this campaign, the facility has been instrumented in order to measure:

- the inducer inlet pressure (Druck, model PMP 1400, 0–1.5 bar operating range, and 0.25% precision class);
- the inducer static pressure rise (Druck, model PMP 4170, 0–1 bar operating range, and 0.08% precision class);
- the volumetric flow rate on the suction and on the discharge lines (model 8732E by Fisher-Rosemount, range 0–100 l/s, 0.5% precision);
- the water temperature inside the main tank;
- the absolute angular position of the driving shaft;
- the absolute angular position of the eccentric shaft that generates the whirl motion.

A rotating dynamometer in the test section allows for the measurement of the forces and moments acting on the impeller. The eccentricity generation is realized by means of a dual-shaft mechanism. The shafts are assembled one inside the other by means of a double eccentric mount and their eccentricity can be finely adjusted from 0 to 2 mm ±0.05 mm by changing the relative angular position of the double eccentric mount before each test. A brushless motor driving the outer shaft generates the whirl motion, while the impeller rotation is imparted by connecting the inner shaft to the main motor by means of an omokinetic coupling. The rotating dynamometer is realized in a solid piece of phase hardening steel AISI 630 H1025 and consists of two flanges connected by four pillars, acting as flexing elements.

The deformation of the dynamometer is measured by 40 semiconductor strain gages suitably installed on the sensing pillars and arranged in ten full Wheatstone bridges, which provide redundant measurements of the forces and moments acting on the impeller ([9]). Each bridge is temperature self-compensated, with separate bipolar excitation and read-out for better reduction of electrical cross-talking. The sizing of the sensing pillars is the result of a trade-off between sensitivity and structural resistance, operational stability, and position control (stiffness). The current design of the dynamometer is optimized for a suspended mass of 4 kg with 70 mm gyration radius, an added mass of about 2 kg (based on the expected magnitude of the rotordynamic forces), a rotational speed of 3000 rpm without eccentricity, and maximum rotational and whirl speeds up to 2000 rpm with 2mm shaft eccentricity. Under these conditions the first torsional mode of the dynamometer has a frequency of about 169 Hz, whereas the first flexural frequency is about 192 Hz, well above the excitation frequency generated by the impeller rotation. The maximum absolute error of the dynamometric force measurements is 1.4 N on the lateral forces and 0.1 N m on the axial torque.

Present experiments have been conducted on the DAPROT3 axial flow inducer and on the VAMPIRE radial turbopump in order to highlight how impeller rotordynamic fluid forces depend on the machine configuration. The VAMPIRE pump is a six-bladed unshrouded centrifugal pump manufactured in 7075-T6 aluminum alloy, with vaneless diffuser and single-spiral volute. The DAPROT3 inducer is a three-bladed, tapered-hub, and variable-pitch inducer, also manufactured in 7075-T6 aluminum alloy. Both the VAMPIRE pump ([24] and [25]) and the DAPROT3 inducer ([26], [27] and [28]) have been designed by means of the reduced order models developed at Alta for preliminary geometric definition and performance prediction of inducers and centrifugal pumps. Their main characteristics are summarized in Table 1 and pictures of their impellers are shown in Figure 2. A moderate value of the blade loading, with a diffusion factor $D = 0.47$ and solidity $\sigma_T = 1.68$, has been chosen for the DAPROT3 inducer in order to reduce the leading-edge cavity and improve the suction performance. The value of the tip incidence-to-blade angle ratio $\alpha/\beta_b < 0.5$ has been selected with the aim of controlling the danger of surge instabilities at design flow under cavitating conditions.

Table 1. Main characteristics of the DAPROT3 inducer and the VAMPIRE pump.

			DAPROT3	VAMPIRE
Design Flow Coefficient	[--]	Φ_D	0.065 [a]	0.092 [b]
Number of Blades	[--]	N	3	6
Inlet Tip Radius	[mm]	r_{T1}	81.0	57.2
Outlet Tip Radius	[mm]	r_{T2}	81.0	105.0
Axial Length [mm]	[mm]	c_a	90.0	46.4
Mean Blade Height	[mm]	h_b	25.95	17.95
Tip Solidity	[--]	σ_T	1.68	2.26
Material	[--]		Anodized AL 7075 T6	Anodized AL 7075 T6

[a] $\Phi = Q/\pi\Omega r_T^3$

[b] $\Phi = Q/\Omega r_T^3$

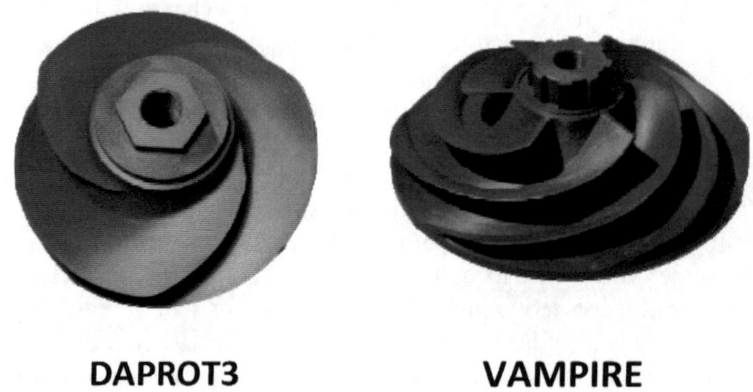

Figure 2. Renderings of the DAPROT3 inducer (left) and the centrifugal impeller of the VAMPIRE pump (right).

Figure 3 shows the test chamber configurations for the experiments on the DAPROT3 inducer and the VAMPIRE pump. The rotating dynamometer is placed between the driving shaft and the impellers. The DAPROT3 inducer has been recessed with respect to the optical access of the test section inlet in order to reduce cantilever effects on the impeller shaft. A nominal radial clearance of 2 mm has been selected for accommodating the imposed whirl eccentricities. It corresponds to nondimensional clearances $c_{\%} = 7.7\%$ for the DAPROT3 inducer and 11.2% for the VAMPIRE pump when referred to the respective mean blade heights. Though relatively large, previous experience ([29]) indicated that this choice of the radial clearance represents an acceptable compromise for generating measurable rotordynamic forces without excessively increasing tip leakage effects.

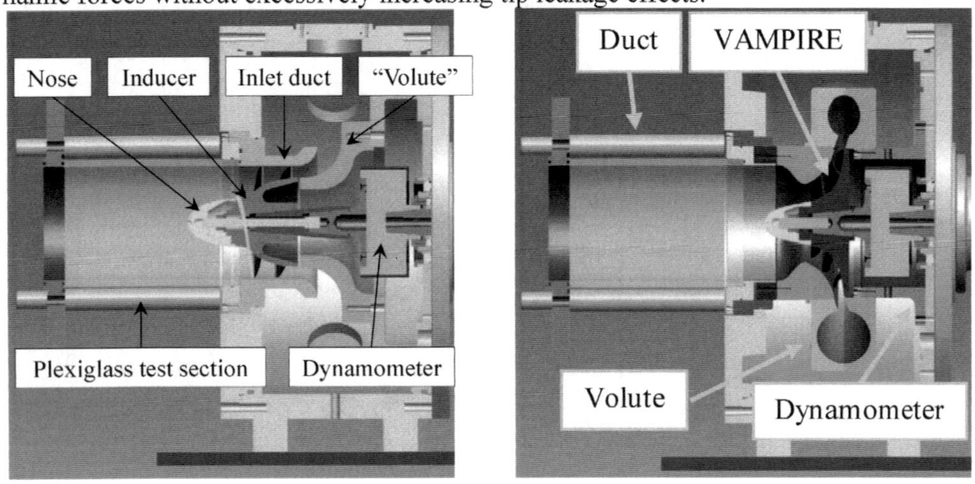

Figure 3. Test configurations of DAPROT3 inducer (left) and the VAMPIRE pump (right).

Experimental Procedure

A circular whirl orbit with constant eccentricity ε has been imposed to the test impellers in order to determine the corresponding fluid-induced rotordynamic forces. The resulting instantaneous fluid force $\boldsymbol{F}$ acting on the impellers can be expressed as the sum of a steady component $\boldsymbol{F}_0$, not depending on the rotor eccentricity, and an unsteady contribution generated by the whirl motion. The latter is the rotordynamic force $\boldsymbol{F}_R$ and is usually a linear function of the eccentricity, which represents small perturbation of the impeller flow.

A schematic of the whirl trajectory in a plane orthogonal to the axis of the machine, together with the decomposition of the total force acting on the impeller into the steady and the rotordynamic forces, are depicted in Figure 4. In the same figure the projections of the rotordynamic force $\boldsymbol{F}_R$ along the normal ($\boldsymbol{F}_N$) and tangential ($\boldsymbol{F}_T$) directions to the whirl orbit are also shown. The normal force is assumed positive if directed outward. The tangential force is assumed positive if it has the same direction of the impeller rotational speed $\boldsymbol{\Omega}$. Therefore, with reference to Figure 5 the normal force is destabilizing when positive (tending to increase the eccentricity) and stabilizing when negative. Similarly, the tangential force is destabilizing when has the same sign as the whirl speed (promoting the whirling motion) and stabilizing in the opposite case.

It is worth noting once more that the rotordynamic forces presented in this paper only refer to the fluid forces induced by the eccentricity of the rotors. The effects of gravity, buoyancy, and tare forces (the submerged weight and the centrifugal force generated by the whirl motion on the rotor mass), as well as the steady fluid force on the impeller (like those induced by azimuthal asymmetries of the flow), have been subtracted from the total force read by the dynamometer. This is typically obtained by performing suitable experiments in air and then under fully-wetted flow conditions, as already shown in [1] and more recently in [30]. The rotordynamic forces have been normalized as follows ([1], [11]):

$$F_R^* = \frac{F_R}{\pi \rho c_a \varepsilon \Omega^2 r_T^2}$$

where ρ is the liquid density, c_a is the axial length of the inducer blades, ε is the radius of the whirl orbit (eccentricity), Ω is the inducer rotational speed, and r_T is the impeller tip radius.

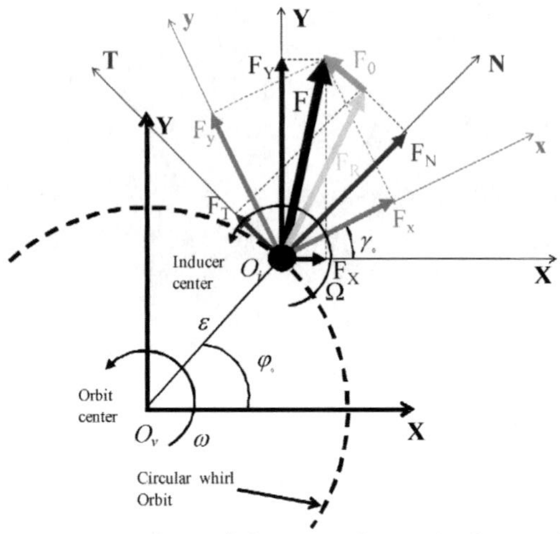

Figure 4. Schematic representation of the rotordynamic forces in the laboratory and rotating reference frames.

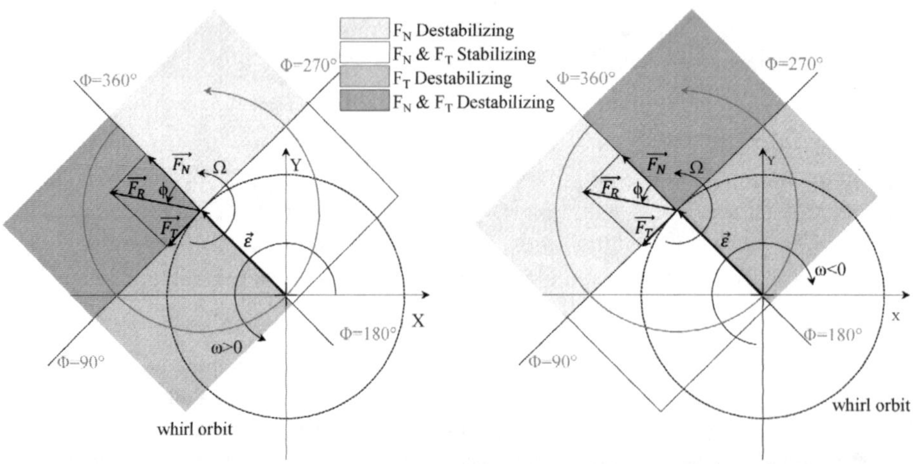

Figure 5. Schematic diagram of the stability/instability regions of the normal and tangential components of the rotordynamic force for positive (left) and negative (right) whirl frequency ratios.

Two different approaches have been used to evaluate the spectral dependence of rotordynamic fluid forces on the whirl speed. The first "traditional" approach consists in the evaluation of the rotordynamic forces from separate experiments carried out at discrete values of the whirl ratio ω/Ω ([20], [31]). The second procedure, illustrated in detail in [21] and [32], yields a continuous measurement of the rotordynamic force spectrum as a function of the whirl ratio from the data obtained in a single experiment by progressively shifting the observation time window throughout the imposed variation of the whirl speed, thus greatly simplifying and speeding up the acquisition and reduction of the experimental data.

The experimental campaign carried out on the DAPROT3 inducer has been aimed at understanding how the rotordynamic forces are influenced by:

- the flow rate (flow coefficient Φ);
- the cavitation regime (cavitation number σ);
- the fluid temperature (inertial/thermal cavitation).

All tests have been carried out with a whirl eccentricity $\varepsilon = 1.13 \pm 0.05$ mm and an impeller rotational speed $\Omega = 1750$ rpm, whose choice represents a suitable compromise between the generation of measurable rotordynamic forces and the structural integrity of the rotating dynamometer under the resulting centrifugal forces on the test impellers. Only subsynchronous whirl ratios have been investigated, with both positive and negative values up to ±0.7 ($\omega/\Omega = \pm 0.1$, ±0.3, ±0.5, ±0.7 for the discrete approach). The combinations of flow rate, cavitating conditions, and water temperature have been considered, as summarized in the test matrix of Table 2.

Table 2. Test matrix for the DAPROT3 and the VAMPIRE pump.

			DAPROT3		VAMPIRE	
Φ/Φ_D	ω/Ω	$T\left[°C\right]$	Φ	σ	Φ	σ
80%		19	0.052	1.015 [a]	0.074	0.60 [a]
100%	$-0.7 \le \omega/\Omega \le 0.7$	70	0.065	0.143 [b]	0.093	0.11 [b]
120%			0.078	0.088 [c]	0.111	0.08 [c]

[a] No cavitation
[b] Limited cavitation
[c] Developed cavitation

Results and Discussion

A series of tests has been conducted on the DAPROT3 inducer in order to characterize its noncavitating pumping performance and hydraulic efficiency at 20 °C in the same nominal configuration used in rotordynamic tests but with zero eccentricity. Figure 6 shows the noncavitating experimental curves in terms of the head coefficient $\psi = \Delta p/\rho\Omega^2 r_T^2$ and hydraulic efficiency $\psi = \Delta p/\rho\Omega^2 r_T^2$ as functions of the flow coefficient $\Phi = Q/\pi\Omega r_T^3$. The volumetric flow rate Q has been measured by means of an electromagnetic flowmeter mounted on the discharge line, the static pressure increase (Δp) by means of a differential pressure transducer. The low pressure tap has been located on the suction line about six inducer diameters upstream of the blade leading edges, in order to reduce the effect of inlet flow prerotation. The high pressure tap has been mounted on the discharge line at about 2.5 duct diameters downstream of the test chamber connection, because of the uncertainties in compensating for the influence of exit flow swirl if mounted on the inducer discharge casing. Hence, head measurements include the (negligible) viscous losses from the upstream pressure tap to the inducer inlet, as well as the diffusion losses from the inducer outlet into the test section and the entrance losses in the discharge line. In order to evaluate the total pressure rise, two main assumptions have been made according to the model explained in detail in a previous paper [31]:

- the static pressure in the discharge line is equal to the bulk static pressure at the exit of the inducer;
- the azimuthal velocity at the inducer exit can be evaluated by means of the Carter's rule.

The torque τ has been directly measured by means of the rotating dynamometer, thereby by-passing the uncertainties associated with seal friction on the inducer shaft. The maximum measured efficiency (about 82%) corresponds to a flow rate close to the design flow coefficient $\Phi_D = 0.065$ that is in good agreement with the typical optimization made for this kind of machine.

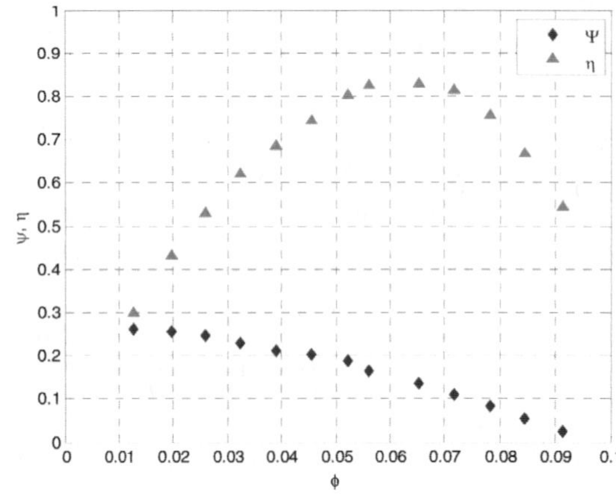

Figure 6. Noncavitating pumping performance and hydraulic efficiency curves of the DAPROT3 inducer with zero imposed eccentricity and whirl motion.

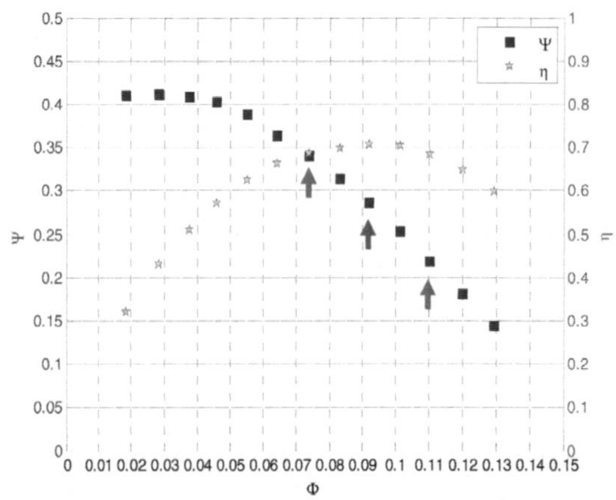

Figure 7. Noncavitating pumping performance and hydraulic efficiency of the VAMPIRE pump with zero imposed eccentricity and whirl motion. The arrows indicate the operating points corresponding to the three values of the flow coefficient tested in the experiments.

The corresponding results for the VAMPIRE pump are presented in Figure 7, which illustrates the noncavitating pumping performance and the hydraulic efficiency characteristics of the machine. The arrows indicate the three flow coefficients used in the experiments.

Rotordynamic Forces on the DAPROT3 Inducer

In the following Figures from 8 to 12 representative results for the rotordynamic forces measured on the DAPROT3 inducer under the conditions indicated in Table 2 are presented in terms of their nondimensional normal and tangential components as functions of the whirl speed ratio ω/Ω. Lines and markers indicate the results of continuous and discrete tests, respectively. The latter approach is likely to be more accurate because the whirl speed is constant over the observation time, while it slowly drifts in continuous tests. On the other hand, the spectral resolution of continuous tests is of special interest as it provides more accurate information on the dependence of rotordynamic forces on the whirl ratio in the regions where they undergo rapid changes and allows for clear identification of the maxima and minima of the curves. As shown in the figures, the results of the two approaches agree almost perfectly, thus confirming the validity of the technique used for the continuous measurement of the rotordynamic force spectra.

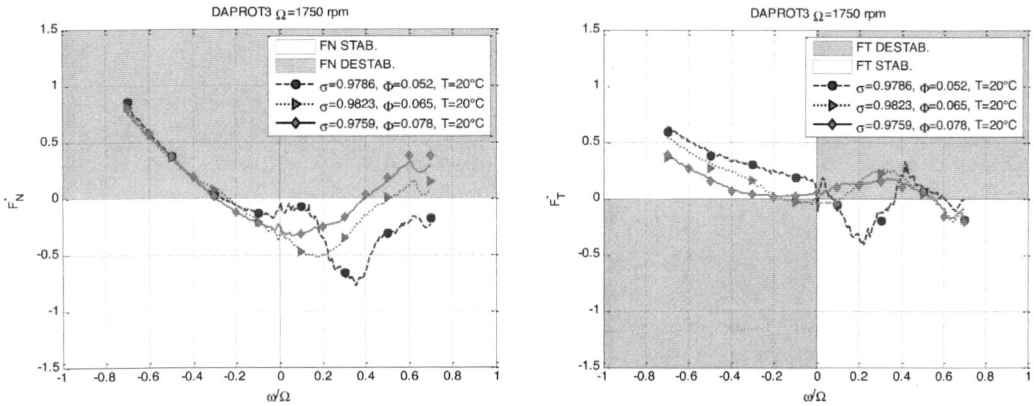

Figure 8. Effect of the flow rate on the rotordynamic force spectra of the DAPROT3 inducer in water tests under fully-wetted flow conditions. Nondimensional normal and tangential components F_N^* and F_T^* as functions of the whirl ratio at 80%, 100% and 120% of the design flow ($\Phi = 0.052, 0.065, 0.078$), $T = 20$ °C and $\sigma \simeq 0.98$.

Figure 8 illustrates the effect of the flow coefficient under noncavitating flow conditions. The results are in a very good agreement with the effects detected in similar inducers tested in previous experiments [19], [32]. The first effect manifest at negative whirl ratios consists in the increase of the rotordynamic force intensity at lower flow coefficients, consistently with the parallel increase of the blade loading. A second effect evident at positive subsynchronous whirl ratios is the presence of maxima and minima in the force spectra, with the tendency to shift towards higher values of the whirl ratio as the flow coefficient decreases.

Previous flow visualizations experiments first conducted at Alta in 2010 ([19], [33]) clearly demonstrated that such peaks in the rotordynamic force spectra for moderately positive whirl ratio are generated by the matching of the impeller whirl speed with the inlet flow prerotation. Under these conditions, the disturbance introduced by the rotor whirl effectively modulates the inlet flow, as especially manifest in cavitation experiments where the cavities tend to concentrate in a single azimuthal lobe co-rotating with the average flow in the inducer eye (see Figure 9). The resulting flow distortion induces more intense lateral forces at the frequency corresponding to the common angular speed of the inlet flow prerotation and the whirl motion. In general, the average rotation of the flow in the inducer eye increases with the load, the azimuthal stagger angle, the leading edge backsweep and the tip leakage of the impeller blades. In particular, for fixed geometry of the inducer, the average prerotation of the inlet flow increases at high blade loads and therefore at lower flow coefficients, consistently with the behavior of the rotordynamic force spectra observed in present and previous experiments both at Alta and Caltech.

For negative whirl speeds, when the interaction with the inlet flow prerotation is weak, the spectral behavior of both components of the rotordynamic forces on the DAPROT3 inducer under fully-wetted flow conditions is rather smooth and regular (Figure 8). In particular, in this range of whirl speeds the normal component is nearly parabolic with upper concavity and the tangential one nearly linear with negative slope, consistently with typical findings in centrifugal impellers ([1] and later Figures 14 – 17). On the other hand, for positive subsynchronous whirl ratios the rotordynamic force spectra are more complex and irregular as a consequence of the interaction with the prerotating inlet flow.

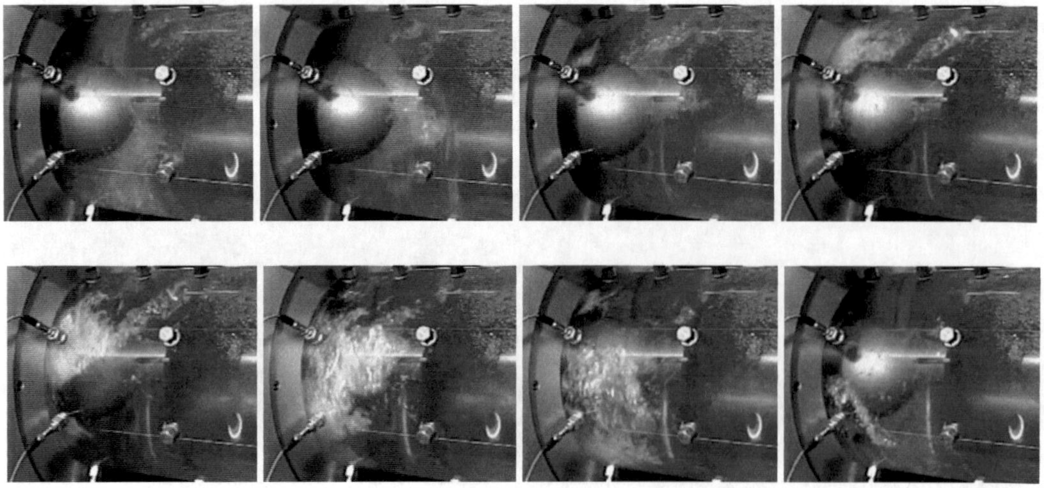

Figure 9. Single-lobed azimuthal cavitation in the eye of the DAPAMITO3 inducer operating in water at whirl speed synchronous with inlet flow prerotation. From left to right and top to bottom the frames correspond to a complete whirl orbit. The flow conditions are: $\Phi = 0.044$, $\sigma = 0.094$, $T = 19.8$ °C, $\omega/\Omega = 0.3$ (from [19]).

At all flow rates, the normal component of the rotordynamic force becomes destabilizing for whirl ratios $\omega/\Omega < 0.3$, is stable for moderate intermediate whirl ratios, and tends to become again destabilizing at higher whirl ratios. The tangential force is significantly less intense that the normal one and is stabilizing at all flow rates for negative whirl, while it exhibits multiple zero crossings in the positive portion of the frequency spectrum. The lowest flow coefficient (higher blade load) is associated to a more stabilizing behavior of both the normal and tangential components of the rotordynamic force. The influence of the flow coefficient on the rotordynamic forces observed in fully-wetted flow conditions has also been confirmed under both limited and more intense cavitation (at $\sigma \simeq 0.14$ and 0.09, respectively), whose results are not shown here for conciseness.

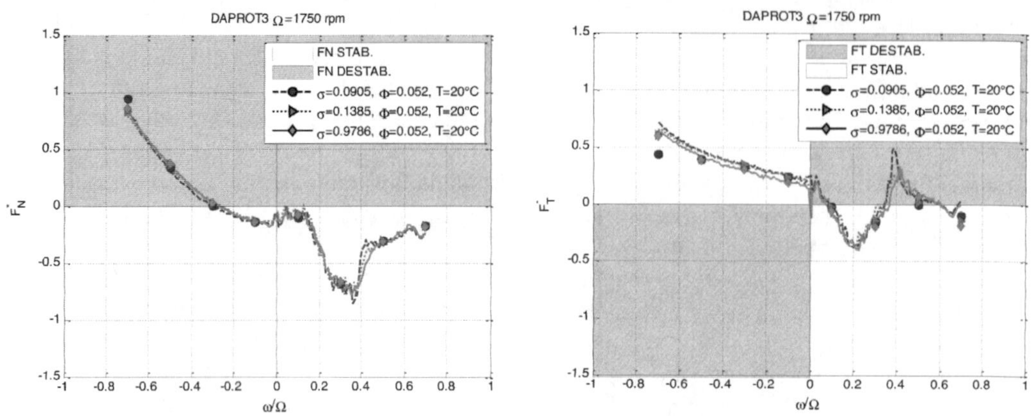

Figure 10. Effect of the cavitation number on the rotordynamic force spectra of the DAPROT3 inducer in cold water tests at 80% of the design flow. Nondimensional normal and tangential components F_N^* and F_T^* as functions of the whirl ratio at $\Phi = 0.052$, $T = 20$ °C (inertial cavitation) and $\sigma \simeq 0.98$, 0.14, 0.09, respectively corresponding to fully-wetted flow, limited cavitation and developed cavitation conditions.

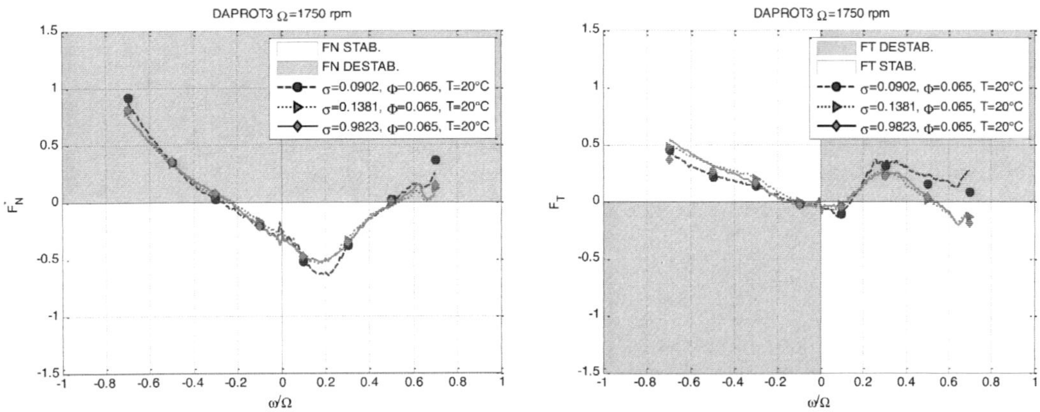

Figure 11. Effect of the cavitation number on the rotordynamic force spectra of the DAPROT3 inducer in cold water tests at 100% of the design flow. Nondimensional normal and tangential components F_N^* and F_T^* as functions of the whirl ratio at $\Phi = 0.065$, $T = 20$ °C (inertial cavitation) and $\sigma \simeq 0.98$, 0.14, 0.09, respectively corresponding to fully-wetted flow, limited cavitation and developed cavitation conditions.

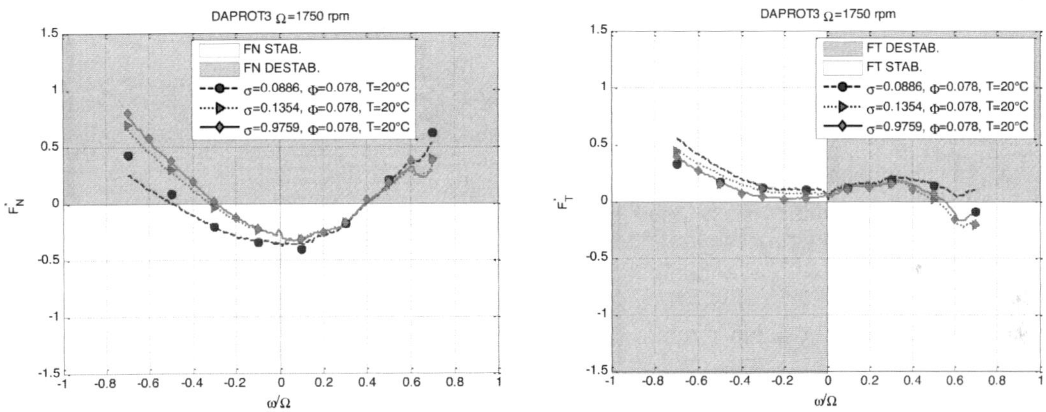

Figure 12. Effect of the cavitation number on the rotordynamic force spectra of the DAPROT3 inducer in cold water tests at 120% of the design flow. Nondimensional normal and tangential components F_N^* and F_T^* as functions of the whirl ratio at $\Phi = 0.078$, $T = 20$ °C (inertial cavitation) and $\sigma \simeq 0.98$, 0.14, 0.09, respectively corresponding to fully-wetted flow, limited cavitation and developed cavitation conditions.

Figures 10, 11 and 12 illustrate the effects of varying the conditions of cavitation in cold water tests at $T = 20$ °C and 80%, 100% and 120% of the design flow coefficient ($\Phi = 0.052$, 0.065 and 0.078). The impact of cavitation on the rotordynamic force is relatively limited, especially at the lowest flow coefficients, and tends to increase only slightly at higher flow rates. In general, the occurrence of cavitation reduces the intensity of the rotordynamic force and delays the destabilizing behavior of the normal force component in the range of whirl ratios from $\omega/\Omega = -0.3$ to $\omega/\Omega = -0.5$ (Fig. 12). The influence of cavitation is most likely the consequence of the general modifications it induces on the impeller blade loading, both steady and unsteady. In fact when cavitation is strong enough to appreciably degrade the pumping performance of the machine, the blade loading is also reduced, and so is the rotordynamic force acting on the impeller.

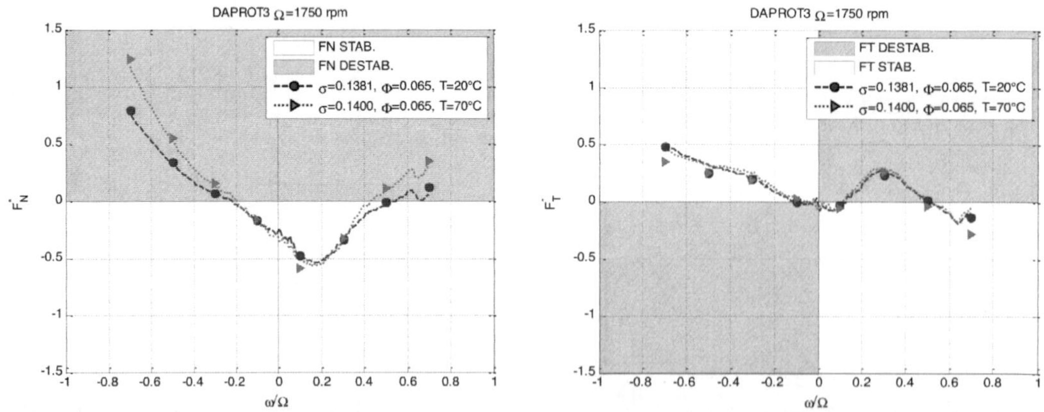

Figure 13. Effect of the water temperature on the rotordynamic force spectra of the DAPROT3 inducer at 100% of the design flow. Nondimensional normal and tangential components F_N^* and F_T^* as functions of the whirl ratio at $\Phi = 0.065$, $\sigma \simeq 0.14$, corresponding to limited cavitation conditions, and $T = 20$ °C (inertial cavitation) and 70 °C (incipient thermal cavitation).

Finally, Figure 13 highlights the effect of the water temperature at design flow rate under slightly cavitating conditions. The occurrence of incipient thermal cavitation in the tests at 70 °C has just a minor influence on the rotordynamic force spectra at the levels of cavitation tested in present experiments. The tangential component is less affected than the normal one which, together with the overall amplitude of the rotordynamic force, tends in general to increase when the whirl ratio is higher than $\omega/\Omega = 0.5$ or lower than $\omega/\Omega = -0.2$. As before, this finding can be readily interpreted if rotordynamic forces are assumed to respond to thermal cavitation like the hydrodynamic loads on the impeller blades. Then the observed effect of the water temperature on the rotordynamic force spectra in the DAPROT3 inducer is coherent with the well-known delay of the pumping performance degradation of hydraulic turbomachinery in the presence of thermal cavitation effects because of the resulting increase of the dynamic actions on the impeller blades.

Rotordynamic Forces on the VAMPIRE Impeller

Representative measurements of the rotordynamic forces acting on the impeller of the VAMPIRE pump under the conditions indicated in Table 2 are presented in the Figures from 14 to 17 for comparison with the results for the DAPROT3 inducer.

The effect of the flow rate on the normal and tangential components of the rotordynamic force in cold-water tests under noncavitating conditions is shown in Figure 14. The influence of the flow rate is larger for the normal force and almost insignificant for the tangential one. In both cases the minimum of the rotordynamic force intensity depends on the flow rate and tends to shift from right to left as the flow rate increases. Besides, the normal force is typically destabilizing at higher whirl speeds, while the tangential force is stabilizing for counter-rotating whirl speeds and typically destabilizing or neutral at positive for $\omega/\Omega > 0$.

However, the most striking difference of the rotordynamic force spectra in the VAMPIRE pump with respect to the results for DAPROT3 inducer is the more regular behavior and the virtual absence of maxima and minima at positive subsynchronous whirl speeds. In particular, the normal force is nearly quadratic and the tangential one roughly linear over the entire frequency spectrum of the experiments, as first documented in earlier investigations ([10]).

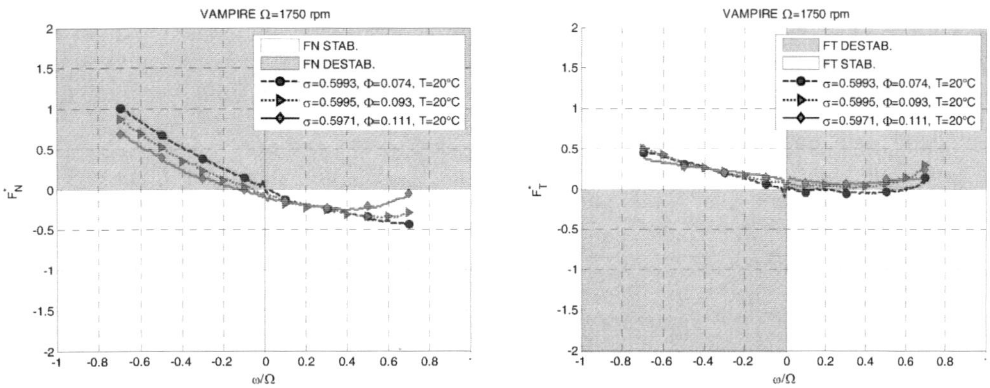

Figure 14. Effect of the flow rate on the rotordynamic force spectra of the VAMPIRE impeller in cold water tests under noncavitating conditions. Nondimensional normal and tangential components F_N^* and F_T^* as functions of the whirl ratio at 80%, 100% and 120% of the design flow ($\Phi = 0.074$, 0.093 and 0.111) at $\sigma \simeq 0.081$ and $T = 20$ °C (incipient thermal cavitation).

As explained earlier, the more complex behavior of the rotordynamic force spectra at moderate subsynchronous whirl speeds is due to the resonant interaction between the inlet flow prerotation and the motion of the impeller eccentricity. This interaction is also occurring in the eye of mixed-flow centrifugal machines like the VAMPIRE pump, whose inlet geometry and physical phenomena responsible for inlet flow prerotation are rather similar to those of inducers. However, the coupling between the inlet flow prerotation and the impeller whirl is proportionally more pronounced in axial machines than in centrifugal impellers, where the inlet radius is comparably smaller. In particular, when (as in the present work) rotordynamic forces are nondimensionalized inversely with the cube of the impeller tip radius, the normalized intensity of the impeller whirl interaction with the inlet flow appears to be greatly reduced in radial machines with respect to axial inducers. This consideration justifies the different behavior observed in present experiments on the VAMPIRE pump and the DAPROT3 inducer.

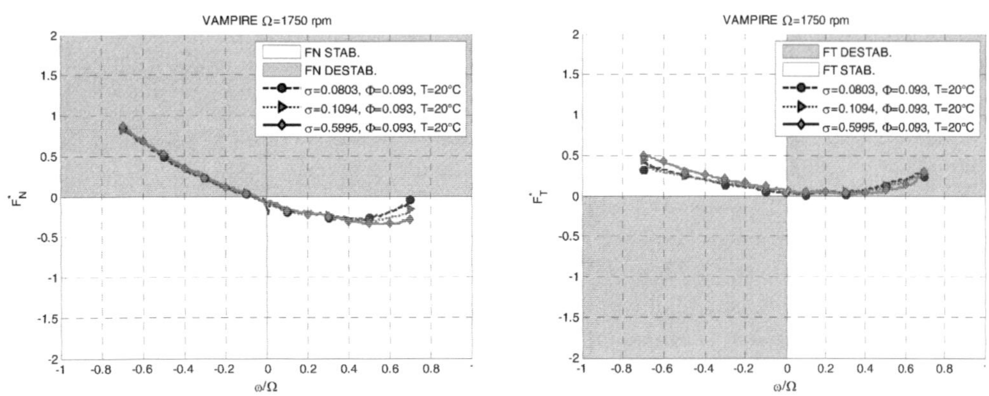

Figure 15. Effect of the cavitation number on the rotordynamic force spectra of the VAMPIRE impeller in cold water tests at 100% of the design flow. Nondimensional normal and tangential components F_N^* and F_T^* as functions of the whirl ratio at $\Phi = 0.111$, $T = 20$ °C and $\sigma \simeq$ 0.60, 0.11, 0.08, respectively corresponding to fully-wetted flow, limited cavitation and developed cavitation conditions.

The effect of cavitation intensity on the rotordynamic forces in the VAMPIRE pump is illustrated in Figure 15. Its impact on the impeller forces is relatively small, as for the DAPRTOT3 inducer. The

IOP Conf. Series: Materials Science and Engineering **129** (2016) 012001 doi:10.1088/1757-899X/129/1/012001

occurrence of cavitation is seen to slightly reduce the intensity of the rotordynamic force, especially at lower negative whirl ratios, and in general acts in favor of the impeller stability with respect to rotordynamic lateral motions. These effects are more pronounced at higher-than-design flow conditions (not shown in the figures) and for the normal force component with respect to the tangential one.

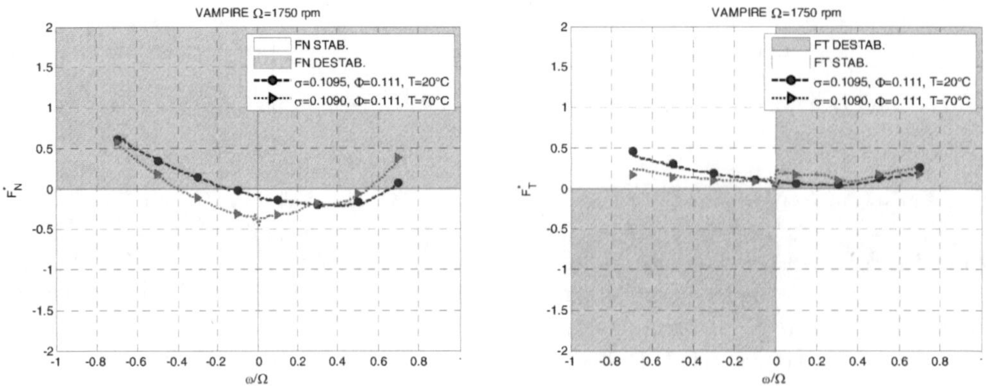

Figure 16. Effect of the water temperature on the rotordynamic force spectra of the VAMPIRE impeller in tests at 100% of the design flow. Nondimensional normal and tangential components F_N^* and F_T^* as functions of the whirl ratio at $\Phi = 0.111$, $\sigma \simeq 0.11$ (limited cavitation) and $T = 20$ °C (inertial cavitation) and 70 °C (incipient thermal cavitation).

The effects of (incipient) thermal cavitation are shown in Figure 16, where the results of tests at room conditions and elevated water temperature are compared. The normal component if more strongly affected, especially at moderate whirl ratios, while the tangential one is almost unchanged. In terms of rotordynamic stability, thermal effects broaden and slightly shift the range of stable whirl frequencies of the normal force component, but have virtually no influence on the stability of the tangential force. In general, therefore, thermal effects tend to have a favorable influence on the rotordynamic stability of the pump impeller.

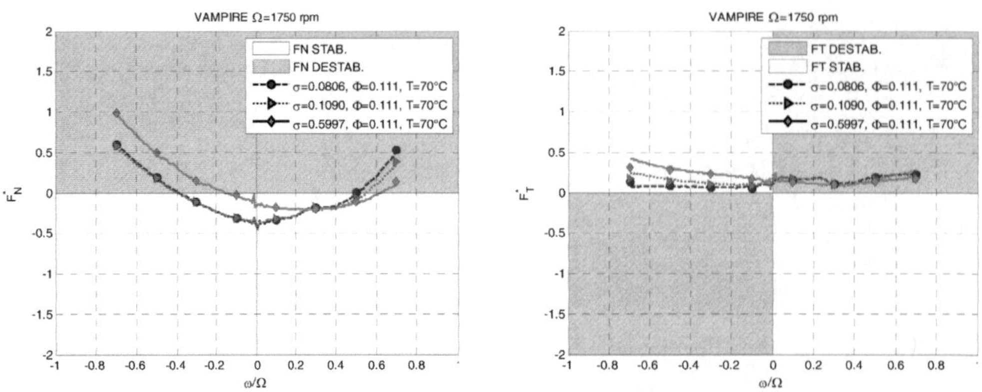

Figure 17. Effect of the cavitation intensity on the rotordynamic force spectra of the VAMPIRE impeller in hot water tests at 120% of the design flow. Nondimensional normal and tangential components F_N^* and F_T^* as functions of the whirl ratio at $\Phi = 0.111$, $T = 70$ °C (incipient thermal cavitation) and $\sigma \simeq 0.60$ (no cavitation), 0.11 and 0.081 (developed cavitation).

Finally, it is worth noting that also the components of the rotordynamic force illustrated in Figures 14, 15 and 16 tend to be smooth functions of the whirl ratio over the whole spectrum covered by the experiments. This result indicates that the effects of the whirl interaction with the inlet flow in the impeller eye of the VAMPIRE pump remain relatively weak also under cavitating conditions. Some evidence of the residual influence of the dynamic coupling between the inlet flow prerotation and the eccentric motion of the impeller can still be recognized in the small subsynchronous irregularities of the rotordynamic force spectra shown in Figure 17, which illustrates the results of hot water tests at 120% of the design flow rate and three values of the cavitation number corresponding to fully-wetted flow, limited cavitation and developed cavitation conditions.

Conclusions

The present series of comparative tests on the DAPROT3 axial inducer and the centrifugal impeller of the VAMPIRE pump confirms in the first place the validity of the procedure recently developed at Alta for the continuous characterization of the spectral dependence of the rotordynamic fluid forces acting on the impellers as functions of the frequency of their eccentric whirl motion. This procedure greatly simplifies and speeds up the acquisition and reduction of the experimental data and proved to be very effective in a) providing accurate, frequency-resolved information on the dependence of these forces on the whirl ratio also in the regions where they undergo rapid changes and b) allowing for clear identification of the maxima and minima of the rotordynamic force spectra.

More importantly, the results of the present research clearly indicate that the complex resonant interaction between the inlet flow prerotation and the impeller whirl motion, which the author and his collaborators recently recognized to be responsible for drastically modifying the behavior of rotordynamic fluid forces on axial inducers in the positive subsynchronous region of the whirl spectrum, is not equally capable of inducing effects of comparable importance also on centrifugal pump impellers. The reasons for this discrepancy are believed to be the relatively weaker intensity of the above interaction in centrifugal machines because of the smaller size of their inlet flow with respect to axial inducers, together with the effect of the nondimesionalization of rotordynamic forces inversely with the cube of the relevant impeller tip radii in the two cases.

Finally, the acquired experimental evidence indicates that, over most of the spectral region covered by present tests, the magnitude of rotordynamic fluid forces acting on both axial and radial impellers tends to decrease with the flow rate, and more weakly with the intensity of cavitation and the occurrence of thermal effects. Besides, in most cases the effect of higher levels of cavitation on the rotordynamic fluid forces is stabilizing. These results are consistent with earlier findings of the author and his research group, as well as with those of previous investigators.

Nomenclature

Symbols

c_a	full-blade axial length
D	diffusion factor
$\boldsymbol{F}$	dimensional force
$\boldsymbol{F}^*$	nondimensional force
h_m	mean blade height
h_m	axial length
N	number of blades
p	static pressure
Q	volumetric flow rate
r_H	inducer hub radius
r_T	inducer tip radius
T	temperature
α	bkade tip incidence angle at design
β	blade tip inlet angle
γ	blade angle
γ_{Tte}	trailing edge tip blade angle

ε	radius of the whirl orbit (eccentricity)
ρ	liquid density
σ	cavitation number ($\sigma = (p_1 - p_v)/0.5\rho\Omega^2 r_T^2$)
σ_T	inducer tip solidity
τ	torque
Φ	flow coefficient ($\Phi = Q/\pi\Omega r_T^3$)
Ω	inducer rotational speed
ω	whirl rotational speed

Subscripts

D	design conditions
H	hub radius
le	blade leading edge
N	normal to the whirl orbit/nominal
R	rotordynamic force modulus
T	tangent to the whirl orbit
te	blade trailing edge
v	vapor pressure
x, y, z	rotating reference frame
X, Y, Z	absolute reference frame
0	initial condition/steady forces
1	inlet station

Acknowledgements
The present work has been supported by the European Space Agency under several TRP contracts and monitored by Dr. Giorgio Saccoccia, ESA-ESTEC, Nordwijk, The Netherlands. His constant attention and encouragement are gratefully acknowledged. The author would like to remember the invaluable work of all present and past members of Alta's Chemical Propulsion Team who collaborated to this exciting experimental research, and in particular of: Angelo Cervone, Lucio Torre, Angelo Pasini, Giovanni Pace, Dario Valentini, and Ruzbeh Hadavandi. Special thanks go to Rafone O. (Oskar) and Laster P. (Paul) of GMYS-Space.

References

[1] Brennen C E 1994 Hydrodynamics of pumps, *Concepts ETI*, (Inc. and Oxford University Press).
[2] Ehrich F and Childs S D 1984 Self-excited vibrations in high performance turbomachinery *Mechanical Engineering*, pp. 66-79.
[3] Hergt P and Krieger P 1969 Paper 10: radial forces in centrifugal pumps with guide vanes, *Proc. Inst. Mech. Eng.*, 184, pp. 101–107.
[4] Ohashi H and Shoji H 1984 Lateral fluid forces acting on a whirling centrifugal impeller in vaneless and vaned diffuser, *Workshop on Rotordynamic Instability Problems in High Performance Turbomachinery*, (NASA Conf. Publ.), pp. 109-122.
[5] Yoshida Y, Tsujimoto Y, Morimoto G, Nishida, H and Morii S 2003 Effects of seal geometry on dynamic impeller fluid forces and moments, *ASME J. Fluids Eng.*, pp. 786-795.
[6] Suzuki T, Prunie`res R, Horiguchi H, Tsukiya T, Taenaka Y and Tsujimoto Y 2007 Measurements of rotordynamic forces on an artificial heart pump impeller, *ASME J. Fluids Eng.*, **129(11)**, pp. 1422–1427.
[7] Hiwata A and Tsujimoto Y 2002 Theoretical analysis of fluid forces on an open-type centrifugal impeller in whirling motion, *ASME J. Fluid Eng.*, **124 (2)**, pp. 342–347.
[8] van Esch B 2009 Performance and radial loading of a mixed-flow pump under non-uniform suction flow, *ASME J. Fluids Eng.*, **131(5)**, p. 051101.

[9] Jery B 1987 Experimental study of unsteady hydrodynamic force matrices on whirling centrifugal pump impellers, *PhD thesis, California Institute of Technology*, Pasadena, California, USA.

[10] Franz R J 1989 Experimental investigation of the effect of cavitation on the rotordynamic forces on a whirling centrifugal pump impeller, *PhD thesis, California Institute of Technology*, Pasadena, California, USA.

[11] Bhattacharyya A 1994 Internal flows and force matrices in axial flow inducers, *Ph.D. thesis, California Institute of Technology*, Pasadena, California, USA.

[12] Bhattacharyya A, Acosta A J, Brennen C E and Caughey T K, 1997 Rotordynamic forces in cavitating inducers, *ASME J. Fluids Eng.*, **119 (4)**, pp. 768–774.

[13] Yoshida Y, Kazami Y, Nagaura K, Shimagaki M, Iga Y and Ikohagi T 2008 Interaction between uneven cavity length and shaft vibration at the inception of synchronous rotating cavitation, *Int. J. Rotating Mach.*, p. 218978.

[14] Yoshida Y, Eguchi M Motomura, T Uchiumi M, Kure H, and Maruta Y 2010 Rotordynamic forces acting on three-bladed inducer under supersynchronous/ synchronous rotating cavitation, *ASME J. Fluids Eng.*, **132 (6)**, p. 061105.

[15] Li C, and van Esch, B P M 2009 Blade interaction forces in a mixed-flow pump with vaned diffuser, *ASME 2009 Fluids Engineering Division Summer Meeting*, pp. 165–173.

[16] van Esch, B, and Li C 2011 Unstable operation of a mixed-flow pump and the influence of tip clearance, *Proceedings of the ASME–JSME–KSME 2011 Joint Fluids Engineering Conference*, pp. 79–87.

[17] Iga, Y, Nohml M, Goto,A, and IkohagiT 2004 Numerical analysis of cavitation instabilities arising in the three-blade cascade, *ASME J. Fluids Eng.*, **126(3)**, pp. 419–429.

[18] Torre L, Pasini A, Cervone A, and d'Agostino L 2011 Continuous spectrum of the rotordynamic forces on a four bladed inducer *In ASME-JSME-KSME Joint Fluids Engineering Conference*, Hamamatsu, Shizuoka, Japan.

[19] Torre L, Pasini A, Cervone A, and d'Agostino L 2011 Experimental characterization of the rotordynamic forces on space rocket axial inducers, *ASME J. Fluids Eng.*, **133(10)**, p. 101102.

[20] Torre L, Pasini A, Cervone A, Pecorari L, and Milani A 2010 Rotordynamic forces on a three bladed inducer, *Proceedings of the Space Propulsion Conference*, San Sebastian, Spain.

[21] Torre L, Pasini A, Cervone A, and d'Agostino L 2011 Continuous spectrum of the rotordynamic forces on a four bladed inducer, *Proceedings of the ASME–JSME–KSME Joint Fluids Engineering Conference*, Hamamatsu, Shizuoka, Japan, pp. 211–222.

[22] Rapposelli E, Cervone A, and d'Agostino L 2002 A new cavitating pump rotordynamic test Facility, *Proceedings of the AIAA Joint Propulsion Conference and Exhibit*, Indianapolis, Indiana, USA, Paper No. AIAA 2002-4285.

[23] Pace G, Pasini A Torre, L, Valentini D, and d'Agostino L 2012 The cavitating pump rotordynamic test facility at ALTA S.p.A.: upgraded capabilities of a unique test rig, *Proceedings of the Space Propulsion Conference*, Bordeaux, France.

[24] d'Agostino L, Pasini, A Valentini D 2011 A reduced order model for preliminary design and performance prediction of radial turbopumps, *47th AIAA/ASME/SAE/ASEE Joint Propulsion Conference*, San Diego, California, USA, August.

[25] Pasini, A, Valentini D Pace G, Torre L, Cervone A, d'Agostino, L 2012 A reduced order model for optimal centrifugal pump design, *14th International Symposium on Transport Phenomena and Dynamics of Rotating Machinery*, Honolulu, Hawaii, USA.

[26] Torre L, Pasini A, Cervone A, and d'Agostino L 2011 Experimental characterization of the rotordynamic forces on space rocket axial inducers, ASME J. Fluids Eng., 133(10), p. 101102.

[27] d'Agostino L, Torre L, Pasini A and Cervone A 2008 On the Preliminary Design and Noncavitating Performance of Tapered Axial Inducers, *ASME J. Fluids Eng.*, **130(11)**, p. 111303.

[28] d'Agostino L, Pasini A, Baccarella D, Cervone A and Milani A 2008 A reduced order model for preliminary design and performance prediction of tapered inducers: comparison with numerical simulations, *Proceedings of the 44th AIAA/ASME/SAE/ASEE Joint Propulsion Conference*, Hartford, CT, pp. 5866–5878.

[29] Torre L, Pasini A, Cervone A and d'Agostino L 2009 Experimental performance of a tapered axial inducer: comparison with analytical predictions, *Proceedings of the 45th AIAA/ASME/SAE/ASEE Joint Propulsion Conference*, Denver, Colorado, USA., pp. 1429–1441.

[30] Pasini A 2010 Pumping performance similarity, cavitation-induced instabilities and fluid-induced rotordynamic forces in tapered inducers, *Ph.D. thesis, University of Pisa*, Pisa, Italy.

[31] Pasini, A, Torre, L, Cervone, A, and d'Agostino L 2010 Rotordynamic forces on a four bladed inducer, *Proceedings of the 46th AIAA/ASME/SAE/ASEE Joint Propulsion Conference*, Nashville, TN, **Vol. 6**, pp. 5390–5405.

[32] Pasini , Torre L, Cervone A, and d'Agostino L 2011 Continuous spectrum of the rotordynamic forces on a four bladed inducer, *ASME J. Fluids Eng.*, **133 (12)**, p. 121101.

[33] Pasini A, Torre L, Cervone A, d'Agostino L 2011 Characterization of the rotordynamic forces on tapered axial inducers by means of a rotating dynamometer and high-speed movies, *WIMRC 3rd International Cavitation Forum 2011*, University of Warwick, UK, 4th-6th July.

Design method of supercavitating pumps

V Kulagin[1, 4], D Likhachev[2] and F C Li[3]

[1] Professor, Siberian Federal University, 26A Kirenskogo Street, building 15, room D3-24, 660074 Krasnoyarsk, Russia
[2] Postdoctor, Zhejiang University, 38 Zheda Road, building 4, room 110A, Hangzhou, Zhejiang Province 310058, P. R. China
[3] Professor, Harbin Institute of Technology, CFHTL, 92 West Dazhi Street, Nan Gang District, Harbin, Heilongjiang Province 150001, P. R. China

E-mail: dima@zju.edu.cn

Abstract. The problem of effective supercavitating (SC) pump is solved, and optimum load distribution along the radius of the blade is found taking into account clearance, degree of cavitation development, influence of finite number of blades, and centrifugal forces. Sufficient accuracy can be obtained using the equivalent flat SC-grid for design of any SC-mechanisms, applying the "grid effect" coefficient and substituting the skewed flow calculated for grids of flat plates with the infinite attached cavitation caverns. This article gives the universal design method and provides an example of SC-pump design.

1. Introduction

The one effective application of the cavitation technology are supercavitating pumps (SC-pumps), which have advantages in variety of industrial processes [1], [2]. The basics for calculation of any cavitating equipment, working in the presence of developed cavitation, are their geometry parameters and hydraulic characteristics, i.e. the solution must contain the flow velocity and pressure distribution, and cavitation bubble's dimensions. For the given flow rate and net head, the design of the SC-pump's impeller requires calculation of its profile's geometry, the number of blades etc., as well as calculation of its rotation velocity, axial velocity etc.; but in addition, the design is complicated by generation of the cavity with required length, the number and size of the cavitation bubbles.

2. Load Distribution

Calculation algorithm of SC-pumps is based on the solution of a supercavitating flow around a blade's element or equivalent flat blades grid [3]. In the first stage of the design of a rotary equipment, the best load allocation along the blade's radii is calculated. This factor would ensure the required net head with minimal energy losses.

The best load allocation along the axis of a real SC-pump's blade is found by the formula obtained by the method of successive approximations:

$$\frac{\mathrm{d}C_P}{\mathrm{d}\bar{r}} = \frac{4L(1-\varepsilon \mathrm{tg}\beta_i)K_Z\bar{r}}{\mathrm{tg}\beta_i^2 + K_\varepsilon + 1}, \tag{1}$$

[4] Author to whom any correspondence should be addressed.

where $C_P = \frac{2\Delta P}{\rho v_0^2}$ – the static pressure coefficient; $H = H_{+\infty} - H_{-\infty}$ – the pressure generated by the SC-pump; $\varepsilon = \frac{C_x}{C_y} = 0{,}075 \div 0{,}100$ – reverse quality of the SC-profile and composition of the grid (first approximation); $K_Z = K_Z(\lambda_1, K_\varepsilon, K_\mu, z, \ldots)$ – amendment to the limited number of blades; $\bar{r} = \frac{r}{R}$ – relative radius of the blade; r, R – the current and the outer radius of the blade respectively; β_i – the angle of entry of the relative speed with influence of the inductive speed ($\text{tg}\beta_i = \frac{v}{\omega_r}$ see figure 1); $K_\varepsilon = \frac{F_{SC}}{F_{CH}}$ – the relative gap; F_{CH}, F_{SC} – square of the channel and SC-impeller respectively; $L = L(K_\varepsilon, \lambda_1, \lambda_0)$ – the Lagrangian coefficient obtained from the solution of variational problem, which computed by the formula:

$$L = \frac{2\left(\frac{\lambda_i}{\lambda_0} - 1\right)}{\frac{1+\lambda_i}{\lambda_i^2 + K_\varepsilon + 1} - \frac{K_\varepsilon \lambda_i^2 ln\left[1 + \frac{(K_\varepsilon + 1)}{\lambda_i^2}\right]}{(K_\varepsilon + 1)^2}}, \tag{2}$$

$$\lambda_i = tg\beta_i \bar{r} = tg\beta_i|_{r=R} = \frac{v_0 + w_{as}^*}{\omega r - w_{\tau s}^*},$$

where $\lambda_0 = \frac{v_0}{\omega_r}$ – the relative pace at the pump's inlet; v_0, ω_r – the axial and circumferential speed of fluid relative to an element of the blade at the pump's inlet; w_{as}^*, $w_{\tau s}^*$ – the axial and circumferential speed at the blades (см. рис. 1). By integrating these values $\frac{dC}{d\bar{r}}$along the radii of the blade, we find the coefficient of static pressure in the first approximation.

The next step in calculation of the SC-pump is to define the spatial characteristics of the blade grid for the given radius, which would ensure maximum value of the quality $k = \frac{C_y}{C_x}$ for the load on the radius, obtained in first stage, and taking into account the conditions of strength.

The study dealt with the SC-grid made of the following profiles: flat plate; low wrap profile $\bar{f} = f/b \leq 0{,}03$; circle's arch profile ($\bar{f} \leq 0{,}04$); concave profile, described by an equation of second, third and fifth order ($\bar{f} \leq 0{,}06$); profile with square load allocation ($\bar{f} \leq 0{,}1$). Ranges of parameters for calculation were chosen as follows: relative length of the cavity $\bar{L}_C = \frac{L_C}{B_C} = 1{,}5 \div 20$ (length to width ration); relative curvature of profiles $\bar{f} = 0 \div 0{,}1$; relative grid density $b/\tau = 0{,}5 \div 1{,}5$; stem angles gratings $\beta = 10 \div 75°$; angles of attack $i = 0 \div 20°$.

In figure 1, 2 and 3 show the hydrodynamic characteristics of SC-grids for flat plates and curved profiles with a rectangular distribution of pressure along the chord $\varepsilon = \varepsilon\left(\frac{b}{\tau}, \beta_0, C_y\right)$. Here $\varepsilon = \frac{C_x}{C_y}, \frac{C_x}{C_y} = f(b/\tau, \beta_0, C_y, \beta_i, \chi)$, $\bar{L}_C = \frac{L_C}{B_C} == f\left(\frac{b}{\tau}, C_y, \beta\right)$.

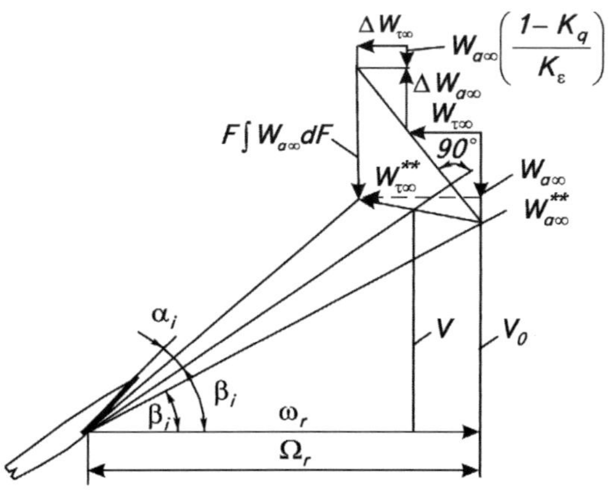

Figure 1. Triangles of flow velocities, flowing on the blades of the SC-grid with an arbitrary gap

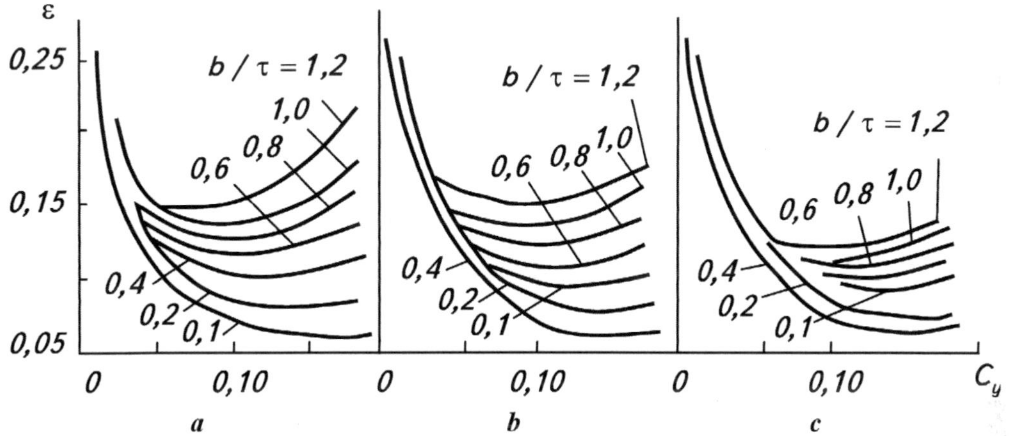

Figure 2. Reverse quality factors $\varepsilon = \varepsilon\left(\frac{b}{\tau}, \beta_0, C_y\right)$ for SC-grids of flat plates with $\bar{L}_K \to \infty$, $\chi \to \chi_{min}$; where a - $\beta_0 = 15°$; b - $\beta_0 = 30°$; c - $\beta_0 = 60°$

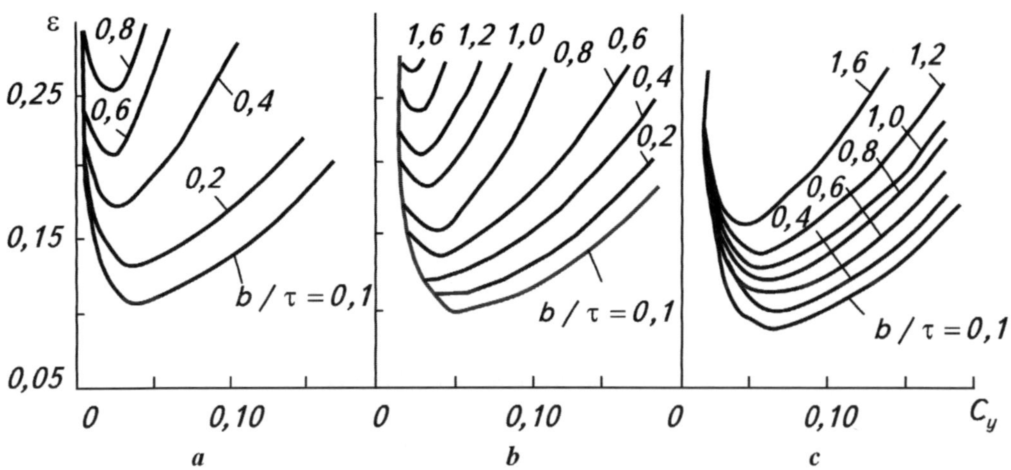

Figure 3. Reverse quality factor $\varepsilon = \varepsilon\left(\frac{b}{\tau}, \beta_0, C_y\right)$ for SC-grid made of curved profiles with the rectangular distribution of pressure along the chord for $\bar{L}_K \to \infty$, $\chi \to \chi_{min}$; where a - $\beta_0 = 15°$; b - $\beta_0 = 30°$; c - $\beta_0 = 60°$

Design calculation of SC-pumps is based on the "base line" vortex theory, the theory of SC-grids, as well as on the method of equivalent SC-grid. The calculation is carried out in two phases. First determine the optimum load distribution along the radius of the blade on the desired characteristics of SC-pump: static pressure of H_S, the flow Q, the rotational speed n, or the diameter blade's outer D (for a given D the unique value n can be evaluated among the several numbers n).

3. Profiling

Once the optimum distribution of circulation (load) is found, one proceed to the choice of structural elements of the blades suitable for this distribution, and profiling in accordance with the strength conditions and optimum for dynamic quality of profiles.

Optimal load distribution along the blade's radius is found by the formula (1). Expressing the profile drag by the multiplier $(1 - \varepsilon \mathrm{tg}\beta_i)$, limit of the number of blades by K_Z, and calculating the Lagrangian L from equation (2) by the method of successive approximations, find the distribution of load factor along the blade's radius of a real SC-pump (1).

On the basis of the calculations one can build diagrams to determine the inductive dynamics λ_i of optimal SC-pump with an arbitrary gap and the degree of cavitation development for the given finite number of blades for the stated values of C_P, λ_0 and strength (parameter t) [1].

In the second stage of calculation chose structural elements of blades that provide found circulation and distribution of the induced speeds. Using the theorem of Zhukovskiy for lifting force on element of the blade $dy = \rho v_i \Gamma dr$, and the following expression

$$dy = C_y \frac{\rho v_i^2}{2} b dr,$$

write down the equation for coupling of the blade and the flow:

$$\Gamma = \frac{1}{2} C_y v_i b. \tag{3}$$

Substituting (3) into the expression for the static head pressure lightly loaded SC-pump

$$QH_S = v_0 \rho \omega \int_{r_S}^{R} r \Gamma K_\mu \, dr,$$

obtain

$$dH_S = \frac{\rho v_{0\omega}}{v_0 \pi R_{CH}^2} z C_y \frac{b}{2} v_i K_\mu r dr. \tag{4}$$

Taking into consideration that

$$v_i = \frac{\omega r - w_{\tau s}^*}{\cos\beta_i} = \frac{v_0 - w_{as}^*}{\sin\beta_i},$$

then from (3) derive the formula to determine the hydro-mechanical parameter of SC-pump's blade:

$$C_y \frac{b}{\tau} = \frac{1}{2} \cdot \frac{\lambda_0}{\bar{r}^2 K_\varepsilon K_\mu} \cdot \frac{\sin\beta_i}{1 + w_{as}^*} \cdot \frac{dC_P}{d\bar{r}}.$$

IOP Conf. Series: Materials Science and Engineering **129** (2016) 012002 doi:10.1088/1757-899X/129/1/012002

It should be noted that in the charts of hydrodynamic characteristics of SC-grids (see figure 2, 3), one can add the curves of equal values of the parameter $C_y\frac{b}{\tau}$. Thus, defining for each radius the value β_i and parameter $C_y\frac{b}{\tau}$ for a set of SC-grids, one can choose the grid that has the lowest inverse quality ε_{min}.

Then, using a method of the equivalent SC-grid [1], for known $\frac{b}{\tau}$, β_i, L_K, find a factor which takes into account the influence of the "grid effect" without the skewed flow $J_P = J_P\left(\frac{b}{\tau}, \beta, L_K\right)$. Similar calculations for grids of curved profiles with arbitrary length of the cavern show that J_P is the factor influencing the "grid effect" (for grid density $\frac{b}{\tau} > 0.6$). But the "grid effect practically does not depends on the curvature *f*, profile's form and the cavern length L_K. The vortex base line theory can be used to find the stream skew $\Delta\alpha_i = \Delta\beta_i$ near the impeller.
By C_y and J_P determine the setup angle of the blade on each radius φ:

$$C_{y_{imp}} = J_P \left(\frac{\partial C_y}{\partial \alpha}\right)_{sap} \alpha_i,$$

where the designation "*imp*" – impeller; "*sap*" – stand-alone profile. Writing $\alpha_i = \varphi - \beta_i$ gives

$$\varphi = \frac{C_y}{J_P}\left(\frac{\partial C_y}{\partial \alpha}\right)_{sap} + \beta_i.$$

Therefore $\frac{H}{D} = \pi\bar{r}\mathrm{tg}\varphi$. For convenience, the calculations of design write down in a table form. Based on the described methods, calculation of the generic diagrams of hydrodynamic characteristics of SC-impellers can be done.

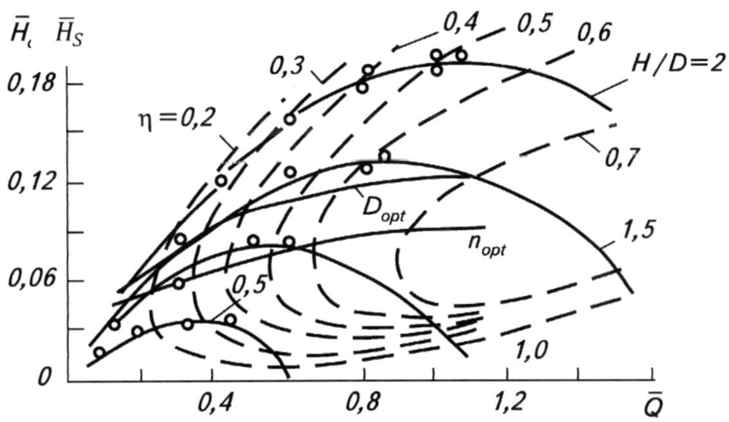

Figure 4. Universal chart of hydrodynamic characteristics for SC-pumps working under developed cavitation with $\bar{H}_S = f(Q)$, $\bar{H}_S = \frac{H_S}{\rho n^2 D^2} = f\left(\frac{H}{D}, \frac{b}{\tau}, \chi, K_\varepsilon\right)$

The figure 4 illustrates one such diagram of supercavitating regimes for SC-pump's impellers ($\frac{b}{\tau} = 1$; $\chi = 0{,}25$; $K_\varepsilon = 0{,}96$; profile is a wedge-shaped flat plate) that are used for simultaneous pumping and processing of the working fluids [1].

4. Conclusion

In its current state, this method allows design of rotary equipment, which simultaneously pumps the medium, and uses effects of cavitation on the pumped medium (SC-pump). The highly turbulent cavitation bubbles collapse, observed under the developed cavitation condition, have numerous applications [1]. For the given conditions and working fluids, designer should be aware of the negative and the positive consequences of cavitation effects, especially for the extremal case – supercavitation.

Acknowledgments
Authors wish to acknowledge Likhachev Dmitriy for preparation of the materials in English.

References

[1] Ivchenko V M, Kulagin V A and Nemchin A F 1990 *The Cavitational Technology* (Krasnoyarsk: KSU Publishing House) p 200
[2] Demidenko N D, Kulagin V A and Shokin Y I 2012 *Modeling and Computational Technologies of Distributed Systems* (Novosibirsk: Nauka) p 424
[3] Kulagin V A, Vil'chenko A P and Kulagin T A 2001 *Modeling of Two-phase Supercavitation Flows* (Krasnoyarsk: KSTU) p 187

IOP Conf. Series: Materials Science and Engineering **129** (2016) 012003 doi:10.1088/1757-899X/129/1/012003

Research on energy conversion mechanism of rotodynamic pump and design of non-overload centrifugal pump

X L Zhang, S B Hu, Z Z Shen, S P Wu and K Li

Hefei General Machinery Research Institute, Hefei 230031, Anhui, China

E-mail: zhangxinglin58@126.com

Abstract. In this paper, an attempt has been made for the calculation of an expression for the intrinsic law of input power which has not yet been given by current theory of Rotodynamic pump. By adequate recognition of the characteristics of non-inertial system within the rotating impeller, it is concluded that the input power consists of two power components, the first power component, whose magnitude increases with the increase of the flow rate, corresponds to radial velocity component, and the second power component, whose magnitude decreases with the increase of the flow rate, corresponds to tangential velocity component, therefore, the law of rise, basic levelness and drop of input power curves of centrifugal pump, mixed-flow pump and axial-flow pump can be explained reasonably. Through further analysis, the main ways for realizing non-overload of centrifugal pump are obtained, and its equivalent design factor is found out, the factor correlates with the outlet angle of leading face and back face of the blade, wrap angle, number of blades, outlet width, area ratio, and the ratio of operating flow rate to specified flow rate and so on. These are verified with actual example.

1. Introduction

Rotodynamic pump (including centrifugal pump, mixed-flow pump and axial-flow pump) is a machinery for realizing energy conversion, which converts the mechanical energy of the prime mover into the liquid energy of the conveyed medium (sum of pressure energy, velocity energy and elevation energy), this involves the energy conversion mechanism, i.e. the relationship of conversion between input power and output power of the pump. Correct understanding of the intrinsic character of input power and obtaining the calculation expressions for the intrinsic law of input power is of important significance for knowing energy conversion mechanism and pump design, etc., for example, effectively controlling the change law of the input power curve so as to make the power curve of centrifugal pump to have maximum value, and thus achieve non-overload design of centrifugal pump; reducing input power as far as possible so as to find out key design factors for improving pump efficiency; realizing prediction of power curve, numerical simulation, etc.

The pump theory and design are still not perfect, many mechanism problems need to be solved, thus it is of important significance to research on input power of pump and relevant problems.

2. Discussion on input power law of pump theories at present

The actual shape of input power (shaft power) curve of rotodynamic pump is shown in Fig. 1(a)[1-7], it

IOP Conf. Series: Materials Science and Engineering **129** (2016) 012003 doi:10.1088/1757-899X/129/1/012003

is characterized that the input power curve of centrifugal pump, mixed-flow pump and axial-flow pump presents ascending, basically level and descending tendency, respectively.

However, the calculation formula $N = \rho g Q_{th} H_{th}$ for input power is given in literature [1,3-5] (neglecting mechanical losses of bearing, etc), their characteristics are: all the input power curves of rotodynamic pump present ascending law, as shown in Fig. 1(b). This viewpoint cannot reasonably explain the actual phenomenon that the shaft power of mixed-flow pump at zero flow rate is approximately equal to rated power and the shaft power of axial-flow pump at zero flow rate is significantly higher than the rated power, etc. Literature [7] points out that the difference between theories and actual value of input power in the small-capacity zone is very obvious, its cause is still not clear.

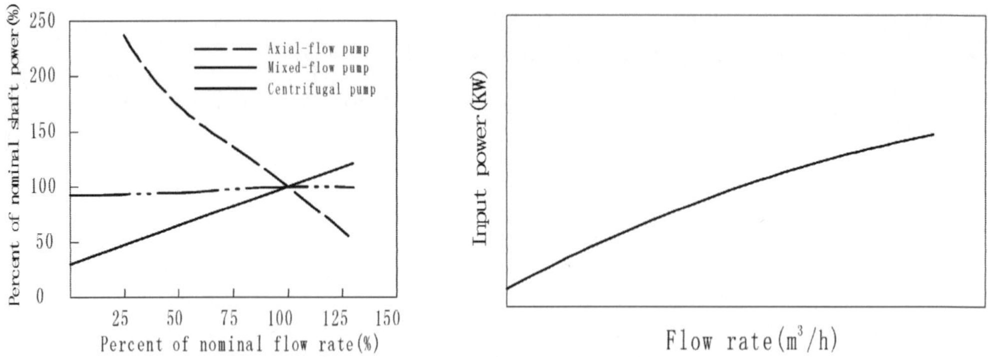

(a) (b)

(a) Actual shape of input power curve (b) theory shape of input power curve

Figure 1. Comparison of input power curve between actual shape and theory shape

3. New knowledge on input power and energy conversion mechanism

Input power is the mechanical energy of the prime mover corresponding to the impelling force of the blades on the medium (neglecting mechanical losses as of bearing, etc.). If the impelling force (unit mass impelling force) of blades on the medium is obtained, the torque of blades on the medium can be obtained, further the input power can be obtained. Therefore, derivation of the impelling force of the blades on the medium is the key to the study of input power and relevant problems.

On the other hand, the head is the energy which is acquired by the medium of unit mass. Therefore, there certainly exists a corresponding relationship between the head and input power, that is, there certainly exists a corresponding relationship between the head and the impelling force of the blades on the medium.

Then, according to the principle that the force can be decomposed and recomposed in the theoretical mechanics, in order to obtain the impelling force of the blades, first the total head needs to be decomposed, and the impelling force of the blades to which each head component corresponds on the medium is to be obtained. Thus the input power component to which each head component corresponds is obtained. Further it is not difficult to obtain the corresponding relationship between input power component and output power component, that is, the relationship between energy conversion mechanisms of input power and output power.

What is worth pointing out is that only the total head is decomposed by the intrinsic characteristics correctly, can the nature of each head component be recognized, and can the correct solution for the impelling force of the blades to which each head component corresponds be found out, thus the action of blades to which each head component corresponds can be obtained correctly, the calculation expression for intrinsic law of input power and the relationship of conversion between input power and output power can be obtained.

3.1 Decomposition and division of the head

In literature [1, 3, 5-7], One of the Equation currently used to calculate theoretical head H_{th} is as follows:

$$H_{th} = (u_2^{\ 2} - u_1^{\ 2})/2g + (w_1^{\ 2} - w_2^{\ 2})/2g + (c_2^{\ 2} - c_1^{\ 2})/2g\,. \tag{1}$$

In literature [1, 3, 5-7], $(w_1^2 - w_2^2)/2g$ is classified into potential head, this is evidently inappropriate. If it (strictly, it should be $(w_{u1}^{\ 2} - w_{u2}^{\ 2})/2g$) is classified into kinetic head, then head classification is very appropriate. The reasons are: ① w is a velocity of movement in a relative coordinates; ② within the impeller there indeed exists friction losses h due to w, and $h \propto w^2$. This is one of the important characteristics of kinetic head; ③the characteristic of w_u is the same as c_u, but it is different from u (refer to the description below); ④because $(w_1^{\ 2} - w_2^{\ 2})/2g$ is converted into pressure, pump theory at present classifies it into the potential head. However, $(c_2^{\ 2} - c_1^{\ 2})/2g$ is finally converted into pressure through pump output, but it undoubtedly belongs to the kinetic head.

Because of $\mathbf{c} = \mathbf{c_u} + \mathbf{c_m}, \mathbf{w} = \mathbf{w_u} + \mathbf{w_m}, \mathbf{c_m} = \mathbf{w_m}$, as shown in Fig. 2, by slight transformation, Equation (1) becomes:

$$H_{th} = (u_2^{\ 2} - u_1^{\ 2})/2g + (c_{u2}^{\ 2} - c_{u1}^{\ 2})/2g + (w_{u1}^{\ 2} - w_{u2}^{\ 2})/2g \tag{2}$$

Equation (2) more clearly reflects the essential law of head because it does not contain meridional velocity c_m and w_m. Therefore, we can classify head as follows:

The right-hand side term of Equation (2) $(u_2^{\ 2} - u_1^{\ 2})/2g$ is potential head H_p, i.e.

$$H_p = (u_2^{\ 2} - u_1^{\ 2})/2g \tag{3}$$

The right-hand side term of Equation (2) $(c_{u2}^{\ 2} - c_{u1}^{\ 2})/2g + (w_{u1}^{\ 2} - w_{u2}^{\ 2})/2g$ is kinetic head H_k, i.e.

$$H_k = (c_{u2}^{\ 2} - c_{u1}^{\ 2})/2g + (w_{u1}^{\ 2} - w_{u2}^{\ 2})/2g \tag{4}$$

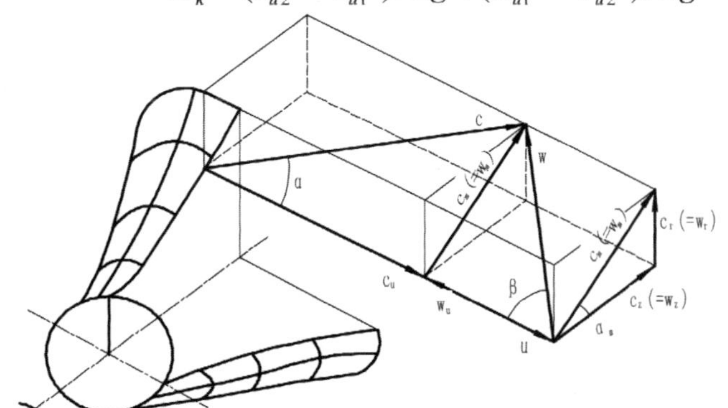

Figure 2. Motion analysis of medium within the impeller, i.e. space velocity triangles

It is not difficult to know that potential head H_p is constant and kinetic head H_k decreases with the increase of the flow rate. The inlet diameter and outlet diameter of axial-flow pump is the same ($D_1 = D_2$), the medium flows along the direction of axis, there is no radial velocity component ($v_r = 0$), therefore, the potential head H_p of axial-flow pump is zero. The law of potential head curve and kinetic head curve is shown in Fig. 3.

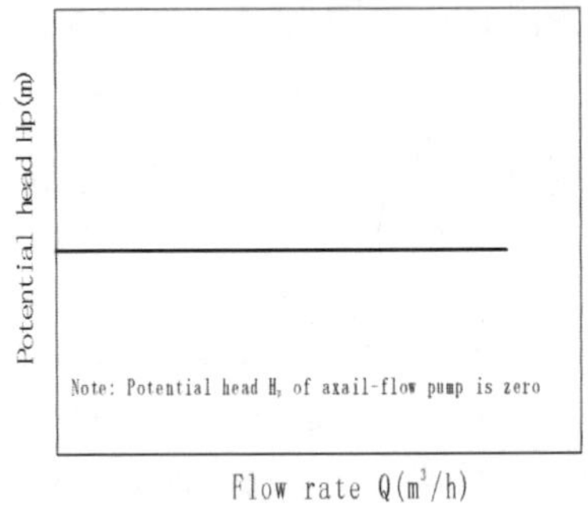

(a)
(a) Law of potential head curve

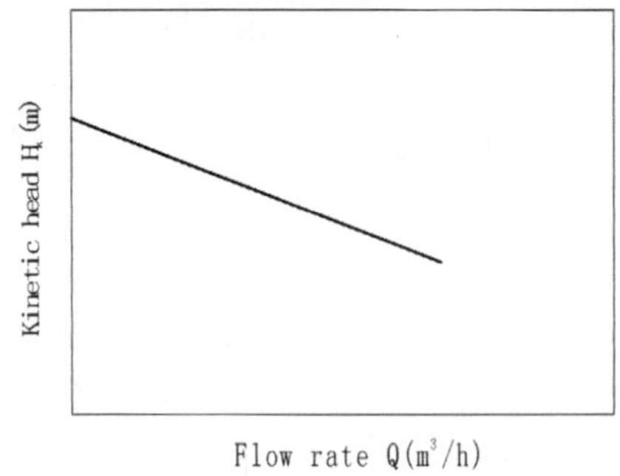

(b)
(b) Law of kinetic head curve

Figure 3. Law of head curve

3.2 Action of unit mass of the blades on the medium

3.2.1 Action of blades to which potential head corresponds If making further analysis on Coriolis force, we will get the calculation formula for the impelling force to which potential head corresponds f_p.

Because of $\mathbf{a}_{cor} = 2\boldsymbol{\omega}\times\mathbf{w}$ and $\mathbf{w} = \mathbf{w_u} + \mathbf{w_m} = \mathbf{w_u} + \mathbf{w_r} + \mathbf{w_z}$, it is not difficult to know that only $\mathbf{w}_u$ and $\mathbf{w}_r$ generates the Coriolis acceleration, that is, $\mathbf{a}_{cor} = 2\boldsymbol{\omega}\times\mathbf{w} = 2\boldsymbol{\omega}\times\mathbf{w}_u + 2\boldsymbol{\omega}\times\mathbf{w}_r = 2\boldsymbol{\omega}\times\mathbf{w}_u + 2\boldsymbol{\omega}\times\mathbf{c}_r$ (due to $\mathbf{w}_r = \mathbf{c}_r$). Therefore, only $\mathbf{w}_u$ and $\mathbf{c_r}$ generates the Coriolis force, their direction is radial and tangential, and their magnitude is $2\omega w_u$ and $2\omega c_r$, respectively. The radial Coriolis force $2\omega w_u$ has no direct relation with torque, whereas the tangential Coriolis force $2\omega c_r$ has direct relation with torque.

The constraining force (also called constraining counterforce) of the tangential Coriolis force $2\omega c_r$ is the potential impelling force f_p, i.e.

$$f_p = 2\omega c_r \tag{5}$$

The potential impelling force f_p corresponds to potential head $(u_2^2 - u_1^2)/2g$. The reason is that from blade entrance to exit edges f_p produces increment of moment of momentum, and from moment of momentum theorem (integral formulation), the increment is

$$\int_{t1}^{t2} mf_u r dt = m\omega\int_{t1}^{t2} 2c_r \cdot rdt = m\omega\int_{r1}^{r2} 2rdr = m(u_2 r_2 - u_1 r_1)$$

It's easy to know that the increment of the moment of momentum is the change of moment of momentum corresponding to u_2 (at radius r_2) and u_1 (at radius r_1), and u_2 (at radius r_2) and u_1 (at radius r_1) correspond to potential head $(u_2^2 - u_1^2)/2g$.

Note: ①the study object here is free body, as shown in Fig. 4(a); ②$c_r dt = dr$; ③ $u_2 = \omega r_2$, $u_1 = \omega r_1$; ④where m is the mass within the free body.

When the medium has no displacement in radial direction, i.e. $c_r = 0$, then $f_p = 0$, and its corresponding energy input is also zero, only when the medium has radial displacement, $c_r \neq 0$, then $f_p \neq 0$, and it have corresponding energy input. Again, in axial flow pump, $(u_2^2 - u_1^2)/2g = 0$, it happens that $c_r = 0$, i.e. $f_p = 0$.

3.2.2 Impelling force to which kinetic head corresponds From Equation (4), the kinetic impelling force

f_k corresponds to kinetic head H_k includes f_{cu} (corresponds to $(c_{u2}{}^2 - c_{u1}{}^2)/2g$) and f_{wu} (corresponds to $(w_{u1}{}^2 - w_{u2}{}^2)/2g$), i.e.

$$f_k = f_{cu} + f_{wu} \tag{6}$$

According to the characteristics of kinetic impelling force, when applying the moment of momentum theorem to obtain f_{cu}, corresponding process is needed. ①Take the control body of the infinite thin annulus as the study object, as shown in Fig. 4(b); ②conventionally, take $c_{u1} = 0$. ③In the classical mechanics, the change in momentum is $m(\mathbf{v}_2 - \mathbf{v}_1)$ or $m\Delta\mathbf{v}$, that is, m is constant, while velocity $\mathbf{v}$ changes. But in the derivation herein, the change of momentum can be considered as $\mathbf{c}_u \Delta m_t$, that is, $\mathbf{c}_u$ is constant ($\mathbf{c}_u$ is regarded as constant when the annulus is developed, similar to the processing on axial flow impeller developed), while Δm_t refers to accumulative mass of medium when it has c_u within the time of dt. From the moment of momentum theorem, then

$$mf_{cu} \cdot rdt = c_u \Delta m_t \cdot r \cdot c_u$$

where r is infinite thin annulus (the control body) radius, c_u is tangential component of absolute velocity c (at radius r), m is actual mass within the control body, and Δm_t is accumulative mass, means the product of number of revolutions of the control body within the time of dt and the actual mass m, i.e. $\Delta m_t = m \cdot \frac{\omega}{2\pi} dt$. Therefore, it is obtained:

$$f_{cu} = \frac{1}{2\pi} \omega c_u \tag{7}$$

Similarly, from the moment of momentum theorem, the calculation formula for f_{wu} can be obtained as follows:

$$f_{wu} \propto -\frac{1}{2\pi} \omega w_u \tag{8}$$

If comparing the calculation formulas of f_{cu} and f_p, an interesting phenomenon can be found: When the centrifugal pump and the mixed flow pump run at the rated flow rate, the value of c_u in the impeller discharge is approximately 10 times of that of c_r (c_r is different from c_m), whereas the ratio of coefficients in the formula of f_{cu} and f_p is 1/(4π), that is, In the impeller discharge, the value of f_{cu} is close to that of f_p. This indicates the value of f_{cu} is not divorced from reality.

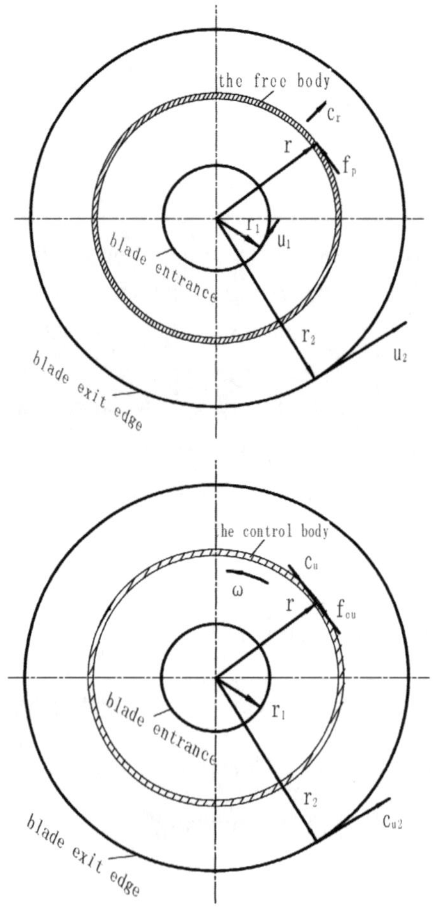

(a) (b)

(a) The free body of action of blades to which potential head corresponds

(b) The controlled body of action of blades to which kinetic head corresponds

Figure 4. Study object for solving the action of blades on medium

3.2.3 Discussion and analysis The two typical kinds of motion in fluid mechanics are shown as in Fig. 5(a) and 5(b).

Due to the impelling force on the medium, the flow within the impeller is a complex non-inertial motion.

The motion to which potential head corresponds within the impeller is analogous to rotary motion as shown in Fig. 5(a).

When the impeller is developed (similar to the processing on axial flow impeller developed), the motion to which kinetic head corresponds within impeller is analogous to accelerating linear motion as shown in Fig. 5(b), the pressure difference between leading face and back face of blades is similar to the pressure difference between two wall faces of the vessel as shown in Fig. 5(b). Note: ①according to the cascade theory, the developed shape of blades of axial-flow pump is cascade, in fact, the motion of cascade after development is analogous to accelerating linear motion; ② Annular cascade of centrifugal pump is changed to cascade by conformal mapping in literature [8].

The non-inertial motion within the impeller contains important scientific problems in the aspects of fluid mechanics, rotodynamic pump, etc.

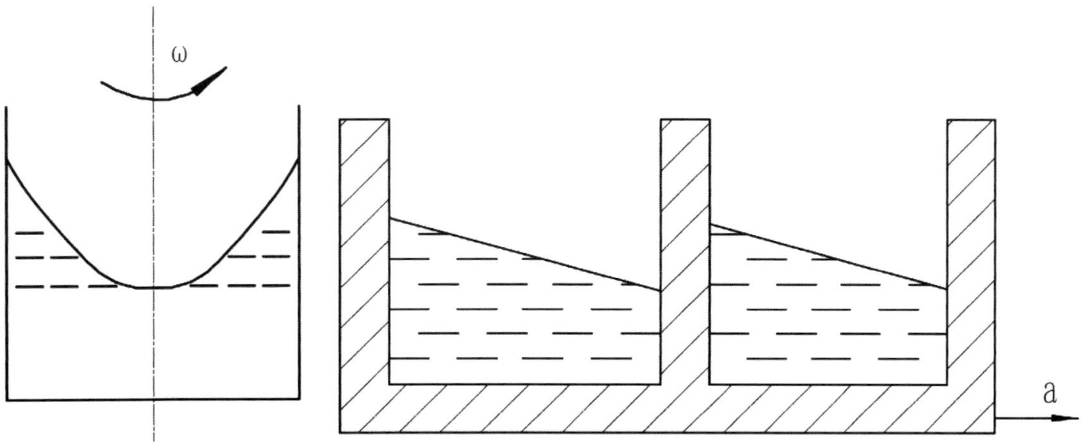

(a)Rotary motion around vertical shaft (b) Accelerating linear motion

Figure 5. Two typical kinds of motion in fluid mechanics

3.3 Input power and energy conversion mechanism

3.3.1 Input power and output power to which potential head corresponds Because the impelling force of blades f_p to which the potential head corresponds is $2\omega c_r$, and v_r is proportional to flow rate Q, therefore, it is not difficult to conclude that the input power to which potential head corresponds is proportional to flow rate. Through further derivation, the calculation formula for input power to which potential head corresponds is obtained, as follows:

$$N_p = K_p Q \tag{9}$$

Where $K_p > 0$, and relates to factors such as geometric parameters of impeller, pump speed, etc.

On the other hand, the output power to which potential head corresponds is apparently proportional to flow rate. The law of input power and output power to which potential head corresponds is shown in Fig. 6(a).

3.3.2 Input power and output power to which kinetic head corresponds According to Equation (7) and Equation (8), the impelling force of blades $f_k\left(= f_{Cu} + f_{Wu}\right)$ to which kinetic head corresponds is proportional to C_u and is inversely proportional to W_u. On the other head, it is easy to know: ① C_u is inversely proportional to flow rate Q, and W_u is proportional to flow rate Q; ② when Q=0, C_u rises to maximum value $\left(= u\right)$, and W_u drops to minimum value (=0). Therefore, input power to which kinetic head corresponds is inversely proportional to flow rate Q. Through further derivation, the calculation formula for input power to which kinetic head corresponds is obtained, as follows:

$$N_k = Const - K_k Q \tag{10}$$

Where $K_k > 0$, and relates to factors such as geometric parameter of impeller, pump speed, ratio of actual flow rate to rated flow rate, area ratio, etc.

The output power to which kinetic head corresponds is apparently proportional to flow rate Q. The law of input power and output power to which kinetic head corresponds is shown in Fig. 6(b).

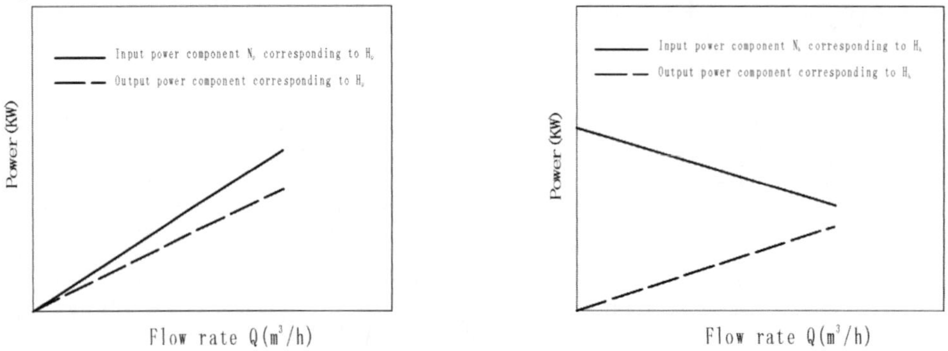

(a) (b)

(a) Law of input power and output power component corresponding to potential head H_p

(b) Law of input power and output power t component corresponding to kinetic head H_k

Figure 6. Component of input power and component of output power

4. Analysis of shaft power curve of rotodynamic pump

Shaft power of rotodynamic pump is composed of two kinds of power components to which potential head and kinetic head correspond, and the power component to which potential head corresponds increases with the increase of flow rate, the increasing amplitude decreases with the increase of specific speed n_s (because the D_2/D_1 decreases with the increase specific speed n_s). The power component of this term in axial-flow pump decreases to zero. The power component to which kinetic head corresponds decreases with the increase of flow rate. Thus it can reasonably be explained that the law which the input power curves of centrifugal pump, mixed-flow pump and axial-flow pump present the law of ascending, approaching to level and descending, respectively. As shown in Fig. 7(a), (b) and (c).

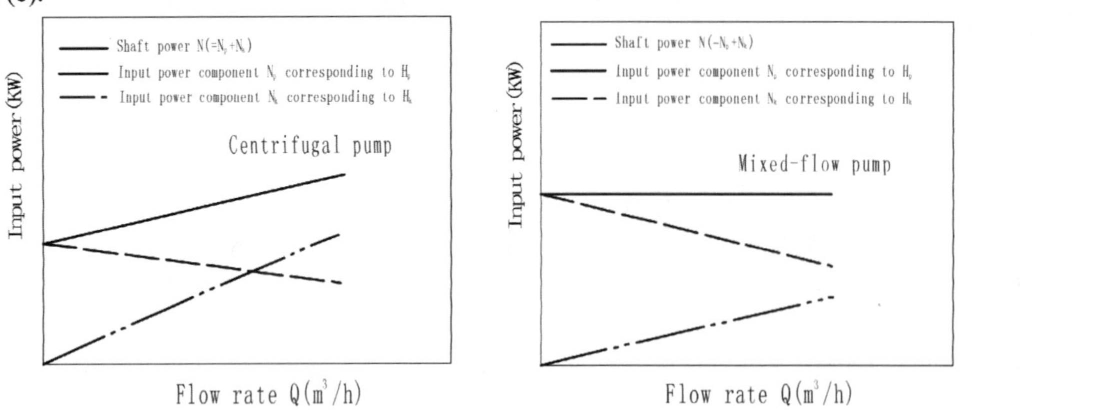

(a) (b)

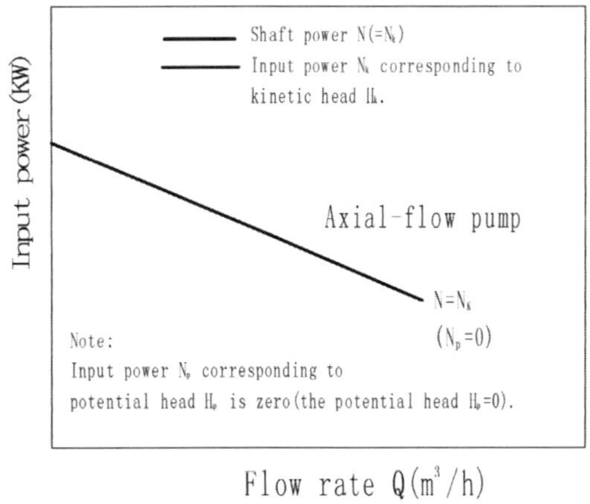

(c)

(a) Analysis of shaft power curve of centrifugal pump (b) Analysis of shaft power curve of mixed-flow pump

(c) Analysis of shaft power curve of axial-flow pump

Figure7. Analysis of shaft power curve of rotodynamic pump

5. Non-overload design of centrifugal pump

In normal case, the shaft power curve of centrifugal pump rises continuously with the increase of flow rate, therefore, it is easy to generate overload in large-capacity zone, and even burn out the prime mover. In some cases, in order to guarantee safe and reliable running of centrifugal pump, the centrifugal pump must be provided with non-overload feature, i.e., the input power curve of centrifugal pump has maximum value, thus no overload problem will be generated over the entire flow rate range.

5.1 Equivalent design factor of non-overload centrifugal pump

From Equation (9) and (10), the shaft power N is $N_p + N_k = K_p Q + (Const - K_k Q)$, i.e.

$$N = Const + \left(K_p - K_k\right)Q \tag{11}$$

In general case, $K_p > K_k$ for centrifugal pump, hence, it cannot realize non-overload feature. However, K_p and K_k are the design factors which depend on multiple aspects of hydraulic model, by some pertinent designs, the change law of shaft power curve can be controlled over a certain flow rate, the rise amplitude of power component of shaft power to which potential head corresponds (K_p) is less than the drop amplitude of power component of shaft power to which kinetic head corresponds (K_k), so as to achieve maximum value of shaft power at a certain flowrate. Thereby, the equivalent design factor for non-overload centrifugal pump obtained is

$$K_p < K_k \tag{12}$$

5.2 Relevant design factors of non-overload centrifugal pump

The hydraulic model of centrifugal pump involves multiple design factors, likewise, the equivalent design factor of non-overload centrifugal pump $K_p < K_k$ also relates to multiple design factors, e.g., the outlet angles of leading face and back face of blades, wrap angle, number of blades, outlet width, the area ratio, etc. Moreover, for a certain centrifugal pump, K_2 has a slight increase as the flow rate increases, therefore, another feature of non-overload centrifugal pump is that maximum value of input power can only appear at a relatively large flow rate point.

5.3 Design example

In a recent subject research, the authors have designed high-efficiency non-overload low specific speed centrifugal pump. By optimizing multiple design factors of hydraulic model (the outlet angles of leading face is 11.5°, the outlet angles of back face is 15.5°, wrap angle is 198°, number of blades is 4, outlet width is 9.5mm, the area ratio is 1.1), the equivalent design factor of non-overload centrifugal pump is satisfied, the shaft power has maximum value, and ideal result is obtained. The centrifugal pump with low specific speed $n_s = 62$, various performance indexes of its prototype (non-overload feature, efficiency, cavitation performance, etc) reach and even are better than the target requirements, among them, the non-overload feature is good, the maximum value of shaft power is 1.1 times of rated shaft power, and the flow rate at maximum shaft power is 1.5 times of rated flow rate, as shown in Fig. 8.

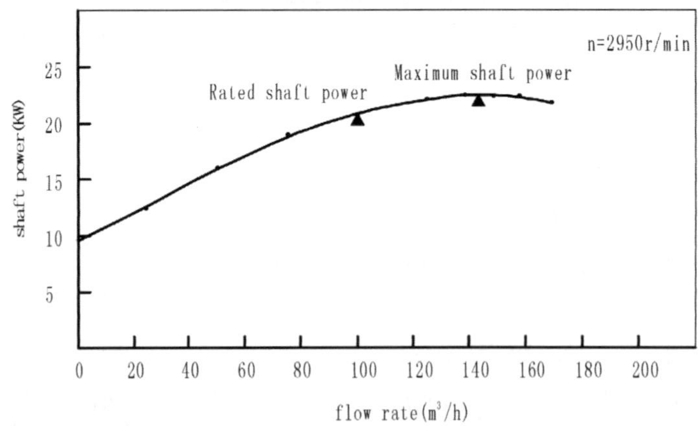

Figure 8. Non-overload power curve of low specific speed centrifugal pump

6. Conclusion

(1) The calculation expression for the intrinsic law of input power which has not yet been given by current theory of Rotodynamic pump;

(2) It is very appropriate that $(w_{u1}^2 - w_{u2}^2)/2g$ is classified into kinetic head, then potential head $H_p = (u_2^2 - u_1^2)/2g$, and kinetic head $H_k = (c_{u2}^2 - c_{u1}^2)/2g + (w_{u1}^2 - w_{u2}^2)/2g$;

(3) The potential impelling force f_p corresponds to potential head H_p, and $f_p = 2\omega c_r$ i.e. the constraining force (also called constraining counterforce) of the tangential Coriolis force;

(4) The kinetic impelling force f_k corresponds to kinetic head H_k includes f_{cu}(corresponds to $(c_{u2}^2 - c_{u1}^2)/2g$) and f_{wu} (corresponds to $(w_{u1}^2 - w_{u2}^2)/2g$). i.e. $f_k = f_{cu} + f_{wu}$, and $f_{cu} = \frac{1}{2\pi}\omega c_u$, $f_{wu} \propto -\frac{1}{2\pi}\omega w_u$;

(5) The input power component to which potential head corresponds is proportional to flow rate Q, and $N_p = K_p Q$;

(6) The input power component to which kinetic head corresponds is inversely proportional to flow

rate Q, and $N_k = Const - K_k Q$;

(7) The shaft power of rotodynamic pump is composed of two kinds of power components to which potential head and kinetic head correspond, i.e. shaft power $N = Const + \left(K_p - K_k\right)Q$,and the power component to which potential head corresponds increases with the increase of flow rate, the increasing amplitude decreases with the increase of specific speed n_s (the power component of this term in axial-flow pump decreases to zero) , the power component to which kinetic head corresponds decreases with the increase of flow rate. Thus it can reasonably be explained that the law which the input power curves of centrifugal pump, mixed-flow pump and axial-flow pump present the law of ascending, approaching to level and descending, respectively;

(8) The equivalent design factor for non-overload centrifugal pump obtained is $K_p < K_k$,and the equivalent design factor correlates with the outlet angle of leading face and back face of the blade, wrap angle, number of blades, outlet width, area ratio, and the ratio of operating flow rate to specified flow rate and so on;

(9) The authors have designed high-efficiency non-overload low specific speed centrifugal pump, and the non-overload feature is good, the maximum value of shaft power is 1.1 times of rated shaft power.

IOP Conf. Series: Materials Science and Engineering **129** (2016) 012003 doi:10.1088/1757-899X/129/1/012003

Nomenclature

$\mathbf{a}_{cor}$	=	Coriolis acceleration, $\mathbf{a}_{cor} = 2\boldsymbol{\omega}\times\mathbf{w} = 2\boldsymbol{\omega}\times\mathbf{w}_u + 2\boldsymbol{\omega}\times\mathbf{w}_r = 2\boldsymbol{\omega}\times\mathbf{w}_u + 2\boldsymbol{\omega}\times\mathbf{c}_r$
$\mathbf{c}$, c	=	absolute velocity of flow, $\mathbf{c} = \mathbf{u} + \mathbf{w}$, $\mathbf{c} = \mathbf{c}_u + \mathbf{c}_m = \mathbf{c}_u + \mathbf{c}_r + \mathbf{c}_z$
D	=	impeller diameter
f	=	impelling force of blade (per unit mass), $f = f_p + (f_{cu} + f_{wu})$
$f_{cu} + f_{wu}$	=	impelling force of kinetic head of blade, f_{cu} corresponds to $(c_{u2}^2 - c_{u1}^2)/2g$, f_{wu} corresponds to $(w_{u1}^2 - w_{u2}^2)/2g$
f_p	=	impelling force of potential head of blade, f_p corresponds to potential head $(u_2^2 - u_1^2)/2g$
H	=	head
m	=	(actual) mass within the free body or the control body of the infinite thin annulus
Δm_t	=	accumulative mass, means the product of number of revolutions of the control body within the time of dt and the actual mass m, i.e. $\Delta m_t = m\cdot\frac{\omega}{2\pi}dt$
N	=	input power, $N = N_p + (N_{cu} + N_{wu})$
$N_{cu} + N_{wu}$	=	N_{cu} corresponds to f_{cu}, N_{wu} corresponds to f_{wu}
N_p	=	corresponding to f_p
Q	=	flow rate
r	=	radius
u	=	peripheral velocity of impeller, $u = \omega r$
w	=	relative velocity of flow, $\mathbf{w} = \mathbf{w}_u + \mathbf{w}_m = \mathbf{w}_u + \mathbf{w}_r + \mathbf{w}_z$, $\mathbf{w}_r = \mathbf{c}_r$, $\mathbf{w}_z = \mathbf{c}_z$
$\boldsymbol{\omega}$, ω	=	angular velocity of impeller

Subscript

1	=	blade entrance
2	=	blade exit edge
k	=	corresponding to kinetic head
p	=	corresponding to potential head
m	=	meridional direction (component of velocity)
r	=	radial direction (component of velocity)

th	=	theoretical
u	=	tangential direction (component of velocity)
z	=	axial direction (component of velocity)

IOP Conf. Series: Materials Science and Engineering **129** (2016) 012003 doi:10.1088/1757-899X/129/1/012003

References

[1] Stepanoff A J 1957 *Centrifugal and Axial Flow Pumps* (Wiley, New York, ed.2)

[2] Karassik I J, et al. 1976 *Pump Handbook* (McGraw-Hill, New York)

[3] ЛОМАКИН А А 1966 ЦЕНТРОЬЕЖНЫЕ И ОСЕВЫЕ НАСОСЫ (ИЗДАНИЕ ВТОРОЕ, ПЕРЕРАБОТАННОЕ И ДОПОЛНЕННОЕ ИЗДАТЕЛЬСТВО " МАШИНОСТРОЕНИЕ "МОСКВАЛЕНИНРРАД)

[4] Guan X F 1995 *Technical Manual of Modern Pumps* (China Astronautics Publishing House, Bejing)

[5] Pfleiderer C 1961 *Die Kreiselpumpen* (Springer-Verlag, Berlin)

[6] Hellmuth S 1977 *Die Pumpen: Arbeitsweise, Berechnung, Konstruktion* (Springer-Verlag)

[7] Troskolanski A T Tazarkiewicz S 1976 *Kreiselpumpen Berechnung und Konstruktion* (Birkhauser Verlag, Basel Und Stuttgart)

[8] Ohashi H 1968 *"Analytical and Experimental Study of Dynamic Characteristics of Turbopumps"* NASA TN D-4298

Numerical analysis of flow interaction of turbine system in two-stage turbocharger of internal combustion engine

Y B Liu[1,3], W L Zhuge[1], Y J Zhang[11], S Y Zhang[2]

1 State Key Laboratory of Automotive Safety and Energy, Tsinghua University Beijing, 100084, China
2 National Key Laboratory of Diesel Engine Turbocharging Technology Tianjin, 037036, China
3 Department of Mechanical Engineering, Academy of Armored Force Engineering Beijing, 100072, China

Abstract: To reach the goal of energy conservation and emission reduction, high intake pressure is needed to meet the demand of high power density and high EGR rate for internal combustion engine. Present power density of diesel engine has reached 90KW/L and intake pressure ratio needed is over 5. Two-stage turbocharging system is an effective way to realize high compression ratio. Because turbocharging system compression work derives from exhaust gas energy. Efficiency of exhaust gas energy influenced by design and matching of turbine system is important to performance of high supercharging engine. Conventional turbine system is assembled by single-stage turbocharger turbines and turbine matching is based on turbine MAP measured on test rig. Flow between turbine system is assumed uniform and value of outlet physical quantities of turbine are regarded as the same as ambient value. However, there are three-dimension flow field distortion and outlet physical quantities value change which will influence performance of turbine system as were demonstrated by some studies. For engine equipped with two-stage turbocharging system, optimization of turbine system design will increase efficiency of exhaust gas energy and thereby increase engine power density. However flow interaction of turbine system will change flow in turbine and influence turbine performance. To recognize the interaction characteristics between high pressure turbine and low pressure turbine, flow in turbine system is modeled and simulated numerically. The calculation results suggested that static pressure field at inlet to low pressure turbine increases back pressure of high pressure turbine, however efficiency of high pressure turbine changes little; distorted velocity field at outlet to high pressure turbine results in swirl at inlet to low pressure turbine. Clockwise swirl results in large negative angle of attack at inlet to rotor which causes flow loss in turbine impeller passages and decreases turbine efficiency. However negative angle of attack decreases when inlet swirl is anti-clockwise and efficiency of low pressure turbine can be increased by 3% compared to inlet condition of clockwise swirl. Consequently flow simulation and analysis are able to aid in figuring out interaction mechanism of turbine system and optimizing turbine system design.

[1] **Corresponding author: Zhang Y J, Professor, yjzhang@tsinghua.edu.cn, +86-010-62797359**

1. Introduction

Energy crisis and CO2 emission problem demands engine raising power density to realize downsizing, and charging system is an effective way to raise power density [1-3]. Besides, high combustion temperature of charged internal combustion engine brings about significant formation of nitrogen oxides [4, 5]. So it is necessary to use high EGR rate to avoid the shortages above. High inlet pressure is needed to meet the demand of high power density and high EGR rate. Present power density of diesel engine has reached 90KW/L and intake pressure ratio needed is over 5[6]. Single-stage turbocharger can realize high pressure ratio, however it will result in flow range of turbocharger narrowed and reliability worsen. Therefore, how to raise engine performance in whole operational conditions becomes the key issue of high charging engine. Reaching the same total intake pressure ratio, pressure ratio of each turbocharger in two-stage turbocharger is lower than single-stage turbocharger. So two-stage turbocharger can broaden flow range and decrease rotation speed of turbocharger to ensure operational reliability.

High pressure ratio will result in compression work increased. When intake pressure ratio reaches 5, ratio of compression work to engine output power is close to 50% [7]. To turbocharging system compression work derives from exhaust gas energy. So efficiency of exhaust gas energy influenced by design and matching of turbine system is important to performance of high charging engine. Conventional turbine system is assembled by single-stage turbocharger turbines and turbine matching is based on turbine MAP measured on test rig. Flow between turbine system is assumed uniform and value of outlet physical quantities of turbine are regarded as the same as ambient value. However, there are three-dimension flow field distortion and outlet physical quantities value change which will influence performance of turbine system as were demonstrated by some studies [8-11]. But because lack of deep analysis for mechanism of interaction between turbine systems, influence caused by flow distortion cannot be considered in process of turbine system design and matching. So in the paper, mechanism of interaction between turbine systems is analyzed by building turbine system simulation model. Based on analysis, flow field matching method can be raised to increase efficiency of turbine system.

2. Models and method

The two-stage turbocharger simulated is based on a type of diesel engine designed for high power density. Because it is difficult to build turbine system model as a whole, calculation model is divided into two parts: HP module and LP module as is shown in figure 1. HP module includes turbine rotor and connection pipe to simulate flow in turbine. LP module includes volute and rotor. To both models, flow control equation uses Reynolds-Averaged Navier-Stokes Equations solved by NUMECA code; turbulence model adopts S-A model which is verified appropriate to simulate turbulent flow in turbine rotor [12]. The flow solver is based on a cell centred finite volume approach, associated with the Jameson's center scheme as space discretization. For HP module, four parameters are given to inlet boundary condition: flow mass rate in accordance with engine operational condition; velocities in three dimensions making angle of attack -20° ; static pressure calculated according to LP module is given to outlet boundary condition. For LP module, flow mass rate and velocities in three dimensions get from HP module are given to inlet boundary condition; and outlet boundary condition of LP module is given static pressure. Rotation speeds of two turbine rotors are given according to actual values measured on two-stage test rig.

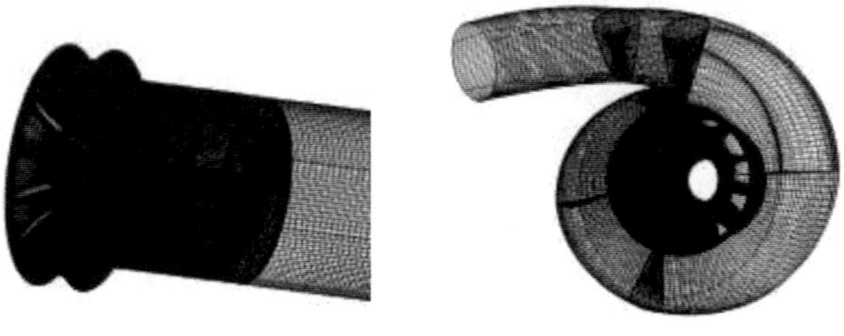

(a) HP module (b) LP module

Figure 1. Simulation model of two-stage turbocharger

In process of calculation of single simulation module, boundary conditions of calculation region are fixed and flow parameters of two modules cannot be exchanged. After convergence is finished, inlet flow field of upstream module is different from outlet flow field of downstream module. So flow in two modules is different from each other which results in simulation credibility decreasing. Therefore, it is necessary to add a module between two calculation modules to transfer flow information. Based on transfer module, outlet boundary conditions of HP module and inlet boundary conditions of LP module are given. The boundary conditions can not only reflect flow interaction between parts but also meet the characteristic compatibility condition for numerical calculation. In transfer module, flow mass rate, static temperature and velocity direction are transferred from upstream to downstream; static pressure is transferred from downstream to upstream. Flow transfer module is expressed by the equations listed below:

$$T_d = \frac{\sum T_u * den}{\sum den} \tag{1}$$

$$\dot{G}_d = \dot{G}_u \tag{2}$$

$$\frac{V_{dx}}{|V_d|} = \frac{V_{ux}}{|V_u|} \tag{3}$$

$$\frac{V_{dy}}{|V_d|} = \frac{V_{uy}}{|V_u|} \tag{4}$$

$$\frac{V_{dz}}{|V_d|} = \frac{V_{uz}}{|V_u|} \tag{5}$$

$$\overline{|V|} = \frac{\sum V * den}{\sum den} \tag{6}$$

$$p_u = \frac{\sum p_d * den}{\sum den} \tag{7}$$

The calculation process of turbine system model is showed in figure 2. Based on boundary conditions, calculation is performed in each module. Then flow characters are exchanged between two modules. By iterating calculation, convergence can be regarded finished when

$$p_u^{n+1} = p_d^n \tag{8}$$

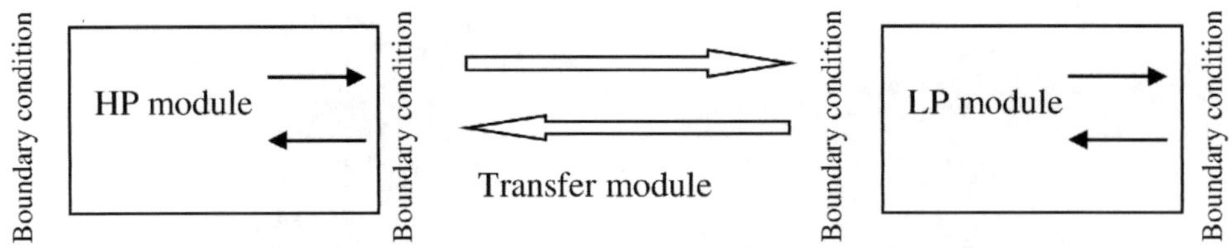

Figure 2. Calculation process of turbine system model

Results and discussion

2.1. analysis of inlet static pressure field of LP module

Because LP module is at downstream, performance of HP module is mainly influenced by static pressure field of inlet to LP module. By numeral calculation inlet static pressure field is simulated. Boundary conditions are shown in table 1. In addition, inlet flow is assumed as uniform.

Table 1. Boundary conditions of LP module

Case number	Flow mass rate(kg/s)	Outlet static pressure(Pa)	Rotation speed(r/min)
1	0.4	101300	75000
2	0.5	101300	80000
3	0.6	101300	85000

Static pressure field is similar among conditions of different inlet flow mass. So the paper analyzed pressure field when flow mass rate is 0.5kg/s. The inlet pressure field is shown in figure 3.

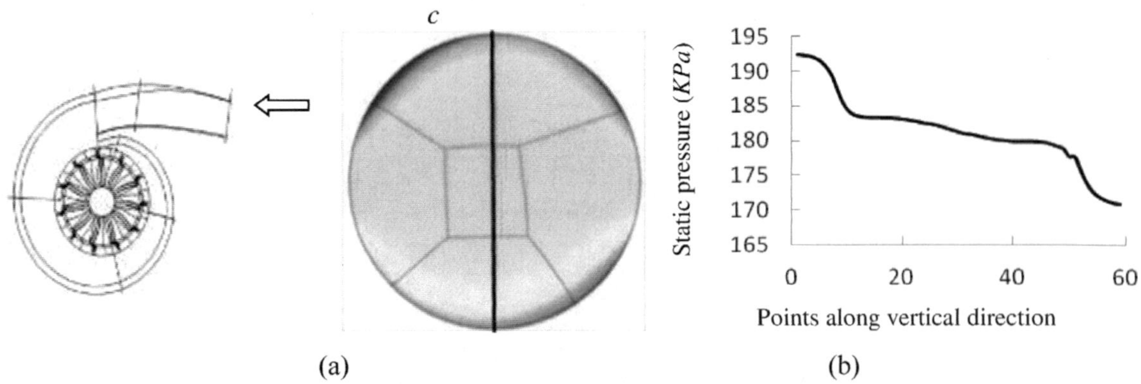

Figure 3. Inlet static pressure field of LP module inlet

In figures above, (a) shows distortion of static pressure field on inlet plane of LP module; (b) is the chart showing law of pressure value along line c from top to bottom in (a).

Figure 3(b) illustrates that static pressure decreases from 195KPa to 170KPa along radial direction on inlet plane. Distortion is large in area close to volute top and bottom; however in middle area pressure field is approximate to uniformity. Besides, static pressure mean value is about 180KPa far greater than ambient pressure value. So outlet pressure of HP turbine in two-stage turbocharger is different from single-stage turbine which will influence flow in rotor and cause performance of turbine changed.

In two-stage turbocharging system, there is connection pipe between HP turbine and LP turbine. Distortion changes in process of pressure field transfers from downstream to upstream. So simulation is done to get static pressure field developing process shown in figure 4.

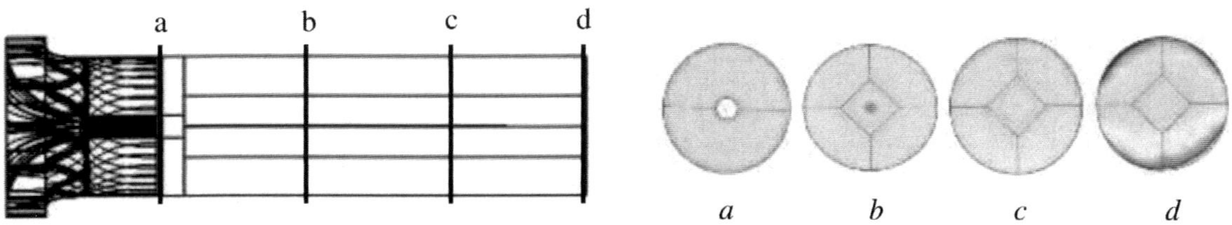

Figure 4. Pressure field in connection pipe

According to figure 4, we can conclude that distortion of pressure field gradually disappears when closing to HP turbine. At outlet to HP turbine static pressure field is uniform. So outlet pressure field of HP module can be regarded as uniform. Moreover, pressure mean values in different planes are almost the same and larger than ambient pressure.

2.2. Analysis of outlet flow field of HP module

By simulation, flow field at outlet to HP module is investigated. Boundary conditions are shown in table 2. In addition, inlet velocities of rotor are given assumed angle of attack is -20° .

Table 2. boundary conditions of HP turbine model

Case number	Flow mass rate(kg/s)	Outlet static pressure(Pa)	Rotation speed(r/min)
1	0.4	150000	80000
2	0.5	170000	85000
3	0.6	190000	87000

Outlet flow field includes pressure field, temperature field and velocity field. Degrees of distortion of pressure and temperature are evaluated by equation following and shown in table 3.

$$\varphi = \frac{V_{max} - V_{min}}{V_{mean}} \tag{9}$$

Table 3. degree of distortion of pressure field and temperature field

Case number	Flow mass(kg/s)	Pressure field (%)	Temperature field (%)
1	0.4	2.1	1.58
2	0.5	2.34	1.74
3	0.6	2.5	1.97

Table 3 illustrates that degrees of distortion of pressure field and temperature field are both below 2%, and pressure and temperature can be regarded uniform at inlet to LP module.

Outlet velocity triangle of turbine rotor analysis shows that outlet velocity has not only axial component but also secondary flow component in plane vertical to axial direction [13] .To turbine rotor, direction of blade curved is opposite to rotor rotation direction to raise turbine efficiency. So relative velocity at outlet inclined and thereby absolute velocity inclined to direction opposite to rotation speed forming outlet swirl. In accordance to velocity triangle analysis and simulation result, swirls at outlet to HP module of different planes are shown in figure 5.

Plane1 Plane2

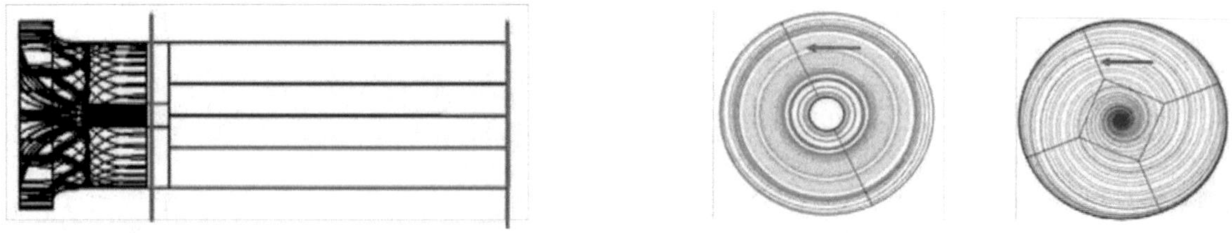

Figure 5. Swirls in different planes

Swirl strength is defined as

$$\emptyset = \frac{v_s}{v_a} \tag{10}$$

Contour of swirl strength for plane 1 in figure 5 is drawn in figure 6 in which positive sign means that direction of swirl is opposite to rotation direction. In plane1, swirl strength increases first and decreases then from rotor hub to rotor shroud. This is because gas expands more fully at hub than at shroud which brings about magnitude of velocity is larger at hub and so does axial component of velocity $\mathbf{v_a}$. Meanwhile, exit angle of blade changes from 24° to 48° with blade height increasing, which results in tangential component of relative velocity increasing with blade height. However, with blade height increasing convected velocity increases which causes tangential component of absolute velocity $\mathbf{v_s}$ goes up first and down then from hub to shroud.

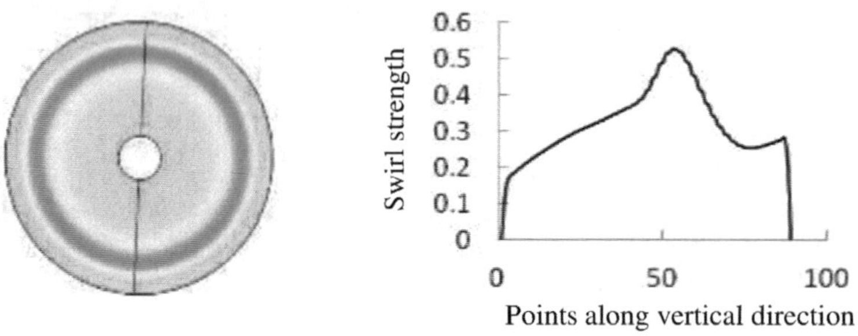

Figure 6. Swirl strength at outlet to HP turbine

2.3. analysis of interaction of LP module to HP module

In two-stage turbocharger, back pressure of HP turbine is higher than that of single-stage turbine because of LP turbine. Greater is expansion ratio of LP turbine, higher is back pressure of HP turbine. To investigate law and mechanism of turbine performance changing with outlet static pressure, flow in rotor is simulated based on HP module. Boundary conditions and rotation speed for calculation is listed in table 4. When flow mass is constant, rotation speed is assumed the same in condition of different back pressure.

Table 4. boundary condition of high pressure turbine model

Case number	Flow mass(kg/s)	Outlet static pressure(KPa)	Rotation speed(r/min)
		100	
1	0.4	190	75000
		250	
		100	
2	0.5	190	82000
		250	
		100	
3	0.6	190	85000
		250	

According to calculation, characteristics of flow capacity and efficiency of turbine in different flow mass and back pressure are shown in figure 7. In figure7 (a), points represent characteristic of turbine flow capacity in condition of different back pressure are close to a line. So it can be deduced that outlet back pressure has little influence on characteristic of turbine flow capacity. Figure 7(b) illustrates that efficiency of turbine in condition of different back pressure are close to a line too. So static pressure at outlet to turbine will not change flow in turbine, and in two-stage turbine system low pressure turbine has little influence on high pressure turbine performance.

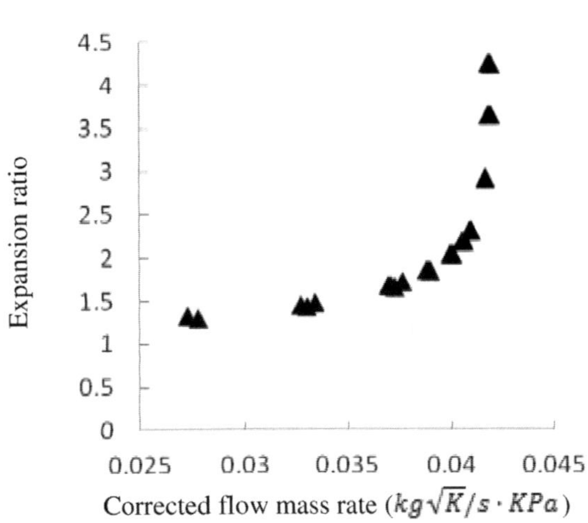

(a) Characteristics of flow capacity

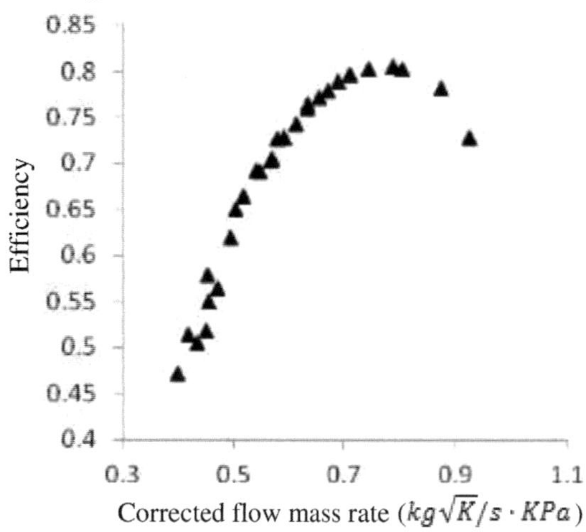

(b) Characteristics of efficiency

Figure 7. Performance of turbine

2.4. Analysis of interaction of HP turbine to LP turbine

Swirl at outlet to HP module causes non-uniform velocity field at inlet to LP module. According to analysis of outlet swirl of HP turbine, operational condition, boundary condition and selecting of two turbines will influence rotation direction and strength of outlet swirl. So inlet swirl of LP module may have different direction and strength. Different inlet condition has different influence on turbine performance. To analyze interaction HP turbine on LP turbine, swirls in clockwise and anti-clockwise direction and with different strength are given to inlet to LP module. Simulation is done to study mechanism of influence inlet condition on turbine. Boundary conditions and rotation speed for calculation is listed in table 5. When flow mass rate is constant, rotation speed is assumed the same in condition of different inlet velocity field.

Table 5. boundary condition of LP turbine model

Case number	Flow mass(kg/s)	Rotation speed(r/min)	Swirl strength
1	0.4	50000	0
			0.173
			0.337
			0.514
2	0.5	55000	0
			0.173
			0.337
			0.514
3	0.6	65000	0
			0.173
			0.337
			0.514

Figure 8 illustrates performance of turbine in different operational condition and in different condition of inlet swirls.

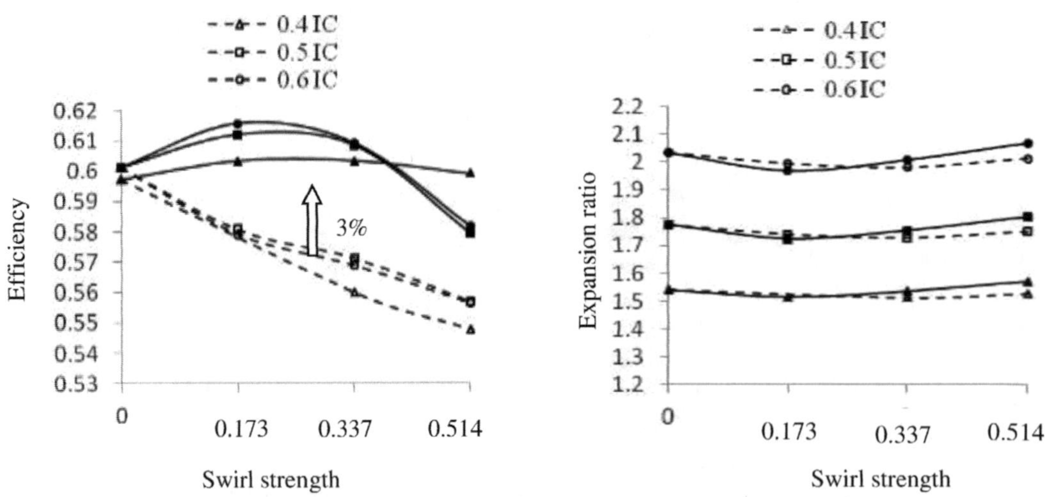

(a) Characteristics of efficiency (b) Characteristics of flow capacity

Figure 8. Performance of turbine

In condition of different inlet swirls, total pressure and total temperature are almost the same when flow mass rate is constant. So flow mass rate can be used to reflect operation condition. In figures above 0.4, 0.5 and 0.6 refer to flow mass rate. X axis is swirl strength and Y axis refers to efficiency of turbine in figure 8(a) and expansion ratio in figure 8(b). It can be deduced that turbine efficiency decreases with swirl strength increasing when flow mass is constant and direction of swirl is clockwise; however, when direction of swirl is anti-clockwise turbine efficiency goes up slightly and down then. Besides, turbine efficiency is higher when direction of swirl is anti-clockwise. Efficiency difference between two inlet swirl directions is about 3% when swirl strength is constant. Because total pressure is almost the same in different inlet condition, expansion ratio almost keeps constant. Taking flow mass rate 0.6kg/s as example, flow in turbine is investigated to study influence inlet swirl on turbine performance.

3.4.1 Flow analysis in volute. Flow in volute influences inlet velocity field of rotor and flow in rotor further. So it is necessary to investigate inlet swirl development in volute. Swirl at inlet is kept and transferred downstream. Because flow in volute is influenced by radial centrifugal force, counter-rotation swirl is formed in volute when inlet velocity field is uniform. When inlet velocity field is non-uniform, inlet swirl will be combined with counter-rotation swirl. At sections close to volute tongue, centrifugal force is small and inlet swirl dominates flow structure. Figure 9 shows outlet angle of volute at section close to volute tongue in three different inlet conditions. Outlet angle in condition of inlet clockwise swirl is about 30° close to rotor hub while close to shroud angle is about 80°. In condition of inlet anti-clockwise swirl, outlet angle is in contrast to the inlet condition of clockwise swirl.

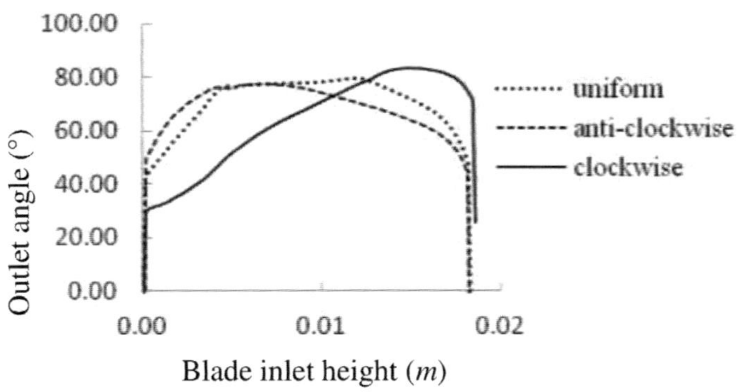

Figure 9. Outlet angle of volute in different inlet conditions

At sections far away from tongue, inlet swirl dissipates and flow structure mainly is counter-rotation swirl. Flow structure is almost the same even if inlet condition is different. So influence inlet swirl on turbine mainly concentrates in sections close to volute tongue.

3.4.2 Flow analysis in rotor. Small incidence angle of rotor means low tangential velocity and large radial velocity which results in great negative angle of attack as is shown in figure 10. Assuming convected velocity u1 and value of absolute c1 is constant, angle of attack of rotor θ is decided by incidence angle θ'[14].

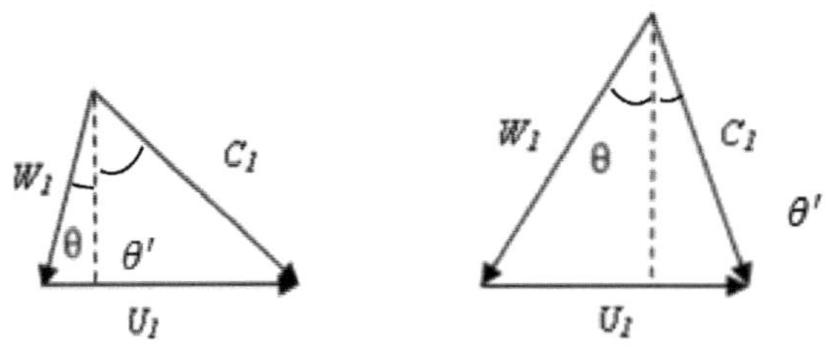

Figure 10. Inlet velocity triangle

So from figure 10 it can be deduced that angle of attack is not uniform along inlet height of blade. When inlet swirl is clockwise, great negative angle of attack is close to shroud and small angle is close to hub. Outlet angle in volute is uniform relatively when inlet swirl is anti-clockwise or inlet velocity field is uniform. So angle of attack of rotor is uniform relatively too in the two inlet conditions. Distribution of angle of attack influence flow in impeller passage and causes turbine performance changed.

Based on simulation results, entropy diagram in blade to blade planes of 10%, 50% and 90% blade height are drawn in the three inlet conditions. In the paper, inlet condition when velocity field is uniform is defined IU and inlet swirl is clockwise and anti-clockwise inlet condition are defined as IC and IAC respectively. Entropy differences between IU and IC and between IU and IAC are calculated and drawn in figure 11.

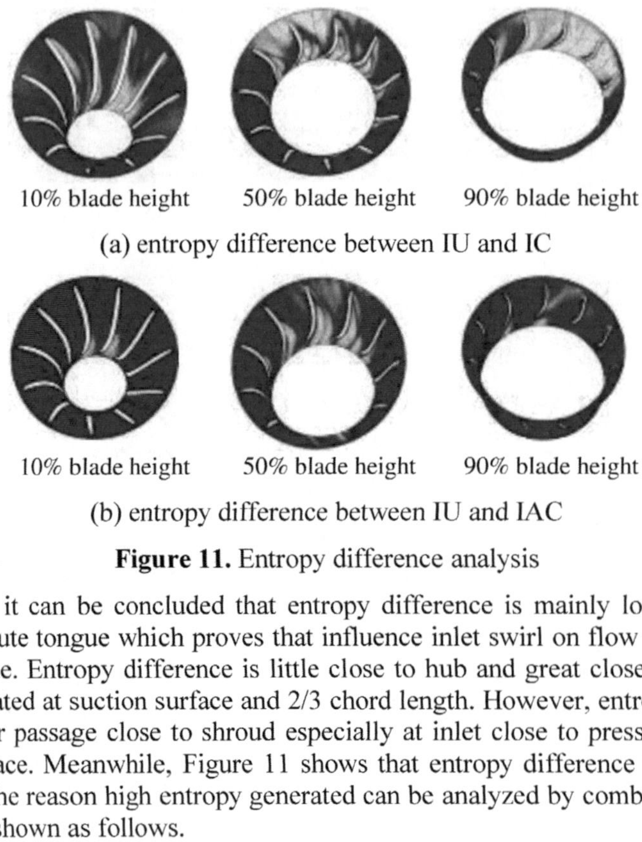

Figure 11. Entropy difference analysis

From figure 11 it can be concluded that entropy difference is mainly located in five impeller passages close to volute tongue which proves that influence inlet swirl on flow in rotor is in passages close to volute tongue. Entropy difference is little close to hub and great close to 50% blade height. The difference is located at suction surface and 2/3 chord length. However, entropy difference is great in the whole impeller passage close to shroud especially at inlet close to pressure surface and outlet close to suction surface. Meanwhile, Figure 11 shows that entropy difference is greater between IU and IAC. When IC, the reason high entropy generated can be analyzed by combination of entropy and streamlines diagram shown as follows.

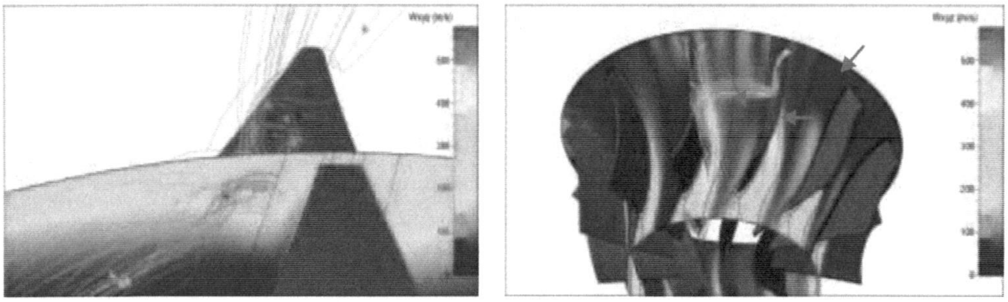

(a) Streamlines at inlet (b) Streamlines in passage

Figure 12. Streamlines in impeller passage

As is analyzed, there is large negative angle of attack at inlet close to shroud when IC. Flow at inlet is wrapped with leakage flow from former passage at inlet and forms vortex as is shown in figure 12 (a). The gas flows along pressure surface from inlet and turn from pressure surface to suction surface at 1/5 chord length of blade. The lateral flow reaches to suction surface and combined with leakage flow from later passage causing high entropy increase. Along blade height, the flow goes down and to outlet. At the zones gas flow passes by high entropy increase is generated as is shown in figure 12 (a) and (b). So negative angle of attack at inlet close to shroud caused by inlet swirl is the reason entropy difference is generated. When inlet swirl is anti-clockwise, angle of attack becomes small as is shown in figure 10. So there is smaller entropy increase which results in little entropy difference between inlet condition IU and IAC as shown in figure11(b).

3. Conclusions

With raising efficiency of exhaust gas energy usage of turbine system as the research target to realize high intake pressure ratio for hydrogen internal combustion engine, the model for flow analysis of turbine system was established and analysis of the mechanism of interaction between two turbines was achieved. The achieved concrete studies and the obtained conclusions are listed below.

1) Using Numeca code, the simulation model for flow interaction among turbine system is established. The calculation model was divided into two parts: LP module and HP module which were modeled respectively.

2) By simulation, the flow field distortion at outlet to HP turbine and inlet to LP turbine is analyzed. Swirl exists at outlet to HP turbine whose strength and rotation direction varies with operation condition and boundary condition; static pressure rises at inlet to LP turbine and distorted pressure field gradually becomes uniform when transfers to outlet to HP turbine.

3) Simulation proves that back pressure has little influence on turbine performance, so in two-stage turbine system HP turbine performance has little relationship with LP turbine selection or design.

4) Inlet swirls caused by HP turbine influence efficiency of LP turbine. To turbine in the paper, large negative angle of attack close to shroud generates strong vortexes in rotor passages and results in great flow losses when direction of inlet swirl is clockwise. On the contrary, anti-clockwise swirl helps to decrease vortex strength and raise turbine efficiency.

References

[1] Lumsden G, OudeNijeweme D, Fraser N and Blaxill H 2009, Development of a Turbocharged Direct Injection Downsizing Demonstrator Engine *SAE paper* **2009-01-1503**

[2] Saulnier S 2004 Computational study of diesel engine downsizing using two-stage turbocharging *SAE Paper* **2004-01-0929**

[3] Lee B C 2009 Dual-Stage Boosting Systems: Modeling of Configurations, Matching and Boost Control Options. *University of Michigan*

[4] Pfluger F 1998 Regulated Two-Stage Turbocharging - 3K-Warner's New Charging System for Commercial Diesel Engines. *The Sixth International Conference on Turbocharging and Air Management Systems, IMechE HQ, London, UK*, November 3-5

[5] Watson N and Janota M S 1982 Turbocharging the internal combustion engine. *Wiley-Interscience*,

[6] Zhang R Y 2003 MT890 heavy duty engine developed by MTU company, *Vehicle Engine*, **1(143)2003**

[7] Baines N C, Fundamentals of Turbocharging. *Concepts NREC*

[8] Vogt M, Frese F, Mai H and Baar R 2011 Improving the Application of Turbine Maps in 1D Engine Process Modelling, *Proceedings of ASME Turbo Expo 2011*, June 6-10, 2011, Vancouver, British Columbia, Canada

[9] Westin F, Burenius R 2010 Measurement of interstage losses of a two-stage turbocharger system in a turbocharger test rig. *SAE Technical Pap*er **2010-01-1221,**

[10] Hellstrom F, Fuchs L 2008 Effects of inlet conditions on the turbine performance of a radial turbine. *Proceedings of ASME Turbo Expo 2008*: Power for Land, Sea and Air, June 9-13, Berlin, Germany

[11] Hellstrom F, Fuchs L 2009 Numerical computation of the pulsatile flow in a turbocharger with realistic inflow conditions from an exhaust manifold. *Proceedings of ASME Turbo Expo 2009*: Power for Land, Sea and Air, June 8-12, Orlando, Florida, USA

[12] Chen L, Zhuge W L, Zhang Y J, Zhang S Y, Zhang J Z 2008 Investigation of the Secondary Flow Structure in the Mixed Flow Turbine for a High Pressure Ratio Turbocharger. *Proceedings of ASME Turbo Expo 2008*: Power for Land, Sea and Air June 9-13, Berlin, Germany

[13] Karamanis N 2000 *Inlet and Exit Flow Characteristics of Mixed Flow Turbines in Advanced Automotive Turbocharging*. Ph.D. Thesis, Imperial College of Science, Technology and Medicine, London, England

[14] Zhu D X 1992 *Turbocharging and Turbocharger*. Beijing Mechnical industry press

Acknowledgments

This work was supported by the National Basic Research Program of China ("973" Program) (Grant No.2011CB707204)

IOP Conf. Series: Materials Science and Engineering **129** (2016) 012004 doi:10.1088/1757-899X/129/1/012004

Appendix A.

Nomenclature

φ	Degree of distortion	θ	Angle of attack of rotor
Φ	Swirl strength	θ'	Incidence angle of rotor
v_s	Tangential component of velocity	u	Upstream characters
v_a	Axial component of velocity	d	Downstream characters
c_1	Inlet absolute velocity of rotor	x	X direction
ω_1	Inlet relative velocity of rotor	y	Y direction
u_1	Inlet convected velocity of rotor	z	Z direction
c_2	Outlet absolute velocity of rotor	den	density
ω_2	Outlet relative velocity of rotor	p	Static pressure
u_2	Outlet convected velocity of rotor	n	Interactions
p^*	Inlet total pressure	G	Flow mass rate
$\Delta\eta$	Turbine efficiency change	V_{max}	Maximum quantity value
$\Delta\eta_R$	Turbine rotor efficiency change	V_{min}	Minimum quantity value
$\Delta\eta_V$	Transmission efficiency of volute change	V_{mean}	Mean quantity value
η_R	Turbine rotor efficiency	V	velocity
$\overline{\lvert V \rvert}$	Mean value of velocity	IAC	Inlet anti-clockwise swirl
T	Static temperature	LP	Low pressure stage
IU	Uniform inlet velocity field	HP	High pressure stage
IC	Inlet clockwise swirl	EGR	Exhaust gas recirculation

IOP Conf. Series: Materials Science and Engineering **129** (2016) 012005 doi:10.1088/1757-899X/129/1/012005

CFD simulation on the turbulent mixing flow performance of the liquid-liquid ejector

W Z An[1], H Y Bie[1,3], C C Liu[1] and Z R Hao[2]

[1] College of Chemistry and Chemical Engineering, Ocean University of China, 238 Songling Road, Qingdao, 266100, China
[2] Institute of Oceanographic Instrument of Shandong Academy of Science, 29 Zhejiang Road, Qingdao, 266001, China

E-mail: haiyanbie@ouc.edu.cn

Abstract. In order to study the flow performance of the liquid-liquid ejector, 3D ejector simulation models were established to investigate the influences of suction angle, suction number and working condition on the ejector performance. The simulation results showed that when the suction angle was 60°, the total pressure was in equilibrium state. The double suction ejector would induced more vortexes in the suction chamber than that of the single suction ejector, and the turbulent intensity of the fluid inside the ejector was bigger, however, it also caused much more loss in energy. When the working pressure was lower than 0.6 MPa, the liquid entrainment ratio increased rapidly. Once the working pressure reached 0.6 MPa, the liquid entrainment ratio basically remained unchanged. The mass flow rate of the suction medium increased with the increasing of suction pressure, and the differential pressure between the suction pressure and the working pressure at the nozzle also increased simultaneously.

Introduction

Ejector realizes the energy and mass transfer between the working medium and the ejector medium making use of the high speed jet shear and turbulent diffusion effect. It is widely used in refrigeration, petrochemical industry, water desalination, aerospace, etc. [1, 2] due to its advantages of no mechanical transmission components, small volume, high efficiency, low noise, low cost and easy to comprehensive utilization [3].

Many studies have been done on the ejector all over the world by experimental and simulation methods. Experimental methods are susceptible to size change and the measurement precision, and difficult to get rid of the impact of the external flow disturbances. Besides, the repeatability is poor and cost is high. With the development of CFD and computer technology, numerical simulation has become one of the important methods on the research of the ejector as it can deeply study the flow characteristics inside the ejector and the influence of various structures on the ejector. Kolhapure et al. [4] carried out the 3-D steady simulation on the single phase turbulence reaction using the standard $\kappa-\varepsilon$ turbulence model. Hassel et al. [5] carried out the simulation study using the Large Eddy Simulation (LES) model and Renault Average Navier-Stokes (RANS) model. The accuracy of the RANS numerical model was verified using the experimental results. Yadav et al. [6] analysed the

[3] To whom any correspondence should be addressed.

influences of area ratio, suction chamber diameter and contraction angle on the liquid entrainment ratio. It was revealed that the liquid entrainment ratio increased with the increasing of the area ratio, but remained constant when the area ratio increased to the critical value. The liquid entrainment ratio had a peak value when the suction chamber diameter varied within a certain range. Yang et al. [7] studied the influence of nozzle structure on the ejector performance. The simulation results revealed that the nozzle structure had great influence on the liquid entrainment ratio and the backpressure. Bi et al. [8] established the turbulence reaction process in the ejector based on the scale theory and PLIF technology. The macroscopic mixing information of the material under different operation conditions was discussed. And the influence of operation parameters and structure sizes on the acid-base reaction conversion rate was revealed using the Eddy-Break-Up model. Then they studied the influence of the structure size on the turbulence mixing performance based on the liquid-liquid ejector by the PLIF technology [9]. It was revealed that the decrease of the entrainment ratio and the increase of the area ratio were good for the mixing effect. Ma et al. [10] studied the influence of structure sizes (area ratio, nozzle location and divergence angle) on the liquid-liquid ejector conversion rate and mixing performance using CFX and taking acid-base reaction for instance. It was revealed that the optimum value for the ratio of nozzle diameters to mixing section was 0.35.

In this paper, 3D ejector simulation models were established. The influences of suction angle and suction number on the internal flow characteristics and mixing effect of liquid-liquid ejector were studied using CFD simulation method. By changing the operating conditions, the influence of operating conditions on the performance of liquid-liquid ejector were obtained. The simulation results were useful for the improving of the ability to adapt different operating conditions.

The simulation model of the ejector

1.1. The structure and conditions of the models

The main structures of established ejector model included nozzle, suction chamber, mixing section and diffuser section. Figure1 showed the schematic of the ejector. The diameter of the import 1 was 40 mm, the diameter of the import 2 was 25 mm, the diameter of the nozzle 7 was 7 mm, the diameter of the throat 4 was 12mm, the mixing section was 60 mm long and the diffusion angle of the diffuser section was 3°. Figure 2 showed the schematic of single suction ejector with the suction angle of 90° and double suction ejector with the suction angle of 60°.

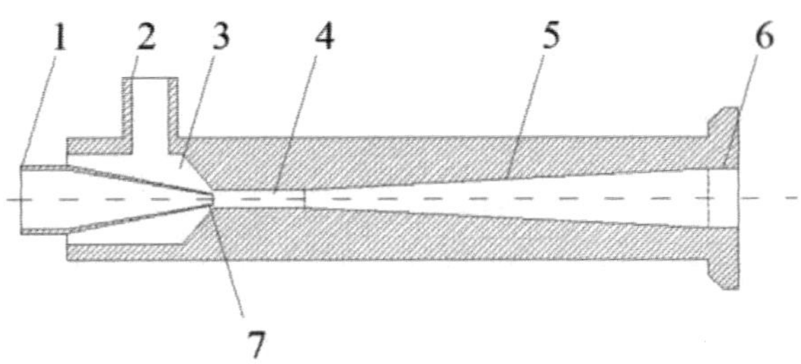

1 - motive flow inlet, 2 - suction flow inlet, 3 - suction section, 4 - mixing section, 5 - diffuser section, 6 - exit, 7 - nozzle

Figure 1. Schematic of the ejector.

a. Single suction ejector with the suction angle of 90°

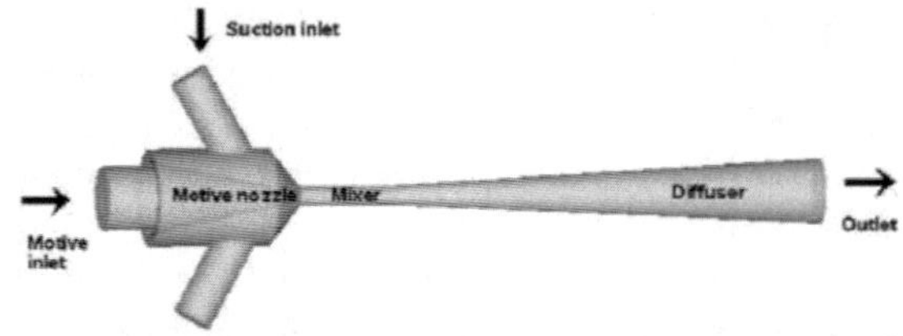

b. Double suctions ejector with the suction angle of 60°

Figure 2. Schematics of the ejector of single and double suctions.

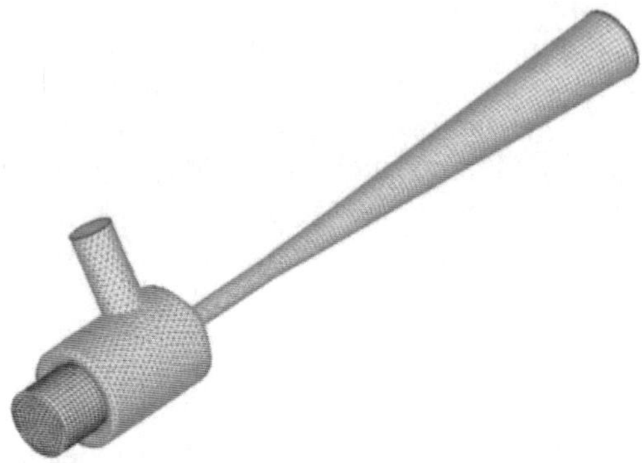

Figure 3. Grids of the single suction ejector.

The ejector model was meshed using structured grid except the suction chamber. The grids of the jet core region and mixing section was refined. The grids of the ejector were shown in figure 3. The two inlets were set to be pressure inlet boundary and the outlet was set to be pressure outlet boundary. The solid wall boundary conditions were dealt with the wall function method. In order to ensure the convergence precision, the momentum equation, the turbulent kinetic energy and turbulence dissipation rate equations were all solved using the second-order upwind difference scheme. The SIMPLE algorithm was adopted in the coupling of pressure and velocity. The density and viscosity of the fluid were 998.2 kg·m^{-3}, 0.001003 Pa·s and 804 kg·m^{-3}, 0.00395 Pa·s, respectively.

1.2. Governing equations

The mass conservation equation [11]:

$$\frac{\partial \rho}{\partial t}+\frac{\partial \rho u_j}{\partial x_j}=0 \tag{1}$$

The momentum conservation equation [11]:

$$\frac{\partial(\rho u_i)}{\partial t}+\frac{\partial(\rho u_i u_j)}{\partial x_j}=-\frac{\partial p}{\partial x_i}+\frac{\partial}{\partial x_j}[\mu(\frac{\partial u_i}{\partial x_j}+\frac{\partial u_j}{\partial x_i})]-\frac{2}{3}\frac{\partial}{\partial x_i}\left[\delta_{ij}\left(\rho k+\mu\frac{\partial u_k}{\partial x_k}\right)\right] \tag{2}$$

The energy conservation equation [11]:

$$\frac{\partial}{\partial t}(\rho E)+\frac{\partial}{\partial x_i}\left[u_i(\rho E+p)\right]=\frac{\partial}{\partial x_j}\left(k_{\text{eff}}\frac{\partial T}{\partial x_j}+u_i(\tau_{ij})_{\text{eff}}\right)+S_h \tag{3}$$

where ρ is density; E is total energy; T is temperature; k_{eff} is effective thermal conductivity; $(\tau_{ij})_{\text{eff}}$ is the deviatoric stress tensor defined as $(\tau_{ij})_{\text{eff}}=\mu_{\text{eff}}\left(\frac{\partial u_j}{\partial x_i}+\frac{\partial u_i}{\partial x_j}\right)-\frac{2}{3}\mu_{\text{eff}}\frac{\partial u_i}{\partial x_i}\delta_{ij}$, and μ_{eff} is the effective dynamic viscosity; S_h is the effect of enthalpy transport caused by species diffusion.

IOP Conf. Series: Materials Science and Engineering **129** (2016) 012005 doi:10.1088/1757-899X/129/1/012005

In the specific system, when the mixing or interaction of two components occurs, each component must abide by the law of conservation of quality components. The component mass conservation equation of component s is shown as

$$\frac{\partial(\rho c_s)}{\partial t}+div(\rho u c_s)=div(D_s\cdot grad(\rho c_s))+S_s \tag{4}$$

where, c_s is the volume concentration of component s, D_s is the diffusion coefficient, and S_s is the productivity within the system. The second item on the left and the first item on the right side of the equation are convection term and diffusion term, respectively. It means that the transmission process includes both convection and diffusion. Therefore, the component mass conservation equation is also known as the concentration equation.

In this model, the turbulence model employs the standard *k-ε* model. The turbulence kinetic energy *k* and the specific dissipation rate *ε* can be obtained from the following transport equations:

$$\mu_t=\rho C_\mu\frac{k^2}{\varepsilon} \tag{5}$$

$$\frac{\partial(\rho k)}{\partial t}+\frac{\partial(\rho k u_i)}{\partial x_i}=\frac{\partial}{\partial x_j}\left[\left(\mu+\frac{\mu_t}{\sigma_k}\right)\frac{\partial k}{\partial x_j}\right]+G_k+G_b-\rho\varepsilon-Y_M+S_k \tag{6}$$

$$\frac{\partial(\rho\varepsilon)}{\partial t}+\frac{\partial(\rho\varepsilon u_i)}{\partial x_i}=\frac{\partial}{\partial x_j}\left[\left(\mu+\frac{\mu_t}{\sigma_k}\right)\frac{\partial\varepsilon}{\partial x_j}\right]+C_{1\varepsilon}\frac{\varepsilon}{k}(G_k+C_{3\varepsilon}G_b)-C_{3\varepsilon}\rho\frac{\varepsilon^2}{k}+S_\varepsilon \tag{7}$$

where, G_b is the generation item of turbulent *k*inetic energy k caused by the buoyancy and for the incompressible fluid, $G_b = 0$. Y_M is the pulsation expansion item of the compressible turbulent flow and for the incompressible fluid, $Y_M = 0$.

The influence of suction structure on the performance of the ejector

1.3. The influence of the suction angle

In order to explore the influence of the ejector fluid feeding angle on the performance of the ejector, 7 suction angles (30°, 45°, 60°, 65°, 70°, 80° and 90°) were selected with the ejector structure size and other operating conditions to be constant in the simulation analyse.

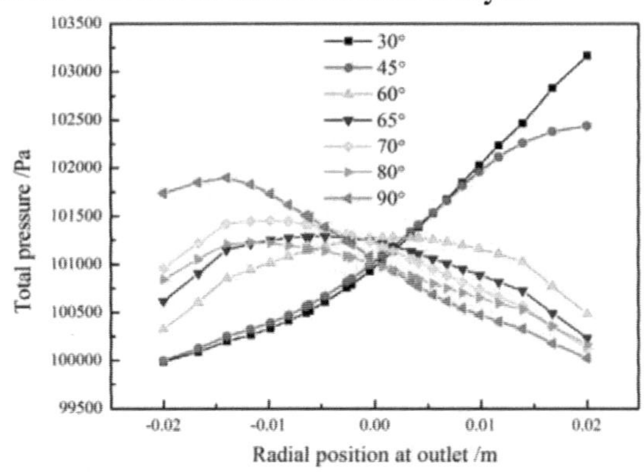

Figure 4. Total pressure distribution in radial at the outlet at different suction angles *θ*.

Figure 4 showed the total pressure distribution in radial at the outlet at different suction angles θ. As the exports static pressures were the same, the offset rule of the total pressure at the outlet would be in accordance with the offset rule of the velocity field. It can be seen that when the suction angle were 30° and 90°, the offsets of the total pressure at the outlet were much more serious. One side was high pressure and the other side was low pressure along the diameter direction of the outlet. With the increasing of the suction angle, the offset of the total pressure decreased firstly then increased,

IOP Conf. Series: Materials Science and Engineering **129** (2016) 012005 doi:10.1088/1757-899X/129/1/012005

reversed from the positive direction to the negative direction at last. When the suction angle was 60°, the offset of the total pressure was not obvious, mostly to be symmetric along the diameter direction, which meaning that the total pressure was in equilibrium state.

1.4. The influence of the suction number

The inside flow state of the ejector was turbulence. Turbulence consisted of wide range scales vortexes which were induced by disturbance, velocity gradient and the boundary. In order to comparative analyse the internal vortex flows of the single and double suction ejectors, slice observation was carried out near suction nozzle. The slice velocity vectors were shown in figure 5 and 6.

It can be seen that the suction flow of the single suction ejector was entrained in high speed at the suction inlet nozzle, and then was shunted into two flows due to the block effect of the outer nozzle wall. The two flows went bypass the nozzle and met at the symmetry bottom of the suction. Because of the reverse interaction forces, counter flows appeared and two vortexes were formed in the suction chamber which were approximate symmetrical.

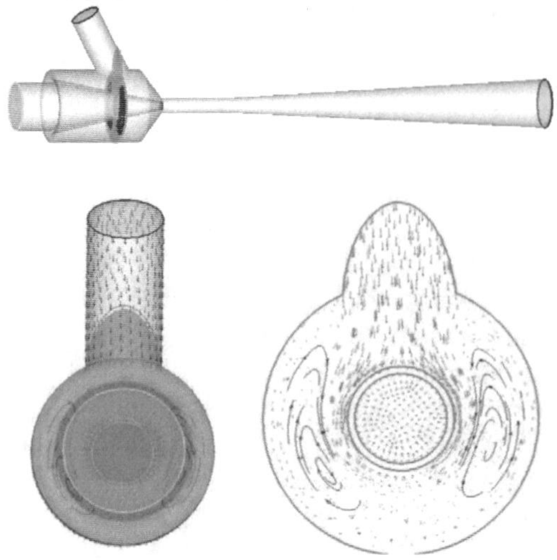

Figure 5. Slices of velocity vector of single suction ejector.

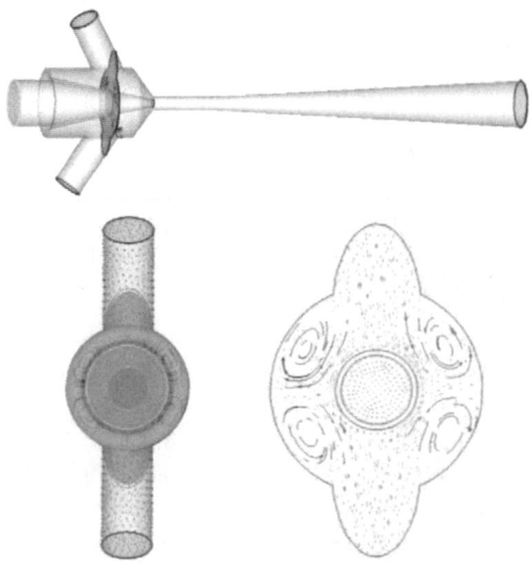

Figure 6. Slices of velocity vector of double suctions ejector.

From figure 6, it can be seen that the vortexes generated in the suction chamber of the double suction ejector were similar to those generated in the single suction ejector. The two suction flows entrained from the inlet set on either side of the suction chamber were shunted into four flows due to the block effect of the outer nozzle wall, and went bypass the nozzle and met at the sides of the nozzle. As a result, four vortexes were formed due to the reverse interaction forces which can be seen clearly in figure 6. It can be concluded that the double suction ejector would induced more vortexes in the suction chamber. The turbulent intensity of the fluid inside the ejector was bigger. From this perspective, the double suction ejector was in favour of the mixing, however, it also caused much more loss in energy.

The influence of the working condition on the performance of the ejector

1.5. The influence of working fluid pressure

In order to study the influence of working pressure on the performance of the ejector, simulations were carried out with the suction pressure p_s =0 MPa and the outlet pressure p_c =0.1 MPa. Figure 7 showed the relationship between the mass flow rates and the working pressure. It can be seen that both the working medium mass flow rate and the suction medium mass flow rate were increasing with the increasing of the working pressure. When the working pressure was in the range of 0.2~0.6 MPa, the working medium mass flow rate was greater than the suction medium mass flow rate, but the growth rate of the suction medium mass flow rate was greater than that of the working medium mass flow rate. When the working pressure was in the range of 0.6~1.0 MPa, the working medium mass flow rate was smaller than the suction medium mass flow rate, and the growth rates of the two mediums gradually become smaller. It meant that the increasing rate of mass flow rate with the increasing of working pressure in the range of 0.6~1.0 MPa was not so strange than that of 0.2~0.6 MPa.

Figure 8 showed the liquid entrainment ratio and pressure drop ΔP under different working pressures. The pressure drop increased with the increasing of working pressure which meaning that the suction ability of the ejector increased. It was because that the increasing of work pressure induced the increasing of fluid velocity at nozzle exit, and formed a larger negative pressure zone. When the working pressure was between 0.2~0.6 MPa, the liquid entrainment ratio increased rapidly, however, the liquid entrainment ratio basically remain unchanged when the working pressure was between 0.6~1.0 MPa. It was in good agreement with the analysis conclusion of the relationship between the mass flow rates and the working pressure as shown in figure 7. Besides, the liquid entrainment ratio was close to zero when the working pressure was 0.2 MPa.

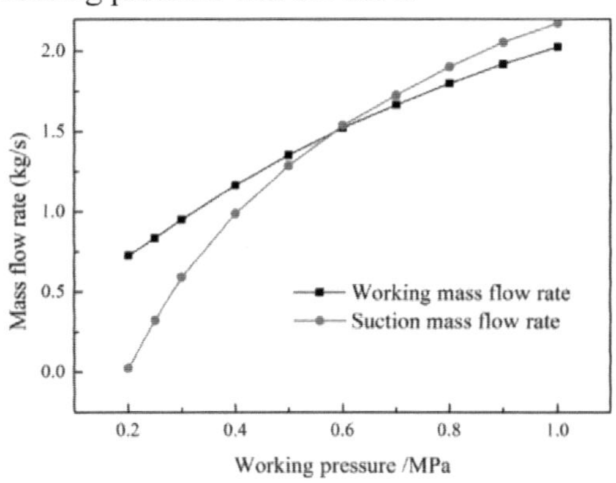

Figure 7. Mass flow rates of the ejector under different working pressures.

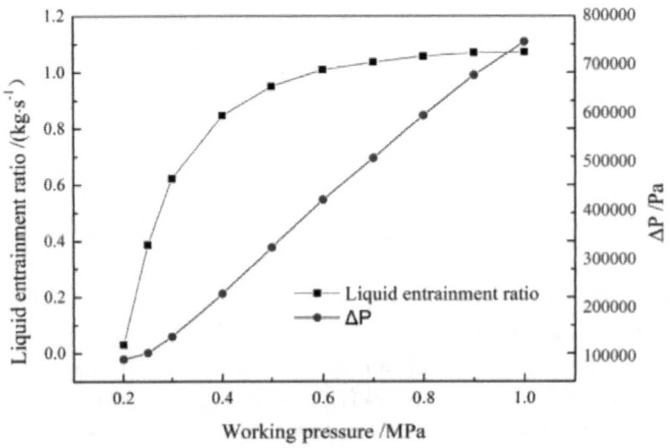

Figure 8. Liquid entrainment ratio and ΔP of the ejector under different working pressures.

1.6. The influence of suction pressure

In order to study the influence of suction pressure on the performance of the ejector, simulations were carried out with the working pressure p_j =0.6 MPa and the outlet pressure p_c =0.1 MPa. The mass flow rates of the ejector at different suction pressures were shown in figure 9. It can be seen that when the suction pressure increased, the mass flow rate of the suction medium increased accordingly, but the mass flow rate of the working medium remained almost the same. It revealed that the influence of the suction pressure on the mass flow rate of the suction medium was great.

Figure 10 showed the liquid entrainment ratio and pressure drop Δp at different suction pressures. It can be seen that the increasing of the suction pressure caused the increasing of the differential pressure between the suction pressure and the working pressure at the nozzle. In other words, the driving force of the suction medium entering into the ejector enhanced, and the mass flow rate of the suction medium would increase simultaneously. On the other hand, the increasing of suction pressure also improved the energy of the suction flow, and furthermore the ability of the mixing flow to overcome the outlet pressure was enhanced, therefore, the liquid entrainment ratio increased.

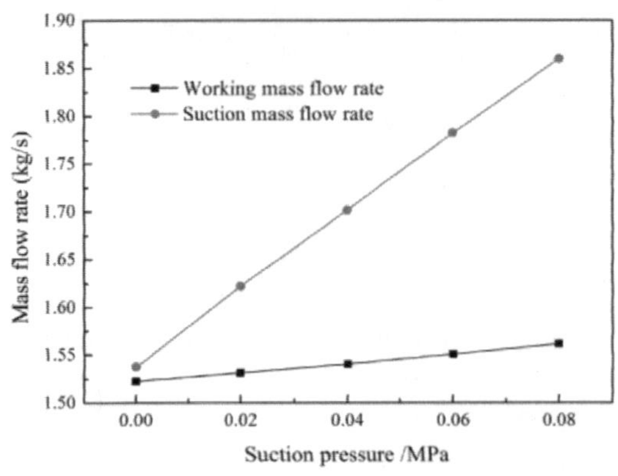

Figure 9. Mass flow rate versus suction pressure under different suction pressures.

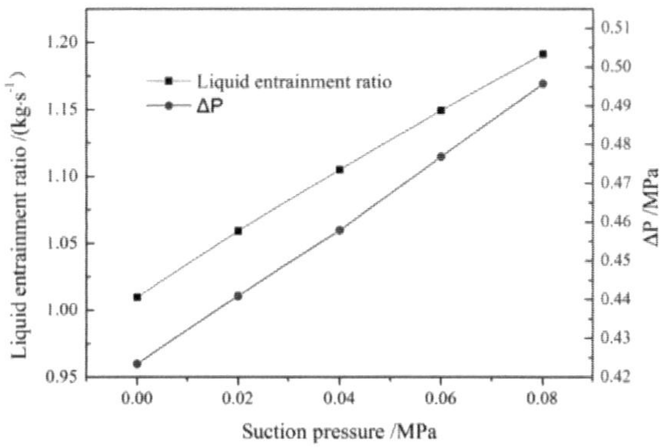

Figure 10. Liquid entrainment ratio and Δp under different suction pressures.

Conclusions

In this paper, simulation study was carried out to analyse the influence factor of the internal performance of the liquid-liquid ejector. The suction angle, suction number and working condition were taken into account. It was revealed that when the suction angle was 60°, the total pressure was in equilibrium state. The double suction ejector would induced more vortexes in the suction chamber than that of the single suction ejector, the turbulent intensity of the fluid inside the ejector was bigger, however, it also caused much more loss in energy. The liquid entrainment ratio increased rapidly when the working pressure was between 0.2~0.6 MPa, but basically remain unchanged after the working pressure reached 0.6 MPa. With the increasing of suction pressure, the mass flow rate of the suction medium increased, and the differential pressure between the suction pressure and the working pressure at the nozzle also increased simultaneously.

Acknowledgments

This research is funded by the National Natural Science Foundation of China (No. 51306166 and 51206101), the Scientific Research Foundation of Shandong Province Outstanding Young Scientist Award (No. BS2013NJ017), the Shandong Province Natural Science Foundation (No. ZR2012EEQ012) and the Qingdao Science and Technology Program of Basic Research Projects (Nos. 13-1-4-210-jch and 13-1-4-248-jch).

References

[1] Luo P C, Cheng Y, Wang Z W, Jin Y and Yang W H 2005 *Chem. Ind. Eng. Prog.* **24** 1319-26
[2] Chong D T, Zhao Q B, Yuan F, Wang W, Chen W X and Yan J J 2015 Exp. *Therm. Fluid Sci.* **64** 134-41
[3] Chen X J, Omer S, Worall M and Riffat S 2013 *Renew. Sust. Energ. Rev.* **19** 629-51
[4] Kolhapure N H, Tilton J N and Pereira C J 2004 *Chem. Eng. Sci.* **59** 5177-84
[5] Hassel E, Jahnke S and Kornev N 2006 *Chem. Eng. Sci.* **61** 2908-12
[6] Yadav R L and Patwardhan A W 2008 *Chem. Eng. Sci.* **63** 3886-97
[7] Yang X, Long X and Yao X 2012 *Int. J Therm. Sci.* **56** 95-106
[8] Bi R S 2008 *Experiment study and multi-scale simulation of turbulent reaction in jet reactors on high Schmidt number* (Qingdao: Qingdao University of Science & Technology)
[9] Bi R S, Tan X S, Lin K L and Zheng S Q 2010 *J.Chem. Eng. Chin. Uni.* **24** 752-7
[10] Ma l X, Bi R S and Zheng S Q 2008 *J. Chem. Ind. Eng.* **36** 33-6
[11] Versteeg H K and Malalasekera W 1995 *An introduction to computational fluid dynamics: the finite volume method* (New York: Wiley)

IOP Conf. Series: Materials Science and Engineering **129** (2016) 012006 doi:10.1088/1757-899X/129/1/012006

Experimental research on pressure loss of rotary pressure exchanger

L Jiao[1], Y Shi[1], Z C Liu[1], L Li, Z M Feng[2], C F Zhao, Y Wei *

[1] Institute of Naval Architecture and Ocean Engineering, Zhejiang University, China
[2] Nanfang Pump Industry Co., Ltd. China
*corresponding author: Tel.:+8613093767519

E-mail address: jiaolei@zju.edu.cn

Abstract. In order to research the phenomenon 'Pressure Loss' of Rotary Pressure Exchanger when it worked in a certain working condition, such as pressure and flow, two different series of experiments were performed respectively in the same experimental apparatus. One is to set pressure as constant when flow is changeable, the other one is to set flow as constant when pressure is changeable. Then, the curves and tables depicting the relationship between the pressure loss rate and working conditions were illustrated. Results from the curves and tables show that flow and pressure has an impact on pressure loss, and the relationship present some of regularity. According to the regularity, an empirical formula which can be used to approximately predict the magnitude of pressure loss for subsequent engineering application was provided by regression analysis on the basis of experimental data. Meanwhile, a 3-Dimensional geometric model of passageway in rotary pressure exchanger were built to verify the accuracy, making a steady calculation on pressure field by Fluent. At last, the feasibility was verified in a field application in a desalination factory.

1. Introduction

With the intention of reducing the cost of water, the rotary pressure exchanger is expected to be playing an important role in the field of seawater desalination. There are three characteristics below: First, simple mechanism, its core part consists of two end covers and a rotor with some passageways [1]. Second, good adaptation, the rotate speed is capable of adjusting itself to flow automatically [2]. Third, high work efficiency, high-pressure fluid and low-pressure fluid meet together directly in the course of pressure exchange [3]. At present, rotary pressure exchanger (PRE) energy recovery device in widespread use on the market is based on the principle of positive displacement [4]. The working principle of rotary pressure exchanger is illustrated by **Figure 1-1**.

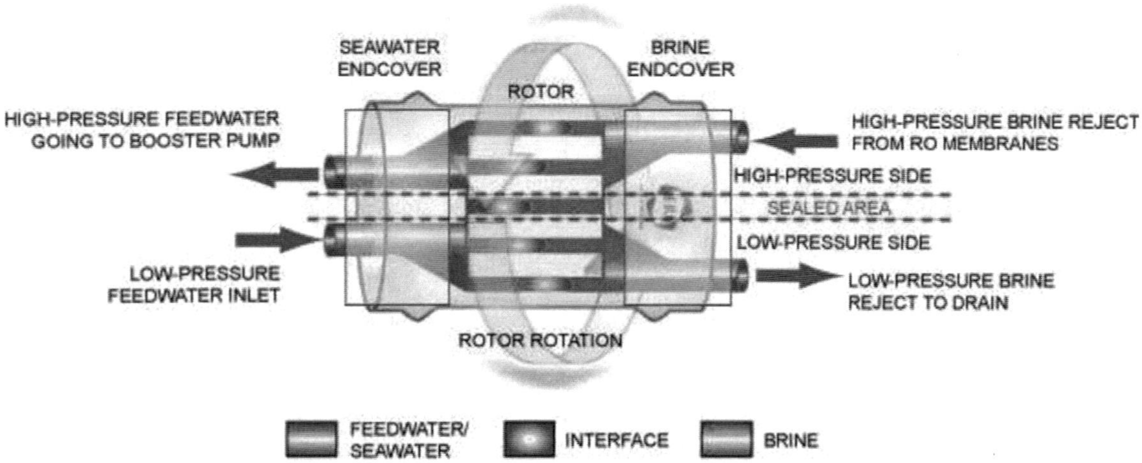

Figure 1-1 Working principle of the rotary pressure exchanger

However, the transmission of power from high-pressure flow to low-pressure flow is not in full force in actual engineering application. The friction, between the wall of rotor and the water film covering the rotor, exhausts a portion of power in the process of rotation of the rotor. The macroscopic performance is that the pressure in the high-pressure outlet is not bigger than the one in the high-pressure inlet, namely, pressure loss [5]. (The research object is the 'pressure loss in high pressure side', which is to be replaced by 'pressure loss' for convenience in this article, because the effect in engineering application brought by the 'pressure loss in high pressure side' is much bigger.) It is the pressure loss that make seawater pressurized by rotary pressure exchanger cannot go to reverse osmosis desalination directly, because the pressure cannot reach the operation standard. Thus, the high-pressure outlet goes immediately to booster device, such as booster pump. Given that pressure loss affects the working conditions of rotary pressure exchanger and the movement of rotor, it is meaningful to research the pressure loss of rotary pressure exchanger. Furthermore, an empirical formula which can be used to approximately predict the magnitude of pressure loss caters to actual demand.

Now, making a number of experiments is the only way which must be passed, in order to research the working conditions of rotary pressure exchanger. For example, Li Gui-Song and his colleagues summarized that there is a positive correlation between the end cover leakage and the flow, the pressure respectively [6]. They also found that the rotor speed has a positive correlation with the flow, and has a negative correlation with the pressure.

To research the pressure loss of rotary pressure exchanger, all the work is in the same way based on two different series experimental programs, which were performed in the experimental apparatus of SWRO system. In a series of experiments, measure the high-pressure pressure loss when the pressure of experiment system was constantly kept on 2.8 MPa, 3.4 MPa, 4 MPa, 4.5 MPa, 4.9 MPa, 5.3 MPa and 6.1 MPa, while the volume flow rate of cycle system was set on 69 m^3/h steadily. In another series of experiments, measure the high-pressure pressure loss when the volume flow rate of cycle system was constantly kept on 50 m^3/h, 55 m^3/h, 60 m^3/h, 65 m^3/h, 70 m^3/h and 75 m^3/h, while the pressure of the experiment system was set on 5.9 MPa steadily. First of all, propose an empirical formula which can be used to approximately predict the magnitude of pressure loss, on the basis of two series of experimental data. Then, take advantage of numerical calculation and field application to verify the accuracy and feasibility of the formula.

2. Experiment

The flow diagram of experimental apparatus is illustrated in **Figure 2-1**.

2.1 Experimental Device

The experimental apparatus contains seven parts, including filter, low-pressure pump, high-pressure plunger pump, circulating pump, electromagnetic flow meter, pressure sensor and data acquisition system.

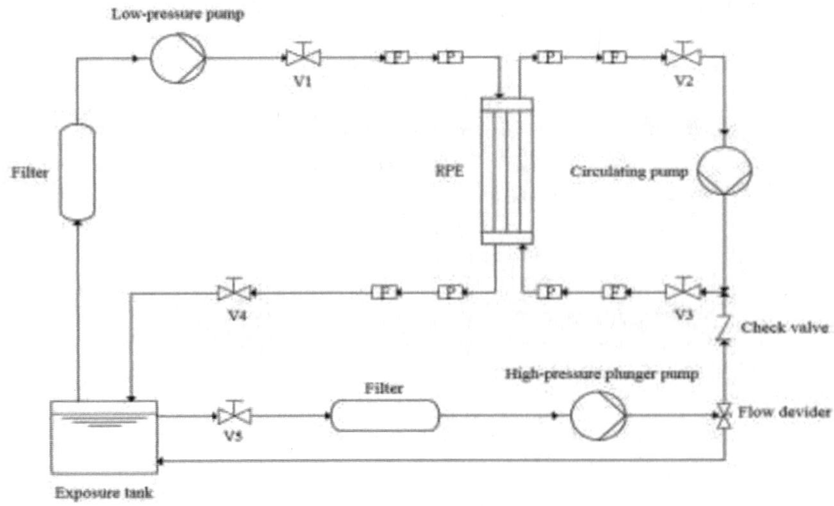

Figure 2-1 Flow diagram of experimental apparatus

2.2 Experimental Process

When there is enough water for the experiment, open the valves V1, V2, V3, V4 and V5. Firstly, open the low-pressure pump to fill the whole pipeline. Secondly, take turns to open the circulating pump and high-pressure plunger pump. Thirdly, control the flow divider and the valves V1, V2 to make sure that the flow in the high-pressure outlet and the pressure in the high-pressure inlet meet the experimental requirements. One series of experimental data includes the flow in the high-pressure outlet and the pressure in the high-pressure inlet in the meantime. Change the pressure or flow to continue next experiments, when one series of experimental data has been noted. All in all, we can reach the goal that experimental conditions are under control by controlling the flow divider and the valves V1, V2.

2.3 Experimental Method

The design flow of experimental apparatus ranges from 50 m^3/h to 90 m^3/h. The design pressure of experimental apparatus is not bigger than 6.3 MPa. The experiment uses orthogonal experimental method: When the pressure of experiment system was constant, make some recording spots of the volume flow rate of cycle system while the value fluctuated from 50 m^3/h to 90 m^3/h; When the volume flow rate of cycle system was constant, make some recording spots of the pressure of experiment system while the value fluctuated from 0 MPa to 6.3 MPa. Then, the data acquisition system captured real time data with the frequency of 12 Hz, to finish the record of the flow and pressure in the high-pressure inlet and outlet.

3. Data processing and analysis

3.1 Pressure loss characteristic

In practical application, there are a lot of factors affecting the pressure loss. Apart from the flow and pressure closely related to the working conditions, the inherent specifications owned by rotary pressure exchanger has something to do with pressure loss [7]. Hence, it is unmeaning to have a research without taking specifications into consideration. The specifications of rotary pressure exchanger is illustrated in Table 3-1.

Symbol	Specification	Parameter
/	Material	Corundum ceramic
R	Rotor outer diameter	190mm
L	Axial length	190mm
T	Outside wall thickness	10mm
t	Wall thickness of passageways	5mm
α	Inclination of end covers	22°
τ	Clearance of end face	30μm
s	Clearance of radial direction	35μm

Table 3-1 Specifications of rotary pressure exchanger

The computational formula of pressure loss $\triangle P$ is:

$$\triangle P = P_{Hin} - P_{Hout}$$

The computational formula of pressure loss rate ρ is:

$$\rho = \frac{\triangle P}{P_{Hin}}$$

Where P_{Hin} denotes the pressure in the high-pressure inlet and P_{Hout} denotes the pressure in the high-pressure outlet.

The relationship between pressure loss, pressure loss rate and the pressure in the high-pressure inlet is illustrated in **Table 3-2**.

The relationship between pressure loss, pressure loss rate and the flow in the high-pressure outlet is illustrated in **Table 3-3**.

Figure 3-1 illustrates the X-Y scatter plot of pressure loss rate and the pressure in the high-pressure inlet. The scatter plot reveals that the bigger the pressure in the high-pressure inlet is, the smaller the pressure loss rate is. That is to say, there is a negative correlation between them. In addition, regression method is used to analyze these data, because the distribution of data points has its regularity, roughly on a curve. Linear function is used to fit these data points shown in Figure 3-1. The regressive curve equation acquired from regressive analysis is:

$$y = -0.0033x + 0.0422$$

Where y denotes pressure loss rate and x denotes the pressure in the high-pressure inlet, the fitting degree index R^2 reaches 0.9816.

Figure 3-2 illustrates the X-Y scatter plot of pressure loss rate and the flow in the high-pressure outlet. The scatter plot reveals that the bigger the flow in the high-pressure outlet is, the bigger the pressure loss rate is. That is to say, there is a positive correlation between them. In addition, regression method is used to analyze these data, because the distribution of data points has its regularity, roughly on a curve. Linear function is used to fit these data points shown in Figure 3-2. The regressive curve equation acquired from regressive analysis is:

$$y = 0.0004x - 0.01$$

Where y denotes pressure loss rate and x denotes the pressure in the high-pressure inlet, the fitting degree index R^2 reaches 0.9728.

No.	P_{Hin}	P_{Hout}	Q_{Hout}	Δ P	ρ
	MPa	MPa	m^3/h	MPa	/
1	2.812	2.719	69.9	0.092	0.0327
2	3.413	3.307	69.9	0.105	0.0309
3	3.996	3.879	69.0	0.116	0.0291
4	4.446	4.320	68.7	0.126	0.0283
5	4.863	4.735	68.8	0.127	0.0262
6	5.335	5.209	68.9	0.126	0.0237
7	6.115	5.978	68.8	0.137	0.0224

Table 3-2 the relationship between pressure loss, pressure loss rate and the pressure in the high-pressure inlet

No.	P_{Hin}	P_{Hout}	Q_{Hout}	Δ P	ρ
	MPa	MPa	m^3/h	MPa	/
1	5.935	5.791	80.1	0.144	0.0243
2	5.810	5.675	75.2	0.135	0.0232
3	5.977	5.847	70.4	0.130	0.0217
4	6.031	5.916	65.7	0.114	0.0189
5	5.804	5.702	60.5	0.103	0.0177
6	5.894	5.817	55.3	0.077	0.01301
7	5.801	5.726	50.5	0.076	0.01302
8	5.872	5.818	45.3	0.054	0.0092

Table 3-3 the relationship between pressure loss, pressure loss rate and the flow in the high-pressure outlet

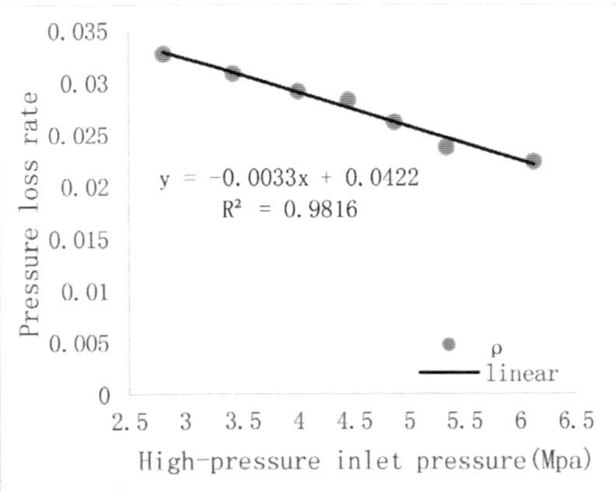

Figure 3-1 the X-Y scatter plot of pressure loss rate and the pressure in the high-pressure inlet

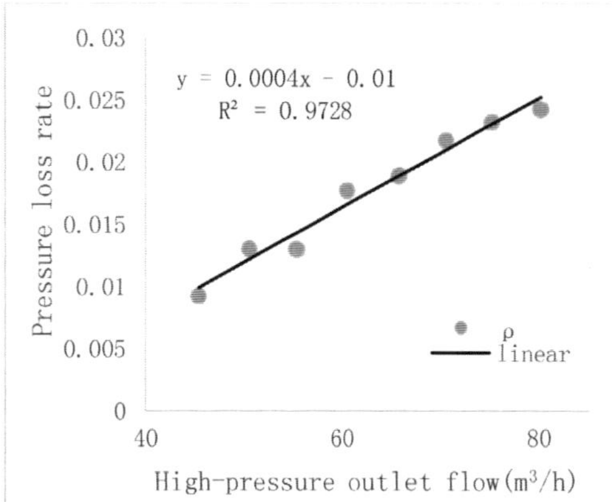

Figure 3-2 the X-Y scatter plot of pressure loss rate and the flow in the high-pressure outlet

3.2 Empirical formula

The relationship summarized in section 3.1 between pressure loss rate and the pressure, as well as pressure loss rate and the flow, indicates that pressure loss rate is not only linearly correlated with the pressure in the high-pressure inlet, but also with the flow in the high-pressure outlet. Now, a general relationship could be acquired when we take two factors, pressure and flow, into consideration together. In consequence, if the pressure and flow is confirmed, we could get the pressure loss rate by calculation to further predict the pressure loss in these working conditions.

Available functions:

$$y = -0.0033x_1 + 0.0422 \tag{1}$$

$$y = 0.0004x_2 - 0.01 \tag{2}$$

Where y denotes pressure loss rate, x_1 denotes pressure in the high-pressure inlet (MPa) and x_2 denotes flow in the high-pressure outlet (m³/h).

Suppose (3) as the general relationship:

$$y = k_1x_1 + k_2x_2 + t \tag{3}$$

Plug the experimental conditions that $x_1 = 5.9$ and $x_2 = 69$ into (3) separately:

$$y = k_1x_1 + 69k_2 + t \tag{4}$$

$$y = k_2x_2 + 5.9k_1 + t \tag{5}$$

In this way, (1) is equivalent to (4), (2) is equivalent to (5).

Thus:

$$k_1 = -0.0033,\ k_2 = 0.0004 \tag{6}$$

$$5.9k_1 + t = 0.0422 \tag{7}$$

$$69k_2 + t = -0.01 \tag{8}$$

Get t_1 and t_2 after plugging (6) into (7) and (8):

$$t_1=0.0146,\ t_2=0.00947$$

Respectively, calculate the standard deviation of the value of flow in table 3-2, as well as the standard deviation of the value of pressure in **table 3-3**. Then, $\sigma_{3\text{-}2} = 0.4865$， $\sigma_{3\text{-}3} = 0.802$. Comparing the two values, the data error in table 3-2 is relatively smaller. Hence, the value derived from (1) is more reliable than (2).In a word, t = 0.0146.

Ultimately:

$$y = -0.0033x_1 + 0.0004x_2 + 0.0146 \tag{9}$$

Empirical formula of pressure loss is:

$$\Delta P = P_{Hin}(-0.0033P_{Hin} + 0.0004Q_{Hout} + 0.0146) \tag{10}$$

4. Verification

4.1 Field application

Figure 4-1. The figure shows that the energy recovery device is in field application. The location of the field application is in Changtu desalination factory in Daishan County in Zhoushan City. The amount of processing of this factory achieves 5000m^3/d [8]. The field application was operated by plunging the energy recovery device into technological process of the desalination factory.

The test began when the device was in regular work. One series of data was record every hour, totally eight times. The data includes the flow and pressure in high-pressure side, either in the inlets or in the outlets. Each series of data results in two values about pressure loss, of which one is the experimental value, another one is the theoretical value calculated by empirical formula. The feasibility of empirical formula would be verified by comparing the two values.

No.	P_{Hin}	P_{Hout}	Q_{Hout}	ΔP_e	ΔP_t	ε
	MPa	MPa	m^3/h	MPa	MPa	%
1	3.3880	3.2773	72.1	0.1107	0.1093	1.239
2	3.3425	3.2331	72.2	0.1094	0.1085	0.834
3	3.3517	3.2450	72.6	0.1067	0.1092	2.371
4	3.3377	3.2309	72.0	0.1068	0.1081	1.194
5	3.3284	3.2237	70.8	0.1047	0.1063	1.490
6	3.3379	3.2307	72.3	0.1071	0.1085	1.264
7	3.3592	3.2529	71.9	0.1063	0.1084	2.039
8	3.3547	3.2442	72.5	0.1105	0.1091	1.266

Table 4-1 Data in field application

Table 4-1 illustrates data in field application. Where ΔP_e denotes the experimental value of pressure loss, ΔP_t denotes the calculated value of pressure loss, ε denotes the data error of two. From the table 4-1, the pressure in the high-pressure inlet was fluctuating on the level of 3.35 MPa, while the flow in the high-pressure outlet was approximately 72 m^3/h. Although there were some small changes in the working conditions, the experimental value of pressure loss was stabilized in the range between 0.1~0.11 MPa. Meanwhile, the theoretical value is identical with the experimental one, data error being under 3%. Hence, the feasibility of empirical formula is verified for this kind of specification.

4.2 Numerical simulation

According to the working conditions and device specification in field application, a steady numerical analysis would be made in the software ANSYS FLUENT, on the basis of the fourth series of data in **table 4-1**.

4.2.1 Numerical process Firstly, establish the geometry model of the energy recovery device with the software—Solidworks, including rotor, sleeve and two end covers. Then, extract the computational domain of passageways to get the geometry model of computational domain. The geometry model of computational domain is shown in **Figure 4-2**.

Secondly, make the mesh generation on the geometry model with the software—ANSYS ICEM. Among all parts, the part of rotor domain is made of structured grid of high qualified, the part of end cover domain is made of unstructured grid for the complexity of details. The mesh quality of whole mesh is more than 0.4, meeting the requirements of numerical calculation.

Thirdly, set the parameter value on computation when the mesh had been imported into Fluent. Concerning the boundary conditions, all the inlets are Velocity Inlet and all the outlets are Pressure Outlet. Namely, V = 7.6358 m/s for the inlet in the high-pressure side, P = 3.23 MPa for the outlet in the high-pressure side, V = 7.4294 m/s for the inlet in the low-pressure side and P = 0.05 MPa for the outlet in the low-pressure side. As to other conditions, the rotation of rotor domain is considered by the use of the multiple rotating reference frame (MRF) method, with the rotation speed set as 700 rpm. At the same time, the effects of turbulence are modelled using the k-ε turbulence with standard wall treatment in the simulation. Numerical convergence absolute criteria is set to a maximum of 1×10^{-4}.
Setting the number of iteration as 1000 with accepting the rest of the default choices.

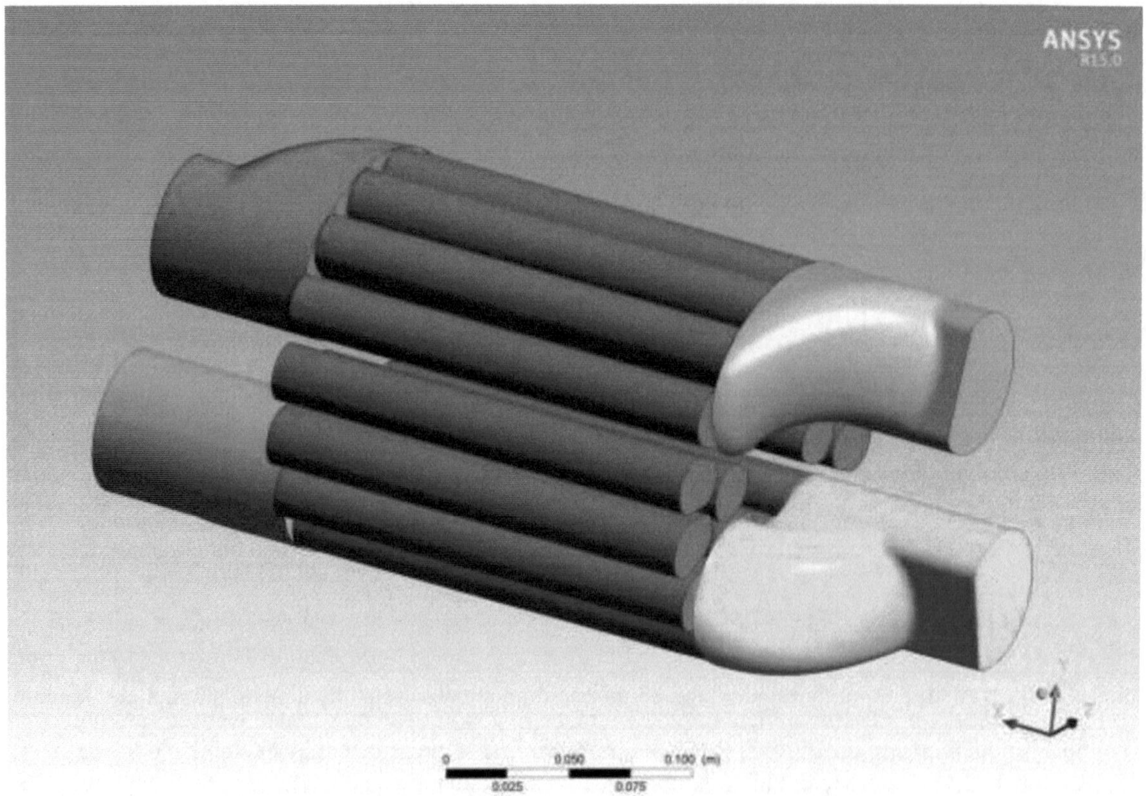

Figure 4-2 Geometry model of computational domain

4.2.2 *Numerical results analysis* After calculation, the computational value on the pressure in the high-pressure inlet came up, $P_{hin} = 3.251$ MPa. Obviously, the error was far beyond prediction value. It is important to note that the model could not simulate the leakage phenomenon. So, it is necessary to compensate the flaw by taking leakage into consideration. A correction term comes to manifest the effect of leakage.

Correction term: $$\gamma = \frac{\Delta Q \times P_{Hin}}{Q_{Hout}}$$

And an extra relationship: $$P_{Hin} = \gamma + P_{hin}$$

Where P_{Hin} denotes the corrected pressure and ΔQ denotes leakage.

Finally, outcome comes that $P_{Hin} = 3.344$ MPa. It means that the real pressure loss $\Delta P_c = 0.114$, in the numerical calculation. The data error is about 5%, so the accuracy of empirical formula is verified for this kind of specification.

5. Conclusion

It is concluded that:

1. With regard to rotary pressure exchangers of the specification, there is a negative linear correlation between pressure loss rate and the pressure in the high-pressure inlet, while there is a positive linear correlation between pressure loss rate and the flow in the high-pressure outlet.
2. An empirical formula coming from the derivation of two linear relationship, is capable of predicting the performance of energy recovery device in a certain condition, especially in the ranges from 50 m^3/h to 75 m^3/h and from 2.8 MPa to 6.1 MPa.

3. The effect of leakage in the clearance surrounding the rotor cannot be ignored, in the numerical calculation about pressure loss in the rotary pressure exchanger. Furthermore, the pressure loss caused by leakage play the backbone, compared with other inherent factors of the device.
4. The correction term γ can offset the effect of leakage, and the correction term is relevant to the value of leakage, the pressure in the high-pressure inlet and the flow in the high-pressure outlet.

Acknowledgements
The research work was funded by National Science and technology support program (No.2013BAB08B03) and the National Natural Science Foundation of China (No. 51276158).

References
[1] Hauge L J 1995 The pressure exchanger—A key to substantial lower desalination cost. *Desalination*, **102** 219-223
[2] Liu Y, Zhou Y H and Bi M S 2012 3D numerical simulation on mixing process in ducts of rotary pressure exchanger. *Desalination and Water Treatment*, **42(1-3)** 269-273
[3] Mambretti S, Orsi E, Gagliardi S and Stover R 2009 Behaviour of energy recovery devices in unsteady flow conditions and application in the modelling of the Hamma desalination plant. *Desalination*, **238** 233-245
[4] Cameron I B 2008 SWRO with ERI's PX pressure exchanger device—a global survey. *Desalination*, **221** 136-142
[5] Lin J Z 2005 *Fluid Dynamics* (Beijing: Qinghua University Press)
[6] Li G S, Feng Z M, Lin H T, Zhao C F and Jiao L 2014 Experimental research on energy recovery efficiency and the leakage characteristic of rotary pressure exchanger. *Proceedings of the 6th International Symposium on Fluid Machinery and Fluid Engineering,* **2014** 868-872
[7] Bross S, Kochanowski W 2004 SWRO core hydraulic module—the right concept decides in terms of energy consumption and reliability. Part II. Advanced pressure exchanger design. *Desalination* **165** 351-361
[8] Ye Z J 2009 Deign of Changtu seawater Desalination Plant in Daishan County, Zhoushan City. *China Water & Wastewater* **25(14)** 40-47

IOP Conf. Series: Materials Science and Engineering **129** (2016) 012007 doi:10.1088/1757-899X/129/1/012007

Investigation on the gas pockets in a rotodynamic multiphase pump

J Y Zhang[1], Y J Li, S J Cai, H W Zhu, Y X Zhang

College of Mechanical and Transportation Engineering, China University of Petroleum, Beijing

E-mail: zhjinya@163.com

Abstract The appearance of gas pockets has an obvious impact on the performance of the rotodynamic multiphase pump. In order to study the formation of gas pockets in the pump and its effects on pump's performance, the unsteady numerical simulation and the visualization experiments were done to investigate gas pockets in a three-stage rotodynamic multiphase pump developed by authors. Meanwhile, the mixture of water and air was selected as the medium. According to the distributions of pressure, gas volume fraction and velocity vector in three compression cells in unsteady flow process, the process of the formation of gas pockets in the pump were analysed generally. The visualization experiments were used to verify the validity of the numerical simulation. The results will be benefit for the hydraulic design of the compression cell of rotodynamic multiphase pump.

1. Introduction

Due to the further demand of energy, the active development of oilfield has become the inevitable trend of society. Because of the overexploitation of the oilfield, the phenomenon that the appearance of outgas from oilfield have restricted the production efficiency [1-3]. In order to prolong the life of the oilfield, the Multiphase Mixed Transportation Technology, which means replacing traditional flow pumps and gas compressors with a multiphase pump, has been developed. The technology has the following features: more simplified devices, low Wellhead Back Pressure, all these are benefit to the production of oil and gas.

But with the application of multiphase pumps, disadvantages emerged: First, due to the difference of density and speed between gas and liquid, the separation of gas-liquid two-phase has appeared; second, the gas pocket forms easily in high gas volume fraction, which causes big flow losses and the sharp drop of the performance of multiphase pumps. Therefore, it is important to study the gas pocket to reduce the separation of gas-liquid. For the further study, a three-stage rotodynamic multiphase pump developed by authors has been selected to do unsteady numerical simulation, visualization experiments were also used to verify the validity of the numerical simulation.

[1] Corresponding author: J Y Zhang

2. the pump model and the numerical simulation

A three-stage rotodynamic multiphase pump developed by authors was adopted in this paper. It contains three compression cells, a diversion cone at the inlet and a diffuser at the outlet of the pump, which is shown in Figure 1. The front and back extension were also set in the pump.

Fig.1 Model of rotodynamic multiphase pump

2.1 The geometry model and the mesh

The software CFX-BladeGen was used to establish the impeller and the diffuser models. The number of the impeller blade was three while the diffuser blade was ten. There was a gap whose distance is 2% times as the span near the rim of the impeller. Large wrap angle was selected to avoid the separation of the gas and the liquid. The software CFX-TurboGrid was used to generate mesh. The number of every impeller mesh was 410448 while every diffuser mesh was 142393, all these are shown in Figure 2. Finally, the hexahedron structured mesh was generated, as is shown in Figure 3.

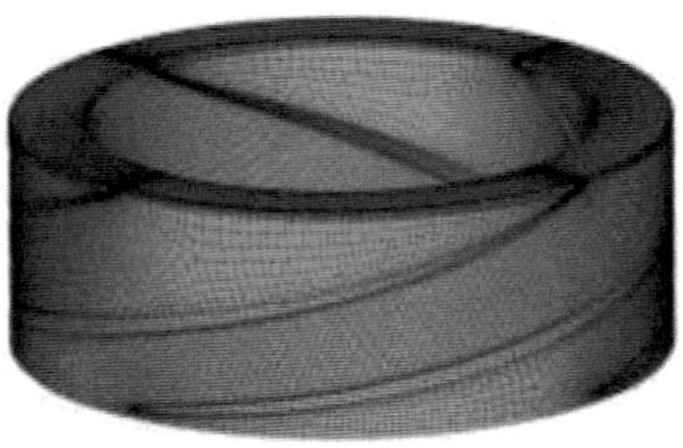

Fig. 2 Mesh of the impeller

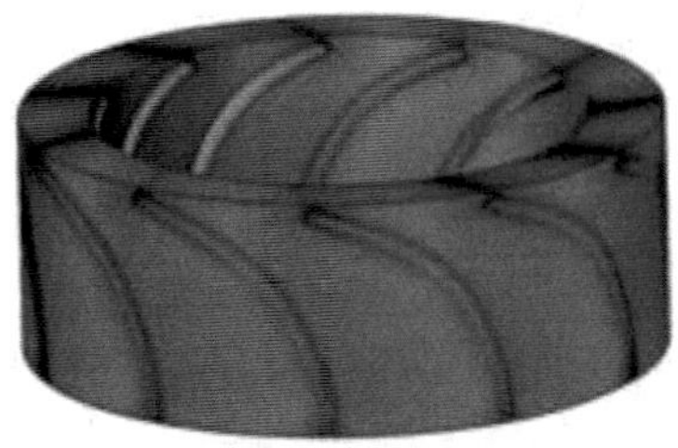

Fig. 3 Mesh of the diffuser

2.2 calculation settings

The commercial software ANSYS-CFX was adopted in this paper. It contains the conservation of Finite Volume Method and the accuracy of Finite Element Method.

Due to the little change of the temperature in gas-liquid two-phase medium[5], the thermodynamic process was considered as an isothermal process, the reference pressure was set at 0.1 MPa, the boundary condition was set at mass flow in the inlet while average static pressure in the outlet, the turbulence model SST which is based on k-ω was applied. The residual convergence value was 1e-6 and with the value condition we can judge the stability of the numerical simulation. The interface of impellers and diffusers was defined as the mixing surface boundary.

The steady numerical simulation has been done when rotational speed was 2700 r/min, the volume flow was 33 m3/h and the IGVF was 30%, the results were the initial file of unsteady numerical simulation, where the time step was set at $\Delta t = 1.85e^{-4}$ s.

3. Results and discussion

3.1 Analysis of the flow field at a certain time

3.1.1 *The distribution of velocity and pressure.* When the calculation reached stable, the velocity vector (Figure 4) has been selected to analyse vortices in different spans of impellers and diffusers. Conclusions can be obtained:

(1) From the span 0.1 to 0.9, the movement of gas-liquid two-phase in impellers and diffusers was more regular. In the span 0.1, different vortices appeared in three stages within impellers and diffusers. In the span 0.9, smaller vortices were only existed in the outlet of the every diffuser passage. The centrifugal force of water was so big that the water moved to the rim while the gas gathered in the hub.

(2) In the span 0.1, the number of the vortices in three compression cells decreased gradually from the first to the third stage in the passages. Vortices in impeller passages stretched from the pressure to suction surface of blades at the first and second stage, but only stayed in pressure surfaces at the third stage. This phenomenon indicated that there was a dynamic and static interference between impellers and diffusers. Vortices prevented the fluid from moving downstream, so that the vortex in impeller passages stretched from the pressure surface to suction surface of blades.

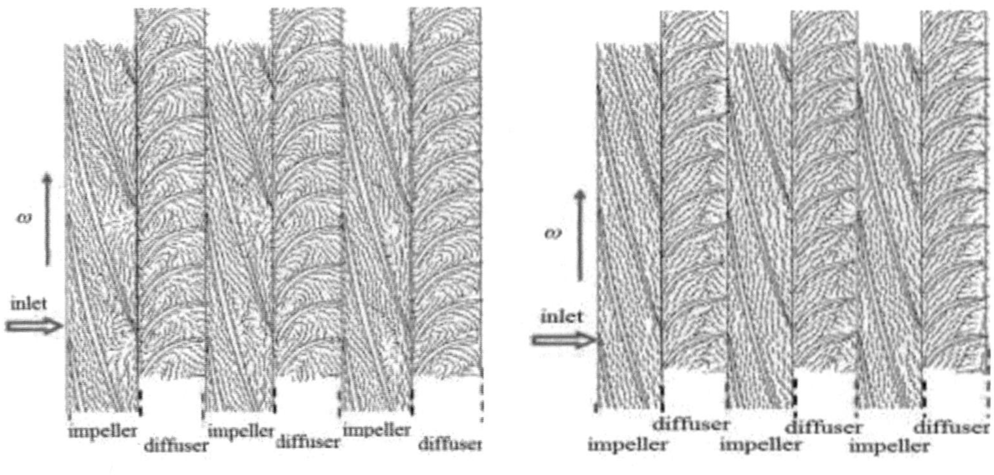

Span 0.1 **Span 0.9**

Fig. 4 Air velocity on the cascade plane

Figure 5 shows the pressure distribution in different spans with the unsteady numerical simulation. We can find that pressure distributions of the gas-liquid two-phase in impeller passages and diffuser passages were uneven. The fluid was not only affected by the rotational impeller that the fluid flowed through, but also affected by the post rotational impeller.

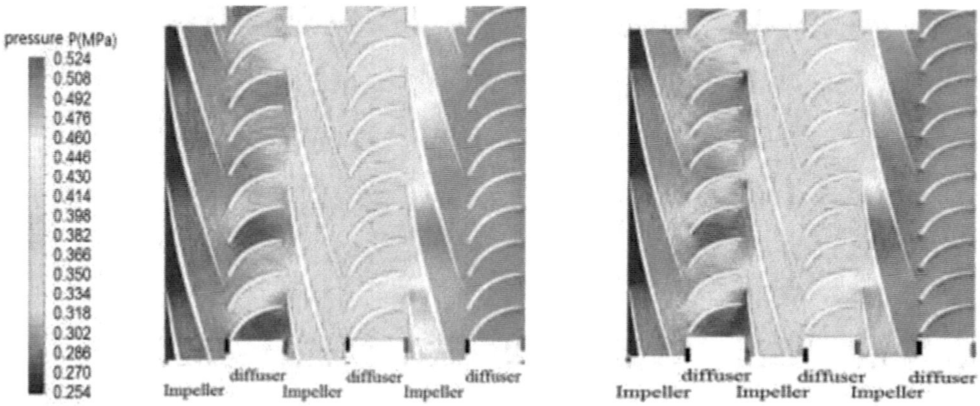

Span 0.1 **Span 0.9**

Fig. 5 Pressure distributions in different spans

3.1.2 *GVF distributions.* Figure 6 shows the GVF distribution in different spans, conclusions can be obtained: In the span 0.1, the high GVF area in impellers of the first stage was so large that it stretched to the suction surface of blades. At the second stage, the high GVF area in impellers was smaller than the first one but still stretched to the suction surface. At the third stage, the high GVF area no longer stretched to the suction surface. The phenomenon showed that GVF was affected by the complex flow situation of the diffuser behind. Vortices formed and pretended the gas from diffusing downstream.

In the span 0.9, the high GVF area almost disappeared in the impeller passage. Unlike the span 0.1, the GVF areas were only existed in suction surface at the first and second stage.

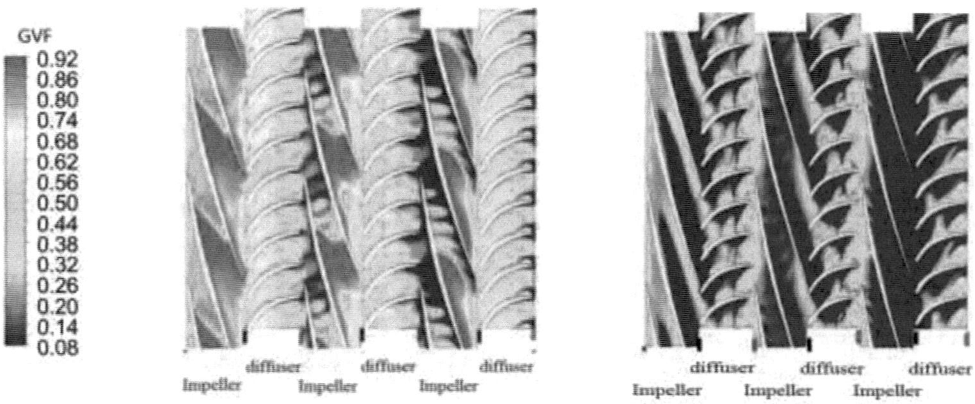

Span 0.1 **Span 0.9**

Fig. 6 *GVF* distributions on the cascade plane of different spans

3.2 Analysis of the flow field with time

3.2.1 *Pressure distributions with time.* From the Figure 7 we can get the analysis. Take four passages of diffusers, label A, B, C and D. At the time t, an impeller blade moved to passage A, the pressure of passage A was higher than passage B; at the time. $t+10\Delta t$, the blade head of the impeller passed passage A, the pressure in passage A decreased obviously while the pressure in passage B increased; at the time $t+20\Delta t$, the impeller blade moved to passage B, the pressure in passage A further decreased and the pressure in passage B increased continuously. The phenomenon can also be observed in passage C and D, at the time $t+30\Delta t$, the impeller blade has not reached passage C, the pressure in passage C and D were uniform; at the time $t+40\Delta t$, the impeller blade moved to passage C, we can find the pressure in passage C was higher than passage D. all these can conclude that the pressure distribution was affected significantly by the rotational impeller downstream, the pressure in diffusers had a periodic change.

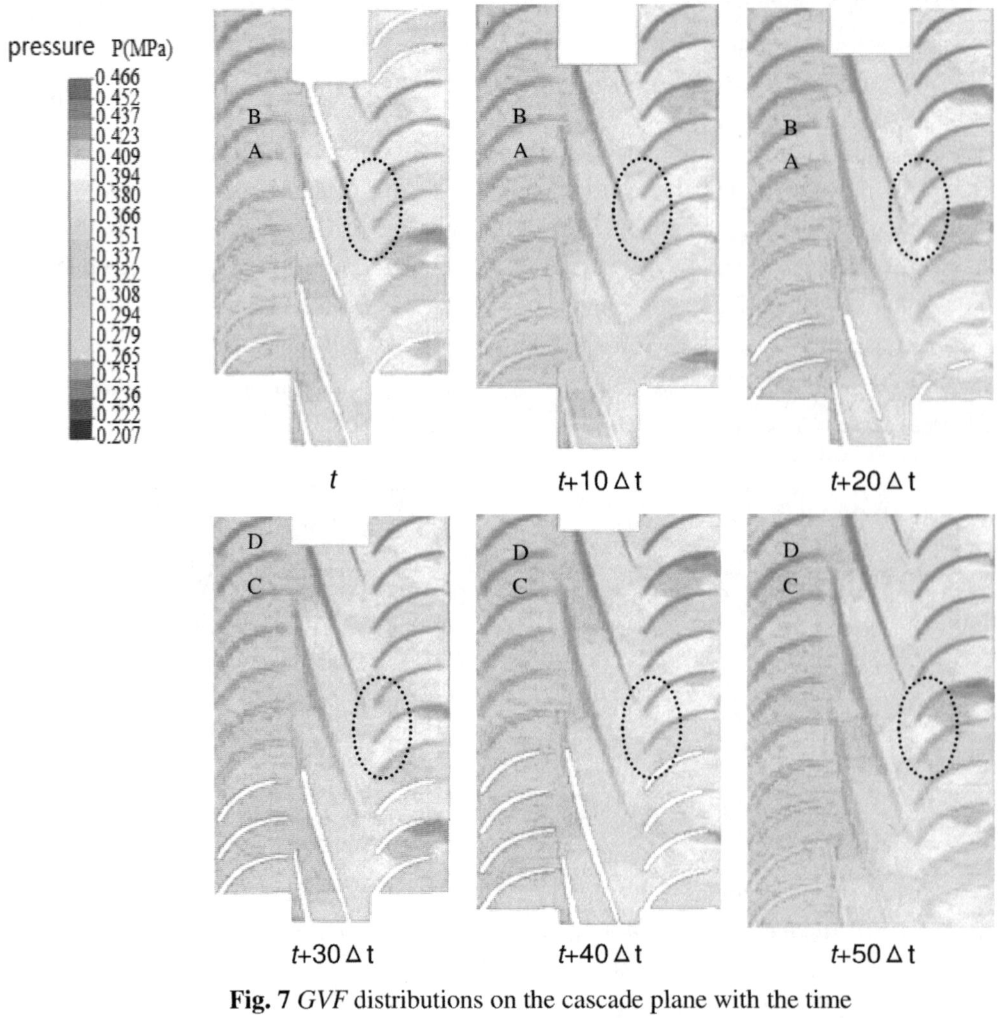

Fig. 7 *GVF* distributions on the cascade plane with the time

3.2.2 *GVF distributions with time.* Figure 8 shows the GVF distribution with time in diffusers at the first and second stage and the impellers at the second stage. Gas pockets formed when the blade head of impellers continuously passed the front diffuser blade tail. A series of discrete gas pockets moved downstream along the impeller passage, and then entered the high GVF area.

From the unsteady calculation results, we find that the transport of gas from the inlet to outlet of impellers was not continuous, but discrete with the formation of gas pocket one after another. At the time *t*, when a blade head in impellers was closed to a blade tail in diffusers at the first stage, the gas pocket was about to appear in the blade head of impellers. At the time $t+10\Delta t$, gas pockets formed when the blade head of impellers passed the blade tail of diffusers, with the further movement of impeller blades, bubbles grew bigger and moved downstream. "Discrete movement of gas pocket", the phenomenon happened when the blade of impellers passed the blade of diffusers, all these showed that the appearance of bubbles were affected by the rotational impellers.

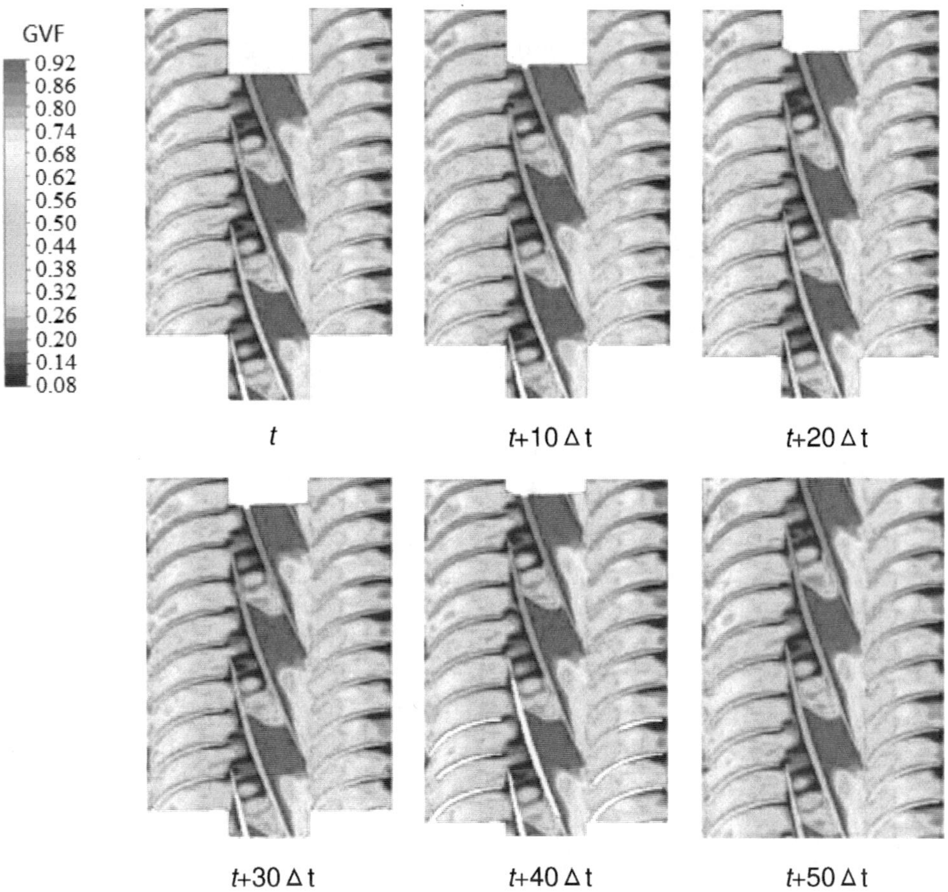

Fig. 8 *GVF* distributions on the cascade plane with the time

4. Experimental comparison

Compared with the experiment, we find that the experimental phenomena were in good agreement with the results of unsteady calculation. As is shown in Figure 9 and Figure 10:

(1) Gas pockets first appeared in impeller passages at the first stage. There were no small bubbles closed to the gas pocket, and the phenomenon meant that bubbles around gathered the gas pocket. Fluid was forced to move along the gap due to the existence of gas pockets.

(2) Compared to the first stage, the position of the gas pocket at the second stage was behind the first one and the scale of the pocket was smaller. Similar to the first stage, the gas pocket was closed to the impeller hub and stretched from the pressure surface to suction surface of blades. The difference was caused by the compressibility of gas.

(3)The position of the gas pocket at the third stage was behind the second one and the scale of the pocket was smallest. Similar to the first and second stage, the gas pocket was closed to the impeller hub and stretched from the pressure surface to suction surface.

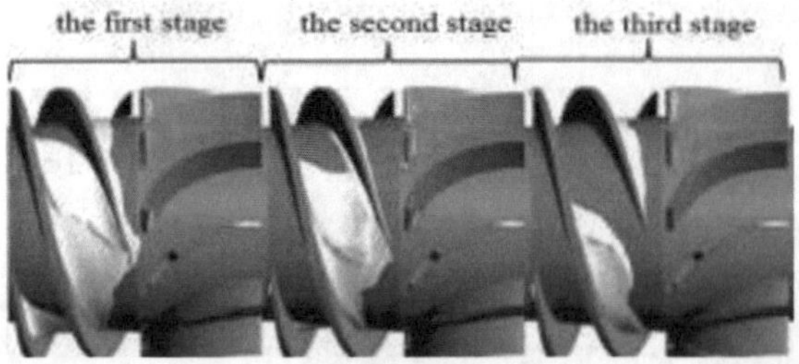

Fig. 9 Gas pockets obtained from transient simulation

Fig. 10 Gas pockets obtained from visualization experiment

5 Calculations

The mixture of water and air was selected as the medium. A three-stage rotodynamic multiphase pump developed by authors has been selected to do unsteady numerical simulation and visualization experiments. With the unsteady calculation, we analysed the velocity distribution, the pressure distribution and the GVF distribution in the second compression cell, as well as the formation of discrete gas pockets.

(1)Gas pockets appeared both in pressure surface and suction surface. It first gathered in pressure surface, then stretched to the suction surface for the rotational impellers, finally be closed to the outlet of impellers.

(2)The gas mainly gathered in pressure surface of the impeller hub. There were a lot of vortices in the hub and the flow was in disorder. The supercharging increased gradually from the first to third compression cell but decreased with the increase of the GVF.

(3)The alternative variation of pressure and GVF at the interface was caused by the dynamic and static interference between impellers and diffusers, which had an effect on flow upstream and downstream; the transport of gas from the inlet to outlet of impellers was not continuous, but discrete with the formation of gas pocket one after another.

IOP Conf. Series: Materials Science and Engineering **129** (2016) 012007 doi:10.1088/1757-899X/129/1/012007

References

[1] Cheng L, Chen T 1999 The Performance and the Present Research Situation of Multiphase Pumps. *Journal of Oil Field Equipment*, **28(5)**:22-25.

[2] Li Q, Xue D 2000 The Research on Multiphase Pumps of Oil and Gas, *Journal of China Offshore Oil and Gas*, **12(1)**: 47-51.

[3] Yves C, Elise A, Sylvain S 2003 Two-Phase Helical Axial Turbines. *SPE Annual Technical Conference and Exhibition*, Colorado, 2003: SPE 84062.

[4] Abelsson C, Busland H, Henden R, et al. 2011 Development and testing of a hybrid boosting pump[C]// *Proceedings of the Offshore Technology Conference. Houston: Offshore Technology Conference*, 2011: OTC 21516

[5] Gao T, Yang D, Cao F, et al. 2011 Temperature and Thermodynamic Deformation Analysis of the Rotors on a Twin Screw Multiphase Pump with High Gas Volume Fractions. *Journal of Zhejiang University Science A*, **12(9)**: 720-730.

[6] Caridad J, Asuaje M, Kenyery F, et al. 2008 Characterization of a Centrifugal Pump Impeller under Two-Phase Flow Conditions. *Journal of Petroleum Science and Engineering*, **63(1)**: 18-22.

[7] Zhang J, Cai S, Zhu H, et al. 2014 Experimental Investigation of The Flow at the Entrance of a Rotodynamic Multiphase Pump by Visualization. *Journal of Petroleum Science and Engineering*, **2(26)**: 254-261.

[8] Yu Z, Zhang Q, Huang R, et al. 2013 Numerical Simulation of Unsteady Flow in Multiphase Rotodynamic Pumps. *Journal of Drainage and Irrigation Machinery Engineering*, **04**: 284-288.

[9] Sooran N, Seyed H, Hashemabadi. 2009 CFD simulation of Inlet Design Effect on Deoiling Hydrocyclones Separation Efficiency. *Chemical Engineering and Technology*, **32(12)**: 1885-1893.

[10] Dogru A H, Hamoud A A, Barlow S G 2004 Multiphase Pump Recovers More Oil in a Mature Carbonate Reservoir. *Journal of Petroleum Technology*, **56(02)**: 64-67.

[11] Humoud A A, Boudi A A, Al-Qahtani S D, et al. 2008 Deployment of a Rotodynamic Multiphase Pump in a Remote Hilly-terrain Oil Field in Saudi Arabia. *Abu Dhabi International Petroleum Exhibition and Conference*, Abu Dhabi, **2008**: SPE 117462.

[12] Scott S L 2003 Multiphase Pumping Addresses a Wide Range of Operating Problems. *Oil & Gas Journal*, **101(37)**: 59-71.

[13] Li Q, 2001 The Performance Research on Multiphase Pump of Oil and Gas. *Journal of China Offshore Oil and Gas*, **13(2)**: 52-56.

IOP Conf. Series: Materials Science and Engineering **129** (2016) 012008 doi:10.1088/1757-899X/129/1/012008

Quasi-three dimensional hydraulic design and performance calculation of high specific speed mixed-flow pump

M Su[1], Y X Zhang[2], J Y Zhang[2] and H C Hou[2]

1. China Petroleum Engineering Co., Ltd., Beijing Company, Beijing 100085, China
2. College of Mechanical and Transportation Engineering, China University of Petroleum, Beijing,102249, China

E-mail: jqsumin@163.com

Abstract. According to the basic parameters of 211-80 high specific speed mixed-flow pump, based on the quasi-three dimensional flow theory, the hydraulic design of impeller and its matching spaced guide vanes for high specific speed mixed flow pump was completed, in which the iterative calculation of S_1, S_2 stream surfaces was employed to obtain meridional flow fields and the point-by-point integration method was employed to draw blade camber lines. Blades are thickened as well as blade leading edges are smoothed in the conformal mapping surface. Subsequently the internal fields of the whole flow passage of the designed pump were simulated by using RANS equations with RNG $k-\varepsilon$ two-equation turbulent model. The results show that, compared with the 211-80 model, the hydraulic efficiency of the designed pump at the optimal flow rate increases 9.1%. The hydraulic efficiency of designed pump in low flow rate condition (78% designed flow rate) increases 6.46%. The hydraulic efficiency in high flow rate areas increases obviously and there is no bad phenomenon of suddenly decrease of hydraulic efficiency in model pump. From the distributions of velocity and pressure fields, it can be seen that the flow in impeller is uniform and the increase of pressure is gentle. There are no obvious impact phenomenon on impeller inlet and obvious wake shedding vortex phenomenon from impeller outlet to guide vanes inlet.

1. Introduction

With more attention of national economic development is paid on energy conservation and emission reduction, the high specific speed mixed-flow pumps with high hydraulic efficiency, strong cavitation erosion resistance, excellent running stability are required urgently. At the present stage, the pump design usually adopts similar conservation method or speed coefficient method based on the similarity theory, or hydraulic design method based on one dimensional or two dimensional theories, but these methods are empirical and design qualities largely depend on experience of designers and improvement after hydraulic performance tests [1]. It is not common for hydraulic design by using quasi-three dimensional or three dimensional methods. Based on the two families of stream surfaces theory, Binghao et al. put forward the direct and inverse problem iterative design method of mixed-flow pump impeller and complete the design of impeller of mixed-flow pump which specific speed is 348[2]. This method can effectively make up for the defect of traditional design method which the calculation of meridional flow only satisfies the fluid continuous equation. Otherwise, the influence of blade shape on the calculation of meridional flow field is also considered. However, there are relatively less researches on high specific speed mixed-flow pump based on quasi-three dimensional theory. On one hand, for the restriction of specific speed, the specific speed is generally from 200 to 500 in the traditional hydraulic design of mixed-flow pump model impellers. With the increase of specific speed, axial flow pumps are for most use and there are not good similar model pumps for reference. In addition, the development of high specific mixed-flow pumps becomes more and more difficult with the uncertainty of empirical parameters selection based on quasi-three dimensional hydraulic design method. On the other hand, due to the complex structure of high specific speed mixed-flow pump, there are obvious bend torsion characteristics of impeller blades. Otherwise, the shape of hub, the clearance between impeller and shroud as well as the clearance between impeller and guide vanes have obvious influences on the three-dimensional unsteady turbulent flow of high specific mixed-flow pump. Up to now, the cognition that people on the inner flow of high specific mixed-flow pump is limited [3, 4].

According to the basic parameters of 211-80 high specific speed mixed-flow pump [5], based on the quasi-three dimensional theory, the self-complied program is adopted to complete the hydraulic design of new mixed-flow pump. The internal flow fields of the designed pump were simulated by using RANS equations with RNG $k-\varepsilon$ two-equation turbulent model. The purpose of this paper is to explore a new rapid development method of high specific mixed-flow pumps.

2. Quasi-three dimensional hydraulic design

The quasi-three dimensional hydraulic design of designed pump is completed by calculation procedure (Program flow chart is shown in figure 1).

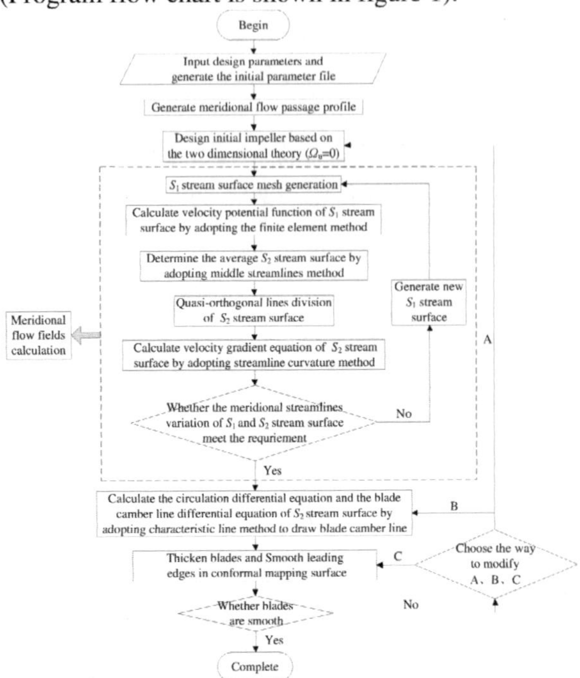

The concrete steps are:

①The initial axial flow passage profile is completed based on the basic design parameters ($Q_d=1440$ m³/h, $H=6.53$ m, $n=1450$ r/min).

②The initial impeller and spaced guide vanes designed based on the two dimensional theory ($\Omega_u=0$).

③ The iterative calculation of S_1 stream surface and S_2 stream surface is conducted.

④ The point-by-point integrate method is applied to draw blade camber line according to the meridional velocity distribution obtained from step ③.

⑤ Blade thickening and smoothing are completed in conformal mapping surface.

Figure 1. The flow chart of quasi-three dimensional hydraulic design

2.1. Initial impeller confirmation

The specific speed of designed mixed-flow pump is $n_s=\dfrac{3.65n\sqrt{Q}}{H^{3/4}}=819$. According to the basic parameters of 211-80 high specific speed mixed-flow pump, combined with traditional speed coefficient method, the initial impeller flow passage profile is determined by referring to the similar specific speed hydraulic model. In order to adapt to the operating condition regulation, the adjustable blade design method is adopted to guarantee the high efficient operating in a wide range of flow rate and head. The blade area is designed as hemispherical and the inlet and outlet of blade are smoothed by arc. The angle between the blade rotational axis and pump axis is normally chosen as $\theta_d = 45° \sim 60°$ and the high specific speed pumps chose the great value. The designed pump chooses as 55°. The range of hub ratio is 0.45~0.65. The designed pump chooses as 0.52. Figure 2 is the profile of meridional flow passage of impeller. The initial impeller is confirmed after the determination of the profile of meridional flow passage based on two dimensional flow theory ($\Omega_u = 0$) [6].

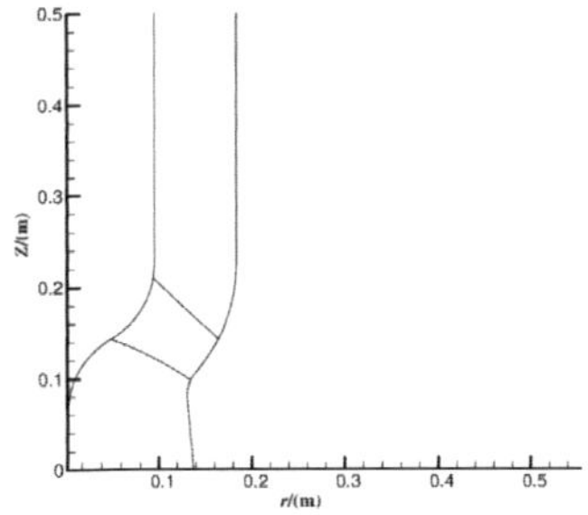

Figure 2. Impeller meridional flow passage profile

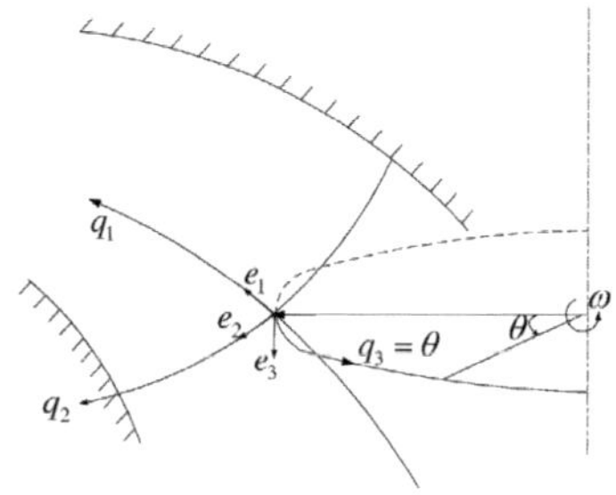

Figure 3. Orthogonal curvilinear coordinate

2.2. Meridional flow field calculation

The calculation of meridional flow field is to use the geometry and flow boundary conditions of whole flow domain formed by the flow passage of initial impeller and guide vanes to achieve. The orthogonal curvilinear is adopted for calculation (shown in figure 3), q_1, q_2 and q_3 represent the direction of meridional streamlines, water section lines and peripheral direction respectively. The basic suppose of steady, incompressible, non-viscous, irrotational and only gravity is introduced. The flow of S_1 stream surface is supposed irrational and the flow of S_2 stream surface is axial symmetry. Where, $H_1 = H_1(q_1, q_2)$, $H_2 = H_2(q_1, q_2)$, $H_3 = r$, $\upsilon_1 = w_1$, $\upsilon_2 = w_2$, $\upsilon_3 = w_3 + r\omega$.

Supposing the absolute motion of the coming flow of impeller is irrational ($\nabla \times \boldsymbol{v} = 0$). There is the corresponding three dimensional potential function ($\nabla \times \boldsymbol{v} = 0$). So, the calculation of the flow of S_1 stream surface transfers to the calculation of velocity potential function φ. The conformal transformation [7] is introduced, so the potential function equation of S_1 stream surface on conformal surface (x-y surface) can be expressed by the follow equation,

$$\begin{cases} \dfrac{\partial}{\partial x}\left(h\dfrac{\partial \varphi}{\partial x}\right) + \dfrac{\partial}{\partial y}\left(h\dfrac{\partial \varphi}{\partial y}\right) = 0 \\ h = h(y) \end{cases} \tag{1}$$

Where, h represents the flow layer thickness of S_1 stream surface.

Ritz variation is introduced to solve potential function (1) by giving the inlet, outlet, surface and periodic boundary conditions. The variation function is as follow,

$$I = -\frac{1}{2}\iint_D h\left[\left(\frac{\partial \varphi}{\partial x}\right)^2 + \left(\frac{\partial \varphi}{\partial y}\right)^2\right]\mathrm{d}x\mathrm{d}y + \oint_{\Gamma\Omega} h\frac{\partial \varphi}{\partial n}\varphi \mathrm{d}S$$

The finite element method is adopted for numerical solution.

For average S_2 stream surface ($q_2 = const$), q_1 is independent variable which represents coordinate surface of streamlines. So there is $\upsilon_2 = w_2 = 0$ (υ_i represents the absolute velocity component, w_i is velocity component of relative coordinate). The flow passage of impeller and spaced guide vans is divided into blade section and non-blade section. The corresponding meridional velocity gradient equation is,

a) The velocity gradient equation in blade section is,

$$\frac{\mathrm{d}\upsilon_1}{\mathrm{d}l} = A\upsilon_1 + B + \frac{C}{\upsilon_1} \tag{2}$$

where,

$$A = \frac{1}{1+\left(r\frac{\partial\theta}{\partial m}\right)^2}\left[-\left(\frac{1}{\tau}\frac{\partial\tau}{\partial m} + \frac{\sin\gamma}{r} + \frac{1}{\cos\delta}\frac{d\gamma}{dl}\right)\left(\sin\delta + r^2\frac{d\theta}{dl}\frac{\partial\theta}{\partial m}\right) + \frac{1}{\cos\gamma}\frac{\partial\gamma}{\partial m}\left(1+r^2\sin\delta\frac{d\theta}{dl}\frac{\partial\theta}{\partial m}\right) + \frac{d\theta}{dl}\frac{\partial}{\partial m}\left(r^2\frac{\partial\theta}{\partial m}\right) - \frac{\partial\theta}{\partial m}\frac{d}{dl}\left(r^2\frac{\partial\theta}{\partial m}\right)\right]$$

$$B = \frac{2\omega r}{1+\left(r\frac{\partial\theta}{\partial m}\right)^2}\left[\frac{d\theta}{dl}\sin\gamma - \frac{\partial\theta}{\partial m}\cos(\gamma-\delta)\right]$$

$$C = \frac{2\omega r}{1+\left(r\frac{\partial\theta}{\partial m}\right)^2}\frac{dE_r}{dl}$$

b) The velocity gradient equation in non-blade section is,

$$\frac{\mathrm{d}\upsilon_1}{\mathrm{d}l} = A\upsilon_1 + \frac{C}{\upsilon_1} \tag{3}$$

where,

$$A = \frac{1}{\cos\delta}\left(\frac{\partial\gamma}{\partial m} - \frac{\mathrm{d}\gamma}{\mathrm{d}l}\sin\delta\right) - \frac{\sin\gamma}{r}\sin\delta \qquad C = -\frac{r^2\omega + r\upsilon_3}{r^2}\frac{\mathrm{d}(r\upsilon_3)}{\mathrm{d}l} + \frac{\mathrm{d}E_r}{\mathrm{d}l}$$

The meaning of variable in the above mentioned equations can be seen from figure 4. Based on two-dimensional flow theory ($\Omega_u = 0$) and iterative calculation of S_1 and S_2 stream surfaces, figure 5 and 6 represent the meridional velocity distribution from hub to shroud along meridional surface streamlines in the whole computational domain of impeller respectively. From the comparison of both, there is a little change of meridional surface velocity distribution in the inlet of domain and there is obvious change of meridional surface velocity distribution in blade area and outlet of impeller.

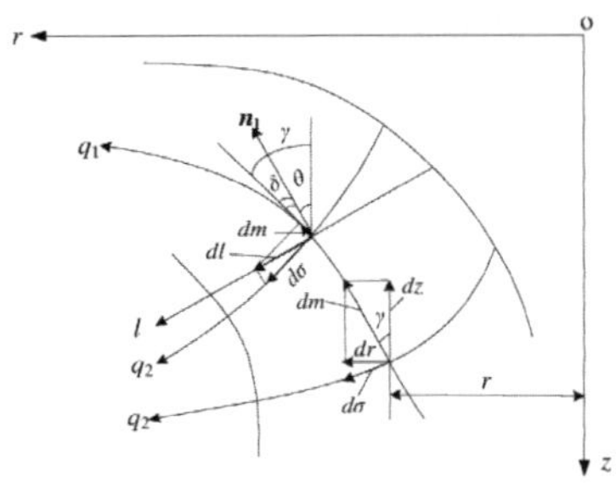

r — radius of meridional surface projection
τ — excretion coefficient
θ — central angle
γ — angle between plumb line and meridional surface streamlines
δ — angle between normal of quasi-orthogonal lines and meridional surface streamlines
ω — rotational angular velocity of impeller
E_r — mechanical energy of relative motion of unit mass fluid

Figure 4. The geometry relationship on S_2 stream surface after introduction of quasi orthogonal line

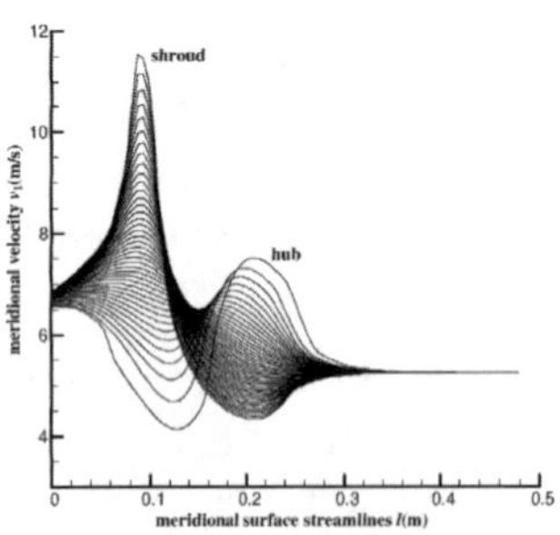

Figure 5. The Meridional velocity distribution based on two-dimensional flow theory($\Omega_u = 0$)

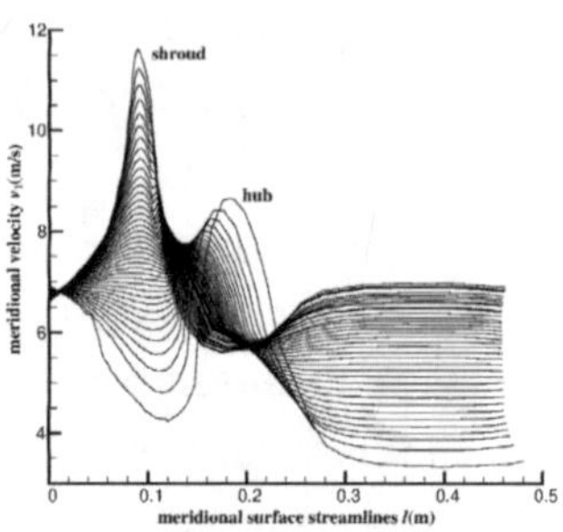

Figure 6. The Meridional velocity distribution obtained by iterative calculation

2.3. Blade camber line confirmation

The circulation differential equation and the blade camber line differential equation are show in equations (4) and (5).

$$\frac{\mathrm{d}(\upsilon_3 r)}{\mathrm{d}q_2} = F(q_1, q_2)\frac{\upsilon_1 r^2}{\upsilon_3 r + \omega r^2} \tag{4}$$

where,

$$F(q_1,q_2)=\left\{\upsilon_1\left[\frac{\partial\gamma}{\partial m}\frac{1}{\cos\delta}-\left(\frac{\partial\gamma}{\mathrm{d}l}\frac{1}{\cos\delta}+\frac{\sin\gamma}{r}+\frac{1}{\tau}\frac{\partial\tau}{\partial m}\right)\sin\delta\right]\right.$$
$$\left.-\frac{\mathrm{d}\upsilon_1}{\mathrm{d}l}+\frac{1}{\upsilon_1}\frac{\mathrm{d}E_r}{\mathrm{d}l}\right\}\frac{\mathrm{d}l}{\mathrm{d}q_2}$$

There is $dq_2 = 0$ on characteristic line $q_2 = const$.

$$\frac{\mathrm{d}\theta}{\mathrm{d}m} = \frac{\upsilon_3 r + \omega r^2}{\upsilon_1 r^2} \tag{5}$$

Normally, the low pressure edge of blade (the inlet of impeller) and the velocity torque on the low pressure edge of blade are given. The low pressure edge is a characteristic line on meridional surface $\theta = const$. Besides, the velocity torque distribution of hub and shroud is given to draw blade camber lines, the wrapping angle of blade θ can be obtained by equation (5). The quartic polynomial is adopted to show the velocity torque distribution [8] along meridional streamlines when the shape of meridional flow passage as well as the position of inlet and outlet of blade are given. Then, in order to reduce the dependence of experience on designed results, one of the parameters a is determined to control the distribution of velocity torque of blade through the theoretical analysis. The simplified quartic distribution function expressed by the follow equation,

$$f(x) = ax^4 + (p-2-2a)x^3 + (3-p+a)x^2 \tag{6}$$

Where, the relative length of meridional streamlines in blade area can be expressed by $x = l / l_0$.

The designed pump chooses p=0, a=1.6. The velocity torque distribution of hub in blade area is shown in figure 7.

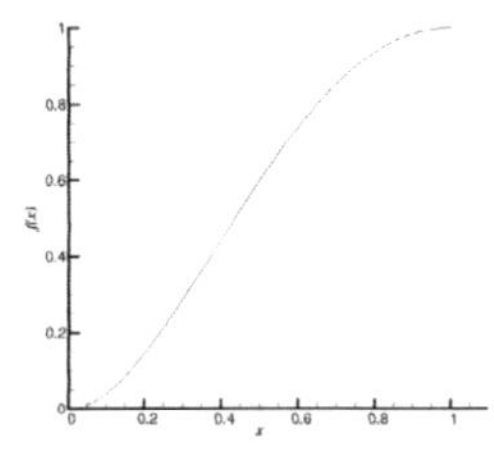

Figure 7. The velocity moment distribution of hub on blade

2.4. Thickening and smoothing of blades

Blades are thickened as well as blade leading edges are smoothed in the conformal mapping surface [9]. The spaced guide vanes adopted the same designed method with impeller. Finally, the inlet and outlet of impeller and spaced guide vanes blade after thickening and smoothing can be obtained, the three dimensional model of impeller and spaced guide vanes can be seen in figure 8.

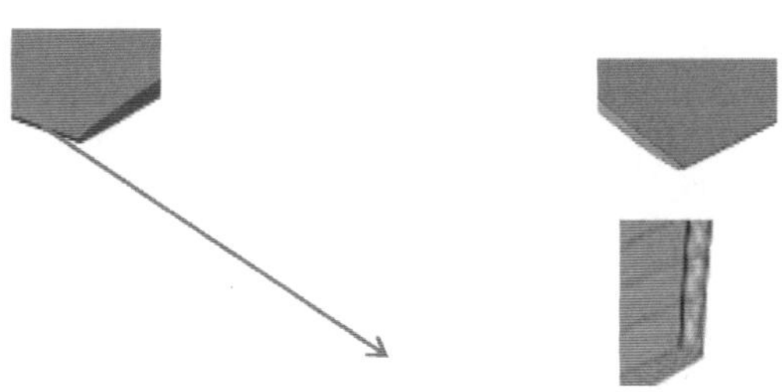

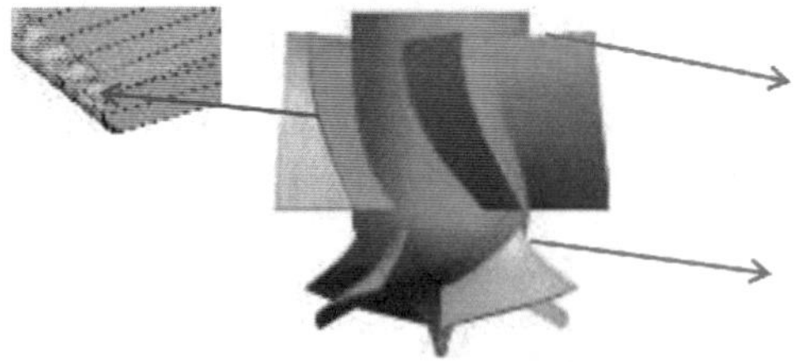

Figure 8. The three dimensional model of impeller and spaced guide vanes

3. Numerical simulation and results analysis

The whole flow passage geometrical model of designed pump is shown in figure 9. In order to compare with the hydraulic performance of 211-80 model pump conveniently, the calculation choose the same eight operating point, the corresponding flow rate are 1728.0, 1671.3, 1551.7, 1499.8, 1457.7, 1347.2, 1224.2 and 1118.8m^3/h.

Unstructured tetrahedral mesh discrete calculation domain is adopted under the comprehensive consideration of time and accuracy. The meshing result is shown in figure 10. Table 1 shows grid independence test results. The influence of grid number on numerical calculation results can be ignored when the correlation of head and efficiency with different grid number is lower than 1%. The finalized total grid number of designed pump is 242 million.

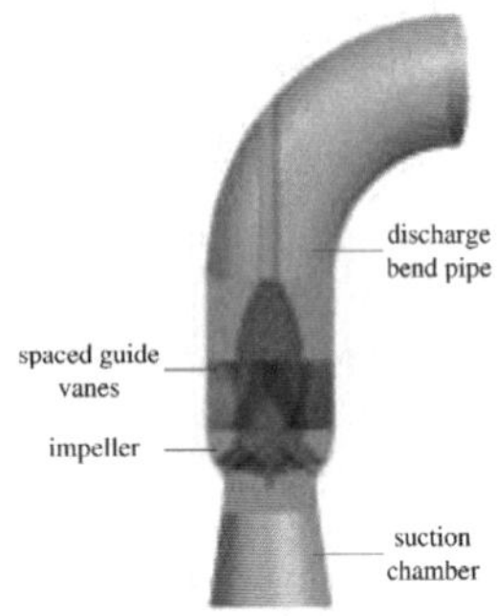

Figure 9. The whole flow passage geometrical model

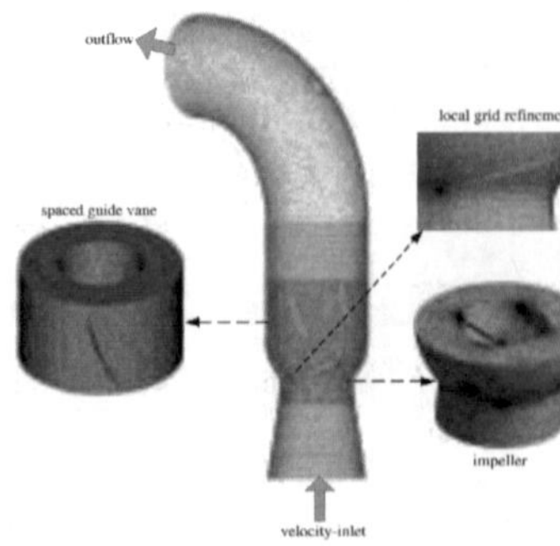

Figure 10. The grid of designed pump

Table 1. The head and efficiency of designed mixed-flow pump with different grids of impeller

numerical order	impeller	entirety	Head H (m)	Efficiency η (%)
1	680487	1707715	5.79	87.47
2	1088790	2116018	5.78	87.36
3	1396984	2424212	5.70	87.73
4	1609834	2637062	5.77	87.38

The RNG turbulent model and SIMPLE algorithm are applied to solve the RANS equations. The Multiple Reference Frame (MRF) model is applied to take into account the interaction between stationary parts and rotating impeller. For such calculations, standard wall functions based on the logarithmic law have been used. Standard Scheme has been used for pressure terms and second-order upwind discretization scheme has been used for convection terms. The boundary conditions are as follows: inlet is the velocity inlet and assumed as a uniform distribution, outflow is given as boundary condition at outlet, and the solid walls with non-slip condition such as blade surface, hub and shroud are given the moving wall.

3.1. The performance curves

The flow rate–head curve and flow rate-efficiency curve of designed pump and model pump can be seen from figure 11. From the figure 11, it can be seen that, the experimental results and CFD simulation results is basically consistent. Thus the reliability of the numerical model was validated. In figure 11, the imaginary line A represents low flow rate condition ($0.78Q_d$), the imaginary line B represents optimal flow rate condition, the imaginary line C represents high flow rate condition ($1.16Q_d$). Compared with the simulation results of model pump, the head of designed pump increases 0.81m and the hydraulic efficiency of designed pump increases 9.1% at B flow rate condition. The hydraulic efficiency of designed pump increases 6.46% at A flow rate condition. In general, the designed pump can restrain the phenomenon of a sudden drop of hydraulic efficiency and the high efficiency areas of designed pump broaden significantly in high flow rate conditions.

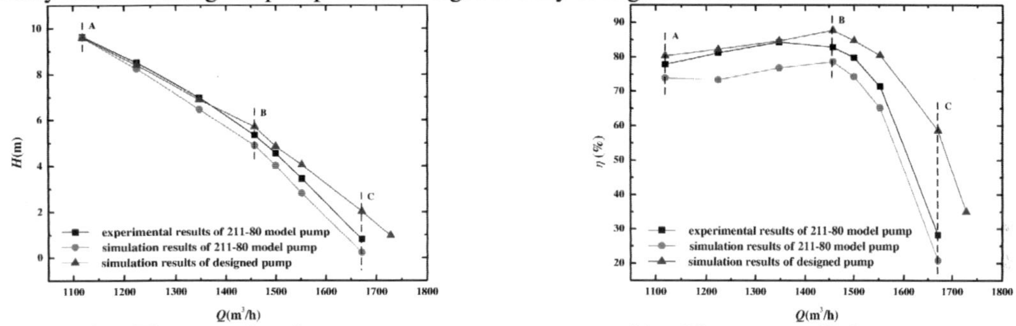

a) Flow rate-head curve b) Flow rate- efficiency curve

Figure 11. The comparison of performance curves between designed pump and 211-80 model

3.2. The analysis of inner flow field

The static pressure distribution is as shown in figure 12. From the distribution of static distribution can be seen that, with the increase of flow rate, static pressure decreases at same location of impeller. The static pressure distributions are relatively uniform. There is no obvious flow separation phenomenon.

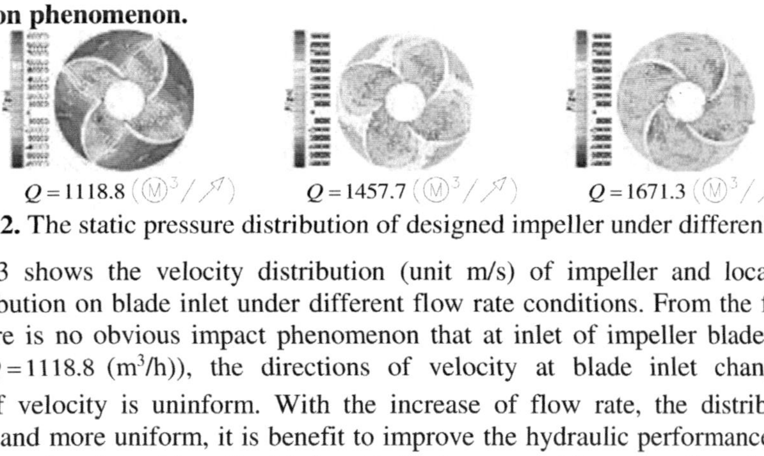

$Q = 1118.8$ (m³/h) $Q = 1457.7$ (m³/h) $Q = 1671.3$ (m³/h)

Figure 12. The static pressure distribution of designed impeller under different flow conditions

Figure 13 shows the velocity distribution (unit m/s) of impeller and local amplification of velocity distribution on blade inlet under different flow rate conditions. From the figure 13, it can be seen that, there is no obvious impact phenomenon that at inlet of impeller blade. At low flow rate condition ($Q = 1118.8$ (m³/h)), the directions of velocity at blade inlet change obviously, the distribution of velocity is uninform. With the increase of flow rate, the distributions of velocity become more and more uniform, it is benefit to improve the hydraulic performance. And the velocity increases at same location of impeller blade with the increase of flow rate, the velocity distribution is still uniform, there is no obvious phenomenon of flow separation.

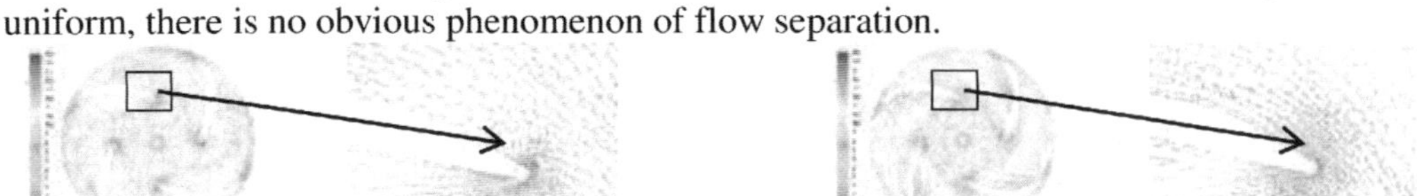

$Q = 1118.8$ **(m³/h)** $Q = 1457.7$ **(m³/h)**

IOP Conf. Series: Materials Science and Engineering **129** (2016) 012008 doi:10.1088/1757-899X/129/1/012008

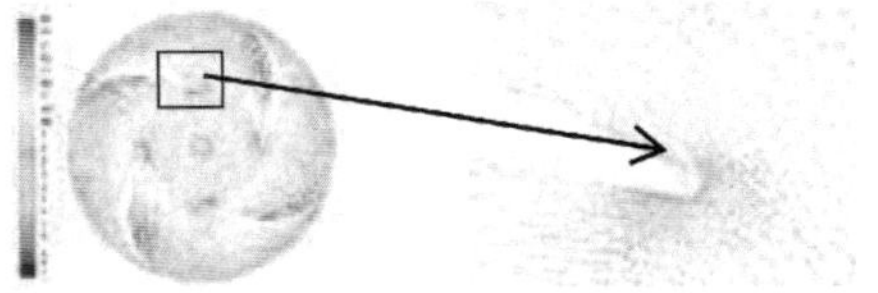

$Q=1671.3$ **(m³/h)**

Figure 13. The velocity distribution (unit m/s) of impeller and local amplification of velocity distribution on blade inlet under different flow rate conditions

In order to observe the streamlines distribution of designed pump, figure 14 show the streamlines distribution of inner flow field of midsection which is parallel to axis of designed pump. From the figure 14, it can be seen that the streamlines distribution is uniform at three different flow rate conditions, and there is no backflow phenomenon caused by flow separation at outlet of pump. There is also no obvious wake shedding vortex phenomenon from impeller outlet to guide vane inlet.

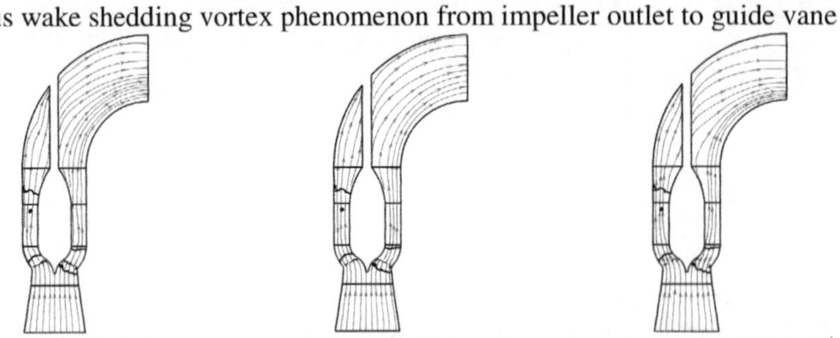

$Q=1118.8$ (m³/h) $Q=1457.7$ (m³/h) $Q=1671.3$ (m³/h)

Figure 14. The Streamlines distribution of inner flow field

4. Conclusion

(1) This paper completed the quasi-three dimensional hydraulic design of impeller and spaced guide vanes of high specific speed mixed-flow pump which specific speed is 819. The meridional speed distribution got by the iterative calculation of S_1 and S_2 stream surfaces is obviously different from the meridional speed distribution got by the traditional two dimensional theory($\Omega_u=0$), and the former is close to the actual flow.

(2) Compared with the 211-80 model, the high specific speed mixed-flow pump designed by the quasi-three dimensional theory has better Hydraulic performance. The hydraulic efficiency of the designed pump on the optimal flow rate condition increases 9.1%. The hydraulic efficiency of designed pump in low flow rate condition (78% designed flow rate) increases 6.46%. The hydraulic efficiency in high flow rate condition areas increases obviously and there is no bad phenomenon of suddenly decrease of hydraulic efficiency in model pump.

(3) The pressure distribution of impeller of designed pump is uniform. There is no obvious impact phenomenon on impeller inlet. The stream lines distribution of whole designed pump is gentle, and there is no obvious wake shedding vortex phenomenon from impeller outlet to guide vane inlet.

Reference

[1] Zhang X T, Ji C J, Gao G H, et al. 2012 Study on hydraulic design method of mixed-flow pump impeller [J]. *Pump technology*, **1**: 23-27.

[2] Bing H, Zhang Q Z, Cao S L, et al. 2013 Iteration design of direct and inverse problems and flow analysis of mixed-flow pump impellers [J]. *J Tsinghua Univ (Sci&Tech)*, **53(2)**:179-183.

[3] Bing H, Cao S L 2013 Multi-parameter optimization design, numerical simulation and performance test of mixed-flow pump impeller. *Science China Technological Sciences* 56.

[4] Kim S, Lee K Y, Kim J H, et al. 2015 High performance hydraulic design techniques of mixed-flow pump impeller and diffuser [J]. *Journal of Mechanical Science and Technology*, **Vol.29 (1)**:227-240.

[5] Liu Q 2006 Numerical Simulation of Interior Flow for High Specific Speed Mixed-flow Pump and Its Characteristics Prediction [D]. *Jiang Su: Jiang Su University Thermal Engineering department*.

[6] Guan X F 2011 *Modern Pumps Theory and Design* [M]. Beijing: China Astronautic Publishing House, 2011:471-472.

[7] Lin K, Cao S L, Zhu B S, et al. 2008 Flow computations for the design of high specific speed mixed-flow pumps [J]. *J Tsinghua Univ (Sci&Tech)*, **48(2)**:219-223.

[8] Zhang Q Z, Cao S L, Wang H, et al. 2011 Effects of velocity moment distribution law on design of mixed-flow pump impeller [J]. *Journal of Drainage and Irrigation Machinery Engineering*, **29(3)**.

[9] Cao S L, Liang L, Zhu B S, et al. 2005 Design method for impeller of high specific mixed-flow pump [J].*Journal of Jiangsu University(Natural Science Edition)*, **26(3)**.

The eddy-eliminating method of guide cone in the closed sump

Y J Wang[1], L Cheng[1], C Z Xia[1], J R Zhou[1], H Q Yan[2], H Y Jiang[2]

1 School of Hydraulic, Energy and power Engineering
Yangzhou University, Yangzhou, Jiangsu, P.R.China, 225009
2 Jiangsu Water Conservancy Engineering & Technology Consulting Co., Ltd.
Nanjing, Jiangsu, P.R.China, 210029

E-mail: chengli@yzu.edu.cn

Abstract. In order to explore the effect on eddy-eliminating method of guide cone in the closed sump, the simple factor analysis and CFD numerical simulation are applied to calculate the flow field of closed sump and select ω-shaped back wall. ω-shaped back wall is consistent with the stream line in the suction sump, on this basis, CFD numerical simulation is conducted with the eddy-eliminating of the triangle guide cone and traditional guide cone. The results show that, for eddy-eliminating measures, with the height of triangular guide cone from 0 to $0.407H_Z/D_L$, the excessive triangle guide cone hinder water into the flared pipe. With the width of triangular guide cone from 0.5 to $1.0B_Z/D_L$, increasing width of triangular guide cone may increase the pumping hydraulic performance and pumping efficiency. However with the width of triangular guide cone from 0.5 to $1.0B_Z/D_L$, too broad traditional guide cone hinder water into the flared pipe. In the design discharge, whether triangle guide cone or traditional guide cone have a little effect on the efficiency of the pumping station. But in terms of the eddy-eliminating on the bottom of suction sump, it is necessary to set up guide cone.

Introduction

Suction sump of the pumping station is the role to guide the water from the pumping station forebay into pump impeller, which may ensure water at the suction sump to finish transition and get into the impeller chamber smoothly. Current scholars focused on the hydrodynamic performance analyses, hydraulic efficiency numerical solution, numerical optimization of model test and analyses of eddy-eliminating method with the structure of the pump device. Cheng [1] applied numerical simulation to get the geometry parameters of the open sump. Qian [2] did channel experimental study and got geometry parameters of the open water pumping station. Wang Bencheng [3] used numerical

[1] Corresponding author: CHENG Li(1975-), male, PhD, Professor

simulation to show the hydrodynamic performance of the bell mouth. Lu Linguang et al. [4] further researched on open sump and completed hydraulic performance optimization. Research on the closed sump is not much and less than the numerical solution and experimental research of box-culvert inlet flow channel [5] ~ [7].Chen [8] and Zhou [9] did experimental study on the height of the box culvert bell dangling. Yang [10] finished water dynamic characteristics analysis of box culvert type inlet axial flow pump system. Different types and sizes of guide cone are studied on the hydraulic characteristics of the closed sump.

2. Calculation model

The closed sump contain suction sump, impeller, guide vane and outlet sump that join with 90o bent pipe. 3D model of vertical axial - flow pump device is shown in Figure 1-1.

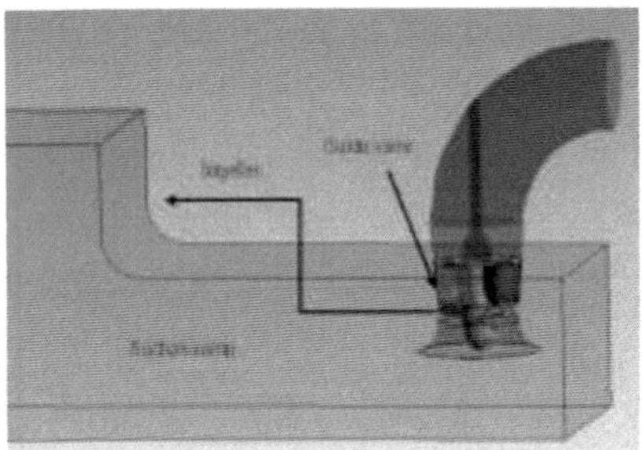

Figure 1-1 3D model of vertical axial-flow pump device

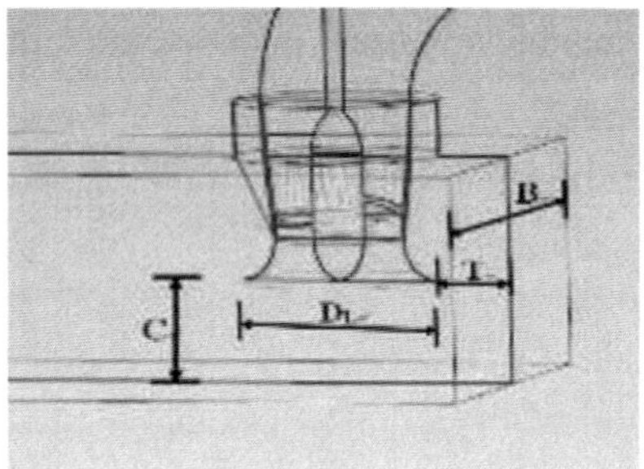

Figure 1-2 Geometric parameters of the closed sump

There are three vanes in impeller. The setting angle of the vane is 0°.The impeller diameter is 1200mm.The speed of the impeller is 370rpm.Guide vane there are seven vanes. The design flow rate of the single pump is $5\,m^3 s^{-1}$.

Typical closed sump structure is rectangular form. The rear wall shape is rectangle. B represents the width of suction sump. T represents the distance from the rear wall. C represents the height of the speaker tube, which can be shown in Figure1-2. D_L represents the diameter of speaker tube. The D_L is 1800mm.The D_L is used as a basic parameter to indicate the geometric parameters of the closed sump.

3. Numerical calculation

3.1 *Governing* equation

Three-dimensional Reynolds N-S equation is applied to describe in-compressible fluid turbulent flow. The turbulence model is used as standard k–ε model. The model by correcting the turbulent viscosity into account the average flow of the rotation and rotational flow, which may deal with high strain rate and streamline the flow of the degree of bend more better.

3.2 *Boundary* conditions and grid meshing

The inlet flow condition is mass flow. The outlet flow is free discharge. The outlet boundary exits the more distant oscula. Due to the pressure or velocity on the boundary are unknown, so the use of free outflows. No-slip conditions be specified in the solid side wall. (Namely $u = v = w = 0$).In the near-wall region of the velocity distribution is determined in accordance with the law of the wall. There is mutual coupling flow existing the interface of suction sump and impeller, the impeller and guide vanes. Dealing the moving and fixed interface play an important influence about computing. The moving and fixed interface with reference system model is adopted to ensure the continuity of the interface.

Taking into account the complexity of the impeller and guide vane, unstructured grid subdivision is adopted .The local grid is refined. Grid-independent verification is completed. Number of grid cells is to 1.9 million. Suction sump is divided 0.6 million grid cells. Impeller is divided 0.21 million grid cells. Guide vanes are divided 0.49 million grid cells. Outlet sump is divided 0.6 million grid cells. Suction sump is conducted grid-independent verification. Grid cells are from 0.2 million to 0.8 million. Efficiency is used as an evaluation Indicators. When number of grid cells from 0.2million to 0.8million, efficiency value exist fluctuation that is from 52.7% to 56.8%.Figure 3-1 indicates that there is a certain number of cells influence on the calculation. Final 0.6 million grid cells is ensured.

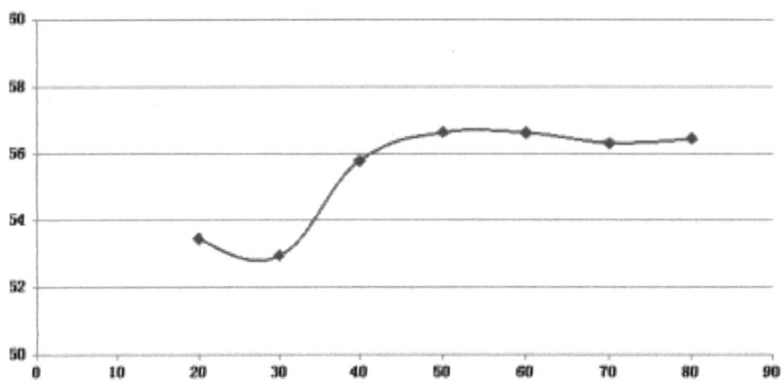

Figure 3-1 Grid-independent verification

4. Numerical Simulation and Analysis

4.1 Case of *guide* cone

In order to eliminate eddy on the bottom of the suction sump and research on the impact guide cone of the pump stations hydraulic performance, nine cases are designed (including not add guide cone program). Table 1 is optimized solution for guide cone. H_z is the height of triangular guide cone.B_z represents the bottom width of triangular guide cone.

Table 1 Optimized solution for guide cone

Case Number	B/D_L	T/D_L	C/D_L	Shape of rear wall	Height H_z/D_L	Width B_z/D_L	Remark

1	2. 222	0. 278	0. 5	ω shape	0	0	No guide cone
2	2. 222	0. 278	0. 5	ω shape	0. 204	1. 0	Triangular guide cone
3	2. 222	0. 278	0. 5	ω shape	0. 305	1. 0	Triangular guide cone
4	2. 222	0. 278	0. 5	ω shape	0. 407	1. 0	Triangular guide cone
5	2. 222	0. 278	0. 5	ω shape	0. 204	0. 5	Triangular guide cone
6	2. 222	0. 278	0. 5	ω shape	0. 204	0. 67	Triangular guide cone
7	2. 222	0. 278	0. 5	ω shape	0	0. 5	Traditional guide cone
8	2. 222	0. 278	0. 5	ω shape	0	0. 67	Traditional guide cone
9	2. 222	0. 278	0. 5	ω shape	0	1. 0	Traditional guide cone

4.2 Triangular guide cone

4.2.1 Height of triangular guide cone Table 2 is comparative efficiency of different height of triangular guide cone. Table 2 shows that increasing height of triangular guide cone may cause the pumping hydraulic performance degradation, which indicate excessive guide cone to some extent hindered the water into the flare, and affect the detour flow below the flare. Several programs streamline are relatively homogeneous. Contrasting case 4 with case 2, we can find hydraulic performance difference is a smaller and efficiency amplitude is 0.27%.Compared with no add guide cone, hydraulic performance of case 4 decrease amplitude by 0.34%.Contrasting case 2 with case 1, hydraulic performance of case 4 decrease 0.07%.Case 2 to case4 show that the pressure increase uniformly at the bell mouth, without negative pressure. On terms of the eddy-eliminating in the suction sump bottom of the bell mouth .If select triangular guide cone, case 2 is recommended as preferable case.

Table 2 Comparative efficiency of triangular guide cone

Case Number	1	2	3	4
Height H_Z/D_L	0	0. 204	0. 305	0. 407
η (%)	56. 633	56. 560	56. 432	56. 296

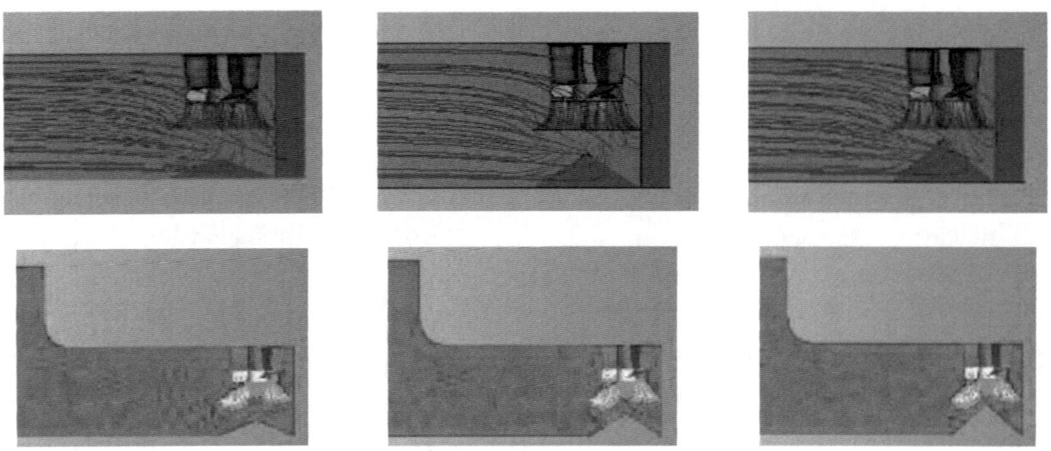

(a) Case 2 (b) Case 3 (c) Case 4

Figure 4-1 Axial pressure distribution of suction sump **Figure** 4-2 Axial pressure distribution of suction sump **Figure** 4-3 Axial pressure distribution of suction sump

4.2.2 Width of triangular *guide cone* Table 3 shows the efficiency comparison about different width of triangular guide cone bottom. Table 3 shows that increasing width of triangular guide cone may increase the pumping hydraulic performance and pumping efficiency .Hydraulic performance of case 2 decrease amplitude by 0.38%.Too small triangular guide cone bottom width is easy to form a large space below the rear flare leaving excessive retention area, which decrease the pumping efficiency. On the axial pressure surface the velocity vector distribute evenly. Flow velocity evenness is 0.856, 0.823, 0.865.If select triangular guide cone, case 2 is recommended as preferable case.

Table 3 Comparative efficiency of triangular guide cone

Case Number	5	6	2
Width B_z/D_L	0.5	0.67	1.0
η (%)	56.176	56.179	56.560

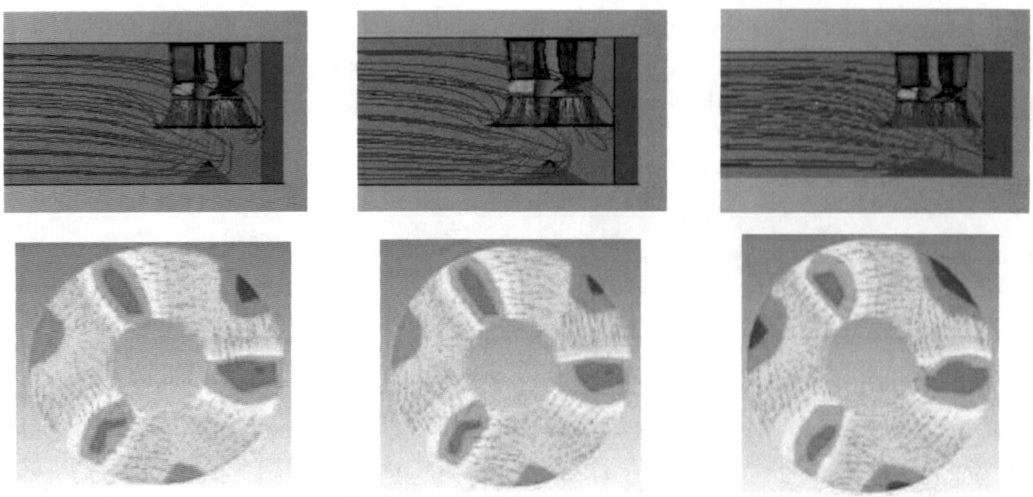

(a) Case 5 (b) Case 6 (c) Case 2

Figure 4-4 Pressure distribution of suction sump exit face **Figure** 4-5 Pressure distribution of suction sump exit face **Figure** 4-6 Pressure distribution of suction sump exit face

4.3 Traditional guide cone

Table 4 is comparative efficiency of different width of traditional guide cone bottom. Table 4 shows that increasing width of triangular guide cone decrease the pumping hydraulic performance and pumping efficiency, which indicates too broad guide cone to some extent hindered the water into the flare. Contrasting case 9 with case 7,hydraulic performance difference is a smaller and efficiency amplitude is 0.21%.Contrasting case 7 with case 1,pumping efficiency increase 0.04%.On terms of the eddy-eliminating in the suction sump bottom of the bell mouth . Axial pressure distribution of suction sump shows that pressure along the flow direction uniform increments on both sides of the guide cone. If select triangular guide cone, case 7 is recommended as preferable case.

Table 4 Comparative efficiency of triangular guide cone

Case Number	1	7	8	9
Width B_Z/D_L	0	0. 50	0. 67	1.00
η (%)	56. 633	56. 671	56. 619	56. 429

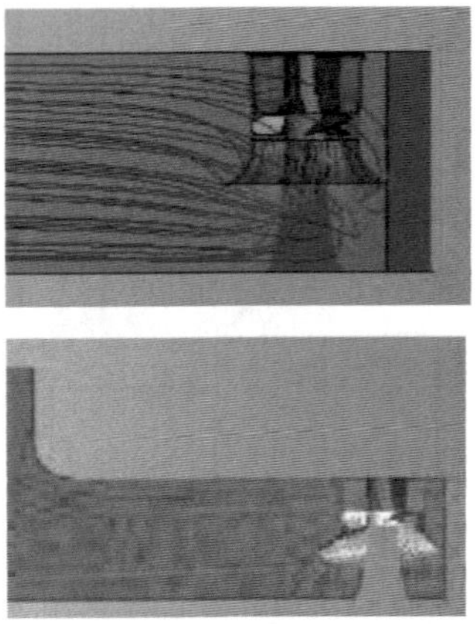

(a) Case 7

Figure 4-7 Axial pressure distribution of suction sump

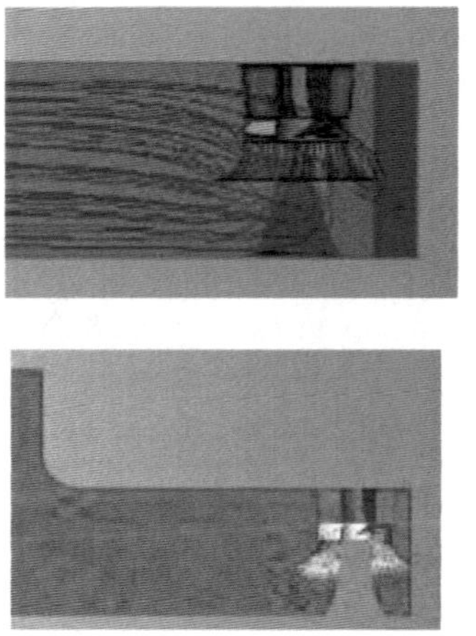

(b) Case 8

Figure 4-8 Axial pressure distribution of suction sump

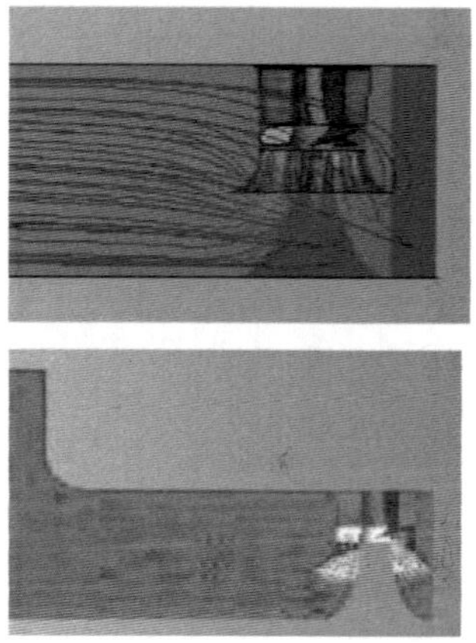

(c) Case 9

Figure4-9 Axial pressure distribution of suction sump

5. Conclusions

On terms of triangular guide cone, increasing height of triangular guide cone may cause the pumping hydraulic performance degradation, which indicates excessive guide cone to some extent hindered the water into the flare, and affect the detour flow below the flare. And as the growth of triangular guide cone's width, it may increase the pumping hydraulic performance and pumping efficiency, which indicates increasing width of triangular guide cone may guide water into the flare.

On terms of traditional guide cone, increasing width of triangular guide cone decrease the pumping hydraulic performance and pumping efficiency, which indicates too broad guide cone to some extent hindered the water into the flare.

Whether select triangular guide cone or traditional guide cone, on terms of pumping efficiency, guide cone have not great effect. However in terms of the eddy-eliminating on the bottom of suction sump, it is necessary to set up guide cone.

Acknowledgments
This work was financially supported by National science and technology plan project in rural areas (Grant No.2012BAD08B03), National Natural Science Foundation of China (Grant No. 51179167), Water conservancy science and technology project of Jiangsu Province (Grant No.2014046), The Young academic Leaders in Blue Project of Jiangsu, Prospective Joint Research Project of Jiangsu (Grant No. BY2015061-12) , A Project Funded by the Priority Academic Program Development of Jiangsu Higher Education Institutions and Major Natural Science Foundation of Universities in Jiangsu(Grant No.12KJA570001).

References

[1] Cheng L, Liu C, Zhou J R 2009 Numerical Simulation on Geometrical Parameters for Open Pump Sump [J]. *Transactions of the Chinese Society for Agricultural Machinery*, **(1)**:58-61

[2] Qian Y D, Yan D F, Liu C et al. 1989 The Experimental Research on The sump of Pumping Station [J]. *Transactions of the Chinese Society of Agricultural Engineering*, **(2)**:47-52

[3] Wang B C, Zhan J M, Yu L H 2012 Numerical simulation for flow characteristics of suspension height in suction of pump station [J].*Advances in Science and Technology of Water Resources*, **(11)**:70-72

[4] Lu L G, Cao Z G, Zhou J 1997 The optimum hydraulic design of pump intakes [J].*Journal of Hydraulic Engineering*, **(3)**:15-20

[5] He T T, Shi Z P, Zhang T 2013 Optimization and design of the after-wall shape about open pump sump on CFD [J]. *China Rural Water and Hydropower*, **(6)**:85-88

[6] Zhu H G 2005 Numerical Simulation and Model Test for the Influence of Pump Intake Design on Pump Flow Conditions [J]. *Transactions of the Chinese Society for Agricultural Machinery*, **(6)**: 57-60.

[7] Shi G P 2012 Numerical simulation and technical transformation for open suction sump of axial-flow pump station [J]. *Advances in Science and Technology of Water Resources*, **(1)**:192-195

[8] Chen S S, Ge Q, Yang G et al. 2004 The experiment on pump setting characteristics with different suction open height [J]. *Journal of Yangzhou University (Natural Science Edition)*, **(3)**:59-62

[9] Zhou J R, Feng X S 1999 Model test on the two-way suction box of lager pumping station [J]. *Journal of Yangzhou University (Natural Science Edition)*, **(4)**: 79-82.

[10] Feng J G 2009 Experimental research on pump setting characteristics with different suction open height [J]. *China Rural Water and Hydropower*, **(11)**:114-117

[11] Yang F, Liu C, Tang et al. 2014 Analysis of hydraulic performance for vertical axial-flow pumping system with cube-type inlet passage [J]. *Transactions of the Chinese Society of Agricultural Engineering*, **(4)**:62-69

[12] He Z N, Chen S S, Zhou Z F 2007 Numerical simulation of the three dimensional turbulent flow of outlet passage [J].*Journal of Yangzhou University (Natural Science Edition)*, **(3)**: 69-73.

[13] Zhou J R, Jin Y, Liu C 2011 Numerical simulation on internal flow of box inlet passage in a large pumping station [J]. *Journal of Water Resources and Architectural Engineering*, **09(6)**:48-50

[14] Ge Q, Chen S S, Wang L S et al. 2006 Experimental study on the characteristic model of pump sets with the bell box passage [J]. *Journal of Hydroelectric Engineering*, **(5)**:129-134

[15] Ansar M, Nakato T, Constantinescu G 2002 Numerical simulations of inviscid three-dimensional flows at single-and dual-pump intakes [J]. *Journal of Hydraulic Research*, **40(4)**: 461-470.

[16] Iwano R, Shibata T, Nagahara T, et al. 2002 Numerical Prediction Method of a Submerged Vortex and its Application to the Flow in Pump Sumps with and without a Baffle Plate[C].*Proceedings of the 9th International Symposium on Transport Phenomena and Dynamics of Rotating Machinery*.

[17] Rajendran V P, Constantinescu S G, Patel V C 1999 Experimental validation of numerical model of flow in pump-intake bays [J]. *Journal of hydraulic Engineering*, **125(11)**: 1119-1125.

[18] Constantinescu G S, Patel V C 1998 Numerical model for simulation of pump-intake flow and vortices [J]. *Journal of Hydraulic Engineering*, **124(2)**: 123-134.

[19] Chuang W L, Hsiao S C 2011 Three-dimensional numerical simulation of intake model with cross flow [J]. *Journal of Hydrodynamics*, Ser. B, **23(3)**: 314-324.

IOP Conf. Series: Materials Science and Engineering **129** (2016) 012010 doi:10.1088/1757-899X/129/1/012010

Numerical simulation on geometrical parameters for closed sump

Y X Wang, L Cheng[1], C Z Xia[1], J R Zhou[1], H Q Yan[2], H Y Jiang[2]

1 School of Hydraulic, Energy and power Engineering
Yangzhou University, Yangzhou, Jiangsu, P.R.China, 225009
2 Jiangsu Water Conservancy Engineering & Technology Consulting Co., Ltd.
Nanjing, Jiangsu, P.R.China, 210029

E-mail:chengli@yzu.edu.cn

Abstract. The closed sump is a typical inlet passage of middle and small pumping station. It has the characteristics of low channel height, small foundation excavation depth, simple structure, a single cross sectional shape changes, ease of construction and other features, so more and more attention and application has been paying on this closed sump in pumping station project. However the flowing pattern within the closed sump is complex, the design is not perfect in some respects, the structure size does not be optimized. Based on the background for renewal and transformation of a pumping station, according to the three-dimensional incompressible fluid Reynolds-averaged N-S equations, the RNG k-e model, the CFD technology. The study on the draught in closed sump might reduce the length of pump shaft to enhance the stability of the pump unit operation. The results reveal the effect of the change of the height of plate. The turbulence in back wall might cause vortex when the height is high. The height of plate had be recommended control in 0.65D-0.85D.The better parameter combination of geometry of closed sump had be given through comparing the results of the orthogonal test and the comprehensive test. The floor clearance should be control in 1.0D. (D is the diameter of flared pipe)

Introduction

The closed sump comes from the open sump with a plate. It have double features of traditional sump and inlet conduit due to the quartet geometric shape and the environment with pressure. So by comparison, it has the unique characteristics of low channel height, small foundation excavation depth, simple structure, a single cross sectional shape changes, ease of construction and other features [1~5], so more and more attention and application has been focused on this closed sump at pumping station project [6~10].

Jiangang Feng [10] put forward that the design of intake sump should make flow smooth-going, no harmful vortex in order to ensure the safe operation of the pump unit. Songshan Chen [11] designed five different suction open height to observe the inlet flow pattern. Charles [12] put forward the velocity in sump should be control near 0.3m/s. Iverson [13] gave the value range of draught for high specific speed pump. Matahel Ansar [14] for the flow pattern in rectangular sump in some inlet conditions. For eliminate vortex bell attached under the bottom, the experiments also designed some vortex suppression program of separator. Flow pattern of the closed sump and hydraulic performance of five different

[1] Corresponding author: CHENG Li(1975-), male, PhD, Professor

heights of the closed sump had be analyzed orthogonal test had be set to study the geometrical parameters of the closed pump.

1.1 Mathematical Model

Calculations are performed using the commercial code CFX®. The numerical model is based on the Reynolds-Averaged Navier-Stokes (RANS) equations with the RNG k-ε model to calculate the Reynolds stresses. First-order upwind discretizations are used for the convective terms and turbulent kinetic energy and dissipation, while a second-order central differencing scheme is used for the diffusive terms. The temporal discretization is second-order implicit. The SIMPLE algorithm is used for the pressure-velocity coupling.

Sliding interfaces up- and downstream of the impeller allow the impeller to rotate with respect to the inlet and stator. The flow in the impeller is solved in the rotating frame of reference. To this end the apparent Coriolis and centrifugal forces are added to the RANS equations as source terms.

Two basic calculation methods are adopted. The first one is a quasi-steady approach in which the flow is assumed steady in its corresponding reference frame. In this multiple reference frame (MRF) method, the transient effect of the rotor-stator interaction is neglected. The convergence criterion for all equations is set to 10^{-4}. A solution normally serves as a good initial solution for the second, truly unsteady, moving mesh method. In this method, the connections between rotating and stationary part are updated each time step. The transient solution is monitored as time progresses until the solution becomes periodic at blade passing frequency.

1.2 numerical simulation

The numerical domain consists of the inlet and outlet passages, the axial flow impeller with three blades and the diffuser with seven vanes. Inflow and outflow boundaries are located sufficiently far away from the pump not to influence the flow characteristics. The extent of the domain in upstream direction is especially important to allow for inlet flow recirculation at part load. Another important part is the tip clearance gap between the impeller blades and the casing. A number of layers of cells are placed in this region to allow for leakage flow over the blade tips.

Using an extrusion method, an O-type grid of hexagonal cells is created around the impeller and stator blades to ensure good mesh quality in terms of size and skewness. Because of the complex topology of the pump, the interior of the domain is filled with an unstructured mesh of tetrahedral cells.

A mesh-sensitivity study was carried out to assess the required mesh density. Several grids were considered, ranging from a total number of cells of 2×10^5 up to 2×10^6, where care was taken that y^+ values at the solid boundaries remained favorable. No further convergence was obtained for grids with more than 1.5×10^6 cells.

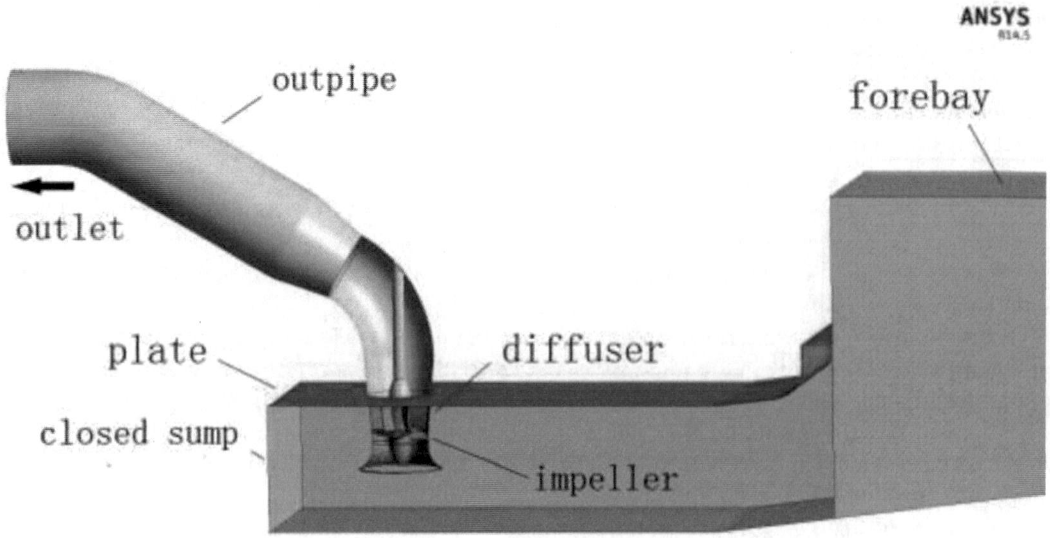

Figure.1 Computational Domain

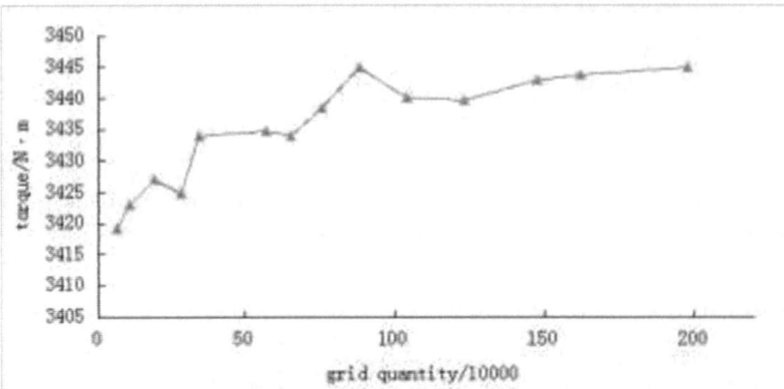

Figure.2 Independence of Grid

The inlet boundary condition is specified as a uniform velocity whereas at the outlet an outflow boundary condition is applied which allows for non-uniformity in both velocity and pressure. No-slip boundary conditions and wall functions are used for the solid walls.

2. Calculation Results

The geometric parameters of sump are shown in Fig3.

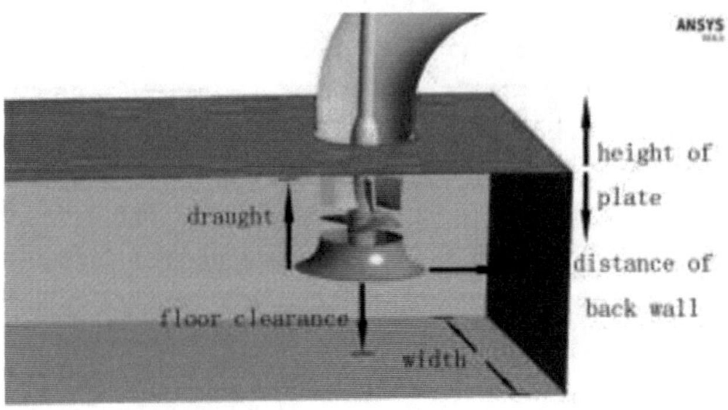

IOP Conf. Series: Materials Science and Engineering **129** (2016) 012010 doi:10.1088/1757-899X/129/1/012010

Figure.3 Geometric Parameters of Sump

2.1. Height of Plate

The closed sump came from the open sump with a plate. The plate changed the flowing law in the sump. So a research about the height of the plate had be taken first. In order to reduce the influence of other geometry parameters, larger geometry parameters of the sump had be set. The calculation schemes are shown in Table 1.Five different heights of plate had be set to analyze the changes of hydraulic performance.

Table 1. Calculation Schemes

NO.	Width	Distance of Back Wall	Floor Clearance	Height of Plate
1				2.05 D
2				1.26 D
3	3 D	0.82 D	0.7 D	1.05 D
4				0.85 D
5				0.65 D
6				0.46 D

Fig.4 shown the distribution of the velocity of the center vertical cross-section and the horizontal cross-section on five heights of plate. On the center section, the height of the area of low velocity increased as the plate height increases. When the plate height is 1.26D, the area of low velocity spread all over the back wall. On the horizontal section, the area of low velocity slightly expended from 0.46D to 0.85D, and the distribution of velocity was uniform. From 1.05D to 1.26D, the flow pattern became turbulence.

The uniformity of velocity and the variety average angle of velocity had be analysed. The uniformity of velocity would get better when the height is 0.65D-0.85D.Also the uniformity of velocity is good in the open sump which the height is 2.05D.When the height is 0.65D-0.85D the average angle of velocity is better. The height of plate should be control in 0.65D-0.85D had be recommend.

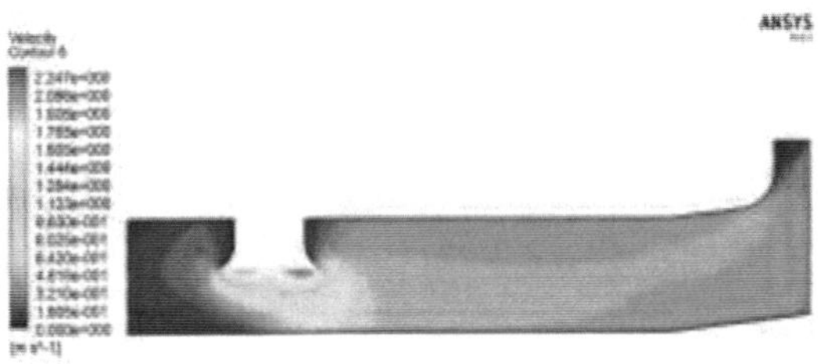

(a)Velocity of Vertical Cross-Section on 0.46D

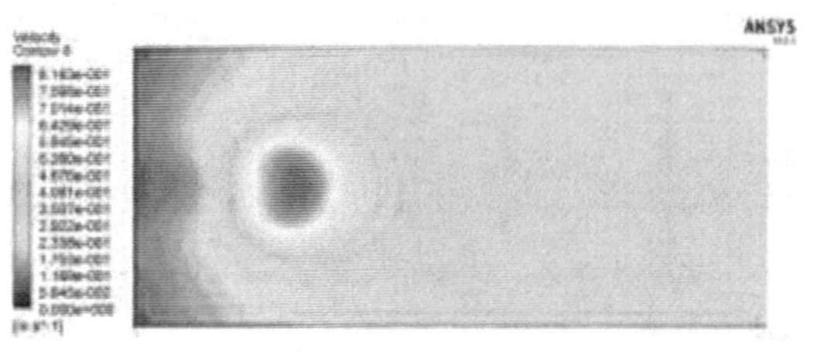

(b)Velocity of Horizontal Cross-Section on 0.46D

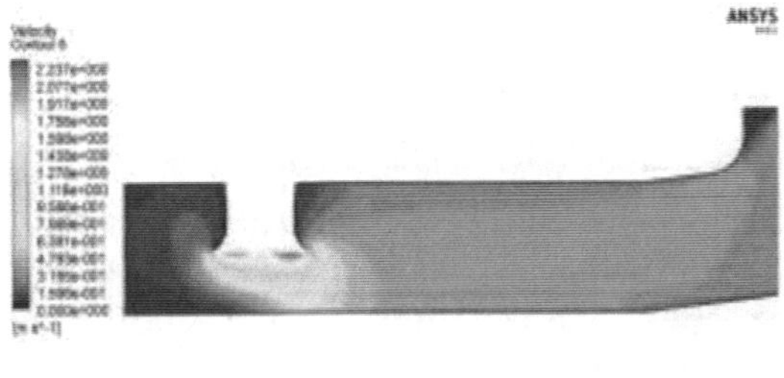

(c)Velocity of Vertical Cross-Section on 0.65D

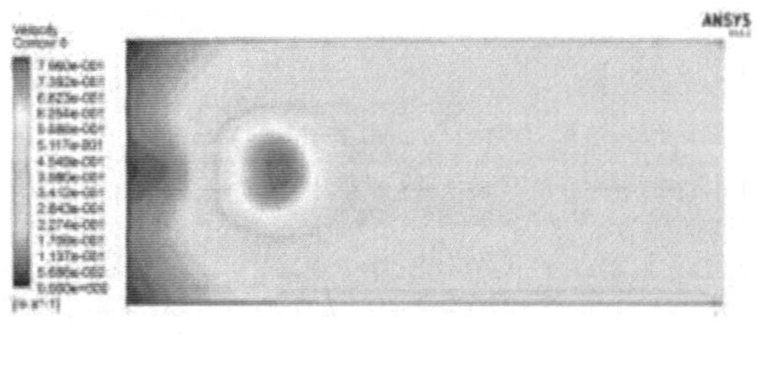

(d)Velocity of Horizontal Cross-Section on 0.65D

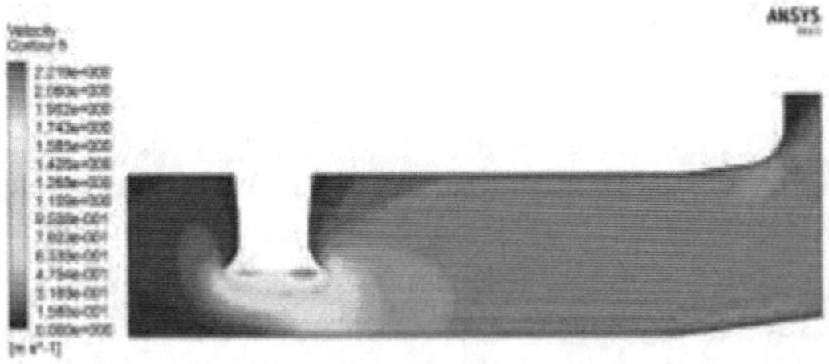

(e)Velocity of Vertical Cross-Section on 0.85D

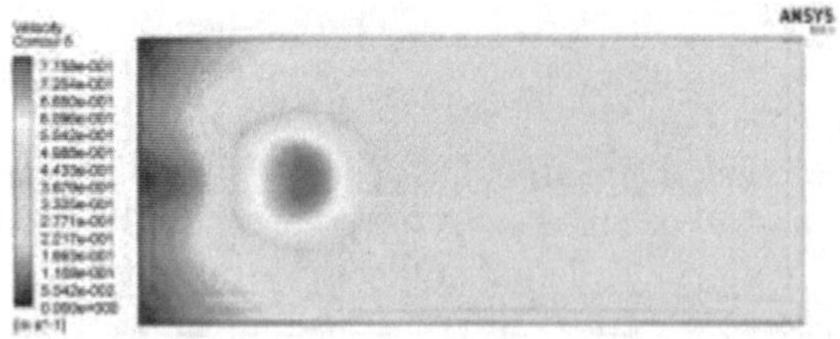

(f)Velocity of Horizontal Cross-Section on 0.85D

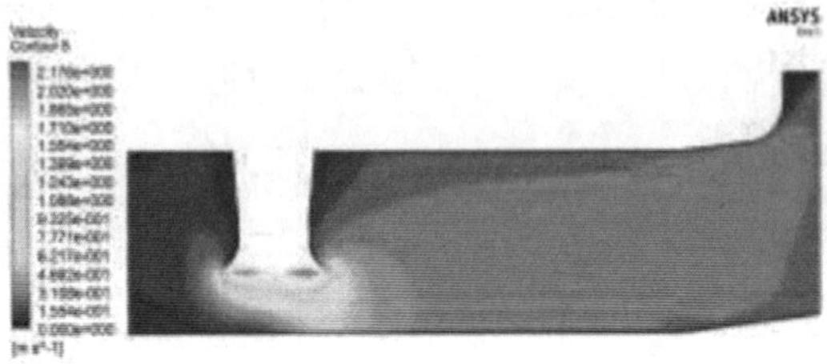

(g)Velocity of Vertical Cross-Section on 1.05D

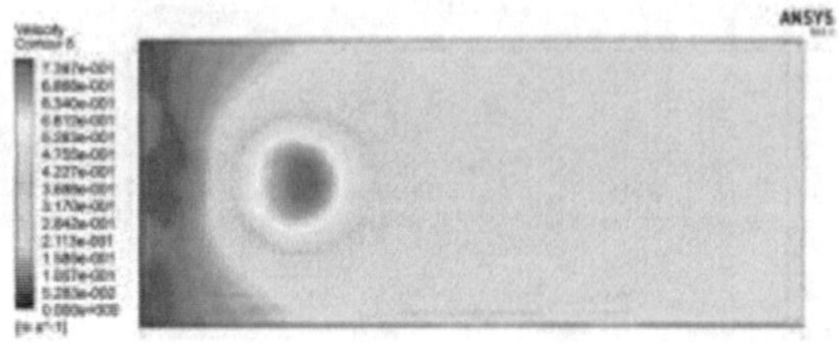

(h)Velocity of Horizontal Cross-Section on 1.05D

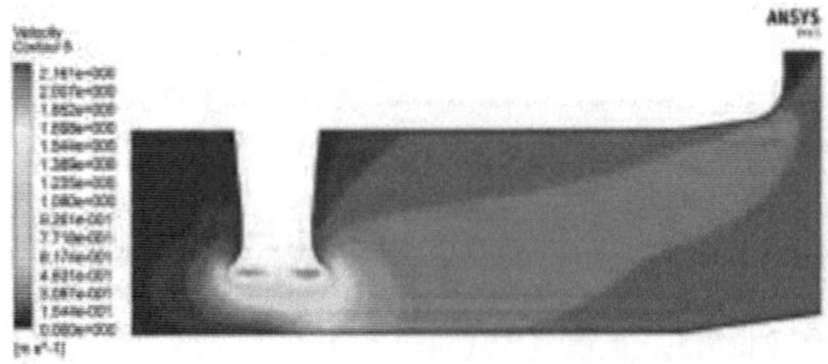

(i)Velocity of Vertical Cross-Section on 1.26D

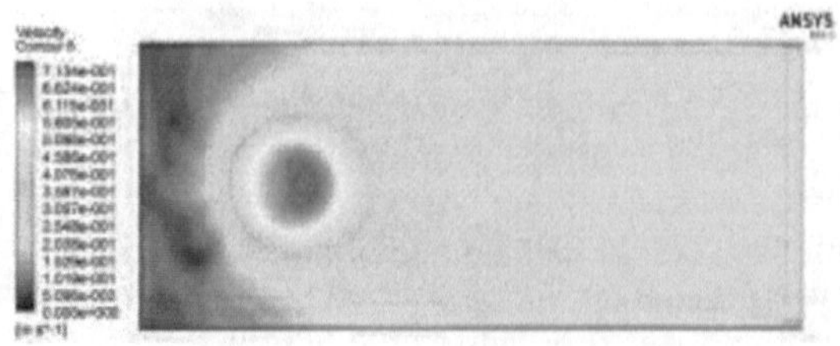

(j)Velocity of Horizontal Cross-Section on 1.26D

Figure.4 Cloud Image of Velocity on Section of Five Heights

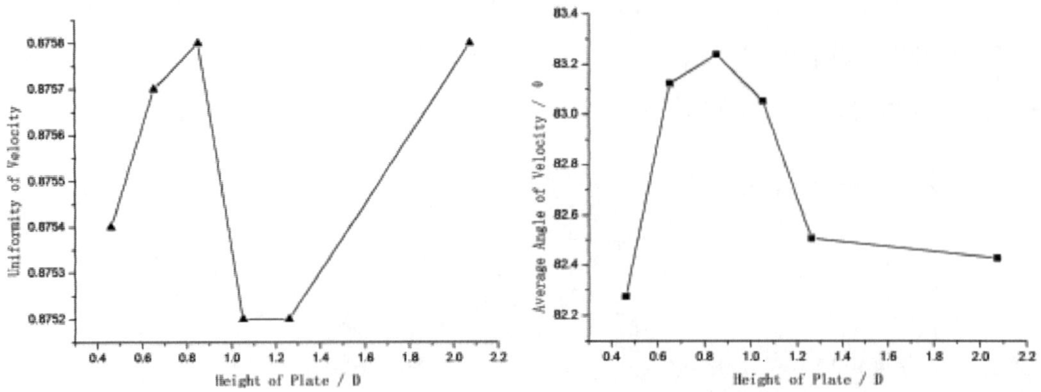

Figure.5 Uniformity of Velocity and Average Angle of Velocity

2.2. Orthogonal Test on Geometrical Parameters

2.2.1 Design of Orthogonal Test Four Geometrical Parameters were chose as the factors for the test. They are width, distance of back wall, floor clearance and draught. Four levels were set in every factor.

The L16 (4^4) table was selected in the test. Otherwise, the pump assembly efficiency, the uniformity of velocity and the average angle of velocity were selected as the index of test.

Table 2. Factors and Levels

Levels	Factors			
	A (Height of Plate)	B (Distance of Back Wall)	C (Floor Clearance)	D (Width)
1	0.46 D	0.4 D	0.4 D	2.25 D
2	0.65 D	0.6 D	0.6 D	2.5 D
3	0.85 D	0.8 D	0.8 D	2.75 D
4	1.05 D	1.0 D	1.0 D	3.0 D

Table 3. Factors and Levels

No.	Factors				Geometrical Parameters			
	A	B	C	D	Height of Plate	Distance of Back Wall	Floor Clearance	Width
1	1	1	1	1	0.45 D	0.4 D	0.4 D	2.25 D
2	1	2	2	2	0.45 D	0.6 D	0.6 D	2.5 D
3	1	3	3	3	0.45 D	0.8 D	0.8 D	2.75 D
4	1	4	4	4	0.45 D	1.0 D	1.0 D	3.0 D
5	2	1	2	3	0.65 D	0.4 D	0.6 D	2.75 D
6	2	2	1	4	0.65 D	0.6 D	0.4 D	3.0 D
7	2	3	4	1	0.65 D	0.8 D	1.0 D	2.25 D
8	2	4	3	2	0.65 D	1.0 D	0.8 D	2.5 D
9	3	1	3	4	0.85 D	0.4 D	0.8 D	3.0 D
10	3	2	4	3	0.85 D	0.6 D	1.0 D	2.75 D
11	3	3	1	2	0.85 D	0.8 D	0.4 D	2.5 D
12	3	4	2	1	0.85 D	1.0 D	0.6 D	2.25 D
13	4	1	4	2	1.05 D	0.4 D	1.0 D	2.5 D
14	4	2	3	1	1.05 D	0.6 D	0.8 D	2.25 D
15	4	3	2	4	1.05 D	0.8 D	0.6 D	3.0 D
16	4	4	1	3	1.05 D	1.0 D	0.4 D	2.75 D

2.2.2 Test Results K_i represents the sum of the test results in every rank when the level is i. $k_i = K_i / s$, s is the quantity of levels. R is range, $R = \max\{K_1, K_2, K_3, K_4\} - \min\{K_1, K_2, K_3, K_4\}$ in every rank。

There are two preferable schemes in the table which are preferable scheme in test and preferable scheme in theoretical analysis. Preferable scheme in test is the preferable parameter combination in this orthogonal test. Preferable scheme in theoretical analysis is the parameter combination which was analyzed through observing the K_i. In pump sets, pump assembly efficiency, uniformity of velocity and average angle of velocity are the index which were bigger is better. So the preferable scheme in theoretical analysis comes from the level which the K is largest. However the preferable scheme in theoretical analysis is not in the orthogonal test which reflected the superiority of the orthogonal test.

Table 4. Test Results

No	Factors				Test Results		
	A	B	C	D	Pump Assembly Efficiency $\eta/\%$	Uniformity of Velocity $\overline{V_u}/\%$	Average Angle of Velocity $\overline{\theta}$
1	1	1	1	1	73.582	87.02	81.3533
2	1	2	2	2	75.449	87.40	83.2057
3	1	3	3	3	75.701	87.60	83.0590
4	1	4	4	4	75.555	87.31	82.1576
5	2	1	2	3	75.396	87.42	83.4115
6	2	2	1	4	74.649	87.33	82.9470
7	2	3	4	1	76.274	87.57	83.4570
8	2	4	3	2	75.284	87.60	83.1019
9	3	1	3	4	75.687	87.54	83.1748
10	3	2	4	3	76.227	87.53	83.6540
11	3	3	1	2	74.164	87.26	82.1396
12	3	4	2	1	75.307	87.45	82.6834
13	4	1	4	2	76.270	87.52	83.6521
14	4	2	3	1	74.861	87.34	81.6463
15	4	3	2	4	75.010	87.48	82.9840
16	4	4	1	3	74.536	87.36	82.6940

Table 5. Anlysis of Results

Index		A	B	C	D
Pump Assembly Efficiency $\eta/\%$	K_1	300.287	300.935	296.931	300.024
	K_2	301.603	301.186	301.162	301.167
	K_3	301.385	301.149	301.533	301.86
	K_4	300.677	300.682	304.326	300.901
	k_1	75.07175	75.23375	74.23275	75.006
	k_2	75.40075	75.2965	75.2905	75.29175
	k_3	75.34625	75.28725	75.38325	75.465
	k_4	75.16925	75.1705	76.0815	75.22525

	Range R	1.316	0.504	7.395	1.836
	Primary-Secondary	C A D B			
	Preferable Scheme in Test	$A_2B_3C_4D_1$ or $A_4B_1C_4D_2$			
	Preferable Scheme in Theoretical Analysis	$A_2B_2C_4D_3$			
Uniformity of Velocity $\overline{V}_u$ / %	K_1	349.33	349.5	348.97	349.38
	K_2	349.92	349.6	349.75	349.78
	K_3	349.78	349.91	350.08	349.91
	K_4	349.7	349.72	349.93	349.66
	k_1	87.3325	87.375	87.2425	87.345
	k_2	87.48	87.4	87.4375	87.445
	k_3	87.445	87.4775	87.52	87.4775
	k_4	87.425	87.43	87.4825	87.415
	Range R	0.59	0.41	0.33	0.53
	Primary-Secondary	A D B C			
	Preferable Scheme in Test	$A_1B_3C_3D_3$ or $A_2B_4C_3D_2$			
	Preferable Scheme in Theoretical Analysis	$A_2B_3C_3D_3$			
Average Angle of Velocity $\overline{\theta}$	K_1	329.7756	331.5917	329.1339	329.14
	K_2	332.9174	331.453	332.2846	332.0993
	K_3	331.6518	331.6396	330.982	332.8185
	K_4	330.9764	330.6369	332.9207	331.2634
	k_1	82.4439	82.897925	82.283475	82.285
	k_2	83.22935	82.86325	83.07115	83.024825
	k_3	82.91295	82.9099	82.7455	83.204625
	k_4	82.7441	82.659225	83.230175	82.81585
	Range R	3.1418	1.0027	3.7868	3.6785
	Primary-Secondary	C D A B			
	Preferable Scheme in Test	$A_3B_2C_4D_3$ or $A_4B_1C_4D_2$			
	Preferable Scheme in Theoretical Analysis	$A_2B_3C_4D_3$			

Interaction might exist in factors. One factor changed may change the influence of other factors. So the preferable scheme in theoretical analysis should be verified. If the results are better than the preferable scheme in test, the preferable scheme in theoretical analysis should be the better scheme. Otherwise the preferable scheme in test might be the better scheme.

2.2.3 Comprehensive test The table given three preferable schemes in theoretical analysis is $A_2B_2C_4D_3$, $A_2B_3C_3D_3$ and $A_2B_3C_4D_3$.A_2 and D_3 is the better level in the three schemes. B and C have two better levels. So a 2×2 comprehensive test was designed to verify the results which shown in Table 6.

Table 6. 2x2 Comprehensive Test Results

Preferable Scheme in Theoretical Analysis	Pump Assembly Efficiency $\eta/\%$	Uniformity of Velocity $\bar{V}_u/\%$	Average Angle of Velocity $\bar{\theta}$
$A_2B_2C_4D_3$	0.761619568	87.49	83.116
$A_2B_2C_3D_3$	0.756011045	87.51	83.2975
$A_2B_3C_3D_3$	0.755241291	87.54	83.0813
$A_2B_3C_4D_3$	0.763619693	87.52	83.4524

The better parameter combination had be given in Table 7. Through comparing the results of the orthogonal test and the comprehensive test. The floor clearance should be control in 1.0D.

Table 7. The Better Parameter Combination

The Better Parameter Combination	Height of Plate	Distance of Back Wall	Floor Clearance	Width
$A_2B_3C_4D_3$	0.65 D	0.8 D	1.0 D	2.75 D
$A_2B_3C_4D_1$	0.65 D	0.8 D	1.0 D	2.25 D
$A_3B_2C_4D_3$	0.85 D	0.6 D	1.0 D	2.75 D
$A_4B_1C_4D_2$	1.05 D	0.4 D	1.0 D	2.5 D

3. Conclusion

Based on the three-dimensional numerical simulation, the effect of the height of plate and the effect of all geometrical parameters were studied. The research results can be applied in the similar closed sump of pumping stations.

The hydraulic performance would be better when the height of plate is lower. The height of plate should be controlled in 0.65D-0.85D.

The better parameter combination of geometry of closed sump had be given through comparing the results of the orthogonal test and the comprehensive test. The floor clearance should be controlled in 1.0D.

Acknowledgments

This work was financially supported by National science and technology plan project in rural areas (Grant No.2012BAD08B03), National Natural Science Foundation of China (Grant No. 51179167), Water conservancy science and technology project of Jiangsu Province (Grant No.2014046), The Young academic Leaders in Blue Project of Jiangsu, Prospective Joint Research Project of Jiangsu (Grant No.

BY2015061-12) , A Project Funded by the Priority Academic Program Development of Jiangsu Higher Education Institutions and Major Natural Science Foundation of Universities in Jiangsu(Grant No.12KJA570001).

Reference

[1] Liu C 2009 *Pump and pumping station* [M].Beijing: China Water Power Press.
[2] Liu C, Ma C S 1994 Experiment and research on improvement measures of forebay flow pattern of pumping stations [J].*Journal of Institute of Architecture (Natural Sciences)*, **2(1)**:9~12.
[3] Cheng L, Liu C, Zhou J R, et al. 2005 Numerical simulation of turbulent flow around sill with RNG k-ε turbulent model[J].*Transactions of the Chinese Society for Agricultural Machinery*,**36(3)**:37~39.
[4] Choi J W, Choi Y D, Kim C G 2010 Flow uniformity in a multi-intake pump sump model. The University of lowa, *Journal of Mechanical Science and Technology*, **24 (7)**:1389~1400.
[5] Gaston E F, Marcela P, Larry W 2009 Numerical simulation of free surface flows on a fish bypass [J]. *Computers& Fluids*, **38(5)**: 997- 1002.
[6] Cai F L, Zhang Z M 2000 Study on guide piers in flank inlet-outlet with double flow directions [J].*Journal Of Hohai University*, **28(2)**:74~77.
[7] Zhou L C 2010 Numerical analysis on improvement of flow conditions in forebay of pumping station by setting separation piers [J]. *Journal of Yangtze River Scientific Research Institute*, **27(2)**:31~33.
[8] Liu C, Han X, Zhou J R, et al. 2009 Numerical simulation of turbulent flow in forebay with side-intake of pumping station [J]. *Drainage and Irrigation Machinery*, **27(5)**:281~286.
[9] Liu C, Cheng L, Tang F P 2001 Numerical simulation of three-dimensional turbulent flow inside a pumping forebay [J]. *Transactions of the Chinese Society of Agricultural Machinery*, **32 (6)**:41~44.
[10] Zhou J R, Liu C, Tang F P 1996 Simulation on the flow pattern in diffusion channel of large pumping station [J]. *Water Resource and Hydropower Engineering*, **1996(9)**:41~43.
[11] Chen S S, Ge Q, Yang G P 2004 The experiment on pump setting characterisics with different open height [J]. *Journal of Yangzhou University (Natural Science Edition)*, **7(3)**:59~62.
[12] Sween C E, Rex A E, Duncan H 1982 Pump sump design experience [J], *J.of Hydraulic Divisin*, **108(3)**:361-377
[13] Iversen H W 1953 Berkeley,Calif. Studies of Submergence Requirements of High-Specific-Speed Pumps [J]. *The American Society*, **1953(75)**:635-641
[14] Matahel A, Tatsuaki N 2001 Experimental study of 3D pump-intake flows with and without cross flow [J]. *Journal of Hydraulic Engineering*, **127(10)**:825-834

IOP Conf. Series: Materials Science and Engineering **129** (2016) 012011 doi:10.1088/1757-899X/129/1/012011

Numerical simulation on the cavitation of waterjet propulsion pump

C Z Xia, L Cheng[1], Y N Shang, J R Zhou, F Yang, Y Jin

School of Hydraulic, Energy and power Engineering
Yangzhou University, Yangzhou, Jiangsu, P.R.China, 225009

E-mail: chengli@yzu.edu.cn

Abstract. Waterjet propulsion system is widely used in high speed vessels with advantages of simple transmission mechanism, low noise underwater and good manoeuvrability. Compared with the propeller, waterjet propulsion can be used flow stamping to increasing cavitation resistance at high speed. But under certain conditions, such as low ship speed or high ship speed, cavitation problem still exists. If water-jet propulsion pump is run in cavitation condition for a long time, then the cavitation will cause a great deal of noise CFD is applied to analysis and predict the process of production and development of cavitation in waterjet propulsion pump.

Based on the cavitation model of Zwart-Gerber-Belamri and a mixture of homogeneous flow model, commercial CFD software CFX was taken for characteristics of cavitation under the three operating conditions. Commercial software ANSYS 14.0 is used to build entity model, mesh and numerical simulation. The grid independence analysis determine the grid number of mixed flow pump model is about 1.6 million and the grid number of water-jet pump system unit is about 2.7 million.

The cavitation characteristics of waterjet pump under three operating conditions are studied. The results show that the cavitation development trend is similar design and small rate of flow condition .Under the design conditions Cavitation bubbles are mainly gathered in suction surface of blade near the inlet side of the hub under the primary stage, and gradually extended to the water side in the direction of the rim with the loss of the inlet total pressure. Cavitation appears in hub before the blade rim, but the maximum value of gas content in blade rim is bigger than that in hub. Under large flow conditions, bubble along the direction of wheel hub extends to the rim gradually. Cavitation is found in the pressure surface of blade near the hub region under the critical point of cavitation nearby. When NPSHa is lower than critical point, the area covering by bubbles is about 40% in the suction surface of blade. It means that the critical point of cavitation of pump system is not the accrue point of install cavitation but cavitation has been developed to a certain stage.

1. Introduction

Along with the high-speed development of mechanical and military technology, high-speed vessels started to use a new way of special ship propulsion waterjet propulsion. Compared with the traditional propeller, waterjet propulsion system not only has the same good performance in acceleration and braking performance, but also has excellent high speed mobility [1]. The noise and vibration in the

[1] Corresponding author: CHENG Li(1975-), male, PhD, Professor of Yangzhou University

cabin of water jet propulsion ship is small. The waterjet system's daily maintenance is also relatively easy. Waterjet propulsion device in high speed ship can use water to flow stamping improve the cavitation resistance device. But in some special conditions, such as turning, speeding, maneuvering in reverse and only part of the pumps work, the waterjet propulsion system will get into the cavitation workspaces [2]. Cavitation will not only lead to sharp decline in pump hydraulic performance, also can make water jet propulsion pump thrust be reduced, the efficiency decline and reduce the service life of the impeller and bring cavitation, vibration and noise. So the research on water jet propulsion pump system has strong engineering application value and academic significance[3-4].

2. Geometric model and numerical method

2.1. Geometric model

By changing the propulsion pump in paper [5-6], new propulsion pump is used in this paper which is about the cavitation on the blade of the impeller. In the model, the inlet is extended so that accuracy and convergence for the simulation is solved, the outlet in connected with optimized nozzle. Figure 1 is the computational domains of propulsion pump.

The computational domains contain impeller, vane, inlet and outlet. In which, impeller has six blades and its rotating speed is 700 r/min, vane has seven blades, diameters of inlet and outlet are D0 and 1.19D0 separately. The straight pipe section shares the same diameter with the inlet. Plane 1 and plane 2 are chosen as the inlet and outlet to calculate the pump head.

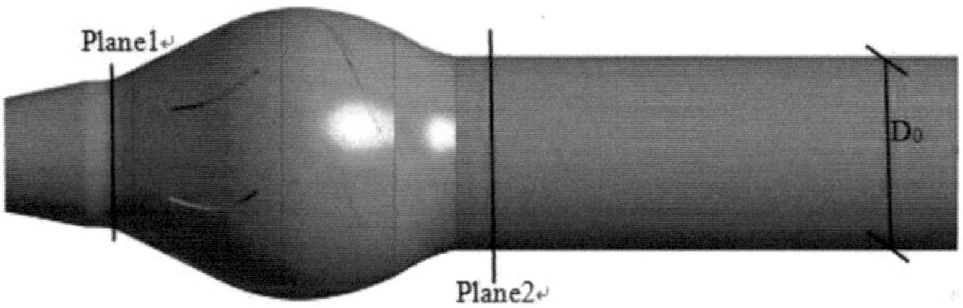

Figure 1. Computational domains

2.2. Mesh generation

Considering structure mesh accelerating CFD calculation speed, the mesh is generated by using ICEM CFD shown in Figure 2. The mesh quantity is 1.6 million.

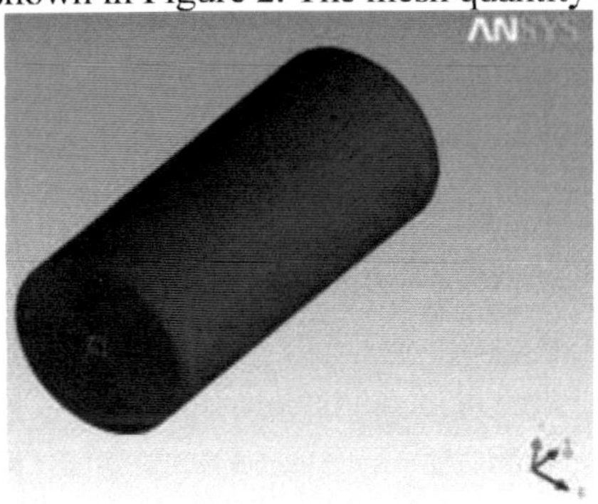

(a) straight pipe section

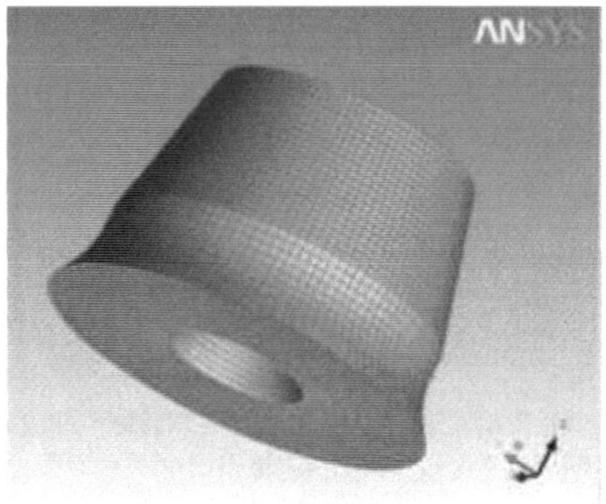

(b) nozzle

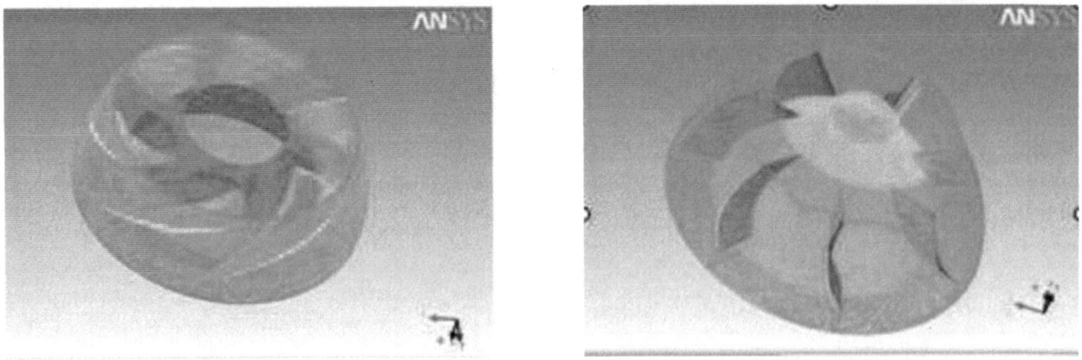

(c) vane (d) impeller

Figure 2. Meshes

In general, mesh quantity affects the accuracy and convergence speed. In theory, more mesh, higher accuracy. Actually, overmuch mesh will increase calculation time and slow convergence speed. The mesh independence discussion is necessary. Impeller is the central domain in the model and its mesh quantity is the key. The result shows that the head and efficiency stay steady when the mesh quantity rises up to 1.6 million. H_d, η_d are head and efficiency separately at design condition.

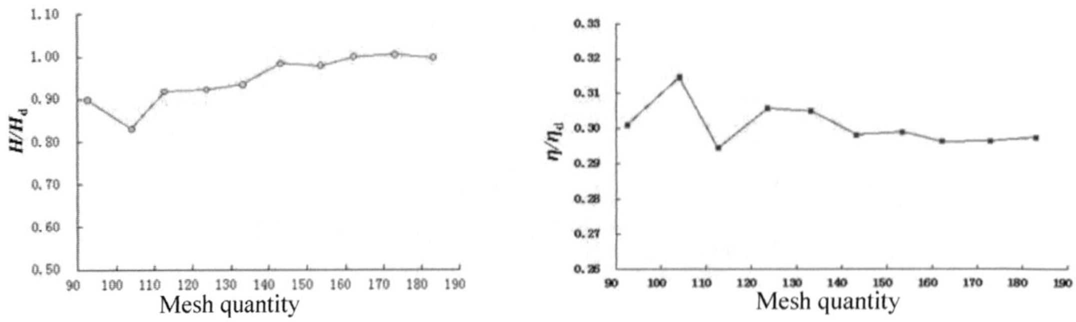

(a) head (b) efficiency

Figure 3. Mesh independence

2.3. Boundary condition

Inlet is set as total pressure stable, supposing no cavitation happens in the water. Outlet is set as mass flow rate. When cavitation model is applied, two physical parameters below are given: water's cavitation pressure P_V equals to 3574Pa at normal atmospheric temperature, cavitation surface tension σ equals to 0.074N/M, Rayleigh—Plesset equation is used to describe bubble motion discipline. No slip wall condition is applied for the wall of hub, rim, blade surface and the other solid walls. The reference pressure is set as 0 atm. RNG k-ε model and mixture model are chosen as turbulent model and multiphase flow model separately [6-9].

Calculated result is more close to the test result at large flow rate coefficient by using Zwart-Gerber-Belamri cavitation model, so Zwart-Gerber-Belamri cavitation model [10-14] is chosen to simulate the cavitation in the impeller.

3. Results and analysis

Three flow rate conditions are simulated and compared which is design flow rate (Q_{bep}), small flow rate (0.7 Q_{bep}) and large flow rate (1.1 Q_{bep}).

3.1. Cavitation performance at design flow rate

IOP Conf. Series: Materials Science and Engineering **129** (2016) 012011 doi:10.1088/1757-899X/129/1/012011

Based on the pump performance, the best efficiency point is chosen as the design flow rate which can be dawn as Q_{bep}. Different total pressures on the inlet are applied which from 0.5atm to 0.1atm, pressure distributions on the pressure side and suction side of the impeller, cavitation area change is observed.

As shown in Figure 4 and Figure 5, there is pressure distribution on the pressure side and suction side of the impeller separately for different cavitation remainders. In the Figure, the blue area is low pressure zone which firstly appears at the rim of suction side inlet. As NPSHa reduces, low pressure zone increases along blade margin. When NPSHa is 0.76m, the entire suction side of the blade is low pressure zone. Meanwhile, there is also low pressure on the pressure side of blade whose shape is almost triangle, pressure at the hub is larger than it at the rim.

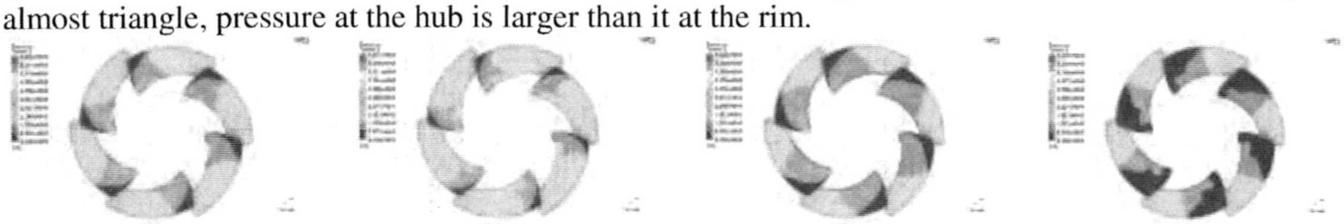

(a) NPSHa=2.41m (b) NPSHa=2.2m (c) NPSHa=1.87m (d) NPSHa=1.62m

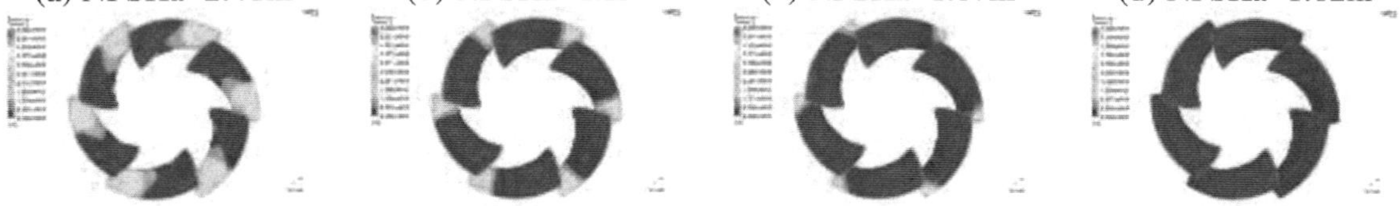

(e) NPSHa=1.38m (f) NPSHa=1.17m (g) NPSHa=0.96m (h) NPSHa=0.76m

Figure 4. Pressure distribution on the suction side of the blade at different cavitation remainders

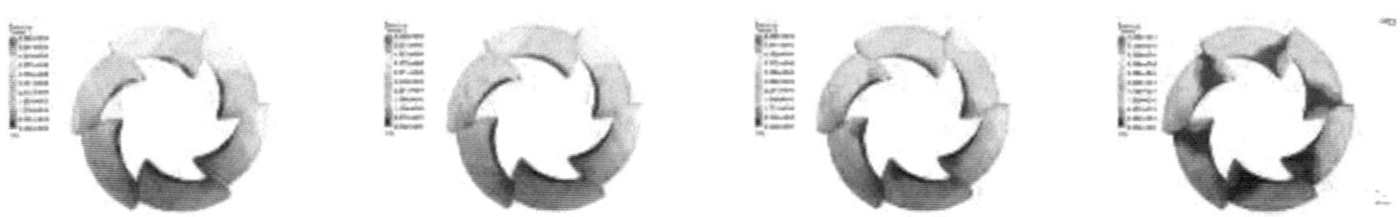

(a) NPSHa=1.38m (b) NPSHa=1.17m (c) NPSHa=0.96m (d) NPSHa=0.76m

Figure 5. Pressure distribution on the pressure side of the blade at different cavitation remainders

Figure 6 and Figure 7 are cavitation volume fraction distribution on the pressure side and suction side of the blade. In the Figure, red zone and blue zone represent large cavitation volume fraction and small cavitation volume fraction. By comparing the pressure distribution on the blade with different cavitation remainders, the same trend is shown. When NPSHa is 0.76m, cavitation on the whole suction pressure has been absolutely developed. The area around the outlet of the blade is affected worst, cavitation will block the entire impeller passage and even result in the cavitation happening on the pressure side of the blade.

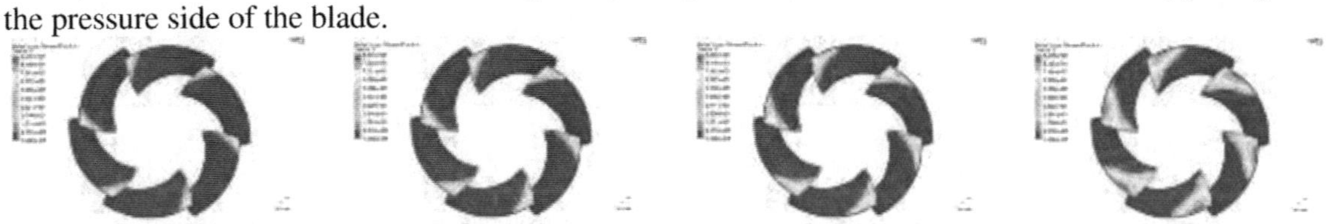

(a) NPSHa=2.41m (b) NPSHa=2.2m (c) NPSHa=1.87m (d)NPSHa=1.62m

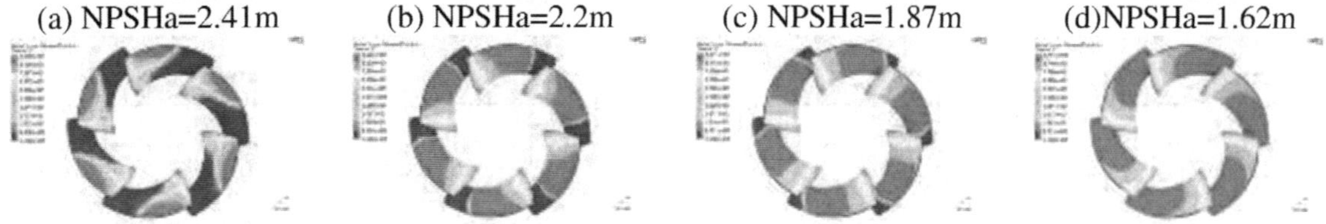

(e) NPSHa=1.38m (f) NPSHa=1.17m (g) NPSHa=0.96 (h) NPSHa=0.76m

Figure 6. Cavitation volume fraction distribution on suction side of the blade

(a) NPSHa=0.96 (b) NPSHa=0.76m

Figure 7. Cavitation volume fraction distribution on the pressure side of the blade

Figure 8 is the static pressure distribution on the blade at different spans. When NPSHa is 2.2m, low static pressure keeps steady. Cavitation happens at the rim of the blade at first. In Figure, on the above and below are the static pressure at the pressure side and suction side respectively. Cavitation will not happen until it happens at the whole suction side. Cavitation at suction side is caused by the low pressure. There is no obvious contact between the cavitation on the pressure side and pressure decrease.

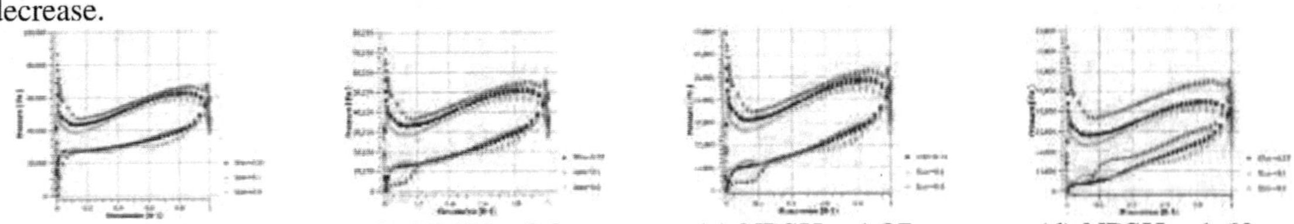

(a)NPSHa=2.41m (b) NPSHa=2.2m (c) NPSHa=1.87m (d) NPSHa=1.62m

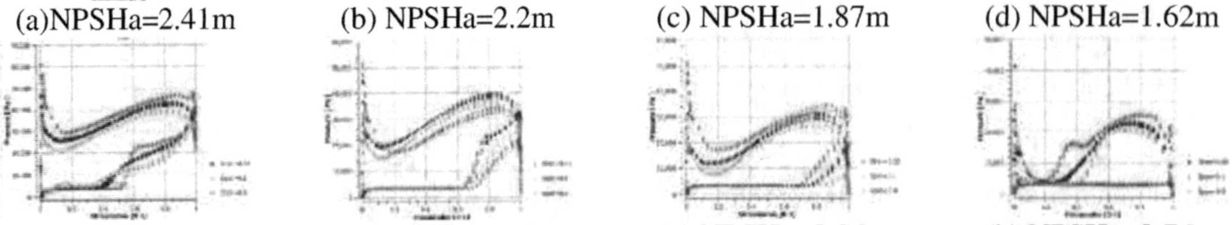

(e) NPSHa=1.38m (f) NPSHa=1.17m (g) NPSHa=0.96m (h) NPSHa=0.76m

Figure 8. Static pressure distribution on the blade at different spans

3.2. Partial conditions of cavitation characteristics

Figure 9 shows the result of numerical simulation on cavitation in small flow rate. The relation between the distributions of water vapour volume fraction on the suction face of blades and the decreases of NPSHa is show in Figure 9. The result shows that the variations tend of the location of bubble volume fraction distribution are consistent with the design condition. The location of bubble on suction face of blade grandly extends to the outside from the inside of blades in same condition.

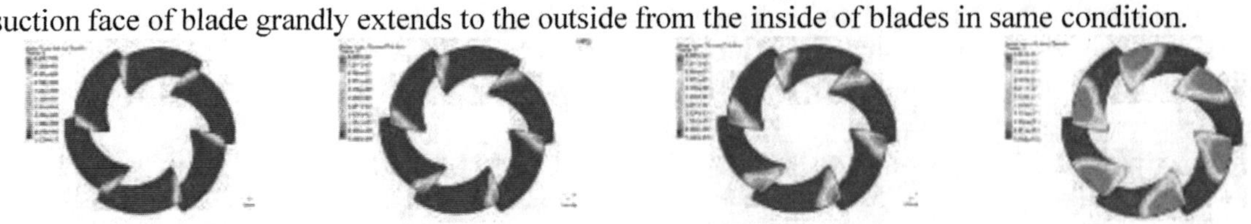

(a)NPSHa=2.2m (b)NPSHa=2.0m (c)NPSHa=1.69m (d)NPSHa=1.28m

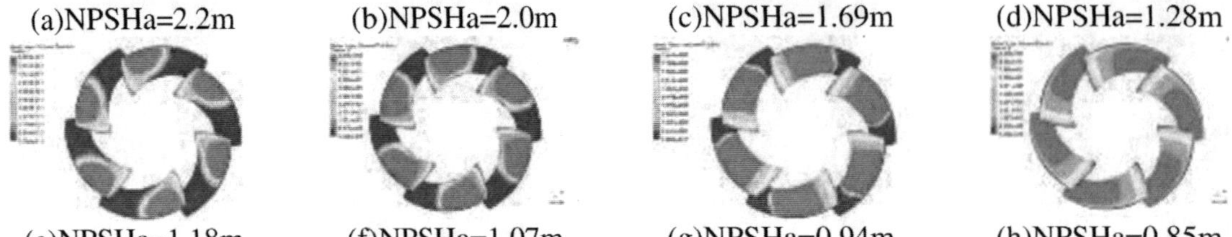

(e)NPSHa=1.18m (f)NPSHa=1.07m (g)NPSHa=0.94m (h)NPSHa=0.85m

Figure 9. the distribution of bubble volume fraction

Figure 10 and Figure 11 show the numerical simulation result of waterjet propulsion pump. The bubbles firstly appear on the edge of the impeller inlet in large flow rate condition, so as in small flow rate and design flow rate. But its variation tendency is greatly different with that in design condition and small flow rate with the decrease of NPSHa. When NPSHa reduces from 1.9m to 1.45m, the region covered by bubble firstly extends along the rim of blade and appears in the middle of blade near the hub. It develops from the hub to the rim of blade. The pressure is higher than the vapour pressure on pressure face of blade. So bubble cannot be found on pressure face of blade.

(a)NPSHa=2.73m (b)NPSHa=2.02m (c)NPSHa=1.9m (d)NPSHa=1.45m

(e) NPSHa=1.45m Static pressure on pressure face (f) NPSHa=1.45m bubble volume distribution on pressure face

Figure 10. NPSHa>1.45m

While NPSHa<1.45m, cavitation is so hard that bubble blocks the passage of impeller. Then the efficiency of pump sharp drop and the pump cannot work. Flow separation is found at the outside edge of suction face of blades for large flow rate and negative angle. Bubble is found on the pressure face and Cavitation volume fraction extends to the rim from the hub at the outside of pressure face. The low pressure area that is blue area showed in Figure 11 is consistent with the distribution of bubble volume fraction. The bubble is found in the large flow rate condition while the development of cavitation is to some extent on the suction face of blade rather than the bubble cover suction face completely in design condition and small flow rate condition. Figure 10 and Figure 11 shows that the critical NPSHa and cavitation area are bigger than that in the other conditions.

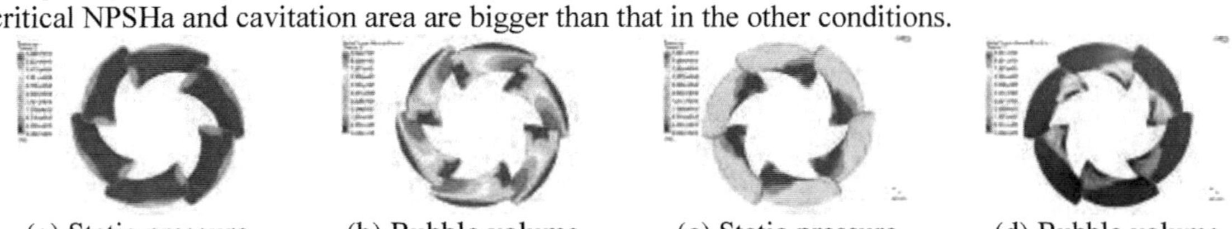

(a) Static pressure distribution on suction face (b) Bubble volume distribution on suction face (c) Static pressure distribution on pressure face (d) Bubble volume distribution on pressure face

Figure 11. NPSHa=1.36m

Figure 12 shows the pump cavitation characteristic curve in three working conditions. The flow rate and efficiency curves are showed little change with the decrease of NPSHa. The reason for that is that the flow is unstable in the pump at the begin of bubble produce. But the flow rate and efficiency curves will sharp drop after NPSHa is greater than critical value. The point of critical value is called critical cavitation point. The critical NPSHa decreases as the flow rate reducing and the dropping extent of pump cavitation characteristic curves is small while the flow rate is small. That means that the pump is hard to get cavitation while the flow rate is small.

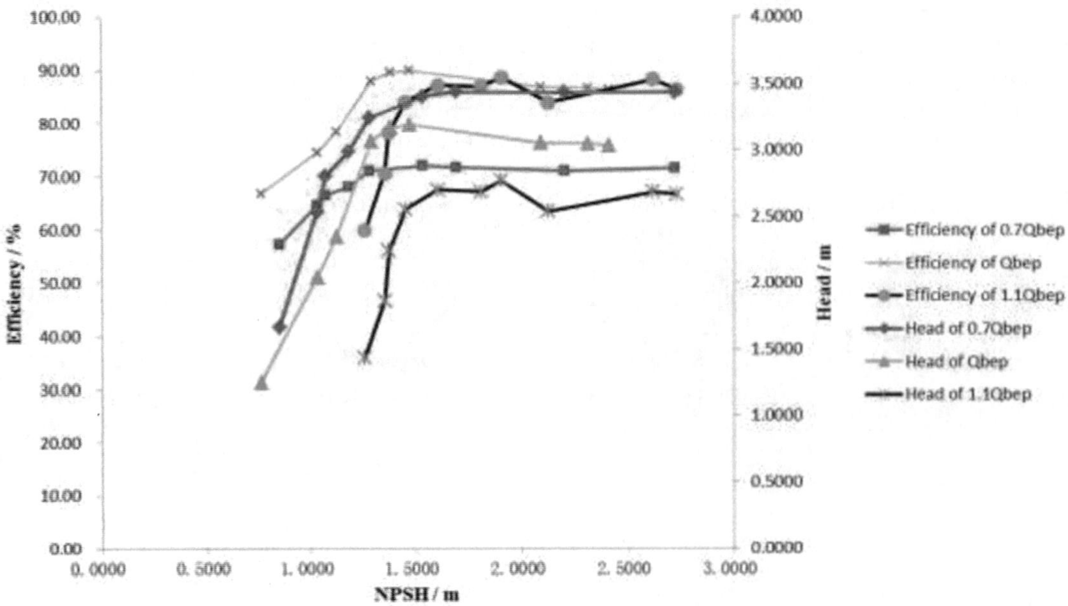

Figure 12 Pump cavitation characteristic curves

4. Conclusions

The paper finds that the cavitation resistances of waterjet propulsion pump are grandly weaken as the critical NPSHa value grandly increase by the numerical simulation on the cavitation flow in waterjet propulsion pump. The harm of cavitation to blades is deepened while deviating from the design operation. The static pressure is bigger in the rim than that in the hub, and the distribution trends of static pressure are almost same in three working condition. Therefore, the bubble volume fraction in the rim is bigger than that in the hub. That means that the cavitation is harder in the rim. And that should be considered in actual design.

Acknowledgment

This work was financially supported by National science and technology plan project in rural areas (Grant No.2012BAD08B03), National Natural Science Foundation of China (Grant No. 51179167), Water conservancy science and technology project of Jiangsu Province (Grant No.2014046), The Young academic Leaders in Blue Project of Jiangsu, Prospective Joint Research Project of Jiangsu (Grant No. BY2015061-12) , A Project Funded by the Priority Academic Program Development of Jiangsu Higher Education Institutions and Major Natural Science Foundation of Universities in Jiangsu(Grant No.12KJA570001).

References

[1] Wang Y S, Ding J M 2004 Research on the relationship between waterjet power absorption and vessel speed[C]// *In-ternational Conference on Waterjet Propulsion*. London: RINA.

[2] Verbeek R, Bulten N W H 1998 Recent development in waterjet design[C]//*Proceedings of international conference on waterjet propulsion II*. Amsterdam: RINA.

[3] Harvey E N, Barnes D K, McElroy W D, et al. 1944 Bubble formation in animals. I. Physical factors[J]. *Journal of Cellular and Comparative Physiology*, **24(1)**: 1-22.

[4] Knapp R T, Daily J W, Hammitt F G 1979 *Cavitation*[J].

[5] Cheng L, Bart P M, Van E, Liu Chao et al. 2012 Radial forces of waterjet propulsion mixed-flow pump[J]. *Journal of Drainage and Irrigation Machinery Engineering*, **30(6)**:636-640.

[6] Bulten N W H, Verbeek R, Van E, Bart P M 2006 CFD simulation of the flow through a waterjet installation[C]//*International Conference on Waterjet Propulsion*. **4**: 11-19.

[7] Gopalan S, Katz J 2000 Flow structure and modeling issues in the closure region of attached cavitation[J]. *Physics of Fluids(1994-present)*, **12(4)**: 895-911.

[8] Pham T M, Larrarte F, Fruman D H 1999 Investigation of unsteady sheet cavitation and cloudcavitation mechanisms[J]. *ASME Journal of Fluids Engineering*, **121(2)**: 289-296.

[9] Pellone C, Peallat J M 1995 Non-linear analysis of three-dimensional partially cavitating hydrofoil[C]//*Proceedings of the International Symposium on Cavitation*, Deauville, France. **1995**: 63-67.

[10] Yang F, Liu C, Tang F P 2012 Cavitation performance prediction of mixed-flow pump based on CFD[C]//*26th IAHR Symposium on Hydraulic Machinery and System*, Auguest 19-23, 2012, Beijing, China

[11] Zwart P J, Gerber A G, Belamri T 2004 A two-phase flow model for predicting cavitation dynamics[C]//*Proceedings of the Fifth International Conference on Multiphase Flow*, Yokohama, Japan.

[12] Orszag S A, Patterson Jr G S 1972 Numerical simulation of three-dimensional homogeneous isotropic turbulence[J]. *Physical Review Letters*, **28(2)**: 76-79.

[13] Merkle C L, Feng J, Buelow P E O 1998 Computational modeling of the dynamics of sheet cavitation[C]//*Proceedings of Third International Symposium on Cavitation* , Grenoble,France, **1998:**307－311.

[14] Senocak I , Shyy W 2002 A pressure-based method for turbulent cavitating flow computations[J]. *Journal of Computational Physics*, **176(2)**:363-383.

IOP Conf. Series: Materials Science and Engineering **129** (2016) 012012 doi:10.1088/1757-899X/129/1/012012

Numerical simulation on macro-instability of coupling flow field structure in jet-stirred tank

D Y Luan[1], J P Lu, Q X Bu, S F Zhang and S X Zheng

School of Mechanical and Electrical Engineering, Qingdao University of Science and Technology, China

E-mail: qddy05@163.com

Abstract. The velocity field macro-instability (MI) can help to improve the mixing efficiency. In this work, the MI features of flow field induced by jet-stirred coupling action is studied by using computational fluid dynamics (CFD) simulations. The numerical simulation method of jet-stirred model was established based on standard turbulent equations, and the impeller rotation was modeled by means of the Sliding Mesh (SM) technology. The numerical results of test fluid (water) power consumption were compared with the data obtained by power test experiments. The effects of jet flow velocity and impeller speed on MI frequency were analyzed thoroughly. The results show that the calculated values of power consumption agree well with the experiment measured data, which validates the turbulent model, and the flow structure and MI frequency distribution are affected by both impeller speed and jet flow rate. The amplitude of MI frequency increases obviously with the increasing rotation speed of impeller and the eccentric jet rate, and it can be enhanced observably by eccentric jet rate, in condition of comparatively high impeller speed. At this time, the MI phenomenon disappears with the overall chaotic mixing.

1. Introduction

The fluid turbulent flow is highly unstable circulation pattern in a stirred tank with the existence of large-scale low-frequency phenomenon termed as macro-instability (MI). The frequency of MI is represented as a distinct peak for frequency oscillations in power spectrum of the region close to the impeller which is usually located between 0.01 and 1 HZ frequency band. The studies have reported that the impeller stream region contains a large number of the symmetry flow field structure, so it is difficult to finish dissipating energy to outward effectively. As a result, near 70% of the mechanical energy will be dissipated in this region [1], which leads to the reducing of mixing efficiency. The energy dissipation concerned with turbulent flow and the large eddy pattern of the flow field may be described well with MI phenomenon [2]. The research on MI phenomenon is of significant value for the better understanding of the flow pattern and mixing mechanism since the existence of MI help to enhance the turbulent flow to an extent, improve the mixing efficiency, moreover, it has the distinct effect on mass and heat transfer and local gas content distribution in a stirred tank [3-4].

[1] Present address: School of Mechanical and Electrical Engineering, Qingdao University of Science and Technology, China

In recent years, a number of studies have pointed to the likely effect of MI phenomenon in stirred tanks in different ways [5-9]. These results indicated that MI frequency is irrelevant to stirred tank size, impeller type and fluid property parameters, whereas MI frequency may produce a certain difference in condition of different Reynolds number and different ratio of kettle and impeller diameter. Up to now, there is few research on MI of coupling flow field in jet-stirred tank [10]. The studies concerned with the jet pattern found that the jet can form a kind of special flow structure different from that of agitation, and the nonlinear coupling flow field in jet-stirred tank inevitably induce the chaotic mixing of fluid. Therefore, it is necessary to analysis the flow field characteristic mixing of jet-stirred coupling flow pattern in more detail. The aim of this paper is to study MI characteristics of jet-stirred coupling flow field and explore the effective way of chaotic mixing induced by coupling flow field using numerical simulation method.

2. Stirred vessel configuration and meshing

The stirred vessel consists of a transparent cylindrical tank of diameter (T) 0.3m. The flat-bottomed tank was fitted with four equally spaced flat baffles, each with a width (w) equal to $T/10$. The fluid height (H) was maintained constant at a height equal to $1.25T$. The dual-impeller system was restricted to the Rushton turbine of impeller and 45° four pitch-blade turbine downflow as upper impeller. The dual-impeller with a diameter (D) of 0.1m and thickness 2mm was mounted on a centrally positioned shaft of diameter 0.016m at an off-bottomed clearance(C) and a layer clearance ($C1$) of 0.1m. The jet nozzle of diameter 6mm was located at the bottom of the tank, the distance was 0.1m off the z axis. The tank was filled with tap water. The impeller speed was fixed at 2 r/s, corresponding to Reynolds number, Re equal to 19960. So the flow was in a turbulent state. The computational domain was separated into two zones: the rotor zone enclosing the impeller and the stator zone with the rest of the vessel. A pre-processor (Gambit 2.3) was used to discretize the flow domain with an unstructured tetrahedral mesh. The density of mesh cells in rotor zone was needed to be five enough to capture the flow details, as shown in Figure 1. Grid independence was verified by demonstrating the additional requirement on mesh cells that did not change the calculated power number and velocity magnitude in the region of high velocity gradients close to the impeller blade by more 3%. First, the original 3D mesh of the model had about 623100 cells for calculation, then the numbers were increased to about 851700 cells. It was found that this increase changed the velocity and power number in regions of high velocity gradients by more than 3%. When the number of the cells further were increased to about 1227400, the velocity and power number changed by less than 3% in the same regions. Therefore, about 1227400 cells were adopted to discretize the computational regions.

Figure 1. Configuration and computational grids of the stirred system.

3. CFD model and simulation methods

The CFD simulation method of flow field in a stirred tank is based on the calculation of the scalar transport equations, considering the stirred tank with the characteristics of the flat bottomed cylindrical vessel, the cylindrical coordinate was used to solve the transport equation in this work as follows:

$$\frac{\partial}{\partial t}(\rho\varphi)+\frac{\partial}{\partial z}(\rho u_z\varphi)+\frac{1}{r}\frac{\partial}{\partial r}(r\rho\rho_r)+\frac{1}{r}\frac{\partial}{\partial \theta}(\rho u_\theta\varphi)=\frac{\partial}{\partial z}\left(\Gamma\frac{\partial\varphi}{\partial z}\right)+\frac{1}{r}\frac{\partial}{\partial r}\left(\Gamma r\frac{\partial\varphi}{\partial r}\right)+\frac{1}{r}\frac{\partial}{\partial\theta}\left(\frac{\Gamma}{r}\frac{\partial\varphi}{\partial\theta}\right)+S \tag{1}$$

The initial condition for CFD simulation was that of still liquid. A flat liquid surface was assumed at the top surface by setting all the shear stress equal to zero. No-slip boundary conditions were adopted at the impeller blades, the shaft and tank walls, and the speed boundary condition was set to the nozzle entrance. The interface between the rotor and stator zone was set as the interior. For modeling the impeller rotation, the sliding mesh(SM) method was used for the $k-\varepsilon$ model computation.

The pressure-based Navier-stoke algorithm was used for the solution of the CFD model with implicit solver formulation. The momentum conservation equation was discretized using the second upwind scheme to obtain a high-precision result. The standard wall function was adopted in near-wall region and the SIMPLE algorithm was performed to couple velocities and pressure terms. Furthermore, the time step of CFD computation is 0.005s. Within each time step, a maximum of 20 iteration was conducted to ensure the accuracy of convergence when the normalized residuals of all variables were less than 1×10^{-4}. All the simulation were carried out using commercial CFD software Fluent v6.3(Fluent Inc.) on a HPXW6200 workstation with 8GB RAM.

The spectrum analysis is a time series method by making Fourier transform for the time series of velocity fluctuation signal. This method may exhibit the periodic time series pulse information based on the analysis of the powers spectral density distribution at the different frequency band. The frequency of MI can be determined using spectrum analysis method according to the CFD simulation results of the time series of velocity signal [11]. In this paper, spectrum analysis is realized by programming with the help of Matlab software, and using the Welch average overlap method and Hanning window function method to improve the resolution of the spectrum map. This may be suitable for the time series spectrum analysis of the flow field in a stirred tank [12]. The literature have reported that the location of the velocity monitoring point has few effect on the MI characteristics of flow field [13]. As a result, the sample point was selected at $r=T/3$, and $Z=0$ in the $\theta=0$ plane, which located at the center height of the impeller. The monitoring frequency at the different windows was 20 HZ, and the monitoring time is 30s.

4. Result and discussions

4.1. Validation of CFD model

The power consumption of the impeller may be conducted based on the shaft torque obtained through numerical simulation. The results of the simulations are compared with the original experimental data and good agreement is obtained in all case, as shown in Figure 2. It can be seen that the power consumption increases as the increasing impeller speed and the same tendency is exhibited between the simulations and the experimental measurement. The maximum deviation is less than 10%, which verifies the turbulent model for simulations in this work.

4.2. Micro structure of the flow field

Figure 3 shows the velocity vector plots within the axial lengthwise section of the stirred tank at impeller speed N=2r/s. It can be seen that the flow between the two impeller can be considered as connected flow without the layer compartment existing. The flow field is almost similar to that of single impeller. However, in condition of no jet, the structure of flow field exhibits the obvious symmetry with double vertex ring shape, which will necessarily result in low mixing efficient because of mixing isolation region existing, as shown in Figure 3 (a). When the jet velocity(v) equals to 2 m/s, the symmetrical structure of flow field is destroyed, accompanied by the asymmetrical flow pattern emergence, and this trend of asymmetrical pattern will increase as the jet velocity increases, as shown in Figure 3 (b) and (c). So the fluid in stirred tank is alternatively stretched and folded under the couple action of agitation and jet, which will induce the chaotic mixing, and furthermore, improve the mixing efficiency.

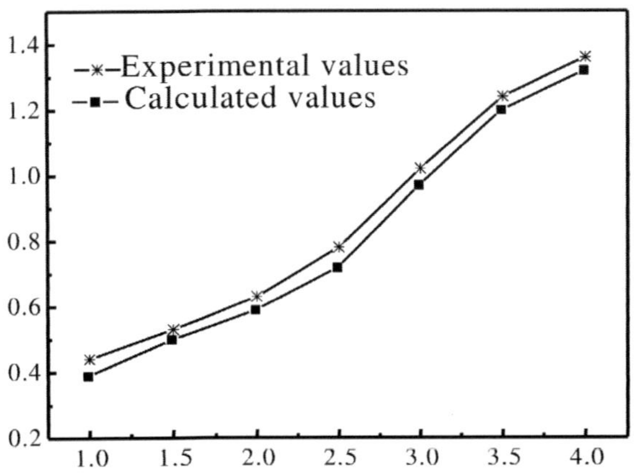

Figure 2. Comparison between calculated powers and experimental values at different rotational speed.

4.3. Action of impeller velocity

MI frequency plots of different impeller speed are shown in Figure 4. It can be seen that MI frequency mainly concentrated on the band of 0~2 Hz, which means the low turbulent intensity at the speed N=2 r/s. The single frequency peak value is found to be 0.040 Hz, corresponding to the dimensionless frequency $f^*=f/N$=0.02. When the speed reaches to 3.5 r/s, the amplitude of MI frequency increases in the whole frequency band. At this time, the fluctuation strength of fluid is enhanced obviously, and two MI frequency peak values are observed, i.e., 0.231 Hz and 0.315 Hz, corresponding to the dimensionless frequency 0.066 and 0.09, respectively, which indicates that the flow is in unstable state obviously. As a result, the distribution of MI frequency may be controlled by the different stirring speed.

4.4. Action of jet velocity

Figure 5 reports the MI frequency plots of different jet velocity at the impeller speed N=2 r/s. It can be found that the fluctuation strength increases and the frequency band of high amplitude becomes a wider distribution as the jet velocity increases. The MI frequency peak value is 0.068 Hz as the jet velocity (v) equals to 2 m/s. When the jet velocity reaches to 2.5 m/s, 3 m/s and 4 m/s, corresponding peak value is 0.097 Hz, 0.113 Hz and 0.187 Hz, respectively. It is seem that the MI frequency diverts to the high frequency band and the MI intensity is improved. This indicates that the jet action can make a big effect on the coherent structure of flow field and MI frequency increases as the jet velocity increases. Therefore, it is concluded that MI intensity of flow field may be improved by adding jet action.

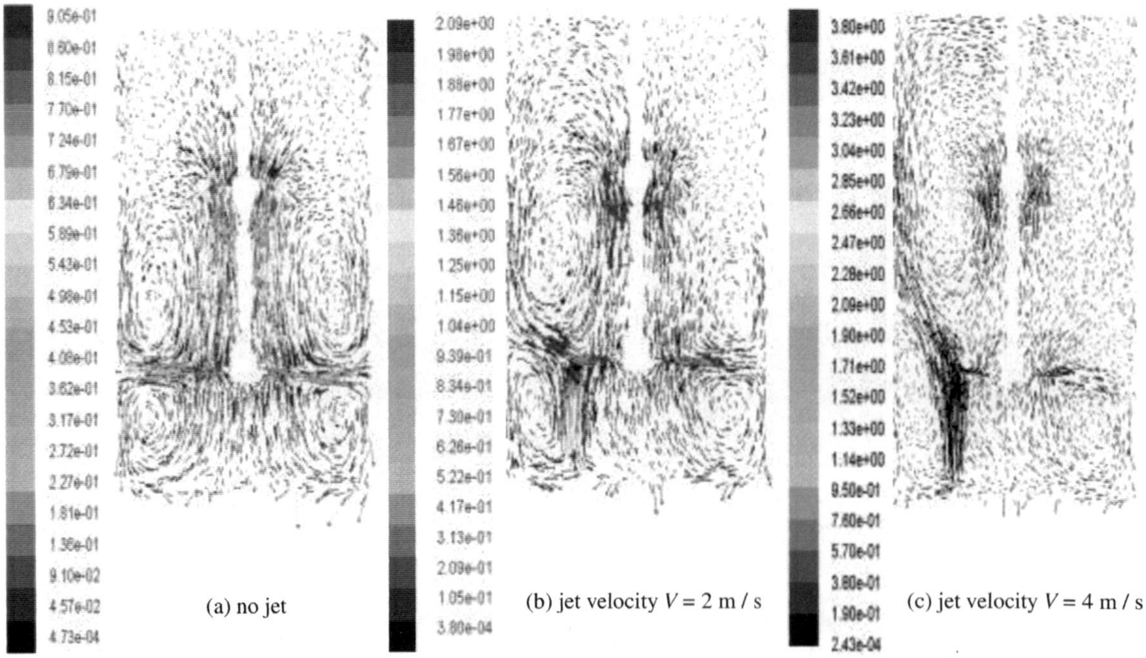

Figure 3. Velocity vector plots within the axial lengthwise section of the stirred tank at $N = 2$ r/s.

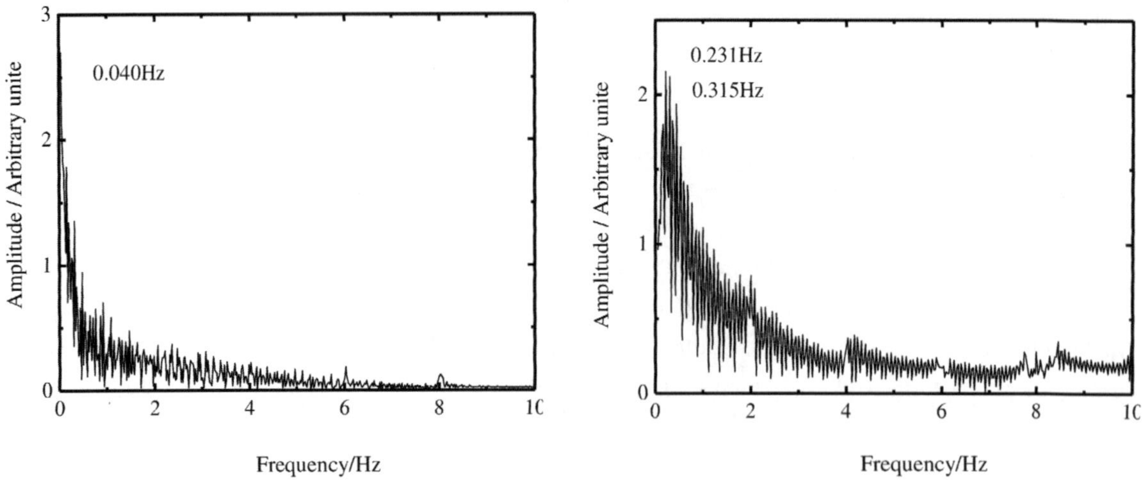

(a) $N = 2$ r/s (b) $N = 3.5$ r/s

Figure 4. MI frequency at different rotational speeds.

Under the high impeller speed N=3.5 r/s, the effect of jet velocity on MI frequency is shown in Figure 6. It is found that the characteristics of MI frequency disappear duo to the jet action. This means that the coherent structure of flow field is thoroughly destroyed and the multi-scale structure characteristics appear, which makes the flow field into whole chaos.

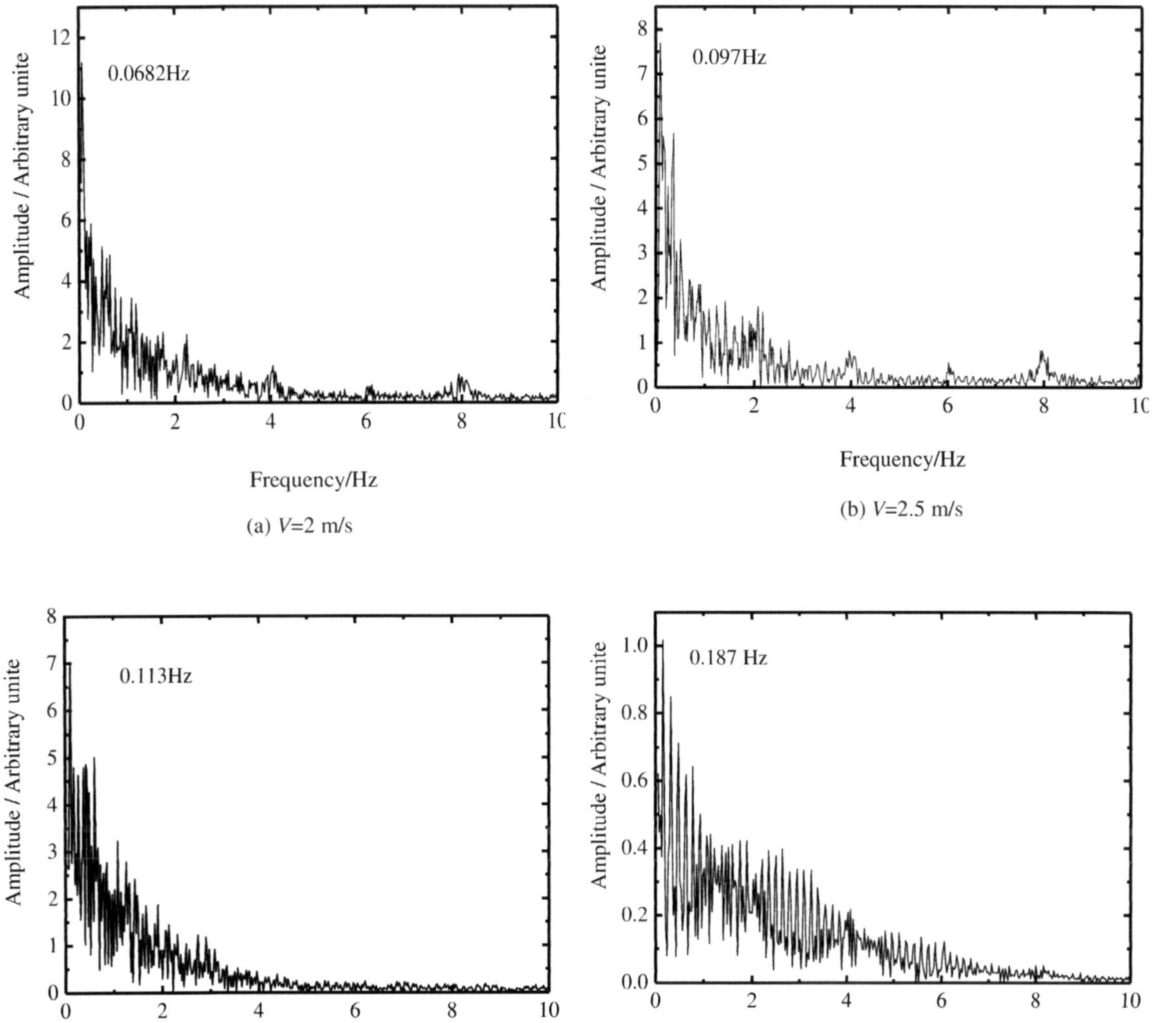

(a) V=2 m/s

(b) V=2.5 m/s

(c) V=3 m/s

(d) V=4 m/s

Figure 5. MI frequency under different jet speeds at N = 2 r/s.

Through the study on the MI frequency variation in the different working condition, it is known that, with the impeller speed increased, the frequency amplitude distribution on the whole band would increase and the fluctuation intensity is also improved corresponding. At this time, the spectrum plot exhibits the characteristics of multiple peak frequency with the high peak values. This indicates that the flow field is of a more apparent random fluctuation character associated with the turbulence intensity of the fluid in stirred tank. Owing to the fact that intensity of MI frequency on the whole band increases as the impeller speed increases, the MI phenomenon can be taken as the weakened signature, which would lead to the rapid mixing. Similarly, Fan Jianhua et al.[7] also obtained such conclusion from the average sequence spectrum plot for the turbine impeller. Moreover, at low impeller speed (*N*=2 r/s), MI frequency and intensity of flow field have been proved to make a better improvement through the eccentrical jet action. When the speed reaches to high(*N*=3.5 r/s), a more dominant effect of jet action on MI may be observed that is a broader distribution band of frequency peak value appears in the spectrum plot, which means that MI frequency disappears and the flow field leads to the chaotic state accompanied by the multi-scale coherent structure feather.

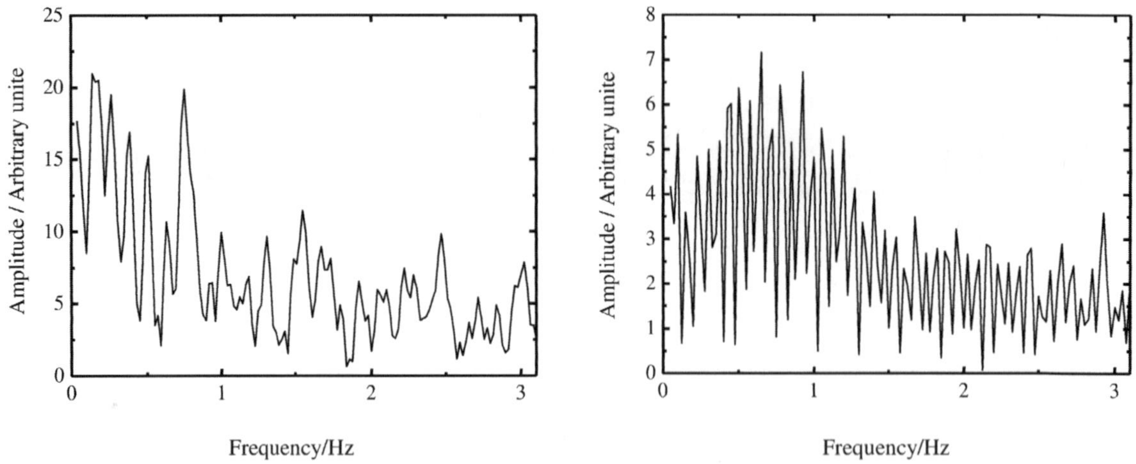

(a) *V*=4 m/s (d) *V*=6 m/s

Figure 6. MI frequency under different jet speeds at *N* = 2 r/s.

5. Conclusions

The eccentric jet-stirring may make an asymmetric flow field structure, therefore, lead to an extensive chaos in the tank and improve the flow and mixing efficiency.

Increasing the impeller speed may improve MI frequency intensity of the flow field, and then the MI phenomenon is weakened. So the random fluctuation level is enhanced in the whole tank, which will finish a rapid mixing progress.

The effect of the eccentric jet action on MI frequency intensity may be observed, and a more dominant improvement can be obtained specially at high impeller speed with MI frequency disappearance. This indicates that the flow field is into the chaotic mixing.

NOMENCLATURE

C	off-bottom clearance, m	C_1	layer clearance
D	impeller diameter, m	H	fluid height, m
T	tank diameter, m	N	impeller speed, r/s
Re	apparent Reynolds number	ρ	density, k_g/m^3
t	time, s	r	radial position, m
Γ	generalized diffusion coefficient	S	Generalized source term
φ	field variable	u_r	radial velocity, m/s
u_θ	tangential velocity, m/s	u_z	axial velocity, m/s

Acknowledgements
The work reported in this paper was supported by the Project Shandong Provincial Science and Technology Development planning Program of China (Grant No. 2013YD09007) and Shandong Provincial Natural Science Foundation of China (Grant No. ZR2014EEM017).

References
[1] Wang K and Feng L F [M] 2001 *The designation of mixed equipment* (Beijing: China Machine Press)
[2] Hartmann H, Derksen J and Van Den Akker H E A [J] 2004 Macro-instability uncovered in a Rushton turbine stirred tank by means of LES *American Institute of Chem Eng* **50** 2383
[3] Hasal P, Montes J L and Boisson H C [J] 2000 Macro-instabilities of velocity field in stirred vessel: detection and analysis *Chem Eng Sci* **55** 391
[4] Ducci A, Doulgerakis Z and Yianneskis M [J] 2008 Decomposition of flow structures in stirred reactors and implications for mixing enhancement *Ind Eng Chem Res* **47** 3664
[5] Paglianti A, Liu Z and Montante G [J] 2008 Effect of macro-instabilities in single-and multiple-impeller stirred tanks *Ind Eng Chem Res* **47** 4944
[6] Fan J, Rao Q and Wang Y [J] 2004 Spatio-temporal stirred vessel via digital particle image velocimetry macro-instability analysis of DPIV *Chem Eng Sci* **59** 1863
[7] Fan J H, Rao Q and Wang Y D [J] 2004 Spectral analysis of the velocity fluctuations in a mechanically stirred tank *J Chem Eng of Chinese Univ* **18** 287
[8] Galletti C, Paglianti A and Yianneskis M [J] 2005 Observations on the significance of instabilities turbulence and intermittent motions on fluid mixing processes in stirred reactors *Chem Eng Sci* **60** 2317
[9] Galletti C, Lee K C and Paglianti A [J] 2004 Reynolds number and impeller diameter effects on instabilities in stirred vessels *Chem Eng Sci* **50** 2050
[10] Liu Z H, Ning W Z and Sun R X [J] 2010 Comparative study on macro-instability of liquid / gas-liquid in stirred tank *J Chem Eng of Chinese Univ* **29** 100
[11] Yang F L, Zhou S J and Wang G C [J] 2012 Detached eddy simulation of the Macro-Instability in eccentrically stirred tanks *J Chem Eng of Chinese Univ* **26** 228
[12] Yin L B [D] 2005 *Experimental study of Macro-Instabilities in a stirred tank* (Beijing: Beijing University of Chemical Technology)
[13] Nikiforaki L, Yu J and Baldi S [J] 2004 On the variation of processional flow instabilities with operational parameters in stirred vessels *Chem Eng J* **102** 217

Experimental research on internal flow in impeller of a low specific speed centrifugal pump by PIV

J F Zhang[1], Y F Wang and S Q Yuan

National Research Centre of Pumps, Jiangsu University,
Zhenjiang, Jiangsu, 212013, China

Abstract. For the purpose of investigating the influence of two different impellers, one is with splitter blades and the other one is without splitter blades, on a low-specific centrifugal pump. The experimental investigation in impellers was conducted at different conditions and phases by means of PIV (Particle Image Velocimetry) to study the internal flow. Meanwhile, the absolute and relative velocity distributions in impellers were obtained. Experimental results show that the head value is higher in the impeller with splitter blades and both two head curves appear hump phenomena at small flow rate. The absolute velocity value increases with radius and from pressure side to suction side at the same radius gradually. The splitter blades can scour the wake, making outlet velocity distribution more uniform and improving the internal flow. The velocity distribution becomes less even in the process of closing to tongue due to reinforced interference of tongue on internal flow.

1. Introduction

Low specific speed centrifugal pump generally refers to the specific speed n_s=30~80, with the characteristics of small flow rate and high head [1]. It is widely used in various fields such as irrigation, urban water supply, petroleum and chemical industry. For low specific speed centrifugal pump, the head curve tends to appear the hump at part load condition as the terrible diffusion in the long and narrow flow path. Therefore, the efficiency of pump is always low and the common way of reducing outlet diameter of impeller $D2$ is applied to decrease impeller disk friction loss, so as to improve the efficiency of pump [2]. However, it is necessary to choose larger outlet setting angle of blade $\beta 2$ and increase blade numbers Z, at the same time reducing $D2$, in order to achieve the required head. Hence, the design method of spaced long blade and splitter blade is taken to alleviate the serious crowding phenomenon at inlet of blade, causing by overmuch blades [3]. Meanwhile, the splitter blades could scour the wake at outlet of impeller to make the velocity more uniform and improve the internal flow. Pump contains many components and the impeller is main flowing and acting component, whose flow field is complex three-dimensional incompressible viscous turbulent rotational flow field, and has an important impact on the pump head, flow rate, efficiency and cavitation performance. Consequently, it is essential to study further the internal flow field in different impellers, with and without splitter blades, and reveal the law of internal flow for improving pump performance and design optimization.

Experimental test measurements are the most basic and reliable research methods, which have an irreplaceable role [4]. PIV technology, with high accuracy, is a non-contact, instantaneous and dynamic measurement and becomes an advanced testing technique for researching flow pattern of

[1] Corresponding author. Tel.:13776476981. Email address: zhangjinfeng@ujs.edu.cn.

turbo-machinery [5-8]. So far, foreign and domestic scholars have conducted numerous studies about internal instability flow phenomena and the development of law by PIV technology for centrifugal impeller [9-16].

In this paper, the internal flow fields in different impellers, with and without splitter blades, of a low specific speed centrifugal pump at different conditions and different phases were tested by PIV technology to obtain internal real flow field. According to experimental dates, the occurrence and development of internal unsteady flow structure were analyzed to reveal the influence of unsteady flow on performance of pump, which could provide reference for improving impeller design methods and experimental study of internal flow.

2. Research model pump

The type of model pump is IS50-32-160 and two different impellers were designed, one is with splitter blades and the other one is without splitter blades. The factors of impacting splitter blade position are as follows, inlet diameter of splitter blade D_{si}, circumferential bias angle of splitter blade θ and deflection angle of splitter angle α, as shown in Figure 1. The main performance and geometry parameters of pump are such as table 1. The distinct difference of two research pumps is the impeller. 4+4 represents the impeller with four long blades and four splitter blades; 6 represents the impeller with six long blades

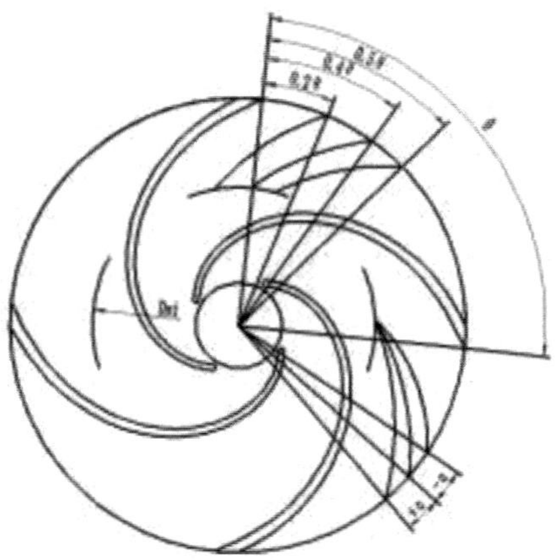

Figure 1. The locating places of splitter blades

Table 1. Main parameters of the research model

performance	Design flow rate Q_d/m^3·h^{-1}	6.3	6.3
	Design head H_d/m	8	8
	Rotational speed n_d/r·min^{-1}	1450	1450
	Specific speed n_s	47	47
geometry	Inlet diameter of pump D_s/mm	50	50
	Outlet diameter of pump D_d/mm	32	32
	Inlet diameter of impeller D_1/mm	50	50
	Outlet diameter of impeller D_2/mm	160	160
	Blade number Z	4+4	6
	Inlet diameter of splitter blade D_{si}/mm	106	/
	Circumferential bias angle of splitter blade θ/°	72	/
	Deflection angle of splitter angle α/°	0	/

The impeller and volute of model pump were made of plexiglass for visual recording easily, and various surfaces were polished, while the non-test side suction of impeller and volute near chamber were blacked to reduce background noise. To reduce the scattering in process of PIV test, the external shape of volute was designed for square. The shape of blade is cylindrical and the volute section is rectangular. Two impellers and volute of model pump are shown in Figure 2.

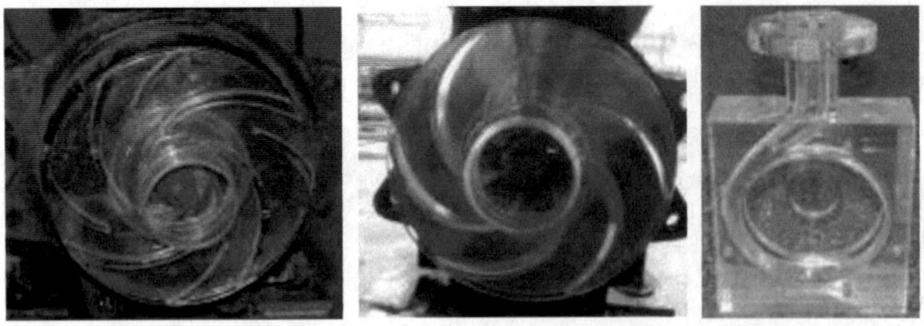

4+4 6 volute

Figure 2. Impellers and volute of model pump

3. Experiment of PIV

3.1 Test system of PIV

Commercial PIV system, made by TSI company in United States, and image acquisition and analysis software platform of Insight 3G were adopted in the test. PIV system are consists of tracer particles, pulsed laser of YAG-NEL, light arm of 610015-SOL, CCD camera of 630059(4MP), time synchronization controller of 610035, shaft encoders and other components, such as Figure 3. Silvered glass balls were chosen as tracer particles, whose diameters are between 10-15μm and the density is 1.6g/cm3, slightly heavier than the water. It has good liquidity and scattering properties. Besides the chemical characteristic is stable, so as to ensure the long-term safety testing requirements. Plane mirror was placed at 45 ° to the axis of the inlet position for facilitating photographing. Meanwhile, the performance of pump was also tested on the PIV test rig.

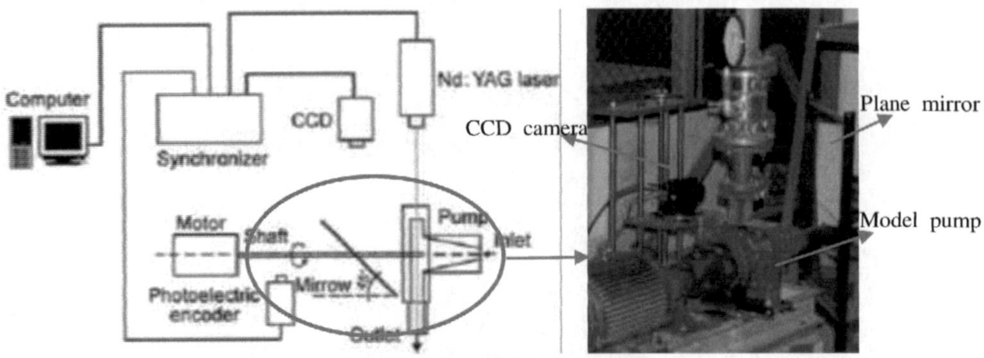

Figure 3. Sketch of PIV test system

3.2 Test scheme of PIV

The test plane is geometric symmetry plane of pump, and parameter θ, the angle between tongue and suction side of long blade in shooting passage, was defined to show the relative position between impeller and tongue, as shown in Figure 4. Two different impellers of 4+4 and 6 were tested under five different conditions (Q=1.8Q_d, 1.4Q_d, 1.0Q_d, 0.8Q_d, 0.4Q_d) at the same phase (θ = 14 °) to reveal the unstable flow and development law, while comparing the difference to study the influence of splitter blades on the pump.

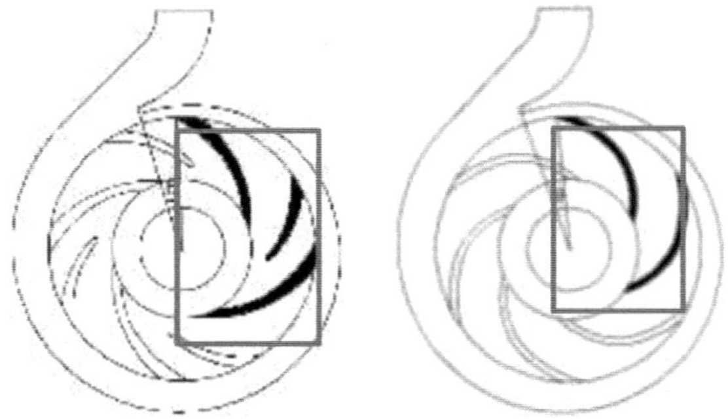

4+4 6

Figure 4. Schematic diagram of testing area

Meanwhile, the internal flow field in impeller of 4+4 was also tested by PIV at various phases position ($\theta = 14$ °, $\theta = 5$ °, $\theta = -4$ °) to study the interference of tongue on the impeller internal flow at the design flow rate, as shown in Figure 5.

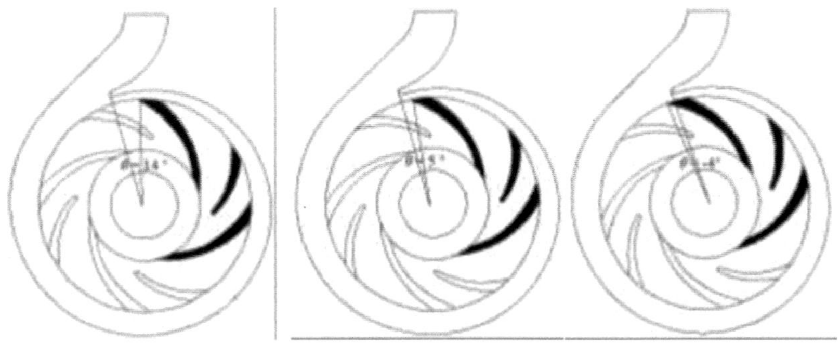

Figure 5. Schematic diagram of blade phase

4. Analysis of experimental results

4.1 Performance curve

Comparison of the performance curves of model pumps between impeller 4+4 and 6 are shown in the Fig.6. Accord to the performance curves of 4+4 and 6, it is obvious that the head value of impeller 4+4 is higher than of impeller 6 and both two head curves appear a certain degree of hump phenomenon at small flow rate around $0.4Q_d$.

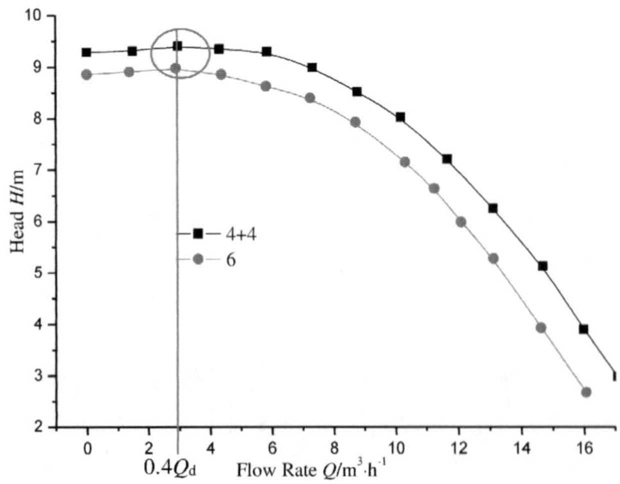

Figure 6. Comparison of the performance curve between 4+4 and 6

4.2 Absolute velocity

The absolute velocity distribution of midsection at different flow rates in impeller 4+4 are shown in Figure 7. It is obvious that the direction of absolute velocity is basically the same with rotational direction of impeller. The value of absolute velocity increases with radius and from pressure side to suction side at the same radius gradually, consistent with powerful flow theory. According to the variation of absolute velocity with flowrate, the value of absolute velocity changes slightly. Therefore, the circumferential velocity is the main component of absolute velocity.

The velocity distribution in channel B is more uniform than that of channel A under different conditions. As the influence of splitter blades, the fluid in the larger channel could not pass the passage in time. This phenomenon results in the fluid gathering in the region near outlet, forming the high-speed zone at the suction side of splitter blade. With the increase of flow rate, the high-speed zone becomes larger and the low-speed zone is formed near the leading edge of splitter blade. Low-speed zone could result in the formation of large velocity gradients, making more unsteady flow phenomenon such as reflow and secondary flow, vortex and cavitation occur in the region.

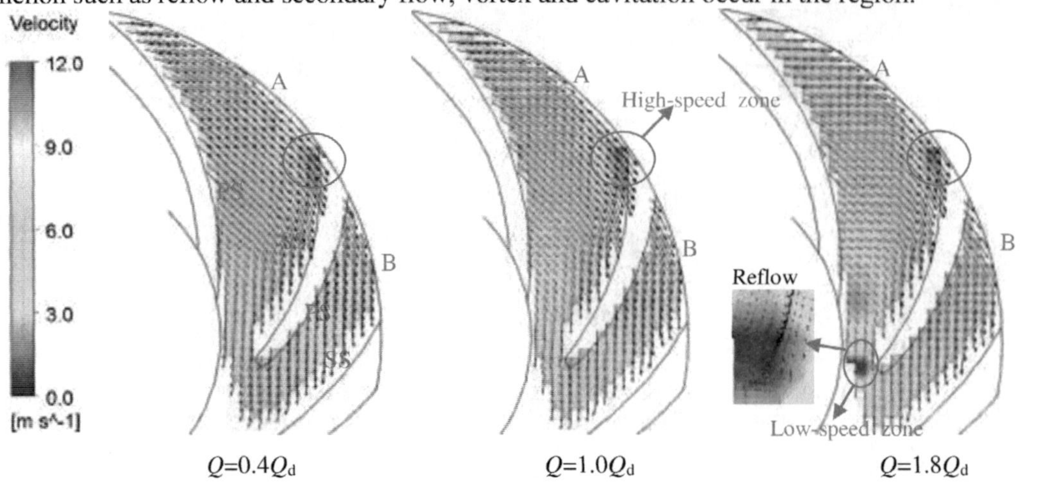

Figure 7. Absolute velocity distribution of impeller 4+4 at different conditions

The absolute velocity distribution of midsection at different flow rates in impeller 6 are shown in Figure 8. Comparing Fig.8 with Figure 7, in can conclude that the absolute velocity distribution in impeller 6 is more uniform than that in impeller 4+4 under over load condition, while more complicated under part condition due to the influence of splitter blades in impeller 4+4.

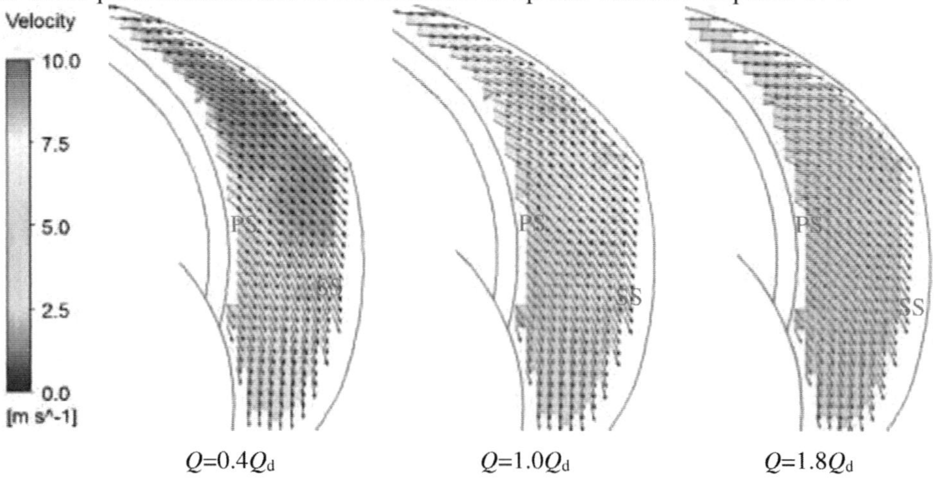

Q=0.4Q_d Q=1.0Q_d Q=1.8Q_d

Figure 8. Absolute velocity distribution of impeller 6 at different conditions

4.3 Relative velocity

The relative velocity distributions in two impellers under design condition are shown in Figure 9. The direction of relative velocity is almost along the blades and the direction of outlet velocity is along the tangential direction of impeller outlet side, which is opposite to the rotational direction of impeller.

Time- averaged relative velocity distribution near outlet of impeller 4+4 and 6 under design condition are shown in Figure 10 and Chanel 1 is the passage where tongue located at and the other channels are labeled along the direction of rotation. It is obvious that the relative velocity is lower close to suction side than that near pressure side of blade at the exit of impeller 6. This flow structure is somewhat of jet–wake flow structure in centrifugal pump. The outlet velocity distribution is improved after adding splitter blades in the impeller 4+4. The splitter blades can scour wake to a certain extent, making outlet velocity distribution more uniform.

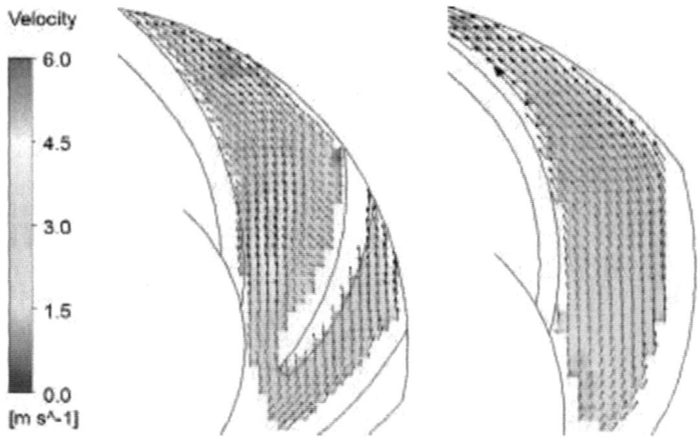

4+4 6

Figure 9. Relative velocity distribution in impeller 4+4 and 6 at design condition

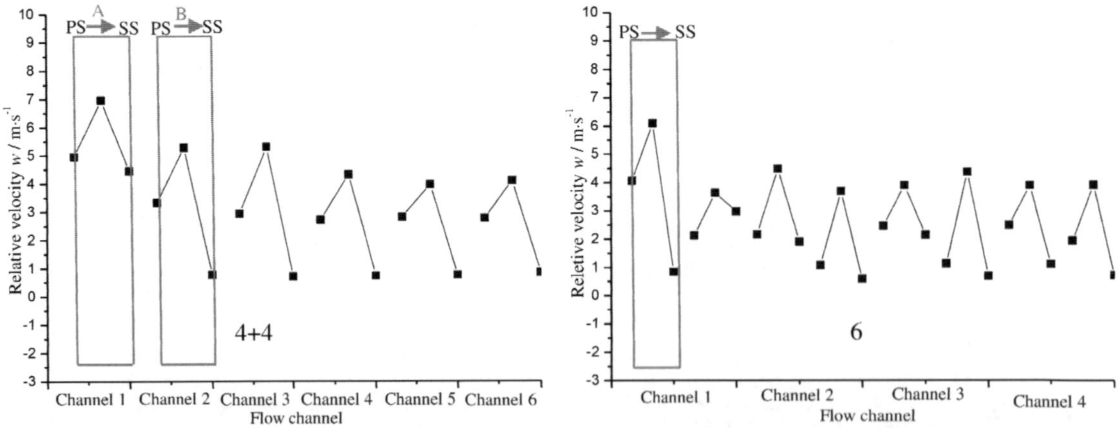

Figure 10. Time- averaged relative velocity distribution at impeller outlet under design condition

Compared to the relative velocity distributions under different flow conditions in Figure 11, the variation of velocity within the channel A is higher than that in channel B. With the increase in flow rate, the value of velocity increases obviously in channel A around the pressure side of long blade. The high-speed region becomes larger to extend outlet when the flow rate continues increasing. At the same time, the velocity of splitter blade leading edge region also increases significantly.

Flow separation exists at the outlet of impeller near suction side of splitter blade and the area of flow separation become larger with the decrease of traffic. When the flow reduce to a certain extent, the long blade pressure suction intermediate region and the splitter blade leading edge near the pressure suction starting to appear flow separation. In conclusion, the flow separation is more terrible with the flowrate decreasing.

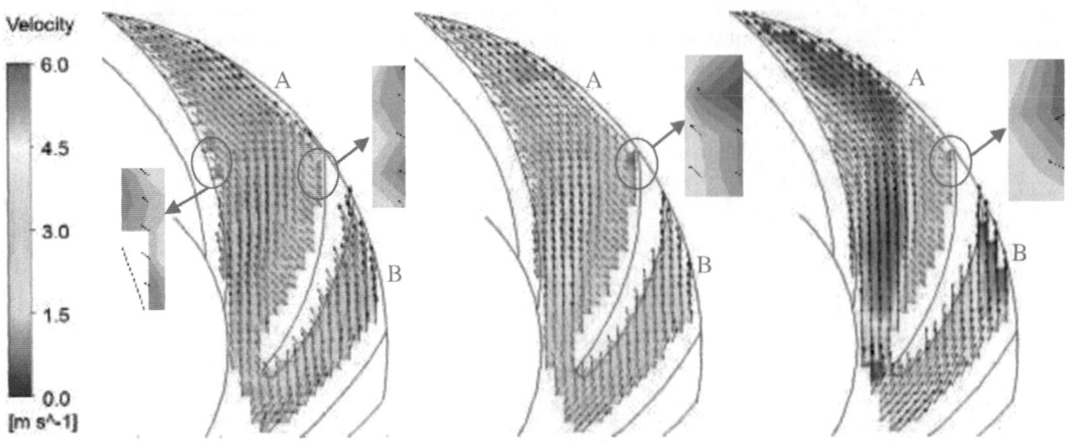

Q=0.4Q_d Q=1.0Q_d Q=1.8Q_d

Figure 11. Relative velocity distribution in impeller 4+4 at different conditions

4.4 Relative velocity distribution at different phases

The relative velocity distributions under design condition at various positions as shown in Figure 12. It shows that the distribution of relative velocity within the flow channel becomes more uneven as the process of near tongue, due to reinforced interference of tongue on the internal flow. The high-speed area continues to expand in the flow passage between pressure side of long blade and suction pressure of splitter blade, as the tongue blocking fluid flowing out of passage. Meanwhile partial pressure gradient becomes larger, resulting in more internal flow disturbance.

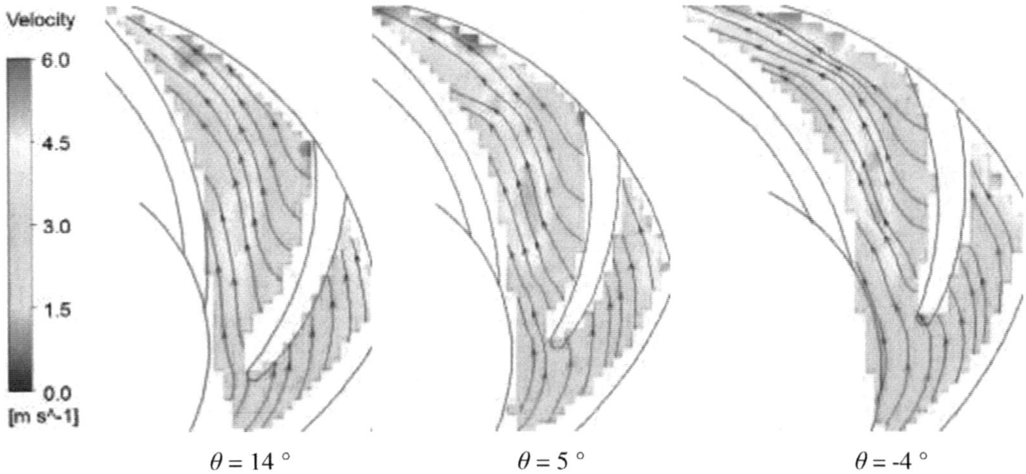

Figure 12. Relative velocity distribution in impeller 4+4 at different phases under design condition

5. Conclusions

In this paper, two different impellers, with and without splitter blades, of low specific speed centrifugal pump were investigated by PIV technology for different conditions and different phases to obtain the velocity distribution within the impeller. The conclusions are as follows through analysis of experimental results.

(1) The head value of impeller 4+4 is higher than that of impeller 6 and both two head curves appear the hump phenomena at small flow rate around $0.4Q_d$.

(2) The absolute velocity value increases with radius and from pressure side to suction side at the same radius gradually. Meanwhile, the value of absolute velocity changes slightly as flow rate declining and the direction is almost consistent with the direction of rotation of impeller.

(3) The outlet relative velocity distribution is improved after adding splitter blades in the impeller 4+4. The splitter blades can scour wake to make outlet velocity distribution more uniform and improve the internal flow. The flow separation is more terrible with the flowrate decreasing in impeller.

(4) The velocity distribution becomes less even in the process of closing to tongue due to reinforced interference of tongue on internal flow.

Acknowledgments

This work was supported by the National Natural Science Foundation of China (No.51009072) and the Foundation for senior person with ability in Jiangsu University of China (No.08JDG040), and a Project Funded by the Priority Academic Program Development of Jiangsu Higher Education Institutions.

Nomenclature

Z	Blade number	D_2	Outlet diameter of splitter blade[mm]
β_2	Outlet setting angle of blade[°]	D_{si}	Inlet diameter of splitter blade[mm]
α	Deflection angle of splitter angle[°]	θ	Circumferential bias angle of splitter blade[°]

References

[1] Yuan S Q 1997 *Theory and design of low specific speed centrifugal pump* [M]. Beijing: China Machine Press.

[2] Guan X F 2011 *Modern pumps theory and design* [M]. Beijing: China Aerospace press.

[3] Zhang J F 2007 Numerical forecast and research on the design method for centrifugal pumps with splitter blades [D]. *Zhen Jiang: Jiangsu University*, 2007. (in Chinese with English abstract)
[4] He Y S, Yuan S Q, Huang L Y, et al. 2004 Development on measurement technology in internal flow field of fluid machinery [J]. *Fluid Machinery*, **32(12)**: 36-40. (in Chinese with English abstract)
[5] Yang H, Tang F P, Liu C, et al. 2011 2-D PIV measurements of unsteady flow field inside the rotating impeller of centrifugal pump [J]. *Transaction of the Chinese Society for Agricultural Machinery*, **42(7)**: 56-60. (in Chinese with English abstract)
[6] Yuan S Q，Li Y L，Tang Y，et al. 2012 Analysis on factors influencing following features of tracer particles in centrifugal pumps[J]. *Journal of Mechanical Engineering*, **48(20)**: 177-184. (in Chinese with English abstract)
[7] Krause N, Zahringer K, Pap E 2005 Time-resolved particle imaging velocimetry for the investigation of rotating stall in a radial pump[J]. *Experiments in Fluids*, **39(2)**:192—201.
[8] Liu H L, Wu X F, Tan M G, et al. 2011 Head calculation and amendments for centrifugal pumps at shut off condition[J]. *Transactions of the Chinese Society of Agricultural Engineering*(Transactions of the CSAE), **27(9)**: 43—47. (in Chinese with English abstract)
[9] Yuan S Q, He Y S, Yuan J P, et al. 2006 PIV measurements and numerical simulations of flow in centrifugal pump impellers with splitting vanes[J]. *Journal of Mechanical Engineering*, **42(5)**: 60-63. (in Chinese with English abstract)
[10] Shao C L, Gu B Q, Huang X L, et al. 2010 Experimental study on flow in the impeller of low specific speed pump using particle image velocimetry[J]. *Journal of Aerospace Power*, **2010 (9)**: 2091-2096. (in Chinese with English abstract)
[11] Choi Y D, Nishino K, Kurokawa J, et al. 2004 PIV measurement of internal flow characteristics of very low specific speed semi-open impeller[J]. *Experiments in Fluids*, **37(5)**: 617-630.
[12] Westra R W, Broersma L, Andel V K, et al. 2010 PIV measurements and CFD computations of secondary flow in a centrifugal pump impeller [J]. *Journal of Fluids Engineering*, **132(6)**: 1-8.
[13] Westra R W, Broersma L, Van Andel K, et al. 2010 PIV measurements and CFD computations of secondary flow in a centrifugal pump impeller [J]. *Journal of Fluids Engineering*, **132(6)**: 0611040-0611048.
[14] Dai C, Dong L, Liu H L, et al. 2013 Unsteady constant value calculation and particle image velocimetry experiment in full passages of centrifugal pump impeller [J]. *Transactions of the Chinese Society of Agricultural Engineering* (Transactions of the CSAE), **29(2)**: 66-72. (in Chinese with English abstract)
[15] Wu X F, Liu H L, Yang H B, et al. 2014 Test and analysis on flow separation in centrifugal pump impeller based on particle image velocimetry [J]. *Transactions of the Chinese Society of Agricultural Engineering* (Transactions of the CSAE), **30(20)**: 51-57. (in Chinese with English abstract)
[16] Ren Y, Wu D H, Liu H L, et al. 2015 PIV Experiment on Flow Instabilities in Centrifugal Pump [J]. *Transactions of the Chinese Society for Agricultural Machinery*, **2**: 008. (in Chinese with English abstract)

Numerical study on the performance of rotational supercavitating evaporator with optimized blade shape

Z Y Zheng[1], Q Li[1], F C Li[1*] and V A Kulagin[2]

[1] School of Energy Science and Engineering, Harbin Institute of Technology, Harbin 150001, China
[2] Department of Heat Technology and Fluid Dynamics, Siberian Federal University, Krasnoyarsk 660041, Russia

E-mail: lifch@hit.edu.cn

Abstract. Based on supercavitation effect, a novel device named Rotational Supercavitating Evaporator (RSCE) has been designed for desalination. In order to improve the performance of RSCE, the optimal design of three-dimensional blade shape is performed by utilizing the empirical formulae which have been validated before. Numerical simulation is then conducted on the supercavitating flow in RSCE with the optimized blade, and the supercavity dimension generated by rotational cavitator is obtained, which is further compared with the above-mentioned empirical formulae obtained by two-dimensional calculations. The results show that the supercavity length obtained by numerical simulation on the optimized blade increases with the increase of radius at first, and then decreases at larger radii, which is much smaller than the result of empirical formula at each radius.

1. Introduction

As a global environmental issue, the shortage of fresh water is becoming increasingly severe. And seawater desalination is considered the most promising solution, which promotes the rapid development of desalination technologies. Desalination methods can be classified into thermal method and membrane method in line with different principles. And the current major large-scale industrial desalination methods include multi-stage flash (MSF), multiple effect distillation (MED) and reverse osmosis (RO). However, the existence of solid walls in MSF and MED induces the generation of scaling and fouling which would result in low heat-mass transfer rate, while rigorous pre-treatment is indispensable in RO process for the mitigation of membrane fouling. In order to overcome above-mentioned shortcomings, supercavitation effect is utilized for desalination.

Cavitation is such a phenomenon that a liquid medium is ruptured by the decrease in local pressure under roughly constant temperature, resulting from the vaporization of liquid phase and the abrupt expansion of nuclei inside the liquid into obvious bubbles. Supercavitation is the most intense level of cavitation under low characteristic pressure or large characteristic velocity. Similar to MSF and MED, phase change is involved in the supercavitation-based desalination. However, the occurrence of heat-mass transfer at the steam-water interface in the supercavitation has the advantages of large energy

[*] To whom any correspondence should be addressed.

intensity and scaling-free [1], that's why supercavitation effect is taken advantage of for seawater desalination.

Most of the studies on the application of supercavitation to desalination are conducted by Russian researchers. In 1984 Machinski [2, 3] introduced stationary conical supercavitating evaporator into desalination, which has the ability to create a relatively stable supercavity and can be connected to vacuum system for steam extraction. However, the entire system included multiple cascading evaporators, and its industrial application for desalination requires continuous high-volume recirculation of the subcooled hot source water through the system. This scheme is metal-intensive and the ratio of supercavity volume to the water bulk volume is very small, and it also uses energy-intensive pump recirculation system. Combining electromagnetic treatment with supercavitation, scientific and technical center "TJEROS-MIFI" in Russia designed and manufactured commercial cavitating desalination device "WATERFALL-1200" with a productivity of 1200 m^3/d [4]. This device consumed electrical energy up to 3 kWh for unit volume of fresh water and ensured more than 99% rejection rate under operation on the source water with TDS (Total Dissolved Solids) up to 65000 mg/L, which was significantly competitive with current major large-scale industrial desalination technologies. Recently, a novel device named Rotational Supercavitating Evaporator (RSCE) was put forward, and preliminary design and research were also performed [1, 3, 5, 6].

The core component of RSCE is a rotational cavitator made of two blades with wedge cross section and alternative thickness of exit edge, which is used for the formation of supercavitation and the production of steam. Therefore, the hydrodynamic characteristic of three-dimensional blade shape is of great importance for the performance of RSCE. In this paper, three-dimensional blades of different sizes are designed based on the empirical formulae of supercavity length which have been obtained by two-dimensional numerical simulations on planar symmetric wedge-shaped cavitators of different wedge angles [7].The calculation method for the blade shape is also improved for better accuracy. The blade shape of optimal performance is then determined based on several geometrical limitations. Numerical simulation is conducted on the supercavitating flow in RSCE with the optimized blade, and the supercavity dimension generated by rotational cavitator is obtained. The supercavity lengths at different radii are also compared with the above-mentioned empirical formulae obtained by two-dimensional calculations.

2. Design and modelling of RSCE with optimized blade shape

All of the empirical formulae of supercavity length as a function of cavitation number have been validated by existing experimental results [7], based on which the blade shape of RSCE can be obtained for each certain wedge angle and rotational speed. Practically, the three-dimensional blade shape is determined by the fixed wedge angles and alternative thicknesses of the exit edge at different radii, and the thickness of the exit edge is determined by the comparison between the supercavity length and arc length at the corresponding radius. Detailed algorithm for the thickness of the exit edge was illustrated in [5].

However, the values of α at different radii in the original calculation method for blade shape didn't strictly equal to the desired wedge angles at corresponding radii. Moreover, the deviation would increase with the decrease of radius r. To address this issue, the improved method is proposed and shown in Figure 1, in which the length of the supercavity L is obtained by the calculation with r_c instead of r. According to the calculation method, there exists a corresponding blade shape for each certain wedge angle and rotational speed as mentioned above. Due to the space limitation and the same qualitative trend, only the variations of half-thickness of the exit edge h with the radius r for different wedge angles at the rotational speed of 5000 r/min and for different rotational speeds at the wedge angle of 45° are extracted for analysis and shown in Figure 2, while the detailed results for other wedge angles and rotational speeds are not provided. It can be obtained from Figure 2 that the half-thickness of the exit edge decreases with the increase of radius for each blade shape. The difference between the half-thicknesses of the exit edge at the blade root and at the blade tip increases with increasing wedge angle for the same rotational speed (Figure 2(a)), thus the blade shape with smaller wedge angle is more suitable for the considerations of the size of the shaft (due to the need of larger shaft length for a certain shaft diameter or larger shaft diameter for a certain shaft length at larger wedge angles) and the blade length (because of the requirement for mechanical strength at larger radius where half-thickness of the exit edge is too small). Besides, half-thickness of the exit edge for each radius decreases with the increase of rotational speed at the same wedge angle (Figure 2(b)), hence the blade shape designed for higher rotational speed is selected on account of the sizes of shaft and blade. With more consideration of blade's mechanical strength, the limitation of h = 1 mm, i.e., the thickness of the exit edge of 2 mm, is set in the design for blade shape. Ultimately, the optimized blade with wedge angle of 45° and design rotational speed of 5000 r/min is designed, whose three-dimensional model is schematically shown in Figure 3. And its detailed parameters of blade shape are tabulated in Table 1. The shaft diameter d_0 and diameter d of this blade are 70 mm and 200 mm, respectively.

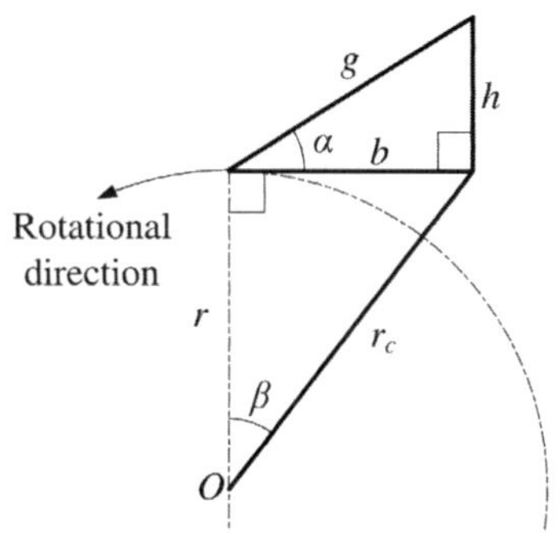

Figure 1. Schematic diagram of improved calculation method for the blade shape.

Table 1. Half-thicknesses of the exit edge (h in Figure 1) at different radii for the optimized blade with wedge angle of 45° and design rotational speed of 5000 r/min.

r (mm)	h (mm)	r (mm)	h (mm)
35	9.89	70	2.72
40	8.63	75	2.25
45	7.33	80	1.88
50	6.09	85	1.58
55	4.98	90	1.34
60	4.06	95	1.15
65	3.31	100	0.99

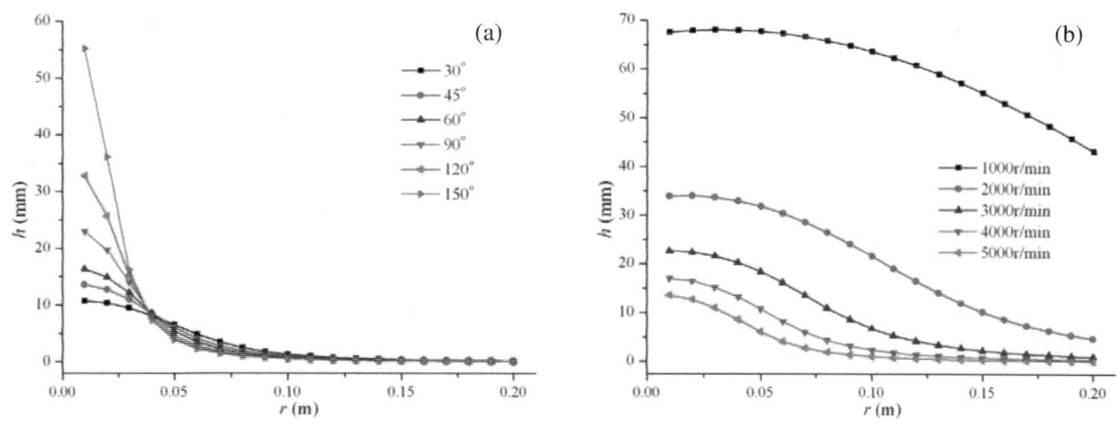

Figure 2. Variations of half-thickness of the exit edge h with the radius r for (a) different wedge angles at the rotational speed of 5000 r/min and for (b) different rotational speeds at the wedge angle of 45°.

3. Numerical simulation on RSCE with optimized blade shape

3.1. Numerical method

Numerical simulation is conducted on the three-dimensional steady supercavitating flow in this paper. And supercavitating flow is a multiphase and turbulent flow. Therefore, multiphase flow and turbulence models should be adopted in the numerical simulations. Herein, Mixture model based on the homogeneous equilibrium multiphase flow theory is selected as the multiphase flow model, in which the mixture of the gas and liquid phases is considered as homogeneous single-phase fluid. And

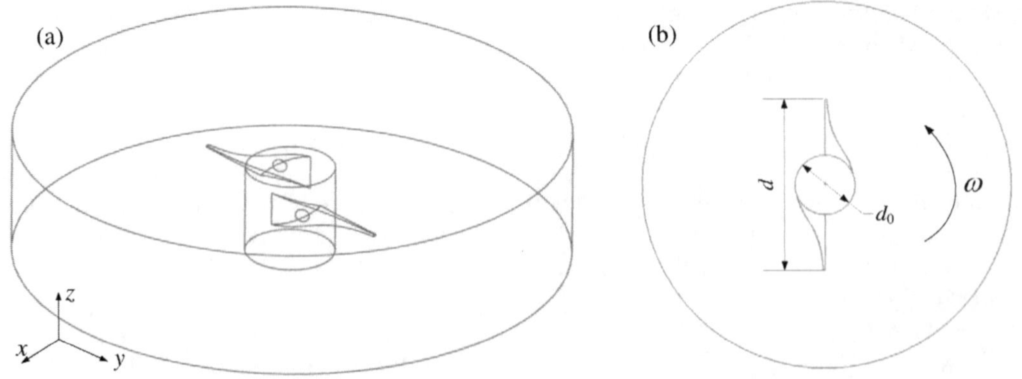

Figure 3. Schematic diagram of the optimized blade. (a) three-dimensional schematic diagram, (b) top view.

the equations in this model are shown below.
Continuity equation of the mixture phase:

$$\frac{\partial}{\partial x_i}(\rho_m u_i) = 0 \tag{1}$$

Momentum equation of the mixture phase:

$$\frac{\partial(\rho_m u_i u_j)}{\partial x_j} = -\frac{\partial p}{\partial x_i} + \frac{\partial}{\partial x_j}\left[(\mu_m + \mu_t)\left(\frac{\partial u_i}{\partial x_j} + \frac{\partial u_j}{\partial x_i} - \frac{2}{3}\frac{\partial u_k}{\partial x_k}\delta_{ij}\right)\right] \tag{2}$$

Transportation equation of the volume fraction of the gas phase:

$$\frac{\partial}{\partial x_i}(\alpha_v \rho_v u_i) = R_e - R_c \tag{3}$$

where u_i or u_j is the velocity of mixture phase, p is the pressure, μ_t is the turbulent viscosity, α_v is the volume fraction of the gas phase, ρ and μ are the density and viscosity, respectively; subscripts m and v represent mixture and vapor phases, respectively. The thermodynamic properties of liquid water and vapor are set in line with the international standard for the thermodynamic properties of water and steam IAPWS95 [8] at 25°C: $\rho_l = 997.00\,\mathrm{kg/m^3}$, $\mu_l = 8.9011\times10^{-4}\,\mathrm{Pa\cdot s}$, $\rho_v = 0.023075\,\mathrm{kg/m^3}$, $\mu_v = 9.8669\times10^{-6}\,\mathrm{Pa\cdot s}$. Herein, the subscript l represents liquid phase. The density and viscosity of mixture phase are defined as $\rho_m = \alpha_v\rho_v + (1-\alpha_v)\rho_l$ and $\mu_m = \alpha_v\mu_v + (1-\alpha_v)\mu_l$. R_e and R_c are the production rate and condensation rate of the gas phase, respectively. In order to depict R_e and R_c, Schnerr-Sauer model [9] is adopted as the cavitation model. Turbulent flow in the numerical simulation is described by $k-\varepsilon-\overline{v'^2}-f$ (V2F) turbulence model, which has been verified for simulating cavitating flows [7]. V2F model is an extension of $k-\varepsilon$ model and includes two additional equations to represent the turbulence anisotropy without damping functions. This turbulence model was proposed by Durbin [10] and several versions have been put forward to improve this model for the enhancement of numerical stability. The adopted version in this paper is derived from the code-friendly one introduced by Lien and Kalitzin [11], and detailed equations are not given due to space limitation.

A cylinder with the height of 100mm and the diameter of 430mm is selected as the computational domain, as shown in Figure 3, as well as the settings of coordinate system in the simulations. The computational domain is meshed by unstructured grids with denser prismatic meshes gathering in the boundary layer around the blades and tetrahedral meshes in the rest space. Both the top and bottom of the computational domain are set to be pressure outlet with the pressure of 101325 Pa. The cylindrical surface is set as fixed no-slip wall, and other boundaries are set to be no-slip wall with constant rotational speed.

3.2. Results and discussions

Numerical simulation is conducted on the supercavitating flow around optimized blade in RSCE at rotational speed of 5000 r/min. The profile of the supercavity is shown in Figure 4, which is defined as the isoline with $\alpha_v = 0.1$ [7]. It can be seen that the supercavity profile is centrosymmetric about the axis. Moreover, re-entrant jet generates at the tail of the supercavity for smaller radii, which can be captured by the depression of the profile. For more intuitive observation, Figure 5 depicts the flow field in the cylindrical surface with the radius of 0.04 m, in which the red line represents the supercavity profile. It can be seen that a jet with larger velocity and the direction towards upstream forms right behind the supercavity, inducing the depression of profile.

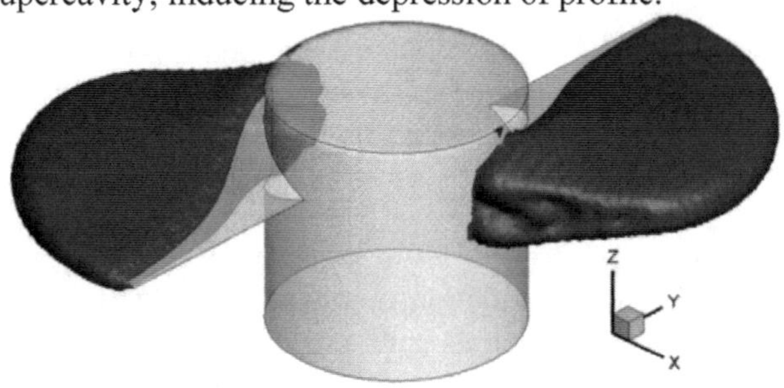

Figure 4. Profile of the supercavity.

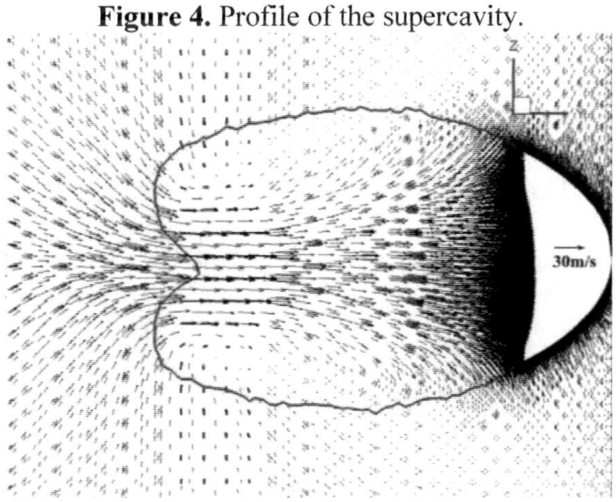

Figure 5. Flow field in the cylindrical surface with the radius of 0.04 m.

However, the supercavity dose not take up most of the space between two blades as expected. The disparity between empirical formula and three-dimensional numerical simulation result is shown in Figure 6. The difference is to a larger extent, which increases with the increase of radius, especially for the larger radii, where supercavity length decreases with increasing radius for three-dimensional numerical simulation result. This significant difference is resulted from the fact that empirical formulae were obtained by numerical simulations on two-dimensional planar cavitating flows around wedge-shaped cavitators with no consideration for the effect of rotation, indicating that empirical formulae for supercavity length should be modified, which can be realized by theoretical derivation or three-dimensional numerical simulation on the blade with constant thickness of the exit edge. Besides, the event that supercavity length decreases with increasing radius at larger radii for numerical simulation result can be possibly explained by tip vortex which has been observed in the experiments on RSCE [3].

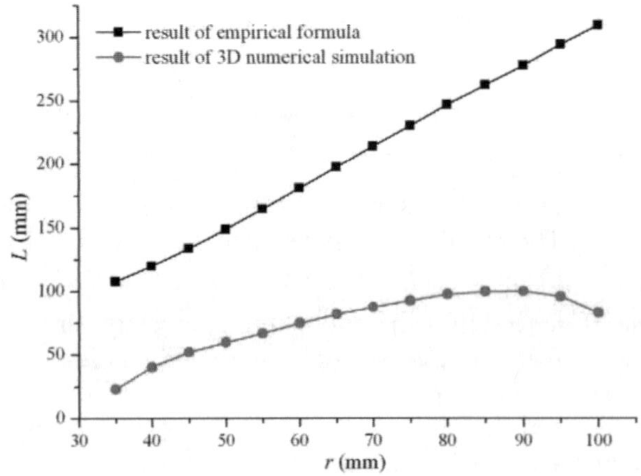

Figure 6. Variation of supercavity length L with the radius r for numerical simulation result and comparison with the result of empirical formula.

4. Conclusions

Optimal design for the blade shape of RSCE is performed by utilizing empirical formulae of supercavity length and improving the calculation method for blade shape. Optimized blade is then determined by several geometrical limitations and mechanical strength. Numerical simulation is conducted on the optimized blade, and the supercavity lengths at different radii are obtained and compared with the result of empirical formulae. The main conclusions are given below.

The blade with wedge angle of 45° and design rotational speed of 5000 r/min is picked out as the optimized blade. The supercavity length obtained by numerical simulation on the optimized blade increases with the increase of radius at first, and then decreases at larger radii, which is much smaller than the result of empirical formula at each radius. And the empirical formulae of supercavity length would be modified in the future research.

Acknowledgments
This work is supported by National Natural Science Foundation of China (51276046). The authors would like to thank the valuable discussions with the members of Complex Flow and Heat Transfer Lab of Harbin Institute of Technology.

References

[1] Likhachev D S and Li F C 2013 *Desalin. Water Treat.* **51**(13–15) 2836–49
[2] Machinski A S 1984 *Hydrodynamics and thermal transfer characteristics of supercavitating evaporators for water desalination* (Moscow: Russian State Library PhD Thesis) (in Russian)
[3] Likhachev D S 2013 *Study on the hydrodynamic characteristics of rotational supercavitating evaporator* (Harbin: Harbin Institute of Technology PhD thesis)
[4] Rudenko B 2007 How to make up water resource *Journal of Science Life* **12** http://www.nkj.ru/archive/articles/12366/ (in Russian)
[5] Likhachev D S and Li F C 2014 *Sci. China-Phys. Mech. Astron.* **57**(3) 541–54
[6] Likhachev D S, Li F C and Kulagin V A 2014 *Sci. China-Technol. Sci.* **57**(11) 2115–30
[7] Zheng Z Y, Li F C, Li Q and Kulagin V A 2015 *Sci. China-Technol. Sci.* **58**(6) 1072–83
[8] Wagner W and Pruß A 2002 *J. Phys. Chem. Ref. Data* **31** 387–535

[9] Schnerr G H and Sauer J 2001 Physical and numerical modeling of unsteady cavitation dynamics *4th International Conference on Multiphase Flow (New Orleans, USA, 2001)*
[10] Durbin P A 1991 *Theor. Comput. Fluid Dyn.* **3**(1) 1–13
[11] Lien F and Kalitzin G 2001 *Int. J. Heat Fluid Flow* **22**(1) 53–61

Design of a new type vapor recovery system nozzle

S H Fu, G J Cao, D S Zhang

Beijing POL Research Institute, Beijing China 102300

Email: fusuhong101@163.com

Abstract. To settle the problem of low-efficiency recovery for Vapor recovery system nozzle, this paper advances a purely mechanical structure of the self-sealing refueling VRS nozzle. The structure, operating principle and controlled process of the nozzle is given. And an application of the nozzle is discussed. All indicated that the nozzle has a reasonable structure, can fuel and vapor recovery simultaneous start and stop. And thus improve the recovery efficiency and reduce oil leakage.

1. Introduction

Gasoline is volatile, which not only causes great economic losses but also causes more environmental pollution and safety hazards especially in the process of the handling and selling. Today more and more vapor emission controlling methods for the vapor recovery in gasoline storing, transporting and filling to handle vapor pollution. [1][2]

Nozzle is the terminal system to the vehicle fuel tank. Nozzle has experienced a normal fuel gun, self-sealing nozzle guns and vapor recovery system nozzle in three stages. Now self-sealing nozzle is the main equipment, which is characterized by automatically shut when full tank to prevent accidents due to human reason. VRS nozzle developed on the basis of the self-sealing nozzle, main difference is an increase of vapor recovery tube and gas-liquid ratio control valve. At the same time, fuel hose is coaxial hose. Its body and tube have both oil and vapor channel.

The vapor recovery system nozzle is the key equipment of the secondary vapor recovery system in gas station, and its role is to get the vapor in the automobile fuel tank into the underground storage tank during refueling. [3][4] In the VRS nozzle application process, there is the problem of low recovery efficiency of vapor. In addition, the check valve of the VRS nozzle opened and closed each time the need for dynamic work, frequency of use, easy to wear. This paper presents a purely mechanical structure of the self-sealing refueling VRS nozzle. The structure, operating principle and controlled process of the nozzle is given. And an application of the nozzle is discussed. All indicated that the nozzle has a reasonable structure, can fuel and vapor recovery simultaneous start and stop. And thus improve the recovery efficiency and reduce oil leakage.

2. Structure and principal

The designed VRS nozzle includes the nozzle body, the tubing assembly and the handle assembly. The structure is shown in Fig.1. Nozzle body with a cavity, the front end of the cavity set up the oil pipeline to connect the tubing assembly and the back end of the cavity set the oil inlet and vapor exports. The handle assembly is located in the bottom of the nozzle body and set the oil switch controller. In the cavity of the nozzle correspondingly set oil and gas separation valve assembly, the return valve assembly, the shaft assembly and the chamber components on the top of the self-styled axis assembly. The self-styled axis connects the oil switch controller. The oil and gas separation valve assembly is the synchronization control valve of the oil inlet and vapor export. The tube assembly includes a tube, airway and vapor recovery unit.

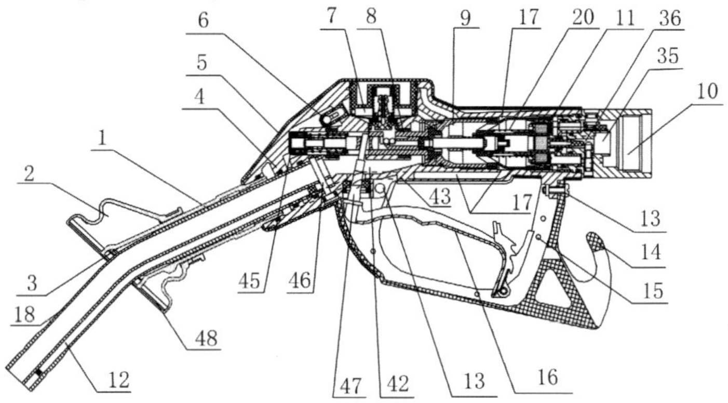

1 oil outlet assembly 2 hood 3 vapor recovery pipe 4 body 5 holster 6 stop valve 7 chamber assembly 8 self-styled axis assembly 9 return valve assembly 10 oil inlet 11oil and gas separation valve assembly 12 airway 13 screw gear 14 butt 15 block 16 handle assembly 17, 36, 43 vapor channel 18 oil tube 19 spring-return 20 outside magnet ring 35 vapor exports 42 oil inlet channel 45 anti-spill plug 46 pin 47 buckle switch pieces 48 side buckle

Figure 1. The structure diagram of the VRS nozzle

Meanwhile, the unique design of the oil and gas dual-channel design to achieve the vapor control valve and oil control valve sync with the action and easy to adjust the gas-liquid ratio while the vapor back to the tank, improve vapor recovery efficiency, Shown in Fig.2. The nozzle body has a dual oil and gas channel, location of the oil inlet with the oil and gas separation valve assembly and location of the oil outlet with the oil tube assembly, which separate the oil and gas. Components in the oil and gas separation valves the wedge-shaped plug, the inside magnet and spool composed of oil and gas discharge valve control valve. The spool and the outside magnet composed of the oil inlet control valve. Under the magnetic attraction of the inside and outside magnet the vapor control valve and oil control valve synchronization linked movements. When the nozzle doesn't work the oil control valve and vapor control valve closed under the pressure of the spring-return.

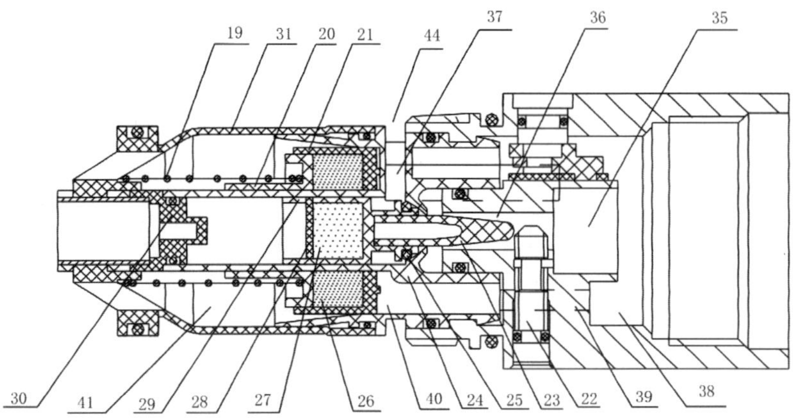

19 return spring 20 outside magnet ring 21 oil inlet control valve spool 22 adjustment screw 23 wedge-shaped plug 24 oil and gas separation joint 25 ring 26 outside magnet 27 inside magnet 28 vapor discharge valve spool 30 plug-return 31 oil and gas separation valve body 35 vapor outlet 36、37、44 vapor channel 38、39、40、41 oil channel

Figure 2. Unique dual channel design of oil and gas

Shown as in Fig.1, the self-styled works as follows:

When lifted the handle assembly 16 to the stopper blocks 15 pieces, the knockout pin push the self-styled axis of the assembly 8 forward movement. The self-styled axis and self-styled sleeve on the shaft with the groove, the two needles of the chamber components 7 stuck in the drive shaft assembly 8 to move. Then the return valve assembly 9 loss of the pressure of the return spring, oil flow out from the return valve.

There is vapor vent at the oil outlet of the return valve assembly body, which is vapor recovery pipe 3, connect the chamber assembly and the oil outlet assembly 1. When the oil goes through the back seat and the back spool of the return valve assembly 9, cause of the oil flow rate, there is an umbrella flow. Thus the umbrella flow and the return the valve arouse negative pressure. The fluid continued to drive the gas from the airway suction of the pipeline components 12. The fluid mixed with the gas outflow from the outlet pipe 18.

When the liquid oil to the airway 12 of the outlet assembly, result in negative pressure of the chamber assembly 7. And then pull up the diaphragm, the diaphragm drive two needle roller bearings made out of self-styled axis of the shaft assembly 8 and self-styled sleeve. At this time under the action of the spring in the self-styled axis and the self-styled sleeve the nozzle self-closed.

The vapor recovery works as follows:

The nozzle body 4 equipped with dual-channel, location of the oil and gas separation valve assembly 11 at the oil inlet and the oil tube at the oil outlet assembly 1, the oil and gas separate. The oil outlet assembly 1 includes a pipe 18 and the vapor recovery pipe 3. The oil from the inlet 10 into the oil channels 38,39,41,42 and out from the pipe 18.

When the nozzle fuel the oil, the role of promoting the flow of oil discharge control valve forward (open), and driving gas discharge control valve is open. Then the space 44 between the external vapor recovery valve 31 and the inner wall of the nozzle body 4, connected with the vapor recovery channel 35, 36 and37. Due to space 44 for the ring structure, the vapor connected with the vapor channel 17. Thus 35, 36, 37, 44, 17, 43 and 3 form an independent vapor recovery channel. Therefore, fuel and gas channel connected to achieve synchronization.

The designed VRS nozzle, fuel oil and vapor recovery start and stop simultaneously, can improve vapor recovery efficiency, reduce oil leakage; The returning valve body material use of more durable materials and advanced molding process, can greatly improve the life of the nozzle. The end of the vapor recovery hood seal in flange design used effectively to ensure that the nozzle and sealing fuel tanks, further reducing the possibility of oil leakage.

3. Application

The designed VRS nozzle reliability test, repeated open and close 200,000 times, no oil leakage. The indicators have higher consistence before and after experiment. Comparison of experimental data is shown in table 1.

Table 1. Comparison of experimental data

item	Oil flow (L/min)	Oil and gas ratio	leakage
Before the reliability test	38-42	1.1-1.2	no
After the reliability test	38-42	1.1-1.2	no

From the table, we can see that the number is the same before and after the test. So we can draw the conclusion that the nozzle has a higher reliability.

The vapor recovery nozzle was installed in the filling machine at a gas station. When a car was refueled, the nozzle started to collect the vapor in the car tank through the gas pipeline to underground storage tanks as shown in Figure 3. And the nozzle has been working well without any failure occurs with very good effect of the actual return air for fuelling 40 tons. The results show that the nozzle can not only fill up the tank with the self-styled features, but also enables fuel and vapor recovery work synchronization.

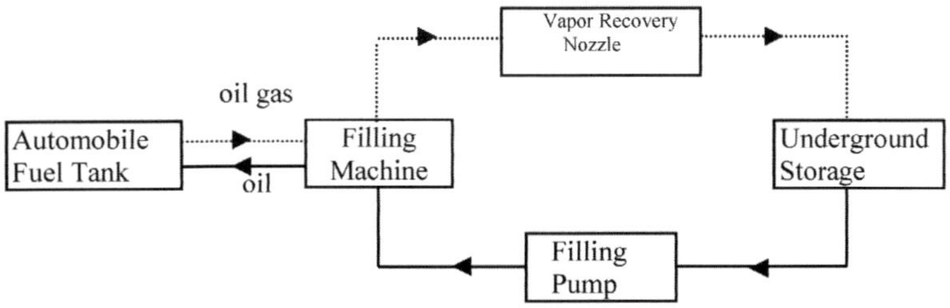

Figure 3. The practical application of the vapor recovery nozzle

According to the vapor recovery nozzle performance test and practical application, the nozzle designed for vapor recovery system has the following three characteristics:

1、The unique dual channel design of oil and gas is synchronous structure for the vapor recovery nozzle filling and recovery. And the outer diameter matches with inlet and exhaust channel.

2、The valve rod on the return valve assembly is made of PTFE (Polytetrafluoroethylene) with self-lubricating properties. When the valve rod moves up and down, it is possible to reduce the frictional force of the diaphragm and the valve chamber. Then heat is reduced.

3、The hood is inflexed on the vapor recovery nozzle body to increase sealing efficiency and reduce the leakage of the vapor.

References

[1] Hao B L 2002 The necessity of carrying out oil gas recovery and exploration on technical plans [J]. *Journal of Sinopec Management Institute*. **27** **(2)**: 41-42.

[2] Zhang S H 2004 Vapor recovery system in the oil depot [J]. *Oil Depot and Gas Station*. **13(3)** :32-34.

[3] Huang Z Q, Wang W Q 2006 The experiment research on the bapor recovery system [J]. *Gas Station*, PP:82-86

[4] Chen J Q, Deng L, Liu X M 2007 The study of the vapor recovery simulation in numerical [J]. *Oil and Gas*, **3**, PP35-37.

Research on liquid sloshing performance in vane type tank under microgravity

Q Hu, Y Li, J T Liu, J Q Liang

Beijing Institute of Control Engineering, Haidian District, Beijing, China

E-mail: hqbuaa03@sohu.com

Abstract. Propellant management device (PMD) in vane type tank mainly comprises of vane type structure parts, whose performance of restraining liquid sloshing should satisfy spacecraft requirements of high stabilization and fast orbital maneuver. Aiming at liquid sloshing performance in vane type tank under microgravity environment, gas-liquid flow model based on the volume of fluid (VOF) method was put forward, and via numerical simulation liquid sloshing performances of vane type PMD with anti-sloshing baffles and without anti-sloshing baffles in microgravity were analyzed and compared. Simulation results reveal that liquid sloshing performance of vane type PMD with anti-sloshing baffles is markedly superior vane type PMD without anti-sloshing baffles and the baffles make liquid surface become stable fast. Then by comparing between results of microgravity experiments and results of numerical simulations, they are very similar. According to present research, vane type PMD with anti-sloshing baffles has better effects on restraining liquid sloshing and is able to restrain observably propellant sloshing in tanks in order to satisfy spacecraft requirements of high stabilization and fast orbital maneuver.

1. Introduction

The vane type tank is a more advanced propellant tank used in spacecraft, and the inner PMD mainly comprises of vane type structure parts and implements orbital liquid management and control initiatively and full range in the tank. With the continuous development of aerospace technology, the propellant weight fraction in spacecraft continuously expands and the requirement for attitude control accuracy and dynamic stability gets higher, but liquid sloshing in the tank of spacecraft has a bad effect on attitude control and dynamic stability because propellant in the tank may shake violently [1], so the effect of liquid sloshing in tank must be considered. Spacecraft works under microgravity in most life time and liquid sloshing performance in microgravity is markedly different from in gravity, so research on liquid sloshing performance in vane type tank under microgravity becomes especially important. So far space equivalent pendulum model and spring-mass model based on linear theory are usually used to characterise liquid sloshing approximately in engineering, and the models can simulate small amplitude liquid sloshing well, but can't accurately simulate large amplitude liquid sloshing [2-4]. So in the paper gas-liquid flow model based on the volume of fluid method was put forward and three-dimensional numerical simulation on liquid sloshing in a vane type tank was carried out by using Reynolds Average Navier-Stokes (RANS) and VOF method, then numerical results about liquid sloshing performances of vane type PMD with anti-sloshing baffles and without anti-sloshing baffles in microgravity were analyzed and compared, and results of microgravity experiments and results of numerical simulations were analyzed and compared also.

2. Research technique

Liquid sloshing process in tank is gas-liquid flow process, so in order to solve liquid sloshing process in vane type tank, Navier-Stokes (N-S) equation of gas-liquid flow process was put forward and three-dimensional numerical simulation on the process was carried out by using Reynolds Average Navier-Stokes (RANS) and VOF method. Gas-liquid interface tracking could be achieved by solving fluid volume transport equation, and the effect of surface tension and wall adhesion force is simulated by adding source item to the momentum equation, and data input and output module is founded by using user defined functions (UDF) macro provided by Fluent soft, and real time input of sloshing velocity and acceleration and output of liquid sloshing characteristics and quantization value could be achieved.

2.1. Gas-liquid flow mathematic model based on VOF method

2.1.1. N-S equation. Two-phase N-S equation about transient, incompressible flow, immiscibility, constant temperature and viscosity, and surface tesion being are given below.

$$\nabla \cdot \vec{u} = 0 \tag{1}$$

$$\rho \frac{\partial \vec{u}}{\partial t} + \rho \nabla \cdot (\vec{u}\vec{u}) = -\nabla p + \nabla \cdot \tau + \vec{F} + \rho \vec{g} \tag{2}$$

$$\tau = 2\mu S = 2\mu \cdot \frac{1}{2}\left(\nabla \vec{u} + (\nabla \vec{u})^T\right) = \mu\left(\nabla \vec{u} + (\nabla \vec{u})^T\right) = \mu \Delta \vec{u} \tag{3}$$

Where $\vec{u}$ is velocity, ρ is fluid density, $\vec{F}$ is surface tension, $\vec{g}$ is acceleration, μ is fluid viscosity, and T means transpose. Equation (1) is the continuity equation, and equation (2) is the momentum conservation equation in which the effect on flow by pressure, turbulent flow, surface tension and acceleration is considered.

2.1.2. VOF method. There is an interface between immiscible gas and liquid on which surface tension comes into being because of the difference of intermolecular force between different fluids. Via VOF method gas-liquid flow and interface could be depicted well just using a set of N-S equation, it simplifies numerical discription. Two phases gas and liquid exist simultaneously in the arithmetic cells which are at the phase interface, and sigle phase exist in the cells which are not at the interface. So k-phase volume fraction ε_k is introduced, and ε_k in a cell can be expressed as:

$$\varepsilon_k(cell) = \frac{\iiint_{cell} \varepsilon_k(x, y, z) dxdydz}{\iiint_{cell} dxdydz} \tag{4}$$

As shown in figure 1, all the cells in the flow can be divided into three areas. If $\varepsilon_k(cell) = 0$ it means k-phase dosen't exist in the cell, and if $\varepsilon_k(cell) = 1$ it means only k-phase exists in the cell, and if $0 < \varepsilon_k(cell) < 1$ it means the interface exists in the cell.

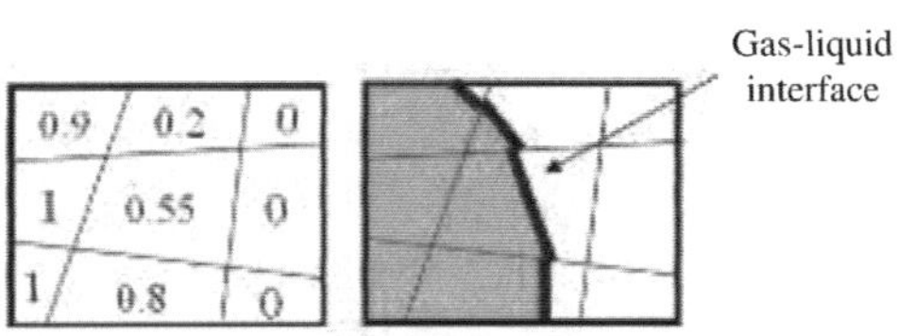

Figure 1. Solution method of gas-liquid interface.

The interface tracking is one of key issues in immiscible multi-phase flow numerical simulation. In order to track the interface, transport equation for each phase volume fraction is put forward below.

$$\frac{\partial \varepsilon_k}{\partial t} + u_j \frac{\partial \varepsilon_k}{\partial x_i} = S_{\varepsilon_k} \tag{5}$$

Where S_{ε_k} is used to depict mass transfer between different phases. Because there isn't mass transfer between different phases in the method, it means $S_{\varepsilon_k} = 0$·

$$\rho = \varepsilon_l \rho_l + (1 - \varepsilon_l)\rho_g \tag{6}$$

$$\mu = \varepsilon_l \mu_l + (1 - \varepsilon_l)\mu_g \tag{7}$$

Where subscripts l and g each separately represent liquid phase and gas phase.

Practical interface structure under the flow is very complicated, especially exact solution about interface curvature and surface tension in three-dimensional space is very difficult, but Brackbill et al solve this difficult problem by using relatively simple continuum surface force (CSF) model [5], and it is most important that CSF model can be easily extended to three-dimensional space and computational quantity increases a little. Then via CSF surface tension as volume force can be add to the momentum conservation equation, and surface tension can be expressed as:

$$\vec{F} = \int_{s(t)} \sigma k' \vec{n}' \delta(\vec{x} - \vec{x}') dS \approx \sigma k \nabla F \tag{8}$$

Where k is surface curvature, k is surface tension coefficient, and F is liquid volume fraction.

2.2. Numerical method

2.2.1. Computational grid model. All the inner space of the tank is used as computational domain for studying liquid sloshing process in vane type tank. In order to simplify calcutation, the inner structures such as propellant acquisition vanes (PAV) and propellant refillable reservoir (PRR) are striked off, and anti-sloshing baffles are preserved. As shown in figure 2, the computing grids of all the domain are hexahedron grids based and five surface body grids as a supplement, and total grid number is approximately one million and the ratio of five surface body grids is less than 5%. In order to increase computational accuracy, hexahedron grids should be used in the computational domain where surface tension effects obviosly, and grid scale near the wall is defined between 0.5mm and 4mm, meanwhile grid independence is validated.

a) Computational domain b) Computational grid c) Grid of anti-sloshing baffles

Figure 2. Computational domain and grid.

2.2.2. Initial condition. The simulation on liquid sloshing process in tank is based on liquid surface steady state, so the liquid steady state of reorientation process in microgravity is used for the initial condition of numerical simulation on liquid sloshing in tank, and liquid filling ratio is 40%, as shown in figure 3. Then liquid sloshing process under the non-along orbit condition in vane type tank is numerically simulated and the effect of anti-sloshing baffles on liquid sloshing performance is analized. Boundary condition of all the walls is infinitely smooth and no slip, and the solid wall adsorption on gas-liquid interface is simulated by using static contact angle which is set as 1° and the flow near solid walls of structure is simulated by using standard wall function. Propellant MON is chosen as the liquid and the microgravity acceleration is $1\times10^{-5}g_0$ and follows negative Y axis direction. Meanwhile the initial point of non-along orbit condition is set as the initial point of liquid sloshing numerical simulation, and the non-along orbit condition makes the tank keep shaking for 15s which is mainly expressed in the changes of angular velocity on X axis and linear velocity on Y axis and Z axis, as shown in figure 4.

a) with anti-sloshing baffles b) without anti-sloshing baffles

Figure 3. Initial distribution of liquid surface in tank.

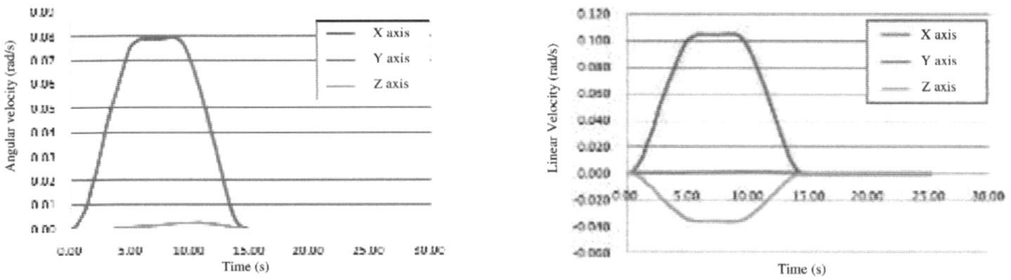

a) angular velocity change b) linear velocity change

Figure 4. The non-along orbit condition.

3. Results and analyses

3.1. Numerical calculation results and analysis

Gas-liquid interface distribution during liquid sloshing process in tank is shown in figure 5, meanwhile at the same time the left side view is about liquid sloshing results with anti-sloshing baffles and the right side view is about liquid sloshing results without anti-sloshing baffles, and velocity on the interface is displayed also. Numerical results reflect that baffles play a role in strengthening the mixing of liquid and make the structure of gas-liquid interface become complicated during tank shaking process under the non-along orbit condition. After tank shaking starts, the velocity in interface central region increases and relatively significant flow arises in the region also, and when they are 4s and 6s the flow velocity in the tank without baffles is significantly higher than the tank with baffles. After tank shaking stops, in the tank with anti-sloshing baffles, the bubble motion is restricted to the area at the top of tank and there is no interface fracture phenomenon on the whole, because anti-sloshing baffles play a role in damping and restraining liquid flow. Meanwhile the holes on the baffles are propitious to liquid flow between baffles, which makes there is no interface fracture phenomenon between baffles, so the holes also play a part in damping. While in the tank with anti-sloshing baffles, under the inertial force the bubble firstly moves from the back side to the bottom and then passes through the front side and finally moves back to the top area, and during the process liquid interface fracture phenomenon comes into being obviously.

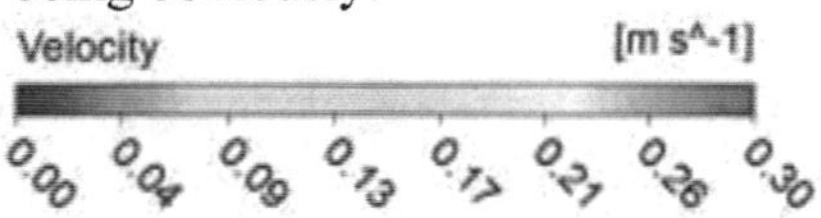

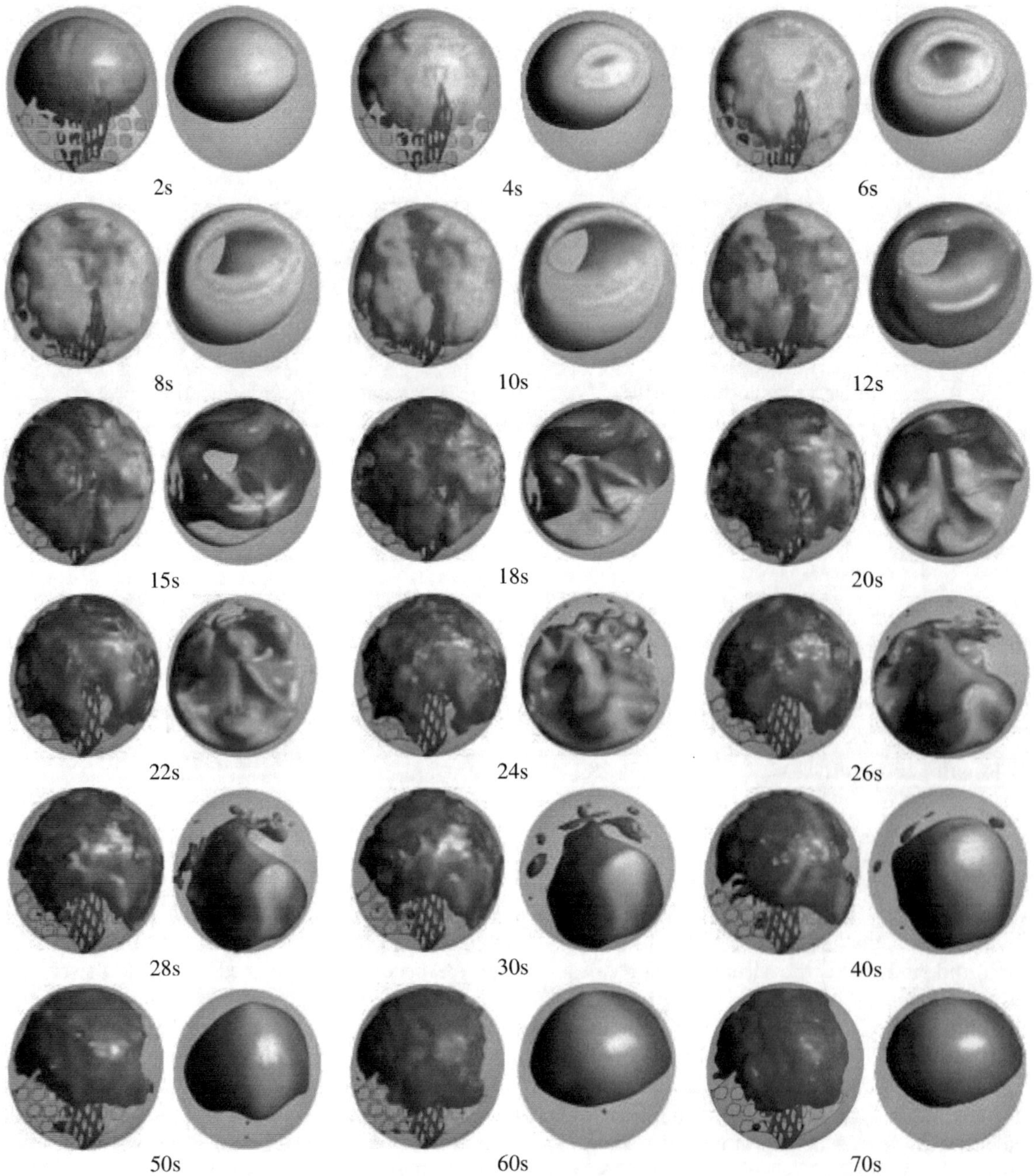

Figure 5. Gas-liquid interface distribution during liquid sloshing process.

The effects with anti-sloshing baffles or not on sloshing mass, sloshing force and slosh torque during liquid sloshing in the tank are shown in figure 6, figure 7 and figure 8 respectively. As shown in figure 6, numerical results of liquid sloshing mass changing process indicate that fluctuation ranges of sloshing mass on Y axis and Z axis are both much greater than on X axis. After tank shaking stops, in the tank with anti-sloshing baffles sloshing mass quickly becomes stable and sloshing mass on Y axis becomes stable after 18s and sloshing mass on Z axis becomes stable after 30s, while in the tank without anti-sloshing baffles sloshing mass on Y axis and Z axis still appears obvious fluctuation. So after tank shaking stops, liquid sloshing could be restrained quickly in the tank with baffles, while the velocity of liquid sloshing attenuation descends significantly in the tank without baffles.

As shown in figure 7, numerical results of liquid sloshing force changing process indicate that fluctuation ranges of sloshing force on Y axis and Z axis are both much greater than on X axis. After tank shaking stops, in the tank with anti-sloshing baffles sloshing force quickly becomes stable and sloshing force on Y axis becomes stable after 20s and sloshing force on Z axis becomes stable after 18s, while in the tank without anti-sloshing baffles sloshing force on Y axis and Z axis still appears obvious fluctuation between 20s and 45s, especially on Z axis. So after tank shaking stops, in the tank with baffles sloshing force rapidly decreases and becomes stable quickly because liquid sloshing could be restrained quickly, while sloshing force amplitude is still relatively large and the velocity of sloshing force attenuation descends significantly in the tank without baffles.

As shown in figure 8, numerical results of liquid sloshing force changing process indicate that fluctuation ranges of sloshing torque on X axis is much greater than both on Y axis and on Z axis. After tank shaking stops, in the tank with anti-sloshing baffles sloshing torque quickly becomes stable and sloshing torque on X axis becomes stable after 20s, while in the tank without anti-sloshing baffles sloshing force on X axis still appears obvious fluctuation between 20s and 45s. So after tank shaking stops, in the tank with baffles sloshing torque rapidly decreases and becomes stable quickly because liquid sloshing could be restrained quickly, while sloshing torque amplitude is still relatively large and the velocity of sloshing torque attenuation descends significantly in the tank without baffles.

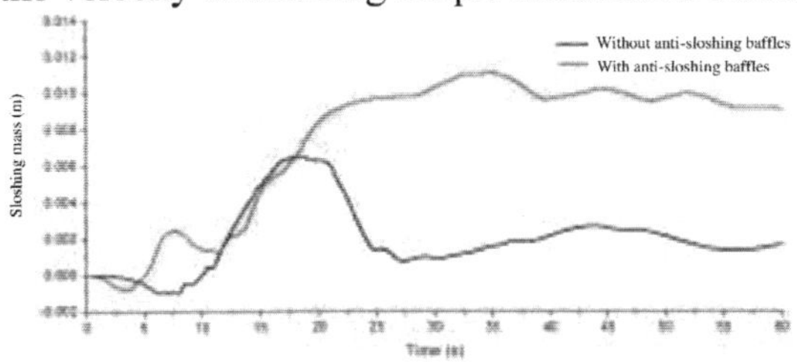

a) sloshing mass on X axis

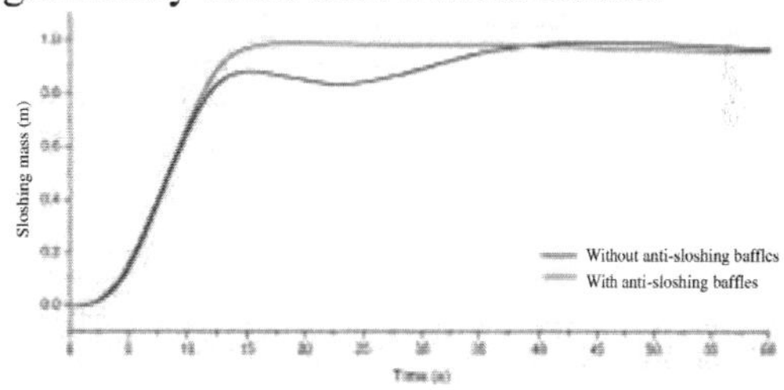

b) sloshing mass on Y axis

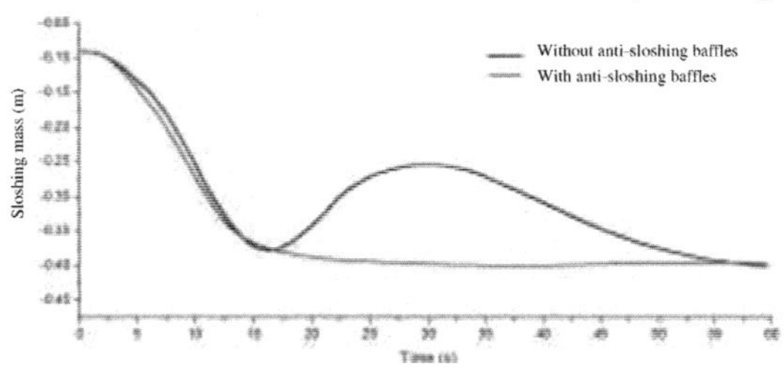

c) sloshing mass on Z axis

Figure 6. The effects with anti-sloshing baffles or not on sloshing mass.

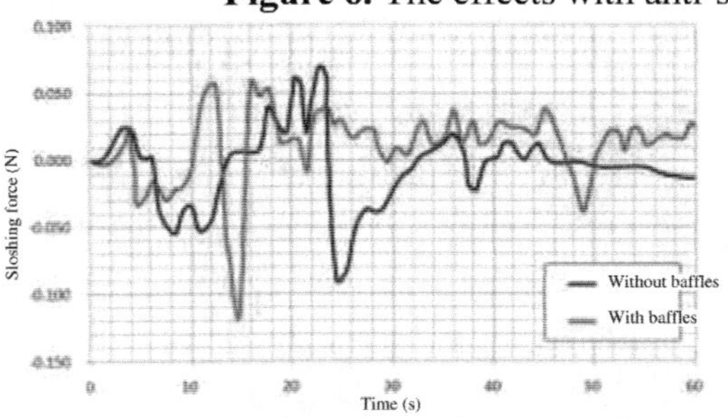

a) sloshing force on X axis

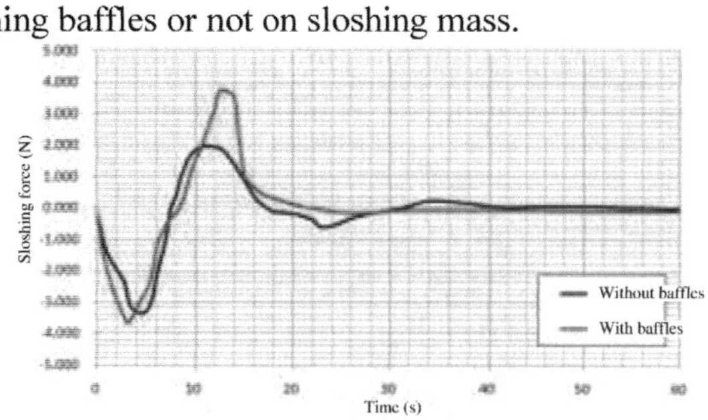

b) sloshing force on Y axis

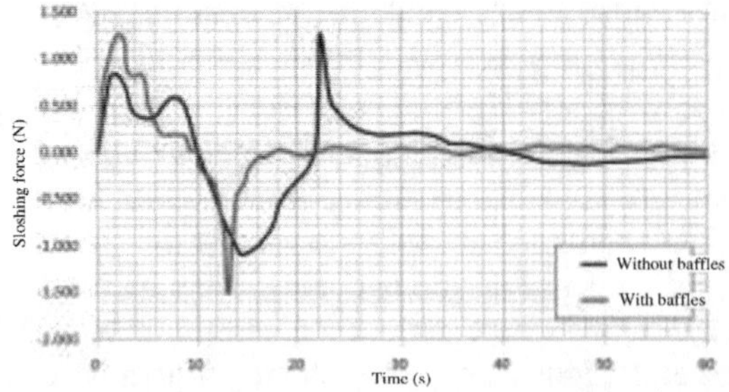

c) sloshing force on Z axis

Figure 7. The effects with anti-sloshing baffles or not on sloshing force.

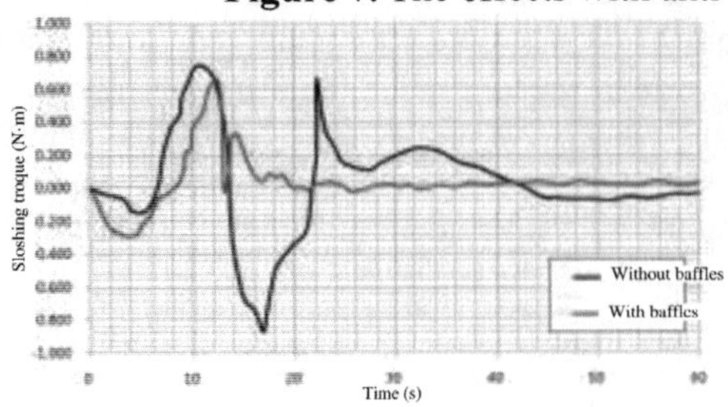

a) sloshing torque on X axis

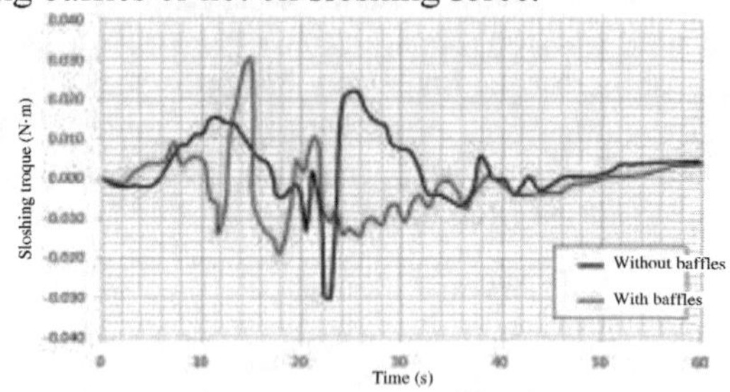

b) sloshing torque on Y axis

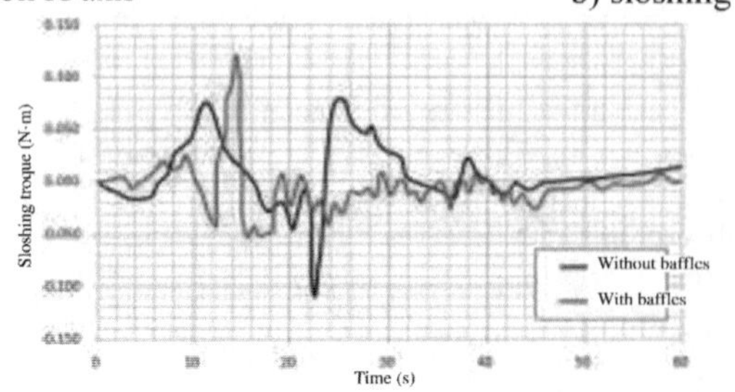

c) sloshing torque on Z axis

Figure 8. The effects with anti-sloshing baffles or not on sloshing torque.

3.2. Comparative analysis between simulation result and experimental result

Figure 9 below shows the contrast between numerical result and experimental result. The comparative analysis indicates that microgravity experimental result and numerical simulation result are very similar, and gas-liquid interface distributoins between the two results during sloshing process are much the same and there are both no interface fracture phenomenons. So the validity and reliability of numerical analysis methods in the paper are verified well.

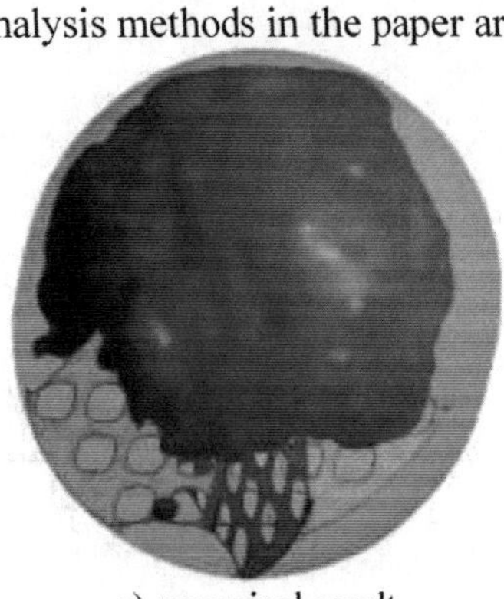

a) numerical result

b) experimental result

Figure 9. Contrast between numerical result and experimental result.

4. Conclusions

Via analysis and research on liquid sloshing performance in vane type tank under microgravity, conclusions below can be gotten.

a) Microgravity experimental result and numerical simulation result are very similar, and the validity and reliability of numerical analysis methods are verified well. Meanwhile numerical calculation can simulate some boundary conditions and microgravity which cannot be simulated by tests and directly give some parameters which can be measured more difficultly in tests, so numerical analysis methods in the paper have great practical value in engineering applications.

b) Vane type PMD with anti-sloshing baffles has better effects on restraining liquid sloshing and is able to restrain observably propellant sloshing in tanks in order to satisfy spacecraft requirements of high stabilization and fast orbital maneuver.

c) Anti-sloshing baffles and these holes on them both effect liquid sloshing in tank significantly, which offer valuable references for the PMD design for restraining liquid sloshing in the subsequent tanks.

Acknowledgments

The authors would like to thank National Natural Science Foundation of China (No. 51406010).

References

[1] Qu G J 2000 *Spacecraft Dynamics Engineering* (Beijing: Chinese Science and Technology Press) pp 288-289

[2] Kana D D 1988 Validated spherical pendulum model or rotary liquid slosh [R]. American Institute of Aeronautics and Astronautics, JSpRo. **26**: 188-195.

[3] Choi J, Sarigul K N 2006 Liquid sloshing in flexible tanks under thruster firing for orbit adjustment [C]. *AIAA Joint Propulsion Conference & Exhibit*, [s. l.] AIAA, 9-12.

[4] Mitra S, Sinhamahpatra K P 2007 Slosh dynamics of liquid-filled container withsubmerged components using pressure-based finite element method [J]. *Journal of Sound and Vibration*, **304**: 361-381

[5] Brackbill J U, Kothe D B, Zemach C 1992 A continuum method for modeling surface tension [J]. *Journal of Computational Physics*, **100**: 335-354.

Effects of radial diffuser hydraulic design on a double-suction centrifugal pump

H C Hou, Y X Zhang[1], C Xu, J Y Zhang, Z L Li

College of Mechanical and Transportation Engineering, China University of Petroleum, Beijing, 102249
Beijing Key Laboratory of Process Fluid Filtration and Separation, Beijing, 102249

E-mail: houhucan@163.com; zhyx@cup.edu.cn

Abstract. In order to study effects of radial diffuser on hydraulic performance of crude oil pump, the steady CFD numerical method is applied and one large double-suction oil pump running in long-distance pipeline is considered. The research focuses on analysing the influence of its diffuser vane profile on hydraulic performance of oil pump. The four different types of cylindrical vane have been designed by in-house codes mainly including double arcs (DA), triple arcs (TA), equiangular spiral line (ES) and linear variable angle spiral line (LVS). During design process diffuser vane angles at inlet and outlet are tentatively given within a certain range and then the wrapping angle of the four types of diffuser vanes can be calculated automatically. Under the given inlet and outlet angles, the linear variable angle spiral line profile has the biggest wrapping angle and profile length which is good to delay channel diffusion but bring more friction hydraulic loss. Finally the vane camber line is thickened at the certain uniform thickness distribution and the 3D diffuser models are generated. The whole flow passage of oil pump with different types of diffusers under various flow rate conditions are numerically simulated based on RNG k-ε turbulent model and SIMPLEC algorithm. The numerical results show that different types of diffusers can bring about great difference on the hydraulic performance of oil pump, of which the ES profile diffuser with its proper setting angle shows the best hydraulic performance and its inner flow field is improved obviously. Compared with the head data from model sample, all designed diffusers can make a certain improvement on head characteristic. At the large flow rate conditions the hydraulic efficiency increases obviously and the best efficiency point shift to the large flow rate range. The ES profile diffuser embodies the better advantages on pump performance which can be explained theoretically that the diffuser actually acts as a diffusion device and is good to transform the dynamic energy to pressure energy. Then through the hydraulic loss analysis of each pump component for all diffusers, it shows that the impeller takes up the biggest part of the whole loss about 8.19% averagely, the radial diffuser about 3.70% and the volute about 1.65%. The hydraulic loss of impeller is dominant at the large flow rate while the radial diffuser is at the small flow rate. Among all diffusers, the ES profile diffuser generates the least loss and combined to the distribution of velocity vector and turbulent kinetic energy for two kinds of diffusers it also shows that ES profile is fit to apply in radial diffuser. This research can offer a significant reference for the radial diffuser hydraulic design of such centrifugal pumps.

[1] Corresponding authors: Yongxue Zhang, e-mail: zhyx@cup.eud.cn;

1. Introduction

Large double-suction oil pump, as one key device in long distance oil pipeline, has widely application for its centrifugal structure with large flow rate, stable running operation and convenient maintenance. Until now, most of the running large double-suction oil pumps online in domestic still depend on import, and not only its high price cost, long delivery time and inconvenient maintenance, but its average efficiency index beyond domestic over 2.5% [1]. In 2015 the ministry of industry and information technology has declared to launch and start the project of Special Action Strong Industrial Base which required improving the control abilities to the key technologies of energy equipment, and thus the study of large double-suction oil pump possesses great value in future.

As the core component in centrifugal pump, the impeller has been researched in many aspects so far. Bi Zhigao [2] numerically studied how blade profile influence centrifugal oil pump and the results indicated that increasing wrapping angle of blade will steepen the head vs. flow rate curve and under the same flow rate condition the head becoming lower, moreover, such characteristic is not effected by medium viscosity. Chen Xiaoling [3] experimentally studied the influence of geometrical parameters of impeller on oil pump in transferring viscous oil and showed that large blade outlet angle can promote transporting ability and less blade number is good to transport high viscous crude oil. By contrast to another important component is radial diffuser, and also as a key component in pump with dual characteristics of suction chamber and pressure chamber, the diffuser generates large hydraulic loss for high velocity in flow channel and thus proper design will increase pump performance and decrease hydraulic loss. Mostly the investigations in open literatures mainly adopt experiment and simulation method to study the influence of radial diffuser on pump performance. Experimental test considers more the hydraulic loss in diffuser and measure its inner flow field and numerical simulation focuses on rotor-stator interaction between impeller and diffuser and rotating stall phenomenon. Ubaldi [4] conducted experiments to study the radial gap between impeller and diffuser on unsteady flow. Shi [5] numerically investigated pressure fluctuations in the entire stage of a diffuser pump due to the interaction between impeller and diffuser vanes. The vane number configuration on the unsteady flow by Feng [6] due to impeller-diffuser interaction in radial diffuser pumps were studied that decreasing diffuser vane number can lead to an increase in pressure fluctuation. Torbergsen [7] and Sano [8] researched the mechanism of rotating stall phenomena in diffuser with numerical simulation.

Inspired by the design methods of cylindrical pump blades [9, 10], the authors in this study totally have designed four kinds of radial diffusers with different camber lines using self-own program and they are double arcs (DA), triple arcs (TA), equiangular spiral line (ES) and linear variable spiral line (LVS). During the design process the inner and outer diameters of radial diffuser still keep unchanged and by changing the inlet angle and outlet angle of vane to make good pump performance. Considering vane number on flow passage diffusion, this study at the last discusses how vane number influence pump hydraulic performance, and in this part the relative circumferential mid-positon between radial diffuser and volute tongue is fixed and the total expelling ratio is constant by adjusting vane thickness distribution.

2. Research model description

The research sample model in this paper is one large double-suction oil main-pump from Shanlan-Lanzhou long distance oil pipeline. This pump is characterized of centrifugal volute consisting of impeller, radial diffuser and volute, and the main characteristic parameter in transferring pure water are rotating speed n of 2980r/min, design flow rate Q_d of 2806m^3/h and head H of 139m. The main geometric sizes of corresponding components are listed in Table 1. According to viscous conversion relationship from KSB company [11] and when dynamic viscosity of crude oil μ is 150mPa • s, the correction ratio of flow rate C_Q is 0.998, of head C_H is 0.975 and of efficiency C_E is 0.88, and therefore the hydraulic performance of pump in transferring pure water can be used to obtain that in transferring crude oil.

Table 1. Main geometrical size of pump components (Units: Length-mm; angle-°)

Impeller	Value	Radial diffuser	Value	Volute	Value
Inlet diameter	255	Inlet diameter	448.4	Base circle diameter	750
Outlet diameter	444.4	Outlet diameter	738	Tongue angle	29
Blade number	5	Inlet width	92		
Outlet Width	87	Outlet width	99		
Wrapping angle	135	Vane number	8		

3. Hydraulic design of radial diffuser

The radial diffuser in double-suction pump usually adopts cylindrical style which is common to see in low specific speed centrifugal pump. The basic element profiles are arc line and spiral line. In this study the DA profile, TA profile, ES profile and LVS profile are generated to construct radial diffusers and by equipped them to investigate how different profiles influence hydraulic performance of double-suction pump. The arc and spiral style of diffuser profiles are designed as follows.

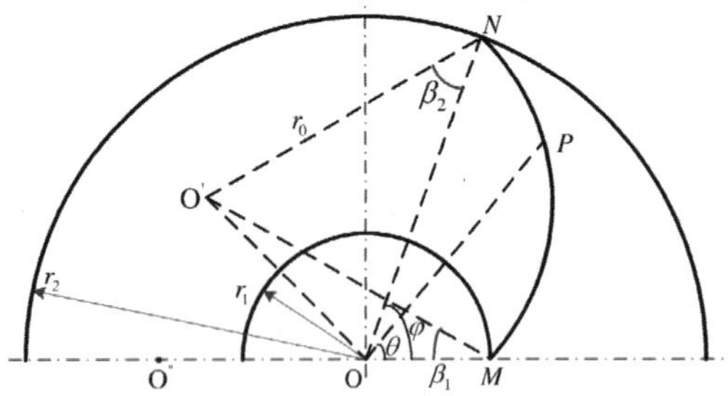

Figure 1. Arc line style profile.

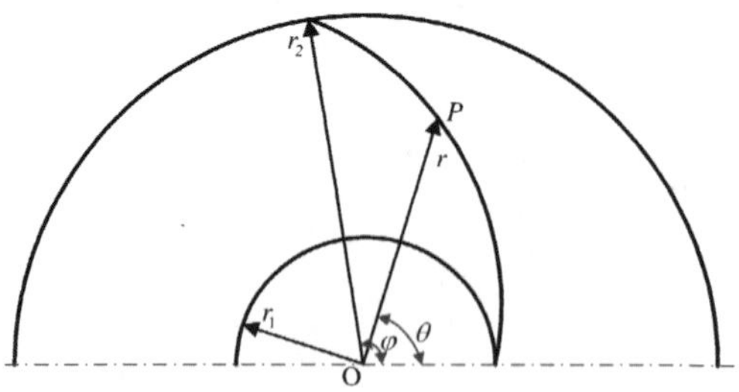

Figure 2. Spiral line style profile.

Figure 1 illustrates the design process of arc line style profile. Taken one arc in multi-arcs as example, *MN* is one basic arc and its centre is point $O^{'}$ and the radius is r_0. The centre of cylindrical diffuser is point O and the inner and outer radius are r_1 and r_2 respectively. Point P is one moving point in profile. β_1 and β_2 are inlet and outlet angle.

$$OO^{'2} = r_0^2 + r_1^2 - 2r_0r_1\cos\beta_1 \tag{1}$$

$$OO^{'2} = r_0^2 + r_2^2 - 2r_0r_2\cos\beta_2 \tag{2}$$

By solving simultaneous equation (1) and equation (2) can obtain $r_0 = \left(r_2^2 - r_1^2\right)/\left(2\left(r_2\cos\beta_2 - r_1\cos\beta_1\right)\right)$. Thus the generating process can be considered as like this: Rotating point $O^{''}$ of angle β_1 in clockwise to get point $O^{'}$ and the distance between $O^{''}$ and M is length of r_0 at symmetrical axis. Point P is obtained by rotating point M in anticlockwise of angle θ, and finally the setting angle β, radius r, profile length and wrapping angle φ can be easily calculated by point-by-point integration method.

Figure 2 shows how spiral line style profile draws. In polar coordinates system the spiral equation can be expressed as in equation (3).

$$r = r_1 e^{\theta\tan\alpha} \tag{3}$$

Here θ is still wrapping angle at point P. ①When α is constant from 0° to 90°, equation (3) represent equiangular spiral line equation and boundary conditions: $\theta=0$, $r=r_1$; $\theta=\varphi$, $r=r_2$. It is to find that the setting angle β is equal to α and in the design process β is set as $\beta=0.5\left(\beta_1+\beta_2\right)$ with the corresponding wrapping angle of $\varphi=\ln\left(r_2/r_1\right)/\tan\beta$. ② When the setting angle is $\beta=a\theta+b$, equation (3) represent linear variable angle spiral line equation and boundary conditions: $\theta=0$, $\beta=\beta_1$; $\theta=\varphi$, $\beta=\beta_2$.

$$\varphi=\left(\ln\left(r_2/r_1\right)/\ln\left(\cos\beta_1/\cos\beta_2\right)\right)\left(\beta_2-\beta_1\right),\ a=\left(\beta_2-\beta_1\right)/\varphi,\ b=\beta_1$$

$$r=r_1\left(\cos\beta_1/\cos\left(a\theta+b\right)\right)^{\frac{1}{a}} \tag{4}$$

It is worth mentioning that as for different profiles of radial diffuser, they must be matched with proper inlet angle and outlet angle which promote good hydraulic performance of pump. That is to say that under the same inlet angle and outlet angle, all designed radial diffuser with its profile cannot match the pump well. Therefore after constant try and error, the proper angle for each kind of diffuser has been obtained as shown in Table 2. The theoretical analysis tells that once the radius and angle at inlet and outlet for each diffuser are given, the setting angle at each point in profile line, profile length and vane wrapping angle are all fixed. From Table 2 it reads that the inlet angle of diffuser is around 18° and the outlet angle differs not much. The ES profile has the smallest wrapping angle and the shortest profile length.

Table 2. Proper parameters for each radial diffuser

Items	DA	TA	ES	LVS
Inlet angle β_1/°	18	18	24.5	18
Outlet angle β_2/°	26	22	24.5	20
Wrapping angle φ/°	69.54	77.73	62.64	82.90
Profile length /m	0.376	0.417	0.320	0.421

Figure 3 and figure 4 show the distribution of vane angle of diffusers and their profile camber lines. At the certain given angle and diameter of each diffuser profile, the vane angle of the ES profile line from inlet to outlet keeps constant while others gradually grow and LVS profile changes very slowly, by contrast, the vane angle of DA profile changes not quite uniformly which may induce flow separation at the exit of its first arc. Figure 4 also tells that from the profile line of DA to LVS, the flow channel length becoming increasing and this is good to delay channel diffusion and good to realize the stable conversion of dynamic energy to pressure energy, however quite long channel length will bring about extra hydraulic loss.

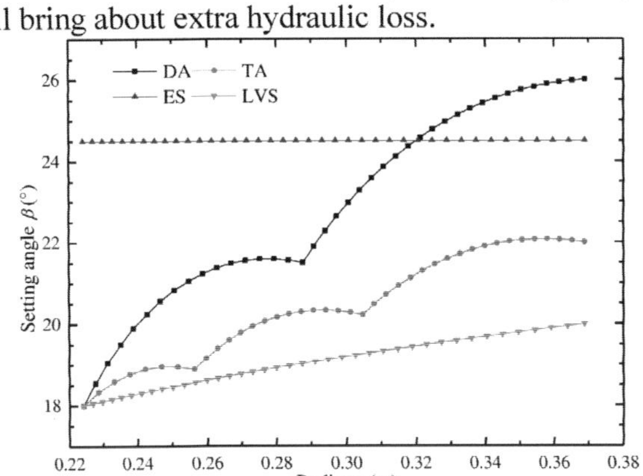

Figure 3. Distribution of setting angle with radius.

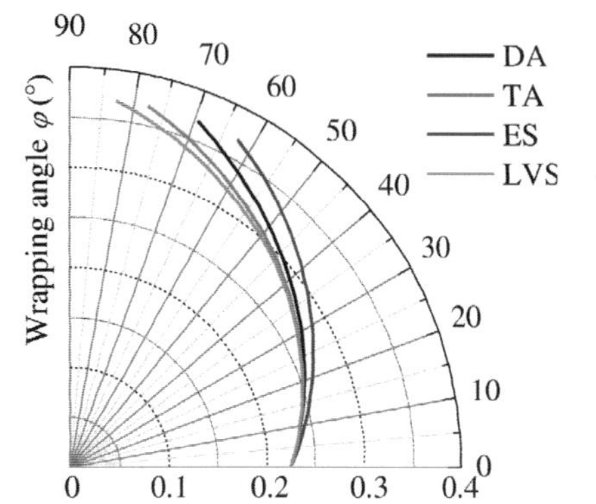

Figure 4. Different diffuser profile lines.

4. Numerical methods and model description

As shown in figure 5, the main computational domain of the pump consists of six parts including inlet extent pipes, impeller, radial diffuser, volute casing and outlet extent pipe. Then the commercial software of ANSYS Meshing is used to generate unstructured tetrahedral grid considering the complexity and skewness of pump components shown in figure 5(b). After this, the solver of ANSYS Fluent 14.0 is used to calculate inner fields for steady, viscous and incompressible flow. The RNG k-ε model and SIMPLEC algorithm were applied to solve RANS equations. The velocity is applied for inlet boundary condition by assuming that velocity at inlet cross-section is uniform. The turbulent intensity and hydraulic diameter are 5% and 145mm respectively. The outflow is used for outlet boundary condition. The non-slip condition is given at solid walls The Multiple Reference Fame model is applied to take into account the interaction between stationary diffuser and rotating impeller with interface pairs. The standard wall function based on the logarithmic law has been used. PRESTO! Scheme is used for pressure term and second order upwind discretization scheme is used for convection terms. The convergence criterion of numerical simulation is set as a residual of 1e-4.

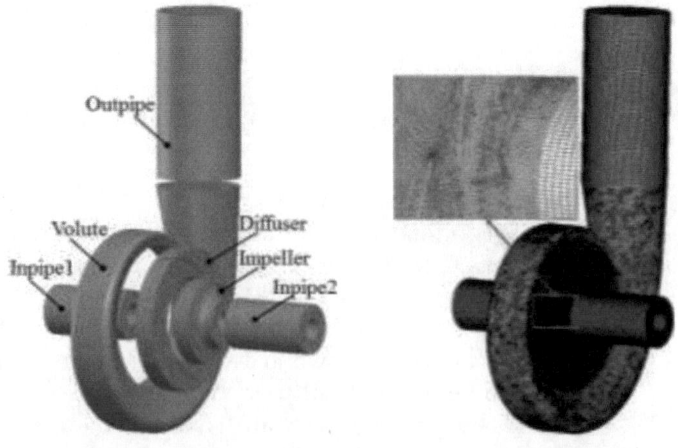

(a)Main computational domains. (b) Grid generation.

Figure 5. Computational domains and grid

Before the real numerical simulation and in order to reduce the computational time and improve the reliability of calculation accuracy, the grid independence under the design operating condition has been investigated. The number of each component keeps at the relative constant proportion compared to the whole grid. And during this process, the characteristic performance of head and hydraulic efficiency are used to evaluate grid effect on final solution. The investigation displays that the head and hydraulic efficiency become stable with the total grid number till 3.04 million cells and thus this refining mesh scheme is considered not to generate value fluctuation. Then the mesh with 3.04 million cells is adopted for the next calculations.

After the inner fields of pump are simulated, the characteristic parameters and the corresponding hydraulic loss efficiency of each component can be defined as follows.

$$H=\left(\left(p_{out}-p_{in}\right)/\left(\rho gQ\right)\right) \tag{5}$$

$$\eta_h=\left(\left(\rho gQ\right)/\left(M\omega\right)\right)\times 100\% \tag{6}$$

$$\Delta\eta_{h-im}=\left(\left(P-\left(p_{im-out}-p_{im-in}\right)Q\right)/P\right)\times 100\% \tag{7}$$

$$\Delta\eta_{h-rd}=\left(\left(p_{rd-in}-p_{rd-out}\right)Q/P\right)\times 100\% \tag{8}$$

$$\Delta\eta_{h-vo}=\left(\left(p_{vo-in}-p_{vo-out}\right)Q/P\right)\times 100\% \tag{9}$$

Here in the equations above, p_{in} and p_{out} are total pressure of pump at inlet and outlet section, P is shaft power, M is torque and ω is angular velocity. p_{im-in}, p_{rd-in} and p_{vo-in} are total pressure at inlet section of impeller, radial diffuser and volute. p_{im-out}, p_{rd-out} and p_{vo-out} are total pressure at outlet section of impeller, radial diffuser and volute. η_h is hydraulic efficiency of pump. $\Delta\eta_{h-im}$, $\Delta\eta_{h-rd}$ and $\Delta\eta_{h-vo}$ are hydraulic loss efficiency of impeller, radial diffuser and volute.

5. Results and discussions

5.1. The influence of diffuser profiles on hydraulic performance of double-suction pump

Figure 6 displays the hydraulic performance of double-suction pump equipped with different profiles of radial diffusers under its own proper inlet angle and outlet angle. Figure 6 (a) shows the head vs. flow rate curves and it reads that compared with the head data from model sample, all designed diffusers can make a certain improvement on head. The head vs. flow rate curve with DA profile diffuser is very close to that with model design diffuser because the measured data from the sample is quite similar to DA profile. The ES profile diffuser contributes more on head vs. flow rate curve and offers the highest head value in all flow rate conditions.

Figure 6 (b) shows the hydraulic efficiency vs. flow rate curves. Totally to say, all designed radial diffusers can improve the hydraulic efficiency of double-suction pump at the large flow rate conditions and the best efficiency point shift to the large flow rate range compared to that the pump data with model sample diffuser. Just like the head vs. flow rate curve, the hydraulic efficiency vs. flow rate curve with DA profile diffuser is very close to that with model design diffuser. In contrast to model design data, the efficiency curves of pump equipped with TA, ES and LVS profile diffusers all performs better and their best efficiency points shift to the left a little but are still inclined to the right side compared with the data of model sample. The ES profile diffuser generate the best hydraulic efficiency as well as head and this can be explained theoretically that the diffuser actually acts as a diffusion device conversing the dynamic energy to pressure energy just like a volute and the fluid flowing through keeps the constant velocity moment under no external force applying on and without considering the viscosity of fluid. Besides, that the high efficiency area of pump hydraulic performance with ES profile diffuser cannot reach and cover that with model sample diffuser may because the unreasonable design of volute casing seen in Ref. [12, 13] and by decreasing the throat area is able to effectively shift the best efficiency point to small flow rate range.

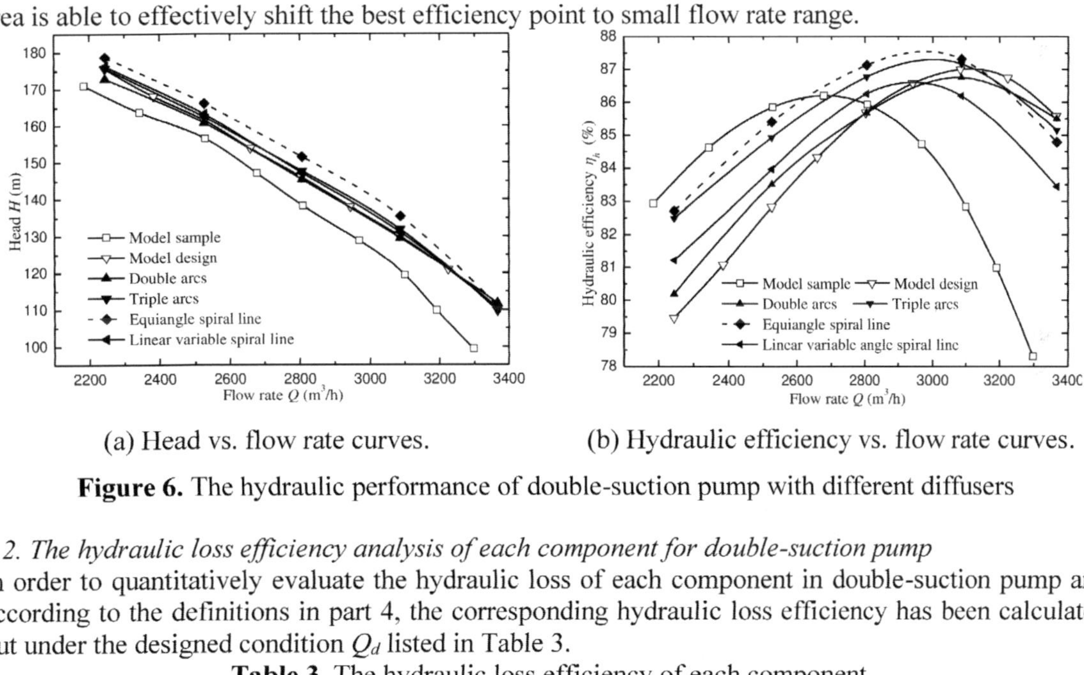

(a) Head vs. flow rate curves. (b) Hydraulic efficiency vs. flow rate curves.

Figure 6. The hydraulic performance of double-suction pump with different diffusers

5.2. The hydraulic loss efficiency analysis of each component for double-suction pump

In order to quantitatively evaluate the hydraulic loss of each component in double-suction pump and according to the definitions in part 4, the corresponding hydraulic loss efficiency has been calculated out under the designed condition Q_d listed in Table 3.

Table 3. The hydraulic loss efficiency of each component

Items	Impeller $\Delta\eta_{h\text{-}im}$ (%)	Radial diffuser $\Delta\eta_{h\text{-}rd}$ (%)	Volute $\Delta\eta_{h\text{-}vo}$ (%)	Total $\Delta\eta$ (%)
DA	8.32	4.20	1.80	14.32
TA	8.13	3.25	1.85	13.23
ES	8.03	3.28	1.56	12.87
LVS	8.27	4.06	1.41	13.74

It reads that among the four pumps with different radial diffusers, generally the hydraulic loss efficiency of the impeller takes up the biggest part of 8.19% averagely, the radial diffuser of 3.70% and the volute takes up about 1.65%. Thus in a sense, the radial diffuser offers an potential ability for reducing hydraulic loss for pump and by improving the geometry of diffuser can realize redistribution of inner flow fields of the total flow passage. Furthermore, among the four different radial diffusers under its own proper inlet and outlet angle, the ES profile diffuser accounts for the least hydraulic loss and the corresponding hydraulic loss efficiencies of each component are also the least of about 8.03%, 3.28% and 1.56. Compared with DA profile diffuser, the component hydraulic loss is decreased obviously by 1.45%, which proves that ES profile is fit to apply in radial pressure diffusion and use in double-suction pump as diffuser profile.

5.3. The inner flow fields of radial diffuser under different flow conditions

The figure 7 presents the distribution of velocity vector at mid-span of DA and ES profile diffusers under three flow rate conditions. It can be seen that the velocity at inlet of diffuser is big and then gradually decrease which indicates that the diffuser can effectively converse the dynamic energy to pressure. With the flow rate increasing, the inner flow improves to some extent. However, there are some obvious flow separations in flow channel of DA profile diffuser while it is rarely to see in ES profile diffuser. Therefore under different flow rate conditions, the ES profile diffuser can better adapt the flow but it is also not so well at the big flow rate condition.

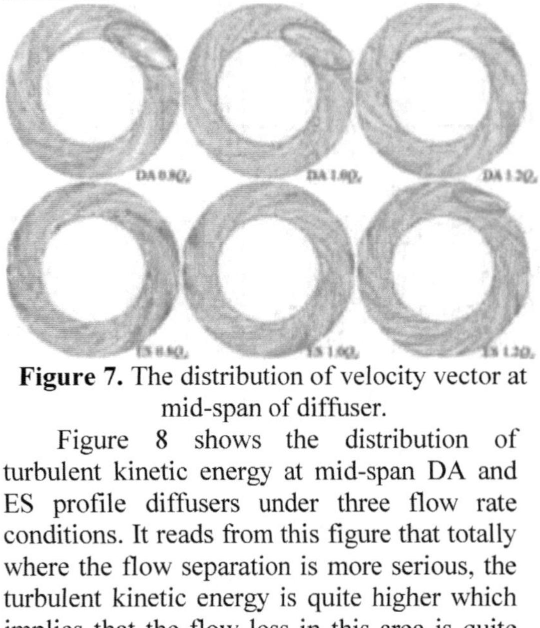

Figure 7. The distribution of velocity vector at mid-span of diffuser.

Figure 8 shows the distribution of turbulent kinetic energy at mid-span DA and ES profile diffusers under three flow rate conditions. It reads from this figure that totally where the flow separation is more serious, the turbulent kinetic energy is quite higher which implies that the flow loss in this area is quite bigger. At the flow rate of 0.8 Q_d condition, the flow separation happens seriously in each channel just like the velocity vector distribution in figure 7. In general, the mass weighted average value of turbulent kinetic energy k for DA profile diffuser is bigger than that for ES profile diffuser and this exactly indicates that the distribution of turbulent kinetic energy for ES profile diffuser is more reasonable. However there is not good under the flow rate of 1.2 Q_d which mainly because the thickness distribution is uniform from leading to trailing edge.

Figure 8. The distribution of turbulent kinetic energy at mid-span of diffuser.

Table 4. The hydraulic loss efficiency of each component with ES diffuser (%)

Items	Impeller $\Delta\eta_{h\text{-}im}$	Radial diffuser $\Delta\eta_{h\text{-}rd}$	Volute $\Delta\eta_{h\text{-}vo}$	Total $\Delta\eta$
$0.8Q_d$	8.96	6.79	1.53	17.28
$1.0Q_d$	8.03	3.28	1.56	12.87
$1.2Q_d$	9.80	3.43	1.62	14.85

What's more, the hydraulic loss efficiency of each component equipped with ES profile diffuser under three different flow rate conditions is presented in Table 4. Under the three flow rate conditions, the hydraulic loss proportion of volute casing is varied not much with the average level of 1.57% while the impeller and radial diffuser changes obviously. The hydraulic loss of impeller is dominant at large flow rate condition and radial diffuser at small flow rate condition. Taking the radial diffuser for granted at the small flow rate condition, the flow separations in flow channels and the inlet incidence phenomenon is quite evident compared with the other two conditions and this is easy to tell in figure7.

6. Conclusions

As the important device in long-distance pipeline for crude oil, the double-suction pump and its component radial diffuser have been focused on in this study by the steady CFD numerical simulation. The diffuser profile is chosen as the key point to investigate and some conclusions can be drawn as follows.

①The four kinds of diffuser profiles are designed by in-house codes consisting of double arcs, triple arcs, equiangular spiral line and linear variable angle spiral line. And after constant try and error, the proper angles for each profile are given.

②Compared with the model pump, all designed diffusers can make improvement on pump hydraulic performance while the best efficiency point shifts to the large flow rate range.

③Through the hydraulic loss analysis of each component, the impeller takes up the biggest part of the whole loss about 8.19%, the radial diffuser about 3.70% and the volute about 1.65. The hydraulic loss of impeller is dominant at the large flow rate while the radial diffuser at the small flow rate.

④Combined to the distribution of velocity vector and turbulent kinetic energy for DA and ES profile diffusers, the ES profile is fit to apply in radial diffuser.

Acknowledgements

This research was supported by Science Foundation of China University of Petroleum, Beijing (No.2462015YQ0411).

References

[1] Wang Y 2012 Analysis of blade inlet edge design parameters variation impact on performance of double suction pump, *MS thesis, Lanzhou University of Technology.*

[2] Bi Z H 2014 Influence of blade profile on performance of centrifugal oil pump, *Journal of Lanzhou University of Technology*. **40** 62.

[3] Chen X H, Zhang W G 1999 Influence of geometric parameters of impeller on centrifugal pump performance in transferring viscous oil, *CPM*. **27**(2) 19.

[4] Ubaldi M 1994 Nonintrusive measurements of the unsteady flow in the radial gap between centrifugal impeller and vaned diffuser, *the 12th Symposium on Measuring Techniques for Transonic and Supersonic Flow in Cascades and Turbomachines*.

[5] Shi F, Tsukamoto H 2001 Numerical study of pressure fluctuations caused by impeller-diffuser interaction in a diffuser pump stage, *Journal of Fluids Engineering*. **123** 466.

[6] Feng J J, Friedrich-Karl B and Hans J D 2007 Numerical investigation on pressure fluctuations for different configurations of vaned diffuser pumps, *International Journal of Rotating Machinery*.

[7] Torbergsen E A 1998 Impeller/Diffuser interaction forces in centrifugal pumps, *Doctor thesis, Norwegian University of Science and Technology*.

[8] Sano T 2000 Alternate blade stall and rotating stall in vaned diffuser-part1: effects of impeller/diffuser clearance, *Trans. Jpn. Soc. Mech*. **66** 650.

[9] Xu H Y 1990 Theoretical research on profile of circular arc blade, *Fluid Machinery*. **8** 7.

[10] Xu H Y 1989 Draw method of variable spiral line for cylindrical blade, *Fluid Machinery*. **7** 6.

[11] KSB Aktiengsllschaft 1990 *Centrifugal pump lexicon, 3rd Edition*. **344**.

[12] Kim J H, Lee H C and Kim J H 2015 Design techniques to improve the performance of a centrifugal pump using CFD, *Journal of Mechanical Science and Technology*. **29** 215.

[13] Yang S, Kong F and Chen B 2011 Research on pump volute design method using CFD, *International Journal of Rotating Machinery*.

IOP Conf. Series: Materials Science and Engineering **129** (2016) 012018 doi:10.1088/1757-899X/129/1/012018

Numerical simulation on the performance of the vortex pump for transporting solid-liquid two-phase with light particles

W Y Mao[1], P Y Song (Correspondence)[1], Q G Deng[1], H J Xu[1]

Faculty of Chemical Engineering,[1]Kunming University of Science and Technology, China

E-mail: maowenyuanyn@163.com, songpengyunkm@sina.com, 971659256@qq.com, why_551@126.com

Abstract. With the purpose of studying performance of the vortex pump for transporting solid-liquid two-phase with light particles whose relative density smaller than 1, the numerical simulation of solid-liquid two phase flowing in the whole channel of a vortex pump with the particle diameter being 0.5 mm, 1 mm, 2 mm, 3 mm and the initial solid phase volume concentrations being 10%, 20% and 30% are respectively carried out by using the commercial software ANSYS Fluent by adopting RNG κ-ε turbulent flow model, Eulerian-Eulerian multi-phase flow model and SIMPLEC algorithm. The simulation results show that in the impeller region, the particles concentrate on the non-working surface of the blades, and the particles are rare on the working surface of the blades. As the initial solid phase volume concentration and particle diameter increase, the pump delivery head of vortex pump decrease. The pump delivery head of vortex pump with different initial solid phase concentrations and different particle diameters are predicted and compared with those obtained by an empirical formula, and they shows good agreement.

1. Introduction

With the advantage of strong passing ability of transporting solid-liquid two-phase particles, vortex pumps are widely used in the fields of transportation of slurry transportation, hydraulic transmission and waste water processing [1]. In the past, the solid particles with relative density larger than 1 was the main research object in vortex pumps, but the research on the solid particles with relative density smaller than 1 was rare. Due to the different relative density of solid particles, the particles in liquid is with different behaviors of settling, as solid particles with relative density larger than 1 or floating, as solid particles with relative density smaller than 1, which will have different influences on the performance of vortex pump. Therefore, it is of great significance to study the performance of solid particles in liquid phase and its influence on internal flowing field and external properties of the vortex pump. The numerical analysis on performance of a vortex pump for transporting solid-liquid two-phase with light particles whose relative density smaller than 1 is carried out based on the commercial software ANSYS Fluent.

IOP Conf. Series: Materials Science and Engineering **129** (2016) 012018 doi:10.1088/1757-899X/129/1/012018

2. Calculation model and method

2.1. Fundamental assumptions

In order to study solid-liquid two-phase flowing pattern inside vortex pump with plastic particles whose relative density smaller than 1, the following assumptions are taken [2]:

(1) Liquid phase (water) is treated as in-compressible fluid, particle as disperse phase, the physical properties of each phase remain constant;

(2) Particles are treated as round, even-diameter solid particle without phase transition;

(3) Internal flow of the pump treated as steady flow with water as principle phase, solid particle as secondary phase;

(4) Axial velocity of inlet is well-distributed, and solid particles and water are evenly mixed with same velocity.

2.2. Calculation model

Numerical simulation was carried out with the type 65-310 vortex pump made by a pump manufacturer where its main design parameter being water flow rate Q=48.78m3/h, delivery head H=38.48 m, rotation speed n=1470 r/min, impeller diameter D=310 mm, density of particle phase ρ=900 kg/m3. The 3-D modelling of flowing region of the vortex pump was built with a CAD software, and the entire calculation domain was divided into two parts, where one being impeller (rotational region), another one being volute (static region), which are shown in figure 1. With the purpose of making solid-fluid two phase of the inlet portion evenly mixed and fluid of outlet fully developed, both inlet pipe and outlet pipe were properly extended [3].

2.3. Mesh Division

For the volute region and impeller region, unstructured tetrahedral meshes were relatively generated with gambit, and the mesh of impeller region was properly refined, which is shown in figure 2. Independence analysis of mesh was conducted in consideration of limitation of computer memory and requirement of time to reach a convergence, as shown in table 1, which turns out that when difference of calculated external property is less than 2%, the requirement of mesh independence is met [4-5], where the total amount of meshes is 603,454, in which the total amount of meshes of volute region is 297,251, whereas the amount of meshes of impeller region is 306,203.

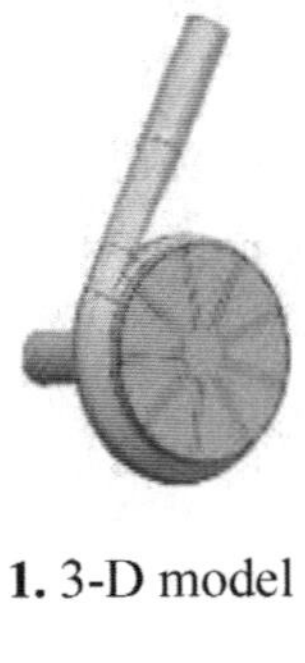

Figure 1. 3-D model of flowing region

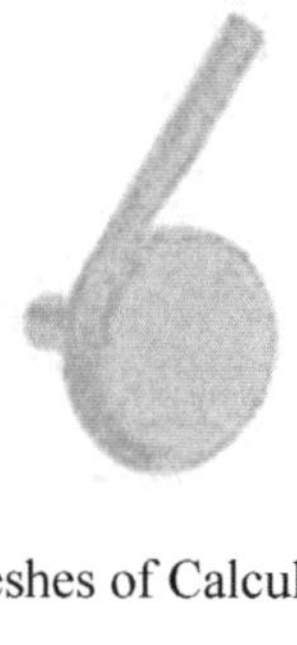

Figure 2. Meshes of Calculation domain

Table 1 Independence analysis of mesh

Mesh amount, N	Pump head, H (m)
603454	42.38
743088	42.20
1131349	42.34
1608276	42.54

2.4. Boundary conditions

Inlet velocity was set as constant flow rate, since the pressure and velocity of the boundary of outlet is unknown, thus free outflow condition was adopted; Eulerian-Eulerian multi-phase model was adopted as multi-phase model, only the special inter-phase drag was retained as interaction between solid phase and liquid phase [6]; velocity and pressure coupling calculation was realized by using SIMPLEC algorithm, the closed RANS equation of RNG κ-ε two-equation turbulence model with curved wall which adapts to intense rotational flow was used; non-slip boundary condition was used at wall, moreover, standard wall function was used at the low Reynolds number zone near wall to deal with problems which caused by high Reynolds number turbulence model; multi-reference model (MRF) method was adopted to conduct data transfer between rotational region and static region in consideration of the rotor-stator interference between impeller region and volute region.

3. Results and Analyses

3.1. The external characteristic of vortex pump transporting clear water

Firstly, the numerical simulation of the external performance of the vortex pump transporting clean water is conducted, and the numerical simulation results are compared with the experiment results, the results are shown in figure 3. It can be seen that the flow curve trend between the simulation delivery head and the experiment delivery head is basically the same, the simulation value of the delivery head is higher than the experiment delivery head, and the reason for this is that the cumulative error caused by the factors such as casting, machining, and not considering the factors such as the roughness of the surface of the impeller and the roughness of volute interior, etc. The trend of simulation efficiency changing with flow rate is in accordance with that of the experiment efficiency that when flow rate is approaching the operating point, and the difference between the simulation efficiency and the experiment efficiency is minor.

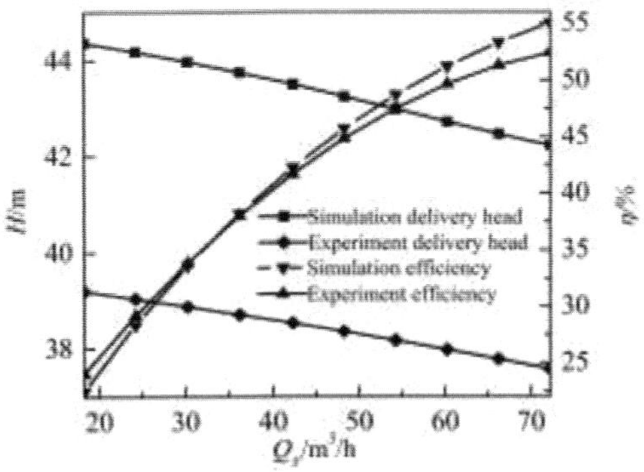

Figure 3. The comparison of vortex pump transporting clean water between simulation results and experiment results

3.2. The distribution of solid phase volume concentration under different particle diameter

Figure 4 shows the solid particle volume concentration distribution in the impeller region with different particle diameter where the initial volume concentration of solid phase being 10%, and the particle diameter respectively being 0.5 mm, 1 mm, 2 mm and 3 mm. It can be seen from the figure 4 that the particle volume concentration distribution trend is consistent with the different particle diameters. Due to the boundary condition in inlet of impeller is set that the velocity of two phases is same, so the volume concentration distribution of solid phase is well distributed under the same particle diameter. When the solid particles enter the impeller, the solids volume concentration changed as the impeller doing work, and there is obvious concentration gradient from non-working surface to working surface of the blade at the same radius. The concentration of solid particles on the blade of non-working surface is significantly higher than that of the working surface. As to the longer and shorter blades, solid particle accumulates more on the longer blades than that of the shorter blades.

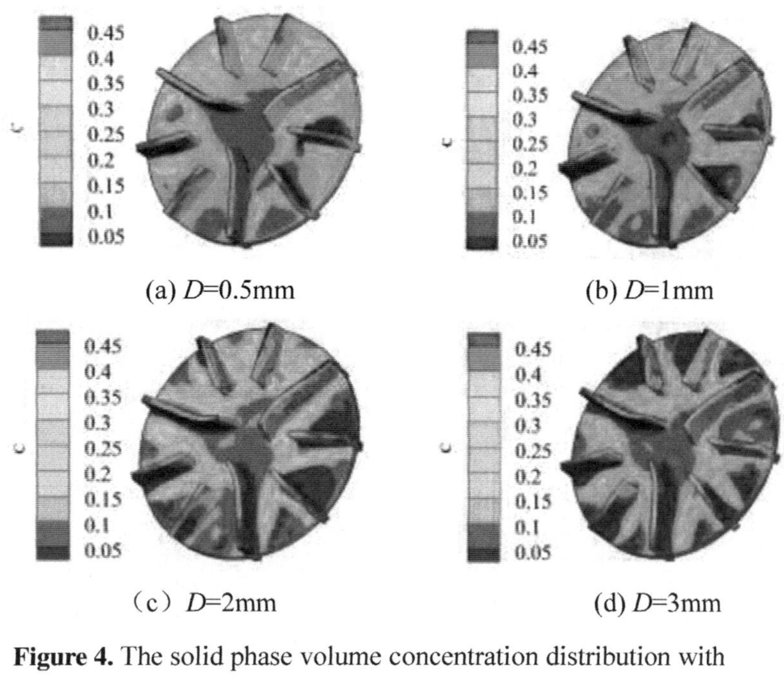

(a) D=0.5mm (b) D=1mm

(c) D=2mm (d) D=3mm

Figure 4. The solid phase volume concentration distribution with different particle diameter(C=10%)

3.3. The distribution of solid phase volume concentration under different initial volume concentration

Figure 5 shows the solid particle volume concentration distribution in the impeller region with different initial volume concentration where the particle diameter is 3 mm, and the initial volume concentration respectively being 10%, 20% and 30%. It can be seen from the figure 5, in the process of solid particles moving from inlet to outlet, the solid volume concentration on non-working surface of the blades is significantly higher than those on the working surface. As the solid phase initial volume concentration increases, the solid phase volume concentration in middle part of the impeller passage increases, and high solid volume concentration area expands, which means it will aggravate the abrasion on the surface of the impeller, and reduce the service life of the vortex pump under the high initial volume concentration.

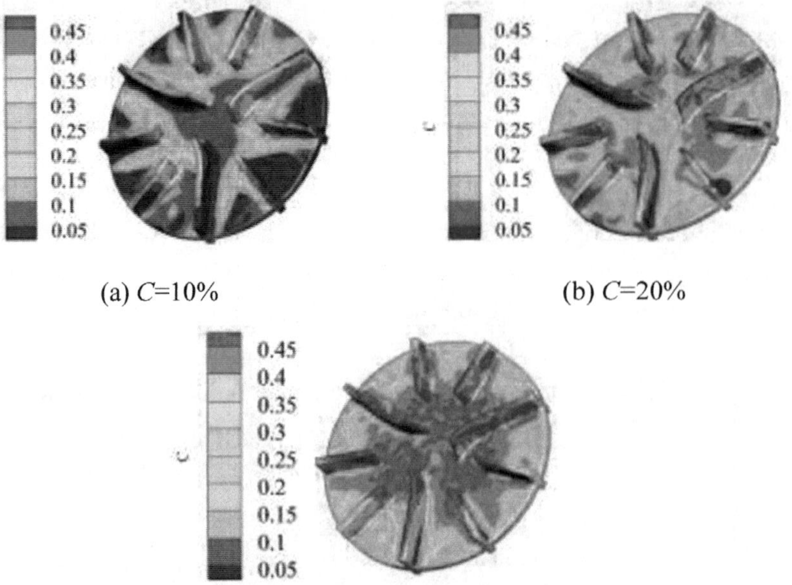

(a) C=10% (b) C=20%

(c) C=30%

Figure 5. The solid phase volume concentration distribution with different initial volume concentration (D=3mm)

3.4. The comparison between the simulation delivery head and the prediction delivery head

In order to verify the simulation delivery head of the vortex pump with different initial volume concentration and different solid phase particle diameter, the experience equation presented by Pengcheng He etc. [7] used to predict the slurry pump delivery head is adopted to predicate the pump delivery head. The formula is shown as:

$$H_s = HR \cdot H_w$$

Where, H_s is the pump head with slurry, H_w is the pump head with clean water; $HR = (1-C_w)^n$; C_w is weight concentration of solid; $n = (0.21 + 0.066\ln d_{50}) \cdot S$, where d_{50} is the mass median diameter, $S = \rho_s / \rho_w$, the relative density, ρ_s is the density of the solid particle, ρ_w is the density of the water.

The numerical simulation pump delivery head and the prediction pump delivery head by the experience equation are shown in figure 6 and figure 7 with different initial volume concentration and different particle diameter respectively.

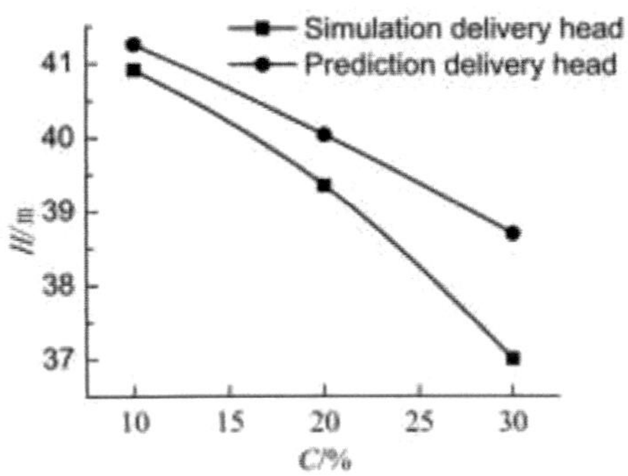

Figure 6. The pump delivery head by numerical simulation and experience equation with different volume concentration, where D=3mm

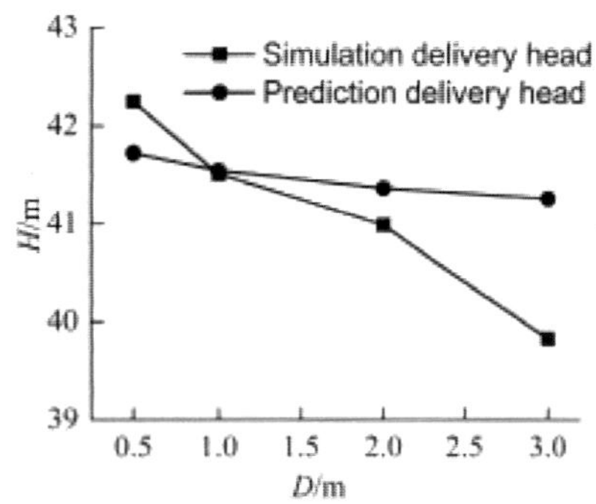

Figure 7. The pump delivery head by numerical simulation and experience equation with different particle diameter, where C=10%

As figure 6 shows, both the pump head predicted by Fluent software and by the experience equation decrease with the initial volume concentration increase. The maximum error between the simulation head and equation prediction head is 4.36%.

As figure 7 shows, both the pump head predicted by Fluent software and by the experience equation decrease with the particle diameter increase. The maximum error between the simulation head and equation prediction head is 3.47%.

4. Conclusions

(1) As to clean water, both the pump head curve and the efficiency curve by numerical simulation are consistent with those by experiment. The simulation head is slightly higher than the experiment head.

(2) The particle volume concentration distribution trend is consistent with the different particle diameters. The volume concentration of light solid particles on the blade non-working surface is significantly higher than on the working surface.

(3) As to different initial volume concentration of light solid, most of the particles are concentrated in the middle and upper part of the impeller passage near the centre. As the solid phase initial volume concentration increases, the solid phase volume concentration in middle part of the impeller passage increases, and high solid volume concentration area expands.

(4) As the solid phase initial volume concentration and particle diameter increase, the pump head and pump efficiency decrease. The numerical simulation pump head are consistent with prediction results by empirical equation.

IOP Conf. Series: Materials Science and Engineering **129** (2016) 012018 doi:10.1088/1757-899X/129/1/012018

References

[1] Sha Y 2011 Experiments on performance and internal flow of a vortex pump *J. Transactions of the CSAE.* **27** 141–14

[2] Wang J Q, Jiang W M, Kong F Y, Qu X Y and Su X H 2013 Numerical simulation of solid-liquid two-phase turbulent flow in centrifugal pump based on particle model *J. Journal of Drainage and Irrigation Machinery Engineering.* **31** 846–850, 878

[3] Shi W D, Long F, Zhang D S, Leng H F and Wang G T 2012 Numerical simulation of solid liquid two phase turbulent flow in submersible axial flow pump *J. Journal of Drainage and Irrigation Machinery Engineering.* **5** 508–512

[4] Zhang Y L, Li Y, Cui B L and Zhu Z C 2012 Computational analysis of hydraulic transmission performance in two-phase flow centrifugal pump *J.* Journal of mechanical engineering. **48** 169–176

[5] Han W, Yue T, Li R N, Jin Y, Jiang L and Hu P L 2015 Numerical analysis of transition on process of solid-liquid two phase in rotor-stator cascades *J. Journal of Drainage and Irrigation Machinery Engineering.* **33** 111–115

[6] Zhang D W, Yang M G, Gao B and Lu B Y 2013 Particle Concentration Distribution and its Effect on Performance in a Vortex Pump *J. Fluid Machinery.* **41** 15–18, 71

[7] He P C, Song P Y 2013 A simple correlation equation to predict pump performance for slurry. *Applied Mechanics and Materials.* **365-366** 365–369

IOP Conf. Series: Materials Science and Engineering **129** (2016) 012019 doi:10.1088/1757-899X/129/1/012019

A numerical study of liquid film distribution in wet natural gas pipelines

X Q Gao[1], Y L Zhao[1], W W Xu[2], X R Guan[1], J J Wang[1], and Y H Jin[1]

[1] College of Chemical Engineering, China University of Petroleum, Qingdao 266580, PR China.

[2] College of pipeline and civil Engineering, China University of Petroleum, Qingdao 266580, PR China.

E-mail: gxqi2003@163.com

Abstract. The software of FLUENT was used to simulate the gas-liquid turbulent flow in wet natural gas pipeline of the Puguang gas field. The RNG k- ε model was used to simulate the turbulent flow, the Mixture model was used to simulate gas-liquid mixed phase, and the Eulerian wall film model was used to simulate the formation and development of liquid film. The gas phase flow field characteristics, the distribution of the axial and circumferential film thickness, and the droplet distribution in the pipeline were studied when the gas Reynolds number is 7.72×10^6(10.8m/s). The results can be concluded as followed: Liquid film distributes unevenly along the circumferential direction and mostly distributes under the pipeline wall because of gravity. The impact of the dean vortex and centrifugal force in the straight section can also influence the liquid film distribution. The wall shear stress distributions in horizontal straight pipeline is concerned with liquid membrane volatility, and consistent with the film volatility period, the wall shear stress reached the maximum value in a certain position of wave front. The influence of the wall shear stress on the film fluctuation in inclined pipeline is weakened by gravity and other factors.

1. Introduction

In wet natural gas pipeline, saturated moisture is easy to produce condensate water in conveying process. Condensate water and the corrosive medium form thin liquid film together which unevenly distributes on the pipe wall. This uneven distribution of liquid film on the metal surface causes the different process of cathode and anode, so metal local corrosion is appeared. There are many kinds of reasons that can influence metal corrosion, such as carbon dioxide, sulfur dioxide, water, corrosion inhibitor, temperature, PH value, medium composition, flow conditions, and so on. Many scholars had gained some good results in this area; however, the study of the distribution of the liquid film is far less for low liquid. The liquid film plays an important role on the corrosion of the pipe, and it lacks further research in this area. So the study of liquid film thickness distribution and how it is changed under the effect of gas fluctuation are very important.

[1] Gao Xueqi, gxqi2003@163.com, 15092115689.

The gas-liquid flow in the wet natural gas pipelines was simulated using the Fluent software, and the simulation results are verified by theory analysis. This paper mainly studied the distribution of liquid film thickness affected by gas fluctuation.

2. Calculation model

The pipeline structure size of the simulation is shown in figure 1. According to the operation data provided by Puguang gas field, the pipe diameter is 0.4636 m, the angle of two curved section is 30.78 °, and the curvature radius is 2.03 m. In order to ensure the turbulence is fully developed, setting the length of two straight pipes to be 9 m, and the inclined pipe to be 9 m, too. The origin of coordinates is located in the center of output section; y-axis negative direction is the direction of gravity. The zoning grid method is adopted in modeling, and the total grid node is 2792250. The meshing model is shown in figure 2.

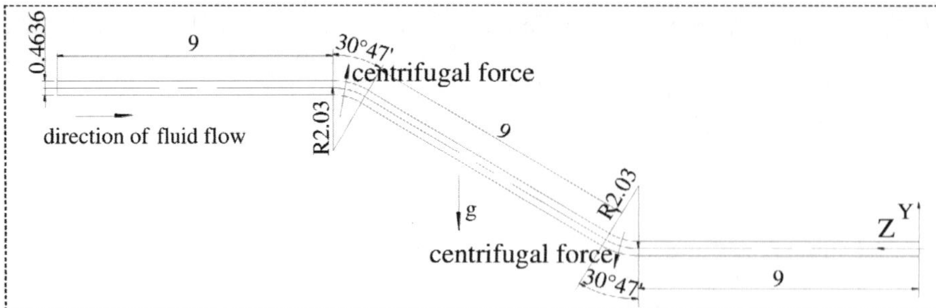

Figure 1. The pipe model structure size (Unit: m).

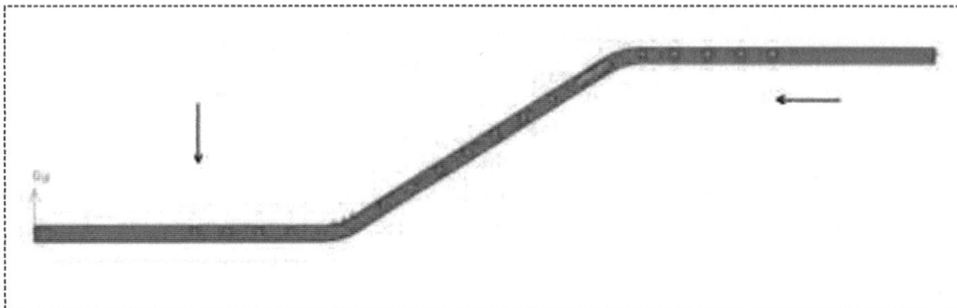

Figure 2. Distribution of meshing in pipe.

3. Calculation method and the boundary conditions

The gas velocity of inlet is 10.8 m/s (7.7×10^6), thus it belongs to the strong turbulent flow, RNG k-ε model was used in numerical calculation. The Mixture and Eulerian wall film model were used to simulate the gas -liquid two phase in the pipeline. The Mixture model was used to simulate gas-liquid mixed phase, and calculate the volume fraction of each phase. While the Eulerian wall film model was used to simulate the formation and development of liquid film. In pulse injection of diesel engine, Wall Film model was validated in the liquid Film thickness and evaporation rate prediction well. Because the formation of the liquid film in the pipeline is similar to the Wall Film model, and satisfies the assumptions and limits as well, therefore, the Wall Film model was used in the simulation.

The inlet boundary condition is 'VELOCITY_INLET'. In order to guarantee the outputs of exhaust pipe meet fully developed flow conditions, pipe length is supposed to be about 10 times of the pipe diameter. The outlet boundary condition was 'OUTFLOW'. The outlet simulation conditions and methods were the same as inlet. Wall boundary conditions was no slip boundary condition and reinforced wall functions was combined in simulation of fluid flow near wall region.

4. The verification of the simulation results

This article respectively verified the influence velocity distribution, wall shear stress and liquid membrane distribution on the circumferential direction in horizontal pipe.

4.1. The verification of velocity

When the gas Reynolds number Re_{sg}=7.72×10^6, comparing the simulation results of average velocity distribution with the theoretical results in horizontal pipe.

Based on a lot of experimental results [1], the distribution of turbulent velocity in circular tube can be calculated as Eq. (4-1).

$$u = u_m \left(\frac{y}{r}\right)^n \tag{4-1}$$

Where u is average velocity of the fluid; u_m is the maximum velocity of the fluid in the tube axis; r is the radius of the pipe; y is radial distance from the wall; n is the index, for the hydraulic smooth tube, when $\mathrm{Re} < 10^5$, $n = 1/7$. When $10^5 < \mathrm{Re} < 4\times10^5$, $n = 1/8$ .and for rough tube, $n = 1/10$.

Figure 3 shows the velocity comparison between simulation and the theory in horizontal pipe. Simulation result shows a good agreement with the theoretical data, and the maximum error is 10.7%.

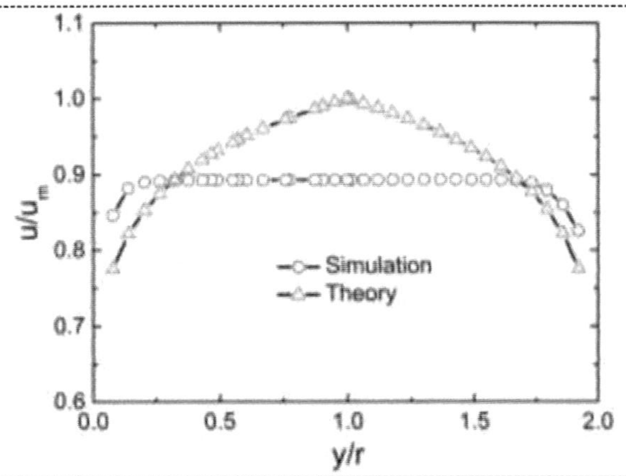

Figure 3. The velocity comparison between simulation and theory.

4.2. The verification of wall shear stress

When the gas Reynolds number Re_{sg}=7.72×10^6, the wall shear stress of simulation compared with the theoretical in horizontal pipe.

The original work shows [2, 3] that when Re_{sg} is more than 10^6, the wall shear stress is as follows:

$$\tau_w = \frac{0.0296\rho U^2}{\mathrm{Re}_x^{\;0.2}} \tag{3-8}$$

where Re_x is the local Reynolds number, U is the velocity in the center of pipe, τ_w is wall shear stress, ρ is the gas density under the condition.

Figure 4 is the comparison of wall shear stress between simulation and theoretical values. The conclusion is present that the simulation results agree well with the theory data. The maximum error is 4.0%.

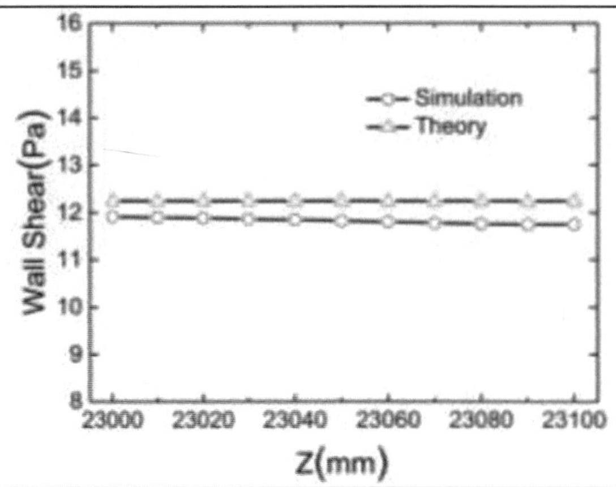

Figure 4. The wall shear stress between simulation theory

5. Phase flow field analysis

The gas phase flow field distribution in wet natural gas pipeline is done to set the foundation of further study of liquid film distribution in the pipeline. Cross sections of the pipe are chosen to be analyzed, which is shown in Figure 5.

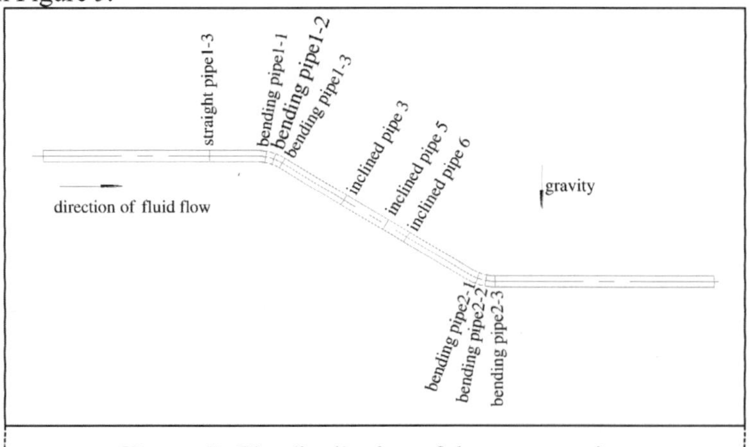

Figure 5. The distribution of the cross section

5.1. Gas streamlines

In this paper, the pipeline bending angles were all set to be 30.78°. The velocity distribution of bend section is chosen to be studied because the flow in bending pipe is complex.

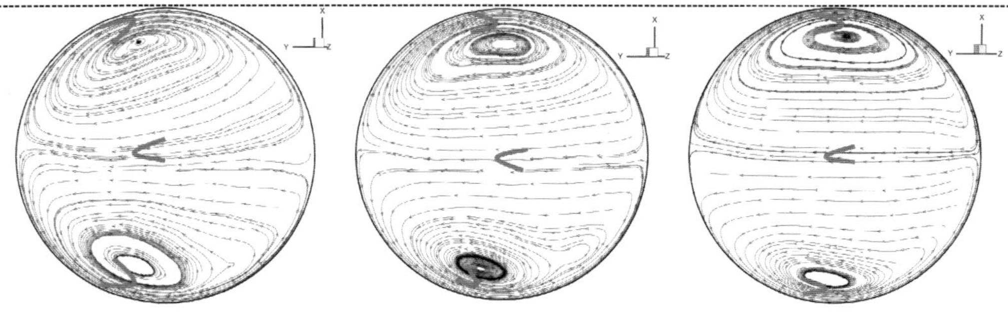

a) The section of 10° b) The section of 20° c) The section of 30.78°

Figure 6. Gas streamlines at the up bending section (Left is the outside lateral, right is the inner side).

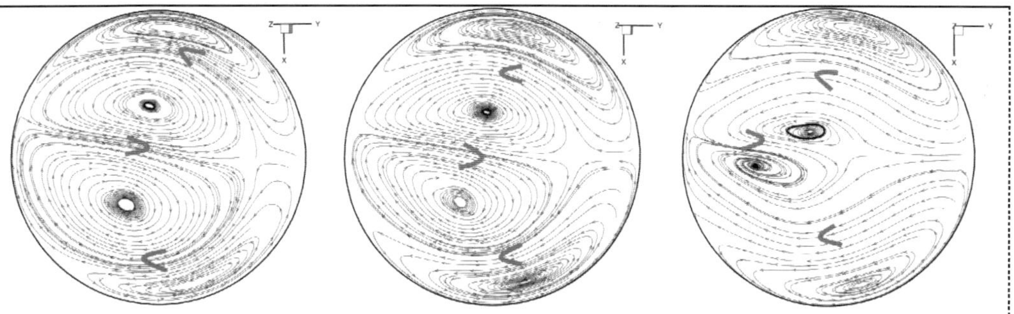

a) The section of 10° b) The section of 20° c) The section of 30.78°

Figure 7. Gas streamlines at the down bending section (Left is the outside lateral, right is the inner side).

As can be seen from above streamline distribution, there are a pair of dean vortex at the bending section of entrance, and the vortex move to the inside gradually. There are two pairs of dean vortex at the bending section of output, one pair at lateral, and other at inner side. The outside vortex moves to the side wall gradually, and the inner side vortex moves close to the center gradually, in the meantime, they became smaller gradually. Dean vortex is a common physical phenomenon [4]. Due to the effects of centrifugal force, viscous fluid produced a pair of symmetric vortex under the mainstream velocity in the bending pipe, and in contrast to the mainstream direction of cross section.

5.2. Velocity distribution curved line

In order to study the speed changes with the Reynolds number, three typical sections in the pipe are analyzed, which are the straight pipe 1-3, bending pipe 1-2, inclined pipe 5(As shown in figure.6).

Figure.8 represents speed curve of straight pipe 1-3, the cross section location of fully turbulent area in straight pipe. There is boundary layer near the wall, and velocity distribution of the boundary layer is more complex. The speed tends to increase along with the increase of radial distance. Velocity is close to the entrance speed in fully turbulent area, and in line with turbulent velocity distribution of the straight pipe.

Figure.9 is speed curve of bending pipe 1-2(the section of up bend 20°). The speed change is bigger at bending pipe, the velocity decreases from inner side to lateral outside in bending pipe. The fluid is squeezed to lateral wall side of pipeline because of the centrifugal factor, in the meantime, the diffusion effects on the outside lateral of bending pipe are generated, and the acceleration channel on inner side is appear.

Figure.10 shows speed curve of inclined pipe 5. The speed has a tendency to increase from down wall side to up wall side in the fully turbulent area. Because of dean vortex moving to down wall side under gravity, fluid is squeezed to the lateral wall, diffusion effect on the outside lateral of bending pipe is generated, and the acceleration channel is formed on inner side. So the velocity near the up wall side is faster than the velocity near the down wall side. Y is radial distance away from the down wall side on diagram.

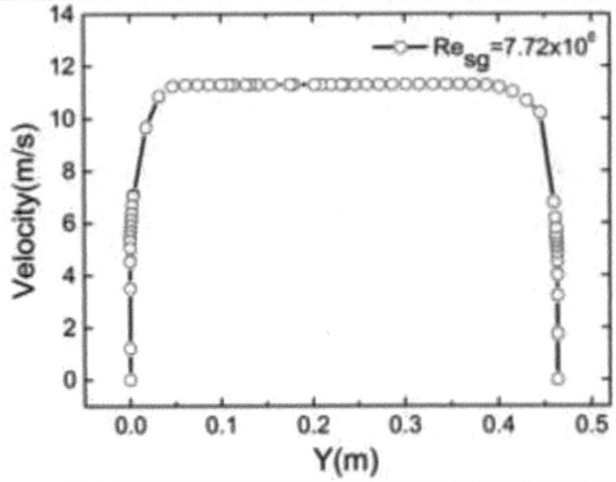

Figure 8. The velocity curve of straight pipe 1-3(Y is radial distance away from the down wall side on diagram)

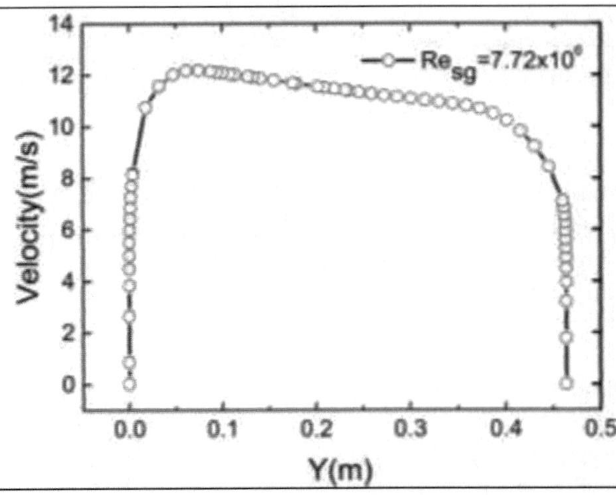

Figure 9. The velocity curve of bending pipe 1-2(Y is radial distance away from the down wall side on diagram)

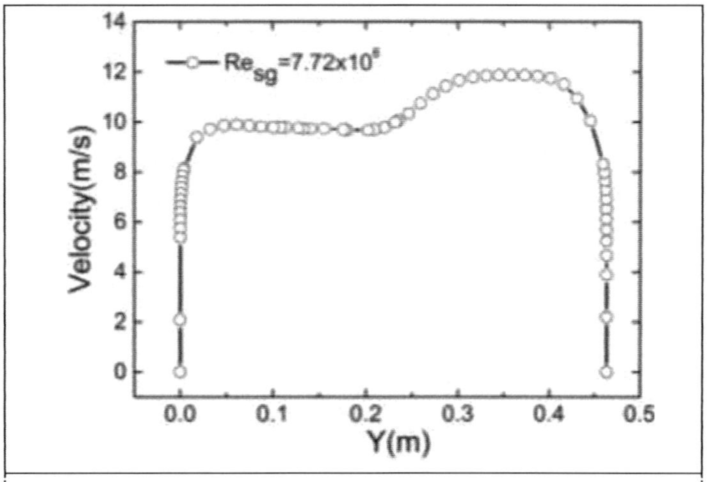

Figure 10. The velocity speed curve of inclined pipe 5(Y is radial distance away from the down wall side on diagram)

6. Phase flow field analysis

6.1. Liquid Film Distribution

Since the liquid film is very thin in the pipe, the liquid film thickness is magnified 1000 times in this article in order to clearly show the liquid film. There are nine characteristic sections (Figure.6) to be studied. They respectively are straight pipe 1-3, bending pipe 1-1, bending pipe 1-2, bending pipe 1-3, inclined pipe 3, inclined pipe 6, bending pipe 2-1, bending pipe 2-2 and bending pipe 2-3.

Figure 12 shows the distribution of liquid film thickness of characteristics section. Liquid film distributed on the circumference of tube wall, because it was influenced by the mechanism of the gas secondary flow, the principle of disturbance wave suction, droplets carrying and settlement mechanism. Figure (a) shows that liquid film is mainly distributed in the under part of pipe because of the influence of the gas phase flow field and the gravity. The liquid film in the upper part of pipe is very thin. The location in Figure (b), (c), (d) is up bending 10°, 20° and 30.78° of cross section. The liquid film is thinner at this time according Figure (b), (c), and there is a big bump in Figure (d). Because of the centrifugal force, liquid moved from inner side to lateral side in bending pipe. In addition, the gas phase flow field driven liquid flow, and the movement of gas fluid squeezed liquid, lead to the liquid film bulged. Fluid went through a tilt angle when the liquid moved from straight pipe to bending pipe. In F.Maroteaux's [5] article, there is a part of the liquid film peeled and forming droplets when the liquid film flows through a certain angle. The droplets made a horizontal cast exercise in bending pipe, and landing in the under part of pipe after peeling off position. Therefore, the liquid film thickness before part of the bending pipe is thinner than the straight pipe. Figure (e) and (f) shows the liquid film thickness in the bending pipe. The liquid film concentrated on the down part of the wall under gravity, and the thickness increases gradually. The location in Figure (g), (h), (i) is down bending 10°, 20° and 30.78° of cross section. The distribution is similar to the up bending pipe. Under the accelerated speed of inclined pipe, the inlet's velocity was much bigger than the one on the up bending pipe. Therefore the liquid film thickness increased gradually.

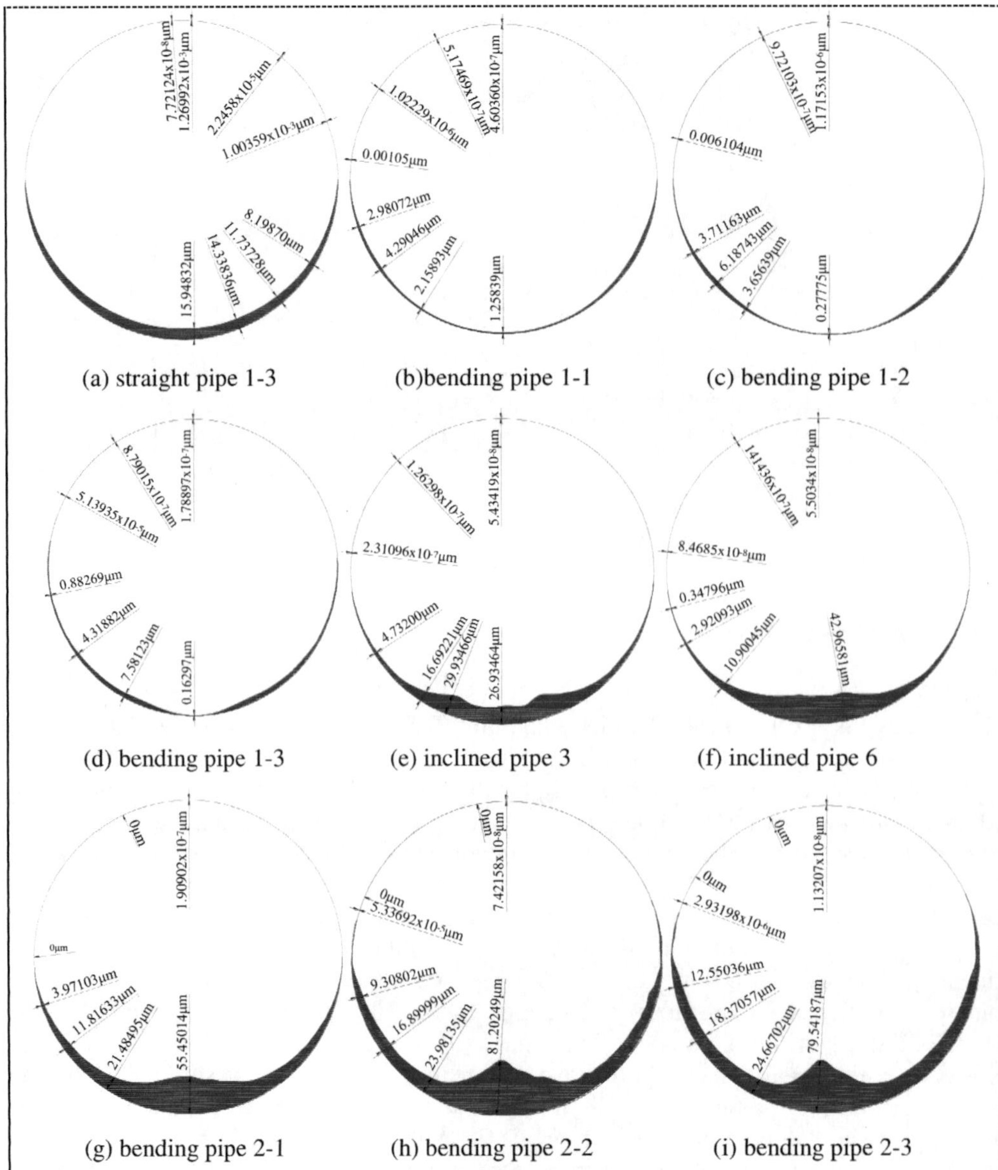

(a) straight pipe 1-3 (b)bending pipe 1-1 (c) bending pipe 1-2

(d) bending pipe 1-3 (e) inclined pipe 3 (f) inclined pipe 6

(g) bending pipe 2-1 (h) bending pipe 2-2 (i) bending pipe 2-3

Figure 11. The distribution of liquid film thickness
(Up is the upper wall, down is the down wall).

6.2. The effects of wall shear stress on liquid film distribution

The wall shear stress was used to verify the simulation results in former part. The conclusion was presented that the wall shear stress increased with the gas Reynolds number. In this section, the relationship between wall shear stress and the distribution of the liquid film was mainly explored.

The curve of wall shear stress and liquid film thickness of straight pipe and inclined pipe were drawn as it is seen in 'Figure 12'.

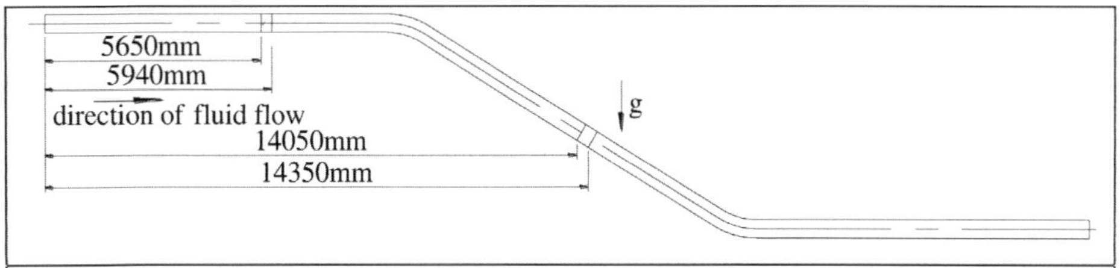

Figure 12. Intercepted section of the straight pipe and inclined pipe.

Figure 13 and Figure 14 are curve of the wall shear stress and the liquid film thickness along the axial direction. Z is the horizontal distance away from the entrance. Figure 13 shows the simulation results is similar to analysis of Xuehu Ma [6] professor, that is the wall shear stress affected the liquid film thickness. The fluctuation curve of liquid film was connected with distribution of wall in straight pipe, and is cyclical. Wall shear stress reached peak in front of wave crest of the liquid film, and then fell rapidly. Figure 14 shows wall shear stress fluctuation in inclined pipe. It hasn't caused too much influence to liquid film thickness. Liquid flow was affected by gas-liquid interface shear stress, weight component along the inclined tube, and wall shear stress.

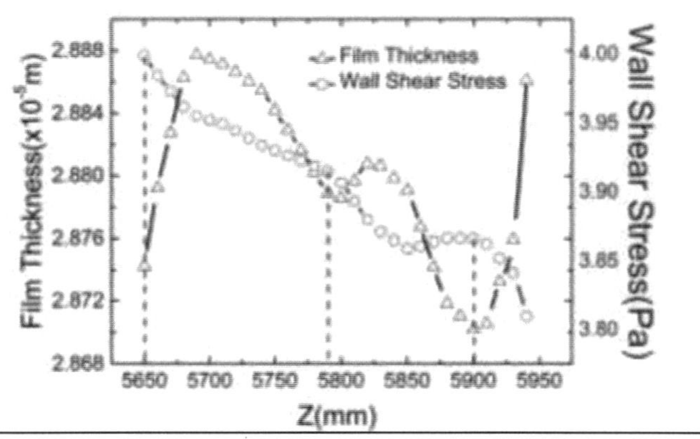

Figure 13. The changing curve of the wall shear stress and the liquid film thickness on axial direction in straight pipe(Z is the horizontal distance away from the entrance)

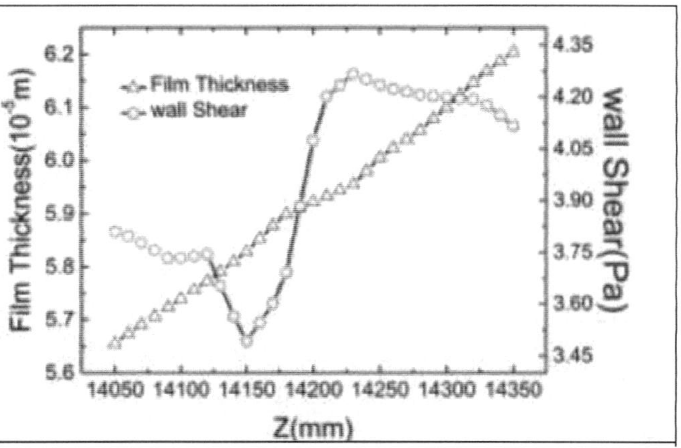

Figure 14. The changing curve of the wall shear stress and the liquid film thickness on axial direction in inclined pipe(Z is the horizontal distance away from the entrance)

7. Conclusion

The main conclusions are as follows:

(1) The dean vortex was caused by centrifugal force in the bending pipe, and gradually moved to the outside with the increase of bend angle.

(2) Liquid film distributed unevenly on circumferential direction of pipe wall. The distribution of liquid film was effected by gravity, the movement of the gas phase flow field, and centrifugal force.

(3) The fluctuation curve of liquid film had some relationship with distribution of wall in straight pipe, and appeared cyclically. Wall shear stress reached peak in front of wave crest of the liquid film. The shear stress didn't cause too much influence on the fluctuation curve of liquid film.

(4) This paper successfully simulated the distribution of the liquid film in large diameter pipe which contained low liquid, supporting the next step of pipeline corrosion research.

References

[1] Yuan X E 2000 *Engineering fluid mechanics* [M]. Peking: Petroleum Industry Press. 2000:120-121.

[2] Zhang Z X, Dong Z N 2011 *Viscous Fluid Mechanics (second edition)* [M]. Peking: Tsinghua University Press. 2011.6.

[3] Schlichting H 1979 *Boundary-Layer Theory* [M]. New York: McGraw-Hill.

[4] Zhu H, Zang H H, Li C and Peng Biao 2009 Numerical studies of movement rule of dean vortex [J]. *Academic journal of Hunan Industry University*. **23**(43):58-62.

[5] Maroteaux F 2002 Liquid film atomization on wall edges-separation criterion and droplets formation model [J]. *Journal of Fluids Engineering*, **124**:565-575.

[6] Ma X H, Bo S S. The influence factors of falling liquid film fluctuation analysis [J]. *Academic journal of chemical engineering of colleges and universities*

Acknowledgment

This work was supported by the National Natural Science Foundation (NSFC) of China (No.51276200, No.41106068), Qingdao Science and Technology Program of China (14-2-4-63-jch) and the Fundamental Research Funds for the Central Universities (15CX06045A).

IOP Conf. Series: Materials Science and Engineering **129** (2016) 012020 doi:10.1088/1757-899X/129/1/012020

Influence of clearance model on numerical simulation of centrifugal pump

Z Wang[1], B Gao, L Yang and W Q Du

Jiangsu University, Zhenjiang, Jiangsu, China

E-mail: initial_wz@sina.com

Abstract. Computing models are always simplified to save the computing resources and time. Particularly, the clearance that between impeller and pump casing is always ignored. But the completer model is, the more precise result of numerical simulation is in theory. This paper study the influence of clearance model on numerical simulation of centrifugal pump. We present such influence via comparing performance, flow characteristic and pressure pulsation of two cases that the one of two cases is the model pump with clearance and the other is not. And the results show that the head decreases and power increases so that efficiency decreases after computing with front and back cavities. Then no-leakage model would improve absolute velocity magnitude in order to reach the rated flow rate. Finally, more disturbance induced by front cavity flow and wear-ring flow would change the pressure pulsation of impeller and volute. The performance of clearance flow is important for the whole pump in performance, flow characteristic, pressure pulsation and other respects.

1. Introduction

Clearance that between pump block and impeller comprises two parts, one is front cavity and the other is back cavity. Mostly, it is convenient for mesh generation and numerical simulation to simplify the 3-D model. And front cavity enable a little of fluid with high energy pass through wear-ring to inlet of impeller, which would generate a disturbance for inlet flow situation. According to some literatures [1,2], The flow conditions in these small axial gaps are of significant importance for a number of effects such as disk friction, leakage losses or hydraulic axial thrust to name but a few. The front cavity between the impeller and the casing is set to achieve an acceptable pump efficiency.

The whole flow field model showed a higher accuracy than the simple model, and the flow pattern of two models were also different [3]. If this clearance is allowed to grow, either through corrosion or erosion, the pump will lose efficiency [4]. The size and shape of the wear-ring clearance directly affects the impeller sealing effect, and have a major impact on the overall performance of the pump [5]. According to [6], the effect of the clearance size change is mainly concentrated in the front and rear chamber, also in the exit of the clearances, while the impact is not significant in other parts of the pump.

[1] E-mail: initial_wz@sina.com

Through theoretical analysis and test, the fluid force of the annular clearance seal of centrifugal pump is studied, which is important for vibration of centrifugal rotor [7]. The head and total efficiency of the centrifugal pump increase as the clearance value of wear-rings narrows, which is induced by three reasons. Such as the energy dissipation decrease within the impeller, the impact of secondary flow at the inlet of impeller on the mainstream weakens slowly, the eccentric whirl of centrifugal pump is dampened; the front shroud leakage diminishes [8].

The size of the clearance would have a greater influence on the distribution of flow pressure and velocity in pump chamber. As the value of the clearance increased, the pressure coefficient would decrease and the velocity increase [9]. Combined with formulas of hydraulic efficiency and volume efficiency, it can be seen that the pump efficiency decreases with the increasing clearances of wear-rings even though the mechanical losses decrease with the increasing of leakage rate [10]. The modification effects manifest in the front and rear chambers, also in the outlet of clearances, while insignificant in other sections of the pump. As the clearance widens, a low pressure area in the chamber tends to expand toward the volute and a high-pressure zone near the clearance outlet has a tendency to expand toward the impeller inlet [11]. There are many reasons for difference of forecasting losing, and especially the simplifying of prototype pump is considered as the dominant factor of large deviation in numerical simulation results [12]. The structure of front and back cavities is very complex on centrifugal pumps, especially the size of front and back wear-rings and balancing hole is tiny so that there are few literatures about numerical simulation of full flow field [13, 14].

2. Numerical simulation

2.1 Model pump

We choose a centrifugal pump with low specific speed, closed impeller and spiral volute to study the effect of clearance to internal and external characters of pump. The part parameters of model pump is shown as follows.

Table 1 part parameters of model pump.

parameters	value
Head (m)	20
Flow rate (m3/h)	55
Speed (r/min)	1450
Impeller type	closed
Blade number Z	6
Specific speed ns	69
Tongue angle ($^{\circ}$)	25
Area ratio of import and export of pump	1

Model pump is built by three dimension software, and there are two cases that case 1 match with the front cavity and back cavity but case 2 is without clearance. The components of model pump is shown in Figure.1 as follow.

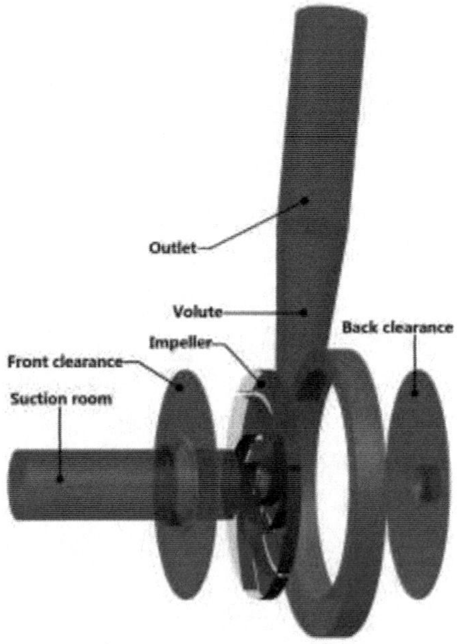

Figure 1. The 3-D computations.

2.2 The method of numerical simulation

The structural meshes of all components are generated by ICEM which is a part of ANSYS software. Based on the non-compressible Reynolds time averaged N-S equations and the RNG k-ε turbulence model, the finite volume method is applied to the equation discretization. And second-order central difference scheme is applied to the diffuse term discretization, second order upwind discretization is used for the convection term. RNG model is a semi-empirical turbulence model, and through solution of two transport equations of the turbulence kinetic energy k and turbulence dissipation ε separately to decide the turbulent velocity and turbulence scale. Separated implicit method is used for the calculation, and SIMPLE algorithm is adopted for the pressure-velocity coupling.

Then LES (large eddy simulation) is used for the transient computing in order to gain pressure pulsation at different monitoring points. Rotor properties is changed to mesh motion, time step is calculated by speed. Four monitoring points are assigned to different place of model pump as follow.

Table 2 Distribution of monitoring points.

Monitoring points	Position
P1	Inlet of impeller
P2	Outlet of impeller
P3	Tongue of volute
P4	Outlet of volute

The midline of span which is from shroud to hub (channel midline) is significant for pressure pulsation to analyse the influence of clearance to impeller, so P1 and P2 monitoring points locate at channel midline. It is one of the dominant factors for impeller operating that the state of inlet flow, which is the reason of placing the monitoring point P1. And the pressure pulsation of impeller outlet is an important basis for the power capacity of impeller. In addition to presenting energy transmitting, the monitoring point P2 could present the rotor-stator interaction. Thus the monitoring point P3 is used for obtained the influence of the front and back cavities to the tongue region. Finally, the monitoring point P4 could present the hydrodynamic characteristic of volute outlet.

3. Analysis of performance

For both case 1 and case 2, nine condition points are computed that range from 0.2 to 1.8 Q/Qv in order to present performance of model pump more completely. Thus the head curve, power curve and efficiency curve of two cases are gained, and they are shown in Figure.2. In this Figure, an obvious disparity could be found in all curves. Hump phenomenon appears in head curve with clearance in the small flow rate condition, but it doesn't appear in head curve without clearance. Because front cavity enable a little of fluid with high energy pass through the wear-ring to inlet of impeller and merge with low energy fluid so that fluid came from outlet of impeller loss a part of energy. The head of both case 1 and case 2 is basically identical in design condition, but case 1's head is obviously lower than case 2's in large flow rate condition. The energy losing phenomenon appears in large flow rate condition too.

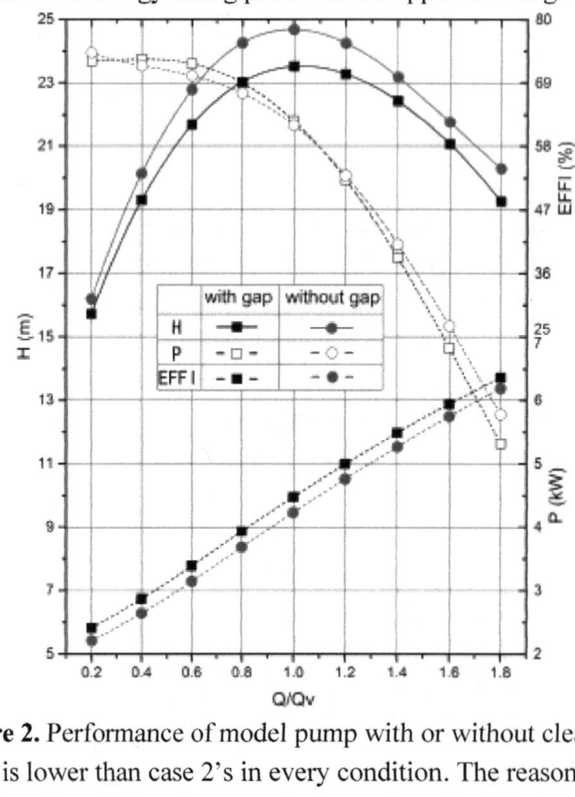

Figure 2. Performance of model pump with or without clearance.

The power of case 1 is lower than case 2's in every condition. The reason similarly is extra loading generated by clearance flow in addition to energy losing at impeller inlet region. Impeller needs more energy to ensure that the fluid of outlet carry enough energy due to the energy losing. And extra loading would consume power so that shaft power of case 1 is higher. The result of these circumstances is the enormous efficiency difference of two cases in all condition. The efficiency of model pump with clearance is obviously lower than another. Thus it is an important factor for numerical simulation that whether or not there is clearance model.

IOP Conf. Series: Materials Science and Engineering **129** (2016) 012020 doi:10.1088/1757-899X/129/1/012020

4. Analysis of flow characteristic

The projection of absolute velocity on meridional plane could demonstrate the difference of flow characteristic and it could present the velocity evolution of fluid from inlet to outlet. As shown in Figure.3, each group of velocity vector is in same flow section, and each group include ten velocity vectors that express the projection of absolute velocity on meridional plane. And we choose three conditions to demonstrate the changing of velocity in different conditions, so the influence of front and back cavity to flow characteristic is presented. Case 1, which is the model pump with clearance, includes Figure. (a), (b), (c) expressing the flow character in 0.6, 1.0 and 1.4 Q/Qv conditions. And case 2 is similar to case 1.

It can be seen in Figure. (a), (b), (c), velocity magnitude increases with the increasing of flow rate in every vector group on meridional plane. The chaos of velocity direction in outlet of impeller in small flow rate condition changes to gather until in large flow rate condition. The first three group, which from inlet to outlet, have similar laws that the velocity magnitude of shroud is bigger than that of central or hub. But beginning from the fourth group, velocity magnitude changes slightly from shroud to hub. The axial velocity is dominant for first three group, but the radial velocity is major component for any other group. It is induced by fluid coming from front wear-ring with high energy that the velocity vector nears to the shroud in group 0-0 is not horizontal. And as the flow rate increases, this velocity vector get horizontal since the increasing of kinetic energy in inlet of impeller.

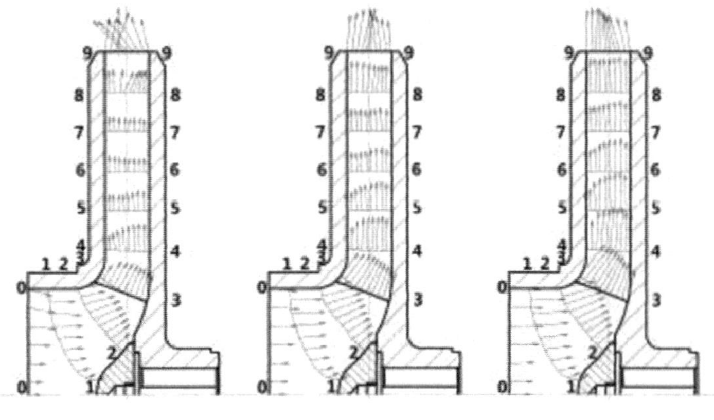

Case 1: (a) 0.6 Q/Qv (b) 1.0 Q/Qv (c) 1.4 Q/Qv

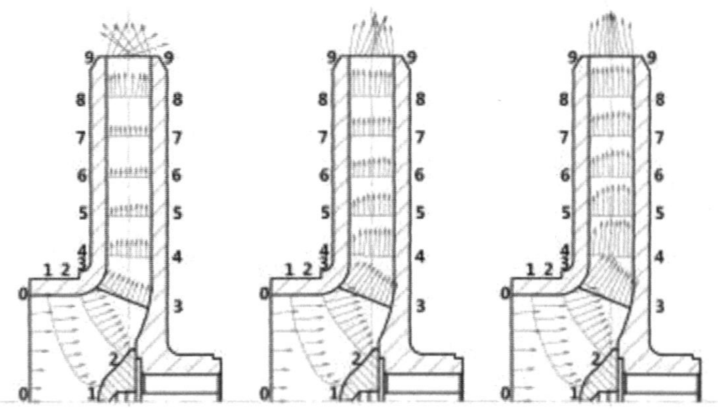

Case 2: (d) 0.6 Q/Qv (e) 1.0 Q/Qv (f) 1.4 Q/Qv

Figure 3. Projection of absolute velocity on meridional plane.

A same law appears in case 2 and the velocity direction in outlet of impeller in small flow rate condition is more disorder. It is obvious that the velocity magnitude in all group, especially in 0.6 Q/Qv condition of case 2 is smaller than that of case 1. Because leakage of wear-ring would reduce flow quantity and it is necessary for fluid in impeller to carry more kinetic energy and guarantee the flow quantity. It could be deduced from velocity triangle that the decrease of the meridional velocity decreases the head.

5. Analysis of pressure pulsation

Pressure pulsation that induced by rotor-stator interaction and other hydrodynamic characteristics is one of the dominant factors for vibration in centrifugal pumps. 4 monitoring points is used for obtaining the pressure pulsation. As shown in Figure.4, there is pressure pulsation at different monitoring points. And it can be seen in Figure. (a), the pressure pulsation amplitude at blade passing frequency (BPF) of case 1 is obviously higher than that of case 2. And at low frequency region, more signals are obtained by numerical simulation in case 1, which demonstrates the pressure pulsation at low frequency region is more complex after add the clearance model. The similar law appears in Figure. (b). As the picture shows, the pressure pulsation amplitude at BPF and its harmonic of case 1 is higher than that of case 2. And the signals of low frequency region is more complicated after add the clearance model. The complex flow region which comprises impeller outlet, volute inlet and front cavity inlet influence the flow characteristic of impeller outlet. So the pressure pulsation of impeller outlet is not affected only by rotor-stator interaction but also by the flow characteristic of front cavity.

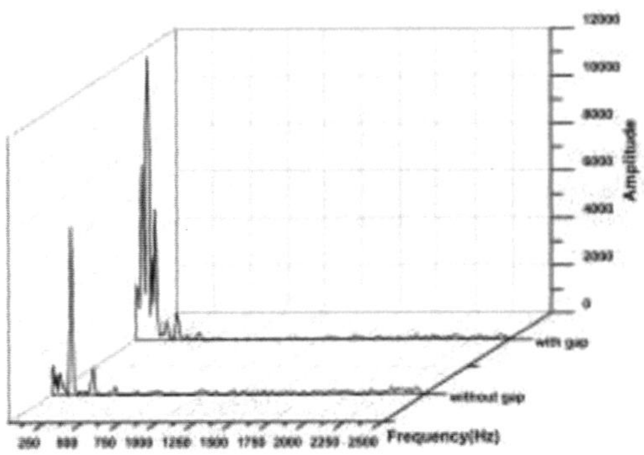

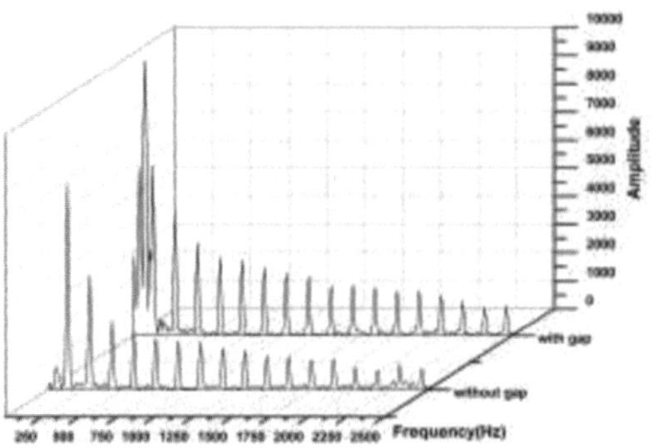

(a) Pressure pulsation at monitoring point P1 (b) Pressure pulsation at monitoring point P2

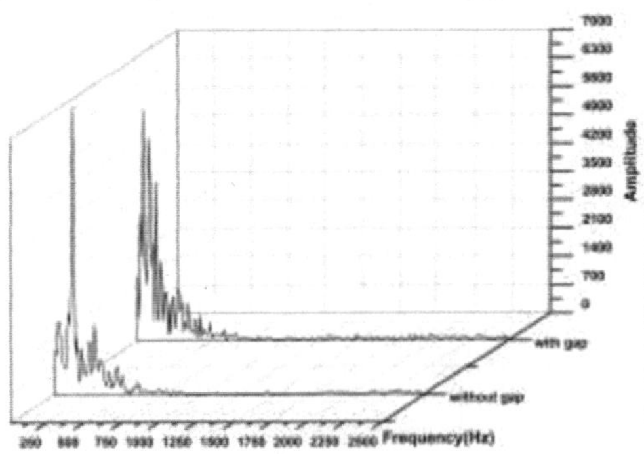

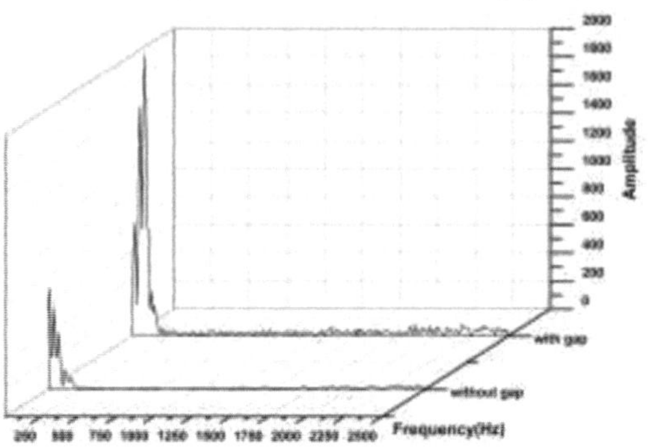

(c) Pressure pulsation at monitoring point P3 (d) Pressure pulsation at monitoring point P4

Figure 4. Pressure pulsation at different monitoring points.

The tongue region is an important area in which the fluid in volute and impeller flow out to diffuser. And impeller-volute interaction generate in tongue so that the pressure pulsation in this place is very complicated and messy. As the Figure.(c) shows, although the amplitude on the BPF of case 2 is higher than that of case 2, the peaks of case 2 is far more than that of case 1 in low frequency region. And in high frequency region which is bigger than 1500 Hz the amplitude of case 1 is bigger than that of case 2. Because of more messy flow in tongue area that computed by full flow field with clearance, the effect of impeller to tongue is weaker.

The flow situation of pump outlet is expected to be uniform, so the pressure pulsation of this region is expected to be small and smooth. The Figure. (d) shows the difference of pressure pulsation of pump outlet between case 1 and case 2. The amplitude on BPF of case 1 is 3 times as big as that of case 2, and the amplitude on high frequency region of case 1 is higher than that of case 2. Therefore the front cavity could influence the flow characteristic of pump outlet. And the energy recovery of case 1 is lower than that of case1, which is one of the reasons for lower head and efficiency.

6. Results

Relevant numerical simulation were performed in comparing performance, flow characteristic on meridional surface and pressure pulsation at different monitoring points. So the conclusions as follow:

(1) The efficiency curve of the model pump with front and back cavities is lower than that of model pump without front and back cavities. And the power curve of the model pump with front and back cavities is higher than that of model pump without front and back cavities. The head gap of two cases is not obvious nearby the design condition, but the model without front and back cavities does not present the hump phenomenon in small flow rate condition.

(2) The projection of absolute velocity on meridional surface in impeller outlet is chaotic in small flow rate condition whether the model pump have or not front and back cavities. With the increasing of flow rate, the velocity become more agglomerate. And the velocity magnitude of the model pump with front and back cavities is lower than that of the model pump without front and back cavities.

(3) The pressure pulsation amplitude of the model pump with front and back cavities is higher than that of the model pump without front and back cavities at impeller inlet, blade outlet and volute outlet. But the law of pressure pulsation amplitude at tongue is opposite to the others.

The clearance flow characteristics would influence the many aspects of centrifugal pump, such as the performance, absolute velocity on several surfaces, pressure pulsation and other internal and external characteristics. It is necessary for numerical simulation of full flow field to compute the front and back cavities, wear-rings and other clearances, which could improve the accuracy. This paper provides a reference for no-clearance numerical simulation to forecast the results via comparing the difference of two cases.

Reference

[1] Will B C, Benra F K and Dohmen H J 2012 Investigation of the flow in the impeller side clearances of a centrifugal pump with volute casing[J], *Journal of Thermal Science*, **Vol.21**, No.3 (2012) 197-208.

[2] Will B C, Benra F K, Dohmen H J 2010 numerical and experimental investigation of the flow in the side cavities of a centrifugal pump, *The 12th International Symposium on Transport Phenomena and Dynamics of Rotating Machinery*, ISROMAC13-2010-0002, Honolulu, Hawaii, USA.

[3] Li X J, Yuan S Q, Pan Z Y, Yuan J P, Si Q R 2013 Numerical simulation of whole flow field for centrifugal pump with structured grid [J]. *Journal of Agricultural Machinery*. **Vol.7**, No.33(2013,7) 49-54.

[4] Shiels S 1997 The importance of running clearances in centrifugal pumps[J], *Centrifugal Pump Academy*. 1997(369):35-37.

[5] Guan X F 2011 *Modern Pumps Theory and Design Beijing* (Beijing, China Astronautics Press).

[6] Zhao W G, Li Y B, Wang X Y, Sun J P and Wu G X 2012 Research on the effect of wear-ring clearances to the performance of centrifugal pump[J], *China Academic journal*. **15 (7)**, p. 072018.

[7] Black H F 1969 Effects of hydraulic forces in annular pressureseals on the vibrations of centrifugal pump rotors [J]. *Journal of Mechanical Engineering Science*, **11(2)** : 206-213.

[8] Pan Z Y, Chen S X, Zhang D Q, Tu Q, Li Y J 2012 Simulation and experiment of the effect of clearance of impeller wear-rings on the performance of centrifugal pump [J]. *Fluid Machinery*. **Vol.40, No.11**: 10-14. (In chinese)

[9] Zhao W Y, Zhao S, Wang L, Ma P F 2013 Influence of seal ring clearance of centrifugal oil pump on flow characteristics of pump chamber [J]. *Journal of LanZhou University of Technology*. **Vol.39 No.1**,2013: 33-36. (In chinese)

[10] Wang Y, Zhang X 2008 Effect of clearance of impeller wear-rings on efficiency of low specific speed centrifugal pump [J]. *Drainage and Irrigation Machinery*. **Vol.26, no.6**, 2008(11): 27-30. (In chinese)

[11] Zhao W G, Wu G X, Li Y B, Wang X Y 2014 Study on effects of wear-rings clearance modifications on performance of centrifugal pump [J]. *Journal of Htdroelectric Engineering*. **Vol.33 No.5**, 2014: 211-215. (In chinese)

[12] Zhou L, Wang Z 2008 Numerical simulation of cavitation around a hydrofoil and evaluation of a RNG k- ε model [J]. *Journal of Fluids Engineering*, **130(1)**: 1～7.

[13] Watanabe S, Ikeda A, Nakamura I, et al. 2010 Observation of unsteady cavitating flow in flat plate cascades [C] *25th IAHR Symposium on Hydraulic Machinery and Systems*, Paper 012047, 2010.

[14] Friedrichs J, Kosyna G 2003 Unsteady PIV flow field analysis of a centrifugal pump impeller under rotating cavitation [C] *Fifth International Symposium on Cavitation*, Cav 03-OS-6-005, 2003.

Analysis of silt abrasion of the impeller ring in a centrifugal pump with J-grooves

Z D Qian, Z Y Wang[1], Z W Guo, J Dong and J Lu

State Key Laboratory of Water Resources and Hydropower Engineering Science, Wuhan University, Wuhan 430072, P. R. China

E-mail: zhiyuan_wang@whu.edu.cn

Abstract: The water flow and movement of silt in a prototype double-suction centrifugal pump was simulated using an Euler-Lagrange multiphase flow model. J-Grooves were adopted to protect the impeller ring from silt abrasion. The influence of J-grooves on the silt concentration and pump efficiency was analyzed. The results show that the radial component of the relative velocity around the impeller ring is too low to move the silt out of the spacing between the impeller plate and the casing. The high silt concentration around the impeller ring is the major contributor to silt abrasion of the impeller ring. The J-grooves induce two strong vortices, which increase the radial component of the relative velocity of water and reduce the silt concentration around the impeller ring, but additional friction losses are introduced and the pump efficiency is decreased. Optimization of the number and shape of J-grooves decreases losses in the efficiency of the pump, and effectively protects the impeller ring. Case 4 was found the most effective configuration in this study.

1. Introduction

The double-suction pump is widely used for pumping irrigation in the Yellow River irrigation district because of its large flow rate and high head, but the silt concentration in the Yellow River is so high that the flow passage components of the pump are easily worn. The impeller ring is usually the most seriously affected by silt abrasion. Its lifetime is only one-third of the impeller. Silt abrasion occurring in hydraulic machinery is a problem for users and designers, and consequently is the subject of considerable research. Liu et al. [1] studied the movement rule of solid particles in a single-suction centrifugal pump using high-speed photography. Yoshiro and Kazuyuki [2] studied silt abrasion of different coating materials. CHEN et al. [3] analyzed the reason for the different appearance of damage at different positions on the hydrofoil. Most research has focused on the silt abrasion of the blade, while neglecting the impeller ring. Kurokawa et al.[4] and Saha et al.[5,6]proposed shallow grooves, named "J-Grooves", mounted on the casing wall of turbo-machinery in the pressure gradient

[1] State Key Laboratory of Water Resources and Hydropower Engineering Science, Wuhan University, Wuhan 430072, P. R. China

direction of the main flow, as a countermeasure to various abnormal flow phenomena. For the present problem, it is believed that the J-groove is an effective method for suppressing silt abrasion of the impeller ring, because the J-groove can increase the radial component of the relative velocity of water, which helps to reduce the silt concentration around the impeller ring.

In recent years, as a new powerful and effective research tool, computational fluid dynamics (CFD) has been widely used in the flow of pumps [7, 8], and it has become an effective and accurate tool in pump design for engineering applications [9-11]. In this study, the water flow and movement of silt in a prototype double-suction centrifugal pump was simulated using an Euler-Lagrange multiphase flow model. J-Grooves were adopted to suppress silt abrasion of the impeller ring in the double-suction centrifugal pump. The influence of the shape and number of J-grooves on silt abrasion of the impeller ring was analyzed.

2. Numerical simulation method

2.1. Mathematical model

The three-dimensional Reynolds-averaged Navier-Stokes equations for incompressible flow are as follows [12]:

$$\frac{\partial \rho}{\partial t}+\frac{\partial}{\partial x_i}(\rho u_i)=0 \tag{1}$$

$$\frac{\partial(\rho u_i)}{\partial t}+\frac{\partial(\rho u_i u_j)}{\partial x_j}=-\frac{\partial p}{\partial x_i}+\frac{\partial}{\partial x_j}\left(\mu\frac{\partial u_i}{\partial x_j}\right)+\frac{\partial \tau_{ij}}{\partial x_j} \tag{2}$$

Considering the rotating flow inside the pump impeller, the RNG $k-\varepsilon$ model, which is effective for strongly rotating flow, was adopted in this simulation [13].

The Euler-Lagrange multiphase flow model was used to simulate silt motion; it can describe the interaction between silt and water and the collision of silt particles and its governing equation takes the form [14]:

$$\frac{d\vec{u}_p}{dt}=F_D\left(\vec{u}-\vec{u}_p\right)+\frac{\vec{g}\left(\rho_p-\rho\right)}{\rho_p}+\vec{F} \tag{3}$$

The silt concentration in the flow passage is obtained when the silt concentration in the inlet and outlet is balanced. The Euler-Lagrange multiphase flow model can describe not only the interaction between silt and liquid but also the collision of silts. The silt concentration around the impeller ring is the major contributor to silt abrasion of the impeller ring. This was chosen as the major performance indicator for optimizing the J-grooves.

All simulations were conducted using Fluent 14.0 software. The SIMPLEC algorithm was used for the pressure-velocity coupling [15]. The second-order upwind difference scheme was used for the momentum, turbulent kinetic energy, and dissipation rate equations. The mass flow rate was defined at the inlet and the static pressure at the outlet. The roughness heights of the hub and casing were also considered [16]. The sliding mesh model was used to simulate the rotor-stator interaction in the pump.

2.2. *Physical model*

The computational domain is the flow passage from the inlet to the outlet, including the impeller and volute as shown in Figure 1. The diameter of the impeller was 0.765 m, the rated speed was 960 rpm, and there were 8 blades. The structures of the impeller are shown in Figure 2. The unstructured mesh was used to form the discrete computational domain for the double-suction centrifugal pump because of their complicated geometry. There were about 3,000,000 mesh cells in total.

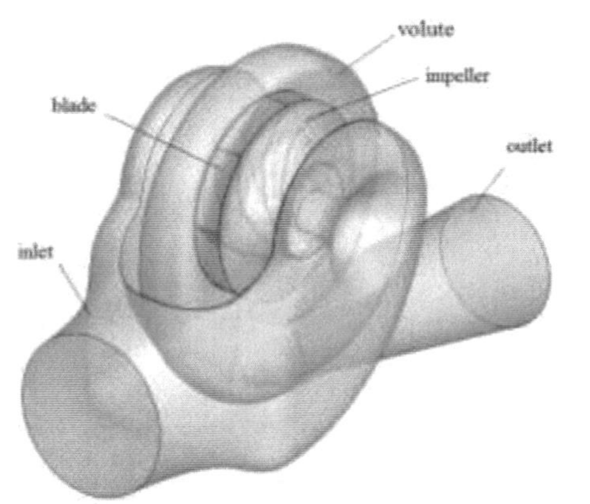

Figure 1. 3D perspective view of the double suction pump

Figure 2. 3D perspective view of the impeller

The operating conditions of the prototype double-suction centrifugal pump and silt are as follows in Table 1.

Table 1. Operating conditions of pump and silt

Discharge	Q=0.95 m^3/s
Total head	H=71.0 m
Rotational speed	n=960 rpm
Silt concentration	C=2.832 kg/m^3
Silt diameter	d=0.031 mm

For the given boundary conditions, the computational head of the prototype pump is H = 71.82 m, and the error was only 1.15 %. Therefore, the mathematical model and boundary conditions were acceptable.

3. Results and discussion

3.1. *Prototype pump*

The outer surface of the impeller plate and a section S1 were chosen to analyze the water flow and movement of silt in the gap between the impeller plate and the casing as shown in Figures 3 and Figure 4 shows the silt concentration on the impeller plate (units: kg/m^3), the silt concentration around the impeller ring was obviously higher than the mean concentration in the water, which was the primary cause for rapid silt abrasion of the impeller ring. Figure 5 shows the velocity of water flow in section S1. It was found that the water flowed into the gap, forming two long vortices. Figure 6 shows the relative velocity on the impeller plate. The radial component of relative velocity on the impeller plate was small. It can be concluded from Figures 5-6 that the silt particles can be easily deposited at the impeller ring and were difficult to extract, which resulted in a high silt concentration around the impeller ring.

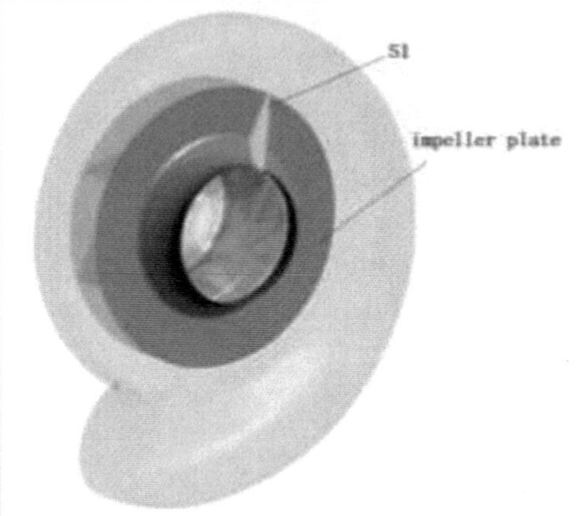

Figure 3. Positions of observations

Figure 4. Distribution of silt on the impeller plate

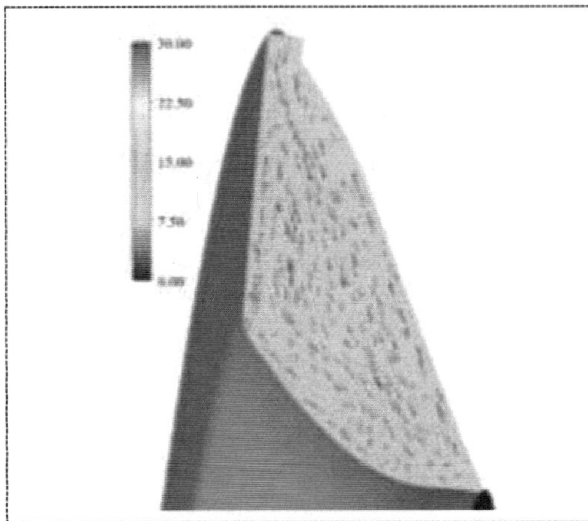

Figure 5. Velocity in section S1

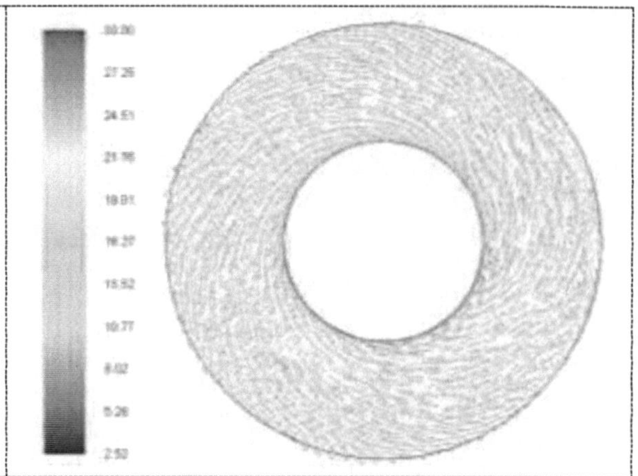

Figure 6. Relative velocity on the impeller plate

3.2. J-Grooves

The main reason for silt abrasion of the impeller ring was the high concentrations of silt particles around the impeller ring. According to the research in Reference 5, as shown in Figure 7(a), J-grooves were designed in the impeller plate. As shown in Figure 7(b), a is the dip angle of the groove, h is the depth, b_1 is the width of lower edge and b_2 is the width of upper edge. Four different cases are listed in Table 2.

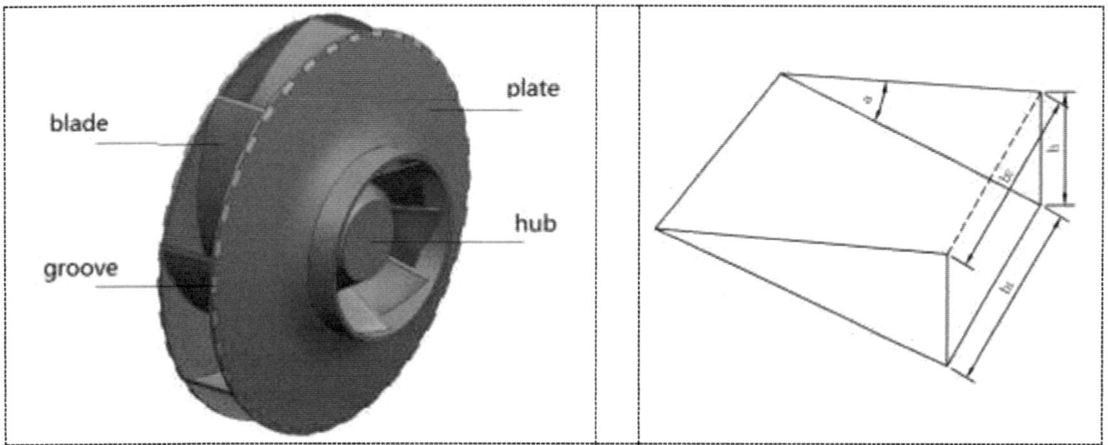

(a) (b)

Figure 7. Schematic view of the J-Groove

Table 2. Specifications of J-Grooves

J-Groove	Case 1	Case 2	Case 3	Case 4
Number	70	35	35	35
a [°]	45	45	30	30
h [mm]	5	5	5	5
b_1 [mm]	20	40	40	40
b_2 [mm]	20	40	40	55

In Case 1, there were 70 J-grooves. Figure 8 shows the silt concentration on the impeller plate. The silt concentration on the impeller plate was relatively higher than that in Figure 4, but the silt concentration around the impeller ring was almost zero. Figure 9 shows the velocity in section S1. Figure 10 shows the relative velocity on the impeller plate.

It can be seen from Figures 9 and 10 that the J-grooves induced two shorter and stronger vortices in the gap between the impeller plate and the casing, and the radial component of the relative velocity was obviously increased. The silt particles were prevented from reaching the area around the impeller ring and the removal of silt particles in the gap between the impeller plate and the casing was easier.

The pump efficiency dropped by 4.21 %. It can be easily understood that the J-grooves increased the roughness of the impeller plate and the friction losses increased accordingly. It can be concluded that the J-grooves had the effect of a dynamic seal and the impeller ring was protected from silt abrasion. The presence of J-grooves also led to additional friction losses and pump efficiency decreased. Therefore, the number and shape of the J-grooves needed to be optimized.

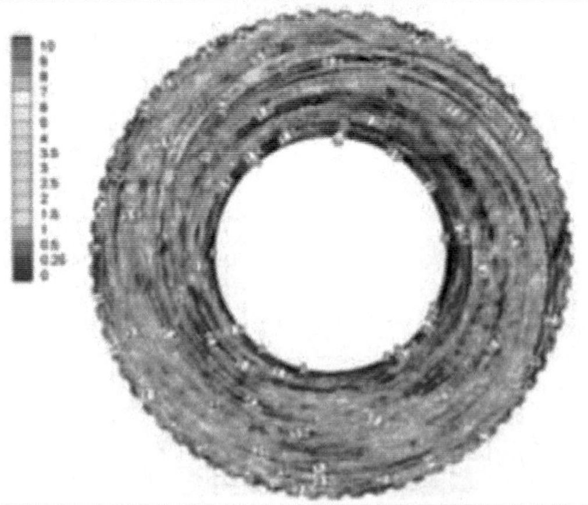

Figure 8. Distribution of silt on the impeller plate (Case 1)

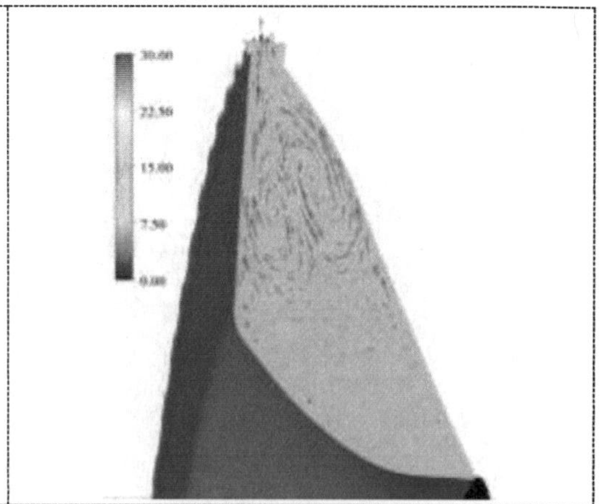

Figure 9. Velocity in section S1 (Case 1)

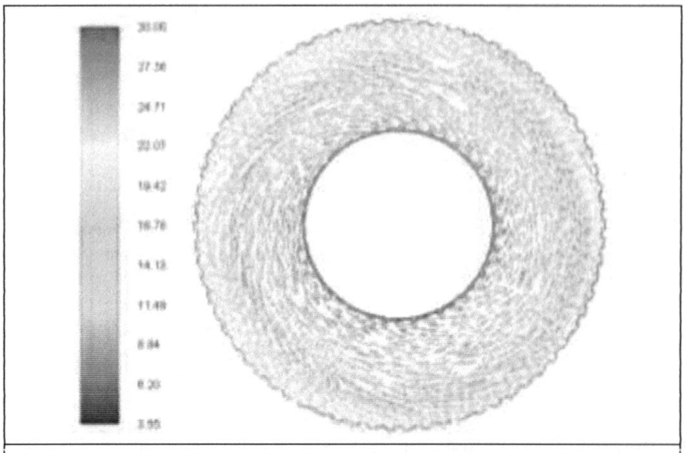

Figure 10. Relative velocity on the impeller plate (Case 1)

In Case 2, the number of J-grooves was reduced to 35 and the width of the lower edge and upper edge of the J-groove was increased to 40 mm. The silt concentration and the relative velocity on the impeller plate for Case 2 are shown in Figures 11 and 12. It can be seen that the silt concentration on the impeller plate changed a little, and the radial component of the relative velocity on the impeller plate was also low. However the pump efficiency for Case 2 dropped by 3.50 %. The silt concentration and the pump efficiency were thus unacceptable. It was concluded that the shape of the J-groove needed to be optimized to achieve acceptable silt concentration and pump efficiency.

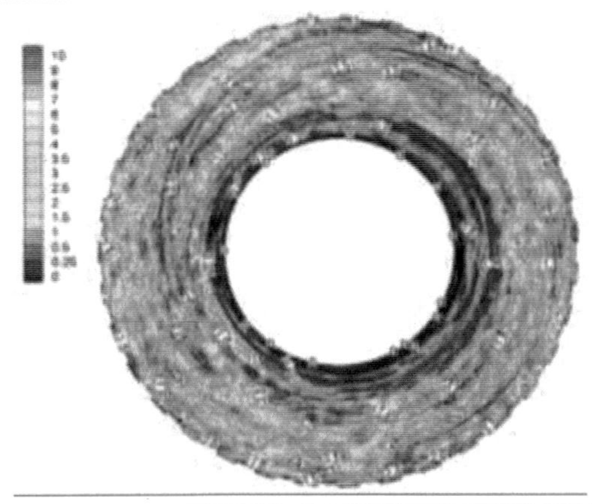

Figure 11. Distribution of silt on the impeller plate (Case 2)

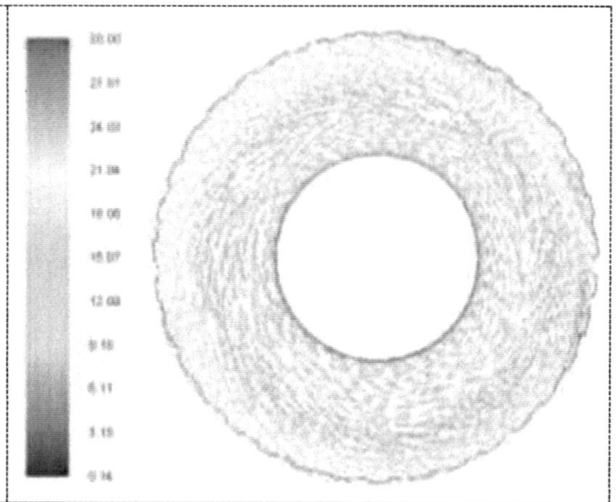

Figure 12. Relative velocity on the impeller plate (Case 2)

In Case 3, the dip angle was decreased to 30°. Figures 13 and 14 show the silt concentration and the relative velocity on the impeller plate for Case 3. In Figure 13, the silt concentration on the impeller plate near the J-grooves was increased and the area where the silt concentration around the impeller ring was nearly zero enlarged. Figure 14 shows that the radial component of the relative velocity obviously increased, which made for efficient silt extraction. The pump efficiency for Case 3 dropped by 3.55 %. The combined effect for Case 3 was an improvement.

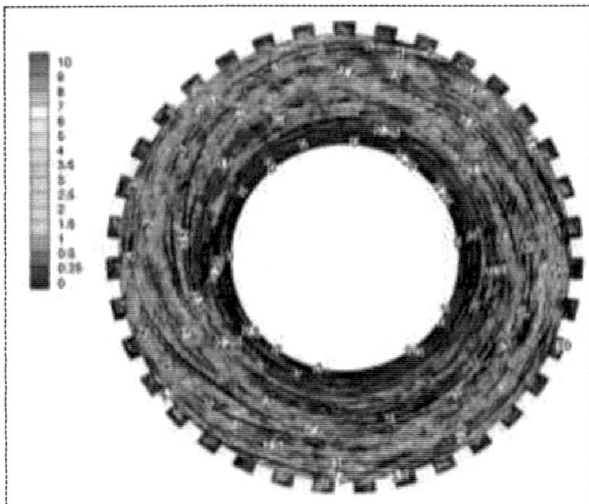

Figure 13. Distribution of silt on the impeller plate (Case 3)

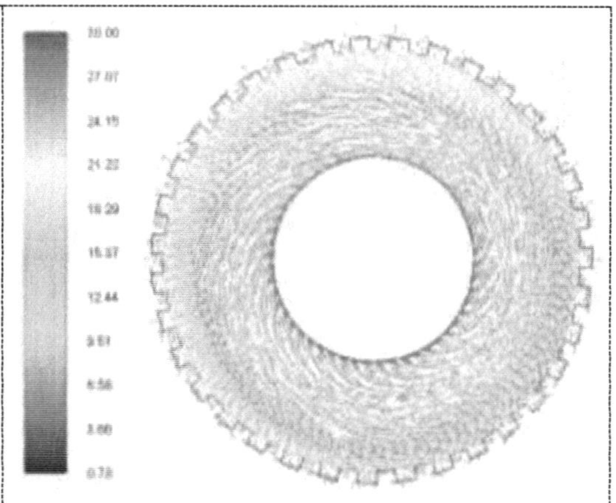

Figure 14. Relative velocity on the impeller plate (Case 3)

To reduce the friction losses caused by the J-grooves, the shape of the J-grooves in Case 4 was optimized on the basis of Case 3. The width of upper edge b_2 was expanded to 55 mm. The results of Case 4 are shown in Figures 15 and 16. It can be seen in Figure 15 that the silt concentration on the impeller plate changed a little but the silt concentration around the impeller ring still remained about zero. In Figure 16, the radial component of the relative velocity only changed a little compared with

Case 3, so the capacity for carrying silts was maintained. The pump efficiency for Case 4 dropped by 3.11 %, which was an improvement on Case 3.

Figure 15. Distribution of silt on the impeller plate (Case 4)

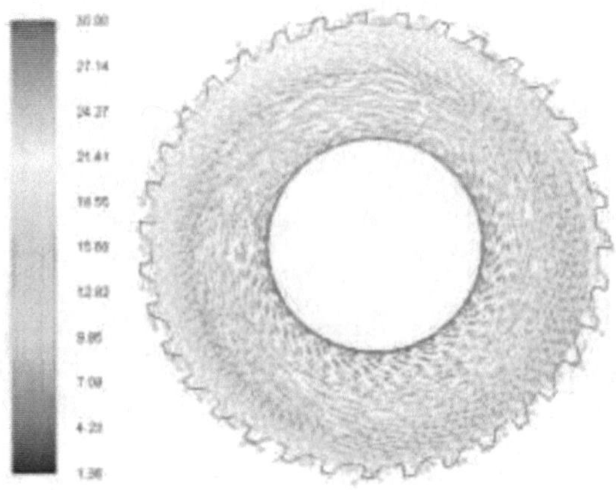

Figure 16. Relative velocity on the impeller plate (Case 4)

The analysis results show that the J-groove was effective in protecting the impeller ring from silt abrasion. The pump efficiency loss can be reduced by reducing the number of J-grooves. In this simulation, optimization of the number and shape of the J-grooves was conducted. Case 4 was considered acceptable in protecting the impeller ring from silt abrasion and keeping friction losses low.

4. Conclusions

The water flow and movement of silt in a prototype double-suction centrifugal pump was simulated using an Euler-Lagrange multiphase flow model. J-Grooves were designed in the impeller plate to protect the impeller ring from silt abrasion. Four different cases were simulated and analyzed. The results show:

(1) The J-Grooves induced two short and strong vortices which formed a dynamic seal and prevented the silt particles from reaching the area around the impeller ring.

(2) The J-groove was effective in increasing the radial component of the relative velocity of water and reducing the silt concentration around the impeller ring, but additional friction losses were also introduced which decreased pump efficiency.

(3) With the optimization of the number and shape of the J-grooves, Case 4 was considered the most effective for protecting the impeller ring from silt abrasion and maintaining pump efficiency.

Acknowledgment

This work was supported by the National Natural Science Foundation of China (No.51422906).

References

[1] Liu J, Xu H Y, Tang S, Lu L and Luo X W 2008 Experimental research on the movement rule of

solid particles in centrifugal pump [J]. *Journal of Hydroelectric Engineering*, **Vol.27** (6):168-173.

[2] Iwai Y, Nambu K 1997 Slurry wear properties of pump lining materials [J]. *Wear*, **Vol. 210**:211-219.

[3] Chen C C, Zhang T, Chen F, Yang C M, Zhang L Q 2007 An experimental study of silt abrasion on the surface of hydrofoil and flow measurement by PIV [J]. *Drainage Irrigation Machinery*, **25**(3):1-5 (in Chinese).

[4] Kurokawa J, Saha S L, Matsui J and Kitahora T 2000 Passive control of rotating stall in a parallel-wall vaneless diffuser by radial grooves, *ASME J. Fluids Eng.*, **122**: 90-96.

[5] Saha S L, Kurokawa J, Matsui J, and Imamura H 2000 Suppression of performance curve instability of a mixed flow pump by use of J-groove, *ASME J. Fluids Eng.*, **122**: 592-597.

[6] Saha S L, Kurokawa J, Matsui J, and Imamura H 2001 Passive control of rotating stall in a parallel-wall vaned diffuser by J-grooves, *ASME J. Fluids Eng.*, **123**:507-515.

[7] Wang F J, Li Y J, Wang Y L 2006 CFD simulation of 3D flow in large-bore axial-flow pump with half-elbow suction sump [J]. *Journal of Hydrodynamics*, **18**(2):243-247.

[8] Blanco M 2000 Numerical flow simulation in a centrifugal pump with impel-volute interaction[C]. *ASME, Fluid Engineering Division Summer Meeting*, Boston, Massachusetts, June, 2000:11-15.

[9] Shukla S N, Kshirsagar J T 2002 Numerical experiments on a centrifugal pump [J]. *American Society of Mechanical Engineers, Fluids Engineering Division (Publication)* FED, 2002, **257**(2B): 709-720.

[10] Chen H X 2004 Research on turbulent flow within the vortex pump [J]. *Journal of Hydrodynamic*: Ser B, **16**(6):701-707.

[11] Liu W C, Zhang K W, Wang J X 2003 The CFD analysis to the dynamic seal of the J-groove[J]. *General Machinery*, **2003**(6):59-62.

[12] Launder B E, Spalding D B 1974 The numerical computation of turbulent flows [J]. *Computer Methods in Applied Mechanics and Engineering*, **No.3**:269-289.

[13] Wang Z, Liu W M 2003 Two modificatory k-ε turbulence models for turbulent swirling flows [J]. *Journal of Hydrodynamics*, **No.2**:51-57.

[14] Habib M A, Badr H M, Ben-Mansour R, et al. 2007 Erosion rate correlations of a pipe protruded in an abrupt pipe contraction [J], *International Journal of Impact Engineering*. **34** (8): 1350-1369.

[15] Patanke S V, Spalding D B 1972 A calculation procedure for heat, mass and momentum transfer in three-dimensional parabolic flows [J]. *Int. J. Heat Mass Transfer*, **15**:1787-1806.

[16] Ishida M, Sakaguchi D, Sun Z 1998 Suppression of rotating stall in vaneless diffuser by wall roughness control [A]. *Proceedings of the international Conference on Pumps and Fans*[C].*ICPE*, Tsinghua, China, Oct 13-16, 1998:232-241. Publisher: Tsinghua Univ. Press, Beijing, China.

Numerical study of particle deposition and scaling in dust exhaust of cyclone separator

W W Xu[1], Q Li[2], Y L Zhao[2], J J Wang[2] and Y H Jin[2]

[1] College of pipeline and civil Engineering, China University of Petroleum, Qingdao 266580, PR China.
[2] College of Chemical Engineering, China University of Petroleum, Qingdao 266580, PR China.

E-mail: xuweiweisydx@126.com

Abstract. The solid particles accumulation in the dust exhaust cone area of the cyclone separator can cause the wall wear. This undoubtedly prevents the flue gas turbine from long period and safe operation. So it is important to study the mechanism how the particles deposited and scale on dust exhaust cone area of the cyclone separator. Numerical simulations of gas-solid flow field have been carried out in a single tube in the third cyclone separator. The three-dimensionally coupled computational fluid dynamic (CFD) technology and the modified Discrete Phase Model (DPM) are adopted to model the gas-solid two-phase flow. The results show that with the increase of the operating temperature and processing capacity, the particle sticking possibility near the cone area will rise. The sticking rates will decrease when the particle diameter becomes bigger.

1. Introduction

The scaling of particles in the gas flue with high temperature relates to the interaction between the particles and the wall surfaces, which is determined by the gas-solid two-phase flow. Two-phase flow is a complicated study field, as the solid particles in the flow interact with the flow around ceaselessly. The transfer of heat and kinetic energy occurs in the interaction. So the particles motion styles in the gas flow differ from each other. So do the interactions between the particles and the wall surfaces. And this can lead to the scaling pattern difference. In the energy recovery system, the speed of the gas is about 2-10 m/s in the exhaust-heat boiler, and about 40-80m/s in the separators. However, it can reach 400-600 m/s in the flue gas turbine. So the apparatus can be divided into two classes, the low slip speed one and the high slip speed one according to the relative moving speed between the particles and the wall surfaces. The paper mainly investigates the scaling formation in the low slip speed equipment, which can provide the theoretical foundation for the study of the scaling mechanism. As the separators are the typical equipment of the low slip speed ones, the single tube in the third cyclone separator (fig.1) has been chosen as the investigation object.

[1] Xu Weiwei, xuweiweisydx@126.com.

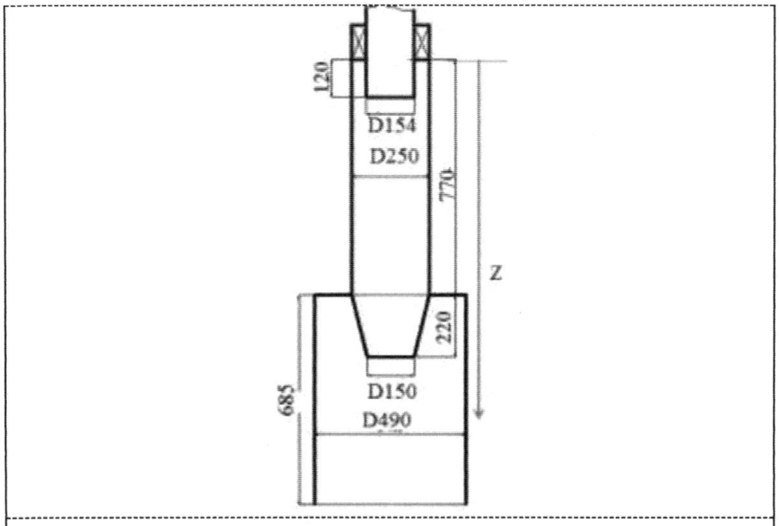

Figure 1. Structure sizes of the calculation model.

Many researchers have done lots of work in the scaling formation or flow field investigation inside the cyclone separators. Jin Ju-hui and Zhang Yang etc. [1] analyzed the operation and low separation efficiency of Fluid Catalytic Cracking (FCC) regenerator's 3rd-stage cyclone, and found that the culprits of low separation efficiency was the higher amount of catalyst fines in flue gas and greater particle sizes, lower flue gas velocity in cyclone separator and back mixing of catalyst fines at cone section of the cyclone separator. Cao Qing-yun [2] found that particle motion was very complex and had a great randomness. For the fine particles, the main factors influencing the efficiency were the air entrainment in the inner vortex region, short-circuit flow near the vortex head and re-entrainment at the bottom of discharge cone. For the larger particles, the main influencing factors were "upper dust ring" near the top plate and short-circuit flow near the vortex head. Hou [3] analyzed the movement of particles in cyclone separators and found that the particle collision and agglomeration mainly existed in the annular space of cyclone. Particle agglomeration due to inertial collision and particle interception was dominant in the cyclone. Wu etc. [4-5] indicated that the precession vortex core (PVC) phenomenon existed in all axial positions of the cyclone separator. And the existence of the PVC would certainly induce the re-entrainment of dust and lower efficiency of the cyclone separators. Tan Hui-min etc. [6] simulated the back mixing of particles in a guide vane separator based on the Reynolds stress model (RSM) and the discrete particle model (DPM). The results showed that the back mixing of particles is serious at the bottom of the dust discharge. And many other researchers [7-11] studied the cyclone separators and gave their explanation for the performance degradation.

2. Physical Model

The geometry data is shown in fig.1. The simulation was carried out with the help of the commercial computational fluid dynamics software Fluent. Based on the finite volume method, the discrete equation was built up. The governing equation was solved with QUICK as its finite difference scheme and Semi-Implicit Method for Pressure Linked Equations (SIMPLE) as the solution method. The Reynolds Stress Model (RSM) was adopted to simulate the unsteady incompressible turbulent flow in the guide vane cyclone. The Reynolds stress transport equation is as follows,

$$\frac{\partial}{\partial t}(\rho \overline{u_i u_j}) + \frac{\partial}{\partial x_k}(\rho U_k \overline{u_i u_j}) = D_{ij} + P_{ij} + \Pi_{ij} - \varepsilon_{ij} \tag{1}$$

In the formula, the two items in the left side is the local times derivative and the convection. The four in the right side are the stress diffusion, the stress production, the pressure strain and the dissipation from left to the right.

As the inlet particle concentration of the third cyclone separator C_i is about 0.5~2.0g/m^3, and the particle density is 2700kg/m^3. The volume fraction of the particle $\varepsilon = C_i / \rho_p = 1.85\times10^{-7} \sim 7.4\times 10^{-7} \ll 1\%$. So it can be dealt as the dilute phase, and the Discrete Phase Model (DPM) can be used to simulate the particles motion.

The particles' density in the simulation in this paper is 2700 kg/m^3, and their mean diameter is 18μm. The minimum diameter of the particles is 1μm, and the maximum diameter is 37μm. The particles are injected to the separator with the same velocity of the gas phase at the inlet.

Boundary conditions at the entrance: the gas enters the equipment is the air at normal atmospheric temperature, and the volume flow rate Q_i=2200 m^3/h; the mass flow rate of the particles $\dot{M}_{in}$=4.4 kg/h. Boundary conditions at the exit: the gas flow at the exit is regarded as fully developed pipe flow. The gradient of all the parameters along the outlet normal direction is zero, viz. $\partial\varphi / \partial z = 0$. To guarantee the fully developed flow at the exit, the outlet pipe was lengthened in the computational model.

Boundary conditions near the wall: the non-slipping shear condition as well as the enhanced wall treatment was adopted. The cyclone separator wall coefficient of elastic recovery (COR) was well set by UDF.

3. Results and Discussions

3.1. Validation

By comparing the simulation results of the cyclone separator and the results obtained by the five holes probe as shown in fig.2, it can be found that the two results fit close to each other. And this indicates that the simulation results are accurate.

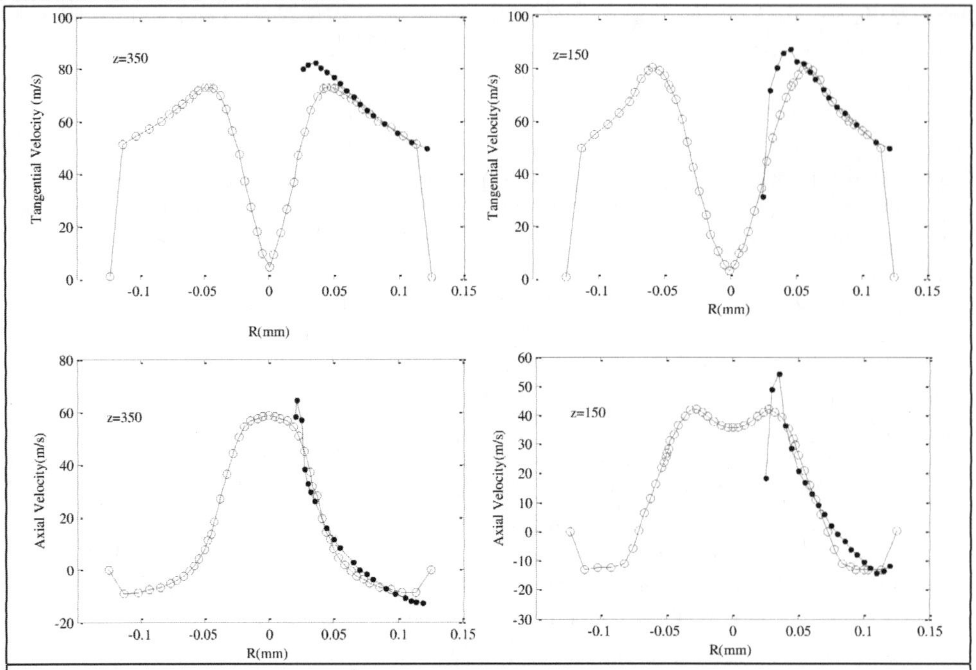

Figure 2. Gas flow verification

3.2. Particle Deposition and Scaling

In fig.3 and the figures below, the particle sticking rates Ra is defined as the mass of particles sticking onto the wall surfaces in a unit interval. The deposition and scaling distribution pattern of the low slip speed equipment is illustrated in fig.6. From the picture, it is obvious that, there is a certain degree deposition on the separator wall surface. And the deposition reached the highest in the dust-exhaust cone zone. This is mainly because of a portrait swirl at the separator's reducing area, which may make the particles wandering for a longer time here. The dust-exhaust cone in the third cyclone separator is the heaviest deterioration area. So the following study is mainly focused on this area.

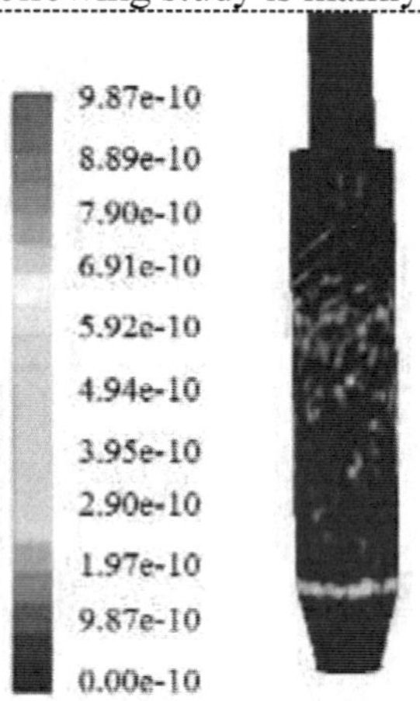

Figure 3. Distribution of particles deposition and adhesion in low sliding speed equipment.

3.3. The effect of the operating parameters on the scaling inside the low slip speed equipment

3.3.1. Processing capacity

(1) Particle concentration near the wall

With the change of the processing capacity, the concentration of particles inside the equipment changes accordingly as summarized in fig.4.

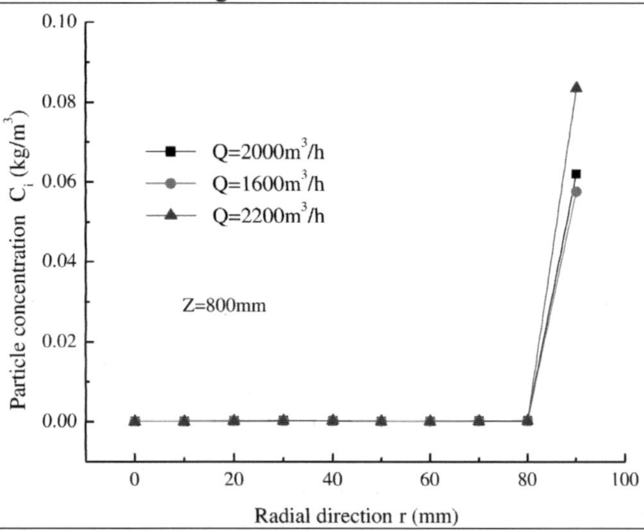

Figure 4. Distribution rules of particle concentration with different inlet flow rates.

IOP Conf. Series: Materials Science and Engineering **129** (2016) 012022 doi:10.1088/1757-899X/129/1/012022

As illustrated in the picture, the gas inlet speed rises with the increase of the processing capacity. This lead to the increase of the gas flow tangential velocity and the centrifugal force exerted on the particles goes up as well. As a result, the particle concentration near the wall surface is high. However, the particle concentration is low and not sensitive to the change of the processing capacity in the central area.

(2) Particle sticking rates

As the processing capacity changes, the particle adherence insider the equipment changes accordingly as shown in fig.5.

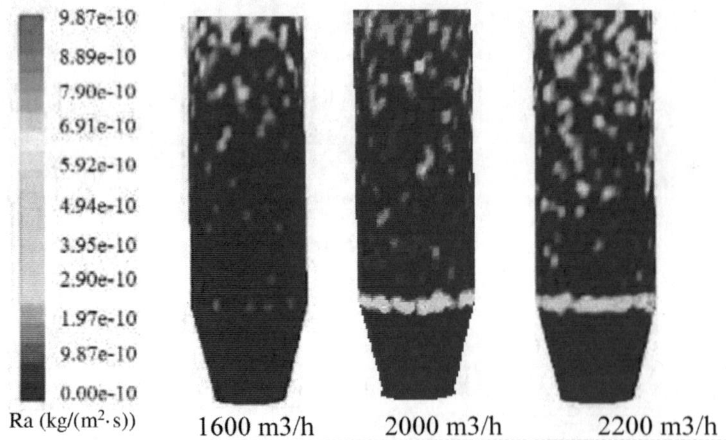

Figure 5. Variations of particle sticking rates with different inlet flow rates.

In fig. 5, with the increase of the processing capacity, the particle sticking rates increases in the dust-exhaust cone. This is mainly because the particle concentration nearby the wall surfaces rises, and the gas shearing velocity adjacent to the wall goes up as well. As the wall surface is arc shaped, there will be a certain degree of centrifugal force exerted to the particles on the wall. The acting force between the particles is enhanced. And this goes against the purpose of elimination of the particles on the wall surfaces and will aggravate the particles adherence to the wall surfaces.

3.3.2. Temperature

(1) Distribution of particle concentration nearby the wall

With the variation of the operating temperature, the particle concentration inside the equipment changes as well as illustrated in fig. 6. In the figure, the particle concentration adjacent to the wall surfaces goes down slightly with the increase of the operating temperature. Through changing the gas density and viscosity, the operating temperature changes the gas flow field and makes the interaction between the gas and particles change. Then the particle trajectories change and the drag force exerted on the particles increases when the operating temperature goes up. And this affects the particle concentration distribution.

(2) Particle sticking rates

In fig.7, the sticking rates get high in the dust exhaust area and goes down in the cylinder shell area with the increase of the operating temperature. The interaction between the temperature and the Young's modulus is adopted in this paper. As the Young's modulus reflects the particles' capacity to resist the deformation, the decrease of Young's modulus with the increase of temperature means it is easier for the particles to get stuck by the wall surfaces with the increase of operating temperature. With the variation of the operating temperature, the particle sticking rates changes as well.

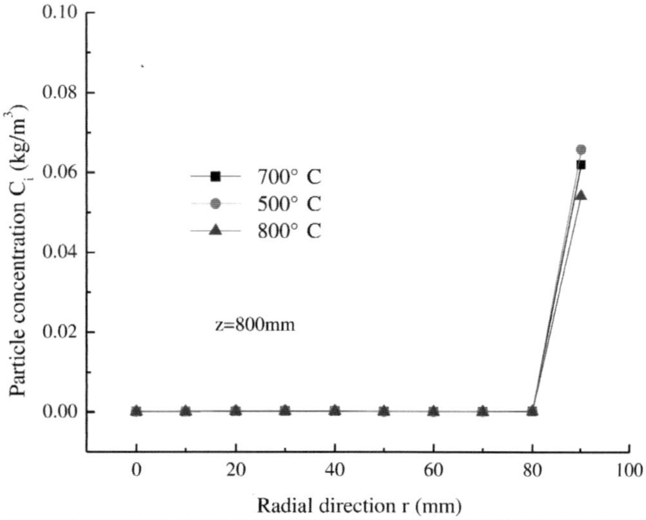

Figure 6. Variation of particles concentrations with different temperatures.

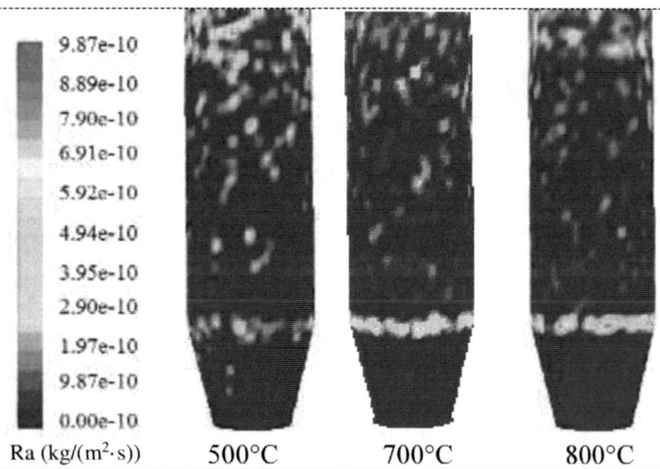

Figure 7. Variation of particle sticking rates with different operating temperatures.

3.3.3. Input concentration. In fig.8, with the increase of the input concentration, the particle concentration nearby the equipment inside wall surfaces is certain to rise. With the sum of particles goes up, the possibility of collision between the particles and the wall surfaces becomes bigger, and the sticking rates in the dust-exhaust cone area is apparently higher.

3.3.4. Particle diameter. Apparently as illustrated in fig.9, with the variation of the particle diameter, the particle sticking rates change.

The maximum sticking rates on the wall surfaces decrease when the particle diameter rises. This indicates that the smaller the particle diameter, the bigger chance the scaling will occur. Particles at 1μm are apt to move with the gas flow, so their separation efficiency is the lowest. And they are less possibly to approach the dust-exhaust cone area, and get captured by the wall surfaces. The larger the particles diameters, the bigger chance they will be captured until the diameter reaches 5μm. After that, the sticking area as well as the sticking rates goes down with the increase of the particle diameter in the dust-exhaust area. When the diameter reaches 10μm, there will be no more particles sticking on the dust-exhaust area. This indicates that it is mainly particles with diameters below 10μm that affect the scaling in the dust-exhaust area, and the sticking rates decreases with the increase of the particle diameter.

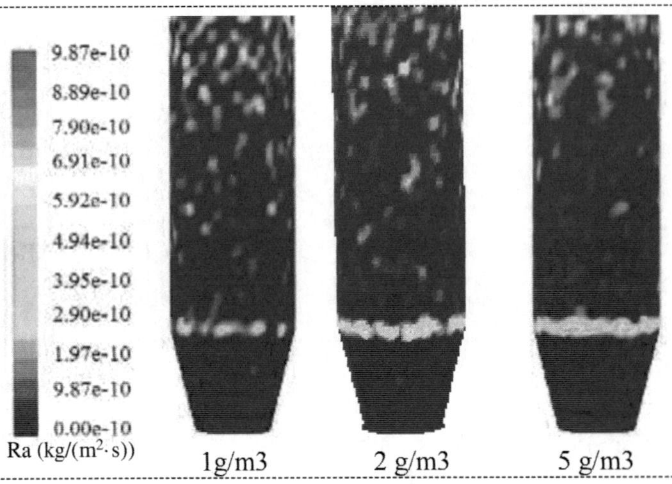

Figure 8. Variation of particle sticking rates with different particle concentrations.

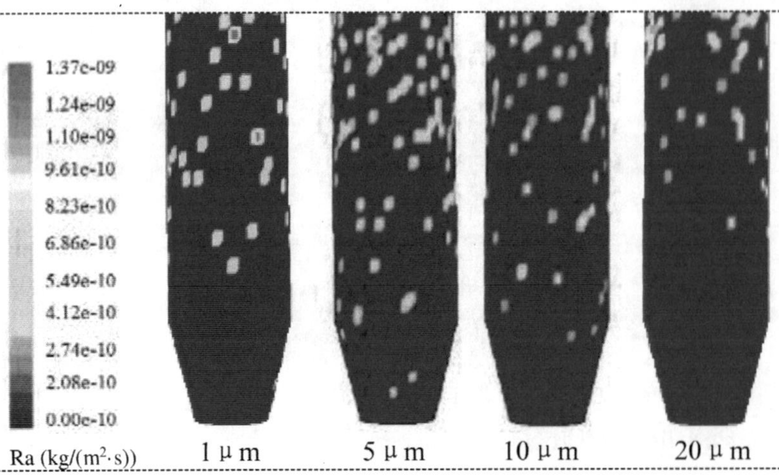

Figure9. Distribution of particle sticking rates with different particle sizes.

4. Conclusions

This paper studies the gas-solid two phase flow inside the low slip speed velocity equipment. The gas shearing velocity and the particle concentration distribution as well as the particle sticking rates near the wall surfaces are analyzed. The study also covers the effects of the operating parameters and physical parameters as well as the structural parameters on the particle sticking rates. The six conclusions are drawn out as follows:

(1) Scaling in the dust exhaust cone area is the severest in the low slip speed equipment.

(2) The high particles concentration makes the possibility of scaling on the dust exhaust cone area get bigger. The particle concentration nearing the wall are not sensitive to the operating temperature. But with the increase of the operating temperature, the particle sticking possibility near the cone area will rise.

(3) When the particle concentration gets higher, the possibility of collision between the particles and the wall surfaces gets bigger. The sticking rates in the cone area go up apparently. The sticking rates will decrease as when the particle diameter is bigger. And the diameters of the particles caused the scaling are mainly under 10μm.

References

[1] Jin J H, Zhang Y, Zheng D Z 2006 Root causes of low-efficiency of FCCU 3rd-stage cyclone [J]. *Petroleum Refinery Engineering*. **2006**(09):21-24.
[2] Cao Q Y 2008 *Numerical simulation of gas-solid flow in cyclone separator of CFB boiler* [D]. Dongying: China University of Petroleum (East China).
[3] Hou G X 2008 *Experimental research on separation model of high temperature cyclone separator* [D]. Dongying: China University of Petroleum (East China).
[4] Wu X L, Xiong Z Y, Ji Z L, Shi M X 2007 Numerical simulation of precession vortex core in cyclone separator [J]. *Journal of Chemical Industry and Engineering (China)*, **58**(2):383-390.
[5] Wu X L, Yan C Y, Shi M X 2002 Research on the precession vortex core in the flow field of cyclone separators with double tangential inlets [J]. *Chemical Engineering & Machinery*, **29**(1):1-5.
[6] Tan H M, Wang J J, Qi C G, Jin Y H 2011 Back mixing of particles in guide vane separator [J]. *CIESC Journal*, (03):716-722.
[7] Wang J Y, Mao Y, Wang J 2008 Numerical simulation of cross-flow back-mixing among multi-pipes in horizontal cyclone separator [J]. *Petroleum Refinery Engineering*, (08): 38-42.
[8] Wang Y, Wang J J, Cui Z N 2011 Numerical simulation of flow field in multi-tube cyclone separator [J]. *Petro-Chemical Equipment*, (06): 24-29.
[9] Tan H M, Wang J J, Ma Y J, Jin Y H 2011 Analysis of the influence of dust discharge cone geometry on the gas-solid flow field in tangential inlet cyclone separator [J]. *Journal of Chemical Engineering of Chinese Universities*, **25**(04): 590-596.
[10] Gao Y W, Guo Q, Li X Y 2001 Study on fine particle entrainment at dust discharge of cyclone tube [J]. *Petro-Chemical Equipment*, **30** (4):8-10.
[11] Hui S L 2008 *The investigation on separation performance of cyclone separator used in the circulating fluidized bed boiler* [D]. Dongying: China University of Petroleum (East China).

Acknowledgments

I would like to extend my sincere gratitude to my friend, Yan Zhao and Xiaowei Gao, for their instruct ive advice on my article. And this paper is supported by The Fundamental Research Funds for the Cen tral Universities (14CX02109A).

The stress analysis of a heavy liquid metal pump impeller

X D Ma, X L Li, Z Q Zhu, C J Li, S Gao*

Key Laboratory of Neutronics and Radiation Safety, Chinese Academy of Sciences, Institute of Nuclear Energy Safety Technology, Chinese Academy of Sciences, Hefei, Anhui, 230031, China

E-mail:sheng.gao@fds.org.cn

Abstract. Lead-based coolant reactor is a promising Generation-IV reactor. In the lead-based coolant reactor, the coolant is liquid lead or lead-bismuth eutectic. The main pump in the reactor is a very important device. It supplies force for the coolant circulation. The liquid metal has a very large density which is about ten times of the water. Also, the viscosity of the coolant is small which is about one sixth of the water. When the pump transports heavy liquid, the blade loading is heavy. The large force can cause the failure of the blade when the fatigue stress exceeds the allowable stress. The impeller fraction is a very serious accident which is strictly prohibited in the nuclear reactor. In this paper, the numerical method is used to simulate the flow field of a heavy liquid metal pump. The SST k-w turbulent model is used in the calculation to get a more precise flow structure. The hydraulic force is obtained with the one way fluid solid coupling. The maximum stress in the impeller is analyzed. The stress in the liquid metal pump is compared with that in the water pump. The calculation results show that the maximum stress of the impeller blade increases with increase of flow rate. In the design of the impeller blade thickness, the impeller strength in large operating condition should be considered. The maximum stress of the impeller blade located in the middle and near the hub of the leading edge. In this position, the blade is easy to fracture. The maximum deformation of the impeller firstly increase with increase of flow rate and then decrease with increase of flow rate. The maximum deformation exists in the middle of the leading edge when in small flow rate and in the out radius of the impeller when in large flow rate. Comparing the stress of the impeller when transporting water and LBE, the maximum stress is almost one-tenth of that in the LBE impeller which is the same ratio of the density. The static stress in different medium is proportional to the pressure distribution because a large percentage of the total stress is caused by hydraulic pressure.

1. Introduction

Fusion and fission energy are the important ways recognized so far to solve the energy crisis problems [1][2][3][4][5]. Lead-based coolant reactor is a promising Generation-IV reactor. In China, Chinese LEAd-based research Reactor (CLEAR) was selected as the reference reactor in the first phase of CAS ADS project ("strategic priority research program" named "the future of advanced nuclear fission energy system-ADS transmutation system")[1][6][7][8][9][10].

In the lead-based coolant reactor, the coolant is liquid lead or lead-bismuth eutectic. The main pump in the reactor is a very important device. It supplies force for the coolant circulation. The liquid metal has a very large density which is about ten times of the water. Also, the viscosity of the coolant is small which is about one sixth of the water. When the pump transports heavy liquid, the blade loading is heavy. So the stress is much higher compared with transporting water. The structure analysis is necessarily.

In the previous studies, CFD method is already successfully used in the stress calculation of hydraulic machineries. Chen Xiangyang et al.[11] calculated the blade stress considering the thermal stress of a coolant reactor pump .Tang Xuelin et al.[12] studied the bulb tubular pump impeller stress and deformation in different flow conditions to check the impeller strength. In Sambhrant Srivastava et al.'s[13] study, the stress analysis was carried out to optimize blade shape. Gao Yongjiang et al.[14] studied the dynamic stress in different flow conditions. It is shown that under small flow condition the dynamic stress is the largest because of the large pressure fluctuation. Li Liuyang et al.'s[15] studies also showed that the dynamic stress has a considerable amplitude compared with the static stress. Xiao Ruofu et al.[16] studied the dynamic stress of a Francis turbine runner at high heads and low loads under which condition the largest dynamic stress occurred. The dynamic stress can not be ignored to the blade's micro-cracks. Wang Zhengwei et al.[17] also found the actual dynamic stress in transient process is much larger than experienced value and is the main reason to the premature cracks. And also there are other valuable studies [18][19]. These studies have important engineering values for the pump design and operation. And also It is proved the numerical method be a valid and economical method. It can be used to analyze the impeller stress of a heavy liquid metal pump.

There are two ways of fluid-solid coupling methods. The first one is the double way coupling. The deformation of the solid will be transferred to the fluid zone. It is usually considered when the large deformation occurred, such as in the wind turbine. For the studied pump, since the size is not so large and the deformation is not so obvious. The influence of the blade deformation to the flow field is ignored in this study. So a one-way coupling method is used in the calculation. In this paper, the flow field in a mixed pump is simulated. The pressure is exerted to the blade and the structural characteristic is analyzed.

2. Computational model

2.1. Physical model

The studied pump is a centrifugal pump with 7 blades. The physical model is shown as follows. The numerical model consists of the impeller, diffuser, casing, front and back clearance, inlet and outlet pipe.

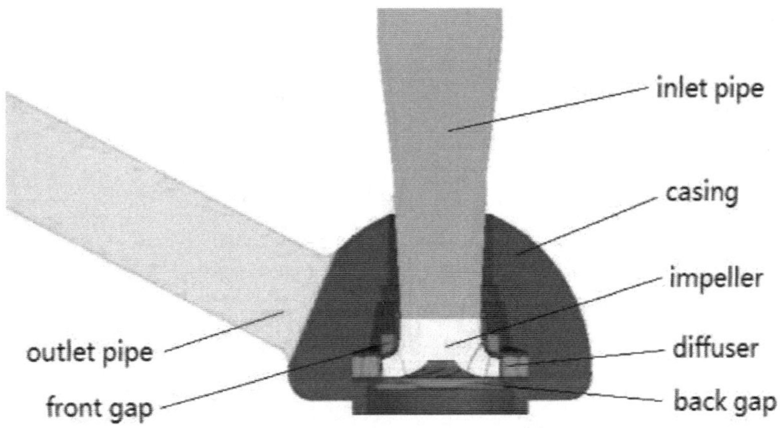

Figure 1. 3-D model of the impeller

2.2. Mesh

Structure mesh was applied to the impeller and the diffuser. Near the blade wall boundary, the mesh was refined to guarantee that the $y^+<30$ near the wall. The total number of the mesh cells is about 2.66 million. For the solid zone of the impeller, unstructured grid with fine size was meshed in workbench.

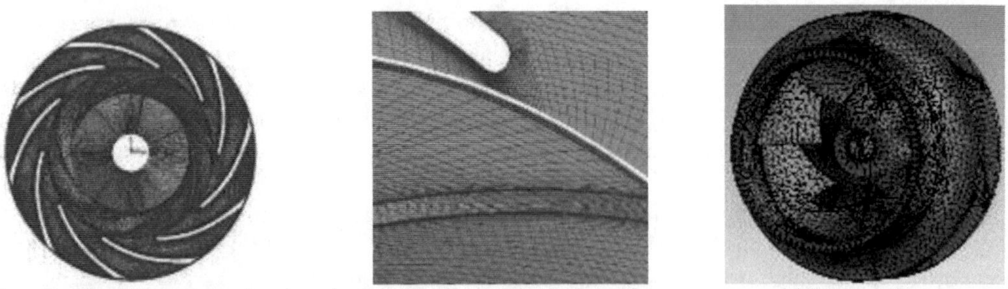

(a) impeller flow channel structural mesh (b) refined mesh in the boundary (c) Solid mesh

Figure 2. Mesh

3. Numerical method

3.1. Boundary conditions
For the fluid domain, the velocity inlet and pressure outlet boundary conditions was set. The blade surface was set as fluid solid interface. The impeller zone was rotational domain. The rotation of the impeller is realized by rotating reference frame method. No slip wall was set for the hub and shroud.

For the solid domain, the fixed support condition is set to the rotating surface. Centrifugal force was applied to the impeller. Pressure in the impeller surfaces from the fluid computational results is exerted.

3.2. Control functions
The continuity and momentum conservation function is shown as follows:

$$\partial u_i / \partial x_i = 0 \tag{1}$$

$$\rho \frac{\partial (u_i u_j)}{\partial x_i} = -\frac{\partial p^*}{\partial x_i} + \mu \frac{\partial^2 u_i}{\partial x_i \partial x_j} \tag{2}$$

The SST *k-w* model is shown as follows:

$$\frac{\partial k}{\partial t} + U_j \frac{\partial k}{\partial x_j} = P_k - \beta^* k w + \frac{\partial}{\partial x_j}\left[(\mu + \sigma_k \mu_t) \frac{\partial k}{\partial x_j} \right] \tag{3}$$

$$\frac{\partial w}{\partial t} + U_j \frac{\partial w}{\partial x_j} = \alpha S^2 - \beta w^2 + \frac{\partial}{\partial x_j}\left[(\mu + \sigma_w \mu_t) \frac{\partial w}{\partial x_j} \right] + 2(1 - F_1)\sigma_{w2} \frac{1}{w} \frac{\partial k}{\partial x_i} \frac{\partial w}{\partial x_i} \tag{4}$$

$$\mu_t = \frac{\alpha_1 k}{\max(\alpha_1 w, SF_2)} \tag{5}$$

The meanings of the parameters can refer to ANSYS Fluent 14 help.

3.3. Numerical methods
The simulation work was conducted with Fluent 14, which contained plenty of physical models. SIMPLEC algorithm was used as a relationship between velocity and pressure corrections to enforce mass conservation. The second order upwind scheme was used to discrete the convective terms. The stress analysis is performed with Workbench static structure analyze. One-way fluid-solid coupling

method is used. The pressure load calculated by hydraulic computation is transferred to the impeller solid domain.

4. Results and discussions

4.1. Steady performance

The following equations defined the dimensionless head and flow rate parameter of the pump. In these equations, the subscript 0 stands for the head and flow rate in nominal condition.

$$H' = \frac{H}{H_0} \tag{6}$$

$$W' = \frac{W}{W_0} \tag{7}$$

Experiments were conducted with water in room temperature. The steady state pump performance was compared between numerical results and experimental results, as shown in Figure 3. In the small flow rate, the experimental results and the numerical results fit well, while in larger flow rate, the numerical results decreases more gentle and slowly. On the contrary, the efficiency curve fits well in large flow rate. The differences are mainly caused by the inlet boundary condition and the physical model. In numerical simulation, the inlet velocity distribution is idealistically uniform.

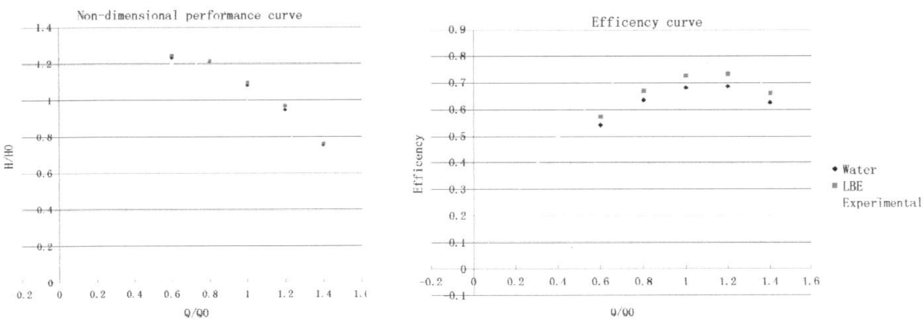

Figure 3. Comparison between numerical and experimental results

4.2. Stress analysis in different flow rate

In the stress analysis, the material properties are shown as follows:

Table 1. Material properties

Material	Structural steel	
Young Modulus	$2 \cdot 10^{11}$	Pa
Poisson Ratio	0.3	
Density	7850	Kg/m^3
Thermal expansion	$1.2 \cdot 10^{-5}$	$/°C$
Yield stress	$2.5 \cdot 10^{8}$	N/m^2

The Von Mises stress is calculated to evaluate the strength of the impeller. Von Mises stress is often used in determining whether an isotropic and ductile metal will yield when subjected to a complex loading condition. The Von Mises stress is directly related to the deviatoric strain energy term. The Von Mises yield criterion suggests that the yielding of material begins when the second deviatoric stress invariant J_2 reaches a critical value.

$$J_2 = k^2 \tag{8}$$

Where k is the yield stress of the material in pure shear.

$$k = \frac{\sigma_y}{\sqrt{3}} \tag{9}$$

Figure 4 shows the stress of the impeller of the LBE pump in different flow rate. The maximum stress located at the middle of the blade inlet edge as well as blade near the hub. The maximum stress increases linearly with increase of the flow rate. But as increase of the flow rate, the stress distribution in the blade is getting more uniform and the stress gradient in the blade is getting smaller. Comparing the stress caused by centrifugal force and hydraulic pressure, the stress caused by centrifugal force is very small compared with that caused by hydraulic pressure. The centrifugal stress keeps the same in different flow conditions while the hydraulic stress increased with increase of flow rate. And the total stress is less than the sum of centrifugal stress and the hydraulic stress.

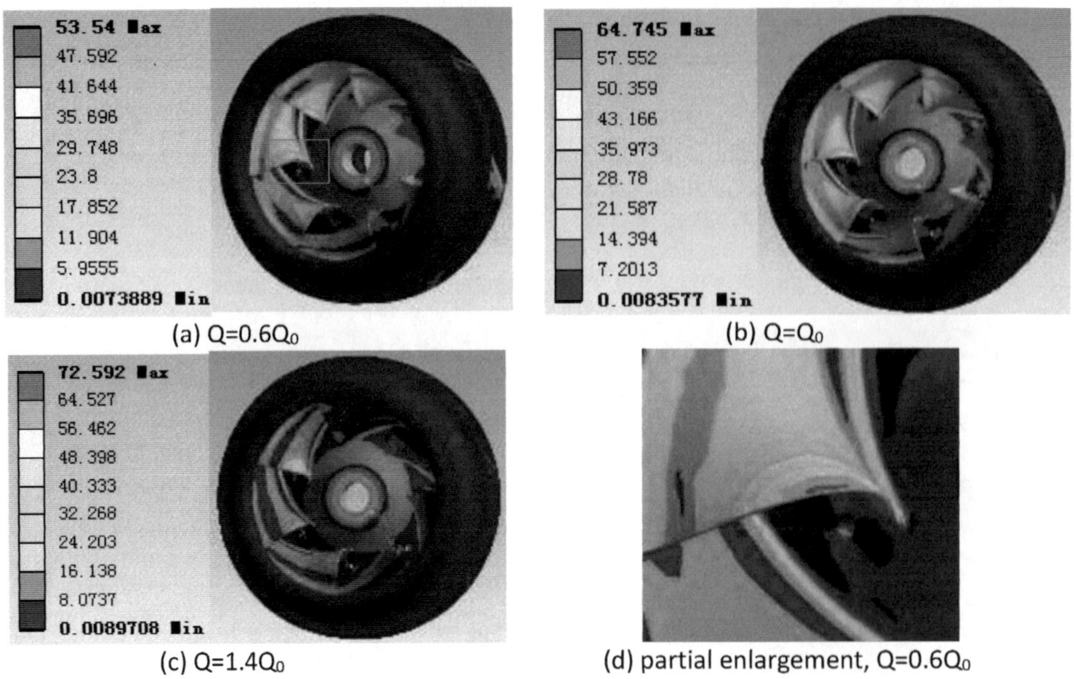

Figure 4. Stress contour in different flow rate (LBE), Unit: MPa

Figure 5 shows the deformation of the impeller of LBE pump in different flow rate. The deformation distribution is different from that of the stress. The maximum deformation changes with flow rate. In small flow rate, the maximum deformation located at the middle of the blade inlet edge, in accordance with the stress distribution. With increase of flow rate, the deformation distribution changes. The deformation increases with increase of the impeller radius. In large flow rate, the maximum deformation located at the out radius of the impeller.

Calculate the deformation in centrifugal force and hydraulic pressure separately. The same conclusion can be drawn with the stress analysis. The deformation caused by hydraulic force is much larger than that cased by centrifugal force. And also the total deformation is less than the sum of centrifugal deformation and the hydraulic deformation.

IOP Conf. Series: Materials Science and Engineering **129** (2016) 012023 doi:10.1088/1757-899X/129/1/012023

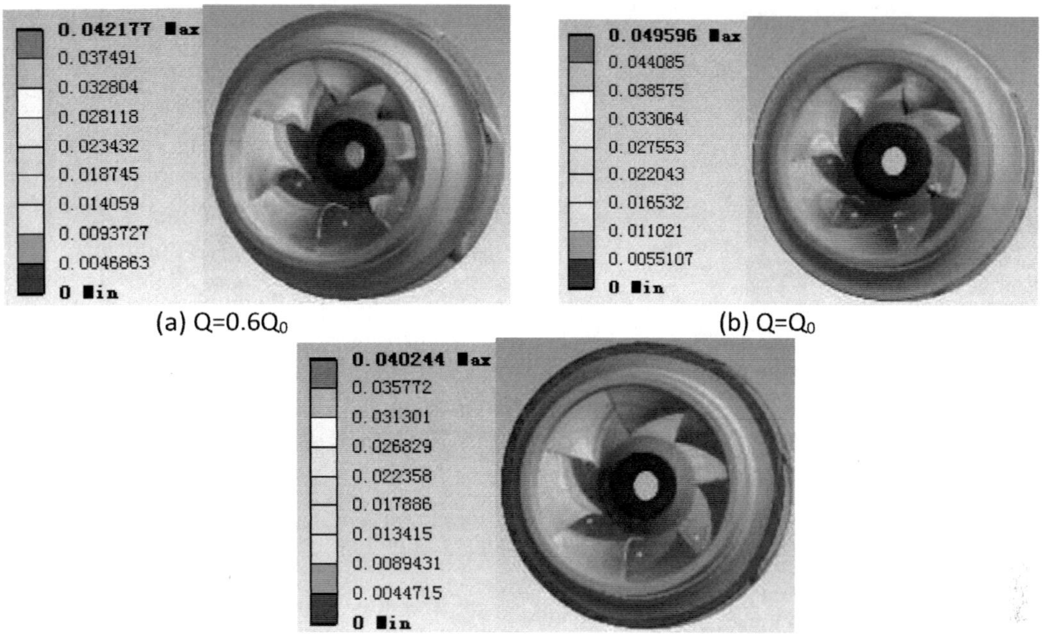

(a) $Q=0.6Q_0$ (b) $Q=Q_0$

(c) $Q=1.4Q_0$

Figure 5. Deformation contour in different flow rate (LBE), Unit: m

4.3. Comparison of different medium

With the same non-dimensional flow rate, as we already know from results shown in Figure 3, the non-dimensional head is very close. Since the density of the lead bismuth is 10 times of that of water, the hydraulic pressure load exerted to the blade of the water pump is also much smaller than that of the LBE pump. From Figure 6(a), Figure 7(a) and Figure 8(a), the maximum stress is almost one-tenth of that in the LBE impeller. The deformation variation with flow rate is also in the same tendency with LBE pump. When transporting LBE, the stress increases in a large range in different flow rate. So when designing a LBE impeller, the maximum operating flow rate should be considered as the design condition to guarantee the impeller strength.

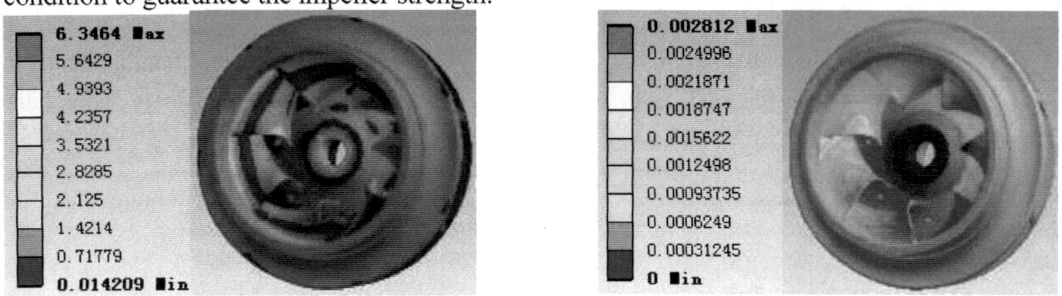

(a) Stress (water), Unit: MPa (b) Deformation (water), Unit: m

Figure 6. Stress and deformation comparison ($0.6Q_0$)

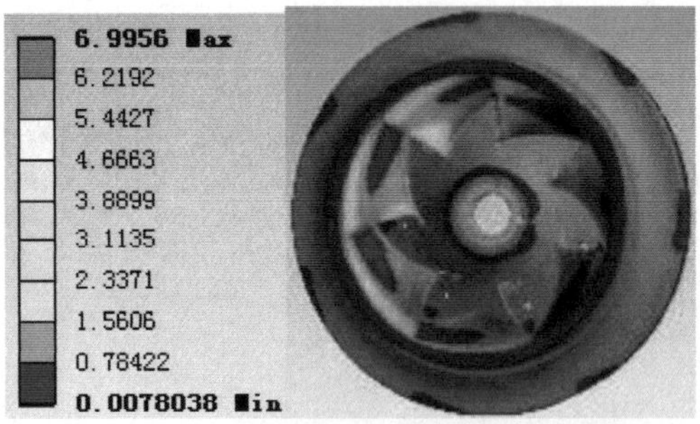

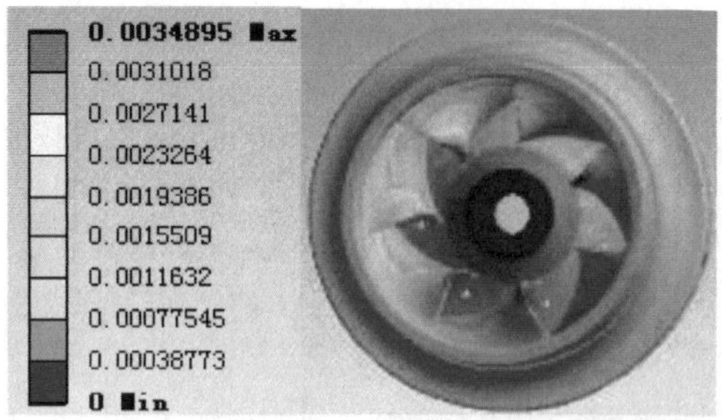

(a) Stress (water), Unit: MPa (b) Deformation (water), Unit: m

Figure 7. Stress and deformation comparison (Q_0)

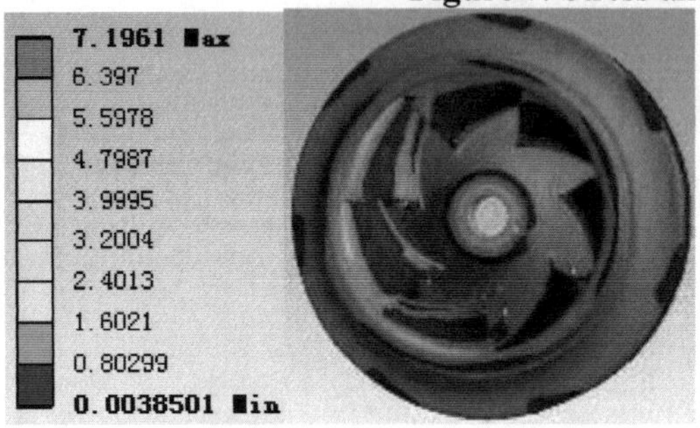

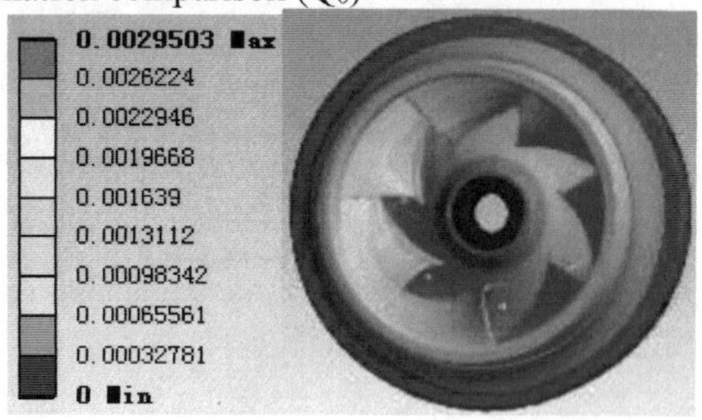

(a) Stress (water), Unit: MPa (b) Deformation (water), Unit: m

Figure 8. Stress and deformation comparison ($1.4Q_0$)

5. Conclusions

In this paper, the impeller stress of a LBE pump is analyzed and compared with a water pump. The following conclusions can be drawn from the study.

1) The maximum stress of the impeller blade increases with increase of flow rate. In the design of the impeller blade thickness, the impeller strength in large operating condition should be considered. The maximum stress of the impeller blade located in the middle and near the hub of the leading edge. In this position, the blade is easy to fracture.
2) The maximum deformation of the impeller firstly increase with increase of flow rate and then decrease with increase of flow rate. The maximum deformation exists in the middle of the leading edge when in small flow rate and in the out radius of the impeller when in large flow rate.
3) Comparing the stress of the impeller when transporting water and LBE, the maximum stress is almost one-tenth of that in the LBE impeller which is the same ratio of the density. The static stress in different medium is proportional to the pressure distribution because a large percentage of the total stress is caused by hydraulic pressure.

For further studies, the dynamic stress in transient and unsteady condition will be studied since the large pressure and pressure fluctuation. The two-way coupling method could be used to evaluate the influence of the structure deformation to the hydraulic and mechanical performance. And also, the impeller stress in multi-physical fields considering thermal stress are also need further study.

6. References

[1] Wu Y C, FDS Team 2006 Conceptual design activities of FDS series fusion power plants in China, *Fusion Eng. Des*, **81** pp.2713-2718.

[2] Wu Y C, Zheng S, Zhu X, et al. 2006 Conceptual design of the fusion-driven subcritical system FDS-I. *Fusion Eng. Des*, **81** Part B: pp.1305-1311.

[3] Wu Y C, FDS Team 2008 Conceptual design of the China fusion power plant FDS-II. *Fusion Eng. Des*, **83**: pp.1683-1689.

[4] Wu Y C, Jiang J Q, Wang M, et al. 2011 A fusion-driven subcritical system concept based on

viable technologies, *Nucl. Fusion*, **51** (10) 103036.

[5] Wang M H, Huang H, Lian C, et al. 2015 Conceptual design of lead cooled reactor for hydrogen production. *Int. J. Hydrogen Energy*, doi: 10.1016/j.ijhydene.04.009.

[6] Wu Y C, Bai Y Q, et al. 2014 Conceptual design of China lead-based research reactor CLEAR-I [J], *Chinese Journal of Nuclear Science and Engineering*, **2**, pp.201-208.

[7] Wu Y C, Bai Y Q, et al. 2012 Overview of China lead alloy cooled reactor development and ADS program in China [C], *NUTHOS-9, Kaohsiung, Taiwan*, Sep. 9-13.

[8] Wu Y C, Bai Y Q, Song Y, et al. 2013 Overview of lead-based reactor design and R&D status in China [C], *International Conference on Fast Reactors and Related Fuel Cycles: Safe Technology and Sustainable Scenarios (FR13), Paris, France*, Mar. 4-7.

[9] Wu Y C, Bai Y Q, et al. Development strategy and conceptual design of China lead-based research reactor, *Ann. Nucl. Energy*. (to be published)

[10] Wang M, Huang H, et al. 2015 Preliminary conceptual design of a LARd-bismuth cooled small reactor (CLEAR-SR), *Int. J. Hydrogen Energy*, doi:10.1016/j.ijhydene. 03.097.

[11] Chen X Y, Yuan D Q, Yang M G, Yuan S Q 2010 Blade Stress of the reactor coolant pump of 300MWe nuclear power plant in China based on fluid-solid coupling method [J]. *Journal of Mechanical Engineering*. **Vol.46** (4), pp.111-115.

[12] Tang X L, Wang X Y, Jia Y X 2014 Strength analysis of bulb tubular pump impeller based on fluid-structure interaction [J]. *Journal of Drainage and Irrigation Machinery Engineering*, **Vol.32** (11), pp.921-926.

[13] Srivastava S, Roy A K and Kumar K 2014 Design of a mixed flow pump impeller blade and its validation using stress analysis [J]. *Procedia Materials Science*. **Vol.6**, pp.417-424.

[14] Gao J Y, Wang F J, Qu L X, Liu Z Q et al. 2012 Dynamic stress of large double-suction centrifugal pump impeller[J], *Journal of Agricultural Machineries*. **Vol.43** (1), pp.42-52.

[15] Li L Y, Wang Y S, Wei Y S, Fu J et al. 2014 Strength calculation and analysis of impeller of mixed flow water jet pump [J]. *Journal of Central South University (Science and Technology)*, **Vol.45** (12), pp.4438-4444.

[16] Xiao R F, Wang Z W and Luo Y Y 2008 Dynamic stresses in a Francis turbine runner based on fluid-structure interaction analysis[J]. *Tsinghua science and technology*. **Vol.13** (5), pp.587-592.

[17] Wang Z W, Luo Y Y, Fang Y, Wen X J 2005 FSI computation of a Francis runner during load adjustment [J]. *Journal of Hydroelectric engineering*. **vol.24** (8), pp.58-61.

[18] Xiao R F, Wang Z W, Luo Y Y 2007 Stress analysis of Francis turbine runners based on FSI [J]. *Journal of Hydroelectrical Engineering*, **26**(3), pp.120-124.

[19] Zheng X B, Luo X Q, Guo P C 2009 Dynamic stress analysis of axial turbine blade based on CFD [J]. *Journal of Hydroelectrical Engineering*, **28**(3), pp.187-192.

Acknowledgments

This work was supported by the Strategic Priority Research Program of the Chinese Academy of Sciences under grant NO.XDA03040300. We further thank the great help from other members of FDS Team in this research.

Performance test of electromagnetic pump on heavy liquid metal in PREKY-I facility

X L li[1], X D Ma[1], Z Q Zhu[1*], Y Li[1], K F Lv[1,2]

1.Key Laboratory of Neutronics and Radiation Safety, Chinese Academy of Sciences, Institute of Nuclear Energy Safety Technology, Chinese Academy of Sciences, Hefei, Anhui 230031, China; 2.University of Science and Technology of China , Hefei, Anhui 230027, China

Email: zhiqiang.zhu@fds.org.cn

Abstract: Pump is a key sub-system which drives the heavy liquid metal circulation in experimental loops. In the paper, the hydraulic and mechanical performances of an electromagnetic pump (EMP) were tested in the liquid metal test facility named PREKY-I. The test results showed that the EMP worked at good state when the working current was up to 170 ampere. In this condition, the flow rate was $5m^3/h$, and pressure head 7.5bar, when the outlet temperature was kept at 380℃ during the test. The performance was close to the expected design parameters. The EMP had run continuously for 200 hours with stable performance. From the test results, the EMP could be used in KYLIN-II loop, which is the upgrade liquid metal test loop of PREKY-I.

1. Introduction

Fusion and fission energy are the important ways recognized so far to solve the energy crisis problems [1-7]. Fusion reactor liquid lithium lead blanket, because of its inherent characteristics and advantages, is widely recognized internationally as one of the fusion power cladding designs with most promising future [8-17]. Lead-based (lead, lead-bismuth, etc.) reactor, because of its good neutronics, thermal hydraulic and security features, has become one of the important options of the fourth generation advanced nuclear energy systems and one of the main reactor candidates for Accelerator Driven-subcritical Systems (ADS) [18-21].

KYLIN-II is a liquid lead-bismuth experimental loop. It is the important platforms to investigate the key technologies of liquid metal reactors. Pump is a key sub-system which drives the heavy liquid metal circulation in experimental loops. The three-phase cylindrical linear inductor ElectroMagnetic Pump (EMP) is chosen for the liquid metal experimental loop circulation. It has the advantages of absence of moving parts, low noise and vibration level, simplicity of flowrate regulation, easy maintenance and no leakage etc. [22-28]

The EMP for the liquid metal has been designed, produced and tested during the recent two years in the liquid Pb-Bi test facility named PREKY-I. Development of an EMP for operation under the liquid metal loop conditions is a difficult task because of the following reasons.

(1) The EMP is submerged in the lead-bismuth, whose temperature fluctuates depending on the proton beam trip in the range 300-400℃. A thin-walled austenitic steel 316L protective hull liable to

endure continuous thermal fatigue cycles is necessary to protect the EMP and flowmeters from direct contact with the Pb-Bi;

(2) The volume of the pump arrangement is very limited. Therefore, loads of electrical and magnetic circuits of the pumps leads to intensive heat release. Minimize the heat release and increase the thermal resistance between the pump inductors and the protective hull (minimal gaps filled with thermal conductive materials, etc.,) are ways for better cooling.

(3) Operating conditions of the EMP (high temperature, irradiation, magnetic field, mechanical stresses) impose strong restrictions to the structural material.

The problem required the development of an original procedure for the EMP calculation in order to optimize the design and technology as well as a comprehensive testing of the system.

2. Design of the EMP

2.1. Working principle of EMP

As shown in Figure 1, EMP consists a pump groove and two deflectors consisting of the excitation coil 1, the outer core 2, the inner core 3, and two guide grooves 4. Three-phase winding in the excitation coil annular groove produces traveling-wave electromagnetic field moving around the axis of the cylinder in the air gap. This magnetic field causes the induced current of in the liquid metal within the pump groove. Interaction between the induced current and the magnetic field generate pressure gradient and then liquid metal flow.

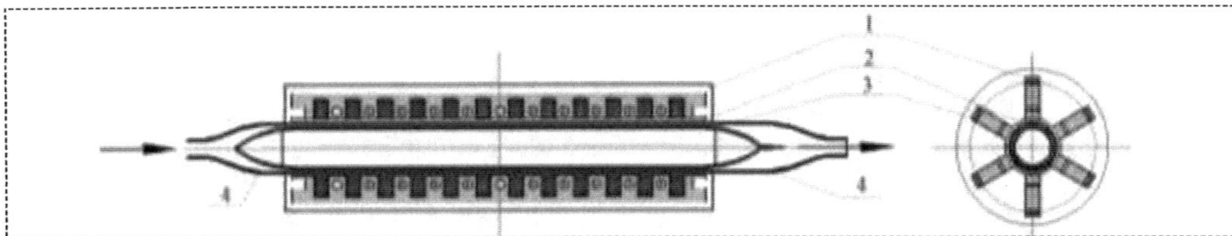

Figure 1. The schematic diagram of EMP working principle

2.2. Design parameters

According KYLIN-II experimental loop design parameters, the main input and design parameters of the EMP is given in Table 1 below:

Table 1. The main parameters of EMP

Parameters	Values
Nominal Flow Rate , m^3/h	5
Pressure , MPa	0.7
Nominal Operating Temperature , ℃	300
Line voltage, V	370
Line current, A	140
Input power, kW	80
Pb-Bi joule heat, kW	30
Winding joule heat, kW	2

3. Experimental set-up

The PREKY-I loop was the pre-study platform of the large high-temperature liquid Pb-Bi loop KYLIN. It was applied to explore the key equipment performance (such as EMP, mechanical pump, heat exchanger etc.,) in liquid Pb-Bi loop, the reliability of measuring instruments and thermal hydraulic measurement technology in Pb-Bi environment. It could provide important experimental data for performance analysis and structural optimization of related measurement instrumentations, including temperature instruments, flow meter, liquid level meter and differential pressure transducers.

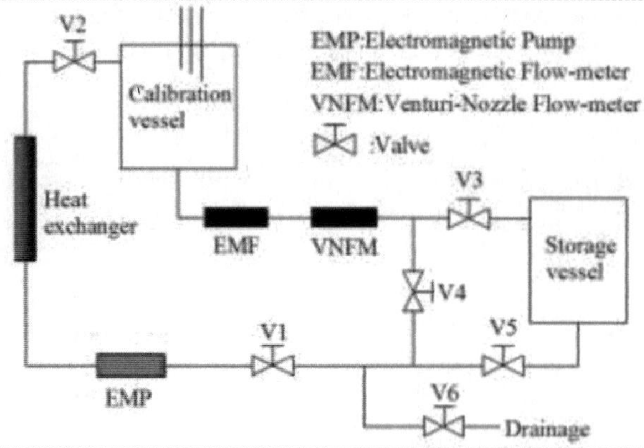

Figure 1. Scheme of PREKY-I

The scheme of PREKY-I was shown in Figure 2. The experimental loop PREKY-I was an isothermal, devious forced Pb-Bi loop, which was composed of main vessel (storage vessel and calibration vessel), EMP, flow meter, temperature sensor, level meter, pressure transducer, liquid metal valves, heaters, heat exchanger, data acquisition and control unit etc.. The main design parameters characterizing the facility were listed in Table 2.

Table 1. Main parameters of PREKY-I

Name	Parameters
Working temperature	200~500°C
Working pressure	0.5~1.0MPa
Velocity	0.5~2.0m/s
Size of flow channel	Φ42(inner)×5mm
Pb-Bi capacity	~1.0t

Several main testing sections were included in the loop under a wide range of operating conditions: (1) flow meter performance test; (2) the temperature comparative experiments both pipe and Pb-Bi; (3) equipment (such as EMP, mechanical pump, heat exchanger etc.,) hydraulic and mechanical performances test; (4) pressure drop testing of valve at different flow rates. Meanwhile, more useful technologies focused on the PREKY itself would provide technical supports for the construction and experiments of KYLIN.

4. Experimental results and discussion

Before test, the PREKY-I experimental needs to be preheated and then filled with liquid Pb-Bi. Flowmeter, thermocouple and pressure sensors which for test using for EMP had been calibrated, and their precisions met the test requirements. The arrangement of PREKY-I test rig are shown in Figure 3.

Figure 3. Appearance of the PREKY-I test rig

4.1. Static test

Static test of EMP is to obtain its static pressures with different phase currents. Firstly, the calibration vessel inlet valve (V2) should be turn off before the EMP was started. Secondly, start the EMP, and adjust the pump input current. Record the EMP inlet pressure (P1) and outlet pressure (P2) before each adjust the pump input current, and the EMP pressure head H= P2-P1. EMP currents relationship with the pressure heads are shown in Figure 4.

From Figure 4, the EMP head with phase current is linearly proportional relationship. With the increase of phase current, the pump head is gradually increased. The greater the phase current, the greater amplitude of pump head increases. The maximum pressure head of EMP is 9.06bar when the working current was up to 170A. When EMP working current is more than 170A, the pump excitation coil temperature is over 200℃, which exceeding the maximum limit of design temperature.

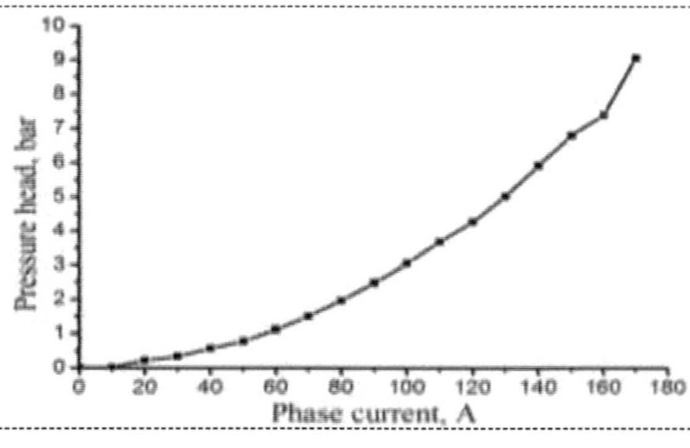

Figure 4. EMP phase current and pressure head Curve (static test)

4.2. Dynamic test

Dynamic test is various hydraulic performances test of the EMP, including the relationship of current, pressure head, flow rate and efficiency, in dynamic testing, calibration tube inlet valve is fully opened. There are two ways to adjust the dynamic test. Firstly, with the EMP inlet valve fully opened, regulate EMP current to get various EMP hydraulic performances; Secondly, with the current of solenoid pump fixed at a certain value, get the hydraulic performance of the EMP by adjusting the EMP inlet valve opening. When the EMP current exceeds 170A, or EMP outlet temperature exceeds 400℃, the test should be stopped.

The following figure 5 (a) is relation curve between static and dynamic pressure head of EMP at different current conditions. From figure 5(a), with current of EMP increases, static and dynamic pressure head both increase. Dynamic pressure head is smaller than the static pressure head, because in the dynamic case, the frictional resistance within device flow path needs to be overcome. When current is 170A, static and dynamic pressure head both reached the maximum value, with static pressure head at 9.06bar, dynamic pressure head at 7.50 bar. As the current increases, the growth rate of the dynamic pressure is less than that of the static pressure head. Definition of difference between static and dynamic pressure head under the same current situation is △P, the following figure 5(b) shown, as variation curve of △P at different currents. As figure 5(b) shows, as the current increases, △P gradually increases. The main reason is that, as the current increases, the driving ability of the electromagnetic pump increases. Liquid Pb-Bi flow rate increases in the loop, and circuit pressure loss increases. According to the pressure loss equation: $\Delta P_h = \frac{1}{2} \times \rho \times v^2$, (Wherein, ρ- density of liquid Pb-Bi, v- average flow rate of liquid Pb-Bi in the loop) as can be seen, the loop pressure loss and flow rate is proportional to the quadratic relationship. The greater the flow rate, the greater the pressure loss in the loop, and the greater the magnitude of the dynamic pressure head of electromagnetic pump decreases.

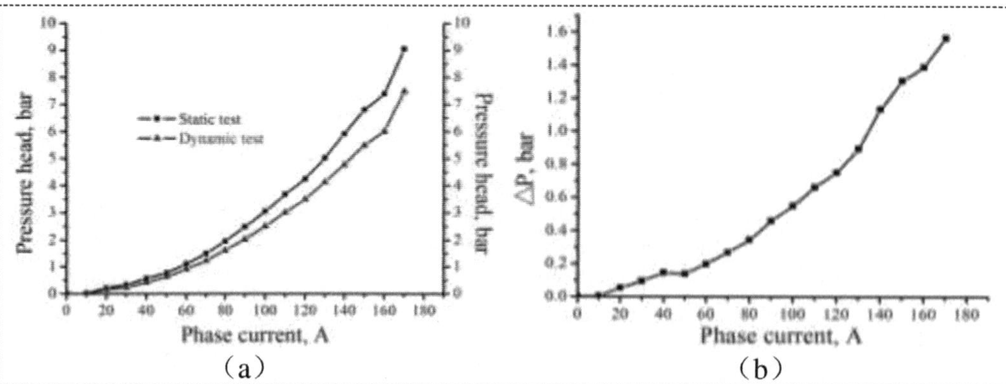

Figure 5. The pressure head (a) and △P(b) curves in static and dynamic test

Figure 6 is performance curve of pressure head of EMP and flow rate under different current conditions. As Figure 6 shows, with the increase of the current, the flow and the pressure head gradually increased; the amplitude of flow growth rate is greater than pressure head growth rate at the beginning. However, as the current increases, the pressure head growth rate is greater than the flow growth rate. When current is 170A, the dynamic pressure head is 8.82bar, corresponding to the flow rate 5.21m³ / h.

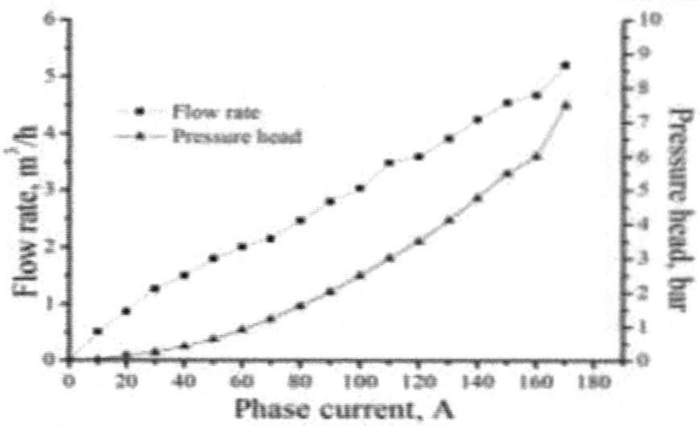

Figure 6. The flow and pressure head curves under different currents

Figure 7 is performance curve between the flow rate of EMP and efficiency under different current conditions. With increase of currents, flow rate and the efficiency gradually increase.

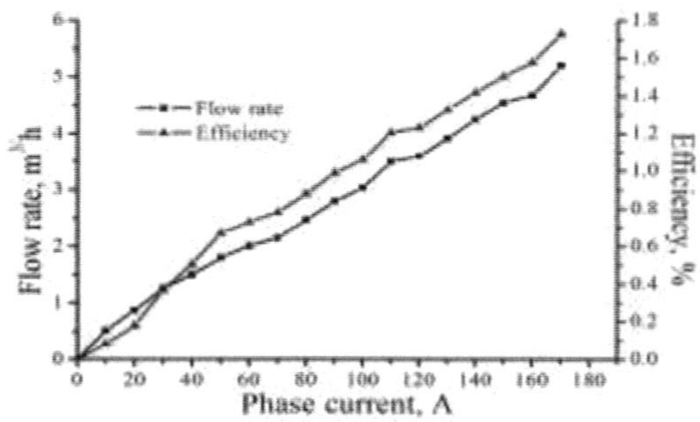

Figure 7. The flow and efficiency curves under different currents

Figure 8 is the relationship between the EMP flow and pressure head which is obtained by adjusting the electromagnetic pump inlet valve opening at fixed electromagnetic pump current. Since the flow is relatively small when the EMP current is <100A, four typical groups of current, respectively 100A, 120A, 150A and 170A, are tested for comparison.

As can be seen from Figure 8, when EMP current is fixed, as the EMP flow rate increases, the pressure head gradually decreases. The main reason is as follows: (1) Under certain current, it's the heat loss fixes. When flow is small, the liquid Pb-Bi temperature within the EMP and the pump coil temperature quickly rise, the performance of the EMP reduced. With increase of flow, temperature of liquid Pb-Bi is gradually reduced, and the temperature of the coil is reduced, thereby the performance of the solenoid pump improves. (2) For EMP in the constant current case, as flow increases, pressure head only increases slightly, which is different from mechanical pump whose pressure head decreased rapidly.

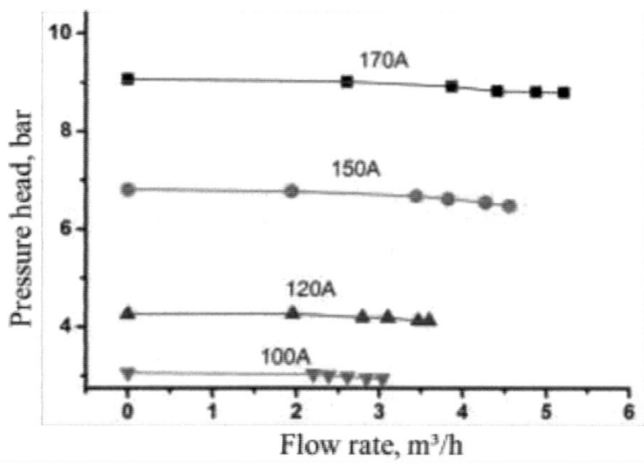

Figure 8. The flow and head curves under different currents

EMP runs on PREKY-I test device for about 200h. During running time various hydraulic performance of the electromagnetic pump is tested. According to test results, the hydraulic performances of EMP are to meet the pre-designed requirements. As is shown in Figure 9, the EMP performance is relatively stable in the entire run-time.

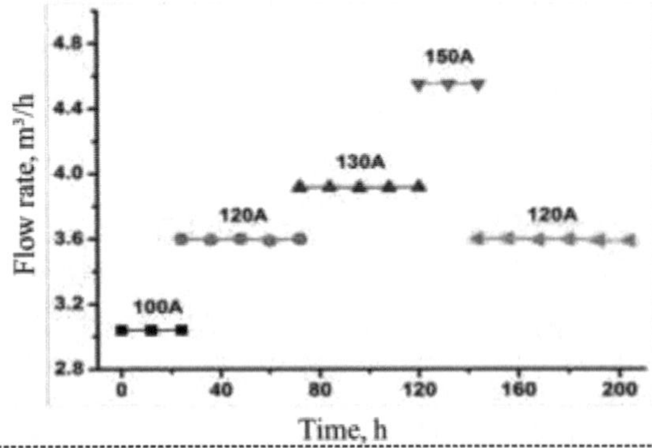

Figure 9. The EMP flow rate and running time during the test

5. Conclusion

In this paper, the EMP performance was tested in the LBE experimental loop. From the test results the following conclusions could be drawn.

(1) The designed EMP reaches a maximum flow of 5m³/h at 170A, maximum pressure head of 7.50bar, which meets the pre-design requirements;

(2) As EMP current increases, the flow rate increases, the circuit pressure loss increases, difference between the dynamic pressure of the pump head and the static pressure head becomes larger;

(3) In the same current, the effect of change of the EMP flow rate on the pressure head is relatively small, mainly because as the flow rate reduces, the pump temperature rises, and the performance degrades.

6. References

[1] Alejaldre C, Marco F, Finzi U et al. 2005 Status report on fusion research. *Nucl Fusion*, **45**(10A):1-28.

[2] Ilhli T, Basu T, Giancarli L et al. 2008 Review of blanket designs for advanced fusion reactors. *Fusion Eng. Des*, **83**(7-9):912-919.

[3] Wu Y, FDS Team 2006 Conceptual design activities of FDS series fusion power plants in China, *Fusion Eng. Des*, **81**:2713-2718.

[4] Wu Y, Zheng S, Zhu X et al. 2006 Conceptual design of the fusion-driven subcritical system FDS-I. *Fusion Eng. Des*, **81**, Part B: 1305-1311.

[5] Wu Y, FDS Team. 2008 Conceptual design of the China fusion power plant FDS-II. *Fusion Eng. Des*, **83**:1683-1689.

[6] Wu Y, Jiang J, Wang M et al. 2011 A fusion-driven subcritical system concept based on viable technologies, *Nucl. Fusion*, **51** (10) 103036.

[7] Wang M, Huang H, Lian C et al. 2015 Conceptual design of lead cooled reactor for hydrogen production. I*nt. J. Hydrogen Energy*, doi: 2015.10.1016/j.ijhydene.04.009.

[8] Antipenkov A, Butko A, Chepovski A et al. 1991 Fusion reactor blanket with Li17-Pb83 eutectic. *Fusion Eng. Des*, **14**(3-4):427-444.

[9] Sardain P, Maisonnier D, Pac L et al. 2006 The European power plant conceptual study: helium-cooled lithium-lead reactor concept. *Fusion Eng. Des*, **81**:2673-2678.

[10] Wu Y 2007 Design status and development strategy of China liquid lithium-lead blankets and related material technology. *Nucl Mater*, **367**:1410-1415.

[11] Wu Y, FDS Team 2007 Conceptual design and testing strategy of a dual functional lithium-lead test blanket module in ITER and EAST. *Nucl. Fusion* **47**(11):1533-1539.

[12] Wu Y, FDS Team. 2007 Design analysis of the China dual-functional lithium lead (DFLL) Test blanket module in ITER. *Fusion Eng. Des*, **82**:1893-1903.

[13] Kumar E, Danan I C, Sandeep I et al. 2008 Preliminary design of Indian test blanket module for ITER. *Fusion Eng. Des*, **83**:1169-1172.
[14] Morely N, Katoh Y, Malang S et al. 2008 Recent research and development for the dual-coolant blanket concept in the US. *Fusion Eng. Des*, **83**:920-927.
[15] Wu Y, FDS Team 2009 Fusion-based hydrogen production reactor and its material selection. *Nucl Mater*, 386-388:122-126.
[16] Wu Y, Huang Q, Zhu Z et al. 2009 Progress in design and development of series liquid lithium-lead experimental loops in China. *Chinese Journal of Nuclear Science and Engineering*, **2**(29):161-169.
[17] Zhu Z, Huang Q, Gao S et al. 2011 Design analysis of DRAGON-IV Pb-Bi loop. *Fusion Eng. Des*, **86**: 2666-2669.
[18] Wu Y, Bai Y, Song Y et al. 2014 Conceptual design of China lead-based research reactor for CLEAR-I. Nuclear Science and Engineering, **Vol, 34, No, 2**:201-208.
[19] Wang M, Huang H, Lian C et al. 2015 Conceptual design of lead cooled reactor for hydrogen production. *Int. J. Hydrogen production. Int. J. Hydrogen Energy*, doi:10,1016/j.ijhydene.2015.04.009.
[20] Wang M, Lian C, Li Y et al. 2015 Preliminary conceptual design of a lead-bismuth cooled small reactor (CLEAR-SR), *Int. J. Hydrogen Energy*, doi:10.1016/j.ijhydene.2015.03.097.
[21] Wu Y, Wang M, Huang Q et al. 2015 Development status and prospects of lead-based reactors. Nuclear Science and Engineering, **Vol.35 No.2**:213-221.
[22] Hideo A, Igor R. Gennady V et al. 2004 Magneto hydrodynamic instability in annular linear induction pump Part I. Experiment and numerical analysis. *Nucl Eng. Des*, **227**:9-50.
[23] Hideo A, Igor R, Gennady V et al. 2006 Magneto hydrodynamic instability in annular linear induction pump Part II. Suppression of instability by phase shift. *Nucl Eng. Des*, **236**,965-974.
[24] Ivanov S, Platacis E, Flerov A et al. 2006 Experience of calculation, design, fabrication and testing of the electromagnetic pump system for the megapie target. *Magneto hydrodynamics*, **Vol. 42, No. 2** - 3, pp. 275 - 280.
[25] Ivanov S, Flerov A 2009 Electromagnetic pump for a liquid metal spallation target: Calculation diagnostics, reliability. *Magneto hydrodynamics*, **Vol. 45, No.2**, pp.239-244.
[26] Dementjev S, Ivanov S, Wohlmuther M 2010 On a concept of electromagnetic pump for the liquid metal target for routine operation in the swish spallation neutron source. *Magneto hydrodynamics*, **Vol. 46, No. 1**, pp. 59 - 67.
[27] Prashant S, Sivakumar L, Rajendra R et al. 2011 Design, development and testing of a large capacity annular linear induction pump. *Science Dierct*, **7**:621-629.
[28] Reyoug H, Bum Y 2012 MHD stability analysis of a liquid sodium flow at the annular gap of an EM pump. *Annals of Nuclear Energy*, **43**:8-12.

Acknowledgements
This work was supported by the Strategic Priority Research Program of the Chinese Academy of Sciences under grant NO.XDA03040300. We further thank the great help from other members of FDS Team in this research.

IOP Conf. Series: Materials Science and Engineering **129** (2016) 012025 doi:10.1088/1757-899X/129/1/012025

Transient thermal-mechanical coupling behavior analysis of mechanical seals during start-up operation

B C Gao[1], X K Meng[1], M X Shen[1] and X D Peng[1]

[1] Engineering Research Center of Process Equipment and its Remanufacturing of Ministry of Education, Zhejiang University of Technology, Hangzhou, 310032, China

E-mail: gaobinchao@163.com

Abstract. A transient thermal-mechanical coupling model for a contacting mechanical seal during start-up has been developed. It takes into consideration the coupling relationship among thermal-mechanical deformation, film thickness, temperature and heat generation. The finite element method and multi-iteration technology are applied to solve the temperature distribution and thermal-mechanical deformation as well as their evolution behavior. Results show that the seal gap transforms from negative coning to positive coning and the contact area of the mechanical seal gradually decreases during start-up. The location of the maximum temperature and maximum contact pressure move from the outer diameter to inside diameter. The heat generation and the friction torque increase sharply at first and then decrease. Meanwhile, the contact force decreases and the fluid film force and leakage rate increase.

1. Introduction

The quality of mechanical seals, which plays a significant role as a vital component in the marine equipment, always directly affects their reliability and safety. Especially under the unsteady conditions, mechanical seals are pressured by instantaneous load, including the mechanical load and the thermal load. Both temperature distribution and face deformation change over time during transient condition, either causing high local temperature so as to heat cracking failure, or oversize leakage. Therefore, it makes a claim for the optimal design of mechanical seals. And accurately predicting the face deformation appears to be particularly important.

Although most of thermal-mechanical models concentrate on steady state conditions [1-3], there are still some theoretical studies dealing with transient conditions. Harp [4] developed a mathematical model to study transient behavior of hydrostatic mechanical seals during the deceleration and acceleration, including fluid film pressure, film thickness and leakage rate variation. Green [5] studied the influence of the balance ratio and thermal time constant on the film thickness, leakage and face deformation etc., under time-varying condition. However, this model overlooked the frictional heat generation and mechanical deformation. Danos [6] developed a numerical model of the THD behavior in transient regime during start-up, which was used to study the viscous heat generation and thermal transfer. It showed that viscous heat generation and axial temperature gradient would result in irreversible thermal instabilities. Tournerie [7] improved a TEHD transient model during start-up, using the method of influence coefficients, to study how the heat generation and minimum film thickness were changing with time. Salant [8] developed an unsteady model of a contact mechanical seal using Duhamel's method, which was used to realize the maximum temperature, average film thickness and leakage rate variation with time.

Most of all the above studies only focused on the time history of the leakage rate, maximum temperature or average film thickness. However, the evolution laws of face deformation and temperature distribution in the interface are neglected, which need further investigation.

This paper presents a transient thermal-mechanical coupling model during start-up, taking into consideration the non-linear and strong coupling relationship among thermal-mechanical deformation, fluid film properties and heat generation. The time history of thermal-mechanical deformation and temperature distribution in the interface are analyzed depending on finite element method and multiple iteration technology.

2. Theoretical Model

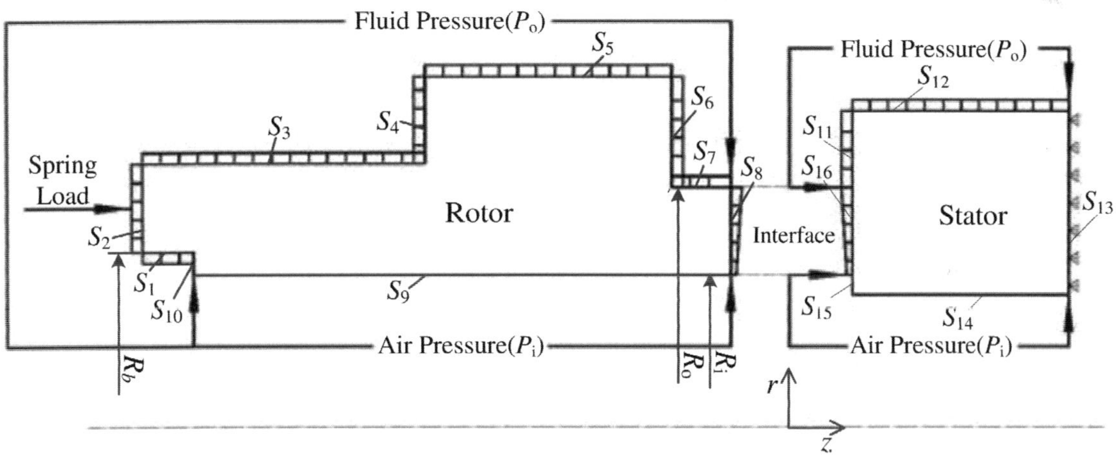

Figure 1. The geometrical configuration and thermal-mechanical boundary conditions of the sealing rings

Basically a mechanical seal is composed of a pair of rings as shown in figure 1. The stator is fixed to the chamber, while the rotor is spring loaded and can float axially. Two rings keep closed by the spring and fluid pressure. It operates in the mixed lubrication regime and the seal gap is full of lubrication film between the sealing faces.

To simplify the calculation process, the following assumptions have been made: (1) the temperature field and stress field of the sealing rings are axially symmetric; (2) the fluid film temperature and viscosity change along the radial direction, which are changeless along the axial direction, and the fluid density remains constant; (3) the fluid film is Newtonian fluid; (4) the influence of thermal lag and inertia is ignored; (5) a quasi-steady-state approach is used.

The character of rotational angular velocity during start-up is

$$\omega = 314.2 \times (1 - e^{-t/0.2}) \tag{1}$$

where ω is rotational angular velocity at any time.

Film temperature distribution and temperature field in the sealing rings are influenced by the heat generation produced by fluid film viscous shear and asperity frictional contact.

$$k_{\mathrm{i}} \frac{\partial^2 T}{\partial r^2} + \frac{k_{\mathrm{i}}}{r} \frac{\partial T}{\partial r} + k_{\mathrm{i}} \frac{\partial^2 T}{\partial z^2} = \rho c \frac{\partial T}{\partial t} \tag{2}$$

where T is temperature of the rings, k_{i} is heat conduction coefficient, i represents the rotor and the stator respectively, which is 1 or 2. Besides, Heat transfer by convection is considered on faces S_1-S_7, S_9-S_{12} and S_{14}-S_{15}[9], as shown in figure 1, governed by the boundary condition,

$$-k_{\mathrm{i}} \frac{\partial T}{\partial n}|_{\Gamma} = h_{\mathrm{ij}} (T - t_{\mathrm{f}}) \tag{3}$$

where the convection coefficient h_{ij} is referenced from Ref. [10], and j represents convection with medium and air respectively which is equal to 1 and 2; t_{f} is the medium temperature entering the seal chamber.

Heat flux generates by friction is transferred by conduction to the rings entirely,

$$-k_{\mathrm{i}} \frac{\partial T}{\partial n}|_{\Gamma} = q \tag{4}$$

where q is heat flux on the sealing face.

The heat flux in the interface is determined by viscous shear q_{v} and asperity frictional contact q_{f},

$$q = q_{\mathrm{v}} + q_{\mathrm{f}} = \mu \frac{r^2 \omega^2}{h} + f p_{\mathrm{c}} r \omega \tag{5}$$

Where μ is dynamic viscosity of the fluid film, and it is modified exponentially with the temperature which is refered to Ref. [11], h is the thickness of the fluid film, f is frictional coefficient, p_{c} is the contact pressure on the sealing face and can be determined by a plastic contact model[1],

$$p_{\mathrm{c}} = H \cdot 2\pi \int_{R_{\mathrm{i}}}^{R_{\mathrm{o}}} \left[\int_{h(r)}^{\infty} \frac{1}{\sigma_{\mathrm{s}} \sqrt{2\pi}} \mathrm{e}^{\left(-z^2/2\sigma_{\mathrm{s}}^2\right)} \mathrm{d}z \right] r \mathrm{d}r \tag{6}$$

where H is yield stress of the softer face material, σ_{s} is combined standard deviation of the roughnesses.

Furthermore, the time domain of the heat conduction equation is discreted by the Galerkin difference scheme, the basic equation is as following:

$$\iint_S \left[2k_{\mathrm{i}} r \left(\frac{\mathrm{d}T}{\mathrm{d}r}\right)\left(\frac{\mathrm{d}N}{\mathrm{d}r}\right) + 2k_{\mathrm{i}} r \left(\frac{\mathrm{d}T}{\mathrm{d}z}\right)\left(\frac{\mathrm{d}N}{\mathrm{d}z}\right) + \frac{3}{\Delta t} r \rho_{\mathrm{i}} c_{\mathrm{i}} T N \right] \mathrm{d}S + \iint_S \left[k_{\mathrm{i}} r \left(\frac{\mathrm{d}T_{\mathrm{old}}}{\mathrm{d}r}\right)\left(\frac{\mathrm{d}N}{\mathrm{d}r}\right) + k_{\mathrm{i}} r \left(\frac{\mathrm{d}T_{\mathrm{old}}}{\mathrm{d}z}\right)\left(\frac{\mathrm{d}N}{\mathrm{d}z}\right) - \frac{3}{\Delta t} r \rho_{\mathrm{i}} c_{\mathrm{i}} T_{\mathrm{old}} N \right] \mathrm{d}S$$
$$+ \int_{l_1} 2h_{\mathrm{ij}} r N (T - T_{\mathrm{w}}) \mathrm{d}l_1 + \int_{l_1} h_{\mathrm{oldij}} r N (T_{\mathrm{old}} - T_{\mathrm{w}}) \mathrm{d}l_1 + \int_{l_2} r N (q_{\mathrm{old}} + 2q) \mathrm{d}l_2 = 0 \tag{7}$$

where N is the interpolation function; Δt is the time step; ρ_i is the density of rings; c_i is the specific heat capacity of rings; q_{old}, h_{oldij}, T_{old} are the heat flux, the convection coefficient and the temperature distribution at previous time, respectively.

Within each time step, two rings are always in force equilibrium,

$$F_{open} - F_{close} = 0 \tag{8}$$

and,

$$F_{open} = 2\pi\int_{R_i}^{R_o} pr\mathrm{d}r + 2\pi\int_{R_i}^{R_o} p_c r\mathrm{d}r \tag{9}$$

$$F_{close} = P_o\pi\left(R_o^2 - R_b^2\right) + P_i\pi\left(R_b^2 - R_i^2\right) + p_{sp}\pi\left(R_o^2 - R_i^2\right) \tag{10}$$

where p is the fluid film pressure, P_o is the sealed pressure, P_i is the atmospheric pressure, p_{sp} is the spring ratio pressure, R_b is the balance radius.

The opening force is affected by the fluid film pressure distribution, which is determined by the Reynolds equation,

$$\frac{\partial}{\partial r}\left(\frac{rh^3}{12\mu}\frac{\partial p}{\partial r}\right) = r\frac{\partial h}{\partial t} \tag{11}$$

Also, the time domain of the Reynolds equation is discreted by the Galerkin difference scheme, the basic equation is as following,

$$2\int_{R_i}^{R_o}\left(\frac{rh^3}{12\mu}\frac{\partial p}{\partial r}\frac{\partial N}{\partial r}\right)\mathrm{d}r + \int_{R_i}^{R_o}\left(\frac{rh_{old}^3}{12\mu_{old}}\frac{\partial p_{old}}{\partial r}\frac{\partial N}{\partial r}\right)\mathrm{d}r - \frac{3}{\Delta t}\int_{R_i}^{R_o}\left(rN\left(h - h_{old}\right)\right)\mathrm{d}r = 0 \tag{12}$$

where h_{old} , μ_{old} and p_{old} are the film thickness, dynamic viscosity and fluid film pressure at previous time.

The fluid film pressure and the contact pressure are decided by the fluid film thickness, which could be described by the face deformation,

$$h(r) = h_m + (\delta_1(r) + \delta_2(r))_{max} - (\delta_1(r) + \delta_2(r)) \tag{13}$$

where $h(r)$ is the film thickness at different radius; h_m is the minimal film thickness; $\delta_i(r)$ is the deformation of sealing faces and compared with original shape, it is positive if the sealing face is stretched, so that negative if compressed; $(\delta_1(r)+\delta_2(r))_{max}$ is the maximum algebraic sum of the face deformation.

Moreover, face deformation is composed of thermal deformation and mechanical deformation. According to the principle of minimum potential energy, the finite element equation of differential equation of elastic mechanics is as follow [12],

$$\frac{1}{2}\iint_S r\{\varepsilon\}^{\mathrm{T}}[D]\{\varepsilon\}\mathrm{d}S-\int_l r\{u\}^{\mathrm{T}}\{Q\}\mathrm{d}r-\iint_S r\{\varepsilon\}^{\mathrm{T}}[D]\{\varepsilon_{\mathrm{T}}\}\mathrm{d}S=0 \tag{14}$$

where $\{\varepsilon\}$ is strain component; $[D]$ is elastic matrix; $\{u\}$ is the displacement vector; $\{Q\}$ is surface pressure including sealed pressure, atmospheric pressure, fluid film pressure and contact pressure; $\{\varepsilon_T\}$ is strain component generated by thermal stress and $\{\varepsilon_T\}=\{\alpha(T-t_b), \alpha(T-t_b), \alpha(T-t_b), 0\}^T$; α is the thermal expansion coefficient; t_b is the room temperature.

3. Numerical procedure

To obtain the sealing performance at each time step, the finite element method and Galerkin difference scheme are used to discrete the spatial domain and time domain respectively and all above equations are solved with each other using multiple iteration technology. The numerical procedure is shown in figure 2 and includes four iteration loops. Firstly, it needs to carry out force equilibrium analysis to obtain the initial film thickness and the temperature filed. Then, the thermal-mechanical coupling analysis under quasi-steady-state would be executed at each instant of time, including the film thickness relaxation iteration, minimal film thickness adjustment and temperature field update[13]. When the fluid film pressure, contact force, thermal-mechanical deformation and temperature distribution in the interface are convergent at the moment, the temperature field at present time will be compared with that at previous time. If it's not convergent, next time step will be done. If it's convergent, the computation is ended and all the parameters changing with time can be obtained.

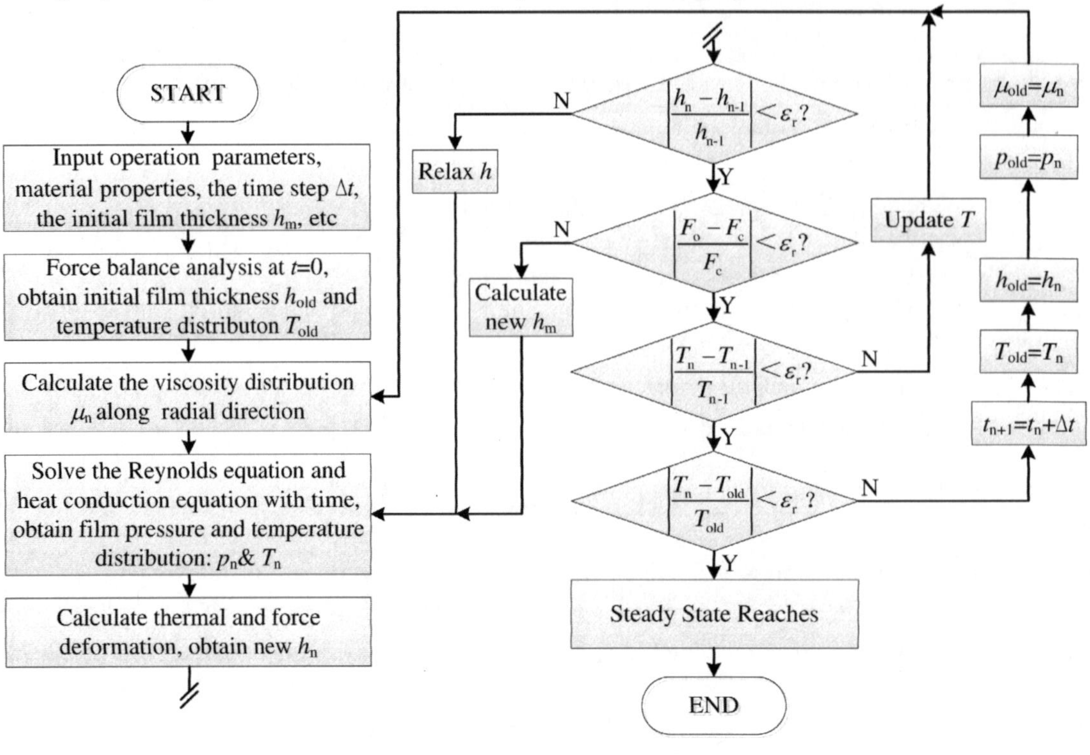

Figure 2. Numerical procedure

4. Results Analysis

The physical parameters and the operating parameters of the computing instance are outlined in table 1 to table 4.

Table 1. material properties of the seal rings

Parameters	Rotor	Stator
Yong's modulus E/ GPa	25	380
Poisson's Ratio υ	0.25	0.27
Thermal conductivity k/ W/(m·℃)	45	150
Thermal expansion coefficient α / 1/℃	6.2×10^{-6}	4.3×10^{-6}
Yield stress H/ MPa	125	——
standard deviation σ_s/ μm	0.15	
friction coefficient f	0.1	

Table 2. fluid properties

Parameters	Values
Thermal conductivity k_f/ W/(m·℃)	0.6256
Density ρ/ kg/m³	994
Dynamic viscosity at 35℃ μ/ Pa·s	0.719×10^{-3}
Specific heat c/ J/(kg·℃)	4178.5
Thermoviscosity coefficient β/ 1/k	0.0175

Table 3. operating parameters

Parameters		Values
Fluid pressure P_o/ MPa		3.45
Atmospheric pressure P_i/ MPa		0.101
Fluid inlet temperature t_f/ ℃		35
Rotation velocity n/ rmp		3000
Spring pressure p_{sp}/ MPa		0.3
convection coefficient h/ W/(m²·℃) ID	$528\omega^{2/3}$	$176\omega^{2/3}$
convection coefficient h/ W/(m²·℃) OD	$1.43\times10^{-3}\omega^{2/3}$	$4.47\times10^{-2}\omega^{2/3}$

Table 4. geometric parameters

Parameters	Values
Balance Ratio B	0.85
External diameter R_o/ mm	53.02
Internal diameter R_i/ mm	48.26
Balance diameter R_b/ mm	49

The film thickness and the contact pressure distribution are illustrated in figure 3 and 5, respectively. And figure 4 shows the maximum and minimum film thickness variation during start-up operation. As shown in figure 3, the seal gap appears to be negative coning and the film thickness distribution is relatively uniform along the radial direction at the beginning of the start-up. Due to the influence of the thermal deformation, the location of the minimum film thickness moves gradually from the outer diameter to the inside one with the increase of the time. In result, positive coning of the seal gap appears to be forming and the film thickness distribution is extremely non-uniform, which is contrary to the initial distribution. Besides, the minimum film thickness increases quickly at the beginning and then quickly reduced to a fixed value which is close to that under steady-state. However, the maximum film thickness is consistently increasing until approaching to a fixed value. It follows that the positive coning is exacerbated, due to the effect of thermal-mechanical coupling. Figure 5 indicates that the location of the maximum contact pressure is gradually moving toward the inside diameter. At the instant of a half second, contact pressure is close to 0 MPa at the outer diameter. Since then, contact area gradually reduced and the maximum contact pressure increases gradually near the inside diameter which can reach 2.5 MPa in the end. As we can see from the film thickness and contact pressure variation, the thermal-mechanical deformation and heat generation also change over time, which will inevitably lead to the higher local temperature and partial wear, even drastically rubbing.

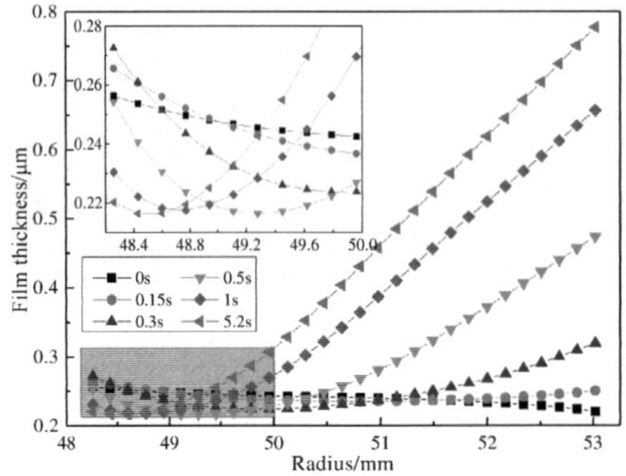

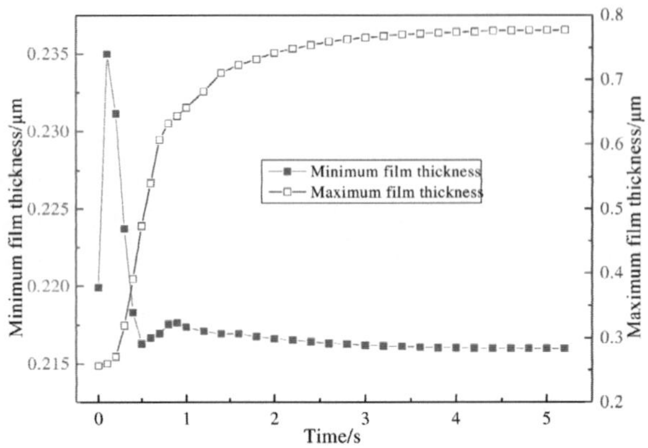

Figure 3. Film thickness distribution variation **Figure 4.** Maximum and minimum film thickness variation

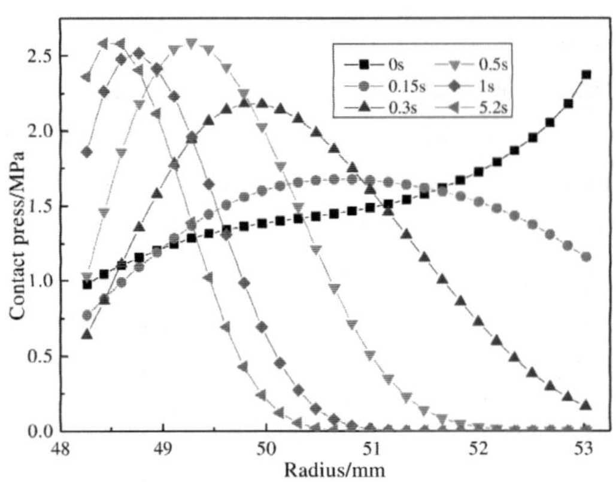

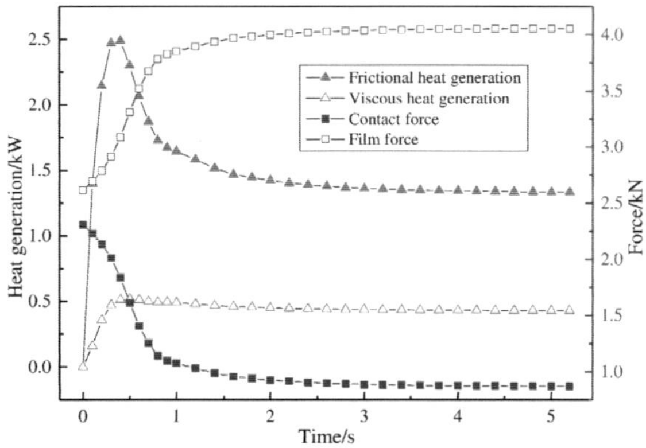

Figure 5. Contact pressure distribution variation **Figure 6.** Force and heat generation variation

Force variation and heat generation variation over time are shown in figure 6. Frictional heat generation and viscous heat generation rapidly increase to a maximum value at the instant of start-up, and frictional heat generation always dominates during start-up. The step change of heat generation is easy to cause large instantaneous temperature and temperature gradient, leading to a higher thermal stress and ultimately resulting in heat cracking failure. In addition, the contact force decreases while the fluid film force increases gradually, and fluid film force always dominates. This is because both the contact area and fluid film pressure drop near the outer diameter gradually decreases with the development of the positive coning.

Figure 7 displays the temperature distribution along the radius direction at the different time. Gradually the face temperature increases and the location of highest temperature moves towards the inside diameter which shows the same change law as the minimum film thickness in figure 4. The main reason lies in the higher heat flux at the location of the minimum film thickness than other locations. Besides, the temperature rise near the outer diameter is much smaller than that near the inside diameter, on account of the better convective heat transfer effect near the outer diameter.

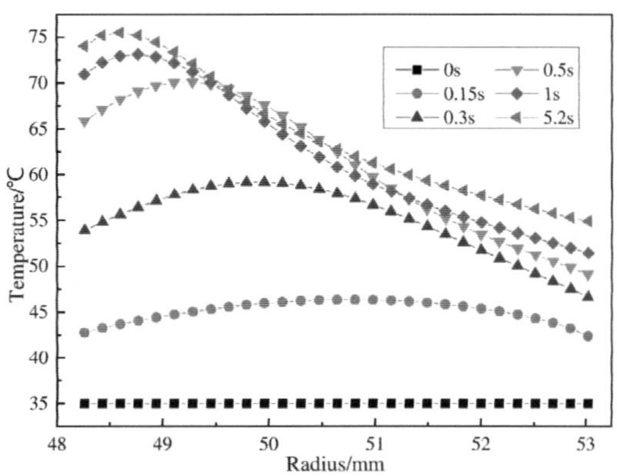

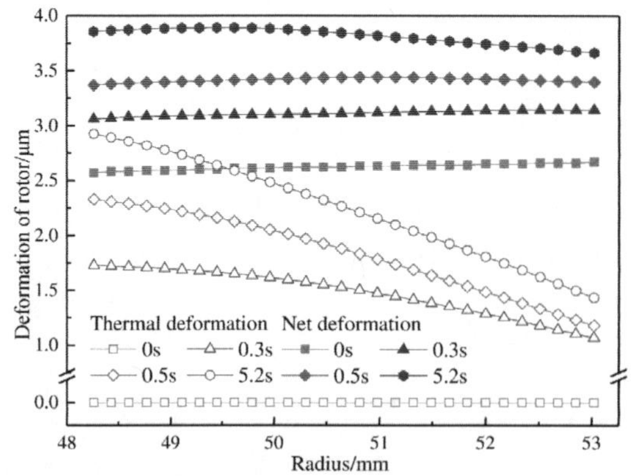

Figure 7. Face temperature distribution variation **Figure 8.** Thermal and total deformation v ariation

Figure 8 shows the thermal deformation and thermal-mechanical deformation of the rotor face in the leakage direction at different instants of time. The thermal deformation increases gradually with time, and causes the seal gap to be positive coning all the time. Similarly, the net deformation increases with time and the seal gap converts from the negative coning into positive coning gradually. This can be explained that the negative coning generated by mechanical deformation is offset by thermal deformation. Thus, the results will have larger deviation when ignoring the influence of temperature field and thermal deformation. Particularly during start-up, seal gap variation caused by the thermal deformation will directly affect the change of film thickness, which gives rise to the redistribution of fluid film pressure, contact pressure and heat generation and finally, the instability of sealing rings.

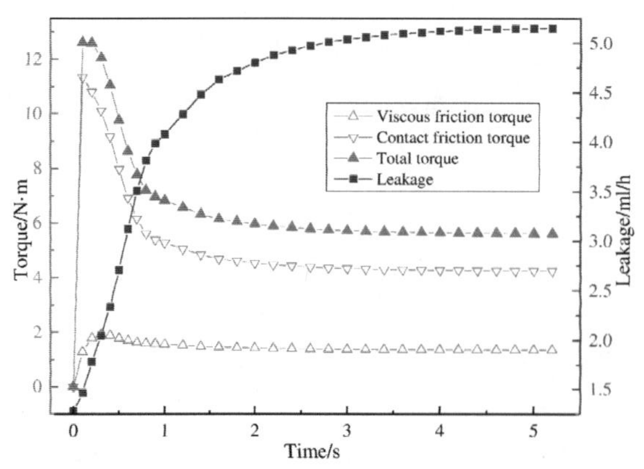

Figure 9. Friction torque and leakage rate variation

The friction torque and leakage rate variation over time are shown in figure 9. Contact friction torque is always greater than viscous friction torque during the start-up and both have a step change and then decrease with time, which is same with the variation in figure 6. The major reasons are that the contact pressure distribution is relatively uniform and the seal gap is negative. The excessive friction torque at the beginning can cause a difficult start-up, and even lead to serious sealing rings fracture. In addition, leakage rate, affected by the seal gap and fluid film viscosity, increases rapidly after the instant of start and then flattens out.

5. Conclusions

(1) The transient thermal-mechanical coupling model of mechanical seals for start-up operation is applies to predict transient behavior during start-up, which can be used to analyze the time history of temperature distribution, thermal and mechanical deformation.

(2) With the increase of the time, the seal gap transforms from the negative coning to the positive coning and the contact area gradually decreases. The location of the maximum temperature and maximum contact pressure move from outer diameter to inside diameter.

(3) The heat generation and torque take on step change and then decrease. Meanwhile, the contact force decreases while fluid film force and leakage rate increase.

Acknowledgments

The work is supported by National Natural Science Foundation (51375449) and Zhejiang Province Natural Science Foundation (LY14E050012).

References

[1] Brunetiere N, Tournerie B and Frene J 2003 TEHD lubrication of mechanical face seals in stable tracking mode: Part 1—Numerical model and experiments *Journal of tribology* **125** 608-616

[2] Brunetiere N, Tournerie B and Frene J 2003 TEHD lubrication of mechanical face seals in stable tracking mode: Part 2—Parametric study *Journal of tribology* **125** 617-627

[3] Galenne E and Pierre-Danos I 2007 Thermo-elasto-hydro-dynamic modeling of hydrostatic seals in reactor coolant pumps *Tribology Transactions* **50** 466-476

[4] Harp S R and Salant R F 1998 Analysis of mechanical seal behavior during transient operation *Journal of tribology* **120** 191-197

[5] Green I 2002 A transient dynamic analysis of mechanical seals including asperity contact and face deformation *Tribology transactions* **45** 284-293

[6] Danos J C, Tournerie B and Frêne J 2003 THD lubrication of mechanical face seals during transient period after start-up: 2D modelisation *Tribology Series* **41** 477-488

[7] Tournerie B, Brunetiere N and Danos J C 2003 2D numerical modelling of the TEHD transient behaviour of mechanical face seals *Sealing Technology* **2003** 10-13

[8] Salant R F and Cao B 2005 Unsteady analysis of a mechanical seal using Duhamel's method *Journal of tribology* **127** 623-631

[9] Thomas S, Brunetière N and Tournerie B 2007 Thermo-elasto-hydro-dynamic behavior of mechanical gas face seals operating at high pressure *Journal of Tribology* **129** 841-850

[10] Lebeck A O 1991 *Principles and design of mechanical face seals* (New Mexico: John Wiley & Sons)

[11] Brunetiere N, Tournerie B and Frene J 2003 A simple and easy-to-use TEHD model for non-contacting liquid face seals *Tribology transactions* **46** 187-192

[12] Cook R D, Malkus D S, Plesha M E and Witt R J 2007 *Concepts and Applications of Finite Element Analysis* (Hoboken: John Wiley & Sons)

[13] Ruan B, Salant R F and Green I 1997 A mixed lubrication model of liquid/gas mechanical face

seals *Tribology transactions* **40** 647-65

IOP Conf. Series: Materials Science and Engineering **129** (2016) 012026 doi:10.1088/1757-899X/129/1/012026

Impeller entrance pre whirl characteristics research

W WU[1], Y Wang[1], Y W Han[1]

[1] National Research Center of Pumps, Jiangsu University, Zhenjiang, Jiangsu 212013, China

Email: Wuwen1122@126.com

Abstract. In order to study the effect of inlet port on the pump performance, the impeller inlet part, should be analyzed for impeller is able to extend the function of water flow to the front of the impeller for a long distance. Impeller flow of pre swirl flow is due to selection of least resistance into the impeller, but the pre swirl in the flow direction according to the impeller blade entrance angle, and the circumferential velocity of flow. The study found that lies in the external characteristic of the pump will be fell when the off-design, but in the case of large flow impeller and impeller in the direction of the front entrance fluid pre whirl steering is on the contrary, when this with little traffic is quite different .this article will study the occurrence, development, and the mechanism of the influence of flow field.

1. Introduction

The earliest research on pre-swirl pump inlet can be traced back to 1909, Steele Watt found the presence of a centrifugal pump suction line pre rotator A.J.Stepanff proposed: At a given speed, the centrifugal liquid can flow along at a certain axially into the working wheel, this flow is called nominal flow. When the flow is less than the rated flow, the direction of the liquid is the same as the rotation direction of the working wheel. When the flow is greater than the nominal flow, the direction of the liquid is opposite to the rotation direction of the working wheel. Lu Linguang, Zhu Honggeng, Chen Wei et al. A new technique for measuring the inlet swirl intensity of the pump inlet is proposed and other research on the inlet flow of the axial flow pump and the optimization of the inlet[1-2];Zhang Yongxue, Song Pengfei, et al. Study the influence of inlet pre-swirl on cavitation. Axial flow pump is a low lift pump[3], with the characteristics of large flow and low head, widely used in the field of farmland drainage, irrigation and large water conservancy projects and other fields. Pump inlet flow is very complex[4-14], the need for in-depth study of the problem is also a lot of. Although there are some studies on the pre-swirl and swirling flow of the inlet fluid of the pump, it is not yet seen in China.

Author: Wen Wu, (1988-), male, Haikou City, Hainan Province, Ph.D., Jiangsu University. Mainly engaged in pump optimization design.

Entrance pre-swirl will lead to the water pump is not working in high efficient area, more serious when can cause the pump cavitation damage. But not with a pre-swirl, there must be bad, sometimes in order to meet the needs of the pump head, is the need to import at some pre-swirl, for example, some in front of the pump to add a wheel in order to induce the want to generate their own pre-swirl. This paper will reveal the generation of the pre-swirl and its different forms in different conditions.

2.Numerical Methods

2.1.Turbulence Model

Similar SST k-ω model and standard k-ω model, but with the following improvements: deformation growth SST k-ω model and k-e model together in the mixing function and dual model. The hybrid function is designed for the near wall region. This region is effective for the standard k-ω model and the free surface, which is effective for the deformation of the k-e modeless k - ω merged model is derived from the cross diffusion equation of ω.Turbulent viscosity took into account the turbulent shear stress of the wave. Model constants of different These improvements make SST k - ω model than the standard k - ω in the field of flow in a wide range of omega has a higher precision and credibility.

Control equation:

Continuity equation: $\frac{\partial \rho}{\partial t}+\nabla.(\rho U)=0$ (1)

Momentum equation: $\frac{\partial \rho U}{\partial t}+\nabla\cdot(\rho U\otimes U)-\nabla\cdot(\mu_{eff}\nabla U)=\nabla\cdot p'+\nabla\cdot(\mu_{eff}\nabla U)^T+B$ (2)

B for volume in the force; μ_{eff} For the effective viscosity; p'For the correct pressure; μ_t For the turbulent viscosity; Its expression is:

$$\mu_{eff}=\mu+\mu_t \quad (3)$$

$$p'=p+\frac{2}{3}\rho k \quad (4)$$

$$\mu_t=C_\mu \rho\frac{k^2}{\varepsilon} \quad (5)$$

Model for both sides, for the turbulent kinetic energy and turbulent kinetic energy dissipation equation

$$\frac{\partial(\rho k)}{\partial t}+\nabla\cdot(\rho U k)=\nabla\cdot[(\mu+\frac{\mu_t}{\sigma_k})\nabla k]+P_k-\rho\varepsilon \quad (6)$$

2.2.Axial flow pump model, meshing

This article select axial flow pump model are studied, Design parameters H=3.2m, Q=392/h, n=1450r/menthe fluid domain as shown in the figure below:

IOP Conf. Series: Materials Science and Engineering **129** (2016) 012026 doi:10.1088/1757-899X/129/1/012026

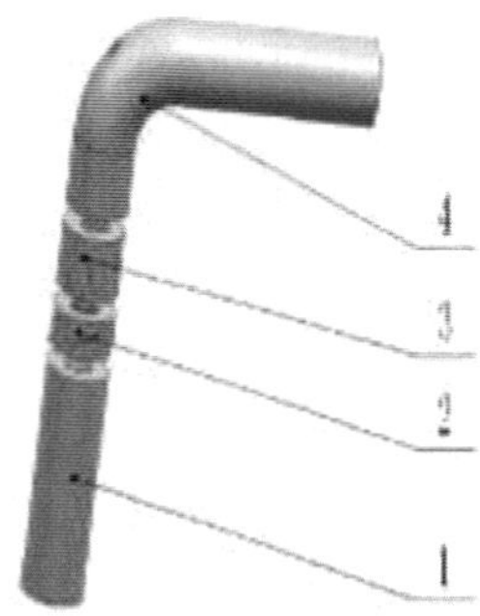

1-Entrance section
2-The rotor section
3-Guide vane section
4-Export section

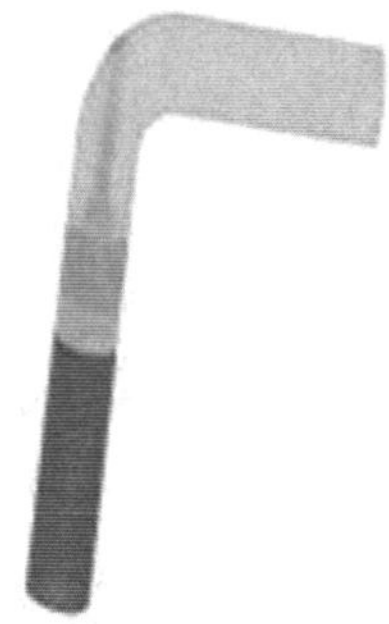

Figure 1. Axial flow pump of the fluid domain **Figure 2.** Mesh

The imported water body diameter, water impeller diameter, the outside diameter of the guide vane water outlet water and the outside diameter is 200mm.In this image, the mesh division of the three dimensional water body of the axial flow pump, the impeller and the guide vanes are used to divide the non-structure tetrahedral mesh. According to the sort of picture, the grid number is shown in the following figure. The import and export have done a proper extension. In the grid independent inspection, the change of the local changes in the velocity gradient is greater.

Entrance section	The rotor section	Guide vane section	Export section
480758	1204493	1107034	991996

Figure 3. Each part of the fluid domain mesh number

2.3.Boundary ***conditions***

The flow field is solved by the governing equations of the rotating coordinate system. The inlet boundary conditions is mass flow, the outlet boundary condition is a standard atmospheric pressure. No-slip wall condition not consider the wall roughness conditions and turbulence model chosen SST k-ω turbulence model.

The finite volume method is used to control the governing equations. In which the velocity and the turbulent kinetic energy viscosity coefficient are divided into two order central difference schemes. The convergence criteria for the calculation are less than the residual value 10^{-4}.

3.The numerical simulation analysis and theoretical analysis

Axial pump suction chamber, also known as Inlet duct, it refers to direct toward the impeller inlet pump pipes. The smooth flow, flow rate and pressure distribution at the impeller inlet, have a great impact on the performance of the pump. Therefore, the design requirements of the suction chamber in the water loss of the minimum conditions of the liquid into the impeller inlet, and ensure that the flow of liquid into the impeller speed distribution. The experimental results show that the hydraulic loss of the suction chamber of axial flow pump is small. But although the water absorption chamber is small, it has a direct influence on the cavitation and safe operation of the pump. Because, if the flow velocity distribution is not uniform, it will greatly reduce the efficiency of the impeller and worsen the cavitation performance of the impeller. In addition, liquid flow in the suction chamber should not produce a vortex, if there is a vortex generated, will cause the impeller inlet flow disorders, resulting in reduced efficiency, resulting in reduced efficiency of the pump, generating vibration. In order to avoid the impact of the pump, the impeller before the entrance to produce a pre rotation. In the inlet of the impeller, the flow of the vortex in the shape of the gradient, which eased the impact. When the liquid flows into the impeller at a large angle, the nature of the hydraulic loss at the entrance of the impeller is caused by the sudden expansion of the liquid or the sudden expansion of the liquid.

In the design of the pump, it is always assumed that the axial distribution of the inlet flow of the impeller is sufficiently uniform. In the derivation of the basic equation of vane pump, it is assumed that the flow is stable and the flow pattern is uniform. If the pump inlet water flow irrational, the absolute velocity of C in the circumferential direction of the Cu1 is zero, and the Euler equation of the vane pump is:

$$H_t = \frac{U_2 C_{U2} - U_1 C_{U1}}{g} = \frac{U_2 C_{U2}}{g} \quad (7)$$

The above equation: H_t-Pump design head; C_{U1}、C_{U2}-Respectively for the impeller import and export of absolute velocity in the circumferential direction of the projection; U_1、U_2-Impeller respectively in the import and export of circumferential velocity of water flow; g-Acceleration of gravity.

Liquid flow to the impeller, and after the outflow of the impeller, is due to the energy gradient to lower than the zero flow of numerical value. Through the impeller, the energy gradient so that the liquid can be gradually increased in the head. According to the principle of energy gradient, liquid flow to choose the path of least resistance impeller and after the outflow of the impeller. The liquid is obtained by the pre rotation so as to enter the impeller flow channel with the smallest obstacle. The inlet angle of the vane of the impeller is related to the flow of the impeller and the circumferential speed of the impeller. All of these three quantities determines the inlet velocity triangles. Obviously, if the liquid is close to the angle of the blade angle, the flow of the liquid is the least resistance. Under the given working wheel speed, only a flow, the liquid can along the axial plane into the wheel. See figure below, when the flow rate of the pump is much smaller than the nominal flow, the direction of the liquid is the same as that of the working wheel, so it may be possible to enter the work in the direction of the β. But when the flow is greater than the nominal flow rate, the pre rotation must be in the opposite direction, in order to meet the requirements of "minimum resistance". Within the actual pump, liquid flow in accordance with the above flow spectrum, just because of the influence of short suction tube and the suction line structure and slightly change. It should be pointed out that, at the entrance of wheel suck liquid prewhirl is not because of the influence of the wheel, it is clear that the wheel can't push liquid to its own direction of rotation in the opposite direction, and when the flow more than nominal flow, often see this kind of situation.

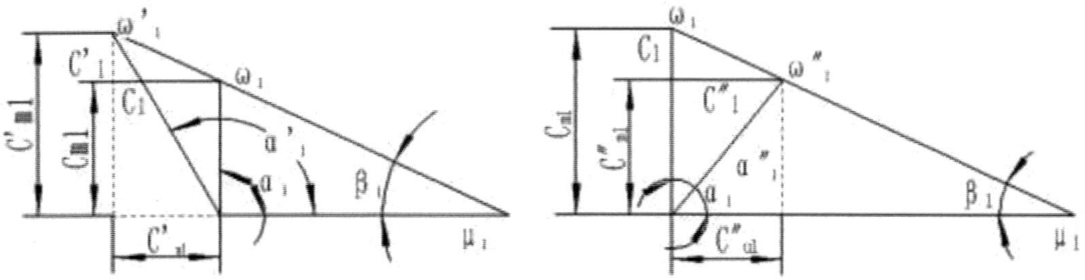

Figure 4. Non-design conditions of the pump inlet velocity triangle

Take two of axial flow pump inlet cross section, respectively to observe two cross section component of circular velocity, respectively.

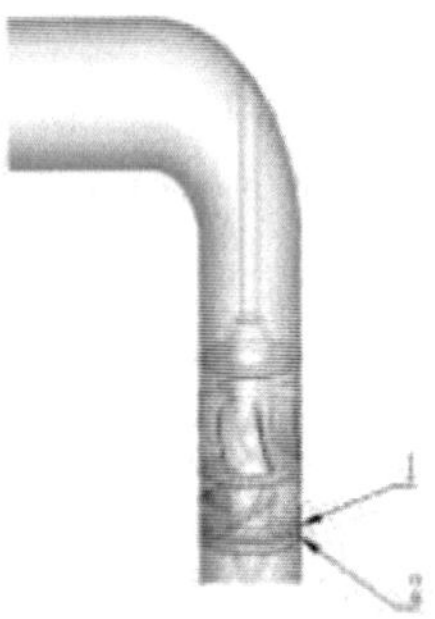

Figure 5. Left for the cross section of two positions

Below for the axial flow pump under different conditions, the entry interface 1 absolute velocity component in the circumferential direction:

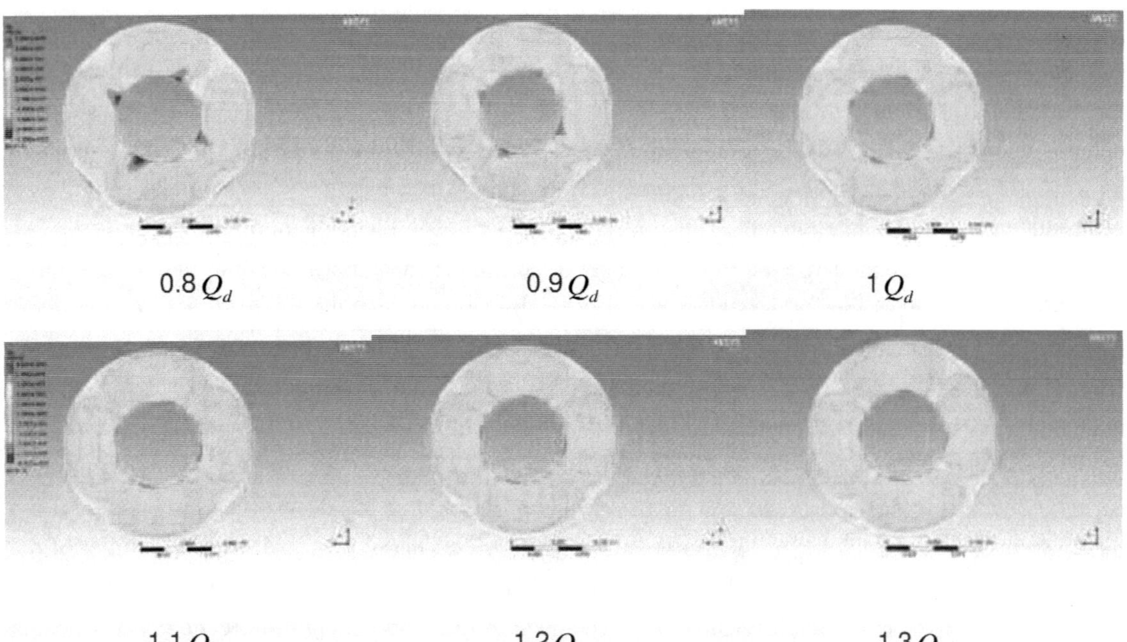

Figure 6.under different conditions, the entry interface 1 absolute velocity component in the circumferential direction;

Below for the axial flow pump under different conditions, the entry interface 2 absolute velocity component in the circumferential direction:

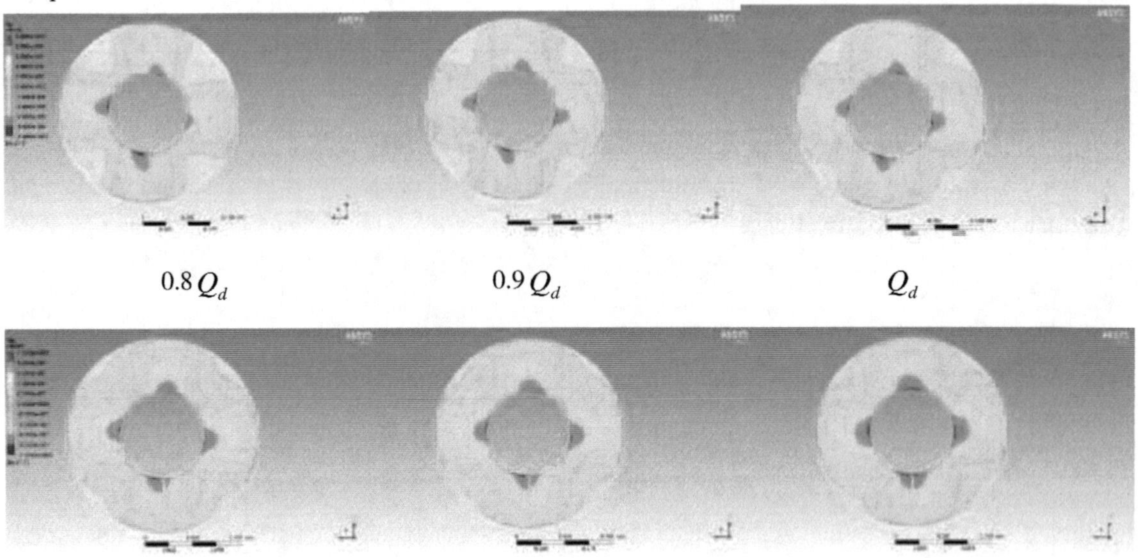

$0.8Q_d$ $0.9Q_d$ Q_d

$1.1Q_d$ $1.2Q_d$ $1.3Q_d$

Figure 7.under different conditions, the entry interface 2 absolute velocity component in the circumferential direction;

From the above we can see from the above, the position of the 1 section of the radius of the impeller wheel position is bigger, relative to the area of cross flow 2 reduced. Axial surface speed increase, according to the velocity triangle, it can be seen in small flow section of Cu value relative to the section 2 Cu value is small, and the big traffic 1 section of Cu is bigger.

People often put the work wheel at the entrance and exit at the loss of the impact. However, this term is not appropriate, because the term "impact" or "crash" in mechanics does not necessarily mean a loss. From the fluid dynamics, it is known that if a collision occurs in the liquid flow direction, most of the energy is recycled (impulse).In the pump, in order to avoid the impact, the liquid flow in the working wheel before the entrance to produce a pre rotation. In the working wheel exit, in the case of a velocity gradient, which will ease the impact. When the liquid flows into the working wheel with a larger impact angle, the nature of the hydraulic loss at the entrance of the working wheel is the loss caused by sudden expansion or sudden diffusion of the liquid after the desulfurization. Wheel at the outlet of the loss is mainly due to the guide leaves lower mean velocity and wheel at the outlet of the high velocity induced in the liquid jet friction loss. It should be noted that, even at maximum efficiency conditions and guide leaves the average speed is still than the circumference of the wheel at the outlet of the absolute velocity points much lower rate. As this is the best working condition, it can not be used to change the area of the guide vane to improve the situation. In addition, the impact damage at the entrance of the guide vane, these losses and the loss of the working wheel inlet, which is the same as the loss of diffusion.

If it is assumed, wheel structure can in flow is rated flow (this time without impact), guarantee in the flow direction and the blade inlet angle, outlet angle consistent, so in this two there is no additional loss. When the flow rate is greater than or less than the nominal flow rate, flow velocity will suddenly change direction and magnitude. Through the above analysis, the axial flow pump is a large flow and the flow of small flow, the entrance and exit of its working wheel will generally have a diffusion loss.

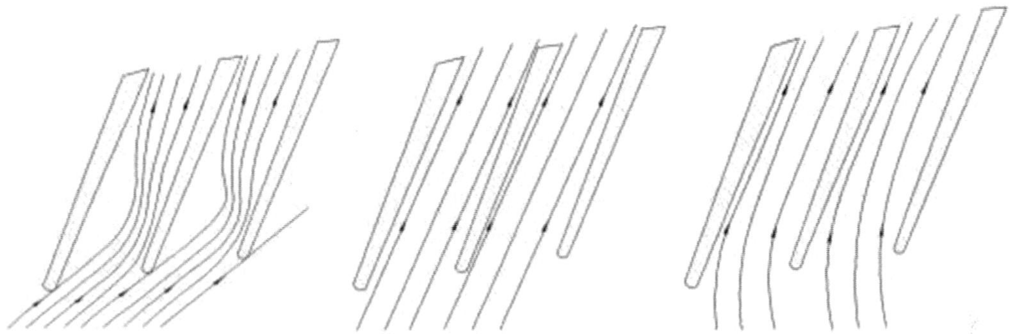

Figure 8. Inlet flow conditions in different conditions

The axial flow pump can reduce the separation of the boundary layer in the large flow rate, and reduce the loss caused by boundary layer separation. The following axial flow pump is given under different operating conditions, the impeller under the different proportions of flow distribution.

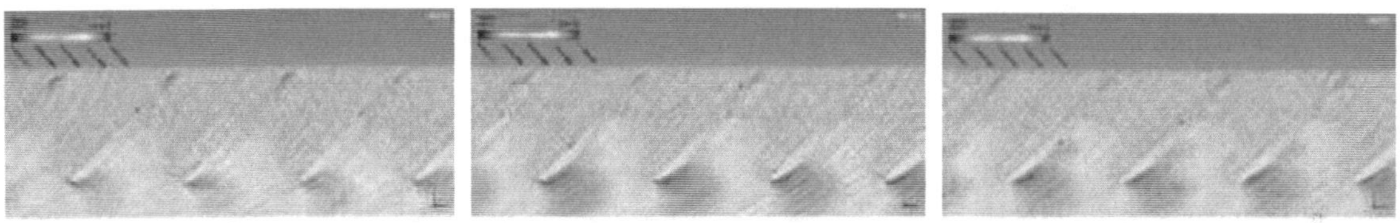

0.8Q 0.9Q Q

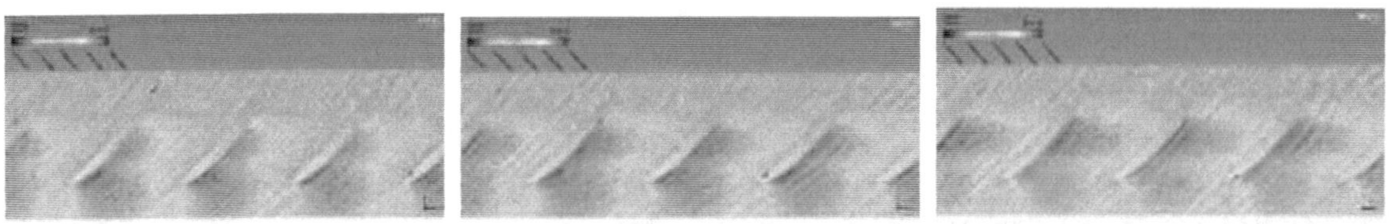

1.1Q 1.2Q 1.3Q

Figure 9. $(r_i - r_h)/(r - r_h) = 0.1$ The radial size ratio of 0.1 on the cylindrical surface of streamline

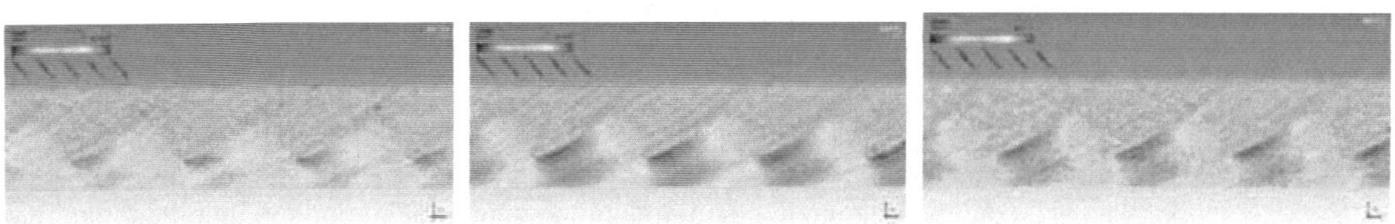

0.8Q 0.9Q Q

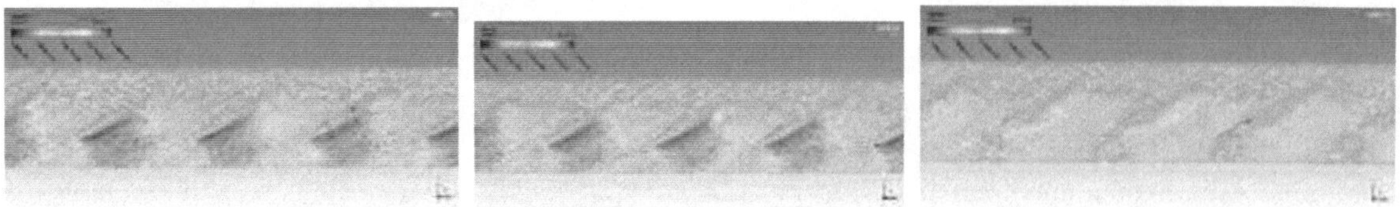

1.1Q 1.2Q 1.3Q

Figure 10. $(r_i - r_h)/(r - r_h) = 0.5$ The radial size ratio of 0.5 on the cylindrical surface of streamline

0.8Q 0.9Q Q

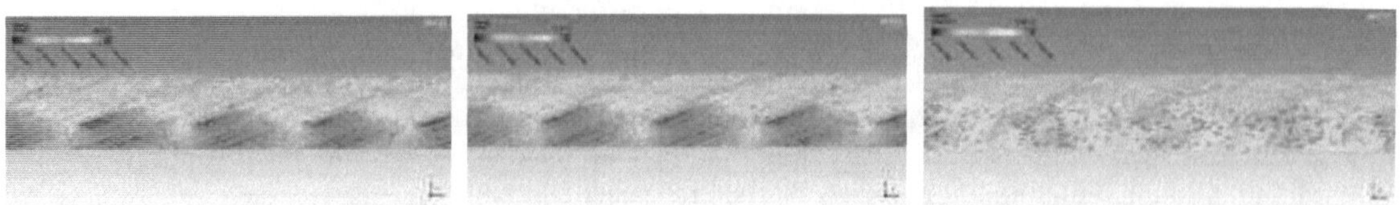

1.1Q 1.2Q 1.3Q

Figure 11. $(r_i - r_h)/(r - r_h) = 0.9$ The radial size ratio of 0.9on the cylindrical surface of streamline

r_i – The radius of the cross - section
r – Working wheel radius
r_h – Wheel hub in radius

When flow is greater than or less than the design flow, the direction of the flow velocity will change, and this can be expressed in the type change caused by the loss.

At the entrance of the loss: $h_{s1} = K_1 \dfrac{\Delta C_{u1}^2}{2g}$ (7)

The exit of the loss: $h_{s2} = K_2 \dfrac{\Delta C_{u2}^2}{2g}$ (8)

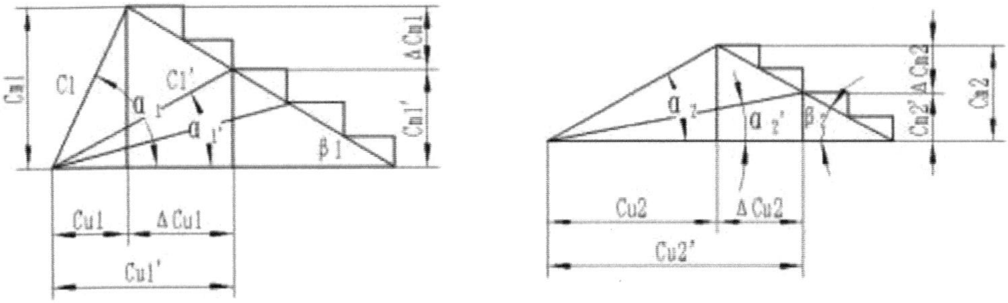

Figure 12.The speed of the pump triangle

$\Delta C_{u1} and \Delta C_{u2}$ are negative values when the flow is greater than the design flow. In the following figure, when the flow rate is designed, the working speed of the working wheel is C_{m1}, And the absolute velocity and the circumference velocity of the liquid flow is the angle of the α.When the flow is reduced （$C_{m1}^{'} < C_{m1}$）, the liquid flow enters the blade angle of the blade, and the circumferential velocity component is $C_{u1}^{'}$ and $\Delta C_{u1} = C_{u1}^{'} - C_{u1}$. Similarly, the flow Q_s, wheel export of axial plane speed C_{m2} ,and absolutely velocity on the circumferential is C_{u2} .When the flow rate is reduced, the circumferential velocity increases to $C_{u2}^{'}$, and the circumferential velocity increment is the difference between the two values of the appeal, that is $\Delta C_{u2}^{'} = C_{u2}^{'} - C_{u2}$.When the flow rate is greater than the Q_s, ΔC_{u2} and ΔC_{u1} are all negative values. It should be pointed out that in the next two graphs, if the C_{m1} increment of the axial velocity and the flow rate is equal to that of the ΔC_{u1} .Similarly, with an increment corresponding to the C_{m2}, ΔC_{u2} increased an equal value. When the flow is greater than or less than Q_s, losses have increased, and are proportional to the square of the flow. In this way, can put the above two losses a formula: the formula of synthesis, this is quadratic parabolic equations.

4. Conclusions

4.1. In heavy traffic and small flow rate, according to the velocity triangle deduction, entrance prewhirl direction is different. The direction of the heavy traffic, prewhirl general contrary to work wheel steering. Prove the import prewhirl is not caused by fluid viscosity

4.2. Reasonable control entry conditions to reduce inlet diffusion loss, improve the efficiency of the pump, provide reference for the design of the pump.

4.3. Import flow caused by flow separation of streamline on the blade surface, and impact on the rotating stall in the pump is critical, this article provides for improving the inlet flow to pump provide reference to some extent avoid rotating stall.

References

[1] Lu L G, Zhu H G et al. 2003 A new technique for measuring the inlet swirl intensity of the pump inlet [J]. *Journal of Irrigation and Drainage*, **22**(3):57~60

[2] Lu L G, Chen W et al. 2012 The influence of axial flow pump operating condition on the inlet of the impeller chamber [J]. *Journal of hydroelectric engineering*, **33**(1):214~219

[3] Zhang Y X, Song T F, Xu C et al. 2014 Experimental and numerical simulation of the influence of the pre swirl adjustment on the cavitation of centrifugal pump [J], *Agricultural Machinery Science*, **45**(9):131~137

[4] Stepanff A J 1980 *Centrifugal and Axial Flow Pumps* [M]. Beijing: China Machine Press, 1980:41~42. (in Chinese)

[5] Sun J Z, Liu D K, Ji S L 1984 Influence of flow pattern at the impeller chamber inlet on axial-flow pump operating [J] *Pump Technology* , (1):30~34. (in Chinese)

[6] Qian J R 1981 Research for flow patterns at the impeller chamber inlet of an axial-flow pump [J]. *Drainage and Irrigation Machinery*, (5):8~12. (in Chinese)

[7] Zhang D S , Shi W D, Guan X F et al. 2009 Measurement on flow field at inlet and outlet of axial-flow pump impeller [J]. *Drainage and Irrigation Machinery*, **27**(4):210~214. (in Chinese)

[8] Vu T C and Shyy W 1990 Navier-Stokes flow Analysis for hydraulic turbine draft tubes [J].*Fluids engineering*, (112):199~204

[9] Li Y J, WANG F J 2007 Numerical investigation of performance of an axial-flow pump with inducer [J]. *Journal of Hydrodynamic*, **19**(6):705~711

[10] Lu L G, Wu K P, Leng Y et al. 2005 Measurement for hydraulic loss of outlet conduit of pumping station[J]. *Drainage and Irrigation Machinery*, **23**(5):23~26. (in Chinese)

[11] Constantinescu G S, Patel V C and Luan Z G, Khonsari M M 2007 Numerical model for simulation of pump-intake flow and vortices [J]. *Journal of Hydraulic Engineering*, (2):123~134

[12] Blazek J 2005 *Computational fluid dynamic: principles and application (Second Edition)* [M]. Oxford: Elsevier Ltd .2005:271~302

[13] Muggii F A, Eisele K, Casey M V 1997 Flow analysis in a pump diffuser-part 2: validation and Limitations of CFD for diffuser flows [J]. *Journal of Hydraulic Engineering*, (11):978~984

[14] Liu Z X, Liu J Z 2008 *Pumps and Pumping Station* [M].Beijing: China and Hydropower press (in Chinese)

Quantitative analysis of backflow of reversible pump-turbine in generating mode

K H Liu[1], Y N Zhang[1]*, J W Li[2] and H Z Xian[1]

[1] Key Laboratory of Condition Monitoring and Control for Power Plant Equipment (Ministry of Education), North China Electric Power University, Beijing China 102206

[2] China Institute of Water Resources and Hydropower Research, Beijing China 100048

*Correspondence author. Email: y.zhang@ncepu.edu.cn

Abstract. Significant vibration and pressure fluctuations are usually observed when pump-turbine is operated during the off-design conditions, especially turbine brake and runaway. The root cause of these instability phenomena is the abnormal unsteady flow (especially the backflow) inside the pump-turbine. In the present paper, numerical simulation method is adopted to investigate the characteristics of the flow inside the whole passage of pump-turbine with two guide vane openings (6° and 21° respectively) and three kinds of operating conditions (turbine, runaway and turbine braking respectively). A quantitative analysis of backflow is performed in both the axial and radial directions and the generation and development of backflow in the pump turbine are revealed with great details.

1. Introduction

The instability problem of pump-turbine caused by abnormal unsteady flow is a big barrier to the development of the pumped storage power plant [1-12]. The safety and efficiency of the units decrease with the increase of pressure fluctuation, vibration, swing of the shaft and so on. For example, rotating stall of pump-turbine could cause significant flow separation and backflow. For further improvement of pump-turbine design, a quantitative description of backflow (e.g. in terms of amplitude, distribution and development stage) is badly needed.

The flow inside the pump-turbine during the off-design conditions (especially turbine brake and runaway modes) is very complex, e.g. generation of significant backflow. The backflow has been investigated by many researchers. Hasmatuchi et al. [7] and Widmer et al. [8] found the backflow phenomenon in the vaneless space during turbine brake. Sun et al. [9] studied the backflow in pump-turbine with misaligned guide vanes. Zhang et al. [10] employed a one-dimensional and three-dimensional coupling approach to analyse backflow in the transient process of pump-turbine. A more detailed study of backflow in pump turbine was performed by Giovanna et al. [11] and Sun et al. [12]. Previous work mainly focused on the backflow in the vaneless space. In fact, the backflow can also be observed in many other domains (e.g. stay vanes or guide vanes) and shows strong non-uniformity in different channels of runner. Those characteristics of backflow in the full domain of pump-turbines have not been revealed yet. Hence, a thorough quantitative analysis of the backflow (e.g. in terms of the distribution and stage of development) is indeed necessary.

In the present paper, a quantitative analysis of backflow in the pump-turbine is performed in the whole passage of the pump-turbine. Distributions of backflow along the axial, radial and circumferential directions in the whole passage of pump-turbines (e.g. stay and guide vanes, vaneless space and runner) are revealed together with their magnitudes. Two typical guide vane openings and three kinds of operating conditions are employed for comparison.

2. Description of method

In the present paper, the inner flow of pump-turbine is numerically studied in generating mode during off-design conditions. Six operating conditions with two different guide vane openings of pump-turbine were quantitatively analysed with a focus on the flow state of the pump-turbine. The simulated domain covers the whole fluid passages of pump-turbine with geometrical parameters shown in table 1.

Table 1. Parameters and corresponding values of simulated pump-turbine

Parameters	Values
Number of blades	9
Number of guide vanes	20
Number of stay vanes	20
Diameter of runner outlet	0.5 m

Two typical guide vane openings (6° and 21° respectively) were selected for the analysis with 6° representing the maximum opening of prototype pump-turbine and 21° demonstrating strong S-shaped characteristics. Three kinds of operating conditions (i.e. turbine, runaway and turbine brake modes) were selected for each guide vane opening. For the simplicity of demonstration, the velocity field was adopted to recognize the backflow. As the runner is a rotating part of the machinery, the relative velocity (or streamline) with the runner blade as the reference frame was adopted for the analysis. In the following analysis, the streamline was colored by the turbulence kinetic energy.

In generating mode, the fluid passes through the spiral case, guide vanes, vaneless space, impeller, and draft tube in sequence. In order to quantify the comparison, velocity in the rotating coordinate can be decomposed into three components (referring to Fig.1): circumferential velocity, axial velocity and radial velocity respectively. Through such kind of decomposition, the radial velocity can be employed to represent the existence and developing stage of backflow in the pump-turbine.

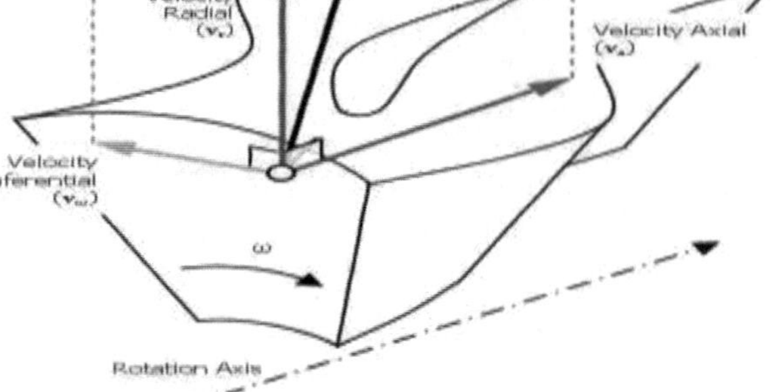

Figure 1. Decomposition of the velocity in rotating coordinates [13]

Backflow could lead to the decrease of total flow entering into the runner, and also cause the generation of vortex. A large amount of backflow occur at off-design conditions of pump-turbine, especially during runaway and turbine brake modes.

In the present paper, a cross-section surface (named as "surface A" in the following analysis) in the vaneless space between the runner and guide vanes was employed for the analysis. The surface was colored by the magnitude of radial velocity. The vector perpendicular to the surface A represents the radial direction with inside being the positive direction and outside being the negative direction. The magnitude of radial velocity is described by the different color and sizes of vectors.

To quantify the distributions of backflow, backflow discharge was calculated in several specific surfaces for all kinds of operating condition and guide vane openings based on the numerical simulations. The total flow rate can be calculated as

$$q=\int_{s} \vec{v} ds, \tag{1}$$

where q is the total flow rate; $\vec{v}$ is the vector of radial velocity. q_{in} and q_{out} are defined as,

$$q_{in}=\int_{s} v_{in} ds, \tag{2}$$

$$q_{out}=\int_{s} v_{out} ds, \tag{3}$$

where v_{in} stands for the radial velocity in the direction from the guide vane to the runner and v_{out} represents the radial velocity in the opposite direction. Then, the total flow rate can be represented as

$$q=q_{in}-q_{out}. \tag{4}$$

The backflow distribution varies with the fluid passing passages significantly along the flow direction. To obtain the variations of backflow along the flow direction or axial direction, the whole passage of pump-turbine was divided to three domains (i.e. guide vane, vanless space and runner).

Figure 2. Selected surfaces and line in runner and guide vane for the analysis

The surfaces or lines employed for quantitative study in the present paper are shown in Figure 2. A line between the hub and shroud in the runner channels was selected for analysis. The backflow of each blade channel was shown using the inlet surfaces of channels (as shown in Figure 2) with "channel 1" representing the inlet surface of the first channel and the other channels numbered counter-clockwise along the circumference direction of the runner. "Surface 1-4" divide "surface A" equally into four parts from top to bottom. Surfaces representing inlet and outlet of guide vanes respectively fit the boundary of the guide vanes.

3. Results and discussions

Firstly, the backflows for different guide vane openings and operating conditions are shown. Figures 3 and 4 display the vector of radial velocity on surface A and relative streamline in the runner. The structure and distribution of vortex can also be observed through the distribution of streamline near the surface A. The color of streamline represents the magnitude of turbulent kinetic energy. The distributions of streamline and turbulent kinetic energy are symmetrical in turbine mode, whereas they are uneven in turbine brake and runaway modes. Duo to a lot of vortex structure blocking the passage during off-design conditions, the internal flow field is distorted significantly and turbulent kinetic energy at runner inlet is higher than one at the other area.

The generation of vortex structure is closely connected with the backflow described by the radial velocity hence the vector distribution on surface A will be emphatically analysed. During off-design conditions (e.g. turbine brake and runaway modes), backflow mainly appears at the runner inlet with strong non-uniformity. The mechanism of the generation of backflow can be explained from two viewpoints. On the one hand, the coming flow collides with the blades when it enters into the runner, leading to the increase of flow resistance and decrease of kinetic energy. On the other hand, the centrifugal force in the flow passage is large for the runner with long channels (e.g. those in the pump turbines) and when the centrifugal force is greater than the flow inertia force, the backflow generates. The backflow disappears when pump-turbine operated at the turbine mode as shown in Figure 3c and also previous work [14]. Therefore, special attention will be paid to the characteristics of backflow during turbine brake and runaway modes.

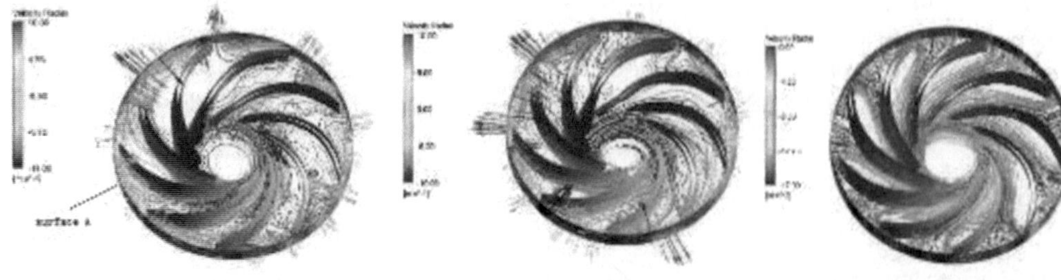

(a) (b) (c)

Figure 3. The backflow at surface A and relative streamlines in the runner during different conditions of pump-turbine with guide vane openings 21°. Surface A was colored with radial velocity and the streamline was colored with turbulence kinetic energy. (a) braking operation; (b) runaway; (c) turbine operation.

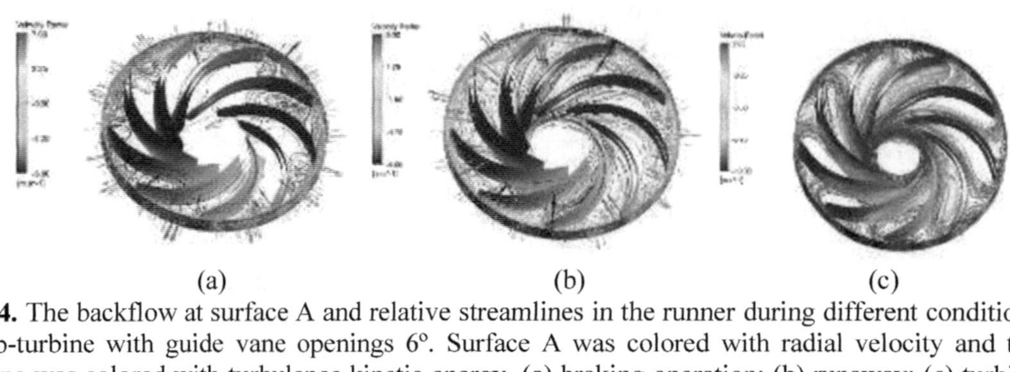

(a) (b) (c)

Figure 4. The backflow at surface A and relative streamlines in the runner during different conditions of pump-turbine with guide vane openings 6°. Surface A was colored with radial velocity and the streamline was colored with turbulence kinetic energy. (a) braking operation; (b) runaway; (c) turbine operation.

Comparing Figures 3 and 4 (with guide vane opening respectively at 21° and 6°), it is obvious that the flow condition gets worse for large guide vane opening in terms of the maximum radial velocity and vortex structures in channel.

The flow state also varies with fluid passing zones. One typical line along the axial direction in the runner and several cross-sectional surfaces in three domains were selected to analyse the distributions of backflow along axial direction and radial direction.

Figure 5 shows the variation of radial velocity from shroud to hub in the runner. Backflow can hardly be observed near the shroud and it occurs at 40% of the total height between the shroud and the hub at the shroud side. After 40% point, the magnitude and direction of radial velocity (V_r) fluctuate with the increase of the height. These phenomena demonstrate that backflow shows strong spatial distribution in the axial direction with significant occurrence near the hub.

As a periodical distribution of backflow in circumferential direction of the runner was observed in Figures 3 and 4, Figure 6 further shows the backflow discharge at each inlet of blade channel. The channels were numbered in Sec. 2. When the guide vane opening is 6°, the backflows in different channels show no observable difference. Whereas, when the guide vane opening is 21°, a remarkable difference among the backflows in different channels can be observed with the biggest difference of discharge even up to 0.16m³/s. To be emphasized, operational modes and guide vane openings are also paramount parameters determining the characteristics of backflow.

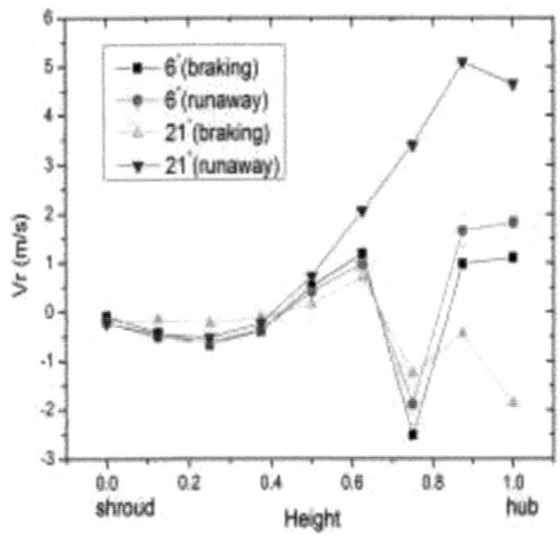

Figure 5. The variations of radial velocity versus the non-dimensional height of runner (from shroud to hub) during different operational conditions.

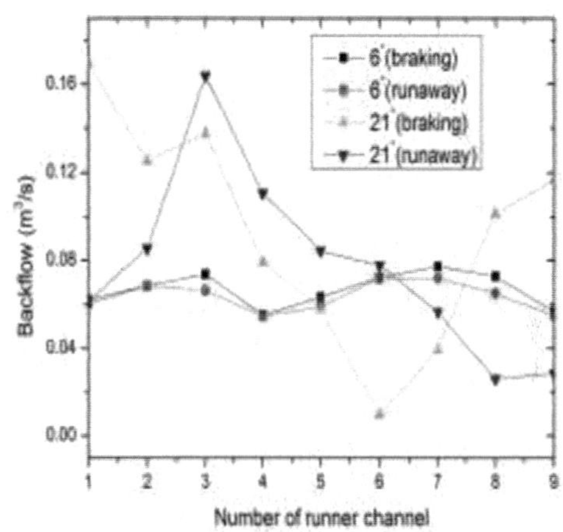

Figure 6. The variations of backflow discharge versus runner channels during different conditions. The runner channels is numbered following the direction of counter-clockwise.

To further demonstrate the distribution of backflow along the axial direction, surface A was divided into four parts with surface 1 representing the top layer near the hub and surface 4 representing the bottom layer near the shroud. The variations of backflow discharge along the axial direction of the surface A is quite different between 21° and 6°.When the guide vane opening is 6°, the backflow discharge is high in the middle part of the surface A but it is small at the hub and shroud. At 21°, the backflow discharge slightly increases with the variations of height from the hub to the shroud.

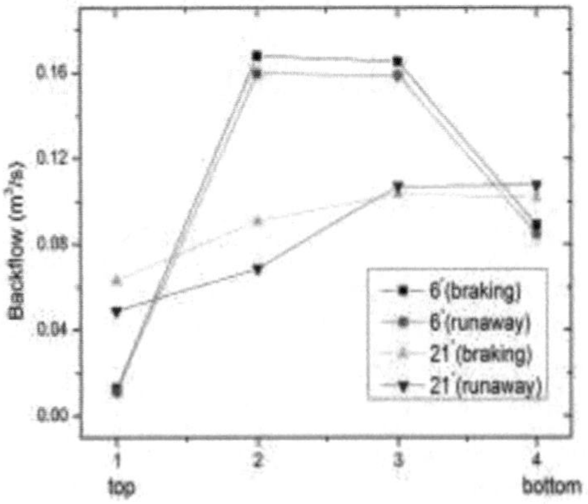

Figure 7. The variation of backflow discharge versus the height of surface A during different operational conditions. 1 represents the top sub-surface of surface A, and so on.

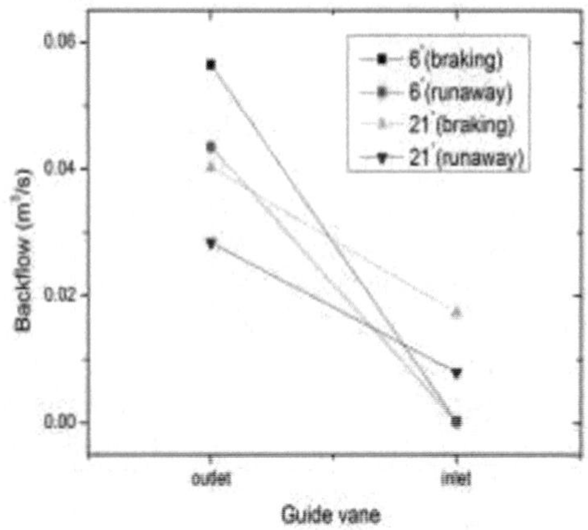

Figure 8. The backflow discharge at the outlet and the inlet of guide vanes during different operational conditions.

Figure 8 represents the backflow discharge at the inlet and outlet of guide vane. The backflow at the outlet is obviously higher than one at the inlet, especially when the guide vane opening is 6°. Figure 8 also demonstrates that backflow is also affected by the operational conditions.

4. Conclusion

The backflow of pump-turbine is strongly influenced by the guide vane opening and operational conditions. Through quantitative analysis of backflow in different zones and directions, the main findings of the present paper can be summarized as following:

1. The backflow occurs during turbine brake and runaway modes of pump-turbine while no significant backflow can be observed for turbine mode.
2. The distribution of backflow shows strong spatial characteristics. For example, at 21°, strong backflow shows near the hub in the runner channel while the backflow is significant near the shroud in the vaneless space in turbine brake mode. In the runner, backflow also distributes non-uniformly in different channels.
3. Guide vane opening is a critical parameter determining the characteristics of backflows.

References

[1] Zeng W, Yang J and Cheng Y 2015 Construction of pump-turbine characteristics at any specific speed by domain-partitioned transformation *Journal of Fluids Engineering* **137(3)** 031103

[2] Zeng W, Yang and Guo W 2015 Runaway Instability of Pump-Turbines in S-Shaped Regions Considering Water Compressibility *Journal of Fluids Engineering* **137(5)** 051401

[3] Zeng W, Yang J D, Cheng Y G and Guo W C 2015 Formulae for the intersecting curves of pump-turbine characteristic curves with coordinate planes in three-dimensional parameter space *Proceedings of the Institution of Mechanical Engineers, Part A: Journal of Power and Energy* 0957650915572840

[4] Yang W and Xiao R 2014 Multiobjective optimization design of a pump–turbine impeller based on an inverse design using a combination optimization strategy *Journal of Fluids Engineering* **136(1)** 014501.

[5] Tao R, Xiao R, Yang W, Wang and Liu W 2014 Optimization for Cavitation Inception Performance of Pump-Turbine in Pump Mode Based on Genetic Algorithm *Mathematical Problems in Engineering* **2014** 234615
[6] Ran H, Luo X, Zhu L 2012 Experimental study of the pressure fluctuations in a pump turbine at large partial flow conditions *Chinese Journal of Mechanical Engineering* **25(6)** 1205-09
[7] Hasmatuchi V, Farhat M, Roth S, Botero F and Avellan F 2011 Experimental evidence of rotating stall in a pump-turbine at off-design conditions in generating mode *Journal of Fluids Engineering* **133(5)** 051104
[8] Widmer C, Staubli T and Ledergerber N 2011 Unstable characteristics and rotating stall in turbine brake operation of pump-turbines *Journal of Fluids Engineering* **133(4)** 041101
[9] Sun H, Xiao R, Liu W and Wang F Analysis of S characteristics and pressure pulsations in a pump-turbine with misaligned guide vanes *Journal of fluids engineering* **135(5)** 051101
[10] Zhang X X and Cheng Y G 2012 Simulation of hydraulic transients in hydropower systems using the 1-D-3-D coupling approach *Journal of Hydrodynamics Ser B* **24(4)** 595-604
[11] Cavazzini G, Covi A, Pavesi G and Ardizzon G 2015 Analysis of the unstable behavior of a pump-turbine in turbine mode: fluid-dynamical and spectral characterization of the S-Shape characteristic *Journal of Fluids Engineering* **138(2)** 021105
[12] Sun H, Xiao R, Wang F, Xiao Y and Liu W 2015 Analysis of the pump-turbine S characteristics using the detached eddy simulation method *Chinese Journal of Mechanical Engine*ering **28(1)** 115-122
[13] 2013 *ANSYS CFD-Post User's Guide* p 306
[14] Wang L Q, Liu J T and Liang C H 2011 *Journal of drainage and irrigation machinery engineering* **29(1)** 1-5

Acknowledgments
This work was financially supported by the Open Research Fund Program of State key Laboratory of Hydroscience and Engineering (Project No.: sklhse-2015-E-01) the National Natural Science Foundation of China (Project No.: 51506051).

Numerical analysis on the cavitation and unsteady flow in a scroll hydraulic pump

S H Sun, P C Guo, Y Huang, J L Zuo, X Q Luo

State Key Laboratory Base of Eco-hydraulic Engineering in Arid Area, Xi'an University of Technology, Xi'an, Shaanxi, 710048, China

E-mail:shh.sun@foxmail.com

Abstract. This paper presents numerical analysis of unsteady flow in a scroll hydraulic pump to discover its flow mechanism. The dynamic mesh model has to be used to simulate the flow field unsteadily. The unsteady flow patterns and pressure distributions in the suction, squeezing and discharge chamber are analysed. The suction process continues until the crank angle reaches the 320 degree. Then the pressure in the chamber rises instantaneously, and the fluid begins to flow out from the chamber. Because of the high pressure difference at the clearance, the jet flow and the vortex appear, and the large flow losses generates with them. In addition, the velocity and static pressure distribution in the two symmetry crescent suction chamber is different remarkably. One reason is that the location of suction port cannot be set symmetrically for the simplification of the pump structure. Another reason for that is the fluid is impelled by different part of the orbiting scroll. The asymmetric pressure distribution will result in the extra force on the scroll. The cavitation generates at the negative pressure region. Therefore, the unsteady simulation shows some important phenomena. The structure of the scroll pump need to be optimized to reduce the maximum pressure, weaken the jet flow, vortex and the uneven pressure distribution to ensure the pump working safely and efficiently.

1. Introduction

Scroll pump was derived from the scroll compressor, which has been used for air-conditioner. The working fluid in scroll pump is liquid, while the fluid in scroll compressor is gas. So the structure of scroll pump is different form the scroll compressor. Because the liquid is almost incompressible, the scroll pump has only a pair of working chamber, as shown in Figure 1. And the number of the scroll profile rotation is 1.5 theoretically. The scroll pump is a kind of positive displacement pump. It is suitable to transport the high viscous fluids, and allows the high gas content in the fluid. The flow rate of the pump is usually small and its lift changes little with the volume flow rate. Compared to other positive pump, the scroll pump operates smoothly, and its noise and vibration are low. Also, the structure of scroll pump is simple, which results in its high reliability.

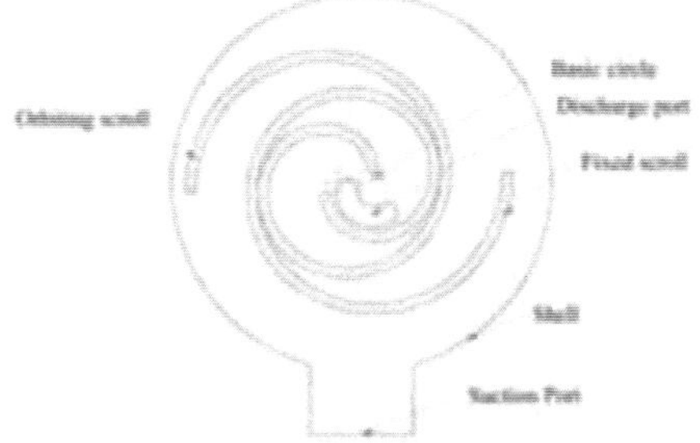

Figure 1. Working principle of scroll pump.

The scroll pump was studied by Zhu Jiang at first[1]. In his research, a mathematic model of a scroll oil pump was developed. With the model, the influence of the discharge port on the scroll pump was analyzed. Then the prototype was manufactured and tested, the results indicated that the high pressure fluctuation existed in the working chamber, which was related to the position angle of discharge groove, the oil viscosity and the leakage clearance[2]. Then, the researcher Jiang Bo and Wang Disheng developed the scroll oil pump SCP-1.08/0.6, which adopted the straight line involute and double acting structure[3]. Based on the pump, the researcher established the mathematic model which included the leakage model and described the working process of the pump[4-7]. With the model, the pressure distribution in the working chamber and the performance of the pump were predicted. The pump was tested. The tested results showed the variation of the pressure with the crank angle in the working chamber, and indicated that the maximum pressure happened at the end of the suction process or the beginning of the discharge process. Moreover, the researcher supported that the maximum pressure can be reduced by enlarging the axial clearance of the pump. Yan and Qu studied the influence of the rotational speed on the pump[8]. At the high speed, the pressure in suction chamber would be lower than the vaporizing pressure and the cavitations happened, which leads to high pressure fluctuation and vibration; at the low speed, the leakage increased and the pump efficiency decreased; so the optimal rotational speed existed. Although the researcher has obtained a lot of achievements, there are many problems unsolved about the scroll pump. One reason is that the present mathematic model was zero-dimension model which is based on the assumption that the pressure and temperature in the working chamber are uniformed. Actually, because of the flow loss and the structure of the working chamber, the fluid parameters in one chamber are inhomogeneous, and it is three-dimensional and unsteady flow in the chamber. So the zero-dimension model can not predicted the pump accurately.

Nowadays, the technology of computer fluid dynamics (CFD) has been applied in scroll compressor[9,10], and can help the researchers to obtain more accurate pressure and temperature values in the working chamber. However, the CFD has merely been used in the scroll pump. In the present paper, numerical analysis of unsteady flow in a scroll hydraulic pump has been done to discover its flow mechanism. The flow field in the scroll pump is figured out, and some special phenomena about the pressure distribution in the chamber are revealed. The reason for the high pressure fluctuation has been analyzed, and the influence of meshing clearance on the pressure field and fluctuation will be studied. And some useful advices about the scroll pump are proposed to improve the performance of the scroll pump.

IOP Conf. Series: Materials Science and Engineering **129** (2016) 012028 doi:10.1088/1757-899X/129/1/012028

2. Mathematic model

The scroll pump consists of many parts, such as the fixed scroll, the orbiting scroll, the main shaft, the motor and the shell. But in this paper, the mathematic model is only relevant to the fluid domain in the scroll chamber, so the main research parts are the fixed scroll and orbiting scroll. As shown in Figure1, the fluid flows into the pump through the suction port, and is inhaled into the working chamber; then, it is squeezed and its pressure reaches at the discharge pressure and flows out by the discharge port. The main scroll profiles geometric parameters are shown in table 1. The basic circle can be seen in the Figure1. The scroll pump has only one pair of working chamber, so the start involutes angle is π. The rotating radius can be changed to regulate the radial clearance.

Table 1. Scroll profiles geometric parameters and operating conditions

Geometric parameters		Operating conditions	
Items	values	Items	values
Radius of basic circle/mm	4.4563	Working fluid	Water
Pitch/mm	28	Rotating speed/mm	1440
Number of circle	1.75	Suction pressure/MPa	0.1
Start angle of involutes	π	Discharge pressure/MPa	0.4
Thickness of scroll vane/mm	4	Suction temperature/℃	25
Suction volume/ml	132		
Discharge port area/mm^2	230		
Rotating radius/mm	10		
Height of scroll vane/mm	30		

Figure 2 shows the computational unstructured mesh which was generated in pre-processing software ICEM. The number of mesh is about 280000, which generally meets the simulation accuracy requirement. Because the orbiting scroll moves with the crank angle, the dynamic mesh technology needs to be used. The triangular mesh was updated by the functions of smoothing and remeshing at appropriate time step to control the mesh skewness. The movement of orbiting scroll was controlled by a file, which is programmed with the user defined function (UDF) code in Fluent. The moving parameters of the scroll pump is shown in table 1.

The flow inside the pump is developed turbulence with curved walls. The *k-e* turbulent model with renormalization group (RNG) is selected as the fluid model. Non-slip boundary condition is employed on the wall. Near wall regions are analyzed using the standard wall functions method. PISO algorithm of the pressure based segregated solver is applied. Pressure boundary conditions are adopted for inlet and outlet, which are given in table1. Because the pressure in the pump will become lower than the vaporized pressure of water at the local temperature, the cavitation model is added in the model to ensure its accuracy. The time step is 1.1574×10^{-4} s, when the orbiting scroll rotates 1 degree ($\pi/180$).

Figure 2. Grid model of the scroll pump.

3. Result and discussion

3.1 *Pressure distribution*

As to the scroll pump, the variation of pressure determines the torque, input power, the axial and radial force on the orbiting scroll and the vibration. Because the water are nearly incompressible, the radial clearance can not be set to 0, or else the pressure in the working chamber will almost reach the infinity. In this paper, the radial clearance is set to be 1mm.

Figure 3 shows the pressure distribution in the scroll pump at different crank angle. At the crank angle 0, it can be seen that a pair working chamber are closed. The suction process finishes and the next suction process will begin at once. The fluid in the working chamber will be discharged at the next discharge process. The pressure in the working chambers is almost uniform, except the two end of the crescent chambers. The pressure is about 0.54MPa, which is larger than the outlet pressure. It is because the volume of working chamber decreases at the crank angle 320°, the water in the working chamber is squeezed, and its pressure increases quickly. So at the two ends of the crescent chambers, the fluid flow through the clearance due to the high pressure difference. It is noticeable that the negative pressure exist at the external end of the crescent chamber. It is caused by the jet flow through the clearance under high pressure difference between chamber and the suction side.

At the crank angle of $\pi/2$, the pressure at the center of pump is about 0.4MPa, which is equal to the outlet pressure. In the external crescent chamber, there are two low pressure regions, the low pressure region at the end of the crescent is cause by the jet flow from the clearance. The velocity of fluid is very large, and the pressure decreases. The high speed fluid decelerates along the narrow passage and the pressure increase. After the low pressure region, a relative high pressure region exists. On the other side, at the entrance of the suction chamber, the fluid flows into the chamber with a high speed because of its volume increment. The high speed flow encounter the high pressure region, and generates the vortex. The energy of fluid is dissipated and a low pressure region forms.

At the crank angle of π, the suction entrance area reaches the maximum. The low pressure region caused by jet flow still exists, but its area decrease largely. As shown in Figure 3. (c), the pressure in the left suction chamber is remarkably different from the right suction chamber. It is caused by the position of the chamber entrance. The entrance of the left chamber faces directly at the suction port, so the fluid flow into the chamber without any block. The velocity of fluid is very high, and the pressure in the chamber is low. The jet flow is stronger, so the area of negative pressure is larger. On the other side, the fluid flows into the suction chamber by a long way and its speed decreases a lot. And the pressure in the suction chamber is almost equal to the suction pressure. The pressure difference at the two sides of clearance is smaller and the negative pressure area is smaller.

At the crank angle of $3\pi/2$, the pressure difference between the left and the right chamber becomes larger. Besides the two low pressure regions in the working chamber, there is another low pressure region on the top of the shell. It is because the orbiting scroll moves up and squeezes fluid between the shell and the orbiting scroll. The fluid velocity increases and the pressure is reduced.

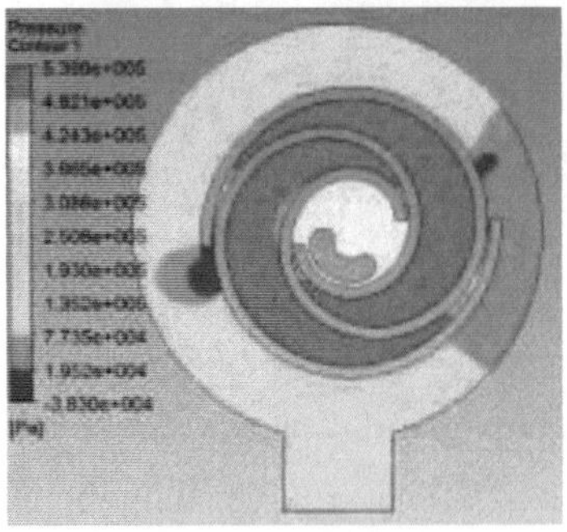

(a) 0

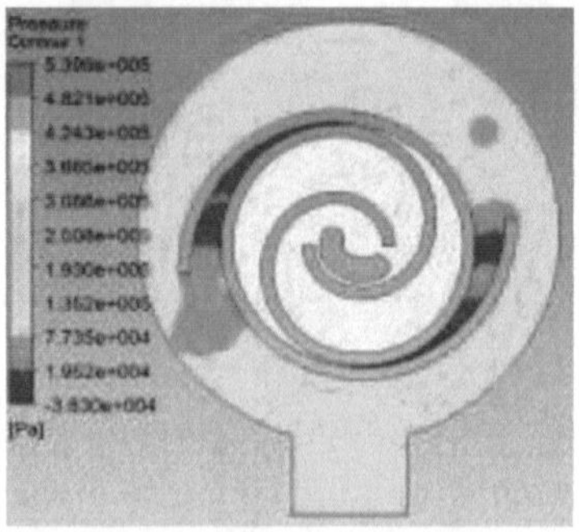

(b) $\pi/2$

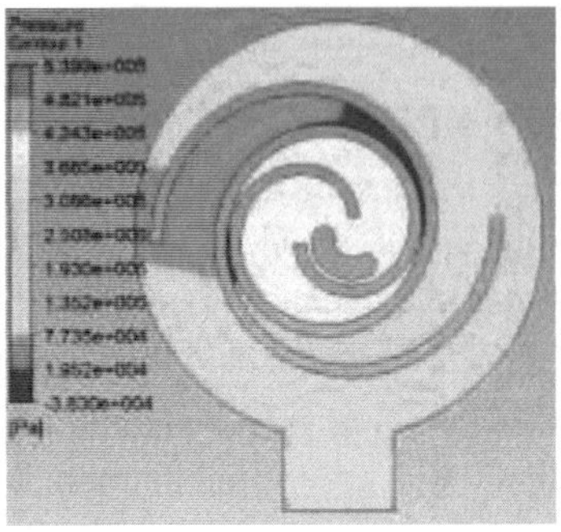

(c)π

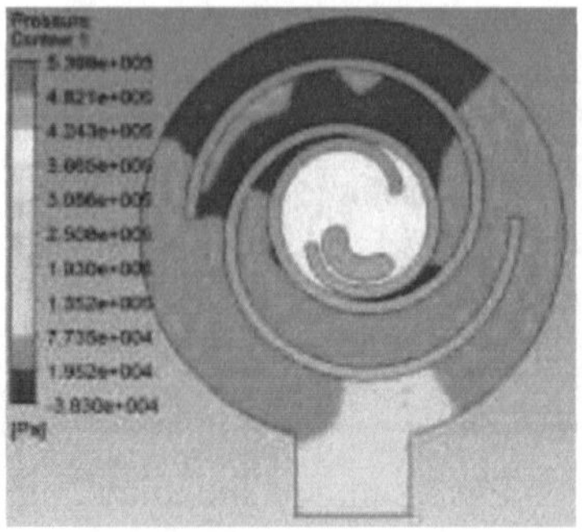

(d) $3\pi/2$

Figure 3. The pressure distribution in the scroll pump.

3.2 *Streamlines and velocity vector*

The Figure 4 shows the streamlines under different crank angle. From the streamline, the vortex can be seen clearly. In the suction chamber, the vortex exists all the time. And the vortex in the left chamber is larger than that in the right suction chamber, which results in the different pressure distribution in the two suction chamber. Between the scroll and the shell, there are many vortex, and the vortex positions change with the crank angle. It can be seen that there are big vortex at the entrance of suction chamber when the crank angle changes between 0 to π. Because the area of entrance is quite small, the high speed fluid is blocked by the small entrance, and the vortex generates. At the top of the shell, the moving orbiting scroll drives the fluid flow with it. So the vortex changes with the position of the orbiting scroll in the four figures.

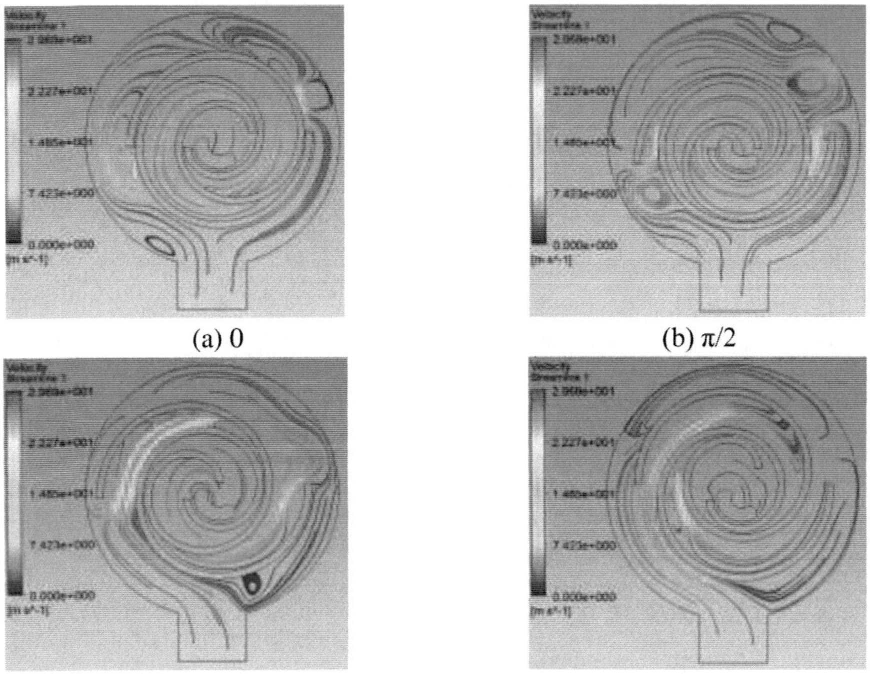

(a) 0 (b) π/2

(c) π (d) 3π/2

Figure 4. The streamline in the scroll pump.

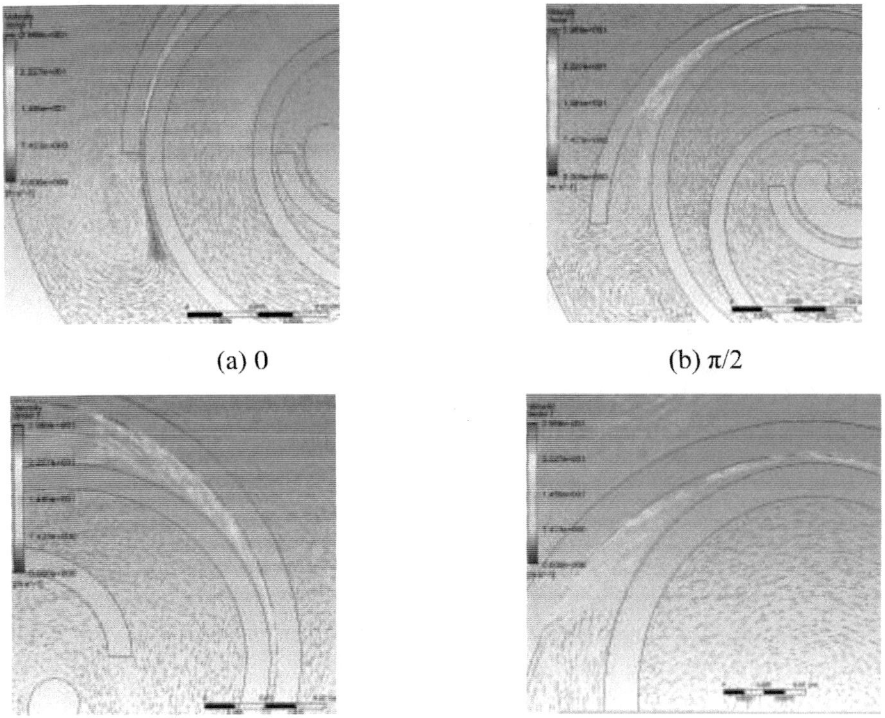

(a) 0 (b) π/2

(c) π (d) 3π/2

Figure 5. The velocity vector at the radial clearance.

Figure 5 displays the velocity vectors at the radial clearances at different crank angles. The clearances is just like a nozzle. The fluid is accelerated through the clearance by the pressure difference. The high speed flow gives rise to the vortex and negative pressure region. When the crank angle is 0, it can be seen that the jet flow from the clearance encounters with the suction flow and generate a big vortex at the entrance of suction chamber. The vortex dissipates the kinetic and pressure energy and the negative pressure region generates, which can be seen in Figure 3. The other three picture shows the vortexes in the suction chamber. It can be seen that the high speed fluid mixes with the suction fluid and change their momentum. The velocity of fluid decreases and the pressure increases. A part of fluid goes back because they get the enough anti-direction momentum.

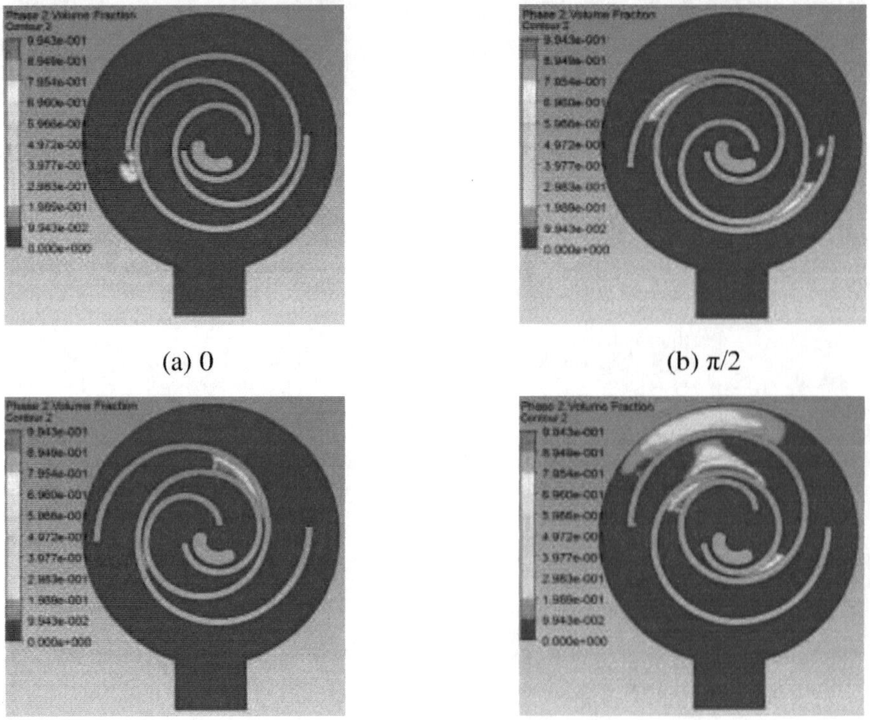

(a) 0 (b) π/2

(c) π (d) 3π/2

Figure 6. The gas phase volume cloud.

3.3 The gas phase volume cloud

At the radius clearance position, the pressure decreases with the increment of the fluid velocity. When the pressure becomes lower than the vaporized pressure of water at the local temperature, the water vaporized. This phenomena is called by cavitation, which is a key problem in the centrifugal pump. If the cavitation model is not applied in the mathematic model, the negative pressure could reaches the -0.6Mpa, which is impossible practically. So the cavitation model is added into the model, and the gas phase volume cloud are shown in Figure 6. It can be seen that the cavitation mainly happens at the low pressure side of the radial clearance. Another places where the cavitation happens are at the vortex, especially in Figure 6(a) and Figure 6(d).Once the cavitation happens, the gas generates and will take the space of liquid. As to the centrifugal pump, the pump efficiency will fall down. However, in the scroll pump, the gas will be compressed again and transported with the liquid. So the cavitation influence the characteristic of pump faintly. But the cavitation will lead to the vibration and cavitation erosion, which reduce the reliability of the scroll pump. Therefore, the structure of scroll pump needs to be optimized to prevent the cavitation from generating.

4. Conclusion

In this paper, a two dimensional transient numerical model of a scroll pump is established. The dynamic mesh technology is applied to solve the problem of moving wall of the fluid domain. The pressure distribution, streamline, velocity vector and gas phase volume cloud at different crank angle are figured out by the model. And, some useful conclusion can be drawn as follows.

(1) The fluid in the working chamber is squeezed after the suction volume reaches maximum, ant its pressure rises quickly and becomes larger than the outlet pressure.

(2) At the radial clearance, the jet flow caused by large pressure difference generates and leads to the negative pressure distribution at the downstream of the clearance.

(3) The asymmetric structure of suction entrance leads to the pressure distribution difference between the two suction chambers, which will lead to the unbalanced force on the orbiting scroll.

(4) At the downstream of the radial clearance, the high speed jet flow encounters with the suction flow and exchange their momentum, the vortex generates and the static pressure decreases.

(5) In the space between the orbiting scroll and the shell, the moving orbiting scroll drives the fluid and the vortex changes with the crank angle, which is also the reason of the imbalance of the pressure between the two suction chambers.

(6) The cavitation generates at the negative pressure region and the vapor volume fraction reaches 0.9; the cavitation will not influence the characteristic of the pump, but the cavitation erosion and the vibration will reduce the reliability of the pump.

Acknowledgment

This work is part of a project supported by the National Natural Science Foundation of China (51406163), National Natural Science Foundation of Shaanxi Province (2015JQ5197), Science Research Project of Education Department of Shaanxi Provincial Government (14JK1541). The supports are gratefully acknowledged.

Reference

[1] Zhu J, Wang D 1992 Research on the discharge port of scroll oil pump[C]. *West Lafayette*, IN, USA: Publ by Purdue Univ, 1992: 611-611.

[2] Zhu J 1992 *Numerical and experimental research on the woring process of scroll oil pump[D]*. Xi'an: Xi'an Jiaotong University, 1992.

[3] Jiang B, Wang D S, Qian L Y 1996 Development of the scroll oil pump SCP-1.08/0.6[J]. *Fluid machinery*, (**02**): 15-17.

[4] Jiang B, Zhu J, Chang Y F et al. 1998 Experimental analysis of the scroll oil pump[J]. *Compressor technology*, (**02**): 16-18.

[5] Qu Z C, Fan L, Yan Y 1998 Leakage research of the scroll oil pump[J]. *Compressor technology*, (**06**): 1-5.

[6] Qu Z C, Jiang B, Wang D S 1998 New type of scroll oil pump[J]. *Compressor technology*, (**02**): 31-32+24.

[7] Qu Z C, Yan Y, Wang D S 1997 Rotational speed optimization of a scroll oil pump[J]. *Comprssor techonology*, (**05**): 28-31.

[8] Cui M M 2006 Numerical study of unsteady flows in a scroll compressor[J]. *Journal of Fluids Engineering*, **128**: 947-955.

[9] Wang J, Zha H B, Zhang X H et al. 2012 Working process study of a novel scroll type multiphase pump for the transportation of gas-liquid mixtures[C]. Beijing, China: *Institute of Physics Publishing*.

IOP Conf. Series: Materials Science and Engineering **129** (2016) 012029 doi:10.1088/1757-899X/129/1/012029

Numerical simulation of effects of sand grain diameters and volume fractions on mass transferring from the water-liquid to the water-vapor

Weiguo Zhao[1, 2], Xiangdong Han[1], Ming Liu[1] and Yingjie Zheng[1]

[1]School of Energy and Power Engineering, Lanzhou University of Technology, Gansu Province, China

[2]Key Laboratory of Fluid Machinery and System, Lanzhou, Gansu Province, China

E-mail: zhaowg@zju.edu.cn

Abstract. The paper analyzed the effects of sand grain diameters and volume fractions on the mass transferring from the water-liquid to the water-vapor in a two-dimensional nozzle. Based on the mixture model, $k-\varepsilon$ turbulence model and *Schnerr-Sauer* cavitation model, the solid-liquid-vapor three phases' cavitation flows were simulated. When the grain mean diameters were defined as constants, volume fractions were changed to investigate the effects of them. The grain mean diameters were 0.013mm, 0.025mm and 0.05mm. Volume fractions were 0.02, 0.04, 0.05, 0.07 and 0.10. Results indicated that cavitation occurred at the beginning spots of the narrow part of the nozzle, low pressure regions. With the different grain mean diameters and volume fractions, effects of the sand on the mass transferring from the water-liquid to the water-vapor were diverse, proved by the curves of the cavitation numbers with the volume fractions of the sand and the curves of the volume fractions of the water-vapor with the volume fractions of the sand, reflecting the distinctions of interactions between the bubbles and the sand grains.

Introduction

Cavitation is that the water-vapor bubbles are created in the low pressure regions in the fluid flow field, causing the destruction of materials, initiating the vibration and the noise and making the performances of the machines decrease. Cavitation also is a complex multiphase flow. Different cavitation models were built by *Singhal*, *Kunz* and *Schnerr&Sauer*, to predict and analyze the features of it. Nowadays, two different methods have been used to study cavitation, including the experiment and numerical simulation. However, exact solutions of some certain parameters can't be gained by experiments, so numerical simulation is applied widely. In this article, this will be followed by a description of the effects of the sand on the mass transferring from the water-liquid to the water-vapor in a 2D flow field.

Nozzle is one kind of significant experiment instrument to simulate the cavitation flow. The study of the cavitation occurring in the nozzle has been done by some prestigious scholars. The *k-ω* SST two-equation turbulence model and *k-kl-ω* turbulence model were employed by Meixin Xue to simulate the non-cavitation and cavitation flow in a diesel nozzle [3]. Some parameters of the cavitation were changed by Hengzhou Wo to research incipient cavitation in orifices [7]. A cavitation water jet flow field of the convergent-divergent nozzle was discussed by Yiyu Lu [4]. The fluid-solid two phases jet flow field in a nozzle was acquired by Mingbo Wang [5]. The effects of different shapes of the nozzle on the cavitation were considered by Liming Yao [6].

To obtain the accurate results of three phases' cavitation flows, the preparation work was model foundation and grid partition. In these simulations, the grain mean diameters of the sand were 0.013mm, 0.025mm and 0.05mm and each volume fraction of the sand was 0.02, 0.04, 0.05, 0.07 and 0.10. With the grain mean diameters and volume fractions of the sand having changed respectively, based on the mixture model, two phases' flows of fluid-solid were numerically simulated in this nozzle. When these flows converged, based on the *Schnerr-Sauer* cavitation model, water-vapor was added to these simulations to investigate the mass transferring from the water-liquid to the water-vapor in the solid-liquid-vapor three phases' cavitation flows.

1. Physical and mathematical equations in multiphase Flow

1.1. Basic equations in multiphase flow [14]

Continuity equation:

$$\frac{\partial \tilde{\rho}_k}{\partial t}+\frac{\tilde{\partial}}{\partial x_j}\left(\tilde{\rho}_k \tilde{u}_{kj}\right)=0 \tag{1}$$

Momentum equation:

$$\frac{\partial}{\partial t}\left(\tilde{\rho}_k \tilde{u}_{kj}\right)+\frac{\partial}{\partial x_j}\left(\tilde{\rho}_k \tilde{u}_{kj} \tilde{u}_{ki}\right)=-\frac{\partial \tilde{p}_k}{\partial x_i}+\frac{\tilde{\partial}}{\partial x_j}\left(\tilde{\tau}_{k,ji}\right)+\tilde{\rho}_{k,ji} \tag{2}$$

Energy equation:

$$\frac{\partial}{\partial t}\left(\tilde{\rho}_k \tilde{C}_k \tilde{T}_k\right)+\frac{\tilde{\partial}}{\partial x_j}\left(\tilde{\rho}_k \tilde{u}_{kj} C_k \tilde{T}_k\right)=\frac{\tilde{\partial}}{\partial x_j}\left(\lambda_k \frac{\tilde{\partial} \tilde{T}_k}{\partial x_j}\right)+\tilde{q}_{\mathrm{c}}-\tilde{q}_{\mathrm{r}} \tag{3}$$

Composition equation:

$$\frac{\partial}{\partial t}\left(\tilde{\rho}_k \tilde{Y}_{ki}\right)+\frac{\tilde{\partial}}{\partial x_j}\left(\tilde{\rho}_k \tilde{u}_{kj} \tilde{Y}_{ki}\right)=\frac{\tilde{\partial}}{\partial x_j}\left(D_\rho \frac{\tilde{\partial} \tilde{Y}_{ki}}{\partial x_j}\right)-\tilde{\omega}_{ks} \tag{4}$$

1.2. Standard $k-\varepsilon$ *two-equation model* [15]

$$\frac{\partial(\rho k)}{\partial t}+\frac{\partial(\rho k u_i)}{\partial x_i}=\frac{\partial}{\partial x_j}\left[\left(\mu+\frac{\mu_t}{\sigma_k}\right)\frac{\partial k}{\partial x_j}\right]+G_k+G_b-\rho\varepsilon-Y_M+S_k \tag{5}$$

$$\frac{\partial(\rho \varepsilon)}{\partial t}+\frac{\partial(\rho \varepsilon u_i)}{\partial x_i}=\frac{\partial}{\partial x_j}\left[\left(\mu+\frac{\mu_t}{\sigma_\varepsilon}\right)\frac{\partial \varepsilon}{\partial x_j}\right]+G_\varepsilon\frac{\varepsilon}{k}(G_k+C_{3\varepsilon}G_B)-C_{2\varepsilon}\rho\frac{\varepsilon^2}{k}+S_\varepsilon \tag{6}$$

1.3. Rayleigh-Porisky equation [14]

$$r_b\frac{d^2 r_b}{dt^2}+\frac{3}{2}\left(\frac{dr_b}{dt}\right)^2=\left(\frac{p_b-p_\infty}{\rho_l}\right)-\frac{4\nu}{r_b}\frac{dr_b}{dt}-\frac{2\sigma}{\rho r_b} \tag{7}$$

1.4. Schnerr-Sauer cavitation model [10]

Bubble radius:

$$r_b=\left(\frac{\alpha}{1-\alpha}\bullet\frac{3}{4\pi}\bullet\frac{1}{n_b}\right)^{\frac{1}{3}} \tag{8}$$

Source terms:

$$R_e=\frac{\rho_v\rho_l}{\rho}\alpha(1-\alpha)\frac{3}{r_b}\left(\frac{2}{3}\frac{p_{vap}-p}{\rho_l}\right)^{\frac{1}{2}}\quad(p\le p_{vap})$$

$$R_c=\frac{\rho_v\rho_l}{\rho}\alpha(1-\alpha)\frac{3}{r_b}\left(\frac{2}{3}\frac{p-p_{vap}}{\rho_l}\right)^{\frac{1}{2}}\quad(p\ge p_{vap}) \tag{9}$$

2. Numerical simulation

2.1. Basic assumption

The water was incompressible and the sand was the continuous medium. Physical properties of every phase were invariant. Shape of the sand grain was regarded as sphere. It was inserted and was not insolated in water. Phase transition of the sand can be neglected. The flowing state was steady.

2.2. Geometric model and grid partition

The geometric model had the same sizes with the nozzle coming from the ANSYS Fluid Dynamic Verification Manual. The geometric parameters were that the inlet diameter was 23mm and the outlet diameter was 8mm; the wide length was 16mm and the narrow length was 32mm. [16]. Quadrilateral structured grids were adopted to discrete the computational domain.

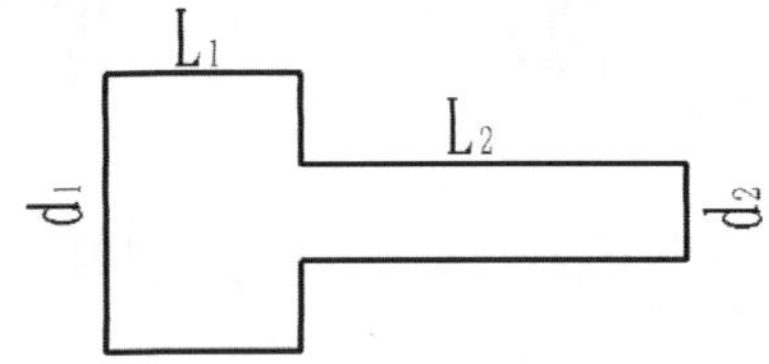

Figure 1. Geometric model of the nozzle.

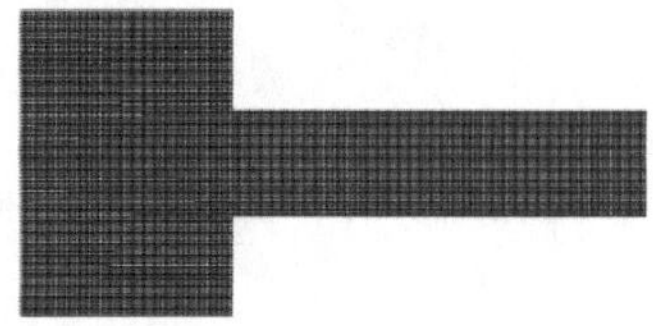

Figure 2. Computational domain and grids.

2.3. Method of numerical simulation and boundary conditions

Pressure was defined at the inlet and the outlet and values were 250,000,000Pa and 95,000Pa. The value of vaporization pressure was 3540Pa. And the temperature was 300K. [16]

Physical parameters used in numerical simulations were that density of the water-liquid was ρ_l=1000 kg/m^3 and the kinematic viscosity of the water-liquid was μ_l=0.001 kg/(m·s); Values of the density and the kinematic viscosity of the water-vapor were ρ_v=0.02558 kg/m^3 and μ_v=1.26×10^{-6} kg/(m·s). Density of the sand was ρ_s=2650 kg/m^3. The bubble diameter was r_b=1.0×10^{-5}m and the bubble density was n_b=1.0×10^{13}. [16]

Commercial Fluent software was used to simulate and analyze the cavitation flow field. Standard wall functions were applied to treat the flow near the wall. "Implicit Body Force" was employed to speed up the convergence. Drag coefficient was set from the sand to the water-liquid and from the water-vapor to the water-liquid. The level of the turbulent intensity was 10%. For the sand and the water-vapor, the backflow volume fraction was zero. All the residuals were less than 1.0×10^{-4}.

Numerical simulations were performed by varying the volume fractions of the sand when the grain mean diameter of the sand was a constant.

3. Results and Analysis

3.1. Contours of volume fraction of water-liquid or water-vapor

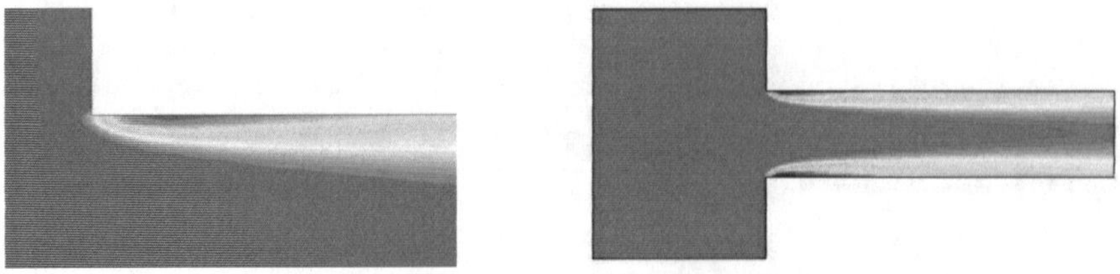

Figure 3. Contour of volume fraction of the water-liquid from the ANSYS Fluid Dynamic Verification Manual and simulated.

Figure 3 exemplified the contour of volume fraction of the water-liquid from the ANSYS Fluid Dynamic Verification Manual. Figure 4 indicated that the contour of volume fraction of the water-liquid simulated under the same boundary conditions. Compared these two contours, regions of the low pressure were the same, verifying that the mass transferring from the water-liquid to the water-vapor occurring at the same place.

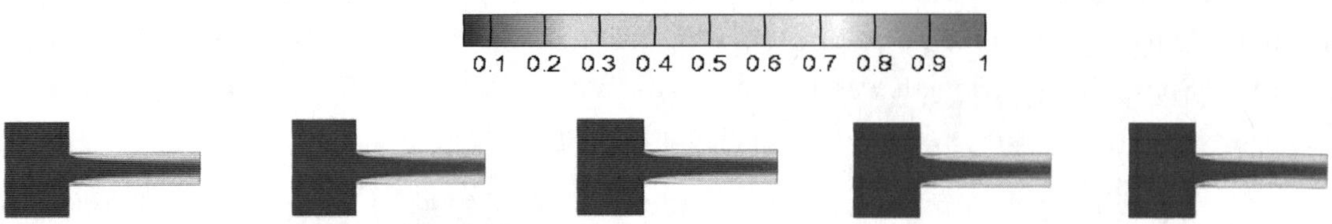

(α=0.02) (α=0.04) (α=0.05) (α=0.07) (α=0.10)

Figure 4. Contours of volume fraction of the water-vapor with mean diameter being 0.013mm

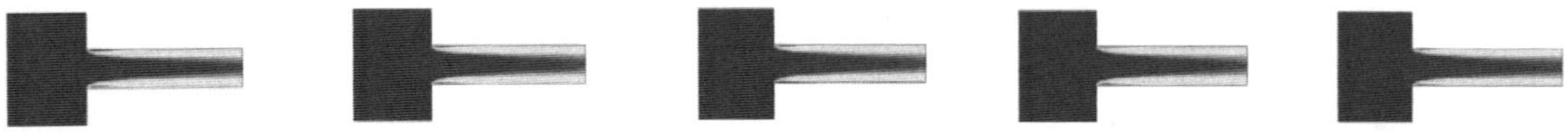

(α=0.02) (α=0.04) (α=0.05) (α=0.07) (α=0.10)

Figure 5. Contours of volume fraction of the water-vapor with mean diameter being 0.025mm

(α=0.02) (α=0.04) (α=0.05) (α=0.07)

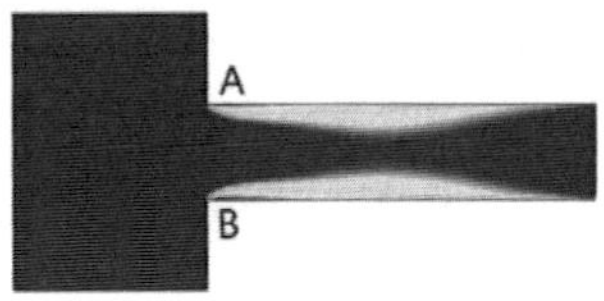

(α=0.10)

Figure 6. Contours of volume fraction of the water-vapor with mean diameter being 0.05mm

Axial symmetrical distribution of the contours of the water-vapor was one obvious feature of these contours. The water-vapor was mainly concentrated on the beginning spots of the narrow part of the nozzle, for example, spot A and B, demonstrating that cavitation mainly occurred at these spots. With the distance increasing gradually, contents of the water-vapor became very little. At the outlet, contents reduced nearly to zero. When the grain mean diameter and volume fraction of the sand was set to 0.05mm and 0.1, changing of the water-vapor was more markedly. A possible explanation was that the decrease of contents of the water-vapor was made by the inter-collisions of sand grains with bubbles. When the grain mean diameters were gotten larger and larger and the volume fractions were grown more massively, inter-collisions developed with far more aggravation and intensification.

3.2. Curves of cavitation numbers with volume fractions of the sand

The cavitation number is

$$\sigma = \frac{p_s - p_v}{\frac{1}{2}\rho v^2} \tag{10}$$

So, curves of the cavitation numbers with the volume fractions of the sand could be got.

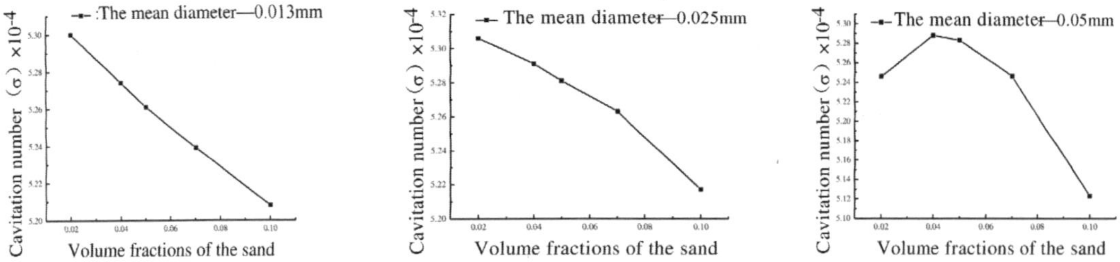

(a) (b) (c)

Figure 7. Curve of cavitation numbers with the volume fractions of the sand.

Figure 7(a) shows that when the sand grain mean diameter was 0.013mm and the volume fractions were grown gradually, cavitation numbers were a steady decline, owning to a linear-change-rate. So equation of a degree could be used to fit this curve approximately.

Figure 7(b) reveals that when the grain mean diameter was set to 0.025mm and the volume fractions were increased gradually, cavitation numbers made a substantial decrease and when the volume fractions were ranged from 0.02 to 0.07, linear-variety-rate could be used to describe this trend approximately. There was a sharp drop of the cavitation numbers when the volume fractions were varied from 0.07 to 0.10.

Figure7(c) indicates that the condition was that the grain mean diameter was 0.05mm. When the volume fractions of the sand were grown constantly, the changing trend of cavitation numbers looked like a parabola. When the volume fraction was 0.04, the cavitation number reached its highest point, reflecting that it had a profound impact on the cavitation flow. So the parabola equation could be utilized to fit this curve approximately.

3.3. *Curves of volume fractions of the water-vapor with volume fractions of the sand*

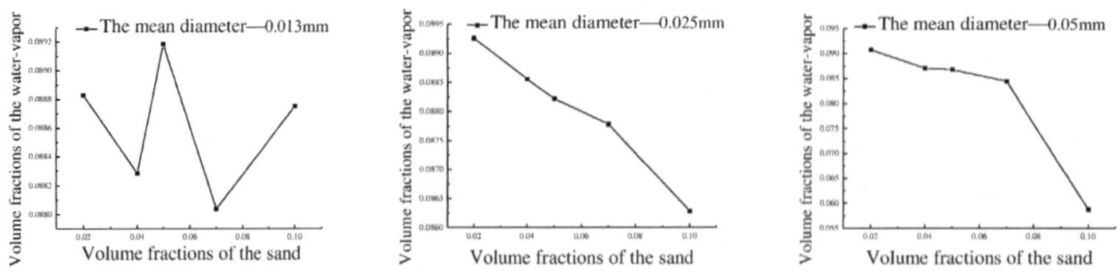

(a) (b) (c)

Figure 8. Curves of the volume fractions of the water-vapor with the volume fractions of the sand.

Figure 8(a) conveys that when the volume fractions of the sand were increased gradually, the whole changing trend of the water-vapor looked like a "W". When the volume fraction of the sand was 0.05, volume fractions of the water-vapor reached its maximum. However, when the volume fraction of the sand was set to 0.07, volume fractions of the water-vapor became its minimum. The fluctuation varied between 0.089186 and 0.088035.

Figure 8(b) illustrates that when the grain mean diameter was set to 0.025mm and the volume fractions of the sand were increased constantly, the substantial decrease of the volume fractions of the water-vapor was one feature of this curve.

Figure 8(c) explains that compared with Figure 11 and Figure 12, this figure showed a completely different trend. The prerequisite was that the grain mean diameter was 0.05mm. When the volume fractions of the sand were changed from 0.02 to 0.07, a slight decline was one character of the volume fractions of the water-vapor. However, when the volume fractions of the sand were ranged from 0.07 to 0.10, there was a slump in the volume fractions of the water-vapor. So, it was palpable that when the volume fraction of the sand reached 0.10, the volume fraction of the water-vapor reached its lowest point.

4. Summary and Conclusions

Based on the *Schnerr-Sauer* cavitation model, the effects of sand grain mean diameters and volume fractions on the mass transferring from the water-liquid to the water-vapor have been explored. It was manifest to comprehend the effects on the cavitation flow and these simulations could be applied to investigate solid-liquid-vapor three phases' cavitation flow in the centrifugal pump.

From the presented results, contours and curves, conclusions are that:

(1) Axial symmetrical distribution was one feature of all contours of the water-vapor.

(2) When the grain mean diameter was 0.05mm and the volume fraction was 0.1, sand had a profound impact on the cavitation flow.

(3) When the grain mean diameters and volume fractions were distinct and diverse, distributions of the water-vapor and values of the cavitation numbers were different, reflecting the distinctions of interactions between the bubbles and the sand grains.

Nomenclature

u = velocity; p = pressure;	p_s = static pressure;	$\tau_{k,ji}$ = component of stress tensor;
g = body force per unit volume;	C = specific heat;	T = temperature;
λ = thermal conductivity;	q_c = radiant heat;	q_r = reaction heat;
~ = instantaneous value;	k = turbulent kinetic energy;	ε = turbulent dissipation rate;
μ = dynamic viscosity;	μ_t = turbulent viscosity;	$C_{1\varepsilon}$ = empirical constant;
$C_{2\varepsilon}$ = empirical constant;	$C_{3\varepsilon}$ = empirical constant;	σ_k = 1.0;
σ_ε = 1.0; b = bubble;	p_∞ = free stream pressure;	r_b = bubble radius;
p_b = bubble wall pressure;	v = kinematic viscosity;	ρ_l = density of water-liquid;
σ = bubble wall tension;	α = fraction of water-vapor;	n_b = bubble density;
ρ_v = density of water-vapor;	ρ = density of mixture;	p_{vap} = vaporization pressure.

Acknowledgment

The authors acknowledge teachers' support from School of Energy and Power Engineering and Key Laboratory of Fluid Machinery and System. And we also thank the support from National Natural Science Foundation of China (Grant No. 51269011 and No. 51269010).

References

[1] Ashok K. Singhal, Mahesh M. Athavale and Huiying Li 2002 Mathematical basis and validation of the full cavitation model *J. Journal of Fluid Engineering* **124** 617-624

[2] Bostjan Gregorc, Matjaz Hribersek and Andrej Predin 2011 The analysis of the impact of particles on cavitation flow development *J. Journal of Fluid Engineering* **133** 1-8

[3] Meixin Xue, Di Wu and Ying Piao 2012 Numerical simulation of internal flow and cavitation of a diesel nozzle *J. Engineer Mechanics* **29(10)** 46-51

[4] Yiyu Lu, Xiaochuan Wang, Yong Kang and Yulong Chen 2009 Numerical simulation for cavitation water jet flow field through convergent-divergent nozzle *J. Journal of China University of Petroleum* **33(6)** 46-49

[5] Mingbo Wang and Ruihe Wang 2005 Numerical simulation on fluid-particle two-phase jet flow in nozzle *J. Journal of the University of Petroleum China* **29(5)** 46-49

[6] Liming Yao, Yi Zhao, Dawei Li, Yu Pei, Hongbin Yao and Mengshi Zhao 2015 Research on the shape of cavitating spray head *J. Energy Conservation Technology* **33(189)** 20-24

[7] Hengzhou Wo, Zhihua Yao, Yafang Zhang, Guofeng Wang, Yufu Xu and Xianguo-Hu 2011

Numerical simulation of cavitation inception in engine orifice *J. China Mechanical Engineering* **22(17)** 2071-75

[8] Zhaojie Li, Songsheng Deng, Yinhua Shen, Jian Feng, Jingbin Liu and Huanou Zhang 2009 Experimental study on cavitation jet flow *J. Journal of Logistical Engineering University* **25(1)** 40-44

[9] Meixin Xue and Ying Piao 2013 Large eddy simulation on cavitation inception of a diesel nozzle *J. Engineering Mechanics* **30(4)** 417-422

[10] Houlin Liu, Dongxi Liu, Yong Wang, Xianfang Wu and Suguo Zhuang 2012 Applicative evaluation of three cavitating models on cavitating flow calculation in cenntrifugal pump *J. Transactions of the Chinese Society of Agricultural Engineering* **28(16)** 54-59

[11] Binjuan Zhao, Shouqi Yuan, Houli Liu, Zhongfu Wu and Minggao Tan 2008 Simulation of solid-liquid two-phase turbulent flow in double-channel pump based on mixture model *J. Transactions of the CSAE* **24(1)** 7-12

[12] Christopher E. Brennen 1995 *Cavitati- on and Bubble Dynamics* (New York: Oxford University Press) pp 47-75

[13] Fujun Wang 2004 *Computational Fluid Dynamics Anasys* (Beijing: TsingHua University Press) pp 74-81

[14] Yulin Wu, Xuelin Tang, Shuhong Liu and Yu Xu 2007 *Cavitation of Hydraulic Machinery and Solid-Liquid Two Phases Flow Dynamics* (Beijing: China Water & Power Press) p 50 and pp 40-41

[15] Hongjun Zhu, Yuanhua Lin and Longhan Xie 2011 *Engineering Simulation and Fluent Analysis* (Beijing: TsingHua University Press) pp 88-89

[16] ANSYS, Inc. *ANSYS Fluid Dynamics Verification Manual* (Canonsburg, PA 15317) pp 75-80

IOP Conf. Series: Materials Science and Engineering **129** (2016) 012030 doi:10.1088/1757-899X/129/1/012030

Numerical research on hydrodynamic characteristics of end cover of pressure exchanger

L Jiao[1]*, H T Lin[1], Y Shi[1], Z C Liu[1], Z M Feng[2], G S Li[1]

[1] Institute of Naval Architecture and Ocean Engineering, Zhejiang University, China
[2] Nanfang Pump Industry Co., Ltd. China
*corresponding author: Tel.:+8613093767519

E-mail address: jiaolei@zju.edu.cn

Abstract. To investigate hydrodynamic performance of the end cover under different inclined angles, a series of 3-Dimensional geometric models of the end cover with different inclined angles were built by Creo. The maximum inclined angle is 32 degrees and the minimum is 6 degrees. Numerical simulations by solving the Navier-Stokes equation, coupled with the "k-ε" turbulence model, were carried out. At last, regressive analysis method is used to deal with the result data. The data obtained by simulation were mainly analysed from three aspects. They are cause of the driving torque, drive efficiency and inclined angle of flow channel. The results show that the driving torque is formed mainly by the positive pressure of the water, and the influence of the viscous force on the driving torque of the rotor is negligible. What's more, both driving torque and pressure difference decrease with the increase of inclined angle in the form of power function. The driving efficiency increases in form of logarithmic function with the increase of inclined angle. This research has a great significance in the control of rotational speed of rotor and revealed the relationship between the driving torque of the rotor and the flow channel of the end cover.

1. Introduction

A rotary pressure exchanger which is based on the positive displacement principle is a kind of fluid energy recovery equipment [1] [2]. It is one of the three core components of seawater desalination industry. Its operating principle is illustrated by Figure 1-1[1]. The key components of RPE include a rotor with axial ducts arranged in a circle around a centre tension rod, two end covers and one sleeve. In the work process, the fluid will generate a term of tangential velocity after it passed through the inclined flow channels and the rotor rotated in the sleeve is driven by this fluid.

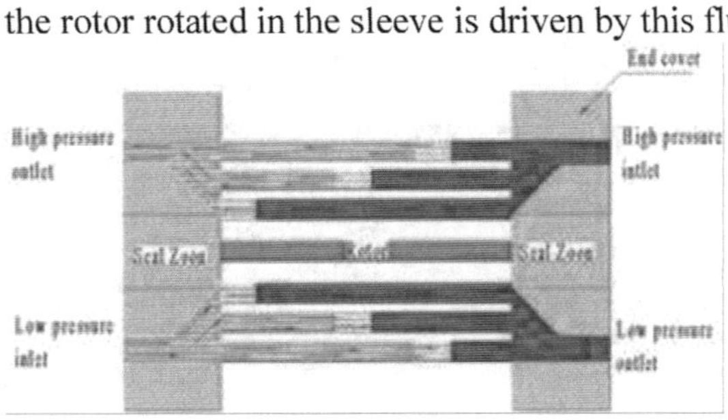

Figure 1. structure view of the rotary pressure exchanger

Low pressure and high pressure brine will directly contact with each other and the pressure energy is transferred directly from the high pressure reject stream to a feed stream with no intervening walls. The pressurized seawater is discharged into lift pump. The pressure of brine decreased and then it is discharged by the feed stream.

The most important three features of pressure exchanger are flow-driven rotation, self-lubricating bearing and pressure transition control [3]. Now the latter two issues have been searched by many people. For example, Zhao Fei and his collaborators have carried out a detailed analysis on the problem of self-lubricating bearing [4]. They researched the support mechanism and stiffness of axial film in rotary pressure exchanger and revealed how much the clearness between cover and rotor should be to guarantee that the rotor can be supported by water and the leakage is maintained at a low level. Another important research was carried out by Yihui Zhou [5] and her partners, which revealed that the rotor speed is very important to control the maximum flow-in length and guarantee that the mixing is maintained at a low level. Many other studies show that the rotor speed has a great effect on the performance of pressure exchanger [6]. So the mechanism and characteristics of flow-driven is a problem we can't avoid to guarantee the rotor rotates at ideal speed. But no detailed research on the problem of flow-driven has been carried out at present. In order to figure out the mechanism and characteristics of the flow-driven of pressure exchanger, a lot of work has been done in this paper.

The structure of the article is arranged as follows. Section 2 gives a detailed introduction of the research problem, in which geometry specifications and operating parameters are listed. Section 3 describes the numerical setup detailed. Section 4 presents the numerical results, including the velocity vectors at different inclined angles as well as the whole hydraulic moment analysis, followed by the conclusion in Section 5.

2. Description of problem

The section at the flow channel of the end cover is shown in figure 2-1. A term of tangential velocity will generate after the fluid passed through the inclined flow channel. Then the fluid produce a tangential impact on rotor so as to drive the rotor rotates in the sleeve

From the qualitative point, bigger inclined angle means that more fluid with tangential velocity component drives the rotor, but the tangential velocity will be smaller. So the ultimate driving torque may be not larger than that in smaller inclined angle. It becomes important to get the change law between driving torque and inclined angle.

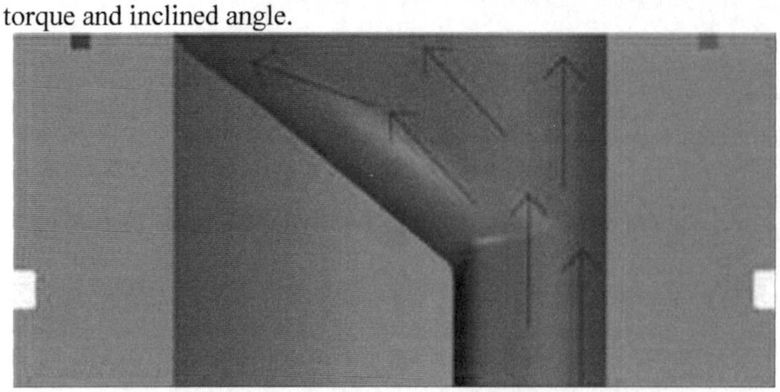

Figure 2. section view at flow channel of the end cover

In order to figure out that problem, a series of flow channels with different inclined angles are build. The inclined angles varies from 6 degrees to 32 degrees with 2 interval degrees. The geometry of flow channel with maximum inclined angle and minimum inclined angle are shown as figure 2-2 and figure 2-3 respectively.

Figure 3. maximum inclined angle with 32 degrees

Figure 4. minimum inclined angle with 6 degrees

In order to study the effect of inclined angle on driving torque, the volume flow rate is set to 70m3/h, pressure at high pressure outlet is set to 6 MPa and that at low pressure outlet is set to 0.4 MPa. In addition, the rotor speed is set to be same.

3. Numerical simulation

3.1. Mesh Generation

The geometry model of the pressure exchanger was established by the commercial software—Creo. It includes two end covers and a rotor. The main geometry parameters are listed in Table 3-1and the whole configuration can be seen in Figure 3-1.

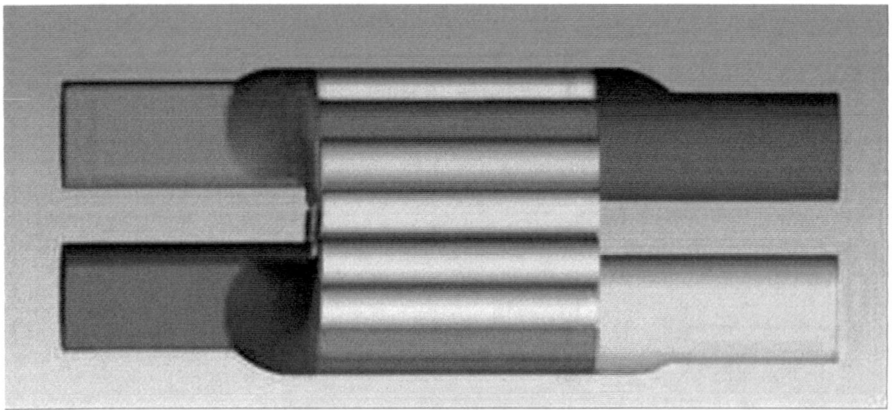

Figure 5. Geometry of whole flow zone

Table 1. Main geometry parameters

Dimension	Parameter
Rotor diameter	170(mm)
Rotor height	190(mm)
Cover height	100(mm)
Inside diameter of cover	50(mm)
Outside diameter of cover	170(mm)
Inclined angle	6~32(degree)

In the present study, FLUENT has been used for calculation, with transport equations solved by finite volume method. The mesh was generated on the platform of ICEM CFD. In practise, the flow zone is divided into three parts (two covers and a rotor). Each part is meshed separately. Tetrahedral elements is used in mesh Generation of rotor and triangular elements is used in mesh Generation of

IOP Conf. Series: Materials Science and Engineering **129** (2016) 012030 doi:10.1088/1757-899X/129/1/012030

end cover. Mesh independent verification is carried out to eliminate the effect of mesh size on computation results. The mesh of rotor and cover is shown in figure 3-2 and figure 3-3.

Figure 6. Mesh of rotor

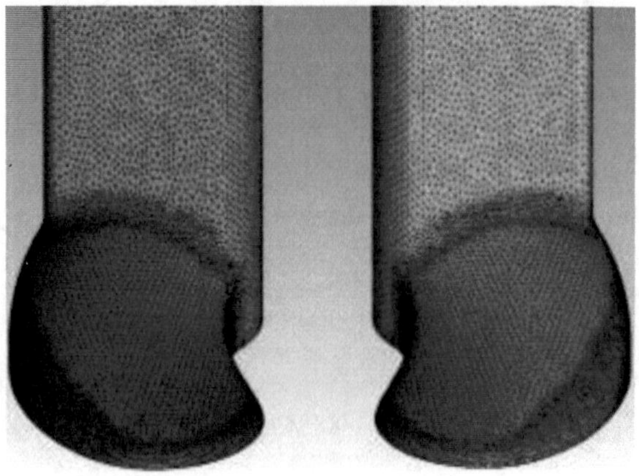

Figure 7. mesh of end cover

The number of mesh element of rotor reached 4 million and that of end cover reached 3.5 million. The quality of mesh of all parts are higher than 0.5. After computational test, it is proved that the mesh system can satisfy the requirement of the k-ε turbulence model and enhanced wall treatment, which can be seen from the y+ values on the walls of rotor and end cover displayed in Figures 3-4 and 3-5, respectively. At last, all the meshes are transferred from ICEMCD to fluent to compose an integral mesh. Interface is used to connect the mesh of different fluid zones.

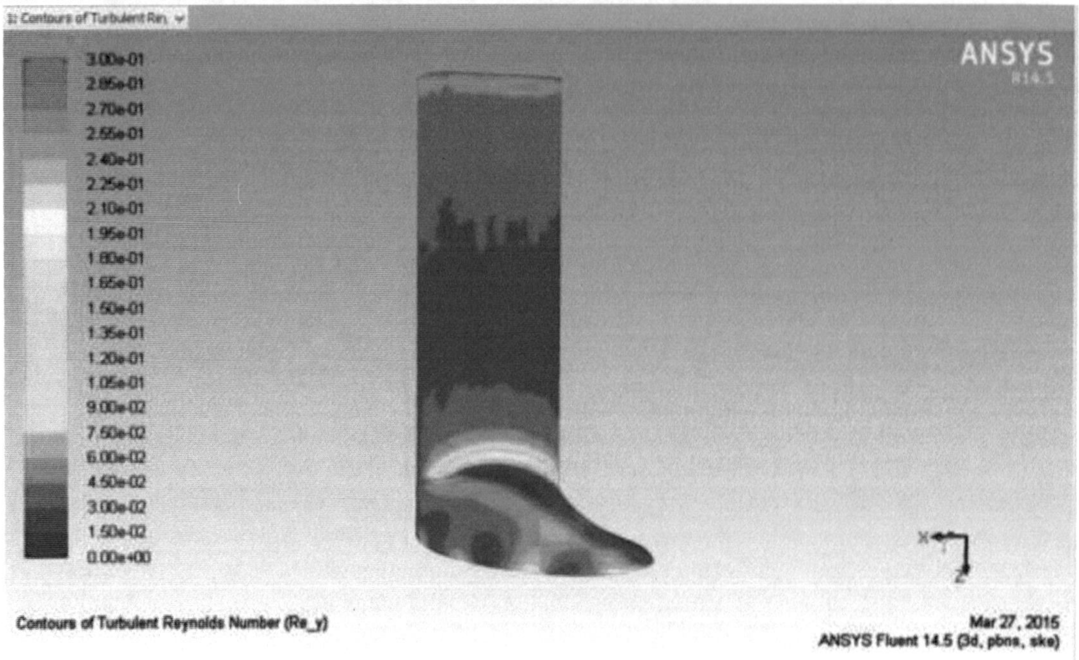

Figure 8. Counter of y+ on the wall of end cover

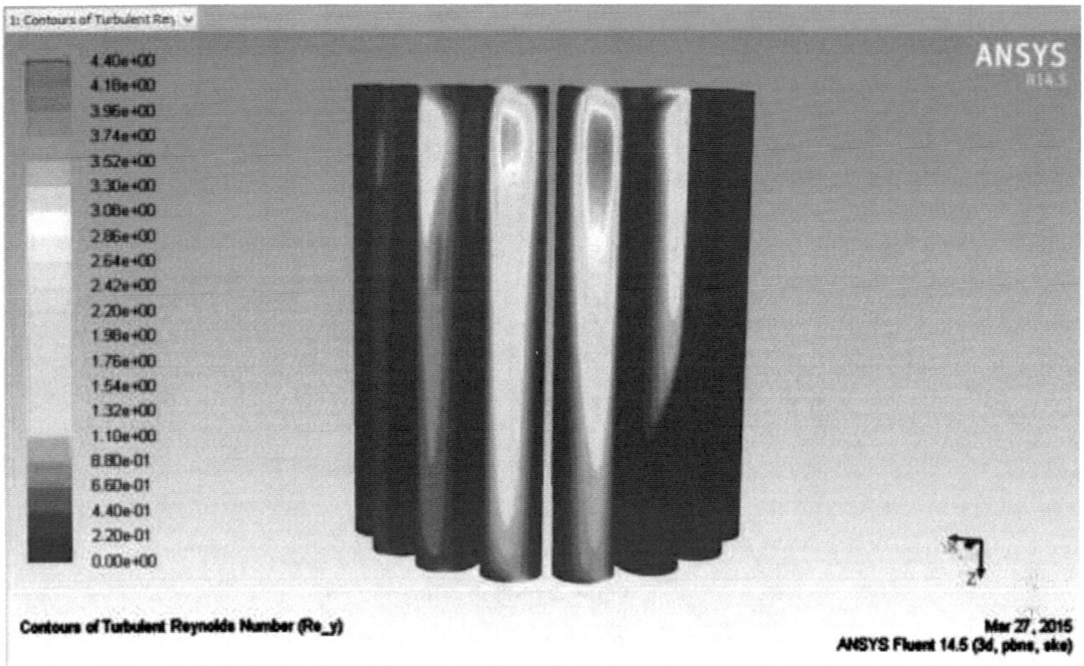

Figure 9. Counter of y+ on the wall of rotor

3.2. Boundary condition

3.2.1. Inlet boundary conditions. The velocity inlet at inlet of end cover is used, whose magnitude is assumed to be uniform and determined by the experiment. The turbulence parameters are specified in terms of turbulence intensity and hydraulic diameter of the inlet.

3.2.2. Outlet boundary conditions. Pressure outlet is used at outlet of end cover, static pressure (p=0.4MPa) is specified for low pressure outlet and static pressure (p=6MPa) is specified for high pressure outlet. The turbulence parameters are specified in terms of turbulence intensity and hydraulic diameter of the inlet.

3.2.3. Other conditions. No-slip condition is assumed on all the solid walls, and enhanced wall treatment function is used to calculate the turbulence kinetic energy and turbulence dispassion frequency near the wall. The rotation of rotor domain is considered by the use of the multiple rotating reference frame (MRF) method.

3.3. Solution strategy
FLUENT is used to carry out the numerical simulations. The code solves the Reynolds averaged Navier–Stokes equations in a primitive variable form. The effects of turbulence are modelled using the k-ε turbulence in the simulation. The second-order upwind scheme is used for discretization of convective term and the second-order central difference scheme for discretization of diffusion term. The separated solver is used to solve the incompressible flow. Numerical convergence absolute criteria is set to a maximum of 1×10^{-4}.

4. Results and discussions

4.1. Driving torque characteristics of pressure exchanger

The rotor rotates clockwise under the driving of the water flow. The Upstream face and Downstream face of rotor are shown in figure 4-1. When the water flow shots to the Upstream face of rotor channels with a certain inclined angle, the kinetic energy of water flow is converted into the static pressure energy. So the pressure of Upstream face will rise, therefore a pressure difference will formed between the Upstream face and Downstream face. It is can be seen from the pressure distribution of the cross section of the rotor shown in figure 4-2.

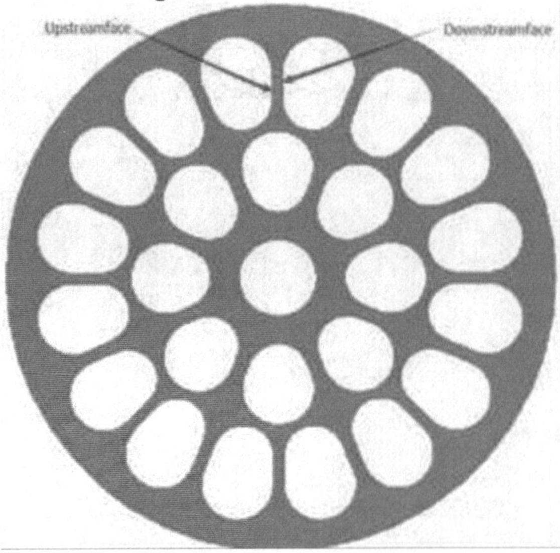

Figure 10. Geometry of rotor

For the sake of clearness, the range of displayed pressure is set as 5.85MPa to6.35MPa, which is suitable for displaying the pressure distribution of high pressure side. It is can be seen from figure 4-2 that the inclined angle is smaller, the pressure difference between the Upstream face and Downstream face is larger. It indicates that the driving torque obtained by rotor is greater at smaller inclined angle.

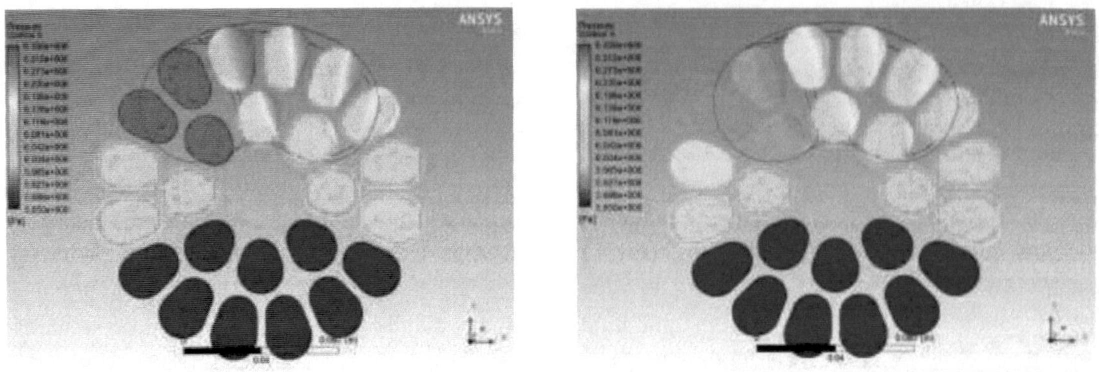

θ=6° θ=8°

IOP Conf. Series: Materials Science and Engineering **129** (2016) 012030 doi:10.1088/1757-899X/129/1/012030

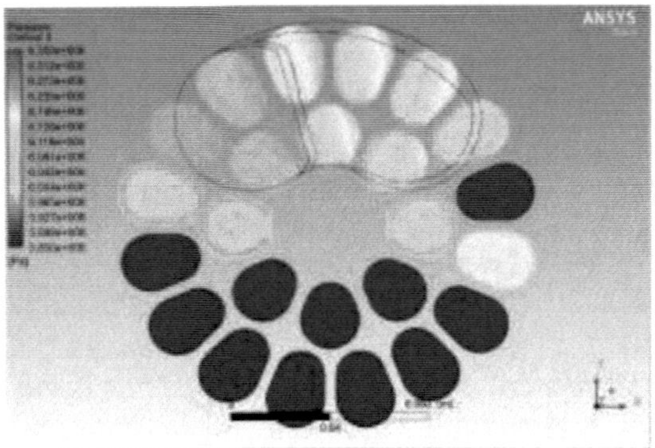

θ=10°

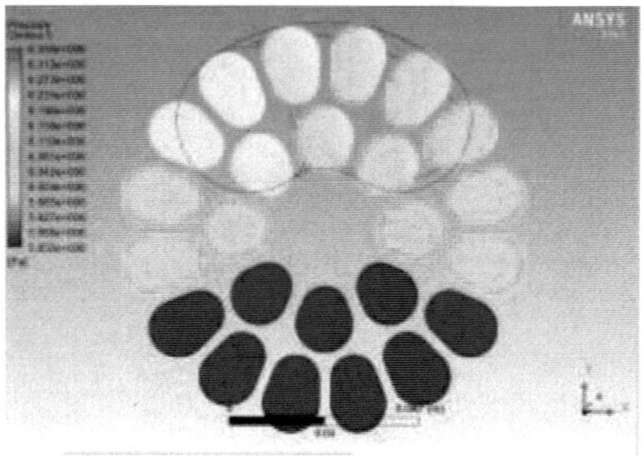

θ=12°

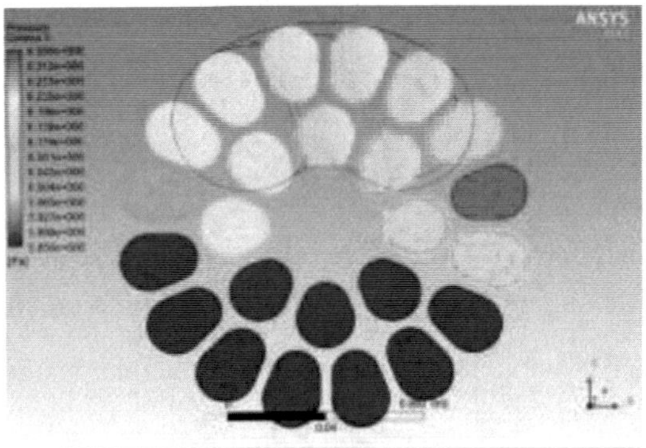

θ=14°

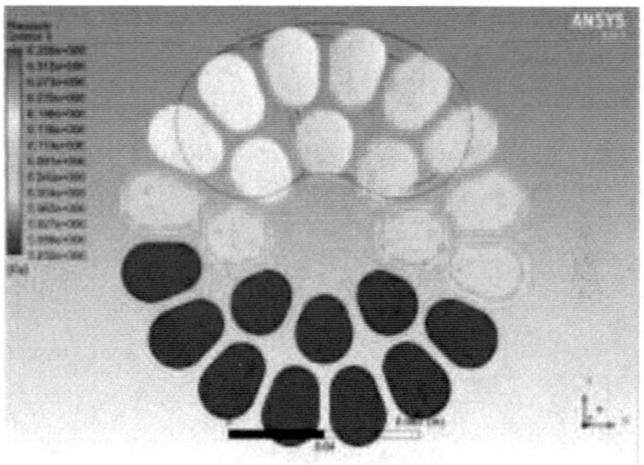

θ=16°

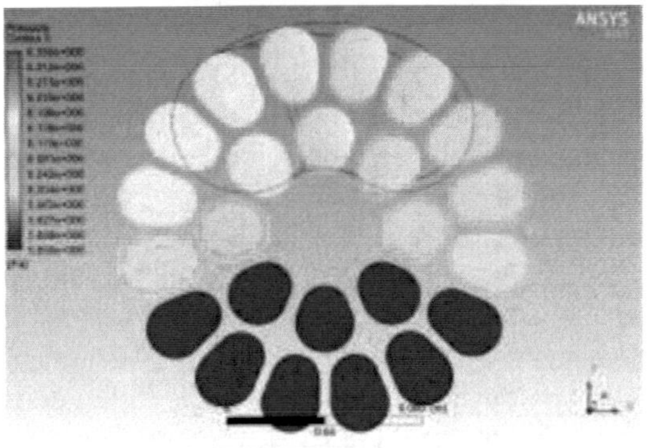

θ=18°

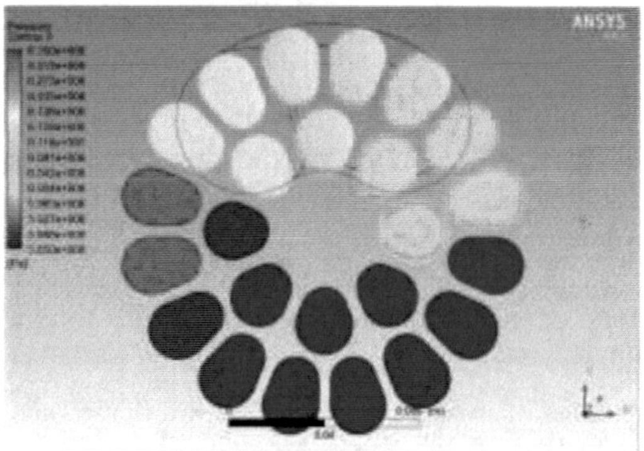

θ=20°

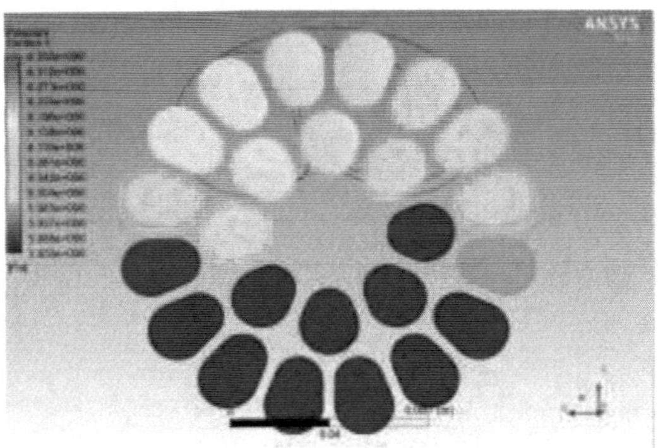

θ=22°

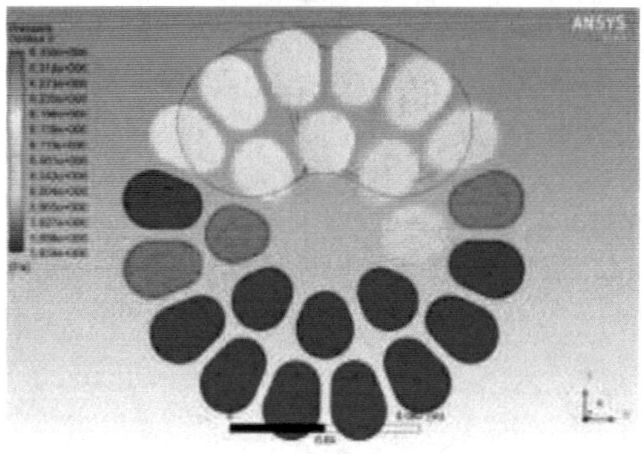

θ=24°

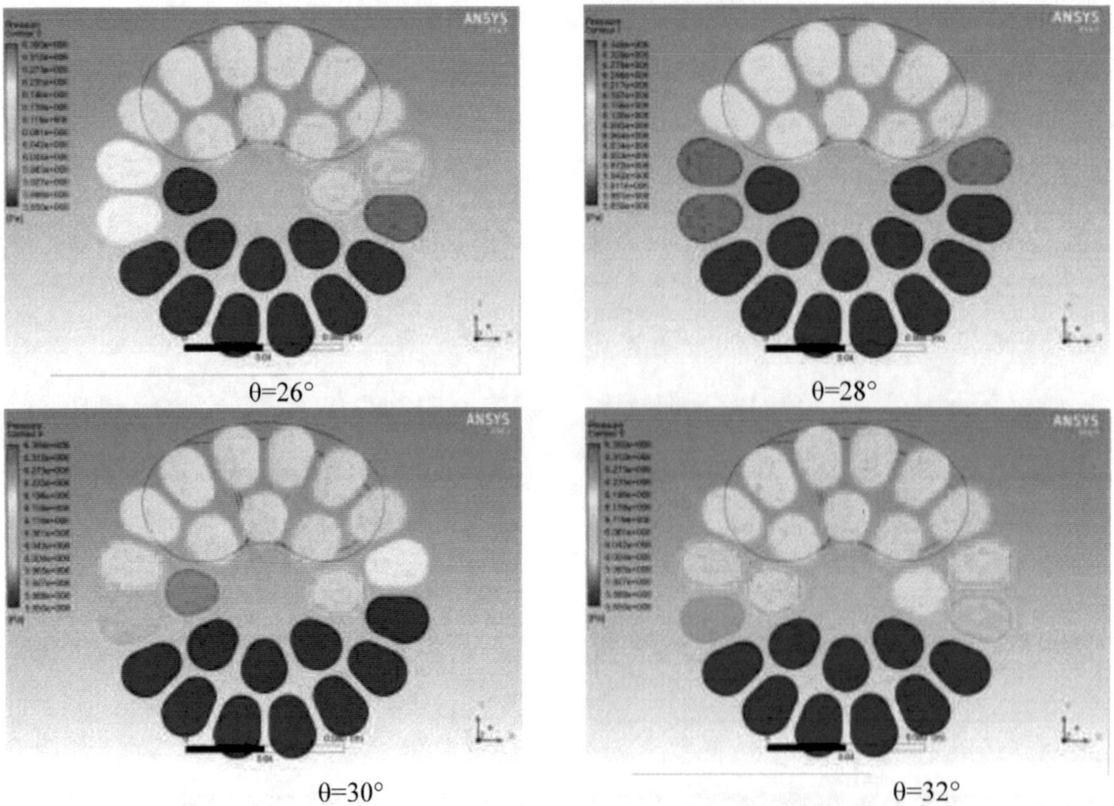

Figure 11. Pressure distribution of the cross section of the rotor

According to the different forming mechanism, the hydrodynamic force acting on any objects can be divided into positive pressure and viscous shear force. The performance characteristics of that two kinds of force is also completely different. So the driving torque of rotor is divided into two parts in this paper. They are respectively called pressure torque caused by positive pressure and viscosity torque caused by viscous shear force. The trend of pressure torque denoted by *Mp* along with the change of inclined angle denoted by θ is shown in figure 4-3. From that figure we can know that the pressure torque at any inclined angle is positive, which suggests that the water' s positive pressure is driving the rotor to rotate. The trend of pressure torque is obviously. The inclined angle is larger, the smaller the pressure torque. Especially when the inclined angle is small than 16°, the pressure torque de-creased rapidly. Another characteristic is that the distribution of data points is regularly, roughly on a curve. Regression method is used to analyse these data and power function is used to fit these data points shown in figure 4-1. The regressive curve equation got from regressive analysis is

$$y = 130 \times x^{-0.787} \tag{1}$$

Where y represents the pressure torque and x represents the inclined angle. The fitting degree index R^2 reached 0.9961.

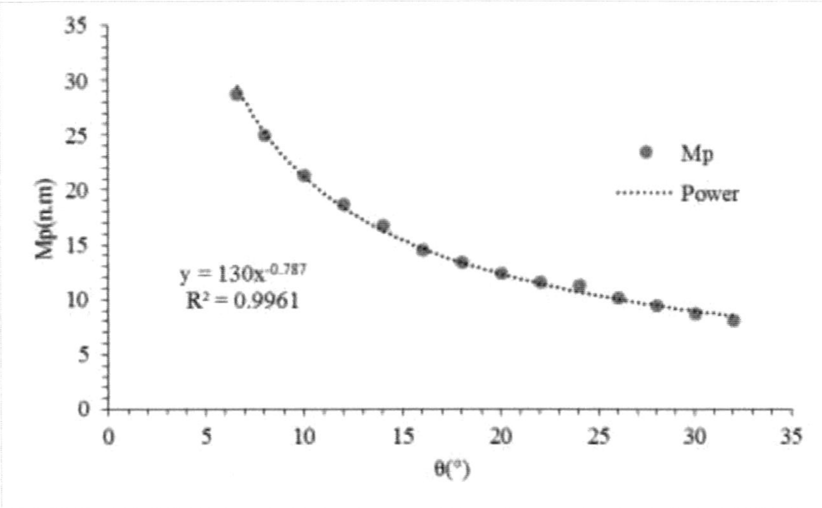

Figure 12. Curve of pressure torque along with the change of inclined angle

The distribution of viscosity torque denoted by *Mv* at every inclined angle is shown in figure 4-4. It is obviously that the viscosity torque at every inclined angle is negative, which suggests that the viscosity torque caused by viscous shear force hinders the rotation of rotor. But the absolute value of viscosity torque is very small. The biggest absolute value of torque caused by viscosity shear force is only 0.108(n*m) at the inclined angle of 6 degree. How-ever the torque caused by positive pressure at that inclined angle is 28.7(n*m), which is much larger than viscosity torque. Although the distribution of date points of viscosity torque is not as regular as pressure torque, regressive analysis is carried out on the viscosity torque and linear function is used to fit these data points shown in figure 4-4. The linear regressive curve equation is

$$y = 0.0014x - 0.1034 \tag{2}$$

Where y represents the viscosity torque and x represents the inclined angle. The fitting degree index R^2 is 0.748.

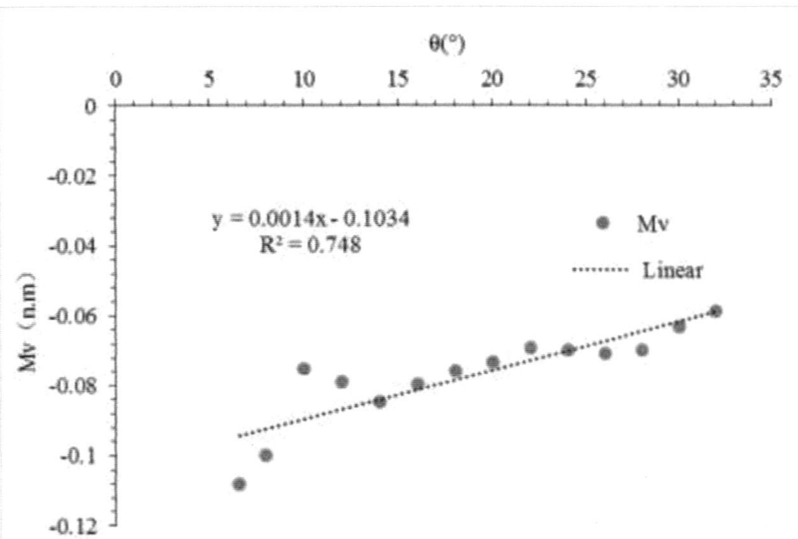

Figure 13. Curve of viscosity torque along with the change of inclined angle

IOP Conf. Series: Materials Science and Engineering **129** (2016) 012030 doi:10.1088/1757-899X/129/1/012030

Total torque denoted by *Mt* is the sum of pressure torque and viscosity torque. Because the absolute value of viscosity torque is too small to influence the variation trend of total torque, the distribution of data points of total torque shown in figure 4-5 is very similar to the distribution of data points of pressure torque. Similarly, regressive analysis is carried out on the total torque and the regressive curve equation is

$$y = 130 \times x^{-0.79} \tag{3}$$

Where y represents the total torque and x represents the inclined angle. The fitting degree index R^2 is 0.996.

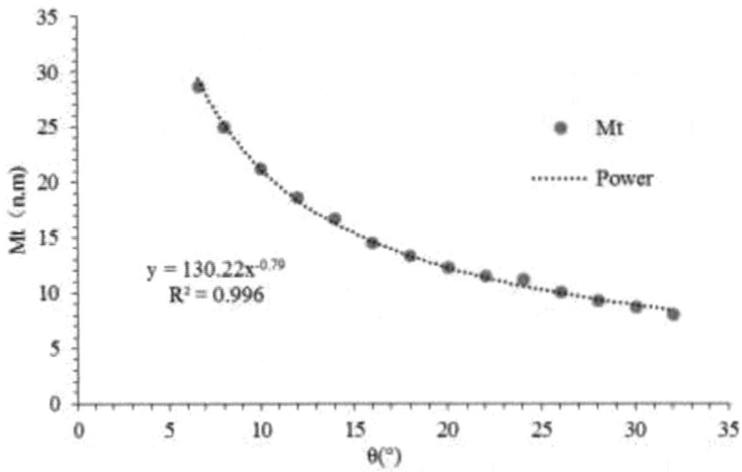

Figure 14. Curve of pressure torque along with the change of inclined angle

4.2. Pressure difference characteristics of pressure exchanger

The previous analysis shows that when water flows through the pressure exchanger, the water will produce a driving torque to drive the rotor to rotate. From the view of conservation of energy law, the water flow must have some loss of energy of itself. This can be confirmed from the pressure difference between the inlet and outlet. The pressure at high pressure inlet, high pressure outlet, low pressure inlet, low pressure outlet are denoted by *HPin, HPout, LPin, LPout* respectively. So the total pressure difference can be described as follows.

$$PD = HP_{in} - HP_{out} + LP_{in} - LP_{out} \tag{4}$$

Where *PD* represents the total pressure difference, which reflects the energy loss of water flow. The distribution of date points of pressure difference is shown in figure 4-6. It is obviously that the distribution of pressure difference and total torque is very similar, which suggests that larger total torque means larger pressure difference. Regressive analysis is also carried out on the pressure difference and the regressive curve equation is

$$y = 57.222 \times x^{-1.314} \tag{5}$$

Where y represents the pressure difference and x represents the inclined angle. The fitting degree index R^2 is 0.9969.

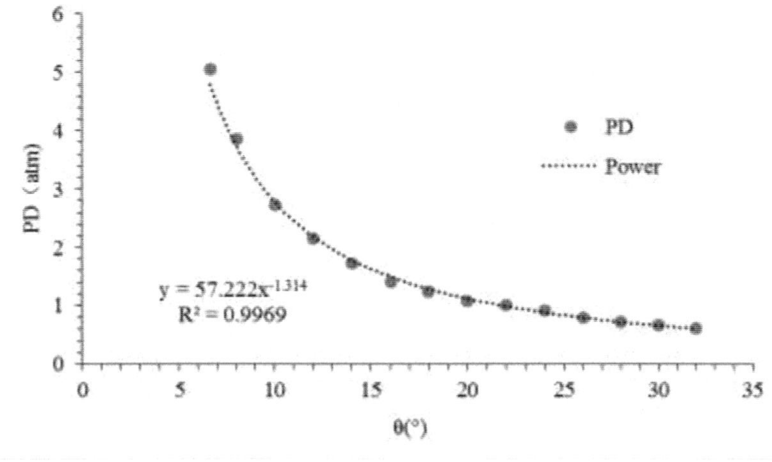

Figure 15. Curve of pressure difference along with the change of inclined angle

4.3. Driving efficiency characteristics of pressure exchanger

Total pressure difference between inlet and outlet represents the energy loss of water flow. Total driving torque is the output of water flow. So the driving efficiency is de-fined as the ratio of total driving torque and total pressure difference, which is described as follows.

$$\eta = \frac{M_t}{PD} \tag{6}$$

Where η represents the driving efficiency, *Mt* represents the pressure torque, *PD* represents the pressure difference. The distribution of driving efficiency denoted by η at every inclined angle is shown in figure 4-7. It is obviously that the driving efficiency increased gradually with the increase of inclined angle, which is contrary to the trend of total torque and pressure difference. Logarithm curve is used to fit the date points of driving efficiency and the regressive curve equation got from regressive analysis is

$$y = 0.485\ln(x) - 0.3378 \tag{7}$$

Where y represents the driving efficiency and x represents the inclined angle. The fitting degree index R^2 is 0.9924.

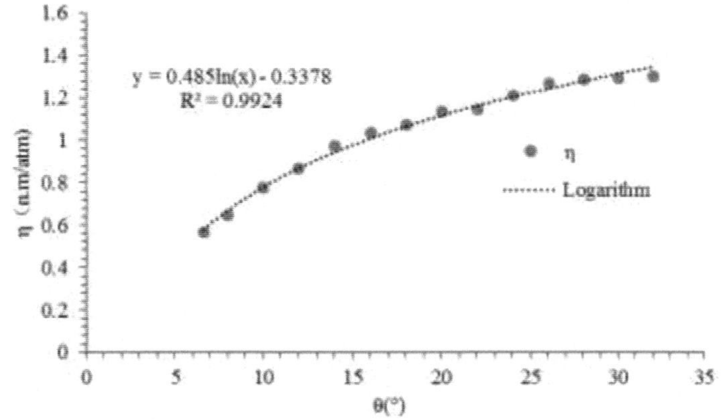

Figure 16. Curve of driving efficiency along with the change of inclined angle

IOP Conf. Series: Materials Science and Engineering **129** (2016) 012030 doi:10.1088/1757-899X/129/1/012030

The isosurface of turbulence kinetic energy is shown in figure 4-8. The value of ISO is *5 J/kg*, which indicates that the turbulence kinetic energy in the region coated by isosurface is Greater than or equal to *5 J/kg*

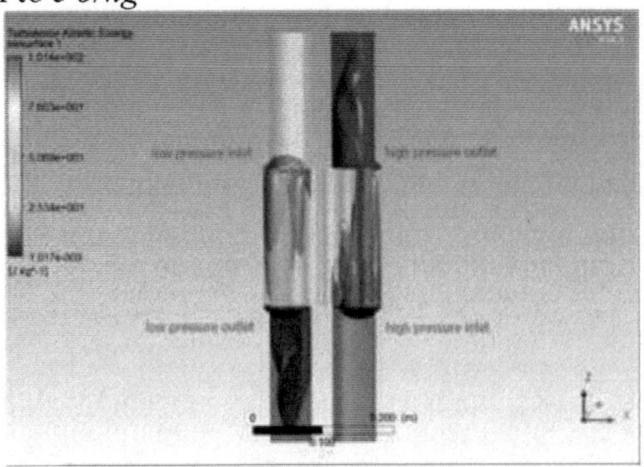
θ=6°

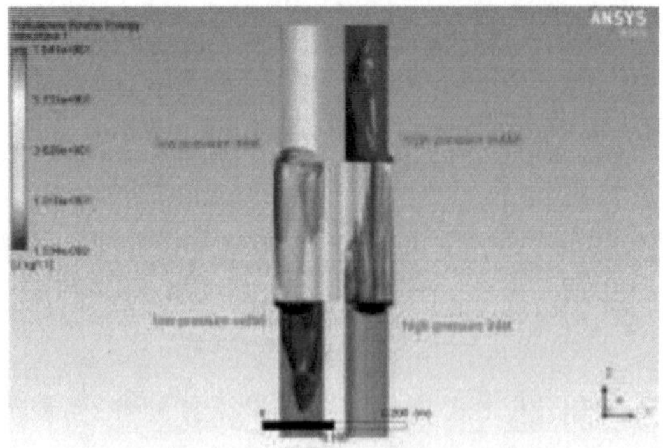
θ=8°

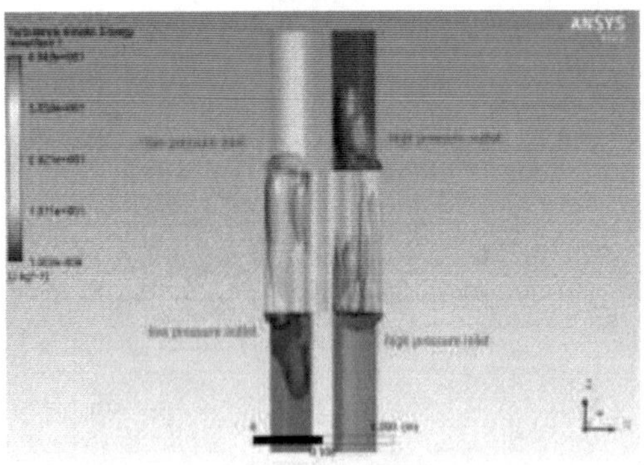
θ=10°

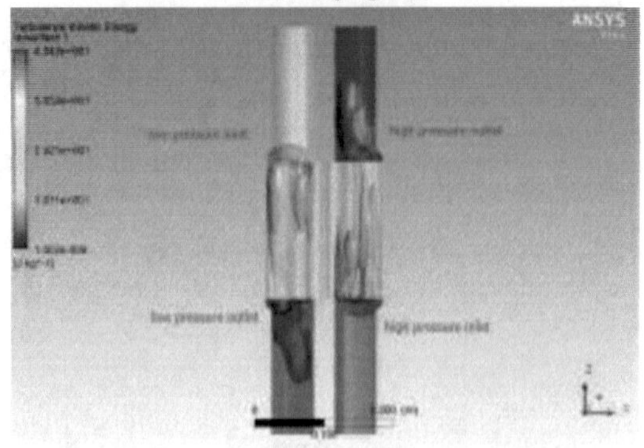
θ=12°

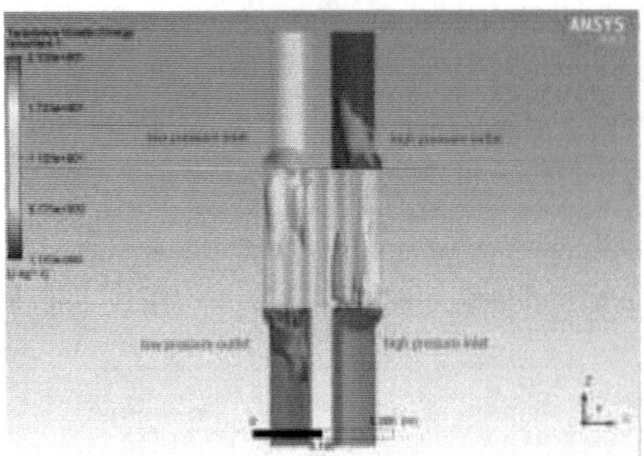
θ=14°

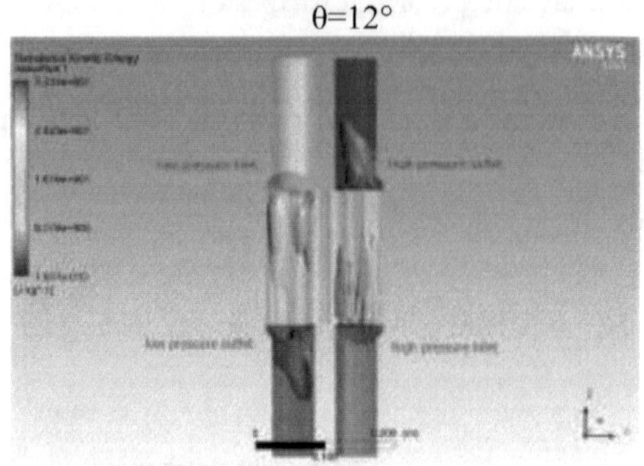
θ=16°

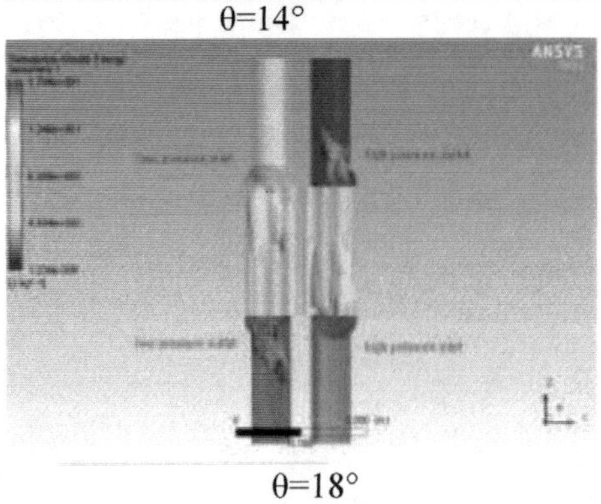
θ=18°

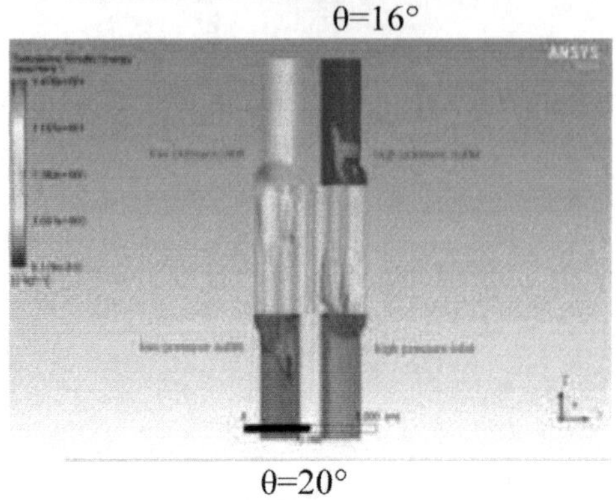
θ=20°

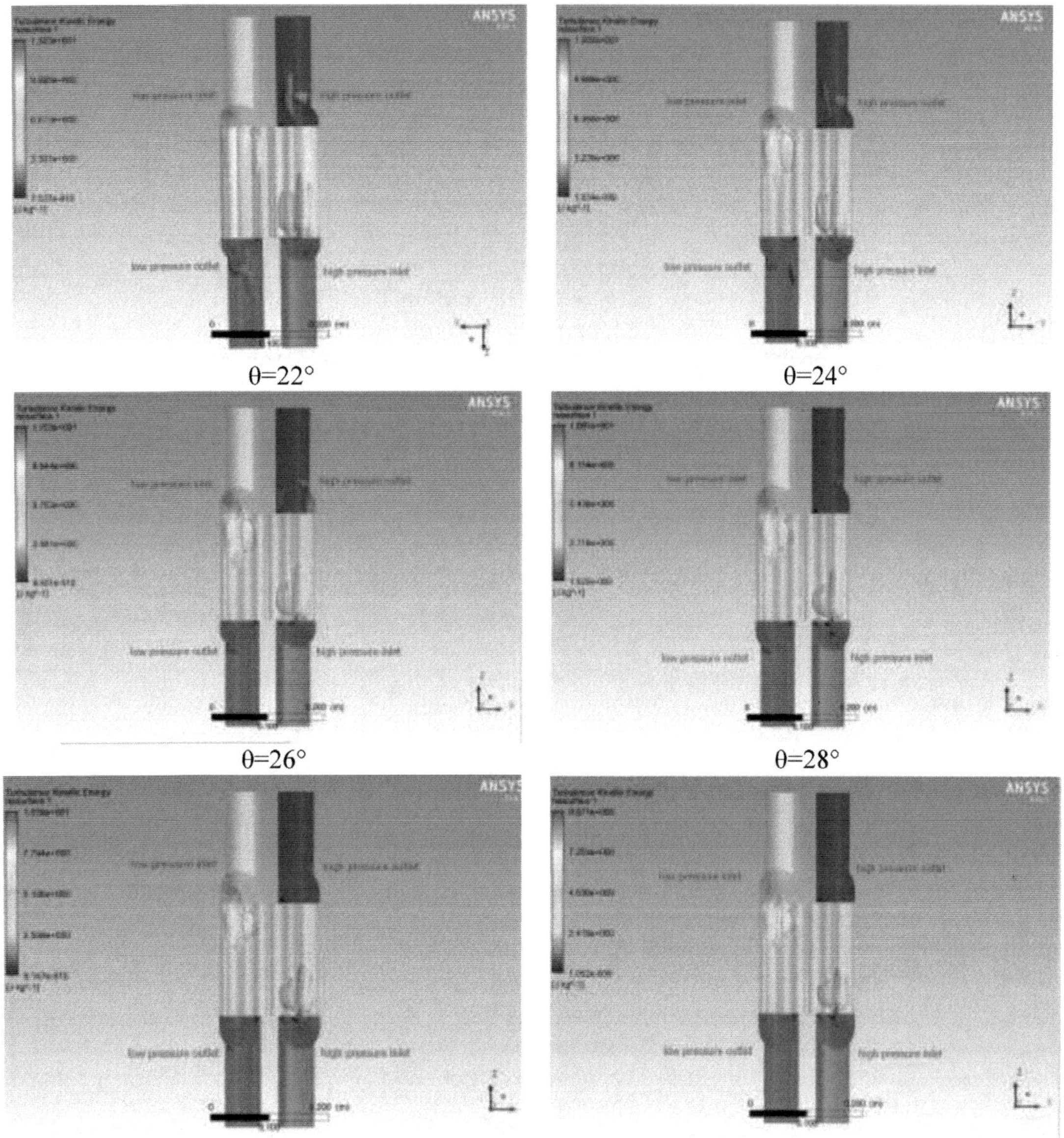

θ=22° θ=24°
θ=26° θ=28°
θ=30° θ=32°

Figure 17. Isosurface of Turbulence kinetic energy

When water flows into channels of rotor, the water flow shots to the walls of rotor channels with a certain inclined angle. So the water flow is disordered by rotor and a large amount of vortex generated in this region. When the water flow into the end cover's channel from the channels of rotor, the direction of water flow is changed. When θ<20° , the water flow is disordered seriously and a large amount of vortex generated in the channels of low pressure outlet and high pressure outlet. It is can be seen from figure 4-8 that the inclined angle θ is larger, the region where the turbulence kinetic energy is greater or equal to 5 J/kg is smaller, which indicates that The greater the inclined angle, the less energy of water flow dissipates into the turbulence. For the sake of clearness, the total energy loss of water flow, energy of driving rotor to rotate, energy dissipates into the turbulence are denoted by *Et*, *Ed* and *Ew*. So the following equation is established

$$E_t = E_d + E_w \tag{8}$$

It is known that the energy of driving rotor to rotate is what we want but the energy dissipates into the turbulence has no use for us. *Ew* is smaller means the less useless energy generated, which suggests that the driving efficiency is higher. It is can be confirmed in figure 4-5.

5. Conclusion

Though flow simulation and analysis on the end cover under different inclined angles, the following conclusions are drawn.

1. k-ε turbulence model is capable of capturing the detailed information of the fluid flow and can be used to predict the driving torque of rotor under different inclined angles.
2. The total driving torque is composed of pressure torque and viscosity torque, in which the pressure torque is positive and the viscous torque is negative. Pressure torque is the main component of total torque. The influence of viscosity torque is negligible.
3. Both driving torque and pressure difference decrease with the increase of inclined angle in the form of power function. The regression equations were shown in equation (3) and (5) respectively.
4. The driving efficiency increases in form of logarithmic function with the increase of inclined angle. The regression equation is shown in equation (7).

Acknowledgements
The research work was funded by National Science and technology support program (No.2013BAB08B03) and the National Natural Science Foundation of China (No. 51276158).

Nomenclature

θ	: Inclined angle
Mv	: Driving torque caused by viscosity
Mp	: Driving torque caused by pressure
Mt	: Total driving torque
HPin	: Pressure at High pressure inlet
HPout	: Pressure at High pressure outlet
LPin	: Pressure at Low pressure inlet
LPout	: Pressure at Low pressure outlet
PD	: Pressure difference
η	: Driving efficiency
Et	: Total energy loss of water flow
Ed	: Energy of driving rotor to rotate
Ew.	: Energy dissipates into the turbulence

References

[1] Stover R L 2004 Development of a fourth generation energy recovery device [J] *Desalination.* **165** 311-321.
[2] Cameron I B and Clemente R B 2008 SWRO with ERI's PX pressure exchanger device a global survey [J]. *Desalination.* **221**(1-3): p. 136-142.
[3] Hauge L J 1995 The pressure exchanger — A key to substantial lower desalination cost [J] *Desalinatio.* **102**(1): p. 219 – 223
[4] Zhao F 2014 Analysis on support mechanism of the axial film in Rotary Pressure Exchanger based on Fluent [J] *Journal of Zhejiang University (Engineering Edition).*p:1528-1533
[5] Yu L 2012 3D numerical simulation on mixing process in ducts of rotary pressure exchanger [J] *Desalination and Water Treatment.* **42**(1-3): p. 269-273.
[6] Yue W 2005 Mixing behaviour of feed and concentration in SWRO energy recovery cylinder and its computer simulation, Membr. Sci. Tech-nol. **25**(6) 36–39.

IOP Conf. Series: Materials Science and Engineering **129** (2016) 012031 doi:10.1088/1757-899X/129/1/012031

Study on water lubricated bearings of high speed pump based on numerical simulation

Y X Bai[1], F Y Kong[2], J R Sun[3]& X Yuan[3]

National Research Center of Pumps,, Jiangsu University, Zhenjiang 212013, China

[1] yxbai1012@163.com, [2] 531264162@qq.com, [3] 2540189073@qq.com

Abstract. A method is presented for calculating and analyzing the performance of water lubricated bearing of high speed pump under different structure. In present work, six kinds of bearings in different radial clearance(C), which are 0.02, 0.04, 0.06, 0.08, 0.10and0.12 respectively, under the same minimum water film thickness, have been designed. The models are built by CREO and numerical simulated by ansys. The main content of the present work is to analyze the relationship between the pressure and the load carrying capacity with different radial clearance(C) by ansys workbench based on Fluid-Solid coupling through ansys workbench.The stress deformations of bearings are also acquired through thermal-structure coupling. From the comparing result among the numerical analysis under the six different model of water lubricated bearing, the relationship between radial clearance(C) and load carrying capacity, as well as the deformation of bearing under different radial clearance(C), are obtained. Further, results indicates that, a proper selection of radial clearance(C) is essential to enhance the bearing performance.

1. Introduction

The capability for water lubricated bearings is critical importance for efficient design and performance of high speed pump due to its good rotation accuracy, excellent damping property, and other advantages. In most cases, the water lubricated bearings determine the life of high speed pump .Over the last several decades, considerable effort from both experimental and analytical fronts has been devoted to understanding the effect of water film and journal bearing.

Singh U studies the influence of rotating speed on the performance of hydrodynamic journal bearing based on two-dimension reynolds equation,energy equation and heat conduction equation [1].Alex introduces a mix elasto-hydrodynamic lubrication model using finite element method based on relaxed iterative method to calculate the performance of water film [2]. .Backar.S researches hydrodynamic journal bearing under different biolubricants through experiment [3]. GERTZOS K P gives a modified boundary condition using dynamic mesh method to obtain the distribution of flow field of lubricated bearing [4]. Gertzos obtains calculation results which is consistent with the experiment results to prove the applicability of the half Sommerfeld's boundary condition [5].Ye xiaoyan expresses the distribution of transient response and the best bearing clearance in water lubricated bearings of seawater-desalination pump[6-7] .But fluid lubricated bearings in high speed condition are studied in few papers.

This paper studies on the optimum of radial clearance(C), which is a key parameter of load carrying capacity of journal bearing, based on fluid-solid coupling using computer fluid dynamics (CFD). In this paper, MFS(multifield solver),which based on the results of the solid domain used transient kinetics and fluid domain used dynamic mesh ,is used to achieve the date exchange through the fluid-solid interface under different radial clearance(C). The control equations of the fluid domain and solid domain are solved in the separation method to simulate the water film model. The purpose of this paper is to provide the theoretical basis and guidance of the journal bearing of the high speed rescue pump.

2. Mathematical Model

2.1. Water lubricated bearing model

The formation of Hydrodynamic Lubrication of water lubricated bearing can be divided into three stages: staring, unsteady operation and steady operation, which is shown in Fig1.With the increasing of rotating speed, the load area of water film is larger, which will lead to the decreasing of friction force. Enough fluid under a certain rotating speed is carried to the axis gap to form water film, which separates the fluid and solid friction surface, to support external load. When steady state is reached, the journal is displaced from the bearing with a center distance (e), which is referred to the journal eccentricity ratio (ε) and the radial clearance(C).

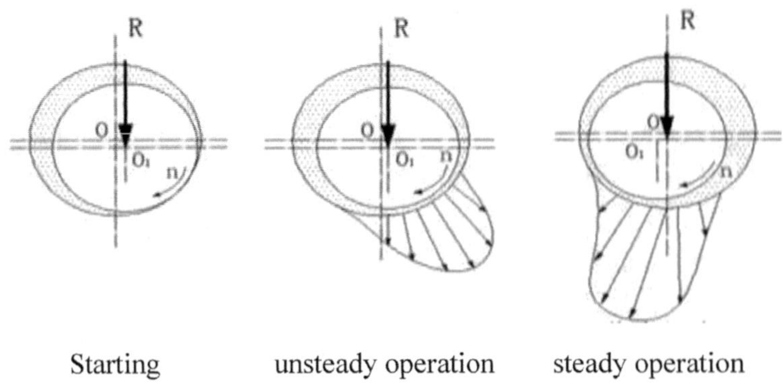

Fig.1 The work process of sliding bearing

2.2. Model of water film

The design parameters of journal bearings based on the require of design are: diameter is 85mm, width is 100mm rotation speed is 8000r/min. For the purpose of acquiring the best radial clearance(C) of the water lubricated bearing, 6 groups of different radial clearance(C) ,which are 0.02, 0.04, 0.06, 0.08, 0.1,0.12 mm respectively, under the same water film thickness, which is 0.05mm, are simulated and compared. Fig 2 shows the mesh of the water film.

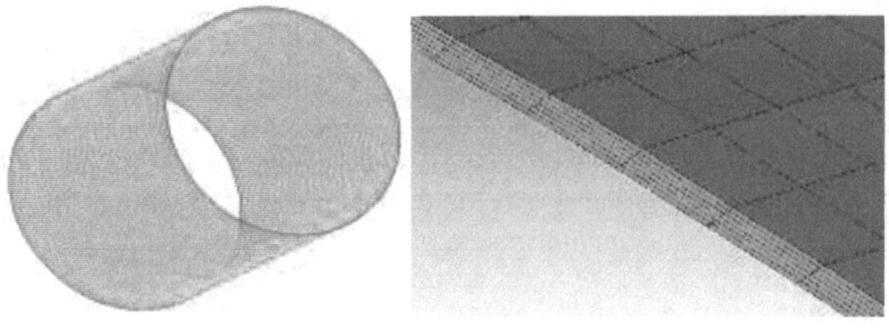

Fig.2 Structured mesh of the water film

2.3. *Basic equation*

2.3.1. Control equation of CFD. The Fluid flow and heat transfer are controlled by physical conservation law, such as law of momentum conservation, mass conservation and energy conservation law[8] .Control equation is the mathematical description of these conservation laws, the continuity equations and the motion equation of newton incompressible fluid.

Any flow question must meet mass conservation, which means the growth of mass of differential fluid in unit time is equal to the net weight of fluid flowed to the micro unit in the same time interval. Based on the mass conservation law, the former of mass conservation equation is:

$$\frac{\partial \rho}{\partial t}+\frac{\partial(\rho\mu)}{\partial x}+\frac{\partial(\rho v)}{\partial y}+\frac{\partial(\rho\omega)}{\partial z}=0 \tag{1}$$

Where ρ is density ,t is time, u, v, w are the velocity component of the x, y and z direction .

Momentum conservation law is the basic law every flow system must need. This law ,which can be called the Newton second law ,should be expressed : the rate of change of momentum with time is equal to the sum of the external force exerted on the micro unit. Based on the law, the Navier-Stokes equation of the x, y and z direction can be deduced.
Induce the vector symbol:

$$div(a)=\frac{\partial a_x}{\partial x}+\frac{\partial a_y}{\partial y}+\frac{\partial a_z}{\partial z}, grad()=\frac{\partial}{\partial x}+\frac{\partial}{\partial y}+\frac{\partial}{\partial z} \tag{2}$$

The former is:

$$\frac{\partial \rho u}{\partial t}+div(uu)=\mu \bullet div(grad(u))-\frac{\partial p}{\partial x}$$

$$\frac{\partial \rho v}{\partial t}+div(vv)=\mu \bullet div(grad(v))-\frac{\partial p}{\partial y}$$

$$\frac{\partial \rho \omega}{\partial t}+div(\omega\omega)=\mu \bullet div(grad(\omega))-\frac{\partial p}{\partial z} \tag{3}$$

Where ρ is density; t is time; u, v and w are the velocity component of the x, y and z direction; p is pressure, μ is dynamic viscosity.

2.3.2. Fluid-Solid coupling equation. The stress formula is:

$$\sigma_{ij}=\lim_{\Delta A_i \to 0}(\frac{\Delta F_i}{\Delta A_i}) \tag{4}$$

Where ΔF_i is force application along the i direction; ΔA_i is the force bearing along the i direction.
The former of constitutive equation is

$$\sigma_{xx}=\frac{E(1-\mu)}{(1+\mu)(1-2\mu)}(\xi_{xx}+\frac{\mu}{1-\mu}\xi_{yy}+\frac{\mu}{1-\mu}\xi_{zz})$$

$$\sigma_{xx}=\frac{E(1-\mu)}{(1+\mu)(1-2\mu)}(\xi_{xx}+\frac{\mu}{1-\mu}\xi_{yy}+\frac{\mu}{1-\mu}\xi_{zz})$$

$$\sigma_{xx}=\frac{E(1-\mu)}{(1+\mu)(1-2\mu)}(\xi_{xx}+\frac{\mu}{1-\mu}\xi_{yy}+\frac{\mu}{1-\mu}\xi_{zz})$$

$$\tau_{xy} = \frac{E}{2(1+\mu)}\gamma_{xy}, \tau_{yz} = \frac{E}{2(1+\mu)}\gamma_{yz}, \tau_{zx} = \frac{E}{2(1+\mu)}\gamma_{zx} \quad (5)$$

Where σ is stress component, τ and ξ are strain components, γ is shear stress component, μ is poisson ratio, E is elastic modulus .

Calculation scheme

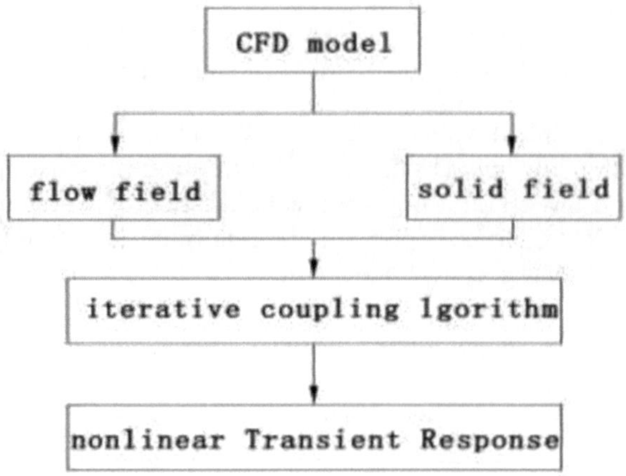

Fig.3 fluid-solid coupling calculation scheme

Fig.3 shows calculation scheme of fluid-solid coupling based on CFD. When calculating transient fluid field of bearing-rotor system, bearing and fluid interact that fluid force causes structure deformation of bearing and the deformation effects fluid field which cause the change of water film pressure distribution and then its carrying capacity that in turn effects bearing structure .Finally, the laws of water film carrying capacity distribution and other parameters iterative coupling operation can be obtained.

3. Analysis and result

The pressure distribution vs. the radial clearance(C) is given in Fig.4.The hydrodynamic water film generates the blue zones which are negative pressure zones, and the red zones which are the positive pressure zones. Hydrodynamic performance of journal bearing depends primarily on the distribution of the positive pressure zones and peak value. In the minimum film thickness of steady state condition, wedge-shaped area of convergence zone and divergence zone makes up the water film. The pressure of the convergence zone drops significantly after the positive pressure arriving the peak value. The pressure increases gradually to positive pressure after the peak of negative pressure.

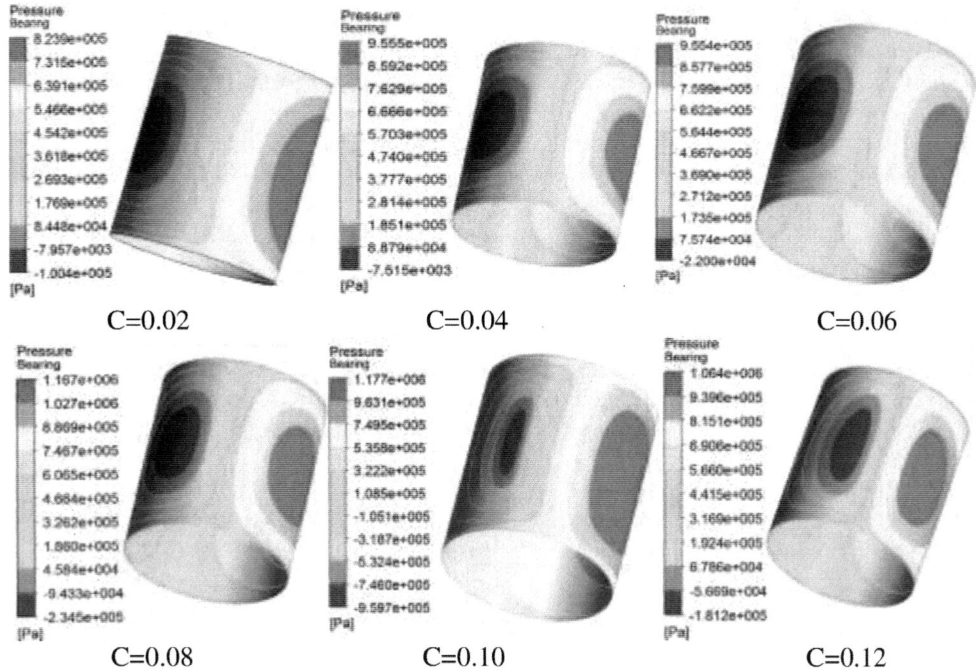

Fig.4 The water film pressure in different radial clearance(C)

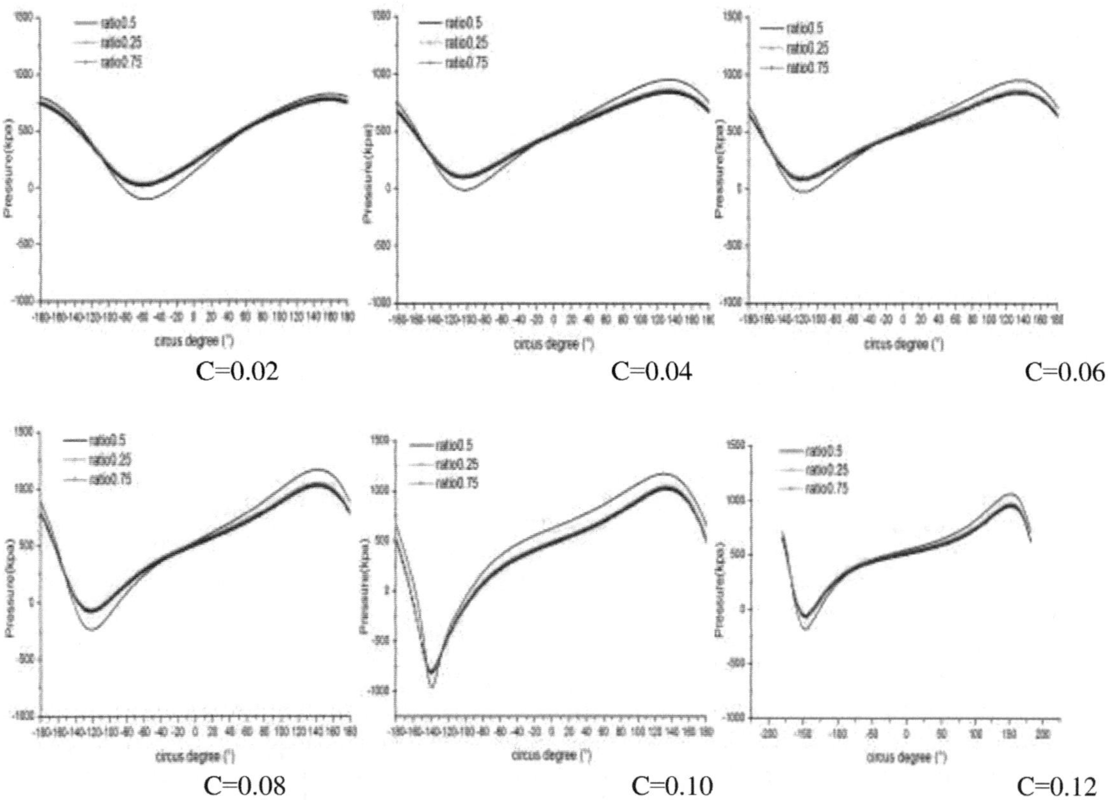

Fig.5 The water film pressure in circus direction in different radial clearance(C)

In order to show the pressure distribution more intuitive and clearly, the pressure distribution of the circus direction is presented in Fig.5. In Fig.5, three circus, which are based on the ratio of the current length vs the whole length of the bearing, are extracted from the pressure of the bearing inner surface. It is obvious that, with radial clearance (C) increasing, the peak of positive pressure is decreasing.

During the hydrodynamic lubrication created by water-lubricated bearing, the deformation of bearing inner surface, which is formed by the pressure of water film, changes the thickness of water film. The change has an influence on the distribution of pressure of water film. This process is fluid-solid coupling. The results obtained from coupling the interaction of flow-solid can predict and simulate the performance of bearing.

To improve the bearing performance, a new graphite alloy F50C is used to make the bearing. The property of F50C is: Rockwell hardness is 26.1; Tensile yield strength is 15.07MPa; Young's modulus is 429.03MPa; Tensile elongation is 94%, Coefficient of expansion is 6.24E-05.

According to the pressure of water film of the wall of bearing in Fig.6, it is inferred to cause the micro-deformation of inner surface, and then the thickness of water film changes. There is no difference on the distribution between the deformation and pressure of water film. The deformation of inner surface of bearing is in direct proportion to the radial clearance(C).

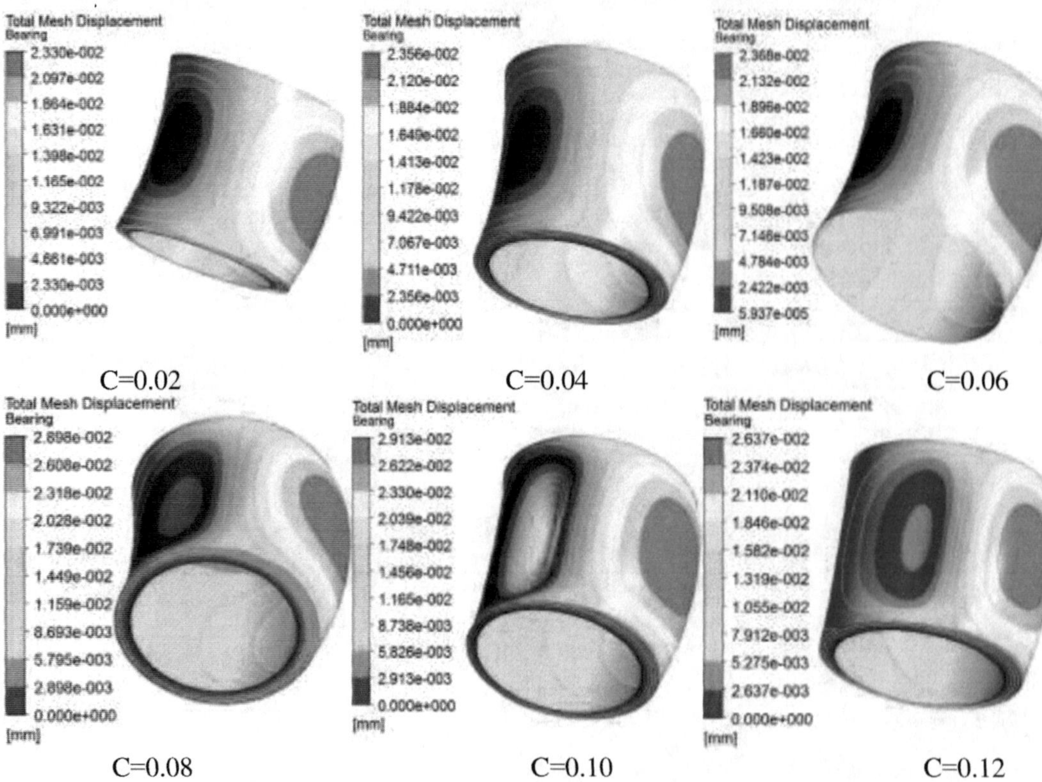

Fig.6 The deformation of bearing inner wall in different radial clearance(C)

The film capacity, ax deformation and max pressure under different radial clearance(C) are plotted in Fig.7.As is shown in Fig.7,the capacity of water film, the max deformation and pressure are in line with the same tendency of the growth of radial clearance(C).Of all the radial clearance(C) points, the values of the three parameters reach the peak at radial clearance(C) 0.10mm .The maximum value obtained at the point(radial clearance(C) 0.10mm) may be caused by the following reason: With the increasing of the radial clearance, the thickness of water film becomes larger which lead to greater load capacity ,which refers to the effect of fluid friction is more significant. Further, when the radial clearance(C) increase after radial clearance(C) 0.10mm, which may be the critical value, the hydrodynamic lubrication of water film may be damaged because of the excessive water film thickness.

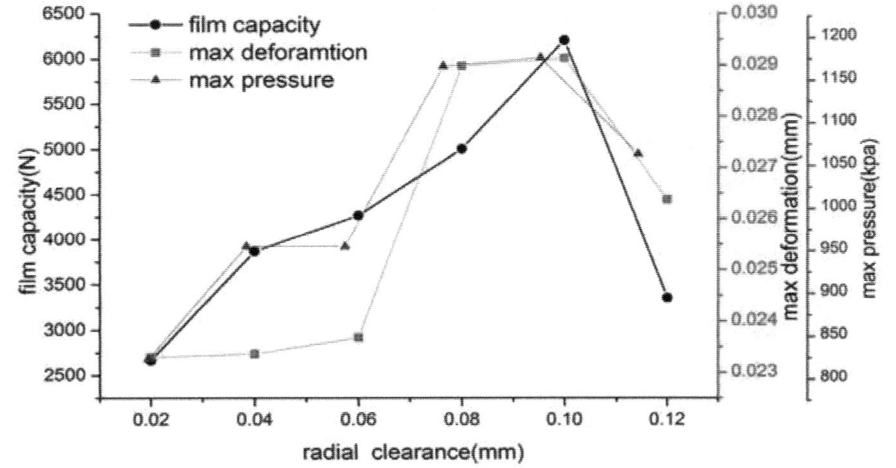

Fig. 7 The water film capacity in different radial clearance(C)

4. Conclusion

In this paper, the relationship between radius clearance and performance of water film was investigated. First, six water film under different radius clearance was numerically simulated based on fluid-solid coupling and its pressure distribution were obtained. Performance of water film consists of film capacity, deformation of bearing and pressure distribution corresponding to different radial clearance(C) reveals the critical effect of the radial clearance(C).

Also, this study can provide theoretical basis for water lubricated bearing designing with higher film capacity to ensure high efficiency of high speed pump

References

[1] Singh U, Roy L, Sahu M 2008 Steady-state thermo-hydrodynamic analysis of cylindrical fluid film journal bearing with an axial groove [J]. *Tribology International*,**2008(41)**: 1135-1144

[2] Kraker A, Ostayen V, Ron A J, Rixen D J 2007 Calculation of stribeck curves for water lubricated journal bearings [J].*Tribology International*, **40(3)**:459-46

[3] Baskar S, Sriram G 2012 Experimental analysis of hydrodynamic journal bearing under different biolubricants [J]. *IEEE-International Conference on Advances in Engineering, Science and Management*, ICAESM-2012, 132-135.

[4] Gertzos K P, Nikolakopoulos P G, Papadopoulos C A 2008 CFD analysis of journal bearing hydrodynamic lubrication by Bingham lubricant [J].*Tribology International*, **41(12)**: 1190-1204.

[5] Gertzos K P, Nikolakopoulos P G, Papadopoulos C A 2008 CFD analysis of journal bearing hydrodynamic lubrication by Bingham lubricant[J]. *Tribology International*. **41(12)**: 1190-1204..

[6] Ye X Y, Zhang J H, Ding Y N 2010 Optimization design for water lubricated bearing clearance of seawater-desalination pump [J]. *Journal of drainage and irrigation machinery engineering*, **28(2)**; 117-121.

[7] Ye X Y, Wang J, Gu R G 2013 Transient response of water lubricated bearing-rotor system [J]. *Journal of drainage and irrigation machinery engineering*, **31(11)**; 934-947

[8] Kong F Y, Jiang W M, Wang W T 2012 Cavtation performance analysis for cryogenic high-speed canned pump based on coupled flow-heat [J]. *Journal of Huazhong University of Science and Technology(Natural Science Edition)*, **40(5)**:24-28.

Theoretical analysis on flow characteristics of melt gear pump

R J Zhao[1], J Q Wang[2], F Y Kong[3]

National Research Center of Pumps, Jiangsu University, Zhenjiang, Jiangsu 212013, China

E-mail: [1]287190436@qq.com, [2]286162100@qq.com, [3]531264162@qq.com

Abstract. The relationship between Geometric parameters and theoretical flow of melt gear pump is revealed, providing a theoretical basis to melt gear pump design. The paper has an analysis of meshing movement of melt gear pump on the condition of four different tooth numbers, stack movement law and flow ripple. The regulation of flow pulsation coefficient is researched by MATLAB software. The modulus formula of melt gear pump is proposed, consistent with actual situation.

1.Introduction

Common external gear pump is hard to be applied for large flow pulsation and large radial force or other defects [1-3] in the situations that require high uniformity such as melt extrusion. The foreign gear pump has compact structure and one import with several export. Simultaneously, it can reduce in-homogeneity of flow largely, improving quality of flow.

In domestic, the research concentrates on common gear pump. Yulong Li etc. have an analysis of relationship among volume oil, the trap pressure and unloading groove parameters of external gear pump [4-5]. Xianliang Xu etc. have an analysis of traffic character and radial force [6-9].

The paper has a research on meshing displacement and superposition motion law about symmetric melt gear pump on the condition of four different teeth numbers by mathematical derivation. The four theoretical flows are derived, according to which, the theoretical flows of the model pumps are calculated.

2. The Related Explain and Definition

2.1. the structure explain of melt gear pump

Melt gear pump principle structure is shown in Figure 1,the driving wheel gear is Z_1=68 ,the pinion gear is Z_2=32,modulus is m=0.75 ,the breadth of tooth is b=12.8 ㎜,the speed is n=60r/min, the displacement is 8×2ml .The first layer is studied assuming the same flow characteristics of each export to simplified analysis. An approximate individual external gear pump is formed between the driving wheel and each driven wheel .The total import of the pump is connected, each sub-pump inlet and outlet independent, thus forming four parallel pinion pump [10].

2.2 .the regulations about the related serial number

As showed in Figure 1,four driven wheel centers O_1, O_2, O_3, O_4 and the driving wheel center are connected at 90°。The pumps with center located O_1, O_2, O_3, O_4 on the driven wheels are defined driven wheels 1,2,3,4. The respective external gear pump is 1,2,3,4 with nodes P_1, P_2, P_3, P_4.

Driving wheel counterclockwise rotation is assumed, and the initial time is defined when a front tooth line medium point and p_1 coincident and the pump is defined Z_1.The other pumps are defined 1、...、(Z_1-1)in turn in counterclockwise rotation .For the driven wheel 1, the tooth which is meshed with the tooth Z_1 of the driving wheel is defined as tooth 1.The other teeth are defined as 2, ... , Z_2 in turn in clockwise[11].

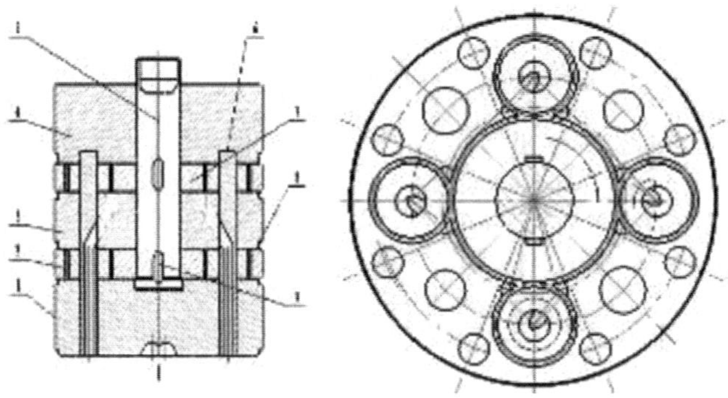

1-bottom cover; 2-meshing cavity; 3-intermediate cover; 4-front shroud
5-driving shaft; 6-driven shaft; 7-driving wheel; 8-driven wheel
Fig.1 Melt gear pump model

3. The analysis of meshing point displacement

According to the above analysis, the 4 meshing points of the driving wheel locates at P_1,P_2,P_3,P_4 respectively at Z_1=4K,t=0.The 4 meshing points move synchronously, the flow pulsation cycle angle is defined as α.According to the gear mesh, there is

$$f_i(\varphi) = f \ \ (-p_b/2 \le f \le p_b/2)$$

Where,

R_{b1}-base circle radius of driving wheel, Rb1=mZ_1cos20°

φ_i-angular displacement that meshing points with respect to nodes P_1、P_2、P_3、P_4

p_b-coxa, p_b=mZ_1cos20°

At Z_1=4K+1 and t=0, we have

$$\varphi_1(0) = 0, \varphi_2(0) = -\alpha/4, \varphi_3(0) = -\alpha/2, \varphi_4(0) = \alpha/4$$
$$f_1(0) = 0, f_2(0) = -p_b/4, f_3(0) = -p_b/2, f_4(0) = p_b/4$$

Considering the meshing discipline and the initial displacement of each meshing point, meshing displacement and alternate of each pinion are obtained.

Meshing displacement can be obtained from the formulas:

$$f_1(\varphi) = f \quad (0 \le f \le p_b/8)$$
$$f_2(\varphi) = f - p_b/4 \quad (0 \le f \le p_b/8)$$
$$f_3(\varphi) = f - p_b/2 \quad (0 \le f \le p_b/8)$$
$$f_4(\varphi) = f + p_b/4 \quad (0 \le f \le p_b/8)$$

By the same method motioned above, if it is Z_1=4K+1 and t=0, the distribution condition meshing displacement absolute values of gear pump 1,2,3,4.

4. The analysis of flow characteristic

4.1. The analysis of instantaneous flow

At Z_1=4k, the 4 meshing points are synchronous, $f_i(\varphi)=f\ (-p_b/2\le f\le p_b/2)$, instantaneous flow is defined as q_{vs} :

$$q_{vs}=4a_1-a_2\sum_{i=1}^{4}f_i^2=b\omega_1(4a_1-4a_2f^2)$$

At Z_1=4k+1、4k+3,the meshing point displacement change at f=0～p_b/8,theory flow is defined as q_E:

$$q_{vs}=b\omega_1[4a_1-a_2(f_1^2+f_2^2+f_3^2+f_4^2)]\ =b\omega_1[4a_1-a_2(4f^2-fp_b+3p_b^2/8)]$$

At Z_1=4k+2,the meshing point displacement change at f=0～p_b/8,theory flow is defined as q_E:

$$q_{vs}=b\omega_1[4a_1-a_2(f_1^2+f_2^2+f_3^2+f_4^2)]\ =b\omega_1[4a_1-a_2(4f^2+3fp_b/2+5p_b^2/16)]$$

$$f_i(\varphi)=f\ (-p_b/2\le f\le p_b/2),$$

$$q_{vs}=4a_1-a_2\sum_{i=1}^{4}f_i^2=b\omega_1(4a_1-4a_2f^2)\quad(1)$$

$$q_{vs}=b\omega_1[4a_1-a_2(f_1^2+f_2^2+f_3^2+f_4^2)]$$

$$=b\omega_1[4a_1-a_2(4f^2-fp_b+3p_b^2/8)]\quad(2)$$

$$q_{vs}=b\omega_1[4a_1-a_2(f_1^2+f_2^2+f_3^2+f_4^2)]$$

$$=b\omega_1[4a_1-a_2(4f^2+3fp_b/2+5p_b^2/16)]\quad(3)$$

Comparative analysis about the four kinds of melt gear pump is conducted(k=17).For the driving wheels,Z_1 ,the number of teeth ,is 68、69、70、71 respectively. The other parameters are defined as:Z_2=32,m=0.75,b=12.8mm,n=60r/min.

The flow characteristics of the melt gear pumps are simulated by MATLAB (Fig.2, φ∈(0, 3α/2)).

According to equations (1)～(3),the flow fluctuation coefficient δ and ripple frequency f are obtained, shown as Tab.1.

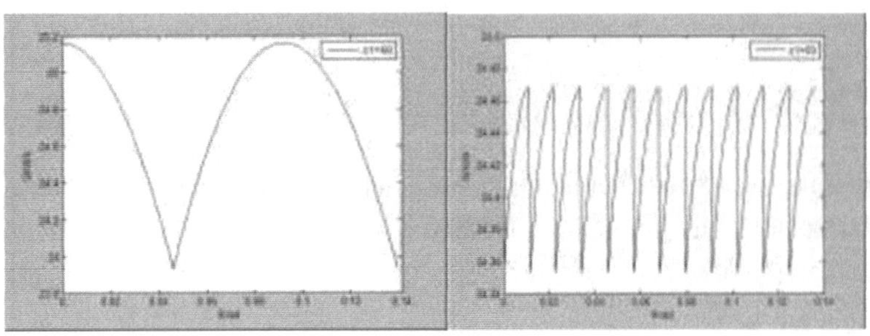

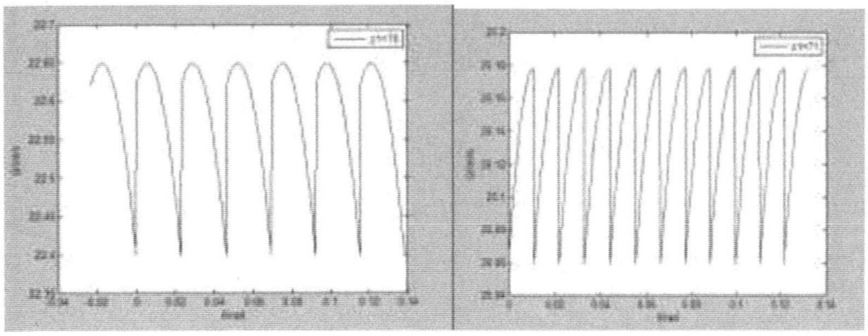

Fig.2 simulation results of four different melt gear pumps

Tab.1 The flow characteristic of different types of melt gear pumps

Category	$\delta(\%)$	Flow pulsation cycle angle	$f(Hz)$
$Z_1=68$	4.885	α	32
$Z_1=69$	0.491	α/8	32*8
$Z_1=70$	1.109	α/4	32*4
$Z_1=71$	0.477	α/8	32*8

Shown as Tab.1 and Fig.2,at $Z_1=68$,the flow fluctuation coefficient (δ)and ripple frequency(f) are same as common external gear pump and these of gear pumps($Z_1=69$ and $Z_1=71$) are basically same, their flow fluctuation coefficients are improved obviously compared with the common one and their ripple frequencies are 8 times that of the common external gear pump; when Z_1 is 70,the flow fluctuation coefficient (δ) is improved obviously and the ripple frequency(f) is 4 times that of the common external gear pump.

3.2. The analysis of theoretical displacement

The analysis of theoretical displacement is as follows:

$$dv = dt * q_{vs}$$

$$dv = dt * q_{vs} = 0.5\omega_1 dtb[4a_1 - 4a_2 f^2]$$

$$= 0.5d\varphi b[4a_1 - 4a_2 f^2]$$

Where

$$d\varphi = df / R_{b1}$$

R_{b1} —base circle radius of driving wheel

$$dv = \frac{0.5dfb[4a_1 - 4a_2 f^2]}{R_{b1}}$$

The first teeth pair starts to discharge fluid at f_1=-0.5pb and ends at f_2=0.5pb, the drainage volume of one teeth pair is:

$$v = \int_{f_1}^{f_2} \frac{0.5b[4a_1 - 4a_2 f^2]}{mz_1 \cos 20°} df$$

The equation (6) depicts the drainage volume of pump with z_2 teeth.

$$V = z_2 \int_{f_1}^{f_2} \frac{0.5b[4a_1 - 4a_2 f^2]}{mz_1 \cos 20°} df$$

$$=\frac{[8p_b a_1 - 2p_b^3 a_2/3]bz_2}{z_1\cos 20^\circ} \quad (4)$$

Similarly, the results of the other model pumps can be obtained:

4k+1、4k+3:

$$V=\frac{[8p_b a_1 - 2p_b^3 a_2/3]bz_2}{z_1\cos 20^\circ} \quad (5)$$

4k+2:

$$V=\frac{[8p_b a_1 - 5p_b^3 a_2/12]bz_2}{z_1\cos 20^\circ} \quad (6)$$

The equations of theoretical displacement deduced provide theoretical basis for the number of teeth and modulus of the melt gear pumps. The more direct relationship can be obtained from the above equations, as follows.

$$V = 50.24m^3 + 3.45m^3\left(\frac{1}{z1}+\frac{1}{z2}\right) \quad (7)$$

at 4k+2,

$$V = 50.24m^3 + 6.86m^3\left(\frac{1}{z1}+\frac{1}{z2}\right) \quad (8)$$

According to (9) 、 (10) ,it can be estimated:

$41.15m^3 < V < 56.86m^3$,

$$m = (0.26-0.29)\sqrt[3]{V} \quad (9)$$

The displacement of the driving wheel in a circle is:2.718×4×68/32=18.513ml

The value range of the modulus (m) is 0.68-0.767, taking it as 0.75.The inverse solution of modulus is consistent with the real value, proving the equation (9) to calculate modulus in designing melt gear pump.

5. Conclusions

(1)At Z1=4k,the four driven wheels mesh with the driving wheel synchronously, flow unevenness coefficient and flow fluctuation frequency are consistent with those of common external gear pumps.

(2)At Z_1=4k+1, Z_1=4k+3, the flow fluctuation coefficient and ripple frequency are basically same, their flow fluctuation coefficients are improved obviously compared with the common one and their ripple frequencies are 8 times that of the common external gear pump; at Z_1=4k+2, the flow fluctuation coefficient is improved obviously and the ripple frequency is 4 times that of the common external gear pump.

(3) The theoretical displacement value deduced is slightly higher than that of real displacement, proving the derivation proper, which provides theoretical basis for the number of teeth and modulus of the melt gear pumps. The calculation equations of modulus of melt gear pump are proposed.

References

[1] Gulich J F 2007 Pump selection and quality considerations [J].*Micro system Technologies*, **2007(10)**:857- 880.

[2] Luan Z H 2001 *Basic Theory of Compound Gear Pump* [M].XUZHOU: China Mining University Press.

[3] Morgu N P, Blesman A I, Kalekin V S, et al. 2007 Nanotechnology incompressor and pump engineering[J]. *Chemical and Petroleum Engineering* **43(5/6)**:275- 279

[4] Li Y L, Wang X J, Gu G H 2004 External gear pump of trapped oil process simulation[J].*Mechanical Transmission*, **28(6)**: 19-21.

[5] Li Y L, Sun F C 2012 Theoretical calculation of the external gear pump high pressure end clearance[J].*Journal of Drainage and Irrigation Machinery Engineering*, **30(2)**:147-152.

[6] Xu X L, Bao H Y, Fan M H 2001 A summary study of the basic theory of compound gear pump[J].*Journal of coal*, **6(4)**: 404-408.

[7] Zhao L C, Xu X L, Sun C J, etc. 2004 The second type of compound gear pump geometric flow characteristic analysis and test [J].*Journal of Agricultural Machinery*, **35(6)** : 82 - 85.

[8] Xu X L, Liu X H, Zhu Y C, etc. 2003 Flow characteristics of corrected tooth gear pump [J]. *Mechanical Engineering*, **39(3)**:44-47.

[9] Xu X L, Zhao L C, Yang Q L, etc. 2004 Balance the theoretical study and experimental analysis of gear pump [J]. *Journal of Coal*, **29(4)**:494-501.

[10] Mao P H, Hou B 2008 Theoretical analysis of three-pole parallel gear pump [J].*Chongqing University (Natural Science)*, **31(10)**:1123-1127.

[11] Wang J Q, Kong F Y, He Y Y 2012 Four planetary gear pump flow characteristics analysis [J]. *Mechanical Transmission*, **36(12)**:80-85.

Numerical simulation and factor analysis of petrochemical pipe erosion-corrosion failure

G F XU[1], G F OU[2*], T Chen[1], P X Li[1], H Z JIN[2]

[1] Hangzhou Special Equipment Inspection and Research Institute, Hangzhou Zhejiang 310018, China;
[2] Institute of Flow Induced Corrosion, Zhejiang Sci-Tech University, Hangzhou Zhejiang 310018, China

E-mail: xugenfu2001@163.com

Abstract. Based on the behavior of carbon steel outlet tube in REAC pipes of Zhenhai Refining & Chemical Company, the mathematical model of fluid-solid interaction was established according to the mechanism of erosion-corrosion damage. The interaction between corrosion products protecting film and multiphase liquid was analyzed by numerical simulation method. The distribution of shearing stress on the inwall of elbow bend, and the distribution of principal displacement, stress and strain of corrosion products protecting film were disclosed, while the erosion-corrosion failure processes was studied. The simulation result coincides with that of the positioned thickness gauging which validated the reliability and feasibility of the finite element analysis software simulation method. The obtained results can be used in the erosion-corrosion failure analysis, structural optimization, in-service testing positioning, life prediction, risk assessment, safety and other security projects for multiphase flow pipeline.

1. Introduction

With the popular pipeline transport and rigorous operating condition, more and more accidents are induced by failure of pipeline, which is harm for manufacture, environment and public safety [1-3]. The failure forms of pipelines are diversified and its mechanism is complicated. Among them, erosion failure is most universal and non-predicted failure, erosion is local, sudden and risk. Erosion failure is interaction between corrosion and multiphase flow [4-7]. Erodent products protecting film is deformed by shearing stress. Based on the deformation of film, it affects fluid flow. Then the shearing stress is augmented. By this way again and again, it accelerates to wash away carbon steel. The substance of process is interaction between corrosion and multiphase flow.

This paper is based on theory of fluid-solid coupling. The elbow made of carbon steel in outlet ductwork of Hydrocracking Reaction Effluent Air Cooler (REAC) is the subject investigated. Fluid-structure numerical simulation is used to analysis erosion failure. Flow quality (velocity), pipe construction (diameter, curvature radius) and film property (Poisson ratio) are impact to erode. All the work will help for REAC tube bundle's on-site monitoring, on-line detection, optimization design and corrosion protection.

*Corresponding author at: The lab of Multi-phase Deposition and Erosion, Zhejiang Sci-Tech University, Hangzhou 310018, China. Tel:+8657186843348; Fax:+8657186843348; E-mail address: ougf@163.com

2. Mathematical model

We can know that interaction between multiphase flow and erodent products protecting film is on the interface [8-9]. Based on above that, during the process of building up the equation and the mechanics model of the flow and the corrosion coupling function, we mainly consider the mechanics condition as following [10]: First, the distortion of coordinated-condition should satisfy the actual distorted condition of the erodent products protecting film. Second, the dynamics of coordinated- condition should satisfy the flow mechanics condition of the corrosive multiphase fluid medium. According to this, we are able to establish the unified coupling function equation as follows.

$$\phi_{fluid} = \phi_{pipe} \tag{1}$$

$$n \cdot \varphi_{fluid} = n \cdot \varphi_{pipe} \tag{2}$$

Among them, ϕ_{fluid} expresses the distorted field of fluid on the coupling interface; ϕ_{pipe} expresses the distorted field of the inner surface of pipeline protecting film on coupling interface; φ_{fluid} expresses the stress field of fluid on the coupling interface; φ_{pipe} expresses the stress field for the inner surface of pipeline protecting film on coupling interface.

Arbitrary Lagrange-Euler (ALE) [11-12] is used to solve the problem of fluid-solid coupling. In fluid region, Euler unit is adopted, and Lagrange unit in solid region, these two units are solved in ALE coordinates which lead that alteration of interface of fluid and solid is accord with the deformation of the film.

According to the fundamental principle of ALE, we can induce N-S equation of viscous multiphase flow:

Component of motion equation:

$$\frac{\partial \sum\limits_{k=1} \alpha_k \rho_k V_i}{\partial t} + \left(V_j - \hat{V}_j\right) \frac{\partial \sum\limits_{k=1} \alpha_k \rho_k V_i}{\partial x_j} = f_i + \frac{\partial[-p\delta_{ij} + \sum\limits_{k=1} \alpha_k \mu_k (\frac{\partial V_i}{\partial x_j} + \frac{\partial V_j}{\partial x_i})]}{\partial x_j} \tag{3}$$

Component of continuity equation:

$$\frac{\partial \sum\limits_{k=1} \alpha_k \rho_k V_i}{\partial x_i} = 0 \tag{4}$$

k equation:

$$\sum_{k=1} \alpha_k \rho_k \frac{\partial}{\partial x_i}(V_i k) = \frac{\partial}{\partial x_i}\left(\frac{\sum\limits_{k=1} \alpha_k \mu_k + C_\mu \rho k^2 / \varepsilon}{\sigma_k} \frac{\partial k}{\partial x_i}\right) + G_k - \sum_{k=1} \alpha_k \rho_k \varepsilon \tag{5}$$

ε equation:

$$\sum_{k=1} \alpha_k \rho_k \frac{\partial}{\partial x_i}(V_i \varepsilon) = \frac{\partial}{\partial x_i}\left(\frac{\sum\limits_{k=1} \alpha_k \mu_k + C_\mu \rho k^2 / \varepsilon}{\sigma_\varepsilon} \frac{\partial \varepsilon}{\partial x_i}\right) + C_{1\varepsilon} \frac{\varepsilon}{k} G_k - \sum_{k=1} \alpha_k \rho_k C_{2\varepsilon} \frac{\varepsilon^2}{k} \tag{6}$$

Among them, α_k expresses the volume fraction of k, ρ_k expresses mass density of k, V_i , V_j expresses component of velocity of fluid, f_i denotes force , P the pressure, μ_k the dynamic viscosity for the fluid, K the turbulent kinetic energy, ε the turbulent dissipation rating, Gk derivative of turbulent kinetic energy causing by velocity gradient.

Main constant of model are:

$C_{1\varepsilon}=1.44$, $C_{2\varepsilon}=1.92$, $C_\mu=0.09$, $\sigma_k=1.0$, $\sigma_\varepsilon=1.3$。

Kinematics of film is described by the way of Lagrange.

Balance equation:

$$\frac{\partial \sigma_{ij}}{\partial x_j}+f_i=\kappa_i \tag{7}$$

Geometric equation:

$$\varepsilon_{ij}=\frac{1}{2}\left(\frac{\partial u_i}{\partial x_j}+\frac{\partial u_j}{\partial x_i}\right) \tag{8}$$

Physical equation:

$$\sigma_{ij}=\lambda\delta_{ij}\varepsilon_{kk}+2\mu\varepsilon_{ij} \tag{9}$$

$$\lambda=\frac{\upsilon E}{(1+\upsilon)(1-2\upsilon)},\mu=\frac{E}{2(1+\upsilon)} \tag{10}$$

Where κ_i is the component pressure of film, ε_{ij} denotes Cauchy strain tensor, λ , μ Lame constant of film, E elasticity modulus of film, υ poison ratios.

3. Numerical method

Erosion failure of petrochemical pipe system involve two fields. One is flow field, the other is structure field. Coupled method of ordinal physical environment is used to analysis erosion failure, which is including establishing geometrical model, meshing, building fluid physical environment and structure physical environment, and so on. The process of numerical calculation is shown in figure 1.

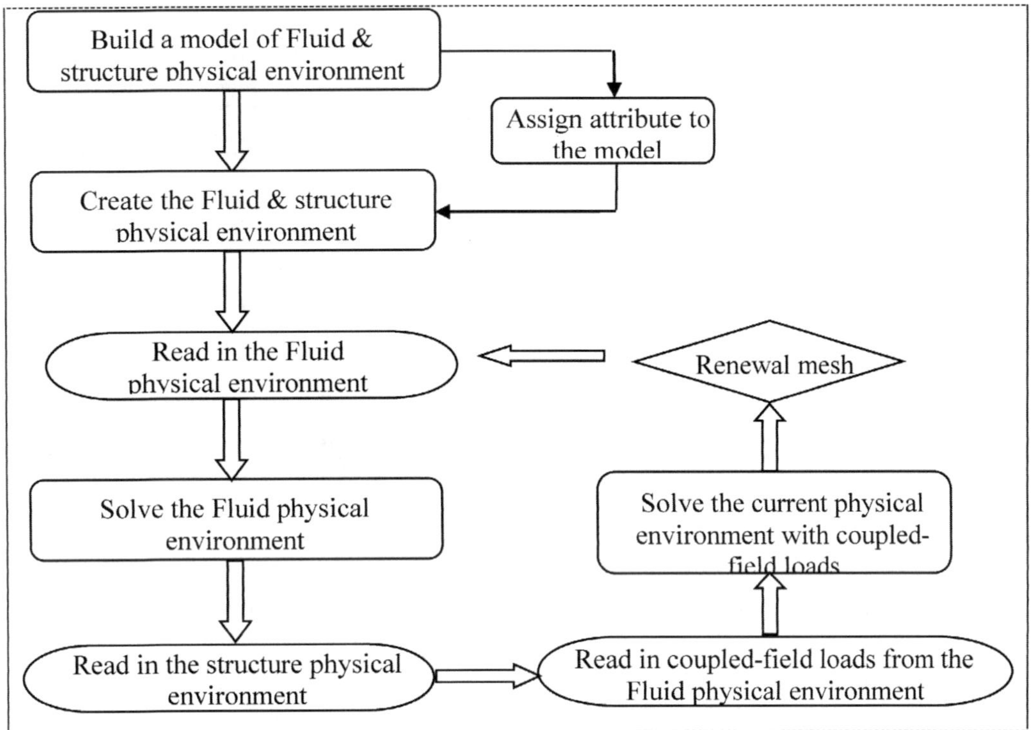

Figure 1. Flow chart of numerical simulation

Fluid 142 three-dimensional fluid element model is chosen as the fluid dynamics analysis model in carbon steel elbow bend. And Solid 186 model is used to analyze the structure stress of protecting film. Meanwhile, mapping mesh is introduced to analysis of fluid dynamics and protecting film structure stress.

Fluid physical environment includes the evaluation of cell type, the infliction of boundary conditions and the setting of FLOTRAN. Speed inlet and pressure outlet are used as boundary conditions (in figure 2), while standard $k-\omega$ turbulence model is chose to the solution of FLOTRAN.

The boundary conditions of structure physical environment are shown in figure 3. And position constraints are the major ways to the solution. Fixed constraint is put on the inner surface of pipeline protecting film and both ends of the pipeline, to restrict the rigid translocation of protecting film. Flow shear stress to the wall is the load of the pipeline. By solving structure physical environment, the influence of multi-phase flow to protecting film can be found.

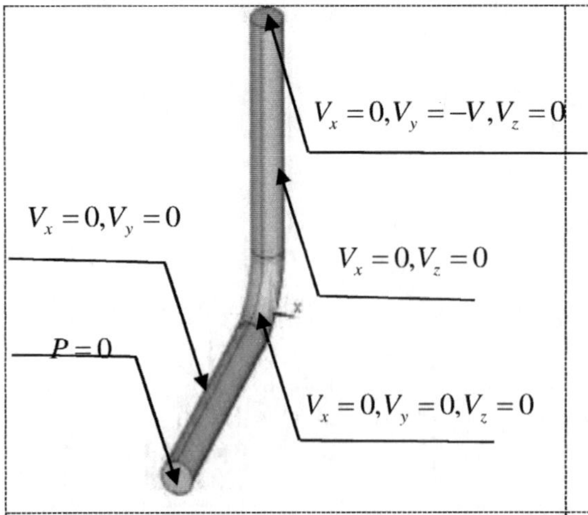

Figure 2. The boundary conditions of fluid physical environment.

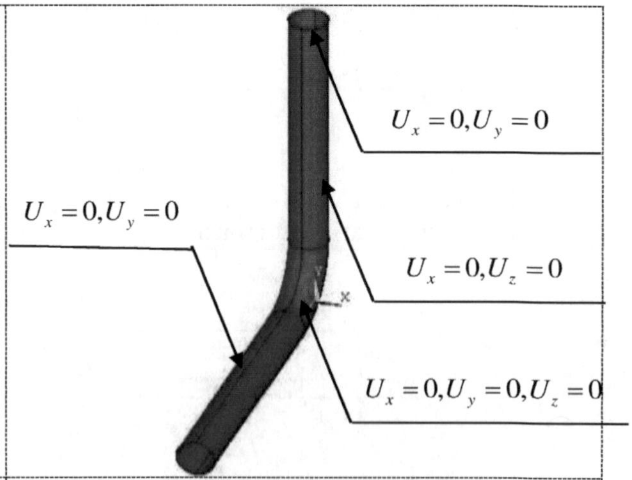

Figure 3. The boundary conditions of structure physical environment.

After reading fluid physical environment, the fluid dynamics analysis of carbon-steel pipeline can be done, and the distribution of fluid dynamics parameters would be obtained. With removing the existing physical information, structure physical environment is read and the analysis load of structure stress is exerted. As a result, the stress analysis of pipeline protecting film structure is done by solving structure physical environment. And then, the influence of multi-phase flow to protecting film is gotten. According to the deformation of protecting film, fluid grid coordinates is updated. Do the calculation of fluid physical field again, and the influence of protecting film deformation to multi-phase flow is found. The shear stress and pressure stress, which are gotten from the process, are used as the load of protecting film again. With that, the solution of structure physical environment field is obtained. Calculate circularly in turn, and the interaction between the two is realized.

4. Analysis of influence factors

4.1. Simulation analysis of actual failure cases

The failure of carbon-steel elbow bend in the outlet pipe system of hydrocracking reactor effluent in Zhenhai, is taken as an example. The dimension is $\phi 114\times 13.5$ mm, the radius of curvature is 1.5 DN, and the flow velocity is 5 $m\cdot s^{-1}$. The physic-chemical natural parameters of working condition medium are shown in table 1. In equation (8), λ and μ are calculated equivalently according to elasticity ways. Meanwhile, nominal elastic modulus is 0.12 GPa, nominal Poisson's ratio is 0.49[13], and the value of thickness is 0.1 mm (gotten by erosion-corrosion simulation experiment).

Table 1. Natural parameter of the causticity multiphase stream in the elbow bend

Parameter	Phase		
	Gas phase	Liquid phase	Water phase
Mass density/ $kg\cdot m^{-3}$	21.8	814.4	995.1
Volume phase fraction/ %	91.924	7.834	0.242
Viscosity/$kg\cdot m^{-1}\cdot s^{-1}$	1.01×10^{-5}	1.89×10^{-3}	6.18×10^{-4}

Based on the fluid-structure simulation analysis of erosion-corrosion failure, several findings are obtained, including the distribution of shear stress on the inwall of elbow bend, the distribution of main displacement, stress and strain of erodent products protecting film. The distribution of shear stress on inwall of elbow bend is shown in figure 4. In this diagram, zone A is on the side face of the exit of elbow bend. The shear stress of position A is larger, with the value of 6.044 Pa. Corresponding to the same area, the main displacement, stress and strain of protecting film also have larger value. And zone B has the maximum. The maximal tensile stress is 15.397 Pa and the maximal elastic stress is 3.8212×10^{-7}. They are shown in figure 5 and figure 6 respectively.

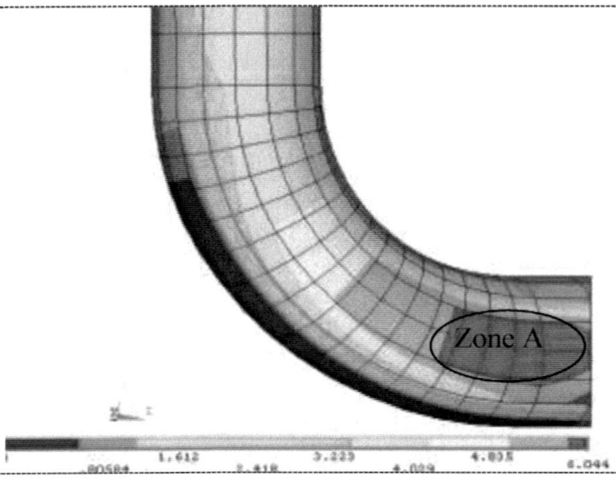

Figure 4. Wall shear stress of elbow bend.

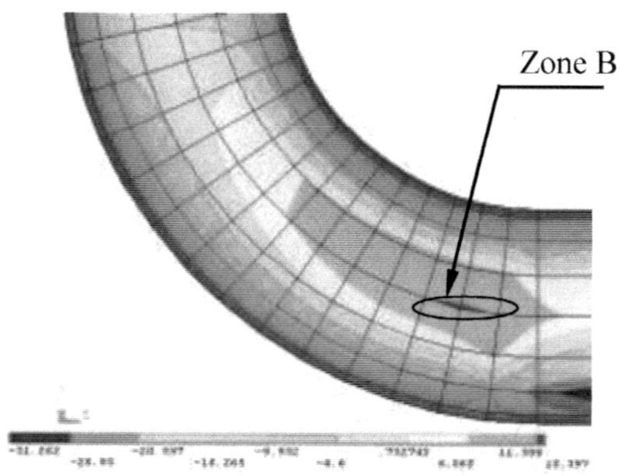

Figure 5. Stress of erodent products protecting film.

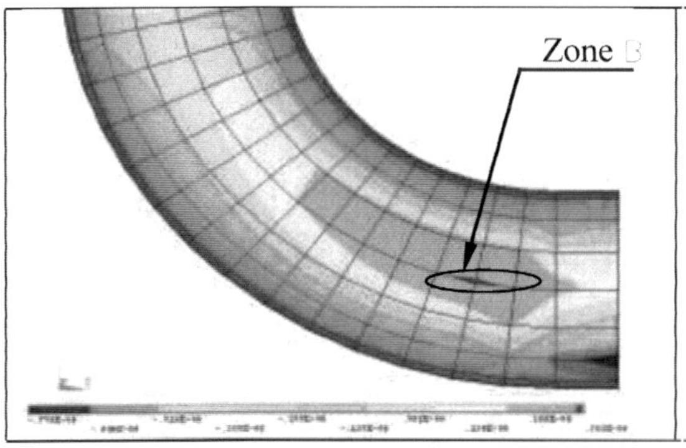

Figure 6. Elastic stress of erodent products protecting film.

4.2. Spot thickness verification

In order to test the result of fluid-solid numerical simulation of erosion-corrosion damage and verify the reliability and feasibility of numerical simulation method, the spot thickness analysis of Zhenhai domestic REAC outlet pipeline system is carried out.

Combining the result of simulation, TT320 ultrasonic thickness gauge is used to measure the spot thickness of erosion-corrosion serious failure area in elbow bend. The position division of thickness points is shown in figure 7, and the specific result is expressed in table 2. From the statistical analysis result of spot thickness data, it can be found that the thinning volume of zone B which is in zone A on the side face of elbow bend exit is the largest, up to 5.9 mm.

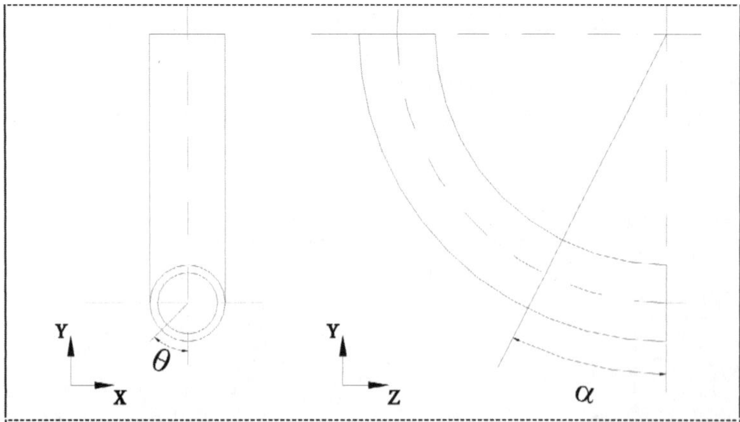

Figure 7. The thickness positions of elbow bend.

Table 2. The thickness thinning volume of elbow bend /mm

α \ θ	180°	157.5°	135°	112.5°	90°	67.5°	45°	22.5°
0°	2.6	1.8	1.8	4.3	4.4	4.0	1.8	2.5
5.625°	2.3	2.0	2.8	5.0	4.8	3.5	2.5	2.1
11.25°	2.3	2.0	2.8	4.5	**5.9**	3.7	2.0	2.3
16.875°	2.3	2.0	3.0	4.7	4.8	3.7	2.1	2.5
22.5°	2.3	2.0	3.0	4.7	5.2	3.8	1.8	2.5
28.125°	2.1	2.3	2.8	4.5	5.2	3.0	1.8	2.3
33.75°	2.0	1.8	2.6	4.5	5.0	2.8	1.8	2.3
39.375°	2.1	1.6	2.5	3.1	4.8	2.6	1.8	2.2

With the contrast of spot thickness data and the simulation result in chapter 3.1, the results show that: the larger that erodent protecting film, the more serious the erosion-corrosion failure, and the closer that the failure position and the numerical simulation. Zone A is the danger zone of erosion-corrosion in elbow bend, and the erosion-corrosion of position B in this zone is the most serious.

5. Conclusions

(1) The numerical model of fluid-solid interaction on multi-phase flow erosion-corrosion failure is set up, which provides theoretical foundation to the deep research of erosion-corrosion failure of multi-phase flow pipe system. According to the result of numerical simulation, the zone of erosion-corrosion failure in elbow bend is determined. And it can be used as the foundation of the location monitor and priority protection of pipe.

(2) In the norms of API932-B[15], the way of limiting flow velocity is adopted to control erosion-corrosion failure of REAC. And it is inappropriate. The influence of some characteristic factors such as the diameter of typical tube, the radius of curvature and the Poisson's ratio of protecting film is also should be considered.

(3) In allowable conditions, the results of fluid-structure interaction numerical simulation of erosion-corrosion failure are used as bases to optimization design. By changing the structure characteristic of such tube with serious erosion-corrosion failure, the erosion-corrosion failure of tube can be reduced effectively. And it is conducive to the security and long-period operation of pipe system.

Acknowledgements
The project is supported by the Natural Science Foundation and ShenHua Group Corporation Limited United Program (U1361107), the Specialized Research Fund for the Doctoral Program of Higher Education (No. 20133318120004) and Scientific Research Projects of Zhejiang province quality and technical supervision system （No.20150237）.

References

[1] Committee of Pipelines for Chinese Pressure Vessel Institution, Heifei General Machinery Research Institute, National Engineering Research Center for Pressure Vessel and Pipelines Safety Technology 2006 *Omnibus of Pressure Pipelines Research Progress on The Third National Conference of Pipeline Technology*[M].Beijing: Professional Engineering Publishing
[2] Zhao M J 2006 Analysis of causes leading to failure of high-pressure hydrogen-adjacent pipelines used for hydrogenated refining of diesel oil [J]. *Journal of Materials Protection*, 2006, **39**(6): 66 ~ 68
[3] Announcement No.148, 2006 of the General Administration of Quality Supervision, Inspection and Quarantine of the People's Republic of China 2006 *Regulations on Safety Supervision of Pressure Pipeline*[S].
[4] Yong X Y, Lin Y Z, Liu J J et al. 2002 Flow-induced corrosion kinetics model of carron steel in flow loop system [J]. *Journal of Chemical Industry and Engineering* , **53**(7): 680 ~ 684
[5] Wu X Q, Jing H M, Zheng Y G et al. 2004 Erosion-corrosion of various oil-refining materials in naphthenic acid [J]. *Wear*. **256**: 133 ~ 144
[6] Nesic S 2006 Using computational fluid dynamics in combating erosion-corrosion [J]. *Chemical Engineering Science*, **61**: 4086 ~ 4097
[7] Malka R, Nesic S, Daniel A et al. 2007 Erosion-corrosion and synergistic effects in disturbed liquid-particle flow [J]. *Wear*, **262**: 791 ~ 799
[8] OU G F, Zhu Z C, Gu W P et al. 2006 Numerical simulation method for erosion failure prediction of hydrocracking REAC Piping[R]. *The 7th International Conference on Frontiers of Design and Manufacturing*, 2006: 97 ~ 102
[9] OU G F 2004 *Study on simulation of multiphase flow and prediction of erosion attack breakage for REAC pipes*: [D]Hangzhou: Zhejiang University, 2004
[10] Xing J T, Zou S, Cui E J 1997 A survey on the fluid solid interaction mechanics [J]. *Advances in Mechanics*, **27**(1): 19 ~ 37
[11] Hu X Y, Li H F, Tao W M et al. 2006 Fluid-structure interaction analysis of valve in pipes [J]. *Journal of Zhejiang University (Engineering Science)*, **40**(6): 971 ~ 976
[12] Pin F D, Idelsohn S, Onate E et al. 2007 The ALE/Lagrangian particle finite element method: a new approach to computation of free-surface flows and fluid - object interactions [J]. *Computers & Fluids*, **36**: 27 ~ 38
[13] Molina F J, Alonso C, Andrade C 1993 Cover cracking as a function of bar corrosion: part 2-Numberical model [J]. *Materials and Structures*, **26**(163): 535 ~ 548

Temperature field study of hot water circulation pump shaft system

Y Y Liu[1], **F Y Kong**[1], **X H Daun**[1], **R J Zhao**[1] and **Q L Hu**[1]

[1] National Research Center of Pumps, Jiangsu University, Zhenjiang, Jiangsu 212013, China

E-mail: 931285515@qq.com

Abstract. In the process of engineering application under the condition of hot water circulation pump, problems of stress concentration caused by the temperature rise may happen. In order to study the temperature field in bearing and electric motor chamber of the hot water circulation pump and optimize the structure, in present paper, the model of the shaft system is created through CREO. The model is analyzed by ANSYS workbench, in which the thermal boundary conditions are applied to calculate, which include the calorific values from the bearings, the thermal loss from electric motor and the temperature from the transporting medium. From the result, the finite element model can reflect the distribution of thermal field in hot water circulation pump. Further, the results show that the maximum temperature locates in the bearing chamber.The theoretical guidance for the electric motor heat dissipation design of the hot water circulation pump can be achieved.

1. Introduction

It is significant to explore the internal temperature distribution of the bearing chamber in the study of hot water circulating pump, because mechanical seals and bearings are most easily going wrong in the process of daily use.

Andrzej Sluialec[1] studied the thermoelectric problems in using process of inductive bearings, and built the heat transfer of electromagnetic field and temperature field equations analysis of inductive bearings. The relationship between bearing speed, frequency and temperature field was obtained by numerical calculation analysis. A. Dadouche et al[2] analyzed the characteristics of thrust bearing in fixed equipment, and studied experiments to verify thermal-fluid model of the thrust bearing, discussed respectively the effects of the load, rotation speed and temperature on the thrust bearing performance. Ryosuke Shiraishi et al[3] estimated the speed decline amplitude of bearing rotor through coupled numerical simulation of the electromagnetic field and heat transfer of superconducting magnetic bearings, according to the temperature increase caused the uneven distribution of the magnetic field so that the superconducting magnetic bearing rotor generated imbalance electromagnetic torque. Gong Enxiang[4] used probability design system(PDS) provided by ANSYS software， and analysed the probabilistic of shaft based on APDL parametric modeling ,and obtained the probability distribution.

National Science Foundation(2013BAK06B02),
Author: Yingying Liu,(1992-),Female, Jiangsu Province, master degree

Aiming at the study of LPm125 type hot water circulating pump bearing chamber structure characteristics, the shafting temperature field and stress field including motor rotor were studied, considering mainly the heat generated in the bearing and motor operates. Hot water circulating pump shaft system studied in this paper includes mainly pump bracket, a pair of deep groove ball bearings, motor cover, motor rotor and the motor shaft.

2. The basic theory to calculate bearing heat[5]

2.1. Bearing heat calculation

Bearing system of LPm125 type hot water circulating pump in the study is a pair of single row radial ball bearings, and model is 6202-2RZ, with thin oil lubrication. The calculation of the bearing heat is particularly important, which is calculated by the following formula:

$$Q = 1.05 \times 10^{-4} M_n \tag{1}$$

where Q is bearing heat(kW), M_n is bearing friction torque(N m).

2.2. Rolling friction torque

Palmgren[6] obtained the empirical formula from the summarized result in the measurement by different loads and different lubricant viscosity, as well as the conditions under different speed of the bearing friction torque. However, the 6202-2RZ deep groove ball bearing in this paper is a standard part. The calculation of the friction torque is given in Eq. (2) as:

$$M = 0.5\mu Pd \tag{2}$$

where M is bearing friction torque(N m), μ is bearing friction coefficient, P is bearing equivalent dynamic load(N), d is bearing nominal bore(mm).

2.3. Equivalent dynamic load calculation of bearing

The 6202-2RZ deep groove ball bearings of the hot water circulation pump in this paper mainly take radial load Fr effect, its equivalent dynamic load P is expressed as:

$$P = f_p F_r \tag{3}$$

Where fp is load factor concern with the machinery smooth operation.

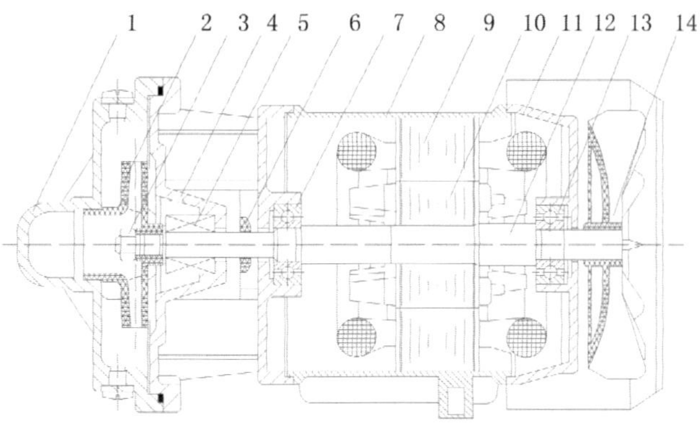

1.pump body 2.impeller 3.impeller nut 4.stent 5.mechanical seal 6.water deflector 7.front bearing 8.motor shell 9.motor stator 10.motor rotor 11.motor rear cover 12.pump shaft 13.back-end bearing 14.blade

Figure 1. structure assembly drawings of LPm125 type hot water circulation pump

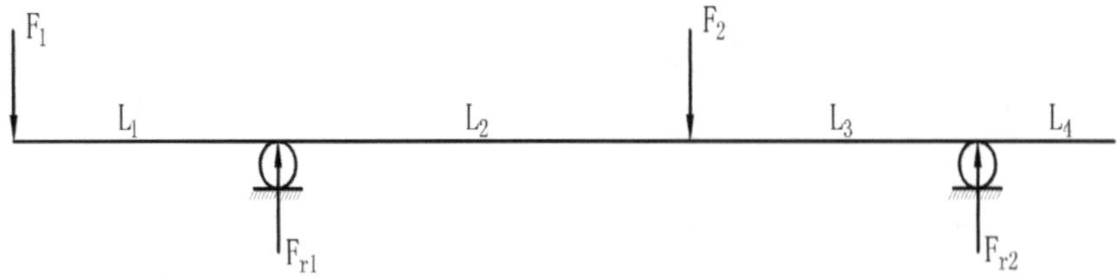

Figure 2. simplified hot water circulation pump rotor system

Figure. 1 shows the structure assembly drawings of LPm125 type hot water circulation pump .The rotor system of hot water circulation pump should be simplified when calculates radial load of bearing. Figure. 2 shows the simplified hot water circulation pump rotor system.

3. Hot water circulation pump shaft system temperature field calculation

3.1. Establishment of three-dimensional model and meshing of hot water circulation pump shaft system

The chamber temperature field which consists of LPm125 type hot water circulating pump shaft system is studied, and the three dimensional model of the hot water circulation pump shaft structure is built by Pro / E software. Using Workbench automatic mesh partition function generates its solid domain grid which resolution is moderate, with a total number of 461,217 grid cells, and the number of nodes is 791,199.Hot water circulating pump shaft structured grid shown in Figure 3.

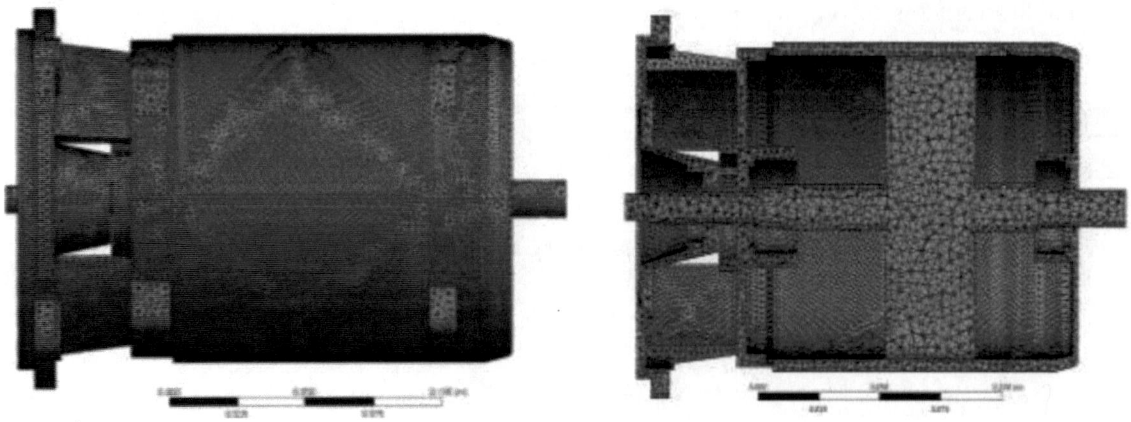

Figure 3. Hot water circulating pump shaft structured grid

3.2. Loading conditions imposed

The load which is in the process of finite element simulation analysis and calculation contains not only the force acting on the finite element model function, but also the constraints imposed by the finite element model. The load which is applied in the finite element calculation, includes two ways entity model load and finite element model load. ANSYS-Workbench finite element analysis software is used to simulate shaft temperature field of hot water circulation pump in this paper and thermal load applied method of those is solid loading. That is the thermal load of the steady temperature field model loaded directly on the physical model (the main load on the face of the solid model unit).

The thermal load of Simulation of hot water circulation pump shaft system temperature field includes the heat flux when bearing working, which is applied in two rolling interior chamber. The process of applying the thermal load on the bearing should pay attention to load respectively heat flux in the two bearing chamber .Due to thermal convection heat generated by the motor during operation, the convection of heat boundary conditions is applied on the motor chamber surface, that the convective heat transfer coefficient h is 27.1573 $W/(m^2 \cdot k)$ .Because some parts of bracket contact with the transport medium ,which possess a certain temperature, the medium temperature is 40 ℃, and the temperature load is applied directly to the contact structure surface of bracket and transport medium.

4. Results and discussion

4.1. Analysis of shaft structure steady temperature field

The shaft structure steady temperature field is analyzed by the steady-state thermal analysis module of ANSYS-Workbench. In calculation, the materials of pump bracket, motor housing, motor cover are gray cast iron, and the materials of shaft and rotor are stainless steel. The most important parameter of structural materials in calculation of steady temperature field is thermal conductivity of materials. The thermal conductivity of gray cast iron and stainless steel is respectively 52 and 15.1 $W/(m \cdot k)$.

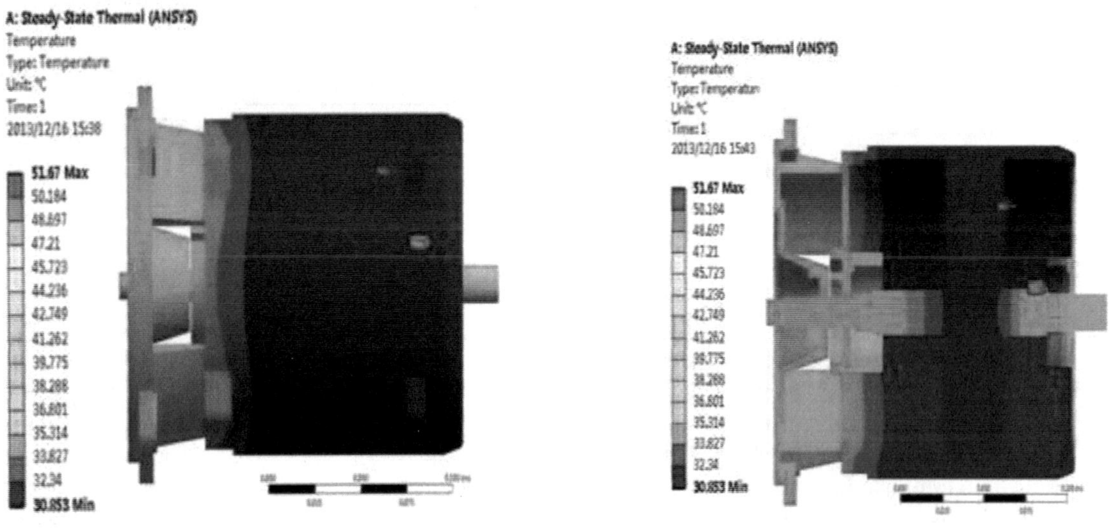

(a) whole structure (b)sectional structure

Figure 4. Temperature field contour of hot water circulating pump shaft system

Figure 4 shows the temperature distribution nephogram of hot water circulation pump shaft structure when medium temperature is 40℃.In Figure 4(a), the temperature gradient contacting with the motor and external environment is small. The temperature of the motor cover and pump bracket junction is the range of 32 ℃ ~ 35 ℃, and away from the pump bracket is 30 ℃ ~ 32 ℃. It is that the temperature close to the motor is slightly higher than away pump bracket, and the temperature difference is the range of 2 ~ 3 ℃.There is a conclusion that the temperature of the hot water circulation pump transmission medium has some influence on the temperature field of the motor.

Figure 4(b) shows that the maximum temperature of the shaft structure is 51.67 ℃, appeared at the back end of the bearing chamber (near the motor blade),therefore the motor ends need to be air-cooled.

The minimum temperature is 30.855 ℃, located at the connecting flange between the motor rear cover and the middle of the post-motor.

In Figure 4, because the temperature value of bearing chamber is larger than other shaft structure, it is important to calculate temperature field of bearing chamber. Although the temperature of motor rear cover changes relatively large after the bearing chamber, the motor rear cover which has an additional air-cooling blade will reduce here temperature in practical application, and here's temperature and changes tend to stable.

4.2. Analysis of shaft structure heat flux

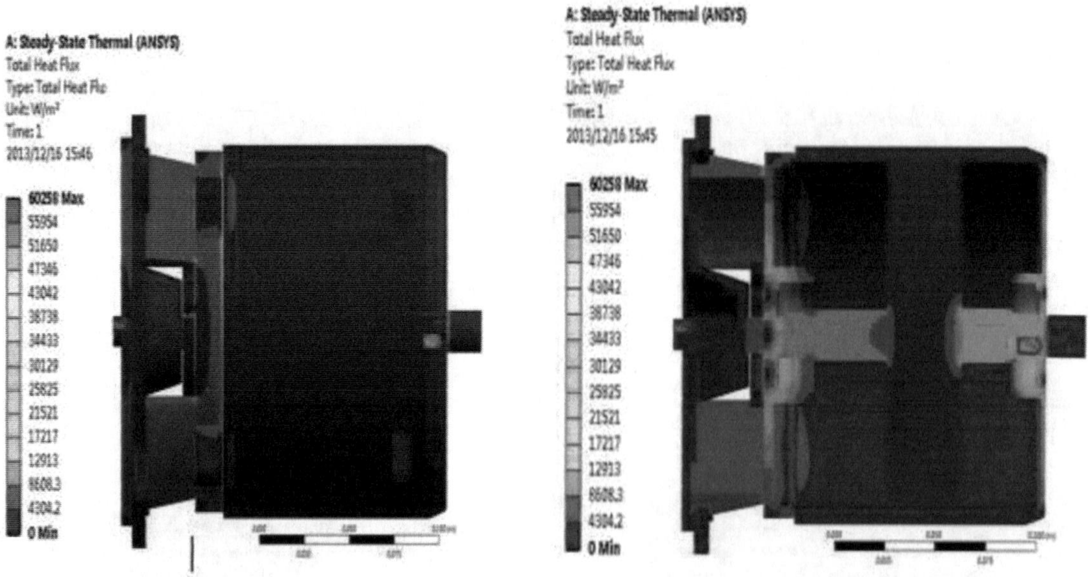

(a) whole structure (b)sectional structure

Figure 5. Heat flux contour of hot water circulating pump shaft system

Figure 5 shows the heat flux density of hot water circulation pump shaft structure when medium temperature is 40℃.In Figure 5(a), the heat flux is well distributed and there is no great mutation. The values of heat flux in stent is greater than motor structure. In Figure 5(b), the values of heat flux in pump shaft is slightly greater than other shaft structure, and there is a gradient values change. The maximum heat flux is 60258W/m2 located on the inner of rear bearing chamber, which is the same position of maximum temperature of the shaft system structure, consistent with the theory of heat transfer. The minimum heat flux is located on the pump shaft of inside mechanical seal chamber in stent. From Figure 4 analysis, the temperature of mechanical shaft seal chamber inner and pump shaft is well distributed, and the values remained at 41.25 ℃, with a small temperature difference, resulting in a smaller heat flux values there.

5. Conclusions

The shafting structure steady temperature field and structure field ,which consist of bearing chamber, stator and rotor, and hot water circulation pump of LPm125 type, were numerically calculated, and the shafting structure temperature field and the heat flux were analyzed in this paper.

It is found from calculation of the shafting structure steady temperature field that the motor temperature closed pump stent is slightly higher than away from stent. The medium temperature of hot water circulation pump transmission has certain effect to the motor temperature field. The temperature values of bearing chamber is greater than shafting structure. The heat flux of shafting structure is well distributed with no significant mutation. The heat flux of pump shaft is generally larger than the other.

Because the influence of the medium temperature, the heat flux values of stent is greater than the motor.

References

[1] Andrzej S 1990 Heating of induction bearings [J]. *Appl. Math. Modelling*, **14**: 386~389

[2] Dadouche M, Fillon J C, Bligoud 2000 Experiments on thermal effects in a hydrodynamic thrust bearing [J]. *Tribology International*, **33**: 167~174

[3] Shiraishi R, Demachi K, Uesaka M 2003 Numerical simulation of coupled problem of electromagnetic field and heat conduction in superconducting magnetic bearing [J]. *Physica C: Superconductivity*, **392(1)**: 734~738

[4] Gong E X, Huang M K, Wu J B, Ding Y N, Pan B H 2010 Probabilistic analyse of shaft based on APDL parametric modeling [J]. *Journal of Drainage and Irrigation Machinery Engineering*, **28 (6)**, 506-509

[5] Harris T A, Kotzalas M N 2007 Rolling *Bearing Analysis FIFTH EDITION: Advanced Concepts of Bearing Technology* [M]. USA： CRC Press

[6] Palmgren A 1959 *Ball and Roller Bearing Engineering* [M], 3rd ed., Burbank, Philadelphia, 34~41

IOP Conf. Series: Materials Science and Engineering **129** (2016) 012035 doi:10.1088/1757-899X/129/1/012035

Characteristic analysis on the pressure fluctuation in the impeller of a low specific speed mixed flow pump

W W Zhang[1], Z Y Yu[1] and B S Zhu[2]

[1] School of Mechanical Engineering, Beijing Institute of Technology, Beijing100081, P. R. China

[2] State Key Laboratory of Hydroscience and Engineering, Department of Thermal Engineering, Tsinghua University, Beijing 100084, P. R. China

E-mail: yuzhiyi@bit.edu.cn

Abstract. To explore the pressure fluctuation characteristics of a low speed specific speed mixed flow pump caused by rotor-stator interaction, the unsteady flow was simulated with CFX for the whole flow passage of a mixed flow pump with a specific speed of 148.8. The structured mesh of the computation domain was generated with ICEM CFD and TurboGrid, and mesh-independent analysis was done in the design condition. Through the comparison with the experiment data, the reliability of the simulation was verified. In different locations of the impeller passage, monitoring points were set. With Fast Fourier Transform (FFT), the characteristic analysis on the pressure fluctuation in the impeller passage was done for three flow rate conditions ($0.75Q_d$, Q_d, $1.25Q_d$). The results show that the pressure fluctuation amplitude increases from the inlet to the outlet. And the maximum values in different flow rates exist near the hub of the outlet; The pressure fluctuation is small in the design condition, but the largest in the small flow rate condition, accompanied by the secondary dominant frequencies with large amplitudes; In the small flow rate condition and design condition, the dominant frequency varies from the inlet to the outlet because the combine action of the impeller and guide vane; while in the large flow rate condition, the pressure fluctuation in the whole impeller passage is affected significantly by the guide vane, and the domain frequency is 8 times the rotational frequency of impeller. In addition, the change of pressure fluctuation from the pressure surface to the suction surface in the off-design conditions is investigated, and the results demonstrates that the intensity of the pressure fluctuation in the impeller passage is closely related with the impeller as well as the distribution of the vorticity and the pressure.

1. Introduction

The performance and structure of the mixed flow pump are between the centrifugal pump and the axial flow pump. Because of the features of compact structure, easy starting and high efficiency, mixed flow pumps are widely used in farmland irrigation, waterlogging drainage, sewage treatment, power plant cooling etc. Usually, low specific speed mixed flow pumps refer to those with a specific speed less than 350. In recent years, domestic and foreign scholars have done a lot of research on the design, experiment and numerical calculation of the mixed flow pump. Shi weidong et al.[1] studied the characteristic on unsteady pressure fluctuation of the high specific speed mixed flow pump in different conditions by using time-average Renoylds equations; Bing hao et al.[2] proposed parametric approach of velocity moment distribution law, designed a series of mixed flow pump impeller and predicted its performance by numerical simulation; Li Yibin et al.[3] analyzed the turbulent flow in a mixed flow pump with the RNG k-ε model and sliding mesh technique; Jin Shuanbao et al.[4] simulated the unsteady flow in a mixed-flow pump based on RANS solver embedded with SST turbulence model and SIMPLEC algorithm; In order to improve hydraulic performance of a mixed-flow pump with guide vanes, Chang Shuping et al.[5] explored the effects of several factors on the performance including the tip clearance, blade number, blade angle and blade thickness; ZHANG Desheng et al.[6] received an abruptly descending characteristic curve by the optimization of the wrap angle and the outlet blade angle, which could meet the need of acceleration at different working conditions; Chisachi Kato et al.[7] investigated the unsteady flow characteristics for a high specific speed mixed flow pump in the small flow rate condition based on LES method; Masahiro Miyabe et al.[8] found that the unsteady characteristic in low specific speed mixed flow pumps is caused by the large scale back flow between the outlet of impeller and the guide vane.

Comparatively speaking, less research is done on the design and the internal flow of low specific speed mixed flow pumps, which limits the scope of such pumps. In some cases, low specific speed mixed flow pumps are replaced unreasonably by centrifugal pumps[9]. In this paper, unsteady simulation was done with CFX for the internal flow of a mixed flow pump with a specific speed of 148.8. Monitoring points were set on different sections in the impeller passage to capture the change of the pressure, then analysis was done with the fast Fourier transform (FFT) to evaluate the characteristics of the pressure fluctuation in the low specific speed mixed flow pump.

2. Structure and parameters of the pump

The main parameters of the mixed flow pump are as below: impeller diameter D is 150mm, blade number Z_1 is 6, guide vane number Z_2 is 8, design flow rate Q_d is 1.39m^3/min, rotational speed n is 2000 r/min, design head H_d is 14.59m, and power P is 5.5kw. The 3-D geometric model of the whole flow passage is constructed with UG NX 6.0 (see figure 1).

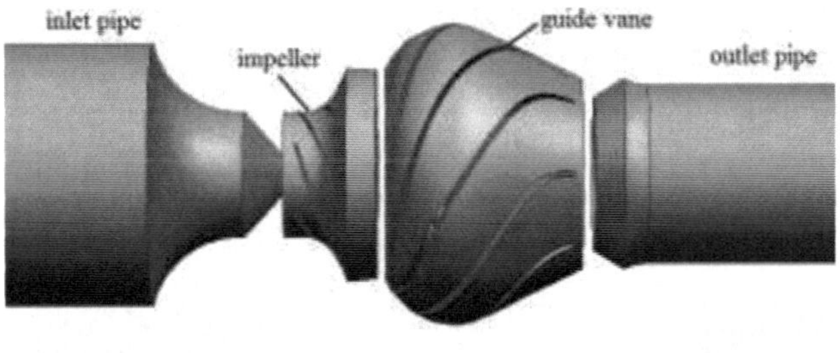

Figure 1. Configuration of the mixed flow pump

3. Numerical methods

3.1. Governing equations

The conservation equations of mass and momentum for incompressible flow are written in Cartesian coordinates as follows:

$$\frac{\partial u_i}{\partial x_i} = 0 \tag{1}$$

$$\frac{\partial u_i}{\partial t} + \frac{\partial (u_i u_j)}{\partial x_j} = -\frac{1}{\rho}\frac{\partial p}{\partial x_i} + \frac{\partial\left[(\mu+\mu_t)\left(\frac{\partial u_i}{\partial x_j}+\frac{\partial u_j}{\partial x_i}\right)\right]}{\partial x_j} + f_i \tag{2}$$

where f_i is the Coriolis force; ρ is the fluid density; u is the relative velocity, p is the pressure;

μ is the viscosity of the fluid; μ_t is the turbulent viscosity. The turbulence is modeled by the shear stress transport (SST) model. This model can give a more accurate prediction for flow separation under adverse pressure gradient. In this model, the eddy viscosity is computed by:

$$\mu_t = \frac{\rho a_1 k}{\max\left(a_1\omega, SF_2\right)} \tag{3}$$

where a_1 is the model constant and $a_1 = 5/9$; S is the invariant measure of strain rate; F_2 is the blending function; k and ω are the turbulent kinetic energy and the turbulence frequency, respectively.

3.2. Mesh and boundary conditions

The computation domain includes four parts, i.e. the inlet pipe, the outlet pipe, the impeller and the guide vane. The structured meshes of the former two parts are generated by ICEM_CFD, while the meshes of the latter two are created with TurboGrid. The y^+ values of the wall mesh of the impeller and the guide vane are between 20 and 30. Through the analysis of mesh independence in the design condition, the mesh numbers for a single passage of the impeller and the guide vane are determined to be 202720 and 200124, while the mesh number for the whole domain is 3979933. The meshes of the four parts are shown in figure 2.

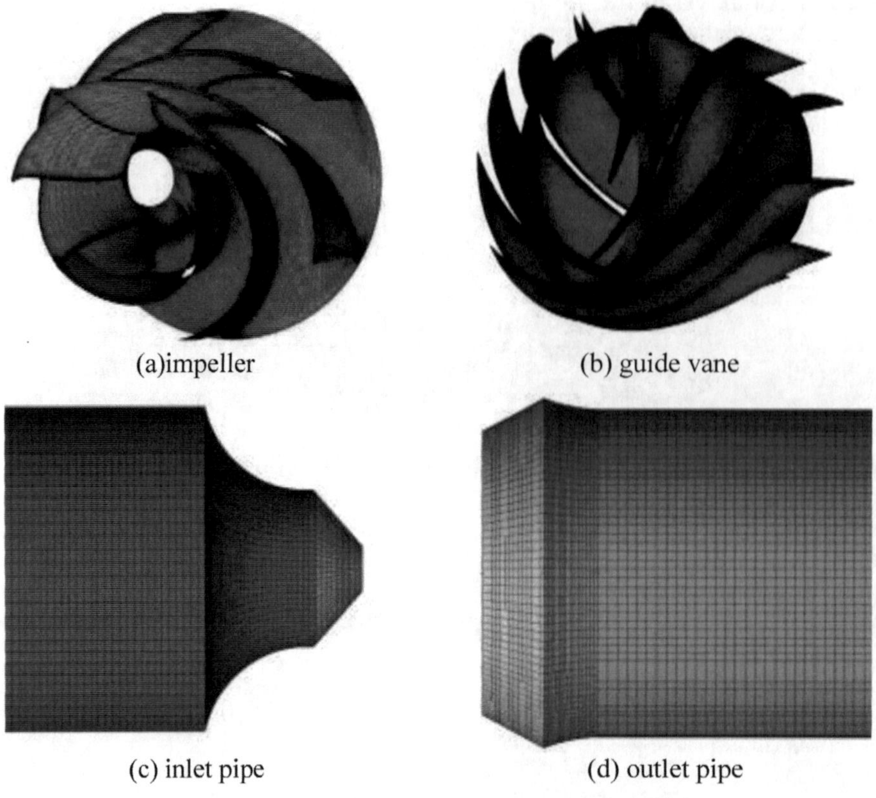

(a)impeller (b) guide vane

(c) inlet pipe (d) outlet pipe

Figure 2. Structured mesh for flow analysis

At the inlet of the computation domain, uniform velocity was specified according to experimental value; at the outlet, free outflow condition was adopted. At all the wall boundaries, the non-slip condition of viscous fluid is used, and logarithmic wall function approach was applied in the near wall region. Transient Rotor-stator approach was adopted for the data exchange in the rotor-stator interaction region (namely the impeller/inlet pipe interface and the impeller/guide vane interface).

3.3. Other numerical settings

In order to achieve the unsteady flow information in a pump with a high resolution, the time step must be small enough[10].As the rotational speed is 2000r/min, the value of the rotational frequency of the impeller fn is 33.3Hz and the period of rotation T is 0.03s . In this study, the time step was set as 0.0002s, thus the impeller rotates for 2.4° in a time step. The total simulation time is 0.45s, which is equivalent to 15 periods.

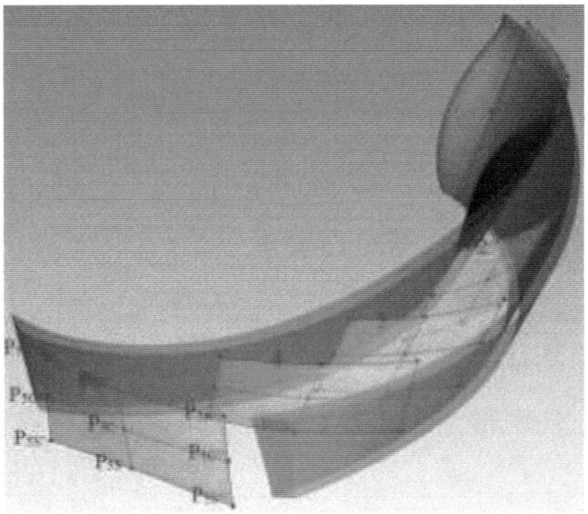

Figure 3. Location of monitoring points

In order to study the characteristics of the pressure fluctuation in the mixed flow pump impeller, 5 monitoring surfaces and 45 monitoring points were set from the inlet to the outlet of the impeller, as shown in figure 3. Here, the monitoring surface 1 is near the inlet of the impeller, while the monitoring surface 5 is near the outlet; the subscripts C, N, S denote that the monitoring points are at the center, near the shroud and near the hub of the monitoring surface, respectively, while the superscripts "'" and """ represent that the monitoring points are near the pressure surface and the suction surface, respectively.

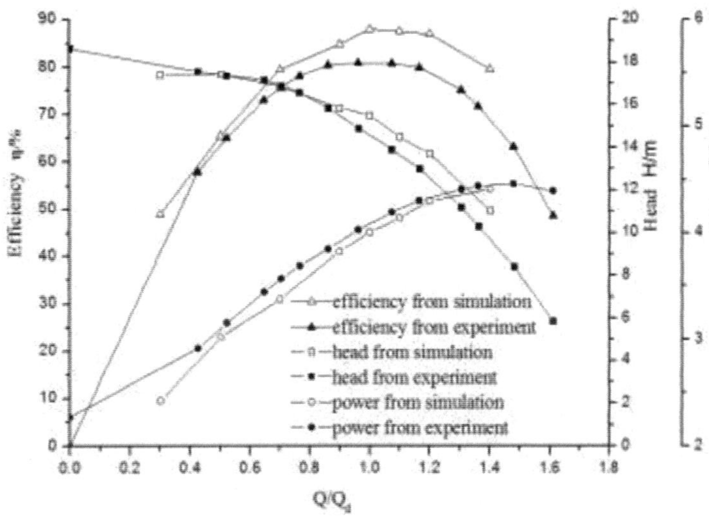

Figure 4. Performance curves from numerical simulation and experiment

4. Results and discussions

4.1. Performance prediction and validation

To verify the reliability of the numerical model, the unsteady internal flow of pump were simulated in eight flow rate conditions, and comparison with the experiment was made for the head, the efficiency and the power, which is shown in figure 4. The calculation errors of the head, the efficiency and the power at each flow rate condition are less than 8.7%, among which the maximum is the efficiency error in the large flow rate condition, and this is because the cavitation is more serious in large flow rate condition. On the whole, the numerical and experimental results agree well with each other, which demonstrates that the numerical approach is reliable.

4.2. Characteristics of the pressure fluctuation in the impeller passage

The pressure fluctuation of the central monitoring points P_{5C} at different working conditions is presented in figure 5. Here, the pressure coefficient is defined as $C_p = \Delta p / (0.5 \rho u^2)$, where Δp is the pressure difference between the monitoring points and the average pressure, u is the circumferential velocity of the impeller. It shows that the pressure fluctuation coefficient in the small flow rate condition ($Q=0.75Q_d$) is much greater than that in the design condition and the large flow rate condition, and the pressure fluctuation is in disorder. In the design condition ($Q=Q_d$) and large flow

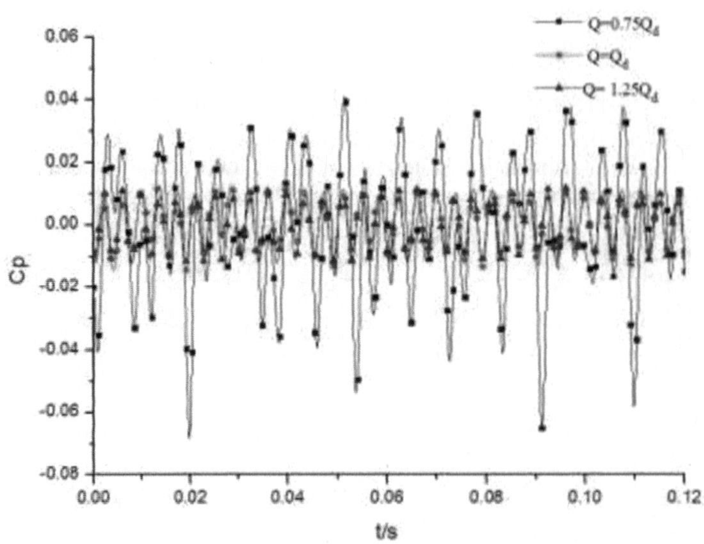

Figure 5. Pressure fluctuation of the point P_{5c}

rate condition ($Q=1.25Q_d$), the pressure fluctuation coefficient is smaller, the fluctuation is more stable and the periodicity is more obvious. Eight peaks and valleys occurs in each cycles of the three flow rate conditions, which coincides with the blade number of the guide vane. Therefore, it can be concluded that the pressure fluctuation of point P_{5C} in the three flow rate conditions is mainly affected by the guide vane.

The maximum values of the relative pressure fluctuation (indicated by $\Delta p/\bar{p}$) of the 9 monitoring points on each monitoring surface are compared in figure 6. It can be seen that the amplitude of pressure fluctuation increase from the inlet to the outlet. The pressure fluctuation in the impeller passage is smaller in the design condition (the fluctuation is greater than 2% only in monitoring surface 5), which illustrates that the design of the low specific speed mixed flow pump is reasonable. The maximum fluctuation occurs in the small flow rate condition, which is consistent with the analysis in figure 5. The reason for this may be that surge phenomenon is easy to happen at small flow rate condition and thus the flow is more unstable. Furthermore, due to the influence of the rotor-stator interaction, the pressure fluctuation in monitoring surfaces 4 and 5 is greater, especially in monitoring surface 5. In the three working conditions ($Q=0.75Q_d$, $Q=Q_d$ and $Q=1.25Q_d$), the maximum values of the pressure fluctuation in monitoring surfaces 5 occur near the hub, with the value of relative pressure fluctuation being 21.8%, 5.5% and 17.1%, respectively.

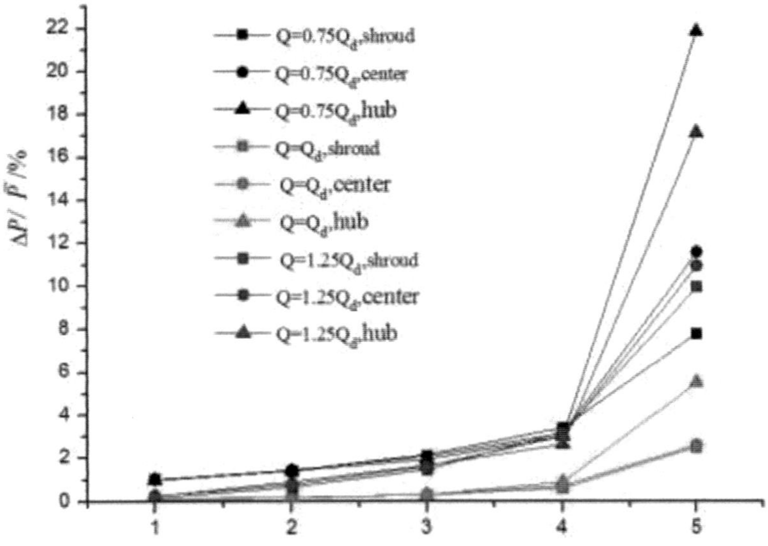

Figure 6. Comparison of the monitoring points with maximum pressure fluctuation

Because the pressure fluctuation occurs mainly in monitoring surface 4 and 5, further analyses for these two monitoring surfaces are carried out. Figure 7 shows the frequency domain of the pressure fluctuation of the central monitoring points (P_{4C}, P_{5C}) in the three flow rate conditions. In the small flow rate condition ($Q=0.75Q_d$), the design condition ($Q=Q_d$) and the large flow rate condition ($Q=1.25Q_d$), the dominant frequency of the point P_{4C} are 3fn, 2fn and 8fn, respectively, while the dominant frequency of the point P_{5C} are all 8fn. On the other hand, according to the calculation

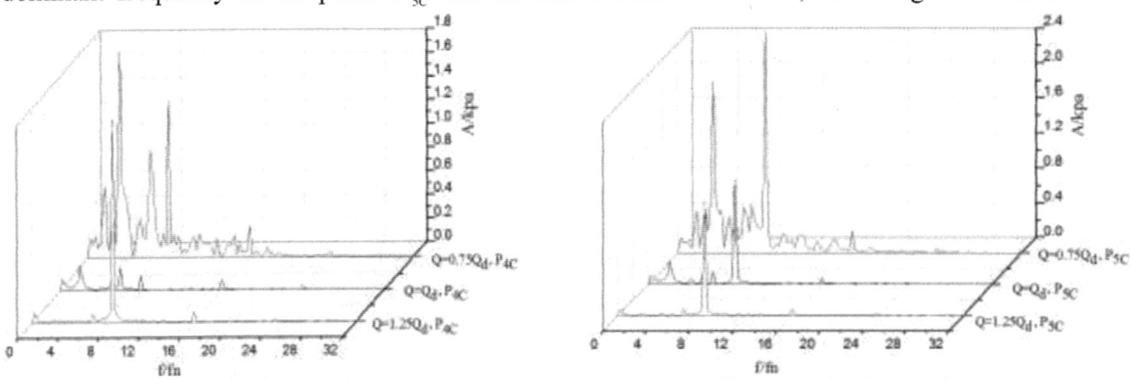

(a) monitoring surface 4 (b) monitoring surface 5

Figure 7. Frequency domain characteristics of the central monitoring points results, all the dominant frequencies of monitoring points $P_{1C} \sim P_{3C}$ are 3fn in the $Q=0.75Q_d$ condition, while the dominant frequency of the same points are 6fn, 6fn and 2fn at $Q=Q_d$ condition, and all of them are 8fn at $Q=1.25Q_d$ condition. Therefore, the pressure fluctuation of the monitoring points $P_{1C} \sim P_{4C}$ is influenced by the interaction of the impeller and the guide vane in the small flow rate condition and the design condition, while the pressure fluctuation of P_{5C} is mainly affected by the guide vane. In large flow rate conditions, the pressure fluctuation of the points $P_{1C} \sim P_{5C}$ is mainly affected by the guide vane.

It can also be found that, the amplitudes of pressure fluctuation of the points P_{4C} and P_{5C} are the largest in the small flow rate condition, which are 8.35 and 2.11 times as great as that in the design condition, respectively. The amplitudes in the large flow rate condition are 8.19 and 1.09 times as great as that in the design condition, respectively. In the design condition, in addition to the dominant frequency, the amplitudes of the secondary dominant frequency of the points P_{4C} and P_{5C} are also great, and many other frequencies are accompanied, which increases the flow instability in the pump. Therefore, off-design conditions, especially the small flow rate conditions, should be avoided for the running of low specific speed mixed flow pumps.

4.3. Pressure fluctuation from the pressure surface to the suction surface

To compare the pressure fluctuation from the pressure surface to the suction surface, we make an average of the root mean square (RMS) for the fluctuation values near the pressure surface, in the center and near the suction surface, as shown in figure 8. On the whole, the pressure fluctuation gradually increases from the inlet to the outlet in the three flow rate conditions. Therefore, the effect of rotor-stator interaction is enhanced in this direction. The fluctuation amplitude of the monitoring surface 1 keeps almost constant from the pressure side to the suction side, and this is because surface 1 locates at the inlet of the impeller, where the impact of guide vane is small for the flow. In general, the pressure fluctuation in surface 2~4 decreases from the pressure side to the suction side, with an exception for the monitoring surface 4 in large flow rate condition ($Q=1.25Q_d$), where the fluctuation at the central point is slightly greater than that near the pressure surface. For the monitoring surface 5

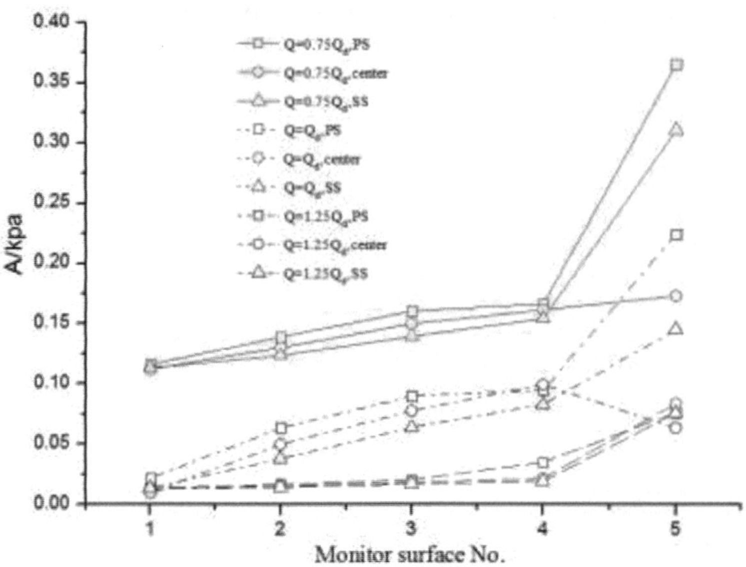

Figure 8. Pressure fluctuation from the pressure surface to the suction surface

in the small flow rate condition ($Q=0.75Q_d$) and the large flow rate condition ($Q=1.25Q_d$), the amplitude of fluctuation near the pressure side is the largest, followed by that near the suction side, and the fluctuation in the middle is the smallest. This may be related to the outlet structure of the low specific speed mixed flow pump.

The vorticity and pressure distribution in the small flow rate condition at the span of 0.5 are shown in figure 9. On the whole, the vorticity changes little in the upstream section of the impeller passage, while at the outlet region, it increses dramatically near the pressure surface and the suction surface and becomes small at the center. This is consistent with the change of the fluctuation amplitude shown in figure 8, indicating that the intensity of the pressure fluctuation in the impeller passage is closely related with the vorticity magnitude. From figure 9(b), it can be seen that the pressure gradient near the outlet of the impeller passage is obviously larger than that in the inlet and the middle section of the passage, which illustrates that the flow herein is affected significantly by the rotor-stator interaction, and thus the steeply increase of pressure fluctuation is aroused in the monitoring surface 5.

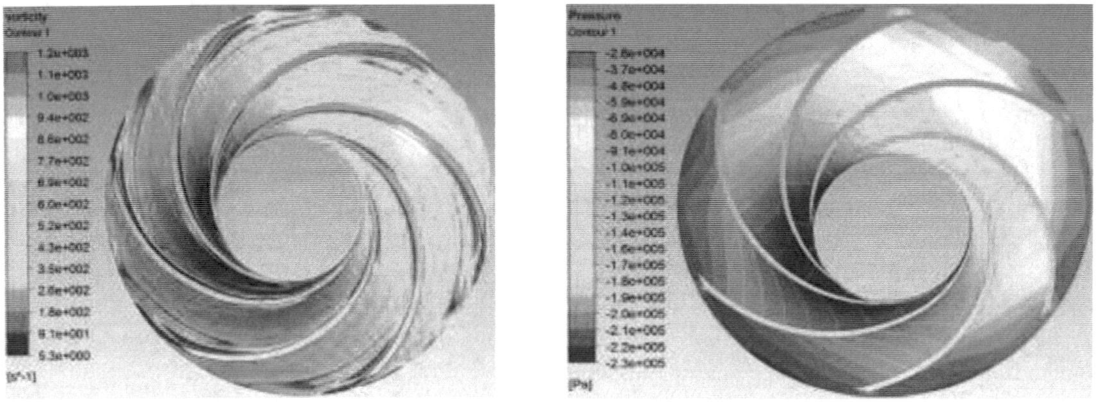

(a) vorticity distribution (b) pressure distribution

Figure 9. Distribution of the vorticity and the pressure in the small flow rate condition (Span=0.5 , t=0.448s)

5. Conclusions

Through numerical calculation, the characteristics of the pressure fluctuation in the impeller of a low specific speed mixed flow pump were analyzed in the design condition and the off-design conditions. The conclusions obtained are as follows:

1) The amplitude of pressure fluctuation increases from the inlet to the outlet of the impeller. In the small flow rate condition, design condition and large flow rate condition, the maximum values of the pressure fluctuation exist near the hub of the impeller outlet, and the values of the relative pressure fluctuation are 21.8%, 5.5%, 17.1%, respectively.

2) The pressure fluctuation in the impeller passage is small in the design condition (exceeds 2% only in the region near the impeller outlet), which means the design of the mixed flow pump is reasonable; while the pressure fluctuation is the largest in the small flow rate condition, accompanied by secondary dominant frequencies with large amplitude, and this will intensify the instability of the flow in the pump.

3) In the small flow rate condition and the design condition, the dominant frequencies varies from the inlet to the outlet because the combined action of the impeller and guide vane; while in the large flow rate condition, the pressure fluctuation in the whole impeller passage is affected significantly by the guide vane, and the domain frequency is 8 times the rotational frequency of impeller.

4) In off-design condition, the pressure fluctuation in the upstream section of the impeller passage tends to decrease from the pressure surface to the suction surface; while in the region near the outlet, the pressure fluctuation near the pressure surface is the largest; followed by that near the suction

surface, and it is smallest in the center. The intensity of the pressure fluctuation in the impeller passage is closely related with the impeller structure as well as the vorticity and the pressure.

Acknowledgments
This work was supported by National Natural Science Foundation of China (51579006, 51009003) and Open Research Fund Program of State key Laboratory of Hydroscience and Engineering (sklhse-2013-E-01).

References

[1] Shi W D, Zou P P, Zhang D S et a1. 2011 Unsteady flow pressure fluctuation of high-specific-speed mixed-flow pump [J]. *Transactions of the CSAE*, **27**(4): 147–152
[2] Bing H, Cao S L, Tan L et al. 2012 Parameterization of velocity moment distribution and its effects on performance of mixed-flow pump [J]. *Transactions of the Chinese Society of Agricultural Engineering (Transactions of the CSAE)*, **28**(13): 100–105
[3] Li Y B, Li R N, Wang X Y, Han W, Gao Q et al. 2013 Numerical analysis of pressure fluctuation in low specific speed mixed-flow pump [J]. *Journal of Drainage and Irrigation Machinery Engineering*, **31**(3): 205–209
[4] Jin S B, Wang Y S, Chang S P, Su Y S et al. 2013 Pressure fluctuation of interior flow in mixed-flow [ump[J]. *Journal of Agricultural Machinery*, **44**(3):64–68
[5] Chang S P, Wang Y S, Su Y S et al. 2012 Effects of some factors on hydraulic performance of mixed-flow pump with guide vanes [J]. *Journal of Drainage and Irrigation Machinery Engineering*, **30**(6): 646–650
[6] Zhang D S, Shi W D, Pan D Z, Cao W D, Li T T et al. 2014 *Journal of Mechanical Engineering*, **50**(6): 177–184
[7] Kato C, Mukai H, Manabe A 2003 Large-eddy simulation of unsteady flow in a mixed-flow pump [J].*International Journal of Rotating Machinery*, **9**: 345–351
[8] Masahiro M, Hideaki M, Isamu U et al. 2005 Unsteady head-flow characteristic generation mechanism of a low specific speed mixed flow pump [J]. *Journal of Thermal Science*, **15**(2): 115–120, 144
[9] Zhang X W 2010 *Design methods and experimental research of a low specific speed mixed flow pump*: [D]. Zhenjiang: Jiangsu University
[10] Cong G H, Wang F J 2008 Numerical investigation of unsteady pressure fluctuations near volute tongue in a double-suction centrifugal pump [J]. *Journal of agricultural machinery*, **39**(6): 60–63

IOP Conf. Series: Materials Science and Engineering **129** (2016) 012036 doi:10.1088/1757-899X/129/1/012036

Characteristics analysis for different water heads on the efficiency hill chart of Francis turbine

Z N Wang[1], P C Guo[1], X Q Luo[1], Y L Wang[2] and S H Sun[1]

[1] State Key Laboratory Base of Eco-hydraulic Engineering in Arid Area, Xi'an University of Technology, Xi'an 710048, Shaanxi Province, China
[2] Dongfang Electric Machinery Co., Ltd., Deyang 618000, Sichuan Province, China

E-mail: guoyicheng@126.com

Abstract. Based on the test results of Francis turbine, the causes and inevitability of various hydraulic phenomena in the model combined characteristic curve for typical water heads were analyzed in this paper. the difference of the model combined characteristic curve from the low water head to the high water head is compared, and the characteristics and commonness of the model combined characteristic curve about different water head are summarized. Further, hydraulic performance and geometric features of Francis turbine are revealed by particularly analyzing model combined characteristic curves, and to provide powerful theoretical basis and definite modification direction for the hydraulic design of hydraulic turbine.

1. Introduction

The hydraulic turbine characteristic curve is used to express the energy conversion, cavitation and other aspects of the hydraulic performance, force properties and other properties at the different operating conditions, these properties is the external characteristic of the internal flow of turbine, known as the external characteristics of hydraulic turbine. The relationship between the parameters of turbine is complex, in order to determine the relationship between some parameters, taking the relationship between two parameters as a simple function though fixing some parameters, which was represented by a curve, this curve was called characteristic curve of hydraulic turbine, we call it the hydraulic turbine combined characteristic curve if the turbine performance curves plotted in a same graph[1, 2].

The main characteristics of the hydraulic turbine can express concentratively on the performance combined characteristic curves or model combined characteristic curve based on the head and flow coordinate or unit parameter coordinate. However, in order to represent the relationship between hydraulic characteristics and hydraulic turbine condition intelligibly, it's a common that applies the basic characteristic curve（equal efficiency curve, blade channel vortex curve, cavitation curve in the pressure and suction surface of the leading edge of the blade）and typical hydraulic phenomenon to describe the hydraulic characteristics of the turbine, such as the critical cavitation, incipient cavitation, Cavitation vortex and pressure fluctuation of typical position[3].

According to the external characteristic expression modes, characteristic curve of turbine can be divided into linear characteristic curve and combined performance curve. And the linear characteristic curve can be represented by operating characteristic curve, rotation speed characteristic curve and water head characteristics curves. (a) Operating characteristic curve is the curve drawn of turbine that work at a fixed speed n and water head H; (b) Rotation speed characteristic curve shows the relationship among the turbine flow Q, output P and efficiency η and speed n when guide vane opening a_0 and head H is a constant;(c) Head curve indicates the relation curve about the output P, flow Q, efficiency η and head H when turbine guide vane opening a_0 and the speed n is a constant. The combined characteristic curve can be divided into the operating combined characteristic curves and model combined characteristic curve according to the prototype operation and model tests. (a)Operating combined characteristic curves uses the operation parameters of head H and output P or flow Q as the abscissa and ordinate, and to express the liner external characteristic curve of the prototype turbine;(b) Model combined characteristic curve uses the unit speed n_{11} and unit flow Q_{11} as the abscissa and ordinate to represent the liner external characteristic curve of the model turbine[4]. Pan xihe et al. [5] summarizes the existing study results towards pump turbine characteristic curves and divide the research results into two dimensional mathematical transformation and three dimensional curved surface fitting. Furthermore, a fitting method for pump characteristic curves of various geometries based on cubic uniform B spline is presented by Zhang lin et al.[6], and proves the method works well by building various pump test data.

In this paper, the typical model combined characteristic curves respectively were selected at the different water heads of 100 meters, 200 meters, 300 meters and 500 meters based on the test results and hydraulic design experience, meanwhile, a longitudinal comparison with the turbine equal efficiency curve, blade channel vortex curve, The cavitation curve in the pressure and suction surface of the leading edge of the blade, Cavitation vortex curve and the output restricted curve was conducted, and the characteristics and differences of hydraulic characteristic curve under different water head were analyzed and summarized, all above provide a powerful theoretical basis and definite direction of modification for the hydraulic design of the turbine.

2. Characteristic of the model combined characteristic curve

It's the turbine model test that is considered still the final and indispensable verification means in hydraulic design of turbine, although the development of CFD technology has provided a new platform for the design of hydraulic turbine. In view of the model test, the model turbine will operate at the head region and output range that the prototype turbine may run, because of the long experiment period and expensive testing costs[7]. With the optimum operating points as the center, test conditions should cover the entire operation region as widely as possible. The final results of the model test of hydraulic turbine is the model combined characteristic curve, the comparison shows the model turbine hydraulic performances, and to provide references for selecting of prototype turbine. Figure 1 shows the Francis turbine model combined characteristic curves of four kinds of typical head respectively 100m, 200m, 300m and 500m, from which the similarities and differences can be seen.

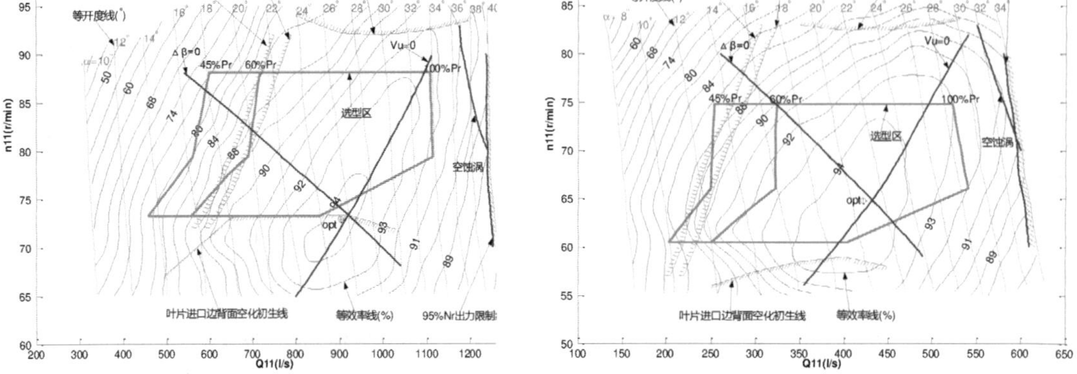

(a) 100m (b) 200m

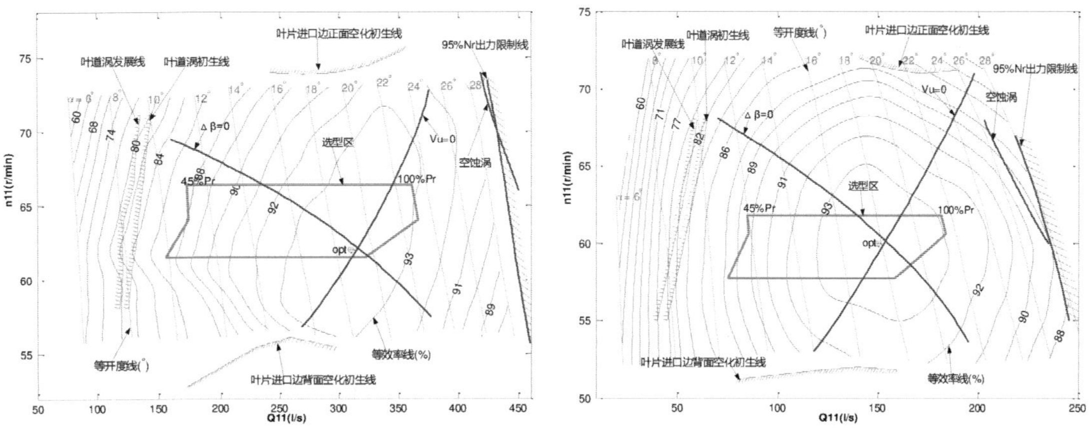

(c) 300m (d) 500m

Figure 1. Francis turbine model combined characteristic curve for different water heads

2.1 Generality analyses of model combined characteristic curve

As shown in Figure 1, the Francis turbine model combined characteristic curve can be plotted when takes the unit speed n_{11} and unit flow Q_{11} as the abscissa and ordinate respectively and marks out the guide vane opening curves and equal efficiency curves at the entire test operating conditions. It can be scaled out the turbine pressure fluctuation isolines, cavitation inception contours and the critical cavitation coefficient curve at different measuring points. Within the test range, it appears a blade vortex inception curve and the development curve if takes the optimal region as central under the small opening and small flow; cavitation inception curves in the pressure surface of the blade inlet edge at high unit speed and large flow region and in the suction surface of the blade inlet edge at low unit speed and low flow region respectively; 95% output limiting curve and Cavitation vortex at the region of large opening and large flow. These curves are the inherent characteristics of Francis turbine; the boundary position can be adjusted in a small range, but cannot be eliminated.

2.2 Cavitation vortex and 95% output limiting curve

The output of Francis turbine increased with the increase of the guide vane opening, but the guide vane opening increases to a certain extent, the output will reach the limit value. Because the prototype turbine can normally achieve power adjustment in maximum output operation, the points of 95% limit output conditions at the same unit speed on the model combined characteristic curve, were connected to draw an output limiting curve. in general, the early hydraulic turbine is designed based on one dimensional or two dimensional flow theory, the manual drawing and semi-mechanization manufacture model turbine output characteristics is poor, it will soon reach the limit output if the guide vane turn around the optimal opening degree, therefore the selection of model turbine was restricted seriously as a result of the output limiting curve often appears in the test range. But now, with the help of powerful computer and CFD technology, in the hydraulic design phase, CFD numerical optimization has been carried out several times before model test. In addition, the advanced processing technology and CNC machining technology have been also applied. So the output limit of the model turbine has passed over far away from the optimum region and rarely appears in the test range. Schematic diagram of the output limiting curve is given in Figure 1. In fact, as shown in the figure, the output limiting curves for four typical water heads in the model combined characteristic curve chart are appear in the test range.

2.3 Cavitation curve in the pressure and suction surface of the blade leading edge

The model combined characteristic curve of Francis turbine has a lot in common and its inevitability from the analysis of the model turbine hydraulic design. In general, on the combined characteristic

IOP Conf. Series: Materials Science and Engineering **129** (2016) 012036 doi:10.1088/1757-899X/129/1/012036

curve chart, the cavitation inception curve in the suction surface of blade inlet is more near the optimum region than the curve in the pressure side and the model selection for the low-head Francis turbine also greatly is limited by the cavitation inception curve in the suction surface of blade inlet. The general opinion is that, it is beneficial for the Francis turbine with slightly positive incidence angle at optimum operating condition to improve the prototype turbine efficiency at low head and rated operating point, accordingly, the stability and the weighted average efficiency could be improved at weight operation condition. The cavitation curve in the pressure and suction surface of the blade inlet is directly related to the runner inlet incidence angle, the cavitation inception in the suction surface of the blade inlet will appear rapidly from the optimum unit speed to the low unit speed on this chart. While the turbine turns from optimal unit speed to high unit speed, the blade leading edge pass zero incidence angle and then turned to negative incidence angle, having reached a certain intensity can appear the cavitation inception in the pressure side of the blade inlet, so it is far from the optimum operating region on the model combined characteristic curve chart.

In some model test institutions, the cavitation inception curve in the pressure and suction surface of the blade inlet and the blade channel curve were accustomed to be connected. Due to the blade channel vortex and cavitation phenomenon belong to observation experiment, it's difficult to identify substantially because of large differences of subjective factors at the operating condition conjunction of the blade channel vortex and the cavitation of the blade inlet.

For the Francis turbine at different water head range, the cavitation inception in the pressure and suction surface of the blade inlet is remarkable difference. In general, for the middle-low water head Francis turbine, the cavitation inception curve in the pressure and suction surface of the blade inlet will appear on the model combined characteristic curve chart because its operating region unit speed is large span. In addition no far away between the cavitation incipient and high head operating region, even it is difficult to eliminate the cavitation incipient curve outside of operation region for the low-head Francis turbine, as shown in figure 1(a). While operating at the high or super high water head, there is few cavitation incipient on the model combined characteristic curve owing to the narrow operation unit speed range and small speed test span.

2.4 blade channel vortex inception and development curve

According to the model test of hydraulic turbine, it is defined as the blade channel vortex inception if the vortex appears simultaneously in three blade channels of the Francis turbine runner, while all channels appear eddy is defined as the blade channel vortex development. In general, the blade channel vortex inception curve and its developing curve will appears in the experimental working condition region, they are parallel from high unit speed to low unit speed inclined down at low flow condition. In comparison, the blade channel vortex of the low water head turbine closes to the optimum region, and the one of the high water head turbine is far away from the optimal region. Many experimental curves showed that, the blade channel vortex of the low head model runner appears in the vicinity of output 60%Pr, and the one of the medium-higher water head runner occurs in the vicinity of output 45%Pr, and it is difficult to exclude it completely outside the stability operation region. The blade channel vortex of the high and super high water head turbine occurs in minimum flow rate condition, eliminated far from the outside output 45%Pr.

2.5 isoefficiency curve

Theoretically, there are two characterization curves to express the operation state of the runner on the model combined characteristic curve chart of the Francis turbine: zero incidence angle ($\Delta\beta$=0) curve and zero circulation (V_u=0) curve. The two curves are orthogonal and are interacted and mutually restricted, which play a controlling role for the trend and gradient of the model efficiency curve trend and gradient. The equivalent efficiency ring on the model combined characteristic curve chart is affected by the interaction between the inlet incidence angle $\Delta\beta$ and the outlet circulation Vu of the runner.

For the iso-efficiency curves on the model characteristic curve chart of the different water heads, their shape and efficiency gradient in all directions shows typical common and obvious difference. It generally believed that the turbine energy performance is the best when the runner operation at zero incidence angle and zero outlet circulation or slightly positive outlet circulation or alternating outlet circulation, therefore, the slope of the iso-efficiency curve is significantly smaller at the direction of curve $\Delta\beta=0$ and Vu=0 on the model characteristic curve chart. Normally, for a prototype turbine, its operation probability at the high water head condition is less than the one at the middle and low water head condition. A positive incidence angle at the blade inlet side is designed to improve the efficiency of a turbine under large opening, but the disadvantage is the rapid decline under small opening and low unit speed.

The guide vane and the draft tube play a decisive role in the overall shape of the efficiency curve on the model characteristic curve chart. For the runner, there are relatively good inflow conditions at zero incidence angle direction, but the hydraulic loss of tandem cascades will increase sharply with the guide vane opening decreasing. This means that the efficiency of the equal efficiency curve will drop sharply under small opening and high unit speed region. the outlet circulation of the runner has a great influence on the energy recovery of the draft tube, the best energy recovery performance of the draft tube will be achieved by the runner outlet with slightly positive circulation or alternating circulation, thereby, the high efficiency region of the turbine shows the characteristics of elliptical partial left distribution along zero outlet circulation curve.

The guide vane and draft tube play different roles for the turbine efficiency improvement in different water head range. For the high and super high water head Francis turbine, the guide vane opening is generally small and the flow velocity is great, so the hydraulic loss of tandem cascades is relatively large. This means that the optimum design of tandem cascades in the high and super high water head turbine is the most priority. For middle and low water head Francis turbine, the energy recovery of draft tube is a very significant. From the model characteristic combined curve perspective, the span of high efficiency curve ring is very large along zero outlet circulation direction. With the increase of water head and the reduction of optimum unit flow, the effects of draft tube gradually weakened and the impacts of tandem cascades are gradually increasing in the hydraulic turbine, the high efficiency curve circle varies from ellipse near zero circulation gradually to a circular by zero incidence angle control, as shown in Figure 1 (a) and (d). In addition, for middle and low water head Francis turbine, due to having the huge momentum in the turbine outlet, success or failure on hydraulic design of low water head draft tube has a significant effect on the low-frequency pressure fluctuation of the turbine.

3 Characteristics of pressure fluctuation in the different quadrants

It is generally known that hydraulic turbine pressure fluctuation is very complicated. Furthermore, the pressure fluctuation in different parts of the hydraulic turbine and different operating conditions shows different characteristics. According to the principle of hydraulic design, hydraulic characteristics of pressure fluctuation can be discriminated approximately in different quadrant in the model combined characteristic curve chart.

In general, the Francis turbine is designed according to the principle of no impact or somewhat positive incidence angle in the inlet and normal flow or somewhat positive circulation or alternating circulation in the outlet for the optimum condition. As shown in figure 2, the circumferential velocity component increases with the decrease of load, then a strong helical off-centered vortex rope appears in the draft tube. It brings out a low-frequency pressure fluctuation and cause the turbine units vibration.

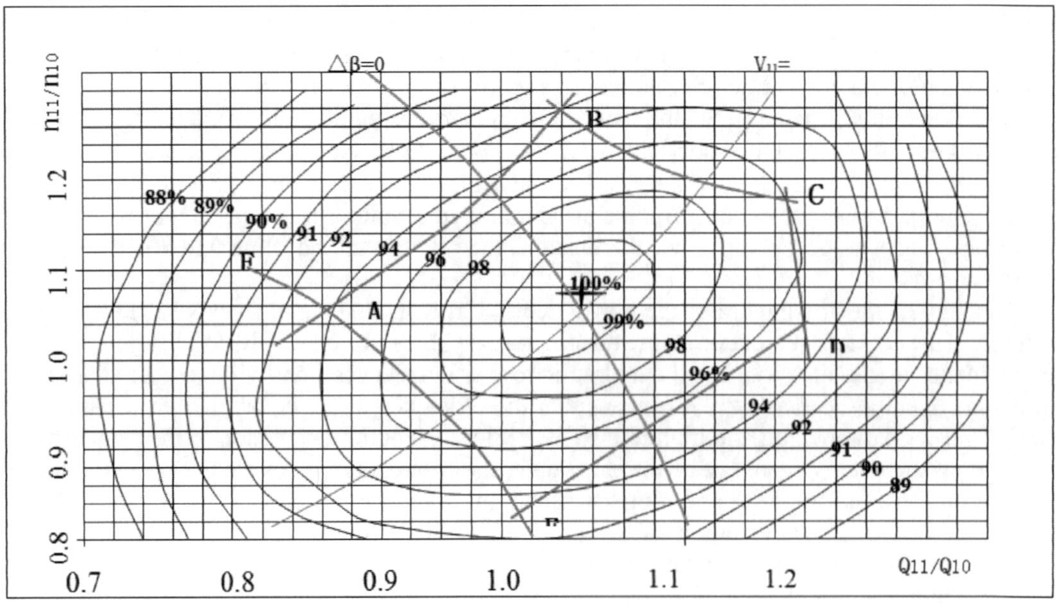

Fig. 2 Schematic diagram of runner combined characteristic curve

In figure 2, the curve AB is the boundary curve of the turbine unstable operation caused by low frequency pressure fluctuation. At low water head and large flow region, negative incidence angle occurs in the runner inlet and flow separation appears on the blade pressure surface as well as cavitation. At this time, several vortex ropes show in the draft tube, which will affect seriously the turbine unit stability. The curve BC is taken as the region boundary curve, and the curve CD is the 95% output limiting curve. In the high load region, the negative circulation of the runner outlet causes the concentric vortex rope in the draft tube, and induces the vibration of the turbine unit, the curve DE is its boundary curve. In the high water head and low flow region, flow separation and cavitation will appear in the suction surface of the blade Due to the existence of large positive incidence angle in the runner inlet. Meantime, the existence of the blade channel vortex will result in the medium high frequency pressure fluctuation and have an effect on the turbine unit stability even blade crack and the curve EA is used as the region boundary curve. The region surrounded by the connecting curves of the point A, B, C, D, E and A is defined as the stable operation region of Francis turbine.

4 Selection of the prototype turbine

Type selection of prototype turbine is influenced by different characteristics of different water head model combined characteristic curve. Their respective selection area also shows distinguishing feature.

Firstly, for low water head hydraulic turbine, the hydraulic turbine customarily operates in a large range of water head variation, and its operating region in the model combined characteristic curve chart is in a wide span, so the type selection is limited by the cavitation inception curve of the pressure and suction surface of the blade inlet, especially the suction surface cavitation inception curve. While the type selection of the high and super high water head hydraulic turbine is rarely influenced by the cavitation inception curve. Secondly, the stability operating region in the model combined characteristic curve chart of the high and super high water head hydraulic turbine is much wider than that of low water head hydraulic turbine, So turbine load adjusting range varies from 45%Pr~100%Pr for the high and super high water head turbine, while load adjusting range for the low water head turbine should be limited within 60%Pr. Thirdly, at low flow condition, low-frequency pressure fluctuation is caused by the draft tube vortex rope, so the steady operation output range of the prototype turbine is not lower than 40%Pr.

5 Hydraulic characteristics of hydraulic turbine

For the low water head Francis turbine (about 100 meters design water head), upper part load pressure fluctuation occurs frequently and is difficult to eliminate in model test, moreover, the more and more difficult to remove it with design water head decreasing. Considering large unit discharge and cavitation coefficient, it needs a larger suction height while low water head Francis turbine operating without cavitation. It is difficult to exclude blade channel vortex below 60% rated output as a result of unit water head containing relatively less energy. The cavitation curve of the suction surface of the blade inlet may enter the turbine operation area when water head variation of prototype turbine is large. The amplitude of the low frequency pressure fluctuation is comparatively fairly high, more than 8% peak to peak value generally.

For the medium and high water head Francis turbine (about 200 meters design water head), its specific speed value range is most suitable to design super giant hydraulic turbine, for instance, the runner diameter of 1000MW Francis turbine in Baihetan Hydropower Plant is about 8.5 meters, while the runner diameter of Three George's 750MW turbine reaches to 10 meters. The blade channel vortex and cavitation curve in the suction surface of the blade inlet can be excluded outside the turbine stability operation region, however, exiting a certain difficulty about hydraulic design. A high suction height and deeper excavation depth is a requirement for turbine operating without cavitation. And the peak to peak value of low frequency pressure fluctuation is below 5%.

For the high and super high water head Francis turbine (above 300 meters design water head), the essential consideration of hydraulic and structural design is silt abrasion problem owing to high flow velocity passing turbine channel. The blade channel vortex and cavitation curve in the pressure and suction surface of blade inlet is far away from turbine stability operation area. For the runner with diameter D_1=1~4 meters, its manufacturing and repairing is very difficult and complex as a result of high synchronous speed, small runner size and more blades, especially is difficult to be welded at the crow and band of the blade midsection, where the number of blade is 17, even 32 splitter blades. Due to small turbine critical cavitation coefficient and cavitation inception coefficient, the higher turbine design water head, the lower its suction height.

6 Conclusions

The model combined characteristic curve can comprehensively reflect hydraulic performance of model hydraulic turbine. The similarities and differences of each model turbine can be explored by detailed analysis of model combined characteristic curve in different water head range. And it can further reveal the curve's change regulation and guide hydraulic design engineers to carry out turbine performance optimization.

Acknowledgement

The research was supported by The National Natural Science Foundation of China (51479166) and the Specialized Research Fund for the Doctoral Program of Higher Education of China (20126118130002).

Reference

[1] Arne K 2001 *Hydropower in Norway: mechanical equipment* (Trondheim: Norwegian University of Science and Technology)

[2] Huang Y F, Liu G N and Fan S Y 2010 *Research on Prototype Hydro-turbine Operation* (Beijing: China Electric Power Press)

[3] Christophe N, Jorge A and François A 2004 Identification and modeling of pressure fluctuations of a Francis turbine scale model at part load operation 22nd IAHR Symp. on Hydraulic Machinery and Systems(Stockholm, Sweden, June 29-July 2)

[4] Dörfler P K 1994 Observation of the pressure pulsation on Francis model turbine with high specific speed *J. The International Journal on Hydropower & Dams.* **(1)** 21-26

[5] Pan X H, Cheng Y C and Wang J F 2014 Research progress of the pump turbine characteristic curve *J. Journal of Yangtze River Science Research Institute* **(31)** 117-123

IOP Conf. Series: Materials Science and Engineering **129** (2016) 012036 doi:10.1088/1757-899X/129/1/012036

[6] Zhang L, Xu H and Yu H H 2008 Fitting method for pump characteristic curve based on optimal knots spline *J. Journal of Hydroelectric Engineering* **(27)** 141-144
[7] Liu Z L and Li Q F 2009 *Testing technology of hydraulic mechanical* (Beijing: China Electric Power Press)

IOP Conf. Series: Materials Science and Engineering **129** (2016) 012037 doi:10.1088/1757-899X/129/1/012037

Cavitation study of a pump-turbine at turbine mode with critical cavitation coefficient condition

J Wang[1], D Yang[1], J W Xu[1], J T Liu[2], L Jiao[*]

[1] Institute of Naval Architecture and Ocean Engineering, Zhejiang University, China
[2]Beijing Institute of Control Engineering, Beijing, China
*Corresponding author: Tel.: +8613093767519;

E-mail: jiaolei@zju.edu.cn

Abstract. To study the cavitation phenomenon of a pump-turbine at turbine mode when it ran at the critical cavitation coefficient condition, a high-head model pump-turbine was disperse using hexahedron grid. Three dimensional, steady cavitating flow was numerically studied using SST k-ω model. It is confirmed that ZGB cavitation model and SST k-ω model are useful ways to study the two-phase cavitation flow in pump-turbine. Mass flow inlet and pressure outlet were specified at the casing inlet and draft tube outlet, respectively. The static pressure was set according to the cavitation coefficient. The steady cavitating flows at critical cavitation coefficient condition were analysed. The cavitation area in the runner was investigated. It was found that the pressure of the suction on the blade surface was decreasing gradually with the decrease of the cavitation coefficient. In addition, the vortex flow in the draft tube was observed at the critical cavitation coefficient. It was found that the vortex flow appeared at the center of the draft tube inlet with the decreasing of the cavitation coefficient. Compared with the experimental data, the simulation results show reasonable agreement with the experimental data.

1. Introduction

The cavitation is one of important phenomenon in a pump-turbine at turbine model. With the development of the cavitation flow, the cavitation bubble would collapse and perish, which would cause material erosion, structure wreck, vibration and noise. But due to hydraulic machine was in low speed state, even if it had cavitation wreck, the wreck was not very critical, that trigger the short of the cavitation research [1]. In recent years, the pump-turbine develop quickly, big capacity, high head and high speed gradually become the main developing aspect gradually. Finally, the condition of the blade wreck, efficiency reduces, device destroys, noise and vibration would aggravate gradually. So the cavitation problem of the pump-turbine has become one of the important research topics of domestic and foreign academia and engineering.

With the rapid development of computing technology in recent years, the simulation by the computational fluid dynamics can get some useful results in the engineering, except experimental method [2, 3]. In recent year, a lot of research about the mechanism of the cavitation form had been done. Wang Ya-yun [4] researched the unsteady hydrodynamic characteristics of a 2D cavitating hyrdrofoil and the evolution of cloud cavition, and compared the phenomenon of shedding and break-off sheet cavitation and behaviour of cloud cavitation by using the Reynolds Averaged Navier-stokers equations, Detached Eddy Simulation and Large Eddy Simulation. The numerical results of 2D LES show that the generation and evolution of the vortex near the wake of sheet cavity in the primary reason of break-off sheet cavitation, and the reentrant jet occurs only after the break of the sheet cavity. This research about the cavitation mechanism Wu [5] studied the unsteady cavitating turbulent flow with two different models and found that the mixture model could fit the experimental data compared with the single-phase model. Nishi [6] analysed the influence of the draft tube flow and the pressure fluctuation of the draft tube caused by the different cavitation coefficient by the experiment, the pressure fluctuation frequency is almost invariant until the cavitation coefficient is less than the certain value that would magnitude the pressure fluctuation.

However, the analysis of the instability of a pump-turbine at critical cavitation coefficient hasn't yet been reported. In present paper, a two-phase cavitation model was used to predict the cavitating turbulent flow in a model Francis pump-turbine by numerical simulation. The distribution of the cavitation area in the runner was investigated.

The main parameters of the model mixed flow water turbine studied are showed in Table.1. D1 denotes the diameter of the runner outlet in turbine mode, Z denotes the number of the blades, Zs and Zg denote the number of the stay vanes and the guide vanes respectively, n denotes the velocity of the runner and H denotes the head of the test. The structure of the turbine is showed in the Figure.1.

Table 1. The main parameters of the model mixed flow water turbine.

Parameter	Value
D(m)	0.26
Zs	20
H(m)	50
Zg	20
Z	9

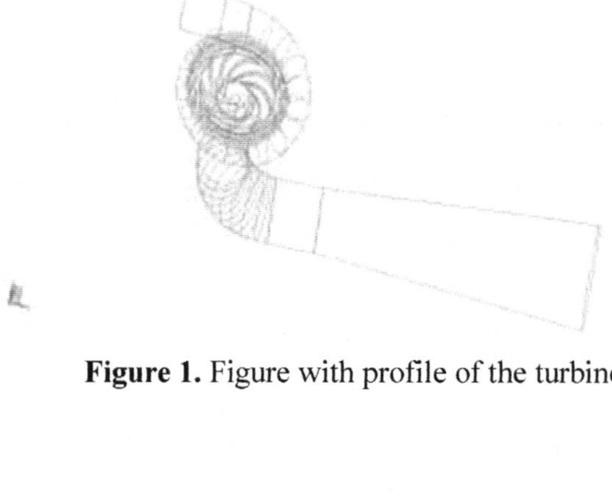

Figure 1. Figure with profile of the turbine.

2. Numerical method

2.1. Cavitation model

Zwart-Gerber-Belamri model was used to model cavitation in the pump-turbine. It was compatible with turbulence model well. It considered the evaporation and the condensation to control the equation. The evaporation and the condensation was included are defined as

When $p \le p_b$,

$$R_e = F_{vap} \frac{3\alpha_V (1-\alpha_V)\rho_v}{R_B} \left[\frac{2}{3} \left(\frac{P_V - P}{\rho_l} \right) \right]^{1/2} \tag{1}$$

When $p > p_b$,

$$R_e = F_{cond} \frac{3\alpha_V \rho_v}{R_B} \left[\frac{2}{3} \left(\frac{P_V - P}{\rho_l} \right) \right]^{1/2} \tag{2}$$

Where the diameter of the bubble was 1μm, $F_{vap} = 50$ and $F_{cond} = 0.01$ which represented factors for vaporization and condensation respectively; ρ_v was the saturated vapour pressure; ρ_l was the density of the fluid; the void fraction of the nuclei was 5×10^{-4}.

2.2. Turbulence model

The SST k-ω model developed by Menter [7] was adopted for the calculation of the unsteady cavitating flow in the pump-turbine. It combined k-ω model in the near-wall region and k-ε model in the far field.

$$\rho_m \frac{\partial k}{\partial t} + \rho_m u_j \frac{\partial k}{\partial x_j} = \frac{\partial}{\partial x_j} \left[(\mu + \sigma_k \mu_T) \frac{\partial k}{\partial x_j} \right] - \beta^* \rho_m k\omega + P_k \tag{3}$$

$$\rho_m \left(\frac{\partial \varepsilon}{\partial t} + u_j \frac{\partial \omega}{\partial x_j} \right) = \frac{\partial}{\partial x_j} \left[(\mu + \sigma_k \mu_T) \frac{\partial k}{\partial x_j} \right] + 2(1 - F_1)\sigma_{\omega 2} \frac{\rho_m}{\omega} \frac{\partial k}{\partial x_j} \frac{\partial \omega}{\partial x_j} - \rho_m \left(\alpha S^2 - \beta \omega^2 \right) \tag{4}$$

Where,

$$P_K = \min\left(\tau_{ij} \frac{\partial u_i}{\partial x_j}, 10\beta^* k\omega \right),$$

$$F_1 = \tanh\left\{ \left\{ \min\left[\max\left(\frac{\sqrt{k}}{\beta^* \omega y}, \frac{500\mu}{\rho_m y^2 \omega} \right), \frac{4\sigma_{m2}}{CD_{km} y^2} \right] \right\}^4 \right\},$$

$$CD_{k\omega} = \max\left(2\rho\sigma_{\omega 2} \frac{1}{\omega} \frac{\partial k}{\partial x_i} \frac{\partial \omega}{\partial x_i}, 10^{-10} \right),$$

Coefficients are chosen as $\alpha = 5/9$, $\beta = 0.075$, $\beta^* = 0.09$, $\sigma_k = 0.85$, $\sigma_{\omega 1} = 0.5$, $\sigma_{\omega 2} = 0.5$.

2.3. Boundary conditions

In the algorithm, the physical timescale was 0.001s. The head was specified 50m. The rotating speed of the runner was set by the function. Mass flow at casing inlet was specified at turbine mode. The outlet condition was used the value of static pressure. The outlet pressure was specified according to the function. The advection scheme was specified high resolution and the turbulence numeric was specified first order.

Three cavitation efficient operation conditions in the optimum operation condition and partial load condition were showed in Table.2, which were set by the cavitation coefficient.

Table 2. Calculation conditions for three cases.

Case	Cavitation coefficient	*n(rpm)*	Mass flow inlet(kg*m3/s)	*Pressure outlet(Pa)*
Optimum	0.12	1061	0.252	632162.0
Optimum	0.08	1061	0.252	42897.0
Optimum	0.06	1061	0.252	32764.5
Partial load (82%)	0.12	1378	0.207	632162.0
Partial load (82%)	0.08	1378	0.207	42897.0
Partial load (82%)	0.06	1378	0.207	32764.5
Partial load (70%)	0.12	1445	0.167	63162.0
Partial load (70%)	0.08	1445	0.167	42897.0
Partial load (70%)	0.06	1445	0.167	32764.5

3. Result and discussion

3.1. Hydraulic performance

The steady state performances under three different rotate speed and mass flow at inlet was simulated and the results were compared with test data, as shown in Figure.2. It can be seen from the Figure.2 that the numerical results meets well with the test results and the numerical method could predict the hydraulic characteristics accurately.

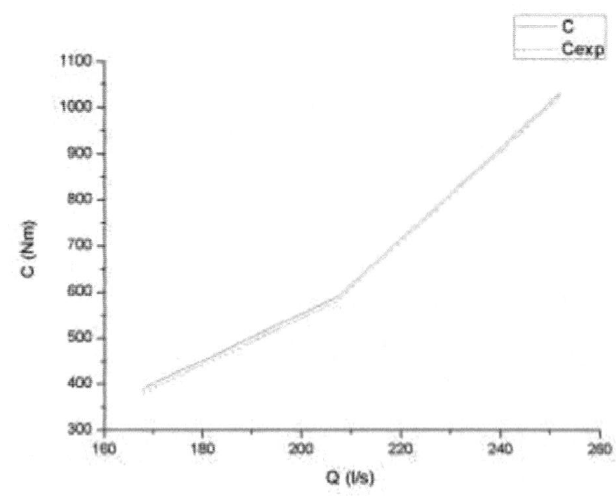

Figure 2. Figure with steady state Q-C performance at different speed.

3.2. Cavitation performance

Figure.3 – Figure.11 show the pressure distribution at the runner blade at different cavitation coefficient and at different load. It can be seen from the figures that, the low-pressure area is obviously exited on the suction side of the blade surface. With the decrease of the cavitation coefficient, the pressure of the low-pressure areas is decreasing gradually. And with the load decreasing, the pressure of the low-pressure areas is increasing at the same cavitation coefficient.

Figure.12 – Figure.20 show the vapour volume on the runner blade at different cavitation coefficient of the optimum load and at different load. It can be learned that, the obvious cavitation area is existed when the cavitation coefficient is at 0.06. Because of the pressure decrease of the low-pressure area, the vapour occurs while the pressure of the area is lowered than the saturate pressure of the water. Compared with the condition at the same cavitation coefficient, the cavitation area at the optimum load is larger than which at the 82% load and the 70% load.

The pressure distribution at the draft tube is showed at the Figure.21 – Figure.29. Figure.21-Figure.23 shows the cavitation coefficient at 0.12, 0.08 and 0.06 when it is at the optimum load, Figure.24 – Figure.26 when at the 80% partial load and Figure.27 – Figure.29 when at the 70% partial load. It can be learned that, compared with the same load, the pressure of the center area of the draft tube inlet is decreased gradually with the decreasing of the cavitation coefficient. And when the cavitation coefficient is lowered to 0.06, whatever the load is, the vortex flow occurs at the draft tube. In addition, compared with the different load when the cavitation coefficient is at 0.06, it show that the cavitation area is bigger at the 80% partial load than which at the 72% partial load and optimum load. And it has occurred a vortex at the 80% partial load. If the vortex frequency is same with the blade rotating frequency, it would lead to the strongly vibration, which will bring about huge damage.

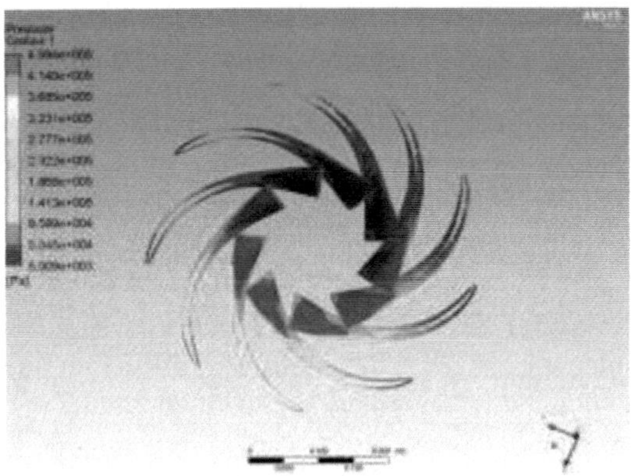

Figure 3. Pressure distribution at runner blade, Optimum, Cavitation coefficient 0.12.

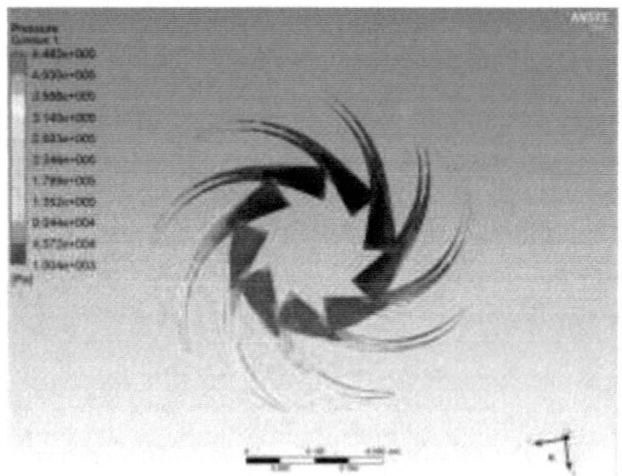

Figure 4. Pressure distribution at runner blade, optimum, Cavitation coefficient 0.08.

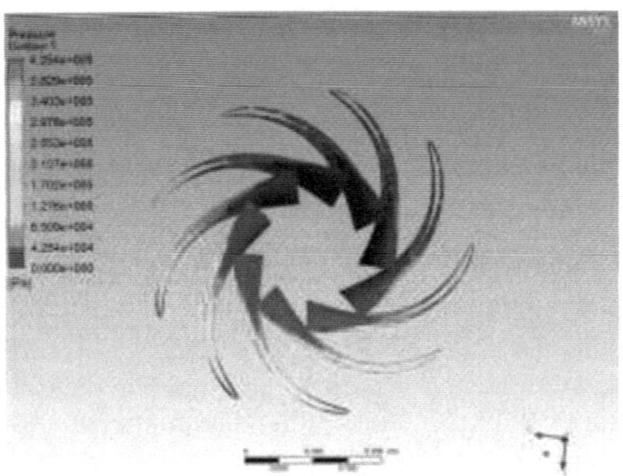

Figure 5. Pressure distribution at runner

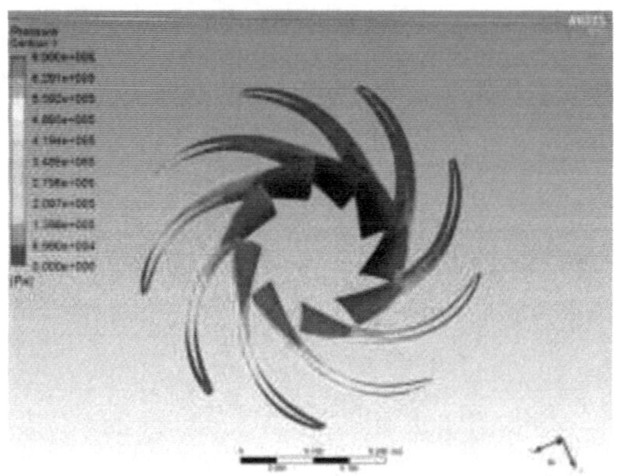

Figure 6. Pressure distribution at runner

blade, Optimum, Cavitation coefficient 0.06.

blade, Load 82%, Cavitation coefficient 0.12.

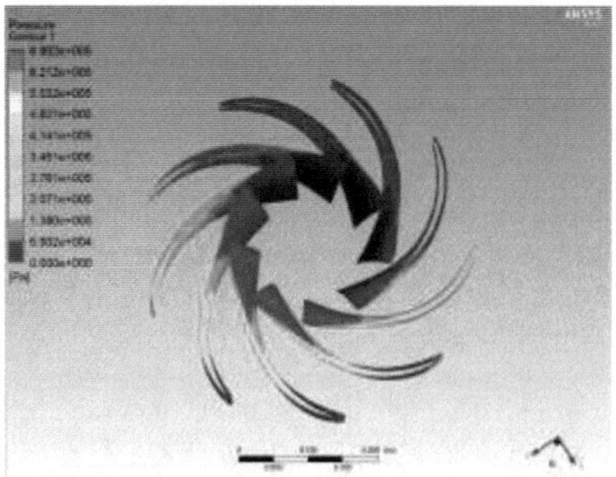

Figure 7. Pressure distribution at runner blade, Load 82%, Cavitation coefficient 0.08.

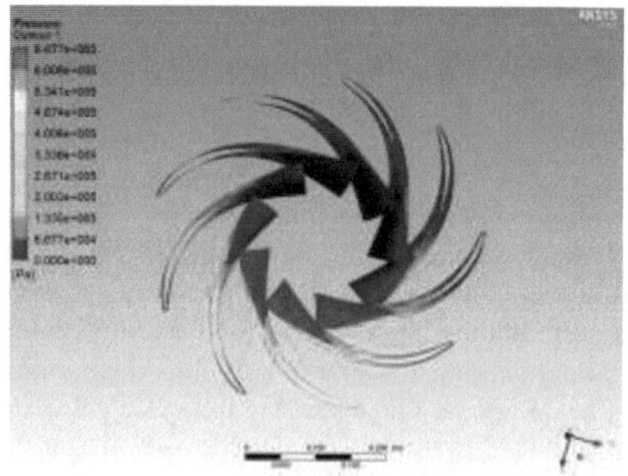

Figure 8. Pressure distribution at runner blade, Load 82%, Cavitation coefficient 0.06.

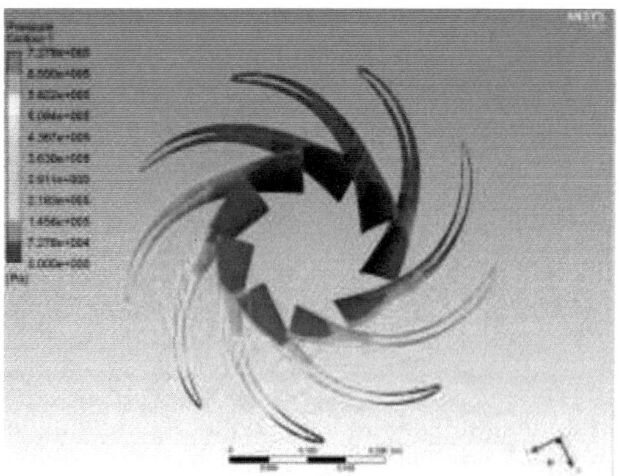

Figure 9. Pressure distribution at runner blade Load 70%, Cavitation coefficient 0.12.

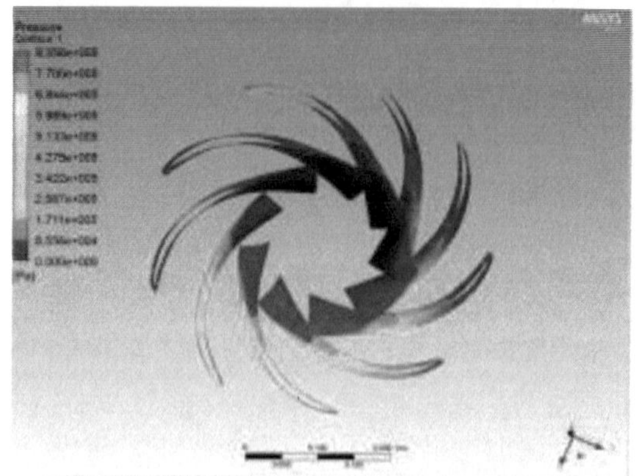

Figure 10. Pressure distribution at runner blade, Load 70%, Cavitation coefficient 0.08.

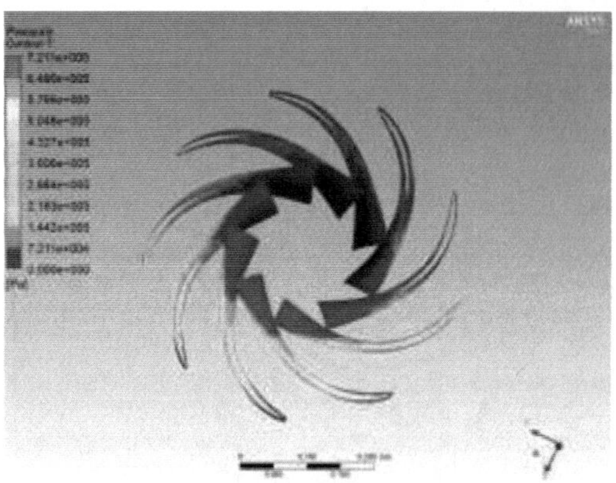

Figure 11. Pressure distribution at runner blade, Load 70%, Cavitation coefficient 0.06.

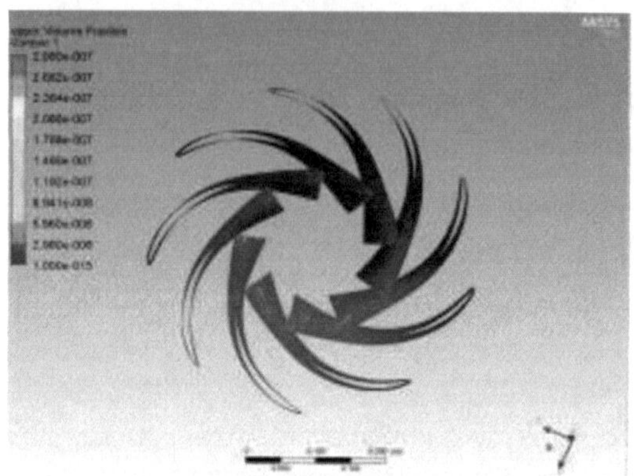

Figure 12. Vapour volume distribution at runner blade, Optimum, Cavitation coefficient 0.12.

IOP Conf. Series: Materials Science and Engineering **129** (2016) 012037 doi:10.1088/1757-899X/129/1/012037

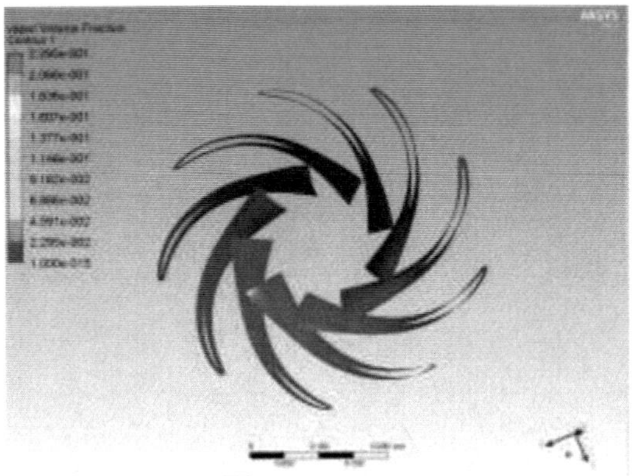

Figure 13. Vapour volume distribution at runner blade, Optimum, Cavitation coefficient 0.08.

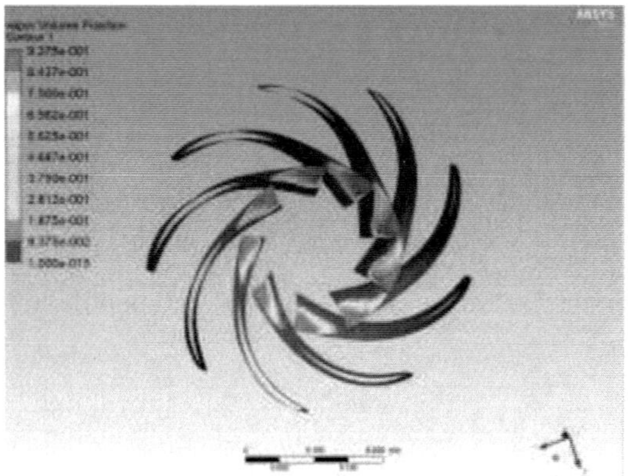

Figure 14. Vapour volume distribution at runner blade, Optimum, Cavitation coefficient 0.06.

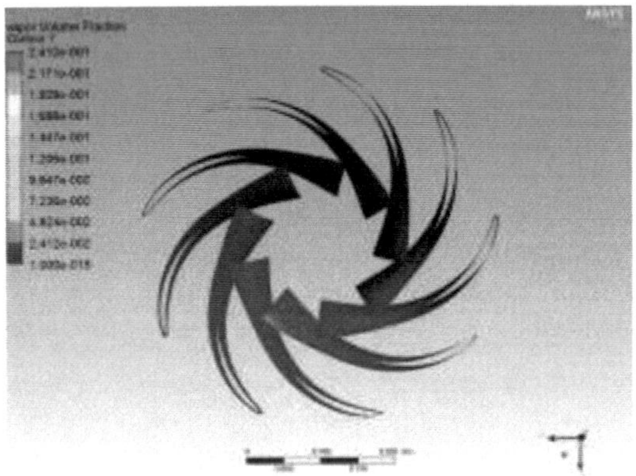

Figure 15. Vapour volume distribution at runner blade, Load 82%, Cavitation coefficient at 0.12).

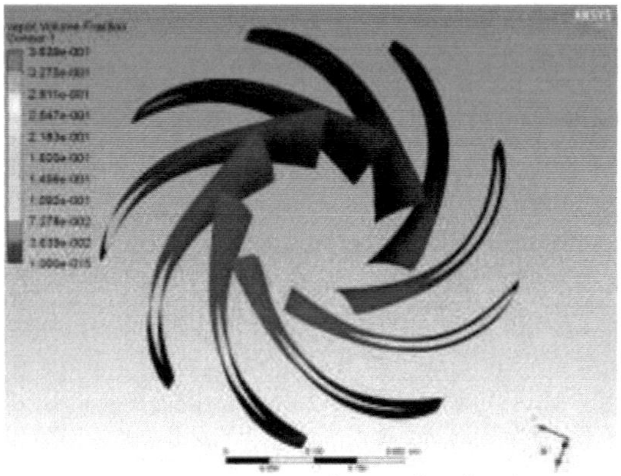

Figure 16. Vapour volume distribution at runner blade, Load 82%, Cavitation coefficient at 0.08.

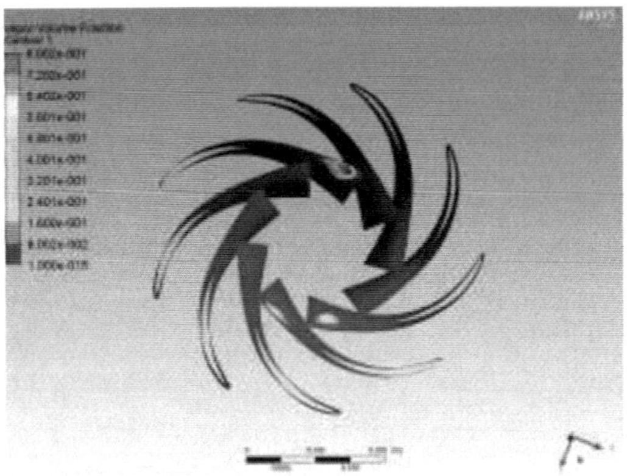

Figure 17. Vapour volume distribution at runner blade, Load 82%, Cavitation coefficient 0.06.

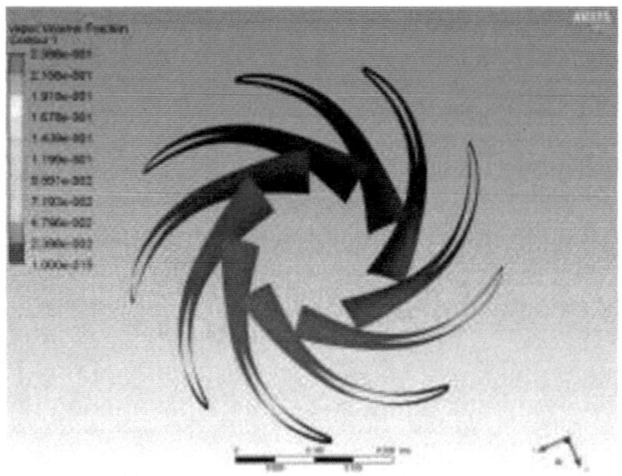

Figure 18. Vapour volume distribution at runner blade, Load 70%, Cavitation coefficient 0.12.

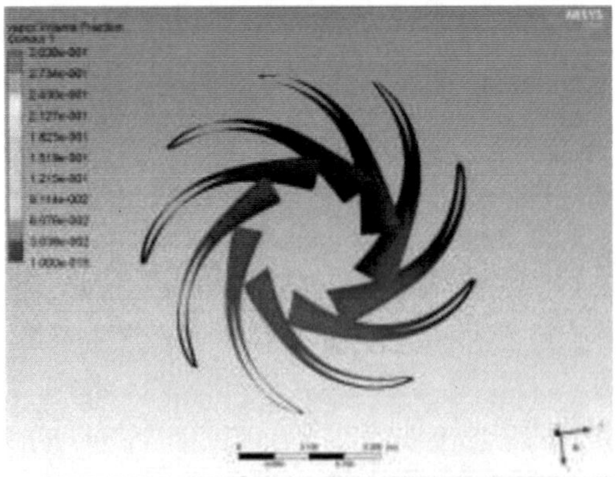

Figure 19. Vapour volume distribution at runner blade, Load 70%, Cavitation coefficient 0.08.

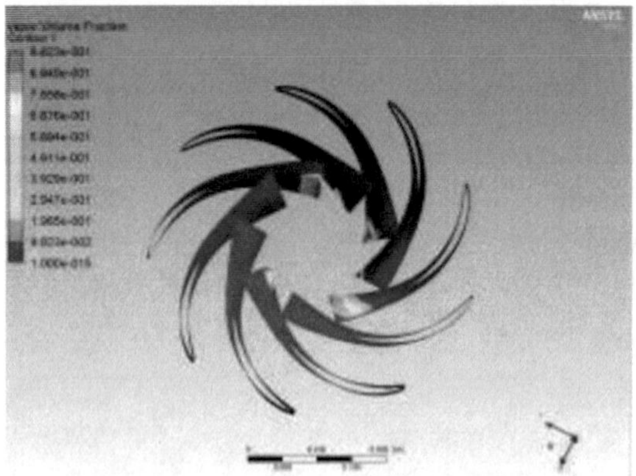

Figure 20. Vapour volume distribution at runner blade, Load 70%, Cavitation coefficient 0.06.

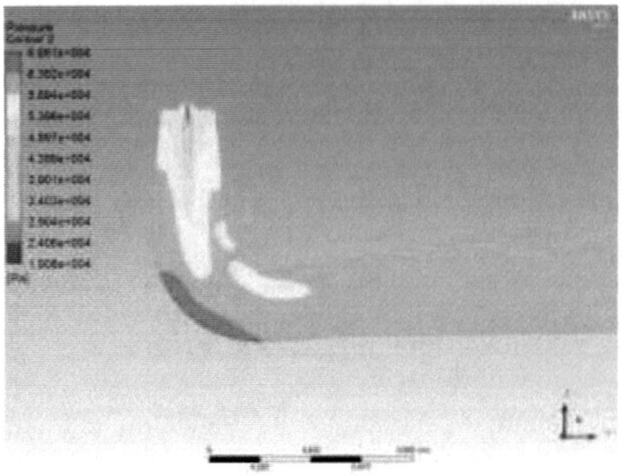

Figure 21. Vapour volume distribution at draft tube, Optimum, Cavitation coefficient 0.12.

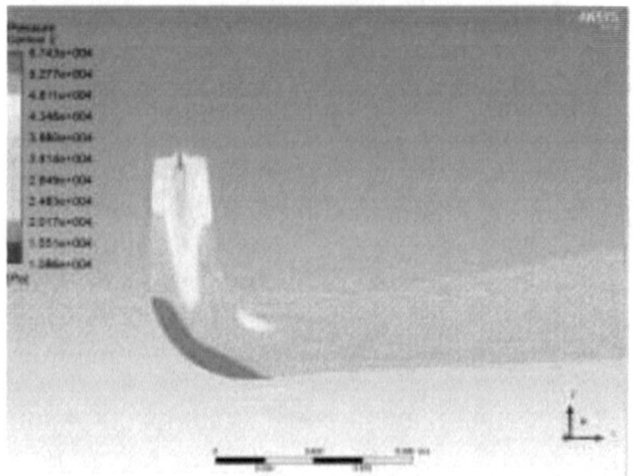

Figure 22. Vapour volume distribution at draft tube, Optimum, Cavitation coefficient 0.08.

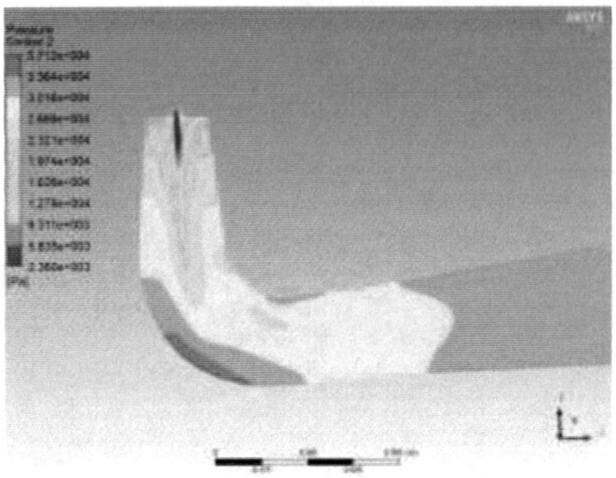

Figure 23. Vapour volume distribution at draft tube, Optimum, Cavitation coefficient 0.06.

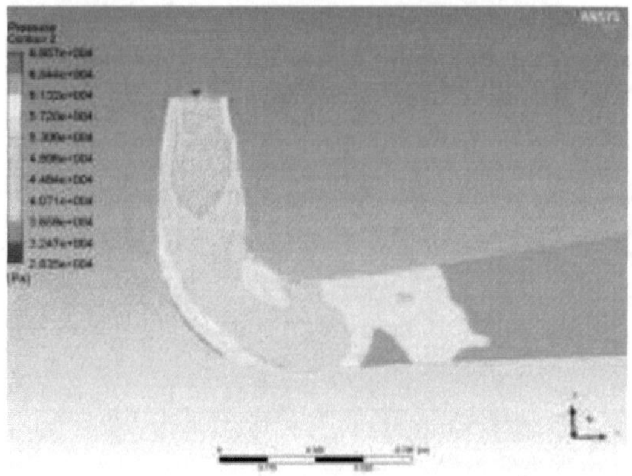

Figure 24. Vapour volume distribution at draft tube Load 82%, Cavitation coefficient 0.12.

IOP Conf. Series: Materials Science and Engineering **129** (2016) 012037 doi:10.1088/1757-899X/129/1/012037

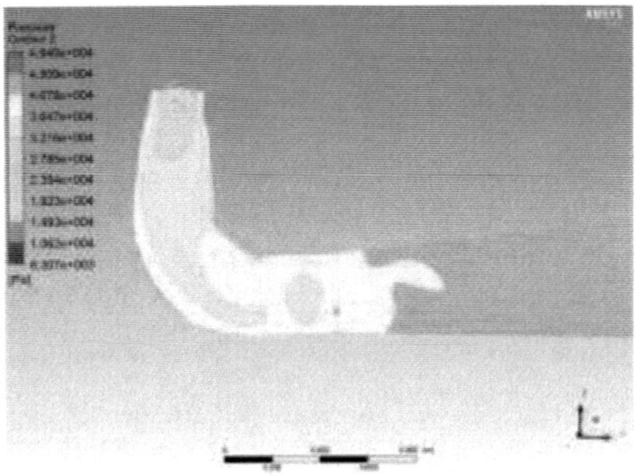

Figure 25. Vapour volume distribution at draft tube, Load 82%, Cavitation coefficient 0.08.

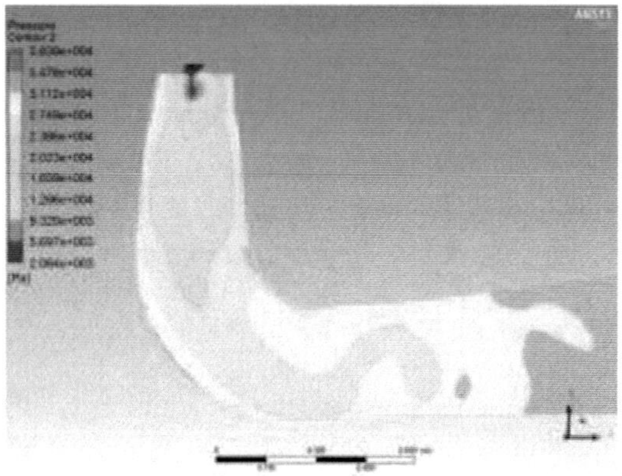

Figure 26. Vapour volume distribution at draft tube, Load 82%, Cavitation coefficient 0.06.

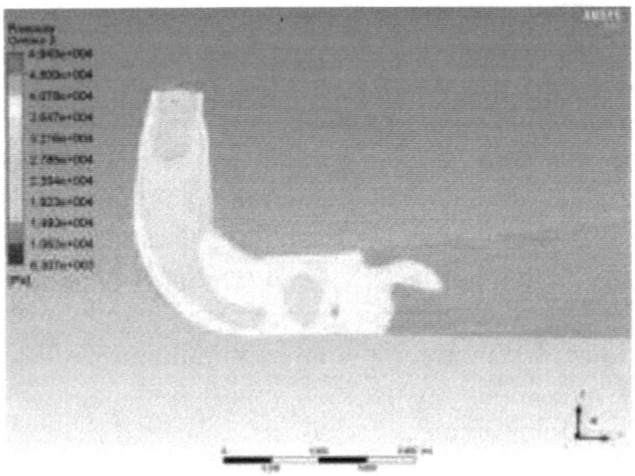

Figure 27. Vapour volume distribution at draft tube, Load 70%, Cavitation coefficient 0.12.

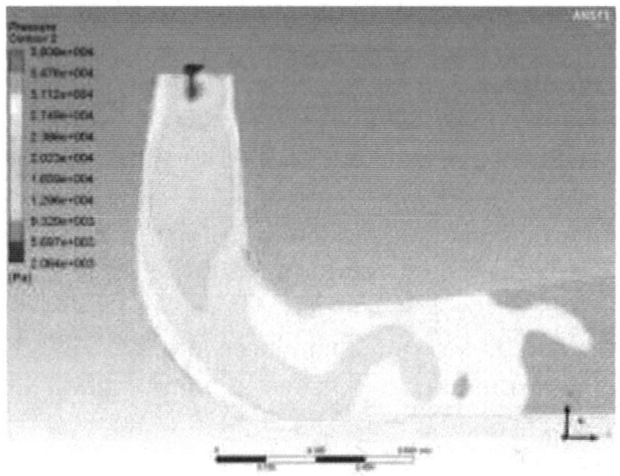

Figure 28. Vapour volume distribution at draft tube, Load 70%, Cavitation coefficient 0.08.

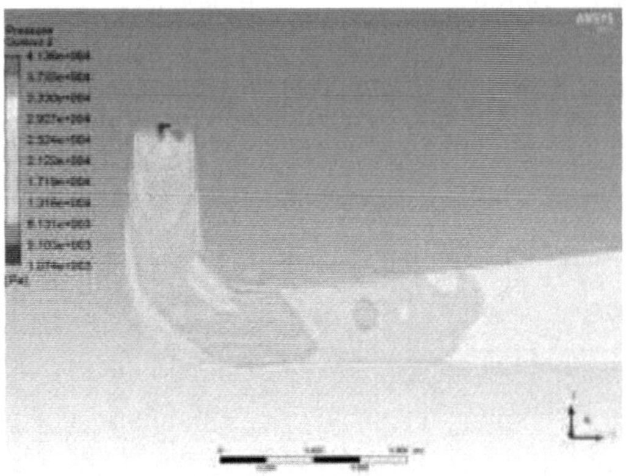

Figure 29. Vapour volume distribution at draft tube, Load 70%, Cavitation coefficient 0.06.

4. Conclusions

In this paper, a kind of numerical method for cavitation performance prediction was presented. In order to verify the accuracy of the simulation, the hydraulic performance was compared with the experiment data. Base on the result of the analysis in the Francis turbine, the following conclusions are drawn.

1. The simulation results by cavitating flow computation based on the mixture model agree well with experimental data.
2. With the decreasing of the cavitation coefficient, the pressure of the suction on the blade surface is decreasing gradually. And when the cavitation coefficient is lowed to a certain extent, the low-pressure area will appear the vapour area. And at the same cavitation coefficient, the vapour area at the optimum condition is largest.
3. With the decreasing of the cavitation coefficient, the vortex flow is appeared at the center of the draft tube inlet.

Acknowledgements

The research work was funded by National Science and technology support program (no.2013BAB08B03) and the National Natural Science Foundation of China (nos. 51276158).

References

[1] Kumar P, Saini R P 2010 Study of cavitation in hydro turbines—A review [J] *Renewable and Sustainable Energy Reviews*. **14(1)** 374-83.
[2] Srinivasan V, Abraham J S, Kozo S 2009 Numerical simulation of cavitation dynamics using a cavitation-induced-momentum-defect (CIMD) correction approach [J] *Applied Mathematical Modelling*. **33(3)** 1529-59.
[3] Zhao J, Wei Y, Zhang J et al 2009 Effect of various turbulence models on simulated results of cavitation flow [J] *Engineering Mechanics*. **8** 042.
[4] Wang Y, Wang B, Liu H 2014 Numerical simulation of sheet cavity shedding and cloud cavitation on a 2D hydrofoil *J. Chinese Journal of Hydrodynamics A*. **2** 006.
[5] Wu Y, Liu S 2011 Simulations of unsteady cavitating turbulent flow in a Francis turbine using the RANS method and the improved mixture model of two-phase flows [J] *Engineering with Computers*. **27(3)** 235-45.
[6] Nishi M, Liu S H 2013 An outlook on the draft-tube-surge study [J] *International Journal of Fluid Machinery and Systems*. **6(1)** 33-48.
[7] Menter F R, Kuntz M, Langtry R 2003 Ten years of Industrial Experience with the SST Turbulence Model [J] *Turbulence, Heat and Mass Transfer*. **4(1)**.

IOP Conf. Series: Materials Science and Engineering **129** (2016) 012038 doi:10.1088/1757-899X/129/1/012038

Flow characteristics on the blade channel vortex in the Francis turbine

P C Guo[1], Z N Wang[1], X Q Luo[1], Y L Wang[2] and J L Zuo[1]

[1] State Key Laboratory Base of Eco-hydraulic Engineering in Arid Area, Xi'an University of Technology, Xi'an 710048, Shaanxi Province, China
[2] Dongfang Electric Machinery Co., Ltd., Deyang 618000, Sichuan Province, China

E-mail: guoyicheng@126.com

Abstract. Depending on the long-term hydraulic development of Francis turbine, the blade channel vortex phenomenon was investigated systematically from hydraulic design, experimental and numerical computation in this paper. The blade channel vortex difference between the high water head and low water head turbine was also analyzed. Meanwhile, the relationship between the blade channel vortex and the operating stability of hydraulic turbine was also investigated. The results show that the phenomenon of blade channel vortex is an intrinsic property for Francis turbine under small flow rate condition, the turning-point of the blade channel vortex inception curve appears at low unit speed region, and the variation trend of the blade channel vortex inception curve is closely related to the blade inlet edge profile. In addition to, the vortex of the high water head turbine can generally be excluded from the stable operation region, while which is more different for the one of the low water head turbine.

1. Introduction

The so-called blade channel vortex is a phenomenon of continuous eddy flow between two blades of a Francis turbine runner, in the hydraulic machinery research fields. It's a natural phenomenon of the Francis turbine, which roots in the room between two blades and disappears in the entrance of the draft tube. Some hydraulic turbine units will suffer serious blade channel vortex at the 60% rated output. Nevertheless, it may also appear at the 30% rated output. The blade channel vortex phenomenon, which is an essential observation and experiment indexes, has been applying to preliminary test and acceptance test in model water turbine.

The general opinion is that the blade channel vortex is a great incentive for the operation stability of a turbine, but not necessary factors. The hydraulic action is a very complex natural phenomenon. a signal factor or the combined effect of many factors together, such as the material properties of metals, the geometry of flow passage components, the frequency characteristics of rigid body, and so on, may induce serious noise, pressure fluctuation, cavitation erosion, resonance etc. In some cases, there is may no serious impact on the turbine unit stability operation although some turbines have existed the blade channel vortex.

When the blade channel vortex affect the stable operation of hydraulic turbines, in some hydropower plants, the stable turbine operation scope is delimited to avoid the unsteady operating region, in others, aeration method is used for reducing noise, pressure fluctuation and improving the operating condition[1]. The third scheme is that the modification or replacement measure of the runner is adopted to achieve stable operation of the hydraulic turbine based on the hydraulic design. Therefore, we must have a correct understanding for the form of the blade channel vortex in order to guide the hydraulic design and engineering practice of hydraulic turbine.

A great lot of related studies have been carried out with a focus on the formation mechanism as well as its influence on the operation characteristic. Xiao Yexiang et al. took a numerical analysis of blade channel vortex in Francis turbine at part load, they investigated the characteristics of the unsteady flow dominant frequencies at different monitored points and concluded the dominant frequencies of blade channel vortex were a low frequency[2]. Zhang Hongming et al. present numerical simulation of blade channel vortex in a low head Francis turbine using OpenFoam code, the simulation results have shown that using cavitation model to analyze blade channel vortex is very effective[3]. A 2D numerical simulations of blade-vortex interaction in a darrieus turbine is given for instance by Amet E. et al.[4], and the results are compared qualitatively with the visualization of the **vortex** shedding of Brochier. The occurrence of the blade channel vortex between the turbine runner blades is discussed depending on the mechanism calculation, and the frequency of vortex cavity is analyzed form Chen jinxia et al.[5]

2. Model test of blade channel vortex phenomenon

A determinant about the blade channel vortex in the model acceptance test is generally believed that, when the vortex occurs in three blade channels of the runner, defined as the blade channel vortex inception, while it is defined as the development of blade channel vortex if all the blade channel vortex phenomenon takes place. The inception curve and development curve are given to express the blade channel vortex characteristics in the model combined characteristic curve chart of hydraulic turbine. Although the manifestation pattern and the overall development trend of the blade channel vortex are similar for different turbines, the model turbine development aims at different water head range, so its characteristics in different model combined characteristic curve is some different.

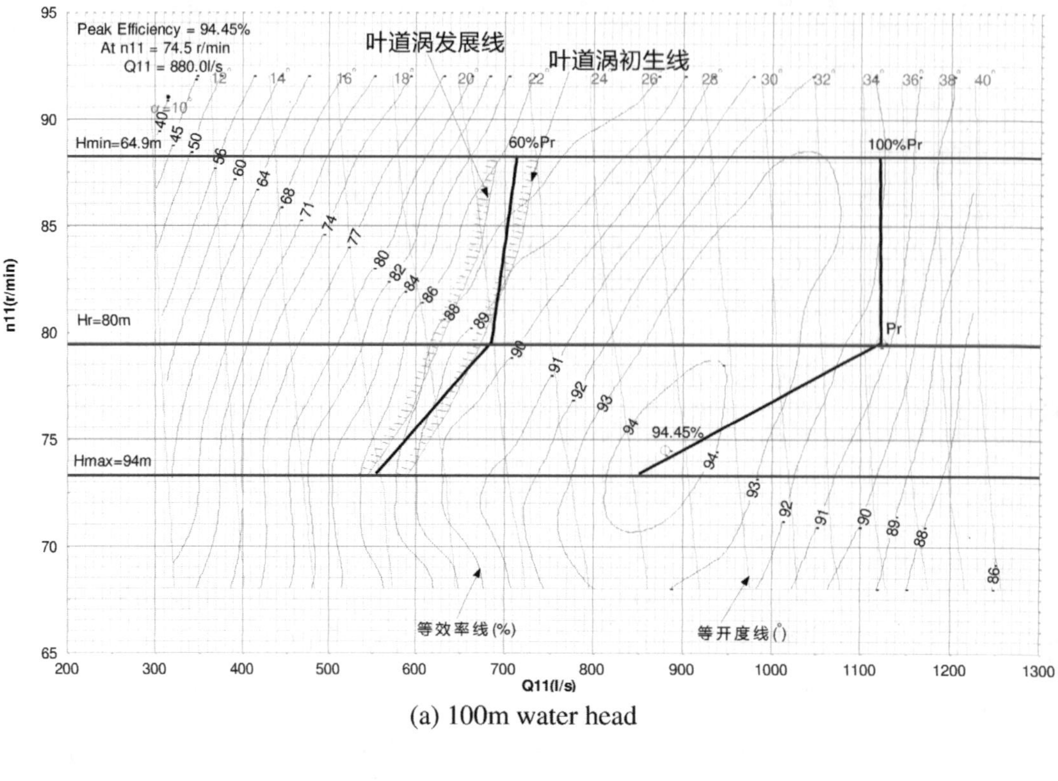

(a) 100m water head

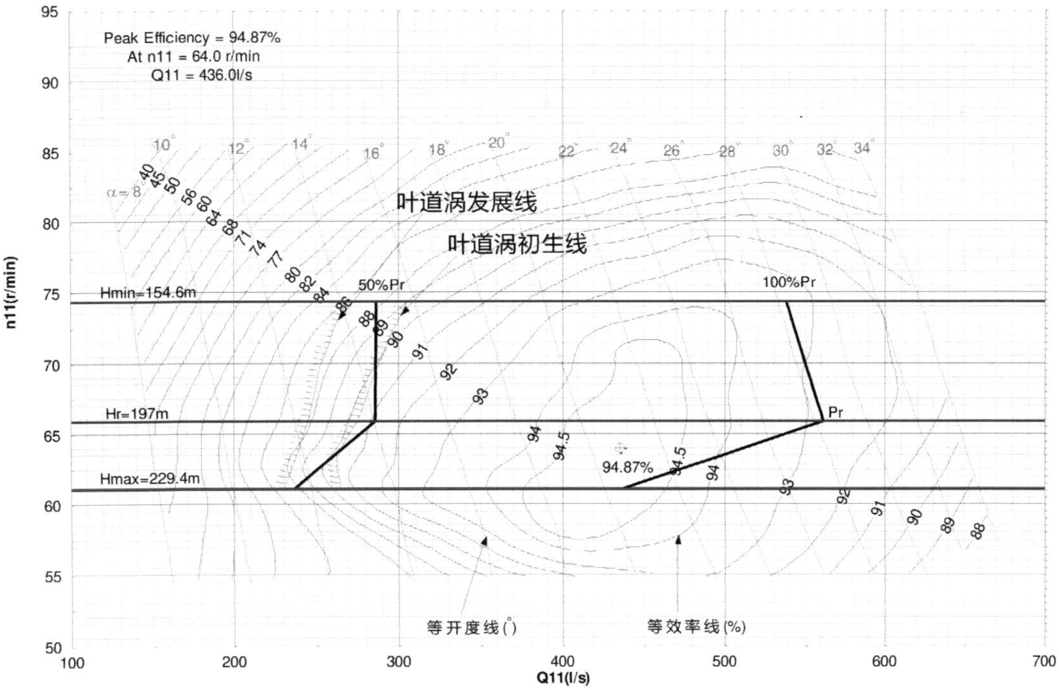

(b) 200m water head

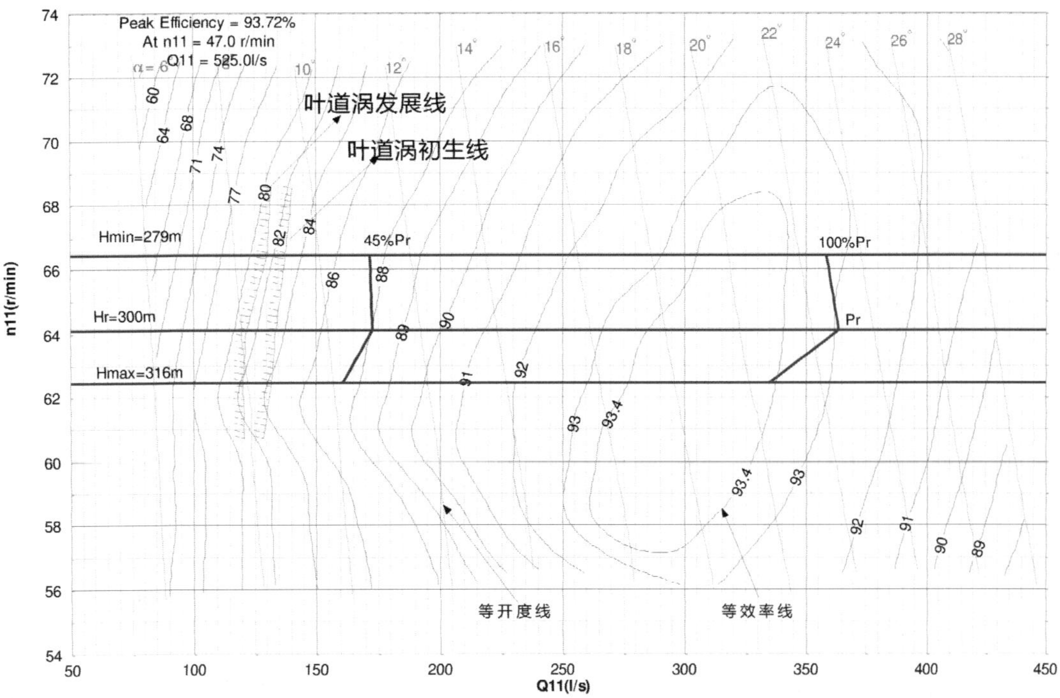

(c) 300m water head

Figure 1. Blade channel vortex curves under different water heads

The blade channel vortex of the low water head turbine occurs near optimum area of the model combined characteristic curve while the blade channel vortex of the high water head turbine is away from optimum area, but they both appear in low flow rate area, that is to say, it emerges in low load condition, below 60% output for the prototype turbine. The representative test curve of the blade channel vortex in different water heads is shown in Figure 1 (a) ~ (c), and from which we can clearly see that similarities and differences of the blade channel vortex in different water heads. The turning-point of the blade channel vortex inception curve appearing at low unit speed region is shown in Figure 1 (b). Almost certainly, the variation trend of the blade channel vortex inception curve is closely related to the blade inlet edge profile.

3. Cause of blade channel vortex formation

About the formation mechanism of the blade channel vortex, some scholars think that blade incidence angle is too large when the guide vane opening is small, form flow separation in the blade leading edge, and result in blade channel vortex generation[6]. Others demonstrate that the flow separation near the crown of runner leading edge gives rise to blade channel vortex generation. All of them agree that flow separation leads to blade channel vortex generation.

A majority of hydraulic plants have adopted the runner with X type blade since emerging of it. As figure 2 shows that one of the hydraulic design characteristics of the X type blade is inlet blade angle increasing gradually from runner crown to band. In order to keep the triangle of inlet velocity basically identical, as showed in Figure 3, the blade inlet edge is also correspondingly increase from the crown to the band.

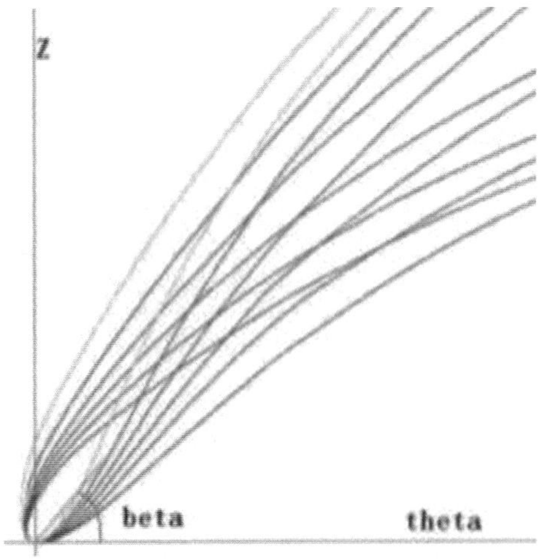

Figure 2. Inlet blade angle beta

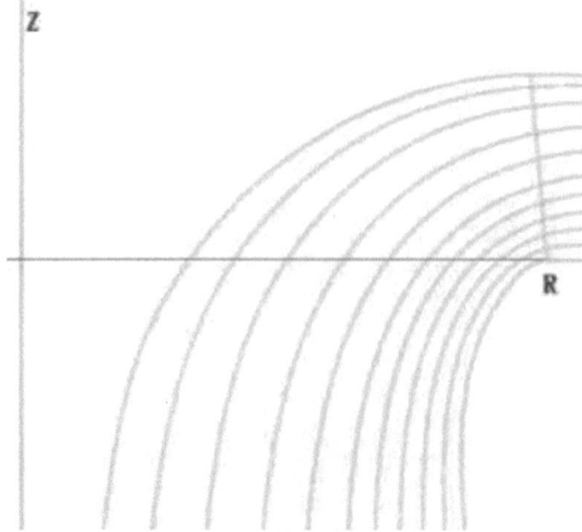

Figure 3. Blade inlet edge in R~Z

According to the results of numerical calculation, for all runners with X type blade, its incidence angle near the band is greater than that near the crown. In theory, if the occurrence of flow separation in the blade suction surface of runner inlet should also be near the band superior to near the crown, however, observation indicates it mostly originates from the runner inlet near the crown position. In addition, theoretically, a large negative incidence angle also leads to flow separation on the blade pressure surface when the guide vane opening is enough large, but experimental results show that eddy flow phenomenon in large flow condition mainly shows the cavitation characteristic, and is defined as cavitation vortex. Cavitation vortex occurs mainly in the low water head and large flow rate condition, as shown in figure 4.

The fundamental reason of blade channel vortex appearing is the turbine operating at the low flow condition with small guide vane opening, and flow separation only incentives channel vortex appearance. Because the blade of Francis runner channel is fixed, the internal flow state of runner will be changed obviously as deviation from the optimum operating condition. Flow velocity increasing result in cavitation occurrence, while a low flow rate lead to blade channel vortex appearing. At the optimum operating condition, steady pressure flow continuously passes through the blade channel of runner, while a marked drop of flow rate, this continuous steady pressure gradient will be broken, partial radial flow separation under the action of the tangential viscous force spins to form blade channel vortex.

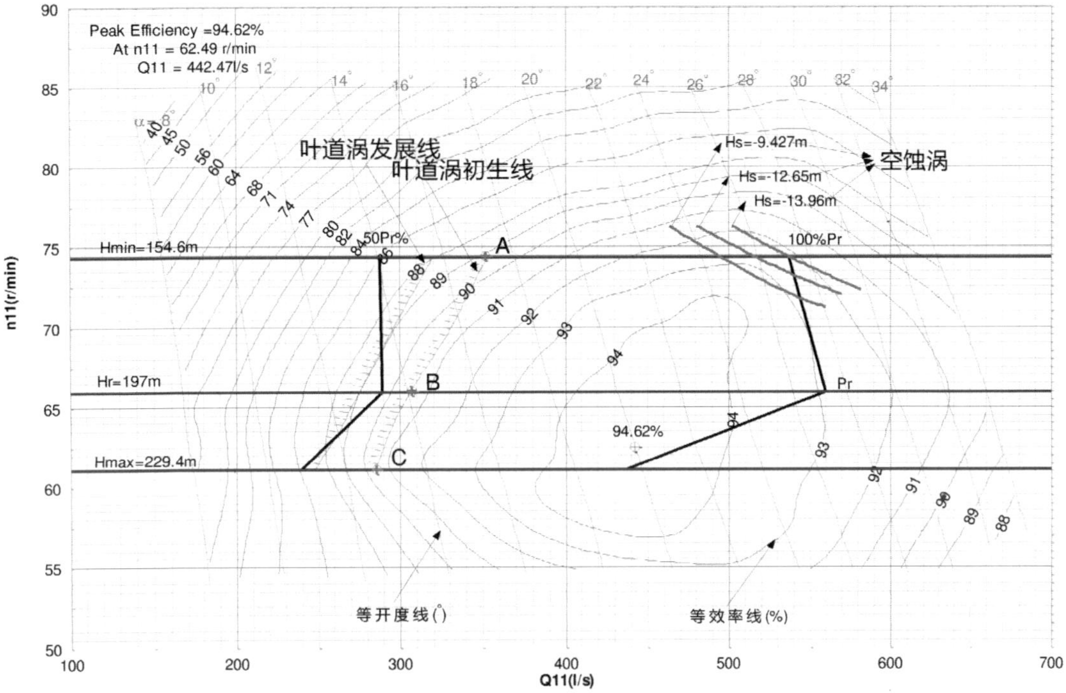

Figure 4. Blade channel vortex and cavitation vortex curve

Characteristic of blade channel vortex can be analyzed by velocity triangle. In figure 4, the curve of variable speed with identical water head marks two operating condition points A and B of the blade channel vortex. Compared with point A, unit speed n_{11}, unit discharge Q_{11} and opening a, decreasing entirely at the point B. As shown in Figure 5 of the velocity triangle, the incidence angle of the point B is larger than that of the point A, and the absolute velocity of the point B is larger than that of the point A. So, two velocity triangles are neither equal nor similar for two operating points A and B.

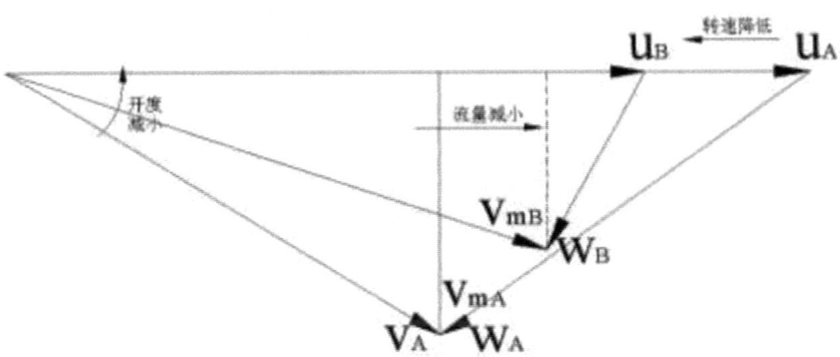

Figure 5. Velocity triangle of blade channel vortex at operating condition A and B

According to test results, the blade channel vortex is expressed by the blade channel vortex inception curve and development curve. No existing absolute symmetric flow in the hydraulic turbine, under the action of the spiral casing and the stay vane relative symmetric flow can occur only in optimum operating condition. For this reason, forming condition of the blade channel vortex is bound to emerge in a circumferential direction under low flow operating condition, and results in blade

IOP Conf. Series: Materials Science and Engineering **129** (2016) 012038 doi:10.1088/1757-899X/129/1/012038

channel vortex inception. With the flow further decrease, all of blade channels of runner will reach blade channel vortex forming condition, which is called as the blade channel vortex development curve.

The author thinks that the blade channel vortex is induced by flow separation only is a superficial phenomenon. When the flow rate drops to a certain extent with an appropriate incidence angle, is the fundamental reason of the blade channel vortex forming.

4. Numerical results of blade channel vortex

Numerical computation was carried out for the three points A, B and C in figure 4, and blade channel vortex appearance and corresponding development was analyzed, as shown in figure 6. Flow pattern of the blade channel vortexes and their evolving rule in the three operating condition points is not the same. In the operating condition point A, the blade channel vortex originates in the flow separation of the runner crown, the blade channel vortex of the point C roots in the flow separation of the blade leading edge, while the blade channel vortex flow state of the point B is not belong to the conventional category.

The common features of the different kinds of blade channel vortexes are obvious streamline drop appearance in the runner passage. In addition, the blade channel vortex flow state in the development stage is more turbulent and disordered than that in the inception stage.

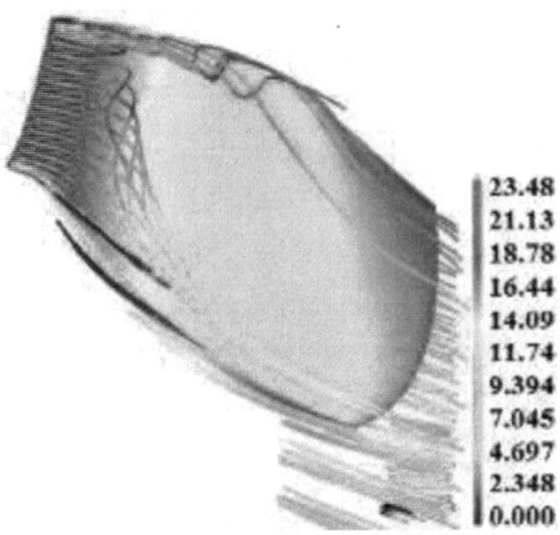

(a) Vortex inception of point A

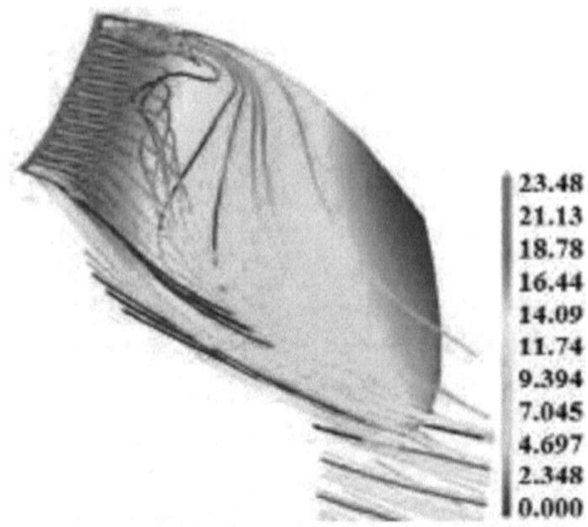

(b) Vortex development of point A

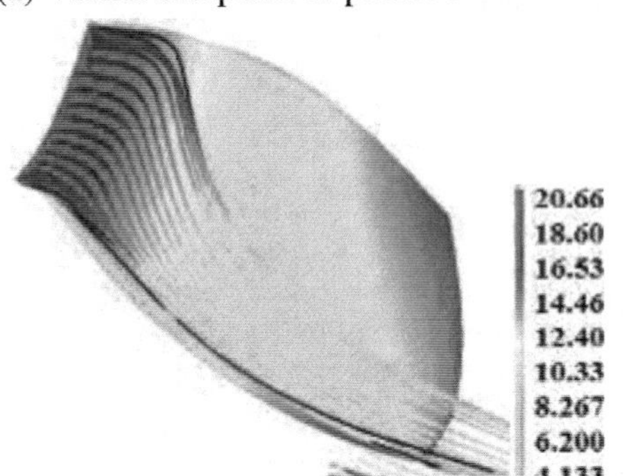

(c) Vortex inception of point B

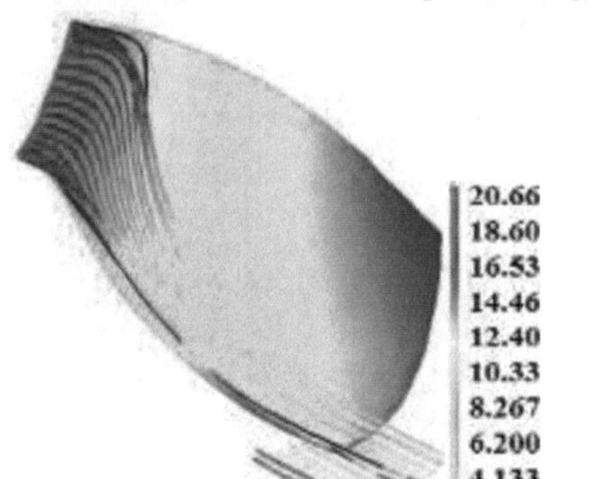

(d) Vortex development of point B

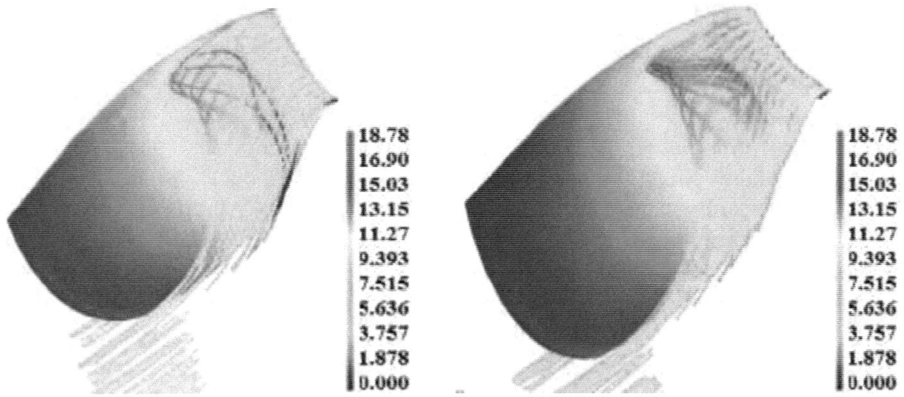

(e) Vortex inception of point C (f) Vortex development of point C

Figure 6. Numerical results of different blade channel vortexes

5. Hydraulic design considerations for blade channel vortex

Although the turbine blade channel vortex cannot be eliminated, in a certain water head range the vortex characteristics can be improved in hydraulic design stage, namely it can be shifted in a certain range towards small flow rate direction.

Two different runner schemes for a power station are shown in figure 1(b) and (c) respectively. From the efficiency and blade channel vortex point of view, the scheme in figure 1(b) is superior to that in figure 1(c), Not only its efficiency is higher, but also the blade channel vortex is shifted outside the 50% Pr, yet part of the blade channel vortex development curve in figure 1(c) is within 50% Pr.

Generally, there is a positive incidence angle in the runner inlet at the design condition. Numerical and experimental results showed that the larger positive incidence angle, the earlier appearance of blade channel vortex, on the contrary, the smaller positive incidence angle, the latter emergence. While the magnitude of positive incidence angle will affect the level of optimum efficiency and restrict the offset of the optimum operating condition point. Besides, the evolution of blade channel vortex is impacted by the runner passage profile, the narrower turbine meridional passage contour, the better adaptability of turbine to low flow rate, and the inception curve of blade channel vortex will shift toward the low flow rate condition, and vice versa. Of course, the decreasing of meridional passage sectional area will give rise to the reduction of the optimum efficiency. So the engineers must conditionally select various hydraulic parameters based on the actual situation of hydropower plant in order to realize the excellent comprehensive performance of hydraulic turbine.

6. Effect of blade channel vortex on the turbine unit stability

Generally, the blade channel vortex doesn't cause serious erosion damage to the blade surface as the cavitation. However the pressure fluctuation amplitude of the turbine will significantly increase in small guide vane opening condition. It's a blind subjective judgment that the blade channel vortex inevitably lead to the turbine unit instability, and it's too stringent in a bidding document that Forced to exclude blade channel vortex from the turbine unit stability operation range[7-9]. In fact, for high water head turbine it's uncomplicated to eliminate blade channel vortex under 40% rated output power, while extremely difficult excluding below 60% rated output power for low water head turbine, so it's essential to make a different treatment for the blade channel vortex in different water head range[10] .

The test and operation experience indicates that at small guide vane opening condition the blade channel vortex will form in the runner passage, and result in high pressure fluctuation amplitude, but its frequency characteristics generally take the forms of typical draft tube vortex rope frequency. As long as the blade vortex did not induce hydraulic resonance or damage, and turbine unit vibration, noise, fatigue and other indicators is in the stability requirements, it is believed that the blade channel vortex is stable, and the units can operate stably in the condition of small guide vane opening and a short period.

7. Conclusion

(1) The blade channel vortex is an intrinsic property for Francis turbine under small flow rate condition.

(2) The blade channel vortex is an inevitable response of runner passage inadaptation for the small flow rate, and the blade leading edge stall in small guide vane opening is just one of the causes of inducing blade channel vortex.

(3) Despite identical regulation and trend of the blade channel vortex for high water head and low water head turbine, the vortex of the high water head turbine can generally be excluded from the stable operation region, while the one of the low water head turbine to achieve this target is more difficult.

(4) Different turbine meridional passage contour has inhibited the inception and development of the blade channel vortex in a certain extent, however, cannot be thoroughly eliminated.

Acknowledgement

The research was supported by The National Natural Science Foundation of China (51479166) and the Specialized Research Fund for the Doctoral Program of Higher Education of China (20126118130002).

Reference

[1] Shi Q H, Xu W W and Li G 2006 Noise reduction in a low head Francis turbine caused by runner inter-blade vortices *Proc. of the 20th IAHR Symp. on Hydraulic Machinery and Systems*(Yokohama, October)

[2] Xiao Y X, Wang Z W, Zhang J, Peng G J, Liu D Y, Ma X H and Xiao M 2009 Numerical Analysis of Blade Channel Vortex in Francis Turbine at Part Load of Middle-Low Head *ASME 2009 Fluids Engineering Division Summer Meeting*(Vail, Colorado, USA, August 2–6)

[3] Zhang H M and Zhang L X 2013 Numerical simulation of blade channel vortex in a low head Francis turbine *International Conf. on Sustainable Energy and Environmental Engineering*(Guangzhou China December 29-30)

[4] Amet E, Maître T, Pellone C and Achard J L 2D numerical simulations of blade vortex interaction in a darrieus turbine *J. Journal of Fluids Engineering* **(31)**

[5] Chen J X, Li G W, Liu S Z and Lv G P *2007* The occurrence and influence of the interblade vortex on the hydraulic turbine stability *J. DA DIANJI JISHU* **(3)** 42-46

[6] Chen J X, Li G W, Liu S Z and Lu G P 2007 The occurrence and the influence of the inter blade vortex on the hydraulic turbine stability *J. Large Electric Machine and Hydraulic Turbine,* **(3)** 42-46

[7] Chu Y T 2006 Exploring effective measures to increase stability of hydro turbines *J. Water Power,* **32(3)** 45-47

[8] Lu L, Zhang L D, Yang J and Xu Z J 2014 Reduction and Elimination of Channel Vortex in Francis Turbine *J. Jinlin Water Resources* **(9)** 50-53

[9] Wu G K, Luo X Q, Feng J J and Li W F 2015 Cracking reason for Francis turbine blades based on transient fluid structure interaction *J. Transactions of the Chinese Society of Agricultural Engineering* (31) 92-98

[10] Wang L H 2011 *Vibration and Analysis of Water Turbine Generating Unit* (Zhengzhou: The Yellow River Water Conservancy)

Numerical investigation of cavitation performance on bulb tubular turbine

L G Sun [1], P C Guo[1], X B Zheng[1] and X Q Luo[1]

[1] State Key Laboratory Base of Eco-hydraulic Engineering in Arid Area, Xi'an University of Technology, Xi'an 710048, Shaanxi Province, China

E-mail: 15191587315@126.com

Abstract. The cavitation flow phenomena may occur in the bulb tubular turbine at some certain operation conditions, which even decrease the performance of units and causes insatiably noise and vibration when it goes worse. A steady cavitating flow numerical simulations study is carried out on the bulb tubular unit with the same blade pitch angle and different guide vane openings by using the commercial code ANSYS CFX in this paper. The phenomena of cavitation induction areas and development process are obtained and draws cavitation performance curves. The numerical results show that the travelling bubble cavity is the main types of cavitation development over a wide operating range of discharge and this type of cavitation begins to sensitive to the value of cavitation number when the discharge exceeding a certain valve, in this condition, it can lead to a severe free bubble formation with the gradually decrement of cavitation number. The reported cavitation performance curves results indicate that the flow blockage incident would happen because of a mount of free bubble formation in the flow passage when the cavity developed to certain extend, which caused head drop behavior and power broken dramatically and influenced the output power.

1. Introduction

Bulb tubular hydro power station is one of the best ways to develop low water head hydropower resources due to some distinguished features of bulb tubular turbine such as excellent technical and economical characteristics and widely applicability, it has become a hot issue and focus for the bulb tubular turbine in the field of development, design and operation technology and experience. Furthermore, the whole flow passage of the power station is basically in the axial direction with less hydraulic loss and convenient for construction, and strong discharge capacity higher specific speed are another particular advantage on the basis of equal water head and power, which also result in a great decrease of the units[1]. Cavitation may be observed in various engineering systems and is usually the main physical phenomenon behind performance alteration in hydraulic machinery, and modern design requirements lead to more compact machines with higher rotation speeds and cavitation risk. However, it's a great difficult and changing task to predict such complex unsteady and two-phase flows as well as its inception and development with an acceptable accuracy [2].

The paper [3] presents the cavitation phenomenon characterized by fluid machinery including type of cavity development related to the specific speed of machines in both pump and turbine mode, the influence of the operation conditions, such as load, head and submergence, also, the influence of cavitation development on the machine efficiency, operation and integrity are discussed. In paper [4], authors has put forward a new "full cavitation model" for performance predictions of engineering equipment under cavitation flow conditions with consideration of effects of turbulent fluctuations and non-condensable gases. In addition, final validation results towards this new cavitation model are presented for flows over hydrofoils, submerged cylindrical bodies, and sharp-edged orifices. Another application of the full cavitation model to pumps and inducers is given for instance by paper [5] by means of simulations of cavitating flows in three types of machines: water jet propulsion axial pump, a centrifugal water pump, and an inducer from a LOX turbo pump. The results show cavitation zones on the leading edge-suction side of each of the machines as expected and all the test cases with cavitation show plausible results. In paper [6], unsteady cavitation turbulent flow around the 3D twisted Delft hydrofoil was simulated by using the Partially-Average Navier-Stokes(PANS) method and a mass transfer cavitation model with the maximum density ratio effect between the liquid and the vapor, the numerical results indicate that the cavity volume fluctuates dramatically as the cavitating flow develops with cavity growth, destabilization and collapse, which presents good agreement with the existing experiment observations. The paper [7] deals with the basic investigation towards the time-dependent cavitating flow around an inclined 2D profile to develop the suitable bubble dynamic model and describes the application of this model to the prediction of the time-average leading edge cavitation and the head drop behavior of a centrifugal pump. Paper [8] employed numerically compare case of a bulb tubular turbine with blade tip gap, with hub gaps and without any gap and presents the impacts of above three types gaps on performance predictions, furthermore, the RANS simulations of the efficiency hill chart of a bulb turbine was investigated. In paper [9], a flow in a horizontal shaft bulb turbine is calculated as a two-phase flow with a commercial Computational Fluid Dynamics code including cavitation model. The results are compared with experimental results achieved at a closed loop test rig for model turbines.

This paper presents steady numerical simulations of cavitation flow study on the model bulb tubular unit with the same blade pitch angle and different guide vane openings by applying the Zwart cavitation model derived from a simplified Rayleigh-Plesset equation using the commercial code ANSYS CFX with the SST turbulence model. The phenomena of cavitation induction areas and development process are obtained and drawn cavitation performance.

2. Calculation model and grid generation

2.1. Mode and computation conditions

Hydraulic losses derived from design of flow passage constitute the majority of weight due to a low water head and large discharge for the bulb tubular turbine. It necessitates a reduction in hydraulic loss with reasonable designing the shape and size of the flow passage, as well as low project cost. In this paper the basic parameters of the investigated model turbine is reported in table 1. The schematic diagram of the flow passage corresponding to table 1 is showed in figure 1, and figure 2 presents the three-dimensional model for the simulation domain, which is the application of the three-dimensional modeling software to generate the calculation domain. Figure 2 shows that the unit is composed of 5 parts, which are respectively the diversion chamber, the bulb body, the guide vane, the runner and the draft tube.

Table 1. Basic parameters of the investigated turbine

Parameters	Remark	Parameters	Remark
Runner diameter(D_1)	350mm	Hub diameter(d_h)	140mm
Text head(He)	4m	Blade numbers(Z_1)	4
Guide Vane numbers(Z_2)	16	Guide vane height(b)	120.7mm

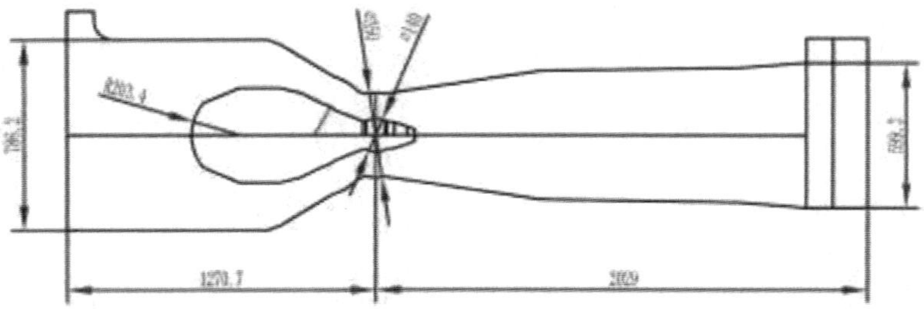

Figure 1. Schematic diagram of the flow passage

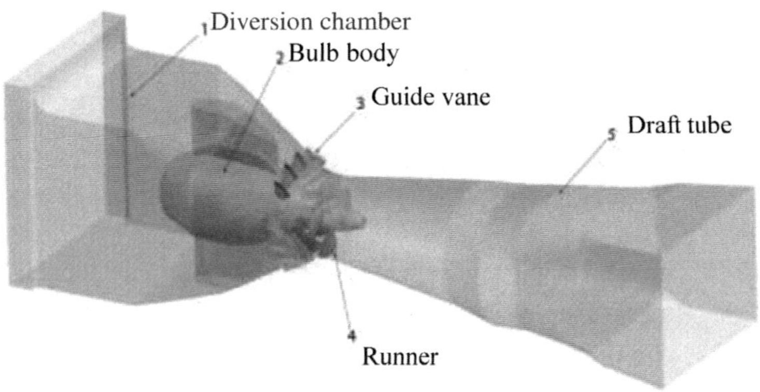

Figure 2. Three-diemnsional model of the investigated turbined

According to the cavitation flow characteristics of tubular turbine, it can be known that the cavitation is more susceptible to be triggered at higher mass flow rates, which is one key factor to select operation conditions, the another is the method of controlling vitiate. In order to comparatively simulate different trends of cavitating flows in the model turbine, two on-cam operation points that is featured by the same blade pitch angle and different guide vane openings are selected to carried out the numerical calculation, corresponding to points A and B in figure 3. The figure indicates that those two operating conditions share the same blade pitch angle of 25 degree, while respectively 90% and 102.5% of the guide vane openings.

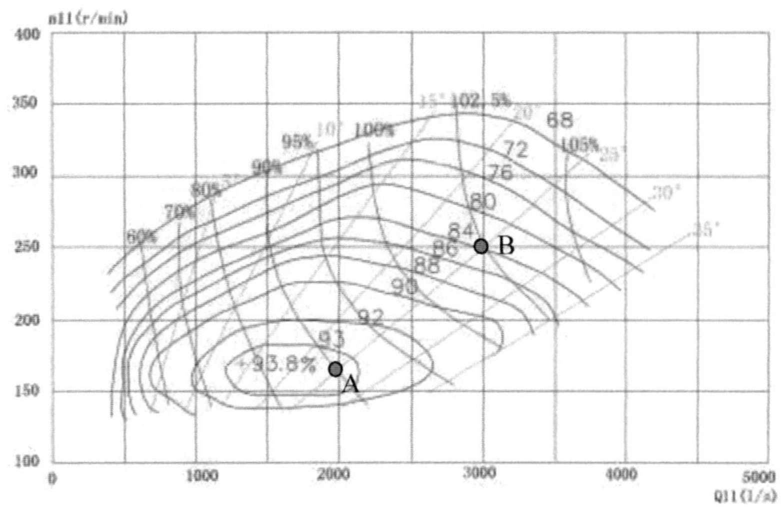

Figure 3. Calculation operation conditions

2.2. Grid generation

One of the most time-consuming tasks in CFD is the generation of grid in the computation domain, particularly for hydraulic machinery, the flow passage components is not only a complex space surface, but also the periodic and non-periodic distribution of complex surfaces. The computation model is discretized into multi-block structured grids based on the grid generation tool ICEM CFD. Figure 4 shows the grid generation of the investigated turbine, an H&L topology is employed in the blade flow passage and an O-grid is created around the blade, which will provide opportunities of excellent boundary layer resolutions. The grids are well refined at the zones around the walls in order to more accurately simulate the viscous flow near the solid wall.

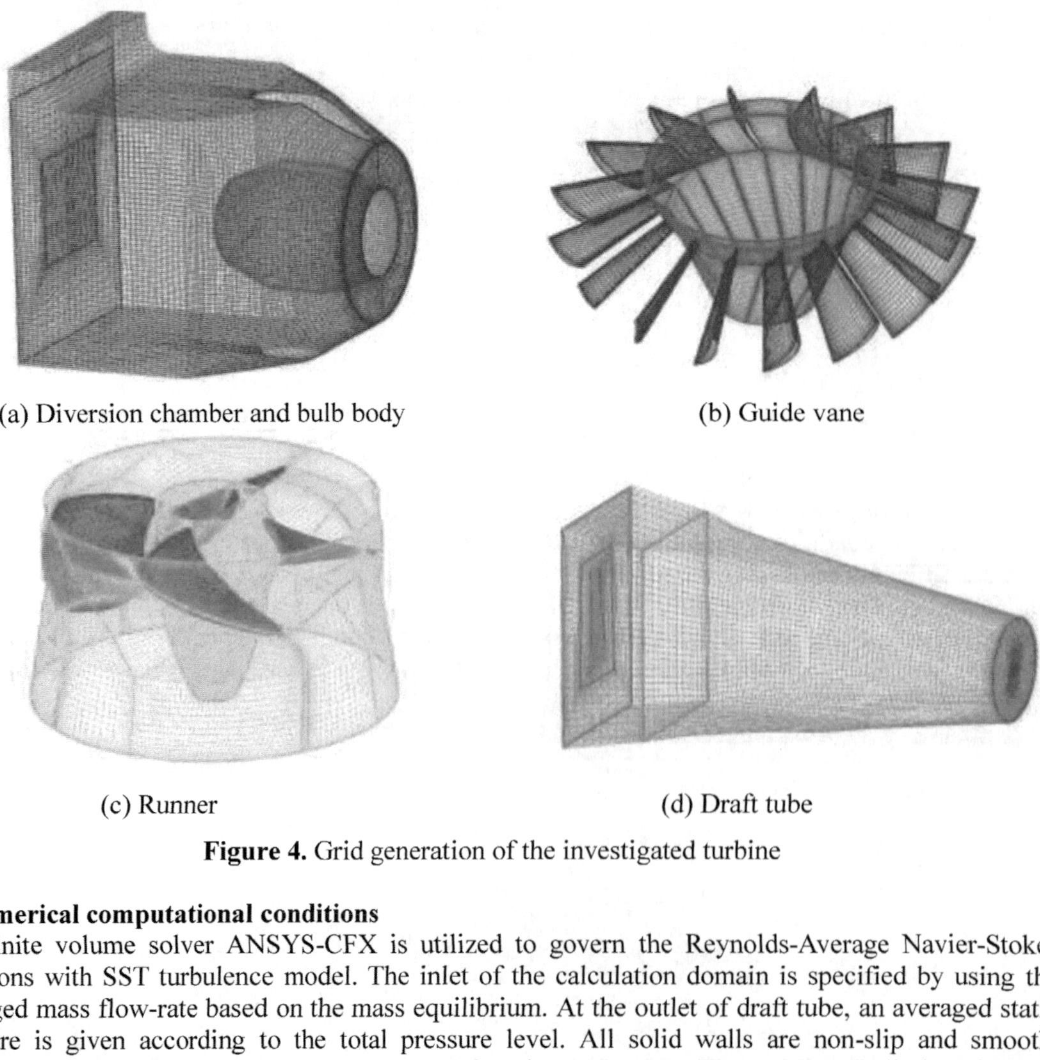

Figure 4. Grid generation of the investigated turbine

3. Numerical computational conditions

The finite volume solver ANSYS-CFX is utilized to govern the Reynolds-Average Navier-Stokes equations with SST turbulence model. The inlet of the calculation domain is specified by using the averaged mass flow-rate based on the mass equilibrium. At the outlet of draft tube, an averaged static pressure is given according to the total pressure level. All solid walls are non-slip and smooth. Defining the interfaces between rotating parts and stationary parts as "Frozen Rotor" slip interfaces at steady-state computation, which treats the flow from one component to the next by changing the frame of reference while maintaining the relative position of the components and has the advantages of being robust, using less computer resources.

In addition, different cavitation development trends in runner already mentioned are observed by means of changing the absolute pressure of draft tube outlet, the "Schnerr-Sauer" cavitation model based on homogeneous muti-phase transport equation is employed for cavitation simulation with vaporization pressure 3540Pa at 25℃, other empirical constants are taken as proposed by ANSYS.

4. Cavitation flow results analysis

Design and operation of hydraulic turbines are strongly related to cavitating flow phenomenon, which may appear in either the rotating runner or the stationary parts. The cavitation development may be the origin of several negative effects, such as noise, vibrations, performance alterations, erosion and structural damages. The cavitation flow numerical results of vapor volume fraction on runner blades and cavitation performance curves are respectively reported from figure 5 to figure 10 at two operation conditions A and B. In figure 5, the cavitation zones are enlarged for better views, and the inlet edge of runner blade on the left in figure 6, figure 7 and figure 8, σ is cavitation number.

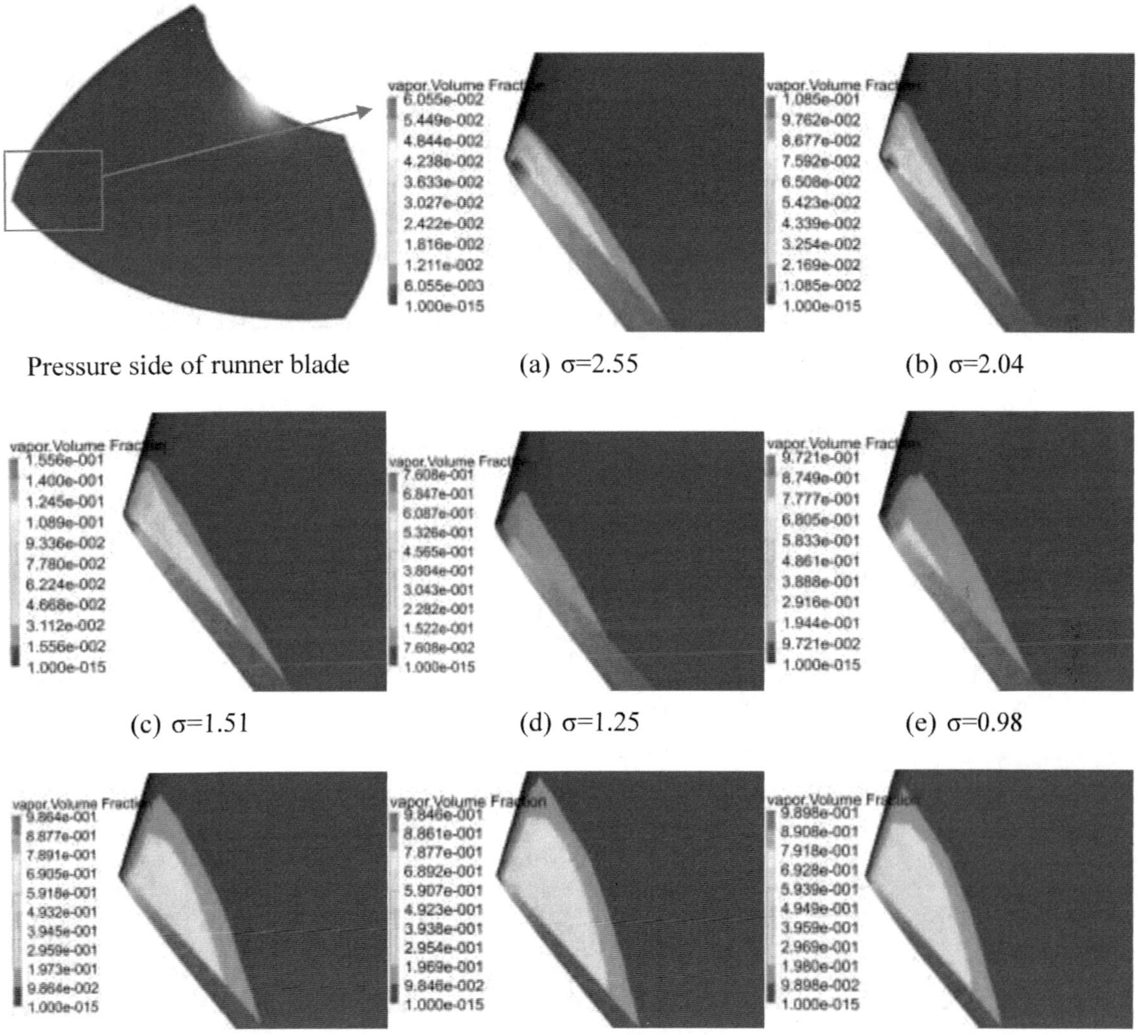

Figure 5. Vapor volume fraction on pressure side at operation condition A with different σ

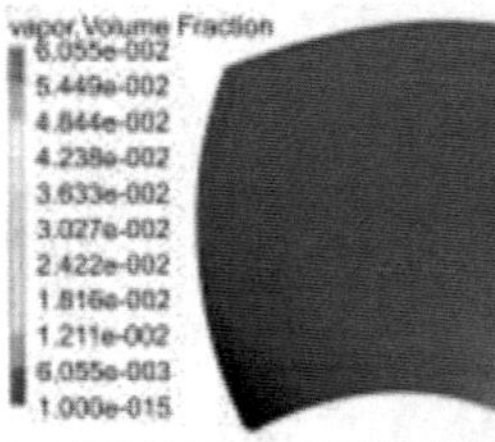

(a) σ=2.55

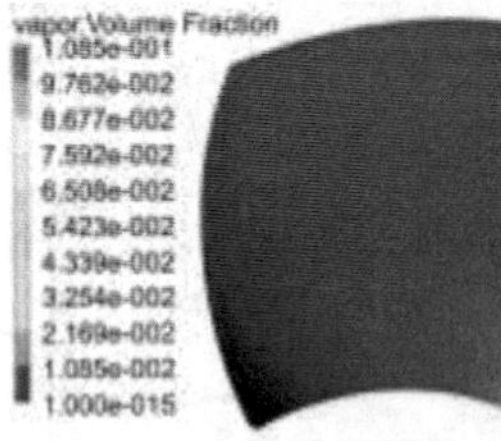

(b) σ=2.04

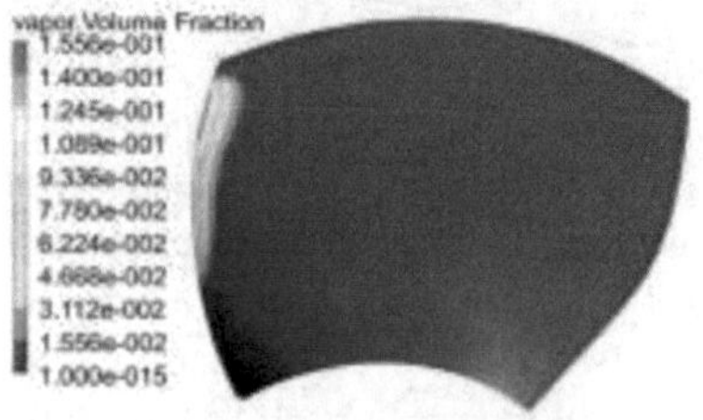

(c) σ=1.51

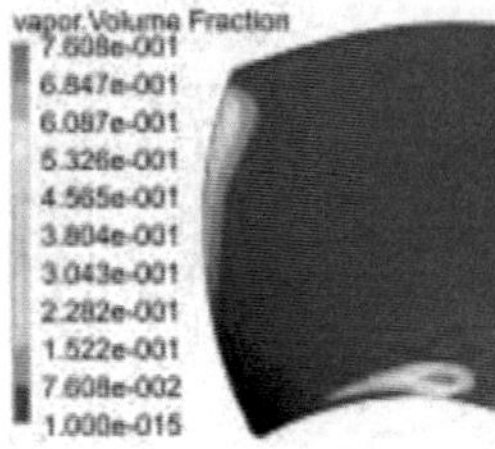

(d) σ=1.25

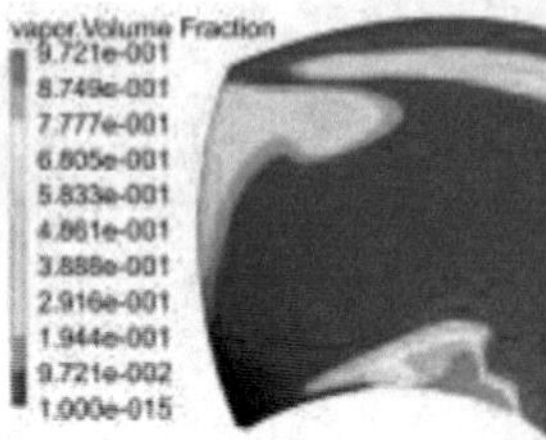

(e) σ=0.98

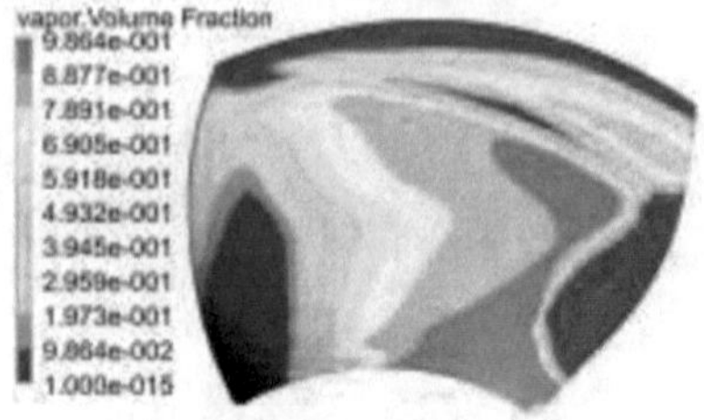

(f) σ=0.86

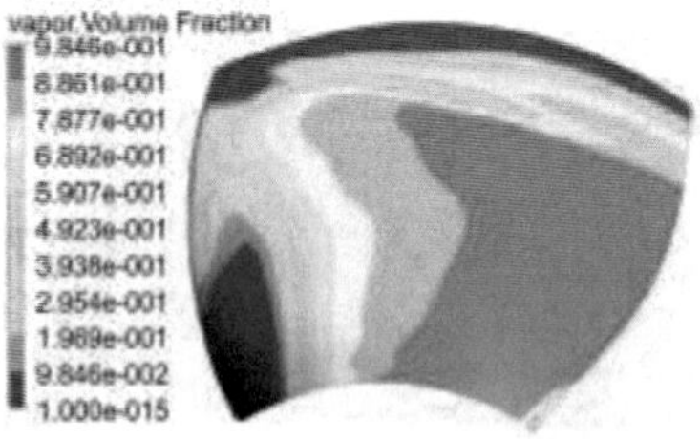

(g) σ=0.78

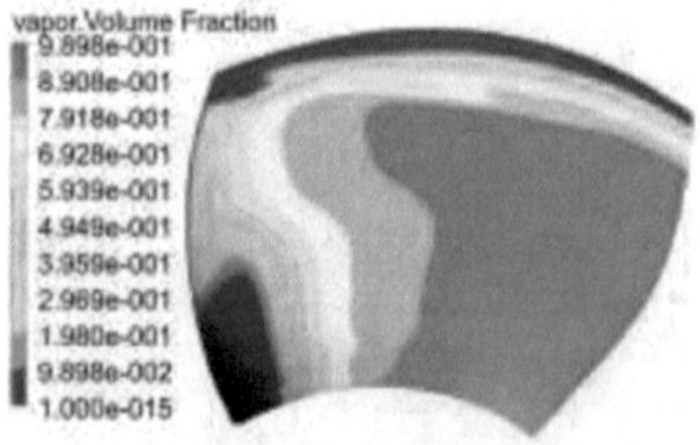

(h) σ=0.72

Figure 6. Vapor volume fraction on suction side at operation condition A with different σ

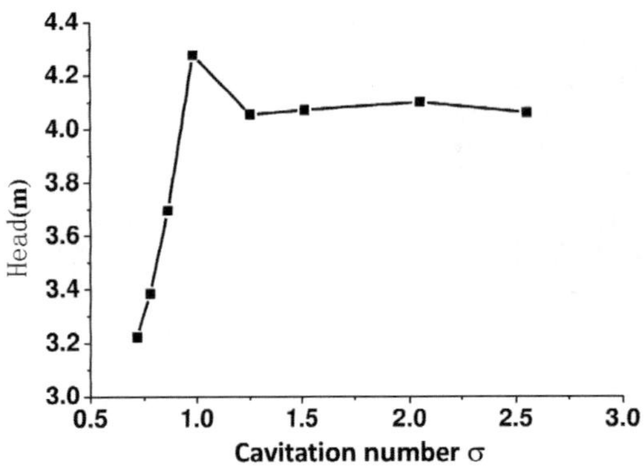

(a) Head drop curves

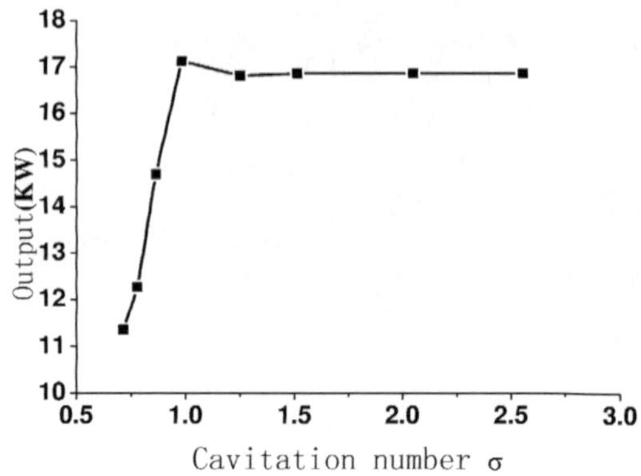

(b) Output broken curve

Figure 7. Cavitation performance curves at at operation condition A

IOP Conf. Series: Materials Science and Engineering **129** (2016) 012039 doi:10.1088/1757-899X/129/1/012039

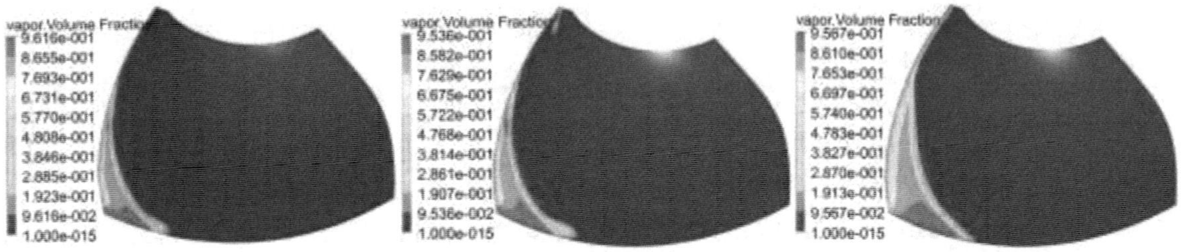

(a) σ=2.66 (b) σ=2.40 (c) σ=2.11

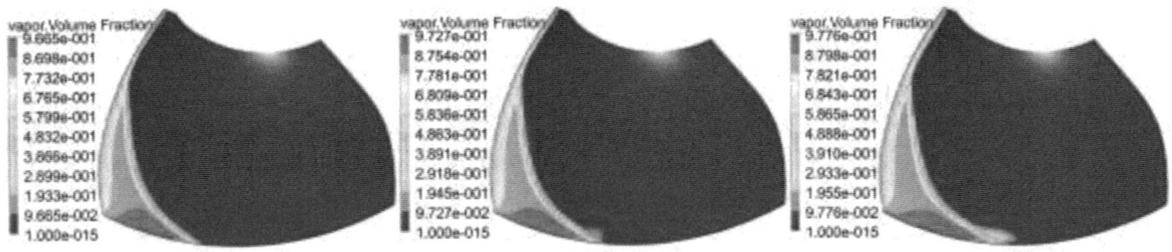

(d) σ=2.04 (e) σ=1.97 (f) σ=1.89

Figure 8. Vapor volume fraction on pressure side at operation condition B with different σ

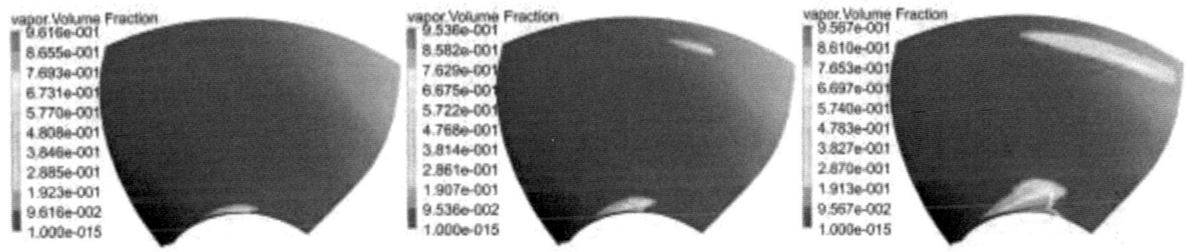

(a) σ=2.66 (b) σ=2.40 (c) σ=2.11

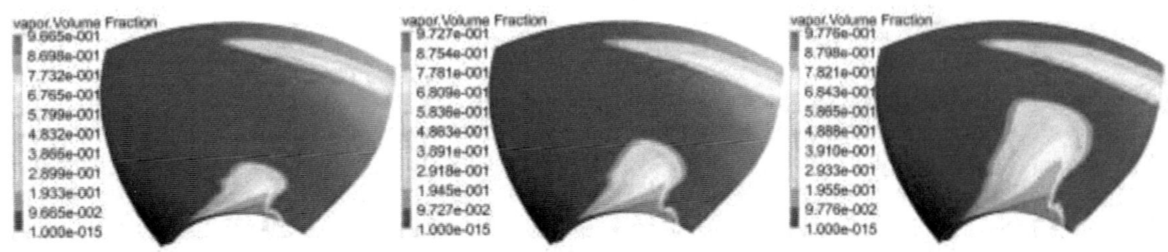

(d) σ=2.04 (e) σ=1.97 (f) σ=1.89

Figure 9. Vapor volume fraction on suction side at operation condition B with different σ

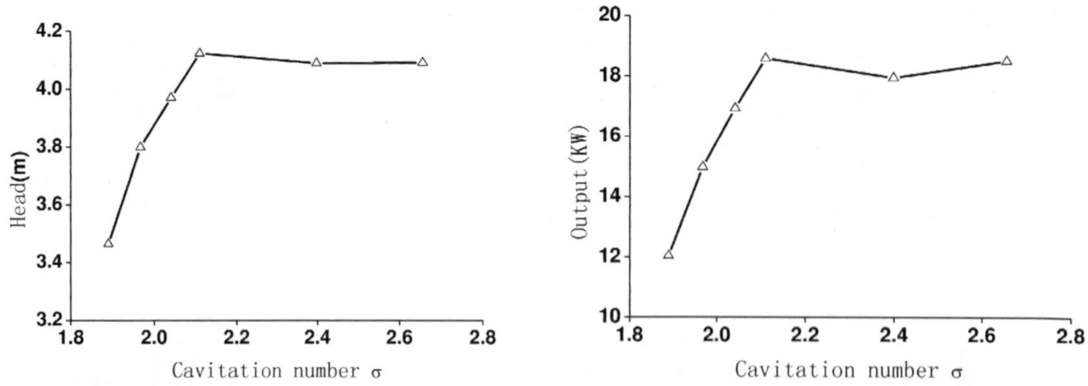

(a) Head drop curve (b) Output broken curve

Figure 10. Cavitation performance curves at at operation condition B

In generally, the majority of vapor can be observed on the suction side of the runner blades although is also occur on the pressure sides at operation condition A, especially towards high σ. It can only observe slight trend cavitation on pressure side due to small negative angle of attack is generated at condition A, while greater negative angle of attack would appear if the flow rate is much higher than the design valve, which results in flow separation and has an obvious influence on cavitation induction. Therefore, the vapor volume fraction on pressure sides at B condition is great than A. The volume of the regions affected by cavitation increases with decreasing cavitation number and increasing flow rates on suction sides a, while it has no significant change on pressure side.

The suction side is the main zones that cavitation occurs and a cavity development takes place at the leading edge of the runner blades firstly, then hub and in the vicinity of the shroud along with the direction of chord length at condition A, and eventually occupied the outlet edge of the runner blade. These leading edge cavities can be avoided by improving the hydraulic design of runner. For operation condition B, the development process of cavitation on the blades suction side can be concluded, the traveling bubble cavity development takes place at the hub of the runner and the cavitation cloud will be more serious at lower outlet pressure. An isolated and banded cavity cloud was observed at the blades tip, and the fully developed cavitation area would connect with the hub cavitation region when the cavitation number declined.

Usually some amount of cavitation is allowed during normal operation for the bulb tubular turbine as the cavitation performance shows in figure 8 and figure 10, which also indicate the influence of cavitation trends to external characteristics of the turbine. The water and output remains relatively constant when σ variates in a certain range, while cavitation performance is very sensitive to σ, which results in a sharp decrease if σ exceeds a certain value. Furthermore, water and output has seriously reduced in spite of light cavitation occuring in suction sides for higher mass flow rates by comparision analysis between figure 9 and figure 10 due to flow blockage incident derived from amount of free bubble formation in the flow passage.

5. Conclusion

In this paper, the three-dimensional numerical calculation of cavitation flow for a model bulb tubular turbine is carried out, the cavitation induction areas and development process are obtained and draws cavitation performance curves. The numerical results show that the travelling bubble cavity is the main types of cavitation development on the suction sides of runner blades and this type of cavitation begins to sensitive to the value of cavitation number when the discharge exceeding a certain valve, which leads to a sharp reduction in performance.

IOP Conf. Series: Materials Science and Engineering **129** (2016) 012039 doi:10.1088/1757-899X/129/1/012039

6. Reference

[1] Tian S T 2003 *Tubular turbine generator set and its selection method* (Beijing: China Electric Power Press)

[2] Pan Z Y and Yuan S Q 2013 *Fundamentals of cavitation in pumps* (Zhenjiang: Jiangsu University Press)

[3] Avellan F 2004 Introduction to cavitation in hydraulic machinery The 6th International *Con. on Hydraulic Machinery and Hydro dynamics*(Timisoara, Romania 21-22 October)

[4] Singhal A K, Athavale M M, Li H Y and Yu J 2002 Mathematical basis and validation of full cavitation model *J Journal of Fluid Engineering.* **124** 617-24

[5] Athavale M M, Li H Y, Yu J and Singhal A K 2002 Application of the full cavitation model to pumps and inducers *J International Journal of Rotating Machinery.* **8(1)** 45-56

[6] Ji B, Luo X W, Wu Y L, Peng X X and Duan Y L 2013 Numerical analysis of unsteady activating turbulent flow and shedding horse-shoe vortex structure around a twisted hydrofoil *J International Journal of Multiphase Flow.* **(51)** 33-43

[7] Frobenius M and Schiling R 2004 Cavitation prediction in hydraulic machinery *22nd IAHR Symp. on Hydraulic Machinery and Systems.* (Stockholm, Sweden June 29-July 2)

[8] Guénette V, Houde S, Ciocan G D, Dumas G, Huang J and Deschênes C 2012 Numerical prediction of a bulb turbine performance hill chart through RANS simulations 26nd IAHR Symp. on Hydraulic Machinery and Systems. (Beijing, China 19-23 August)

[9] Necker J and Aschenbrenner T 2010 Model test and CFD calculation of a cavitating bulb turbine. *25th IAHR Symp. on Hydraulic Machinery and Systems.* (Timişoara, Romania 20-24 September)

Numerical simulation analysis and optimum design for combined type pressure reducing valves

D M Gou[1], P C Guo[1], X B Zheng[1], X Q Luo[1] and L G Sun[1]

[1] State Key Laboratory Base of Eco-hydraulic Engineering in Arid Area, Xi'an University of Technology, Xi'an 710048, Shaanxi Province, China

Email:guoyicheng@126.com

Abstract. Pressure reducing valve is an extremely significant equipment of energy dissipation for the water supply by gravity with pressure reducing technology in hydropower stations, and which has a pronounced effect on the normal technical water supply even safety operation for the hydropower units. A three-dimensional numerical calculation of flow field and cavitation characteristics towards a combined type pressure reducing valves was carried out based on the system of technical water supply in this paper. The numerical results show that the investigated valve could meet the requirements of technological supply water pressure and great pressure loss was caused when the water flow was accelerated by narrow overflowing section between throttling cone and valve seat. At working operation, obvious cavitation phenomenon was observed on the surface of throttling cone, and the maximum volume fraction of vapor reached 0.537%. Based on above researches, this paper introduces an optimization model for profile line design of throttling cone. The optimal results show that the cavitation performance is effectively improved with identical pressure drop compared with original results.

1. Introduction

The supplied water of the water supply in hydropower stations is for various equipment which is working, including air cooler and bearing cooler for hydroelectric generators, water cooling transformer cooler and air compressor cooler for hydraulic turbines and so forth. The main function and role of technical water supply is applied to the cooling and lubrication of all kinds of facilities consumed water, which have some requirements for water, such as water quantity, water temperature, water pressure and water quality. The application of water supply by gravity with pressure reducing technology to hydraulic station with high head is becoming a new trend because of its advantages of simple equipment, no rotating parts, stable operation, convenient maintenance and less investment, more importantly, this way shows reliable technique and better economic benefit, thus is increasingly enjoyed by engineering designers in high-head hydro plants[1].

For the water supply by gravity with pressure reducing technology of high head stations, the water intake is generally located in both sides of the steel penstock with an inclination angle of 45 degree. Given water pressure of water intake is far higher than the technical water supply required, it is easy to see it is essential to install a reliable depressurizing facilities, that is the pressure reducing valve, to release excess pressure. The selection and application of pressure reducing valves turn out to have an even more important bearing on the water supply system of gravity flow for reducing pressure, which will provide appropriate cooling water with lower maintenance costs and labor intensity in a very long time.

At present, numerical simulation analysis technology has gradually become an important auxiliary tool for engineering design since the rapid development of computer technology. Performance characteristic of the valve can be designed and optimized by engineers with this technology, which can help avoid blind experiment and make the design more reasonable. In this way, the manufacturer could fully utilize limited development resources to develop high quality valve products and improve the competitiveness of product[2]. In the literature[3], an internal flow numerical simulation analysis was carried out towards a pressure reducing valve with high pressure difference, the results obtained show that numerical simulation technology can be used to stimulate the complex flow of the valve, furthermore, an optimization scheme was posed on the basis of numerical results, which expected to lengthen the lifespan of valve. According to the structure and working principle of the pilot diaphragm valve, a mathematical model of this valve is established, and the hydraulic transient process of the water pipeline with pressure reducing valve is calculated in literature [4], the results show that the maximum water hammer pressure of the relief valve is significantly reduced, and the pressure valve on the inlet and outlet of the valve and its change process is obviously different. Literature[5] took a three-dimensional transient flow computation in the annular orifice of pressure reducing valve applied in the rocket attitude control system and obtained some parametric drawings about the flow fields. Literature[6] applied the finite volume method of nonstructural and non staggered grid to simulate separated flows towards three kinds of common valves in the pipeline system, the results show that with the decrease of the valve opening, two vortices were formed in the back of the butterfly valve and two sides of ball valve, further experiments indicate the downstream noise induced by water flow is greater than the upstream and the noise mainly originated from the vortex sound. Based on two-phase cavitating flow control equation and RNG k-ε turbulence model, literature[7] numerically analyzed cavitation erosion and cavitation characteristics of a liquid control valve, the numerical investigation is in agreement with the experimental ones, the flow velocity increase sharply when the fluid flows through the position between the throttling core and valve seat, which lead to a sudden pressure lowering and induced cavitation phenomenon. Literature[8] qualitatively presented force situation of important parts of a butterfly valve under different opening by using the commercial software Fluent, the numerical results show that the flow pattern is better at the position of fully open and large opening, the subnormal pressure areas can be observed in the posterior of valve plate , where the velocity gradient increase evidently. In literature[9], the generation model of valve flow noise in gas transmission pipeline is established, and the steady and unsteady flow field of gas flow through the valve is simulated, the results show that the fluid turbulent fluctuations is the fundamental reason of the fluid noise, in addition, the simulation results are verified by experiments, which show that the feasibility of application of Fluent to simulate the flow field and acoustic field. The authors study the fully automatic performance and life test system with a special interest on the comparatively analysis of internal flow fields before and after retrofitting the profile line of the valve cone in literature [10], the optimization results indicate that modified curved surface type cone can achieve steady flow with a reduction in drag losses.

In this paper, a three-dimensional numerical calculation of the pressure reducing valve of the technical water supply system is carried out, and the internal flow characteristics of the valve are analyzed. Based on the steady calculation results, the cavitation performance is investigated by FLUENT software. According to the analysis results of cavitation performance of the valve, the throttle cone profile that vitally influence the valve performance is modified by profile line optimizing. Furthermore, a comparative analysis of the flow field characteristics and cavitation characteristics between the original results and optimized numerical results are performed.

2. Calculation model and grid generation

The combined pressure reducing valve are highly thought by users owing to their high ratio of reduced pressure, large flow, strong adaptability to inlet pressure , low noise and long service life. This type valves are comprised of main valve and its corresponding feedback and realize automatically regulation to keep the outlet pressure always the formerly adjusted low-pressure valve.

Figure 1 shows the overall assembly of the combined type pressure reducing valve, which consists of body, valve seat, spindle and throttling cone, and whose basic parameters are listed in table 1. Three-dimensional modeling software is utilized to generate calculation model by means of surface rotation, stretching, fillet and so on. Figure 2 presents the three-dimensional sectional drawing of the investigated valve and the assembly relationship of throttle cone and valve seat. The working principle of the valve also shows in figure 2, the flow rate and outlet pressure were regulated by changing the openings of throttle cone, which would reduce the outlet pressure with decreasing area of passage. The valve would realize automatically adjustment of rated pressure and flow rate by the energy of the flow medium itself when variation of pressure or flow rate in outlet were detected, so as to ensure the outlet pressure an original preset value.

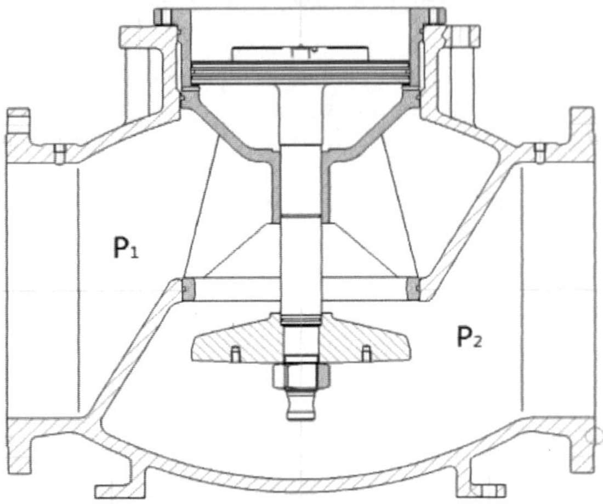

Figure 1. The assembly view of the model

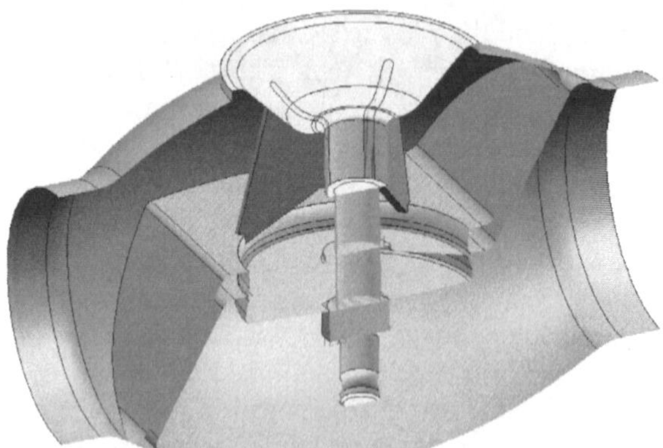

Figure 2. The 3-d sectional drawing

Table 1. The basic parameters of pressure reducing valve

Parameters	Remark	Parameters	Remark
Nominal diameter(DN)	450mm	Nominal pressure(PN)	2.5MPa
Inlet pressure(P_1)	1.56-2Mpa	Outlet pressure(P_2)	0.6MPa
Flow rate(Q)	1600m^3/h	Opening(H)	28mm

The generation of computational grid is the most crucial and time-consuming tasks, which accounts for the majority of the workload for the numerical calculation. The quality of the grid has direct affect on the numerical calculation accuracy and the reliability of solution results. Also, the accuracy and convergence rate of numerical calculation have a good advance by using a set of high quality grid. For the computational model of the pressure reducing valve, the grid generation tool ICEM-CFD is used to create the unstructured grids for the calculation domains because of good adaptability to complex model. In the addition, the method of local mesh refinement of critical region has been introduced in this paper for more credible flow calculation results, for example the end surface of the throttle cone, which seriously influence the valve flow pattern. Figure 3 shows the entire schematic diagram of the computational grid and the internal grid. The total number of grid nodes is 314,000, and the tetrahedral mesh is 1,800,000.

(a) Entire gird (b) Internal grid

Figure 3. Grid generation of the investigated pressure reducing valve

3. Numerical computational conditions

In order to simulate the water current movement more realistically, it is particularly essential to keep the flow pattern in inlet and outlet of valve as much as steady, therefore the valve were equipped with two sufficiently long pipeline on inlet and outlet. As a general rule, the length of the pipelines are setup as 5~10 times of the DN(nominal diameter=450mm) on the basis of the engineering experiment we have learned, in this paper the length of the pipes is 5 meters respectively for the investigated domain. The numerical model of ANSYS CFX is shown in figure 4.

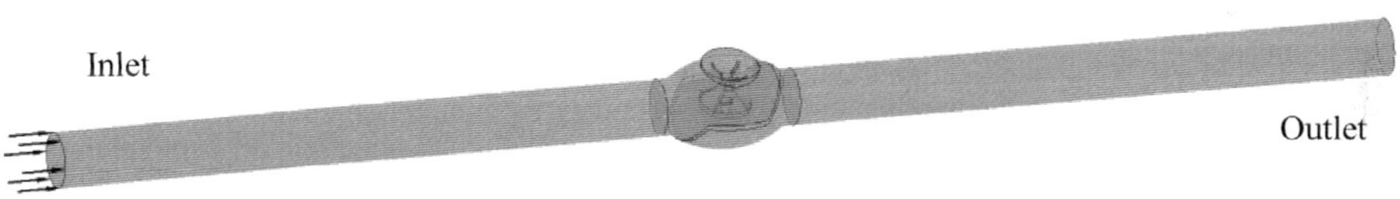

Figure 4. Numerical model of the investigated valve

The flow field and cavitation numerical computation are carried out using ANSYS Fluent, given an inlet total pressure and a mass flow rate at outlet, the average Reynolds equation is closed with the standard k-ε turbulence model. In the near wall region automatic near wall treatments were employed, at the same time, no-slip and smooth conditions for solid boundary, and GGI(General Grid Interface) connections for the interfaces between the valve and pipes.

On the other hand, the "Schnerr-Sauer" cavitation model based on homogeneous muti-phase transport equation was employed for cavitation simulation with vaporization pressure 3540Pa, and bubble number density 1e+13, as well as zero water vapor volume fraction of inlet and outlet.

3. Numerical results analysis

3.1. Flow field characteristic analysis

Figure 5 shows the valve pressure distribution and velocity distribution on the symmetry plane at the working condition of opening 28mm, as is shown in figure, inlet pressure of the valve uniformly distributes and outlet pressure is set around 0.6Mpa. The pressure drop of pressure reducing valve mainly concentrate in the narrow flow passage between the throttle cone and valve seat, where observes obvious pressure gradient, which shows that the flow pattern of the region is relatively not stable, and presents significant pressure decrement.

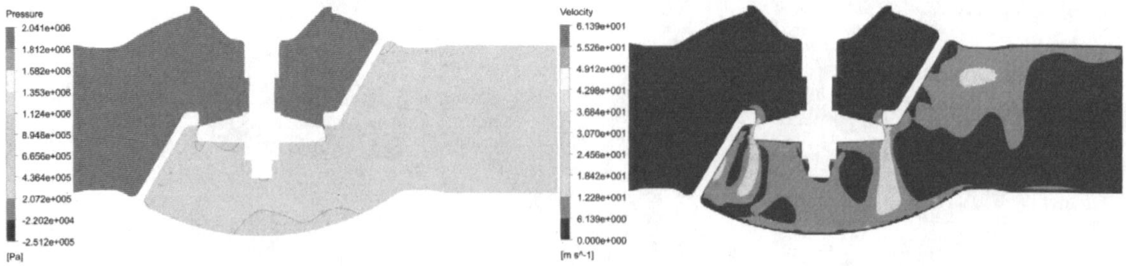

(a) Pressure distribution (b) Velocity distribution

Figure 5. Pressure and velocity distributions on the symmetry plane at H=28mm

For the velocity distribution, the influence of incoming originated form upstream on the inlet velocity seems to be neglected, water flow velocity increased gradually with decreasing area of passage, and it would reach its extreme points finally. The spindle serves as the boundary for the velocity distribution of upstream and downstream as shown in figure 5, high speed water flow directly flows into steel pipe on the right side of the spindle, while it is distributed as vortices at the upstream region. The low-pressure water flows along the superior wall of pipe in a great extent after flow through throttle cone, which leads to a remarkable pressure reduction in this region. Poor operation stability may be caused by the unequilibrium condition of forces acting on throttle cone because of appearance of larger vortices between the region of cone and valve seat, even induces body vibration and damage to the valve parts especially operating at several worst working situations.

High pressure distribution can be observed on the upper surface of cone owing to the impact action form upstream water, as reported in figure 6 form different views, and the extreme low pressure lies in end face of the cone. The minimum value of pressure is as low as 2550Pa, lower than the vaporization pressure, as a consequence, cavitation phenomenon may occur in the low-pressure region.

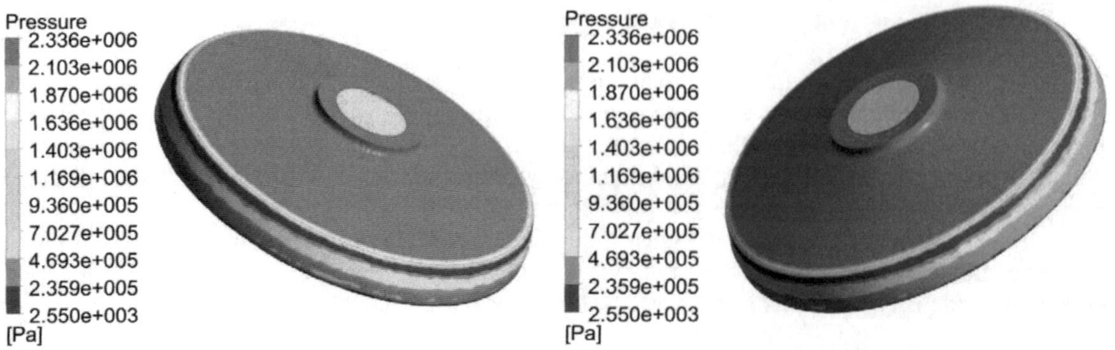

(a) Inlet direction (b) Outlet direction

Figure 6. Pressure distribution of the cone from different views

3.2. Cavitation characteristic analysis

Cavitation is the formation of vapor or gas cavities within a given liquid due to pressure drop, whose inception and development may be the origin of several negative effects, such as noise, vibrations, performance alterations, erosion and structural damages. Cavitation generally cause material breakage because of the impact action of collapsed bubble. The cavitation phenomenon of vapor volume fraction can be extracted form numerical results, see figure 7.

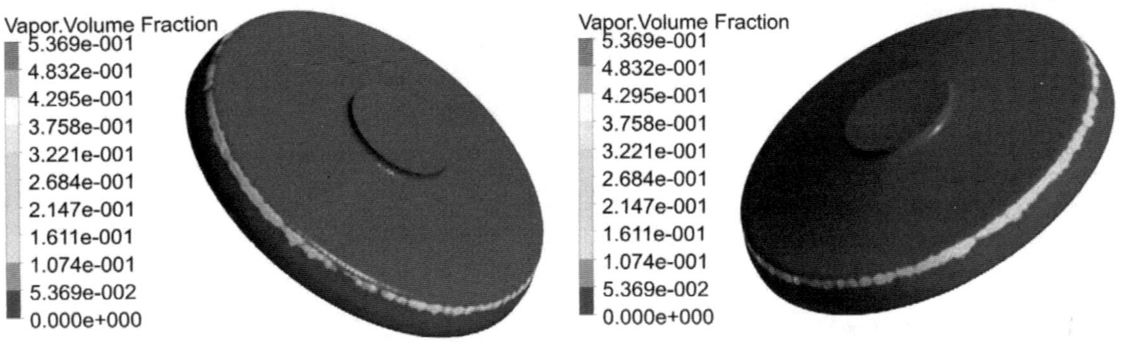

(a) Inlet direction (b) Outlet direction

Figure 7. The vapor volume fraction on the surface of cone from different views

The figure 7 implies the almost all vapors is distributed on the end surface of the cone, and the vapor volume fraction from outlet direction is larger than the inlet direction largely due to greater pressure drop, which has resulted from the high-speed water flow discharging into from upstream to downstream, all of cavitation results present good agreement with that of flow characteristics results. It is apparent that increasingly pressure sink induces more serious cavitation phenomenon. In addition, it's a failure to observe vapor in the rest of the computational domain.

4. Optimization design of throttle cone profile line

Cavitation phenomenon can be observed in the region where pressure drop to below the vaporization pressure, the cavity destruction which mainly occurs on the solid wall highly cause structure alteration by erosion conversely due to the cumulative effect of impact action where travelling bubbles collapse, which may results in seriously noise and vibrations of the valve, as well as performance degradation, even can't meet the requirement for technical water supply and increases cost of operation and maintenance. It is expected that to reduce the incidence ratio of cavitation in the premise of meeting the requirement of reducing the pressure depending on the analysis of above flow field and cavitation performance of the valve. The reason of cavitation development on the end surface of the cone consists in significant pressure drop with drastic changes of passage area, therefore, a modification scheme towards profile lines of the cone is given in figure 8 and identical numerical computation about flow field and cavitation is performed at the same level of pressure drop.

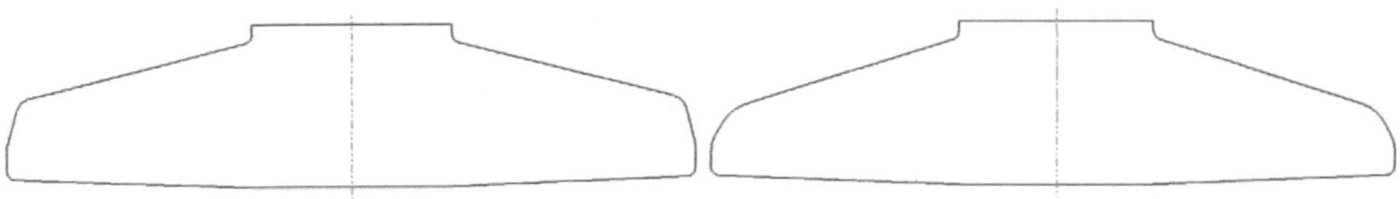

(a) original (b) modified

Figure 8. Comparison views of the original and modified cone cross-section

4.1. Comparative analysis of flow characteristics

Pressure distribution comparative contours on the symmetric plane in the region of upstream and downstream are respectively reported in figure 9 and figure 10.

(a) Original (b) Modified

Figure 9. Pressure contour at upstream

(a) Original (b) Modified

Figure 10. Pressure contour at downstream

It's evident that there is a significantly decrease in the areas of low pressure in figure 9 and figure 10 for the modified throttle cone, which is favorable to inhibit the development of cavitation. As regard the original cone, pressure drop mainly concentrated in the vertical area between the throttling cone and valve seat, where a large number of negative consequences occur such as vortices formation, uneven pressure gradient, unstable flow field and pressure falling rapidly. Contrary to the original results, the improved throttle cone shows better performance in pressure distribution and flow characteristics, and the pressure drop across the region is relatively uniform. All above analysis indicates the optimization design for the throttle cone seems to be reasonable and acceptable.

4.2 Comparative analysis of cavitation characteristics

The vapor volume fraction on the surface of the throttle cone are reported in figures 11 and figure 12 for the original model and optimized one form different views, which illustrates improvement of the cavitation performance with a drop of 60% maximum vapor volume fraction from 0.537% to 0.214%.

The numerical results show that it has attain the expected purpose to improve cavitation performance of the pressure reducing valve and raised reliability of technology water supply.

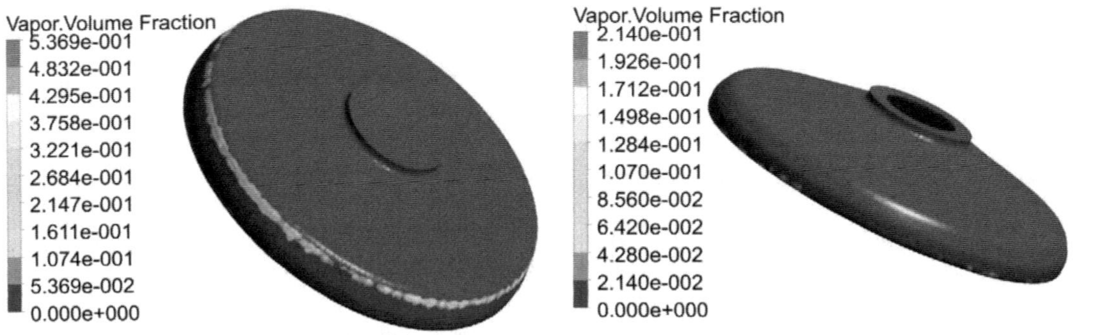

(a) Original (b) Modified

Figure 11. The vapor volume fraction on the surface of cone at inlet direction

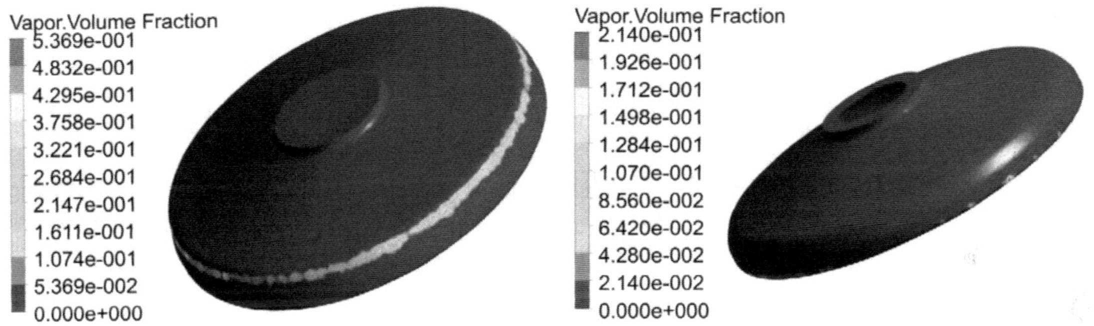

(a) Original (b) Modified

Figure 12. The vapor volume fraction on the surface of cone at outlet direction

Compared with the original calculation model, inception and development of cavitation that strongly related to the operation stability of the valve significantly decreases for the modified model and no vapor on the end surface of the cone, only a slight degree of vapor in the vicinity of lower surface of the cone, which indicates that the investigated valve performance is very sensitive to the profile line of the cone and more smooth flow section effectively inhibits the occurrence and development of cavitation. It turns out that the optimization approach towards the throttle cone is successful and can be for reference.

5. Conclusion

In this paper, the three-dimensional numerical calculation of the pressure reducing valve of the technical water supply system is carried out and flow fields and cavitation characteristics are analyzed. The paper finally emphasizes the importance of the optimization approach to the throttle cone and conducts a comparative analysis between the original model and optimized one and draw some conclusions, which should include:

(1) Great pressure loss is caused when the water flow was accelerated by narrow overflowing section between throttling cone and valve seat, it forms serious vortices in the bottom of the valve. In terms of the cavitation cases, almost all vapors is distributed on the end surface of the cone, and the

IOP Conf. Series: Materials Science and Engineering **129** (2016) 012040 doi:10.1088/1757-899X/129/1/012040

vapor volume fraction from outlet direction is larger than the inlet direction.

(2) Profile optimization design and performance analysis on the improved model show inception and development of cavitation on throttle cone significantly decreases and offers valuable reference for the optimization design of this type of pressure reducing valve.

6. Reference

[1] Fan X H 1987 *Auxiliary equipment of hydraulic units* (Beijing: China WaterPower Press)

[2] Sun Q and Zheng J L 2013 Application of numerical simulation analysis technology to valve design *J GM in Electric Power.* **(9)** 67-69

[3] Zhang X W, Lan J H and He J H 2011 High-pressure difference pressure reducing valve numerical simulation *J Chemical Equipment Technology.* **32(9)** 1-3

[4] Zhang Y H, Zhou X D, Zhu M L and Wang T 2007 Numerical analysis of pressure reducing valve in the process of hydraulic transient *J Journal of Xi'an University of Technology.* **23(4)** 360-364

[5] Wang D J, Song H L, Bai S Q and Wei C 2009 Simulation and analysis of the regulator orifice flow field *J Journal of Rocket Propulsion.* **36(3)** 37-46

[6] Wu S and Zhang W P 2005 Investigated numerically on flow field of valves and experimental study of valve noise *J Valve.* **(1)** 7-10

[7] Wang L, Ou G F and Zheng Z J 2013 Numerical analysis on the cavitation and cavitation erosion of the high temperature and large pressure differential fluid control valve *J Hydraulic Pneumatics & Seals* **(6)** 40-43

[8] Huang G Q and Cao Z W 2011 The research on numerical simulation of centric type buttery valve flow field *J Machinery design & Manufacture.* **(7)** 186-188

[9] Liu C W, Li Y X, Li X J and Cao J 2012 Flow noise simulation of gas transmission pipeline based on CFD *J Oil & Gas Storage and Transportation* **31(9)** 657-662

[10] Yu X M, Mao Z M and Kong B L 2009 Computer assistant test and numerical simulation analysis for pressure reducing valve *J J. University of Shanghai for Science and Technology.* **31(2)** 183-189

IOP Conf. Series: Materials Science and Engineering **129** (2016) 012041 doi:10.1088/1757-899X/129/1/012041

Study on the effect of the impeller and diffuser blade number on reactor coolant pump performances

Y Long [1] **J L Yin** [1] **D Z Wang**[1*] **T B Li** [2]

[1]School of Mechanical Engineering, Shanghai Jiao Tong University, 800 Dongchuan Road, Shanghai, 200240, China

[2] Shanghai Electric KSB Nuclear Pumps and Valves Co., Ltd, Shanghai, 201306, China

E-mail: longyunjs@sjtu.edu.cn

Abstract. In this paper, CFD approach was employed to study how the blade number of impeller and diffuser influences reactor coolant pump performances. The three-dimensional pump internal flow channel was modelled by pro/E software, Reynolds-averaged Naiver-Stokes equations with the k-ε turbulence model were solved by the computational fluid dynamics software CFX. By post-processing on the numerical results, the performance curves of reactor coolant pump were obtained. The results are as follows, with the blade number of the impeller increasing, the head of the pump with different diffuser universally increases in the $0.8Q_n$~$1.2Q_n$ conditions, and at different blade number of the diffuser, the head increases with the blade number of the impeller increasing. In $1.0Q_n$ condition, when the blades number combination of impeller and diffuser chooses 4+16, 7+14 and 6+18, the head curves exist singular points. In $1.2Q_n$ condition, the head curve still exists singular point in 6+18. With the blade number of the impeller increasing, the efficiency of the pump with different diffuser universally decreases in the $0.8Q_n$ and $1.0Q_n$ conditions, but in $1.2Q_n$ condition, the efficiency of the pump with different diffuser universally increases. In $1.0Q_n$ condition, the impellers of 4 and 5 blades are better. When the blade number combination of impeller and diffuser choose 4+11, 4+17, 4+18, 5+12, 5+17 and 5+18, the efficiencies relatively have higher values. With the blade number of the impeller increasing, the hydraulic shaft power of the pump with different diffuser universally increases in the $0.8Q_n$~$1.2Q_n$ conditions, and with the blade number of the diffuser increasing, the power of different impeller overall has small fluctuation, but tends to be uniform. This means the increase of the diffuser blade number has less influence on shaft power.The influence on the head and flow by the matching relationship of the blades number between impeller and diffuser is very complicated, which still need further research.This paper provides a reference for exploring the match relationship between the impeller and diffuser blade number of reactor coolant pump.

1. Introduction

Reactor coolant pump is one of the most important equipment in Nuclear Power Plant. There are many scientific issues which need to be researched in hydraulic components.

Previous researchers have done a lot of work in the pump optimization and the effect of blade number on the pump performance by the numerical simulation. Li Ying [1] used Andritz reactor

* Dezhong Wang, professor, Email: dzwang@sjtu.edu.cn.

coolant pump of Pakistan's Chashma Nuclear Power Plant as the research object, and got effective access to internal flow characteristics like pressures, streamlines, velocity distributions by numerical simulation. Kim Jin Hyuk [2, 3] adopted numerical simulation method to optimize mixed-flow pump hydraulic components, and finally obtained best internal flow and high hydraulic efficiency. Qin Jie [4] comparatively analysed reactor coolant pump with twisted and diffusion diffuser by numerical simulation, and the results showed that static pressure distribution of twisted diffuser is more uniform. Zhang Dongjun [5] studied the influence of orbicular volute outflow pipe structure on performance of reactor coolant pumps based on CFD method and found out that the conical outlet pipe has better flow characteristics than the cylindrical outlet pipe. Zhu Rongsheng [6] designed 13 kinds of pump chamber outlet shrinking angle to study the influence on pump performance and revealed that the contraction angle has a significant effect on the connection area between pumping chamber and outlet, and as the angle is between 12° and 16°, which the efficiency of the RCP is better. Yan Bipeng [7] researched the influence on pump performance by changing the number of axial-flow pump blades and it turned out that the number of blades affects very little on the efficiency of axial-flow pump but much on its cavitation characteristics. Yang Qiongfang [8] took CFD method principally to investigate effects on the characteristics of flow pattern and water-jet propeller performance when the number of pump blades of both rotor and stator is changed. Eventually, the water-jet optimum performance can be obtained when 5 blades of rotor match 9 blades of stator. Bing Hao [9] simulated the distribution of static pressure and total pressure in the mixed-flow pumps and analysed the consequences of the diffuser vane number, the impeller blade number and the blade thickness on the hydraulic performance of mixed-flow pumps. Wang Xiuli [10] used CFX in the numerical simulation for the radial force variation with the number of blades or diffuser changed and it proved that when the number of blades is 5 and the diffuser is 11, the radial force on the impeller gets the minimum.

This paper summarized the results of previous studies at first, and then used the computational fluid dynamics software CFX to conduct the numerical simulation for reactor coolant pump, and thus analysed the influence on reactor coolant pump performance of different impeller and diffuser blade numbers, and finally provides a reference for the optimize design of the hydraulic model.

2. Numerical Simulation

2.1. The pump basic parameters

According to the parameters of reactor coolant pump, the rated flow is 17886m^3/h, the rated head is 111.3m, the design speed is 1480 r/min, the specific speed n_s is 352.25 ℃ water at 1 atm pressure, volume flow design point Q_n=17886m^3/h converted into mass flow Q_m=4968.3kg/s. Hydraulic components include mixed flow impeller, twisted radial diffuser, and pump casing, which resembles a spherical considering the pressure and operational safety [11]. The geometry model of the impeller, diffuser and pump casing can be established with Pro/E software. The pump three-dimensional structure is shown in figure 1.

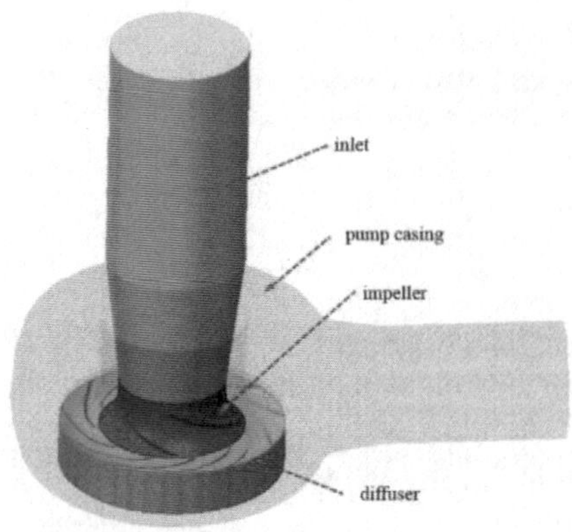

Figure 1. Reactor coolant pump hydraulic component three-dimensional model.

2.2. Match program

Table 1 shows matching relationships of the blade number between impeller and diffuser. The impeller blade number varies from 4 to 7， and the diffuser blade number varies from 11 to18. And 32 different blade number matching relationships between the diffuser and the impeller at a given condition are calculated and the corresponding hydrodynamic performance data is obtained.

Table 1. Matching relationships of the blades number between impeller and diffuser.

Impeller blade number	Diffuser blade number							
4	11	12	13	14	15	16	17	18
5	11	12	13	14	15	16	17	18
6	11	12	13	14	15	16	17	18
7	11	12	13	14	15	16	17	18

2.3. Meshing and grid unrelated test

Use ANSYS-CFX14.5 pre-processing meshing software Workbench to mesh for the model of this study, and take automatic meshing division method to ensure the accuracy of the calculation results. In order to make the number of meshes, the mesh density and the overall quality of grid meet the actual computing requirements, and so that the grid-independent model can be established and researched. The curve graph in figure 2 shows the relationship between the reactor coolant pump head and the number of grids. And it can be obviously seen from the figure, when the number of grid is much more than 3.3 million, the range of variation of the pump head at rated operating point is less than 0.5%, in order to ensure the accuracy of the calculation and the economy, the number of the grids nearly 4.5 million is appropriate.

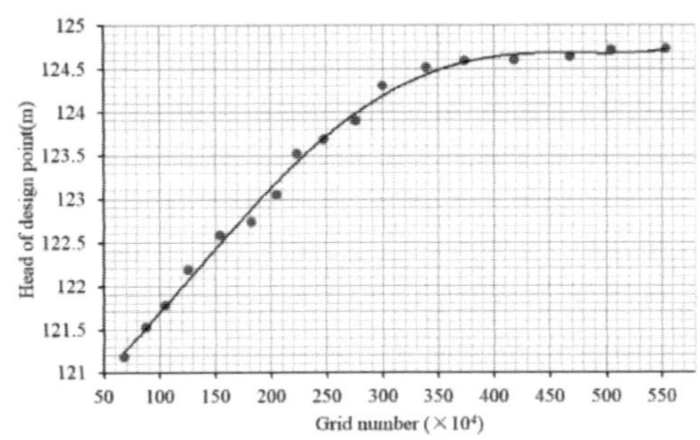

Figure 2. Grid-independent curve.

Figure 3 shows the calculation domain grids of each component. The grid number of the inlet section is 280,612, impeller grid number is 1,638,962, the diffuser grid number is 903,385, the spherical pressurized chamber grid is number 1,930,696, and the total number of grid is 4,753,655.

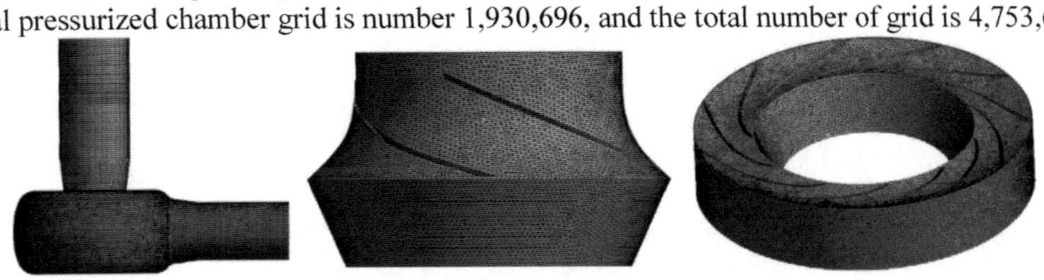

(a) Assembly Grid (d) Impeller (e) Diffuser

Figure 3. Mesh of each flow component water body and the assembly.

2.4. Boundary Conditions

Reynolds-averaged Naiver-Stokes equations with the *k-ε* turbulence model were solved by the computational fluid dynamics software CFX. When the flow rate is constant, in order to get more accurate velocity and pressure gradient, pressure-inlet condition is used at inlet, meanwhile using mass-outflow condition at outlet, the mass value determined by the flow, using no-slip wall boundary conditions, movement-static coupling surfaces of the impeller and diffuser, impeller and import section using Frozen Rotor interface. To better handle the flow of the boundary layer, use the standard wall function in the near wall region. Transportation medium is water [12]. Choose default values of CFX software as under-relaxation factors in the calculation process, and the residual convergence precision is set to 10^{-4}.

3. Results and analysis

3.1. The performance curves variation with different impeller and diffuser blade number in different conditions

As shown in figure 4, the head curves variation regular patterns of 4 blades impeller in different conditions are nearly unanimous while diffuser blade number is 11~18. With the flow rate increasing, the head decreases, but the absolute value of the head slope firstly decreases, and then increases, like S-type. The efficiency curves variation regular patterns of 4 blades impeller in different conditions are nearly unanimous while diffuser blade number is 11~18. With the flow rate increasing, firstly the efficiency increases, and then decreases. When Q_m is nearly 5500kg/s, the efficiency reaches the maximum. The power curves variation regular patterns of 4 blades impeller in different conditions are nearly unanimous while diffuser blade number is 11~18. With the flow rate increasing, the power firstly decreases sharply, and then increases slowly until the power curve is smooth, and the power curve is like saddle type. When Q_m is less 2500kg/s, the power reaches to the maximum, and overload phenomenon occurs. That is because the reactor coolant pump hydraulic model is unstable in low flow conditions, and numerical instability leads to the shaft power increasing when the pump operates in small flow conditions. It is obviously in the figures that while diffuser blade number is 11~18, the head curves exist a little difference, which is not clear, and the efficiency and power curses are also like this.

When blade number of impeller changes to 5, 6, 7, variation regular patterns are the same as 4 blade impeller, which are not shown in the paper. But the little differences between the performance curves are researched deeply in following chapters.

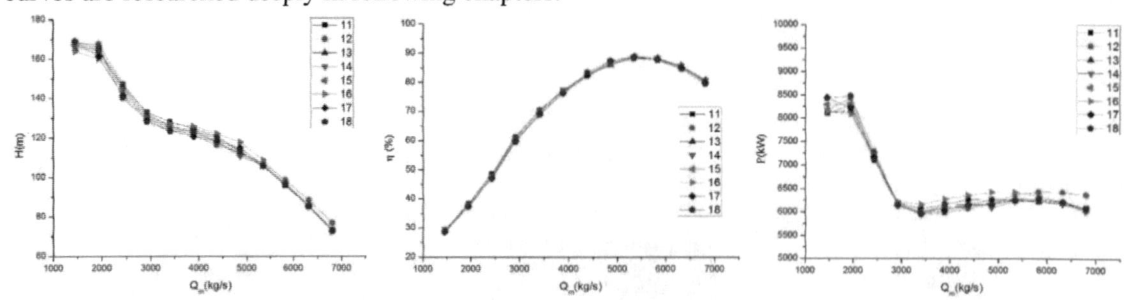

(a) The head (b) The efficiency (c) The power

Figure 4. The performance curves of 4 blade number impeller in different conditions.

3.2. The effect on the head when only change the blade number of the impeller

As shown in figure 5, with the blade number of the impeller increasing, the head of the pump with different diffuser universally increases in the $0.8Q_n$~$1.2Q_n$ condition, but the head increasing trend slows down. This is because the theoretical head is built by Stodala finite number of blades modified formula. In the formula, H_t is

$$\mathrm{H_t} = \frac{u_2}{g}\left\{u_2\left(1-\frac{\pi}{z}\psi_2 sin\beta_{2b}\right)-\frac{Q_t}{D_2\pi b_2\psi_2 tan\beta_{2b}}\right\} \tag{11}$$

$$\psi_2 = 1-\frac{Z\delta_2}{D_2\pi}\sqrt{1+\left(\frac{cot\beta_{2b}}{sin\lambda_2}\right)^2} \tag{12}$$

Where H_t is the finite number blade hypothesis theoretical head, u_2 is the blade outlet circumferential velocity, g is acceleration of gravity, Q_t is the flow rate, D_2 is the impeller diameter, b_2 is the blade outlet width, ψ_2 is the crowding coefficient of blade outlet, Z is the number of blade, δ_2 is the thickness of blade outlet, λ_2 is the included angle of axis section transversal and flow line, β_{2b} is the blade outlet angle.

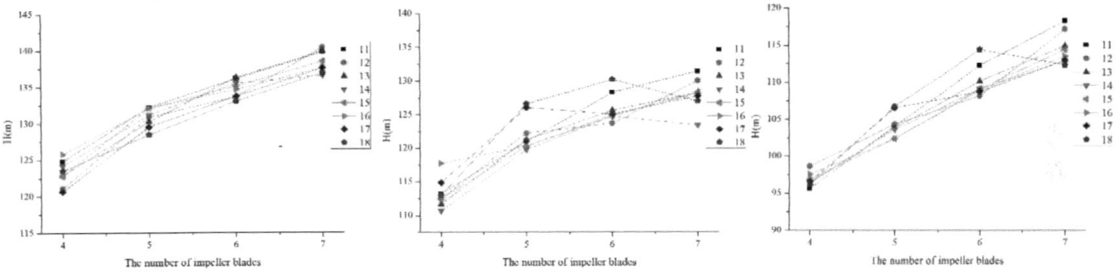

(a) $0.8Q_n$ (b) $1.0Q_n$ (c) $1.2Q_n$

Figure 5. The head curves with the blade number of the impeller increasing.

In Equation 11, substituted into designed impeller geometric parameters, the relation between H_t and Z is shown in figure 6, the theoretical head increases with the blades number increasing, but the head increasing trend slows down. The head variation trend of numerical simulation is consistent with theoretical analysis. But in each condition, there are some differences in the head of different diffuser blade number, and the head changing law is not obvious. This is mainly because the crowding and surface friction at different blades number of impeller and diffuser have great effect on the pump head, and even small changes of the machine geometry may generate big response to the flow and performance.

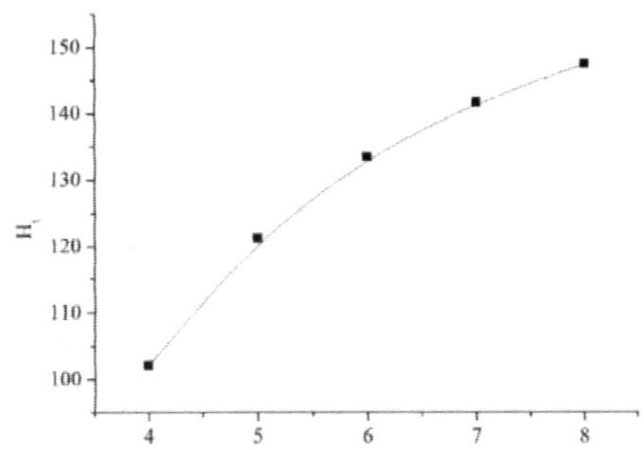

Figure 6. The relation between H_t and Z.

3.3. The effect on the efficiency when only change the blade number of the impeller

As shown in figure 7, with the blade number of the impeller increasing, the efficiency of the pump with different diffuser universally decreases in the $0.8Q_n$ and $1.0Q_n$ condition, and the efficiency changing law is not obvious at different impeller blade number. But in $1.2Q_n$ condition, the efficiency of the pump with different diffuser universally increases with the blade number of the impeller increasing, and the efficiency changing law at different impeller blade number is also not obvious.

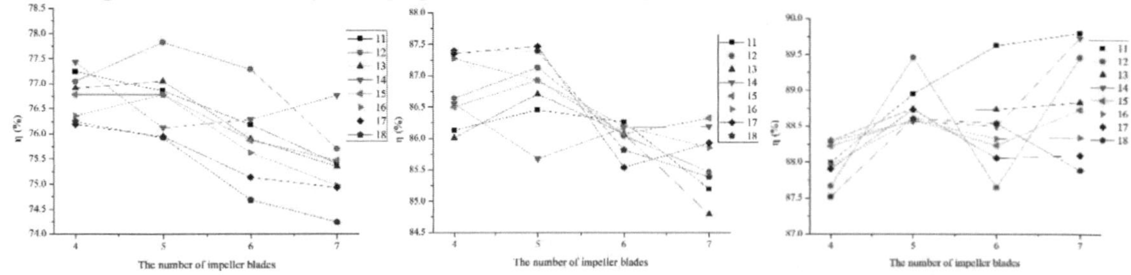

(a) $0.8Q_n$ (b) $1.0Q_n$ (c) $1.2Q_n$

Figure 7. The efficiency curves with the blade number of the impeller increasing.

3.4. The effect on the power when only change the blade number of the impeller

As shown in figure 8, with the blade number of the impeller increasing, the power of the pump with different diffuser universally increases in the $0.8Q_n$~$1.2Q_n$ conditions, and the power changing law at different impeller blade number is more obvious than the head and efficiency.

Figure 5~8 macroscopically reflects the variations of the head, efficiency and power with different blade number of impeller and diffuser, but its performance variation still needs deep analysis.

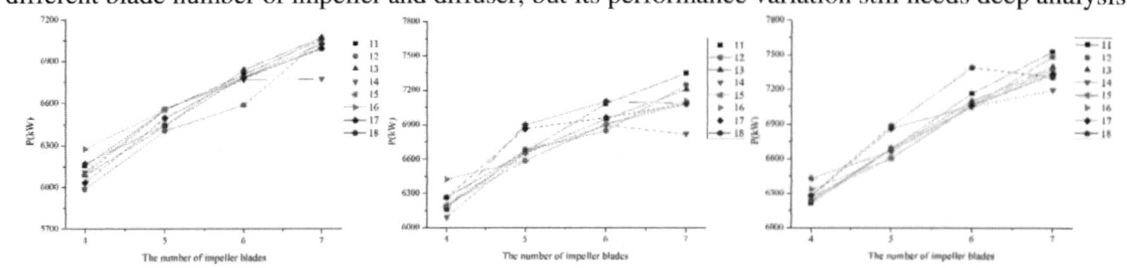

(a) $0.8Q_n$ (b) $1.0Q_n$ (c) $1.2Q_n$

Figure 8. The power curves with the blade number of the impeller increasing.

3.5. The effect on the head when only change the blade number of the diffuser

As shown in figure 9, in $0.8Q_n$ condition, at different blade number of the diffuser, the head increases with the blade number of the impeller increasing, and the head with 7 blades impeller is maximum, the head with 4 blades impeller is the minimum. In $1.0Q_n$ condition, the head changing law is basically same with $0.8Q_n$, but when the blades number combination of impeller and diffuser choose 4+16, 7+14 and 6+18, the head curves exist singular points. In $1.2Q_n$ condition, the head changing law is basically the same as $0.8Q_n$. But when the blades number combination of the impeller and diffuser choose 6+18, the head curve still exists singular point. From the figure 11, it can also be seen that with the increase of the diffuser blades number, the head curve is not linear, nor is a regular curve. In summary, the impact on the head and flow by the matching relationship of the blades number between impeller and diffuser is very complex, which still needs further study.

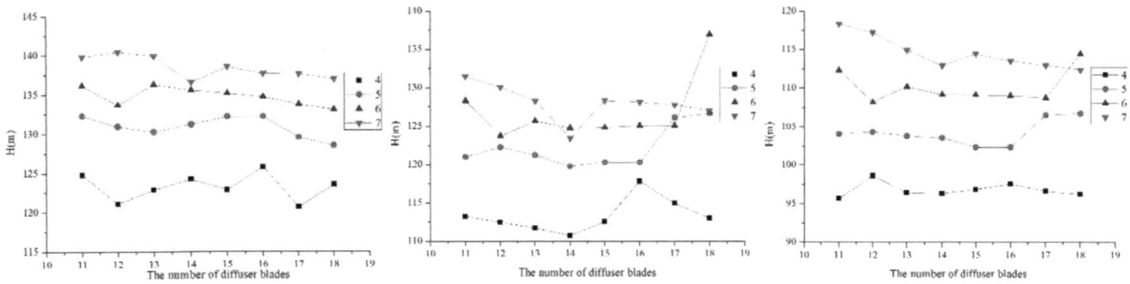

(a) $0.8Q_n$ (b) $1.0Q_n$ (c) $1.2Q_n$

Figure 9. The head curves with the blade number of the diffuser increasing.

3.6. The effect on the efficiency when only change the blade number of the diffuser

As shown in figure 10, in $0.8Q_n$ condition, at different blade number of the diffuser, the efficiency decreases with the blade number of the impeller increasing. And with the blade number of the diffuser increasing, the efficiencies of different impeller overall decrease. When the blades number combination of impeller and diffuser choose 4+11, 5+12 and 4+14, the efficiencies relatively have larger values. In $1.0Q_n$ condition, at different blade number of the diffuser, the efficiency decreases with the blade number of the impeller increasing, the efficiencies of 4 and 5 blades impeller overall show an increasing trend, but the efficiencies of 6 and 7 blades impeller overall show a decreasing trend. When the blades number combination of impeller and diffuser choose 4+11, 4+17, 4+18, 5+12, 5+17 and 5+18, the efficiencies relatively have larger values. In $1.2Q_n$ condition, with the blade number of the diffuser increasing, the efficiencies of different impeller overall decrease. The efficiency changing law is not obvious at different impeller blade number, and the efficiency overall increases with the blade number of the impeller increasing.

At different blades number of the impeller, pump efficiencies fluctuates with changes in the blades number of the diffuser. This is mainly because the efficiency is determined by the head and power, which also fluctuate.

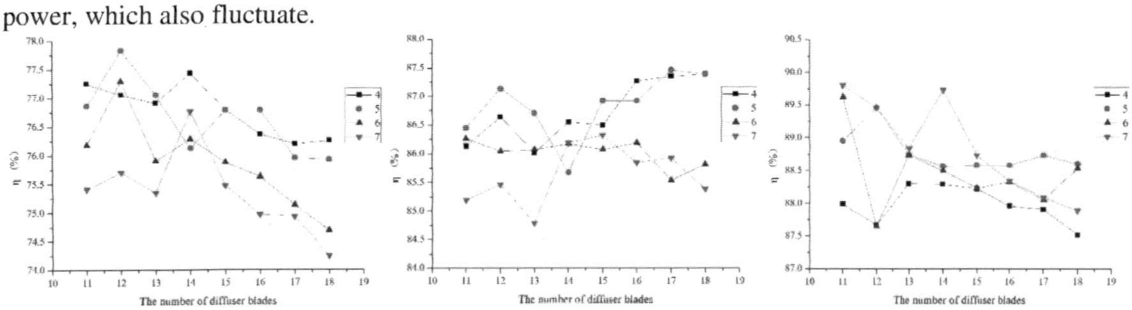

(a) $0.8Q_n$ (b) $1.0Q_n$ (c) $1.2Q_n$

Figure 10. The efficiency curves with the blade number of the diffuser increasing.

3.7. The effect on the power when only change the blade number of the diffuser

As shown in Figure 11, in $0.8Q_n$~$1.2Q_n$ conditions, at different blade number of the diffuser, the power increases with the blade number of the impeller increasing. And with the blade number of the diffuser increasing, the powers of different impellers overall have small fluctuations. This means the increase of the diffuser blade number has less impact on shaft power.

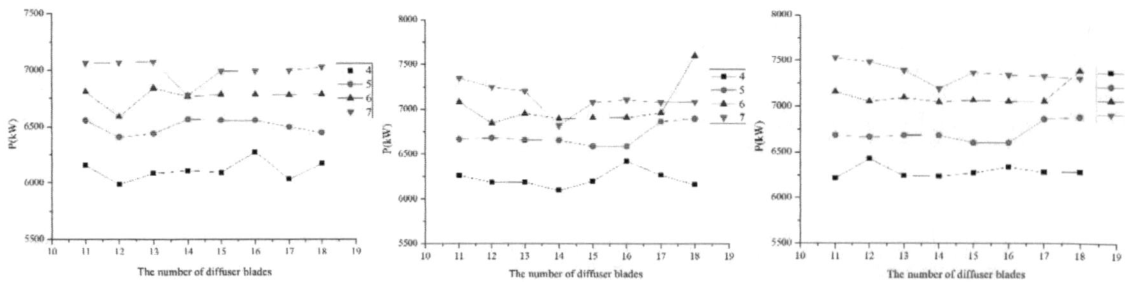

(a) $0.8Q_n$ (b) $1.0Q_n$ (c) $1.2Q_n$

Figure 11. The power curves with the blade number of the diffuser increasing.

4. Conclusion

In this work, the effects of the impeller and diffuser blade number on reactor coolant pump performances are studied with CFD techniques. The following conclusions could be summarized.

(1) With the blade number of the impeller increasing, the head of the pump with different diffuser universally increases in the $0.8Q_n$~$1.2Q_n$ conditions, and at different blade number of the diffuser, the head increases with the blade number of the impeller increasing. In $1.0Q_n$ condition, when the blades number combination of impeller and diffuser choose 4+16, 7+14 and 6+18, the head curves exist singular points. In $1.2Q_n$ condition, the head curve still exists singular point in 6+18.

(2) With the blade number of the impeller increasing, the efficiency of the pump with different diffuser universally decreases in the $0.8Q_n$ and $1.0Q_n$ condition, but in $1.2Q_n$ condition, the efficiency of the pump with different diffuser universally increases. In the design condition, 4 and 5 blades impeller are better. When the blades number combination of impeller and diffuser choose 4+11, 4+17, 4+18, 5+12, 5+17 and 5+18, the efficiencies relatively have larger values.

(3) With the blade number of the impeller increasing, the power of the pump with different diffuser universally increases in the $0.8Q_n$~$1.2Q_n$ conditions, and with the blade number of the diffuser increasing, the power of different impeller overall has small fluctuation, but tends to be uniform. This means the increase of the diffuser blade number has less impact on shaft power.

Acknowledgments

This work is funded by the National Natural Science Foundation of China (No.51406114).

References

[1] Li Y , Zhou W X , Zhang J G, Wang D Z 2009 Numerical simulation of three-dimensional flow through ful l passage and performance prediction of nuclear reactor coolant pump *Atomic Energy Science and Technology* .**43(10)** 898-902(in Chinese)

[2] Kim J H,Kim K Y.2011 Optimization of vane diffuser in a mixed-flow pump for high efficiency design *International Journal of Fluid Machinery and Systems.* **4(1)**:172-178.

[3] KimJ H 2010 High-efficiency design of a mixed-flow pump *Science China*.**53(1)** 24-27.

[4] Qin J,Xu S M. 2010 Influence of diffuser structure on performance of reactor coolant pumps *Power Equipment* 2010(5):315-318(in Chinese)

[5] Zhang D J, Xu S M 2010 Influence of obicular vlute outflow pipe Structure on Performance of Reactor Coolant Pumps *Pump Technology*.**(1)**: 21-30. (in Chinese)

[6] Zhu R S, Li X L, Yuan S Q, Fu Q, Wang X L 2012 Effect of pumping chamber outlet contraction angle on hydraulic performance of main nuclear reactor pump *Nuclear Power Engineering*.**33(2)**:97-103(in Chinese)

[7] Yan B P, Tang F P 1998 Study on the performances of axial-flow pump by changing the number of its blades *Journal of Yangzhou University (Natural Science Edition)*, **3(1)**: 53-55(in Chinese)

IOP Conf. Series: Materials Science and Engineering **129** (2016) 012041 doi:10.1088/1757-899X/129/1/012041

[8] Yang Q F, Wang Y S, Zhang Z H, et al 2009 Computational Fluid Dynamics Analysis of Effects of Number of Pump Blades on Water-jet Propeller Performance *Journal of Mechanical Engineering*, **45(6)**: 222-228(in Chinese)

[9] Bing H,Tan L,Cao S L 2013 Effects of blade number and thickness on performance of mixed-flow pumps *Journal of Hydroelectric Engineering*, **32(6)**: 250-255 (in Chinese)

[10] Wang X L, Yuan S Q, Zhu R S, et al 2014 Effect of number of blades on reactor coolant pump radial force under variable flow transition conditions *Journal of Vibration and Shock*, **33(21)**, 51-59 (in Chinese)

[11] Wang C L, Peng N, Kang C, et al 2009 Numerical simulation of interior flow field of nuclear model pump *Nuclear Power Engineering*, **30(4)**: 81-85 (in Chinese)

[12] Zhang Y, Wang X F, Jie H E 2011 Effect of boric acid concentration on reactor coolant pump performance in PWRs *Nuclear Power Engineering*, **32(4)** (in Chinese)

The numerical simulation based on CFD of hydraulic turbine pump

X H Duan[1] ,F Y Kong[1] ,Y Y Liu[1],R J Zhao[1] ,Q L Hu[1]

[1] National Research Center of Pumps, Jiangsu University, Zhenjiang, Jiangsu 212013, China

E-mail: 694140559@qq.com

Abstract. As the functions of hydraulic turbine pump including self-adjusting and compensation with each other,it is far-reaching to analyze its internal flow by the numerical simulation based on CFD, mainly including the pressure field and the velocity field in hydraulic turbine and pump.The three-dimensional models of hydraulic turbine pump are made by Pro/Engineer software;the internal flow fields in hydraulic turbine and pump are simulated numerically by CFX ANSYS software. According to the results of the numerical simulation in design condition , the pressure field and the velocity field in hydraulic turbine and pump are analyzed respectively .The findings show that the static pressure decreases systematically and the pressure gradient is obvious in flow area of hydraulic turbine ; the static pressure increases gradually in pump . The flow trace is regular in suction chamber and flume without spiral trace . However , there are irregular traces in the turbine runner channels which contrary to that in flow area of impeller . Most of traces in the flow area of draft tube are spiral .

1. Introduction

Computational fluid dynamics (CFD), which is theoretical and practical, describes the numerical solution of the flow field quantitatively on the time and space by computer numerical calculations and a method of image displayed[1]. The relationship of performance parameters between pumps and PAT was predicted by theoretical analysis and the empirical formulas based on the experimental data[2]. However, turbine Pump has some advantages， such as low cost and the ability of simulating complicated or desirable fluid motion. As the functions of hydraulic turbine pump including self-adjusting and compensation with each other , it is far-reaching to analyze its internal flow.

2. Computation of parameters and conditions

Table 1. The main parameters of the pump and the turbine

Pump flow $/m^3 \cdot h^{-1}$	Pump head /m	N $/r \cdot min^{-1}$	T /℃	Turbine flow $/m^3 \cdot h^{-1}$	Turbine head /m	P /KW	ρ $/kg \cdot m^{-3}$
60	11	1475	20	90	41.5	1.8	998.2

National Science Foundation(2013BAK06B02), Project supported by Postgraduate Innovation Foundation of Jiangsu Province(CXZZ13_0678)
Author: Xiaohui Duan,(1990-),Male, Jiangsu Province, master degree

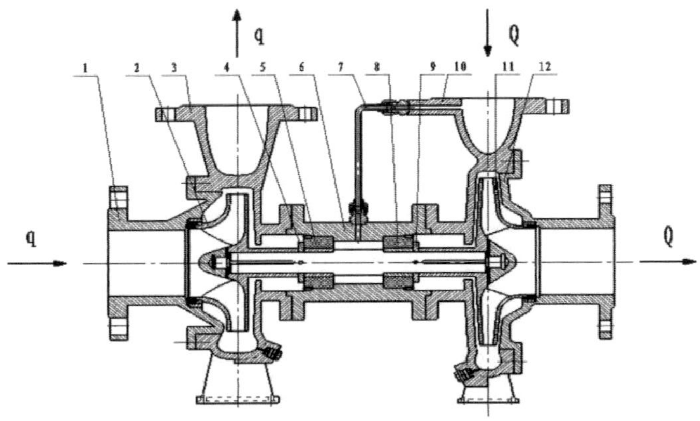

1.pump lid 2.impeller 3.Pump body 4.Balance disc 5.Sliding bearing 6.Connector 7.Rushing pipette 8.Sliding bearing 9.Balance disc 10.Turbine body 11.Runner 12.Turbine lid

Figure 1. Structure diagram of hydraulic turbine

3. The geometric model and numerical simulation

3.1. Solid model of fluid region

Using Pro/Engineer software, the three-dimensional solid models are made. The models are presented in figure2 to figure3[3].

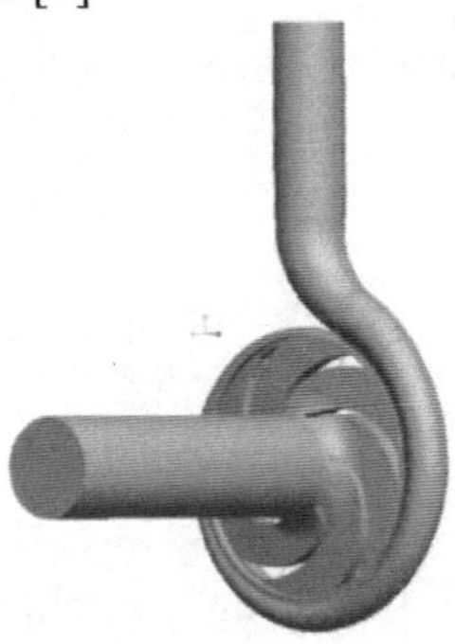

Figure 2. Hydraulic Turbine

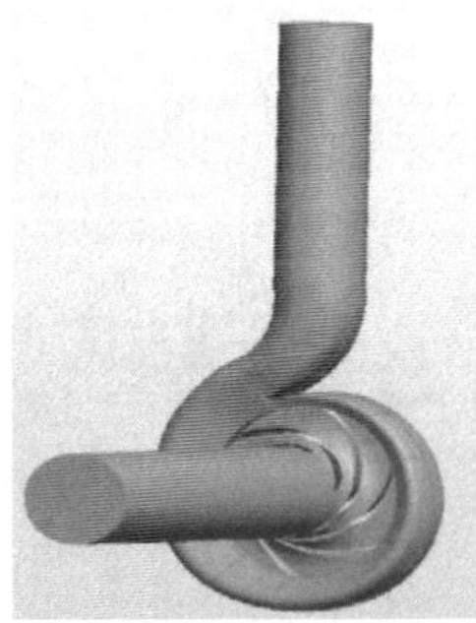

Figure 3. Pump

3.2. Mesh generation of fluid region

The number of grid of hydraulic turbine is 1158877 and pump is 1458943. The grids are presented in figure4 to figure5[4].

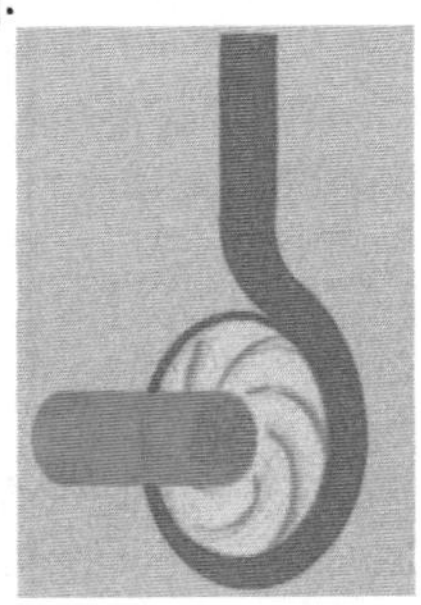

Figure 4. Meshes in Hydraulic Turbine

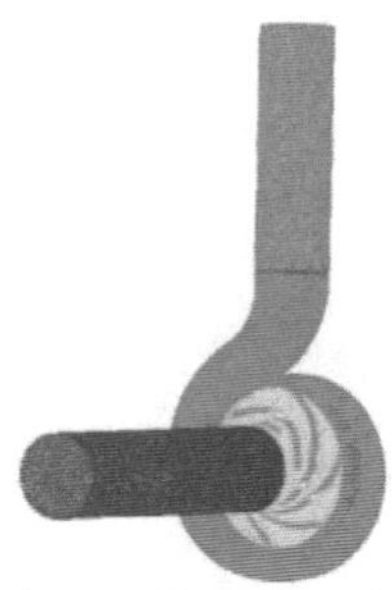

Figure 5. Meshes in Pump

3.3. Turbulence Model

At present , the standard k- ε model is the most basic one of the double equations model being the most widely used the eddy viscosity model. The equations[5] are as follows :

$$\frac{\partial}{\partial t}(\rho k)+\frac{\partial}{\partial x_i}(\rho k u_i)=\frac{\partial}{\partial x_j}\left[\left(\mu+\frac{\mu_t}{\sigma_k}\right)\frac{\partial k}{\partial x_j}\right]+G_k-\rho\varepsilon-S_k \tag{1}$$

$$\frac{\partial}{\partial t}(\rho\varepsilon)+\frac{\partial}{\partial x_i}(\rho\varepsilon u_i)=\frac{\partial}{\partial x_j}\left[\left(\mu+\frac{\mu_t}{\sigma_\varepsilon}\right)\frac{\partial\varepsilon}{\partial x_j}\right]+C_{1\varepsilon}\frac{\varepsilon}{k}G_k-C_{2\varepsilon}\rho\frac{\varepsilon^2}{k}+S_\varepsilon \tag{2}$$

Where: μ_t —turbulent viscosity, $\mu_t=\rho C_\mu\frac{k^2}{\varepsilon}$; G_k —turbulent kinetic energy, $G_K=\mu_T\left(\frac{\partial u_i}{\partial x_j}+\frac{\partial\mu_j}{\partial x_i}\right)\frac{\partial u_i}{\partial x_j}$; $\sigma_k,\sigma_\varepsilon$ —turbulence Prandtl constant of k -ε equation ; S_k —vicious stress S_k =1.0 ; S_ε —turbulence stress S_ε =1.3; $C_\varpi=0.09;C_1=1.44;C_2=1.92$

3.4. The simulation and analysis of internal flow field in pump and hydraulic turbine

The inner flow field in pump and hydraulic turbine are simulated by ANSYS CFX ,and then the velocity profile and pressure profile are analyzed to get some conclusions through the pictures from the simulation.

3.5. The analysis of pressure profile in hydraulic turbine

The pressure value compared to 101325 Pa is a relative value in this model[6].

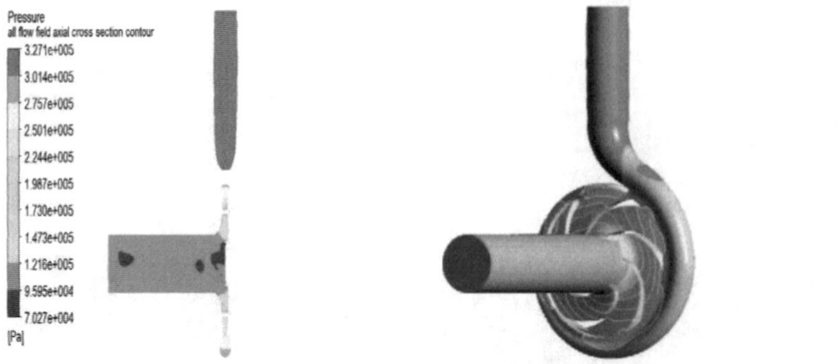

Figure 6. Static pressure distribution diagram of hydraulic turbine

From Figure6 ,the static pressure is lower and lower as the order of intake pipe , turbine flume ,impeller and draft tube. The dividing line is obvious among the different pressure fluid ,and the static pressure value is rather low in the draft tube.

The static pressure contours of the whole flow passage of hydraulic turbine are shown in Figure6. As can be seen from the picture, the static pressure of the water diversion chamber, the runner and the draft tube is decreasing in order .The indoor static pressure of the water diversion chamber begins to decrease from the IX section, which is decreasing as the diversion chamber flow section area reducing. Some regions appears high-pressure. The static pressure of the runner is decreasing from the inlet to the outlet and the pressure gradient is significant. The static pressure of draft tube is basically constant.

3.5.1. The analysis of pressure field of the internal flow region of the pump. The pressure values in the picture is relative to the reference value of the 101325Pa of the model. And the values in the following static pressure contours are relative.

The static pressure contours of the axial section of the whole passage of pump is shown in Figure7.As can be seen from the picture, the static pressure of the water inlet pipe, the water diversion chamber, the impeller, the water pressing chamber and the water outlet pipe is gradually decreasing in order. And the pressure gradient is significant.

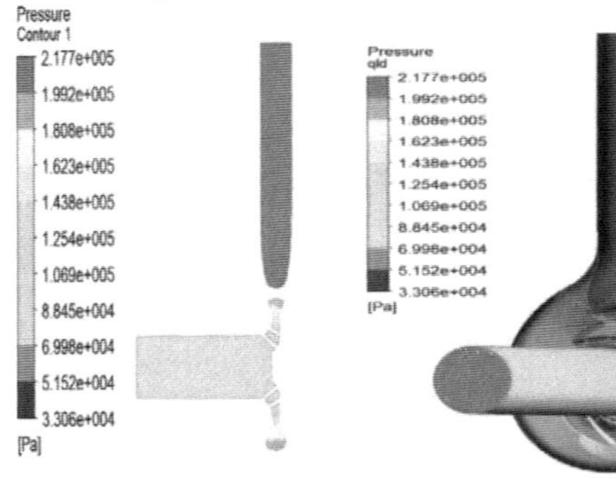

Figure 7. Static pressure contours of the whole passage of pump

The static pressure contours of the whole passage of pump is shown in Figure7. As can be seen from the picture, the pressure of the suction chamber, the impeller, the pump chamber and the outlet pipe is gradually increasing .The the static pressure gradient of liquid impeller is significant, which is increasing gradually.

The indoor static pressure of the pumping chamber is decreasing from I section to XIV section. Some regions appeared high-pressure. The static pressure of impeller passage increased gradually from the inlet to the outlet direction, the adverse pressure gradient. And the pressure gradient is significant. The static pressure of outlet pipe is basically constant.

3.5.2. The analysis of velocity field of the internal flow region of the hydraulic turbine. The velocity vector distribution of the whole flow passage of pump is shown in Figure8.The velocity vector distribution of the axial section of the whole flow passage is shown in Figure8. As can be seen from the picture, the velocity of the inlet pipe, the suction chamber, the impeller and the pumping chamber increase first ,then decrease. The liquid velocity value reached the maximum in the impeller. And it reached the minimum in the draft tube. The velocity field of inlet liquid in the tube is regular, which is spiral in the tail pipe and the outlet pipe.

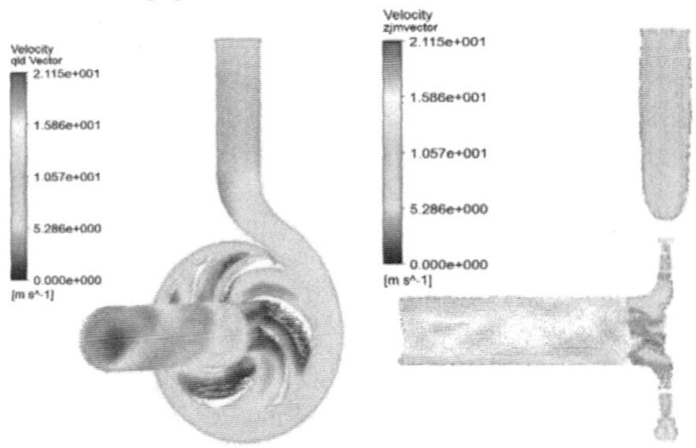

Figure 8. Velocity vectors contours of hydraulic turbine

The trace diagram of the whole flow passage of pump is shown in Figure9.As can be seen from the picture, the trace in the inlet pipe and the water diversion chamber is regular. There is no spiral trace appearing. There is irregular trace appearing in the flow passage of impeller. Most of trace in the tail pipe is spiral.

3.5.3. The analysis of velocity field of the internal flow region of pump. The velocity vector distribution of the whole flow passage of pump is shown in Figure10.As can be seen from the picture, the trace in the suction chamber and the impeller is regular. And the trace of the outlet pipe is bending up.

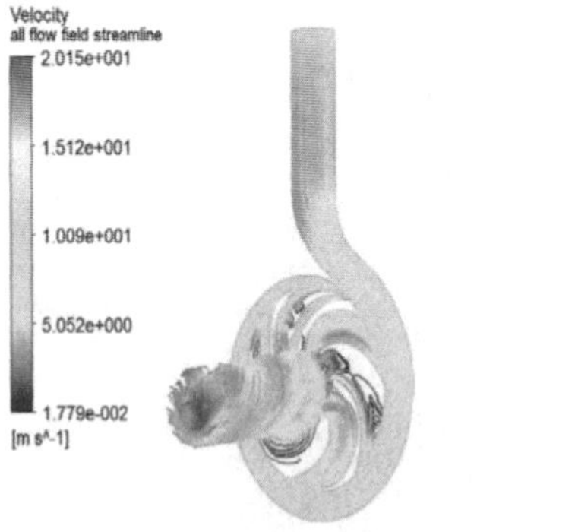

Figure 9. Trace diagram of the whole passages in hydraulic turbine

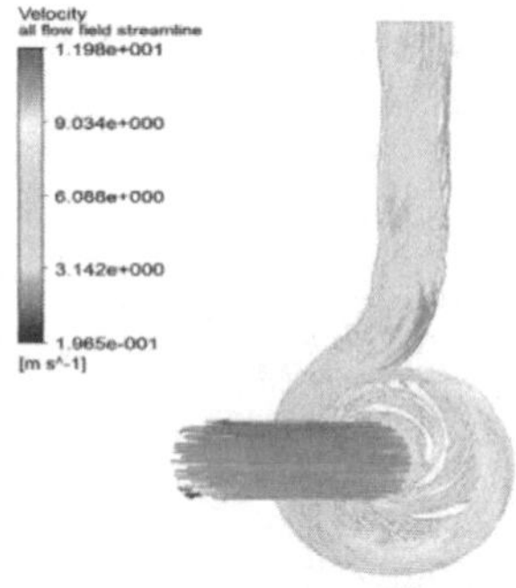

Figure 10. Trace diagram of the whole passage in pump

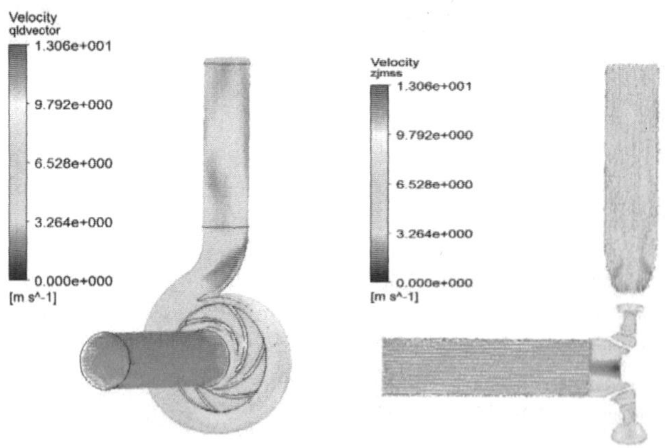

Figure 11. Velocity vectors contours of pump

The velocity vector distribution of pump is shown in Figure11.As can be seen from the picture, the relative velocity of the suction chamber, the impeller, the volute and the outlet pipe increases first ,then decreases.

The liquid relative velocity value reached the maximum in the flow passage of the impeller. And it reached the minimum in the center of back cover plate. The velocity field in the suction chamber and the impeller is regular. And there is vortex of the velocity field in the volute. The velocity field in the outlet pipe is regular.

4. Conclusion

In the hydraulic turbine, the static pressure of the suction chamber, the impeller, the tail pipe is lower and lower, the pressure gradient is significant. In the pump, the pressure of the suction chamber, the impeller, the pumping chamber and the outlet pipe is gradually increasing .

The trace in the inlet pipe and the diversion chamber is regular, There is no spiral trace appearing. There is irregular trace appearing in the flow passage of the turbine impeller which contrary to in the follow passage of pump impeller. Most of trace in the tail pipe is spiral.

5. References

[1] Wang F J 2004 *The Analysis of Computer Fluid Dynamics* (Beijing：Tsinghua University Press)

[2] Wang T,Kong F Y,He Y Y,Yang S S 2013 The current research status of centrifugal pump for turbine *Journal of Drainage and Irrigation Machinery Engineering*, **08**:674-680

IOP Conf. Series: Materials Science and Engineering **129** (2016) 012042 doi:10.1088/1757-899X/129/1/012042

[3] Yang S S, Kong F Y, Su X H , etc 2012 Numerical simulation and external characteristic of pumps and pump as turbine *Academic Journal of Xi'an Jiaotong University* , **3(46)**: 36-41.f
[4] Yang S S, Kong F Y, Shao F, et al. 2010 Numerical simulation and comparison of pump and pump as turbine *ASME Fluids Engineering Summer meeting*(Montreal:Canada) 1-10
[5] Xie L H, Zhao Xinyu 2013 *ANSYS CFX analysis and simulation of fluid* (Beijing: Electronic Industry Press)
[6] Yang S S, Shahram Derakhshan, Kong F Y 2012 Theoretical numerical and experimental prediction of pump as turbine performance *Renewable Energy*, 507-513

IOP Conf. Series: Materials Science and Engineering **129** (2016) 012043 doi:10.1088/1757-899X/129/1/012043

Suitability research on the cavitation model and numerical simulation of the unsteady pulsed cavitation jet flow

S Y Chen[1], X F Yu[1], D Y Luan[2], Y P Qu[1] and C Zhou[3]

[1] Key Laboratory of High-efficiency and Clean Mechanical Manufacture, School of Mechanical Engineering, Shandong University, Jinan, Shandong 250062, P. R. China
[2] School of Mechanical and Electrical Engineering, Qingdao University of Science and Technology, Qingdao, Shandong 266061, P. R. China
[3] Qingdao Special Equipment Inspection and Testing Institute, Qingdao, Shandong 266071, P. R. China

E-mail: chensy66@sdu.edu.cn

Abstract. In order to explore the cavitation jet mechanism, it can first study its critical state of single-phase flow before cavity occurrence to explore the trend of pulsed cavitation jet. Then select the cavitation model to simulate the complex multiphase flow state. Such a step-by-step approach is beneficial to advance research reliably and steady, relying on the foundation for further solving the problem.

Three turbulence models such as Euler Hybrid Model, Euler Two Phase Model and Euler Lagrange Model are discussed on their suitability. In this paper, it states only RNG k- ε turbulent model can simulate small scale vortex of jet in the transient simulation. Grid independent verification and the effect of time step is presented. The simulation results show that a large scale vortex ring surrounding jet flow in the nozzle, the pressure of vortex core is slightly lower than the upstream nozzle pressure.

Considering the capture ability of small scale eddies, an equivalent pressure is established. The single-phase flow turbulence model is modified to simulate the turbulence flow in the self-excited pulsed cavitation after the cavitation occurs. Through different results comparison of not modified cavitation model and the modified cavitation model to the experimental results, it proves that the latter simulation results are relatively accurate.

1. Introduction

Lots of invisible gaseous kernels will grow up and become macroscopic visible bubbles when the local pressure of liquid is lower than its saturated vapor pressure. If the liquid with bubbles pass through a relative high pressure area, the bubbles will break down under the action of pressure and change into microscopic size of the invisible kernel. The phenomenon of partial vaporization in the liquid is called the cavitation. There are some important factors that influence cavitation occurrence and development such as the size and quantity of the gaseous kernel, the amount of undissolved gas contents, the relative velocity between gas and liquid, the pressure distribution, turbulence intensity and velocity gradient[1-3].

First, the greater the size of the gaseous kernel, the shorter time it takes from generation to visible size. According to the stability theory of gas nucleus, when the gas nucleus size is greater than the critical size, gas nuclei will unlimited expansion just below the saturated vapor pressure, the cavitation will occur. When the pressure is lower than the saturated vapor pressure, gas nucleus can be in a stable condition under a corresponding tensile stress state, the cavitation will be suppressed. Second, the greater the number of gas nucleus contained in liquid, cavitation phenomenon will be more obvious. Again, gas nucleus in the flow field will affected by drag force, Saffman force and viscous drag force, etc. The power of the force are related to the size of the gas nucleus. In some cases, the large size of gas nucleus can't get through the low pressure area under the action of those forces[4].

Chahine and Hsiao[5] deduced bubble dynamics equation under the condition of the relative velocity between the gas nucleus and water. They found that the effects of bubble expansion and compression caused by the relative speed between gas nuclear and water is equivalent to added equivalent pressure to the saturated vapor pressure. They got the modified Rayleigh-Plesset equation.

Cavitation occurs is not sudden, it need to experience a period of time. Gas nucleus needed to experience a period of time in the low pressure area before developed into a critical radius and generate cavitation .So, it needed a scope of low pressure area in the flow field before the formation of cavitation. If the scope of low pressure area is too small even it lower than the saturated vapor pressure also cannot generate cavitation. Usually backflow is easy to form on the surface of the object in the low pressure area. Some gas nucleus with backflow will return back. This will prolong the time of gas nucleus experience, the cavitation will occur. In addition, when the gas nucleus with liquid entered the stagnation point of near the surface of the object, the gas nucleus will be deviated from the original streamline under the action of large fluid acceleration and pressure gradient. And the large gas nucleus are more likely to deviating from the original streamline. Johnson[6] called it filtering effect. Recent studies have shown the importance of filtering effect in the aspects of cavitation inception.

2. Analysis of cavitation model

There is a big challenge to establish the physical model and the steady numerical method for the numerical simulation of the cavitation flow. Delannoy and Kueny[7] pioneered to use single-phase mixing model which combines with the experience of positive pressure model to simulate the cavitation. This model can calculate the density by pressure and velocity and only get qualitative consistent with experimental results in cavitation number. In the positive pressure model which was established by Schmidt, Rutland and Corradini[8], cavitation was simulated by a compressible water - steam mixing method which the sound velocity based on homogeneous equilibrium flow model (HEM). While considering the compressibility to improve the accuracy of the simulation, but it not used for occasions of small pressure gradient and great density change, and this model didn't consider the effect of turbulent on the cavitation. Dumont, Simonin and Habchi[9] expanded this model to three dimensional problems. Marcer and LeGouez[10] proposed a cavitation model which applied to the simulation of the diesel nozzle. In this model, the volume of fluid method (VOF) was modified and combined with a mass transfer model which was driven by energy. The basic hypothesis of this model is that the cavitation area can use a large-scale interface to approximate, this will exclude the possibility of some small bubbles seeing in the experiments which dispersed in the liquid. This model also took the compressibility into account and ignored the effect of turbulent on the cavitation. Kubota, Kato and Yamaguchi[11] proposed a model which assumed that the cavitation is caused by grew up of tiny gas nuclear which existed in the liquid. The pure liquid and pure steam is treated as incompressible medium, and the overall density of cavitation area determined by the content of steam. All above models are roughly divided into two kinds of model. The first model based on Euler method. The second based on the Euler - Lagrange method.

2.1 Euler - hybrid model

Hybrid Singhal model[12] assumed that the presence of large amounts of gas nucleus in water, gas nuclear size and quantity is a constant value, there is no relative velocity between gas and water. By raising the threshold pressure to control the cavitation production. This model takes the turbulent, not dissolved gas into account on the influence of cavitation, but it also ignored many factors that can affect the cavitation, such as the nuclear spectrum of gas nuclear, the relative velocity between liquid and gas nuclear, the viscosity of liquid, surface tension and velocity gradient, etc. But, when this simplification simulated the problem of the rapid flow of the steady-state cavitation, the accuracy of simulation was also acceptable.

2.2 Euler two-fluid model

Advantages of Euler two-fluid model compared with the hybrid model is solving continuity equation and momentum equation of liquid and gas phases, and can also set turbulent models for the liquid and the gas phases respectively, we can improve the accuracy of simulation, this will certainly increase the computer memory and calculation time for the simulation. Two kinds of cavitation model based on Euler two fluid models are Zwart-Gerber-Belamri[13] model and Schnerr-Sauer[14] model.

2.3 Eulerian - Lagrangian method

Liquid phase is treated as continuous phase which solved by N-S equation. Bubbles are treated as discrete phase which dispersed in the continuous phase, getting its position at a certain moment in the continuous phase through integrating the force equilibrium equation in the flow field of continuous phase, so its trajectory is obtained. It also can achieve the coupling transfer of momentum, mass and energy between discrete phase and continuous phase. After analysis of the essence of the cavitation occurs, we can find that the Eulerian-Lagrangian method compared with the former two models can more precisely describe the process of cavitation happening. Because the Eulerian-Lagrangian method can research the state of a single bubble in the flow field, the individual characteristics of many bubbles will be considered, such as the size and number of bubbles and the trajectory, this will be conducive to improve the accuracy of cavitation simulation.

3. Modification of the Euler model

Based on the analysis of factors which have effect on cavitation, using the UDF function in Fluent, we can make appropriate modifications to above three models as following.

Considering the previous analysis of turbulent influence on cavitation, in order to measure the effect of turbulent on cavitation, the final threshold pressure P_v is:

$$P_v = P_{sat} + \frac{1}{2}\left(\frac{C_1}{n\Delta t} + C_2\right)\rho k + \frac{\rho U^2}{4} \tag{1}$$

Where P_v is the vapor pressure, P_{sat} is the saturated vapor pressure, ρ is the density of liquid, k is the turbulent kinetic energy, n is the number of per unit area grids. ∇t is the time step, C_1 and C_2 are coefficient. Some effects of cavitation from turbulent is due to the relatively large structure vortex. In the numerical simulation, the capture precision of vortex in the turbulent depends on the number of grids and the time interval. In other words, the capture of vortex is different. Different capture precision of vortex will directly affect vortex impact on cavitation, so add $C_1/(n\nabla t)$ in Eq. (1). In some occasions, some small vortex which can't be simulated by the numerical simulation, people used to think that as a kind of nonlinear pressure pulsation. So add C_2 to Eq. (1) in order to consider the effect of simulating cavitation caused by pressure fluctuation.

4. The numerical simulation of the unsteady flow field of cavitation jet

In this article, we first study its critical state of single-phase flow before pulsed cavitation occurring, then select the cavitation model to simulate the complicated multiphase flow state. Such a step-by-step approach is beneficial to advance research reliably and steady, laying the foundation for further to solve the problem.

4.1 Geometric modelling, time step, mesh generation and boundary condition

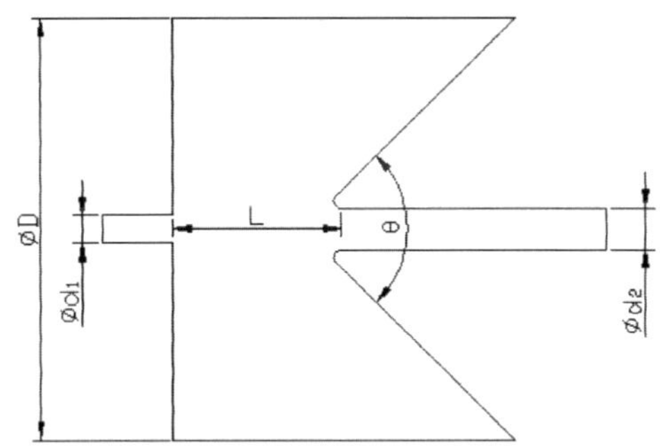

Fig. 1. The geometric model of nozzle.

In Fig 1, there are five major parameters of nozzle structure such as the upper nozzle diameter d_1=6mm, the cavity length L=45mm and the cavity diameter D=78mm, the collision angle θ=120 and the down nozzle diameter d_2=11mm. The cavity is simplified as a two-dimensional axisymmetric model which is divided into five parts in Fig 2: The upper and down nozzle parts respectively uses fined grids of quadrilateral structure. The middle cavity is divided into three parts and they all adopt quadrilateral grids.

In order to verify the effect of grid independence and the time step, simulations are carried on the same geometric model with different mesh density and time step, the specific settings are shown in table 1.

Due to the axial symmetry, we take its half as the simulation area as shown in figure 2. The boundary conditions of inlet $\overline{AB}$ is set as pressure inlet and the value is 0.2 MPa, $\overline{CD}$ is

IOP Conf. Series: Materials Science and Engineering **129** (2016) 012043 doi:10.1088/1757-899X/129/1/012043

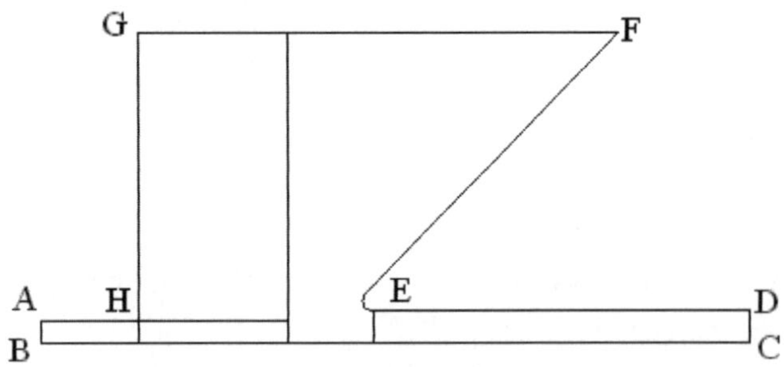

Fig. 2. The partitioning of nozzle grid area

Table 1. The setting of the grid number and the time step

The number of grid cells n	Time step Δt
108087	infinite（steady state）
182545	0.00005
270945	0.000005

pressure outlet and set the value as 101325 Pa. Others are defined as the solid wall. We use turbulence model of RNG k- ε, simulations are performed on steady and transient state with grids number 182545, the time step 0.00005 s and 0.000005 s, respectively.

Table 2. The comparison of different time step

Time step（s）	Vortex core pressure（Pa）	Inlet pressure（Pa）	Inlet velocity（m/s）
0.00005	32000	39200	18.0
0.000005	27200	39000	18.0

Under steady state, the residual fluctuates greatly and does not reach a stable solution. Under transient simulation with time step 0.00005s, within a time step, residual fluctuates significantly, but eventually gets a relatively stable solution. Meanwhile with 0.000005s, the convergence speed of residual is so quickly. Vortex cores appears in both transient states, the vortex intensity is different. As shown in table 2, the inlet pressure and velocity are basically the same, but there exists difference of vortex core pressure. The transient flow simulated with 0.000005 s appears obvious vortex structures in the shear layer than 0.00005s. Besides, the small scale vortex core pressure is low, this will have a great influence on cavitation inception.

Performing transient numerical simulation respectively on three different number of mesh grids with time step 0.000005s. Convergence of three grids is good, and they all can converge as soon as possible. With the increase of grid number, the vortex core pressure and inlet pressure increase, and the inlet velocity is reduced. The results of the grid number 182545 have been closed to the number

270945, we think the grid number 182545 state has certain reliability.

For the transient numerical simulation with the time step 0.000005s, the RNG k- ε ,the realizable k- ε , the standard k- ω and SST k- ω turbulence model is employed, respectively. From the simulation, we know that small scale eddies can be captured by RNG k- ε model except other three models.

4.2 The numerical simulation after cavitation

In the FBM[6] model, the turbulent viscosity μ_t [14]can be written as:

$$\mu_t = C_\mu \frac{\rho_m k^2}{\varepsilon} \cdot f_{FBM}, \quad f_{FBM} = \min\left[1, C_3 \frac{\Delta \cdot \varepsilon}{k^{3/2}}\right] \tag{2}$$

Where C_μ = 0.85, σ_k and σ_ε is Prandtl number 0.7179, G_k is the generation term of turbulent kinetic energy, f_{FBM} is filtering function which was determined by the ratio of the filter scale to turbulence scale. Considering the vapor-liquid density was influenced by the turbulent viscosity, the filtering function is modified as follows:

$$f_{FBM} = \min\left[f(\rho), C_3 \frac{\Delta \cdot \varepsilon}{k^{3/2}}\right], f(\rho) = \frac{\rho_v + \alpha_l^n (\rho_l - \rho_v)}{\rho_v + \alpha_l (\rho_l - \rho_v)} \tag{3}$$

Where C_3 = 1.0， n =10.

The choice of filter scale is particularly important for the smooth realization of the filtering process. Johansen[15] suggested the filtering scale Δ was,

$$\Delta = \max(\Delta_{select}, \Delta_{grid}), \Delta_{grid} = (\Delta x \cdot \Delta y \cdot \Delta z)^{1/3} \tag{4}$$

Where $\triangle x$、$\triangle y$ and $\triangle z$ are the grid space The filtering scale should be larger than grid scale in computational area. We carried on UDF programming on the above modified turbulent model.

We perform simulation on the internal flow field of low pressure self-excited pulsed cavitation jet nozzle with the inlet total pressure is 2MPa. The absolute pressure inside the cavity is below 0.058MPa which was measured by pressure sensor.

In Fig. 3, 4 and 5, we performed the numerical simulation on the nozzle internal flow field by using Singhal and the single-phase RNG k-ε model, Singhal and the modified multiphase RNG k- ε model, modified Singhal and the modified multiphase RNG k- ε model, velocity, pressure and the volume fraction contours of water vapor in a period of pulse were shown, respectively.

From figure 3, the flow contours seems more reasonable, bubbles distribution in the nozzle cavity is more uniform, gas-liquid mixing area is not visible. From figure 4, The volume fraction of gas in the cavity shear layer center is lower compared with single-phase RNG k- ε turbulence model. This should be due to the modified multiphase RNG k- ε model reduces the turbulence intensity so that the threshold pressure of cavitation inception is reduced too.

From figure 5, the volume fraction near the cavity center line increases, this should be caused by the modified Singhal model which considers the effect of large velocity gradient and the

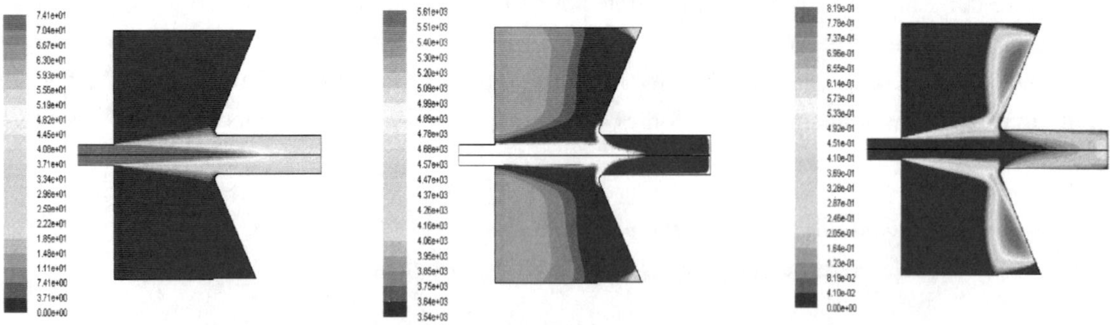

a) Velocity vector b) Pressure contours c) Volume fraction contours of vapor

Fig. 3 The flow field contours simulated by Singhal-single phase RNG k-ε model

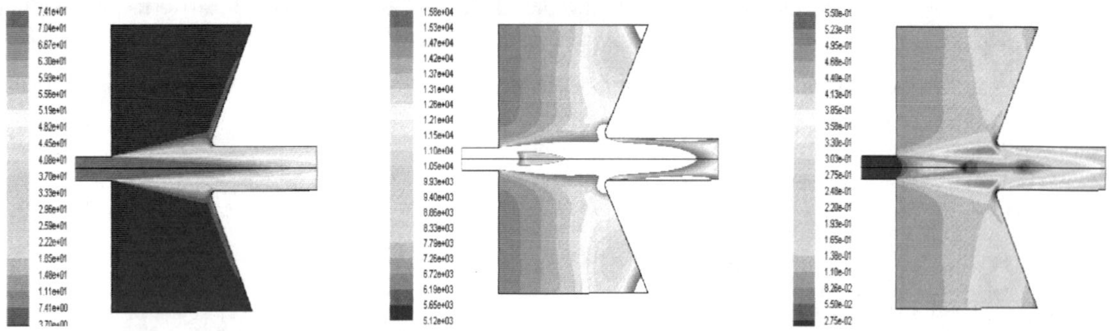

a) Velocity vector b) Pressure contours c) Volume fraction of vapor

Fig. 4 The flow field contours simulated by Singhal- modified multiphase RNG k-ε model

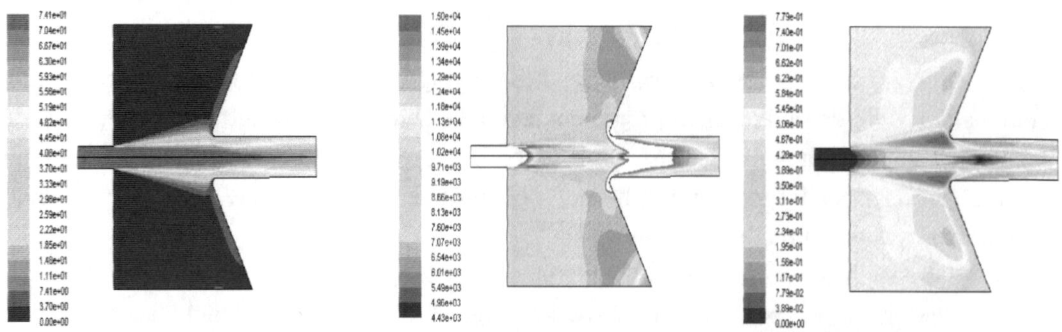

a) Velocity vector b) Pressure contours c) Volume fraction of vapor

Fig. 5 The flow field contours simulated by modified Singhal- modified multiphase RNG k-ε model

relative velocity between bubbles and water. The simulation results of modified Singhal model is close to the experimental data, gas content and chamber pressure value is also closes to the experimental results. Therefore, the modified Singhal model has good accuracy in the simulation of internal cavitation jet flow field.

5. Conclusions

Conclusions are drawn as following.

(1) Suitability of cavitation model such as Euler Hybrid Model, Euler Two Phase Model and Euler

Lagrange Model is discussed. Those models do not consider the effect of the breakage and merger of bubble on inception, development and breaking of cavitation. Turbulence intensity, larger velocity gradient in cavitation chamber, relatively velocity between bubble and liquid are merged to modify the Euler cavitation model.

(2) Through the numerical simulation on the self-excited pulsed cavitation jet. we find that only RNG k- ε turbulence model can capture the small scale eddies of jet by establishing an equivalent pressure in the transient condition. In the cavity chamber, large scale vortex ring around the jet center line is produced, the pressure in vortex core is slightly lower than that in the upper nozzle.

(3) The original Singhal model and modified one, single-phase flow turbulence model and modified multiphase model are combined to simulate the turbulence flow in the self-excited pulsed cavitation after the cavitation occurs. Through different simulation comparison to the experimental results, it proves that the simulation results by modified Singhal model are relatively accurate.

Acknowledgement
The authors acknowledge the National Natural Science Foundation of China (51176102). Deyu also thanks the support of Shandong Provincial Science and Technology Development planning Program (2013YD09007) and Shandong Provincial Natural Science Foundation (ZR2014EEM017).

Reference

[1] Huang J T, Chen J F, Ding T et al. 1996 *Journal of Hydraulic Engineering* **12**: 1-7.
[2] Huang J T, Li Z M 1987 *Journal of Hydraulic Engineering*. **8**: 44-48.
[3] Huang J T, Tian L Y 1988 *Journal of Hydraulic Engineering*. **12**: 41-45.
[4] Christopher E B 1995 (London, Oxford university press) 54-55.
[5] Chao-Tsung, H Georges, L Chahine, et al. 2000 *Fluid Mech*, **11**:123-135.
[6] Johnsson C A. *Proc 12th Int. Towing Tank Conf*, Rome, 381-392.
[7] Delannoy J K. 1980 *Journal of computational physics*, **35**: 229-253.
[8] Avva R K, Singhal A, Gibson D H 1995 *Journal of Fluids Engineering*, **24**:134-143.
[9] Schmidt D P, Corradini M L 2001 *International journal of engine research*, **2**:1-22.
[10] Dumont N, Simonin O, Habchi C 2001 *Fourth international symposium on cavitation*(California) 7.
[11] Marcer R, Le Cottier P, Chaves H, et al.2000 *Journal of engines*, **10**:23-35.
[12] Kubota A, Kato H, Yamaguchi H. 1992 *Journal of fluid mechanism*. **240**:59-96.
[13] Singhal A K, Li H Y, Athavale M M, et al..2001 *ASME fedsm01*.(New Orleans)
[14] Zwart P J, Godin P G, et al.2004 *Fifth international conference on multiphase flow* (Yokohama)
[15] Johansen H, WU J, Shyy W 2004 *International Journal of Heat and Fluid Flow*, **25(1)**:10-21.

IOP Conf. Series: Materials Science and Engineering **129** (2016) 012044 doi:10.1088/1757-899X/129/1/012044

Numerical study on thermodynamic characteristics of rotational supercavitating evaporator

Q Li[1], Z Y Zheng[1], F C Li[1*] and V A Kulagin[2]

[1] School of Energy Science and Engineering, Harbin Institute of Technology, Harbin 150001, China
[2] Department of Heat Technology and Fluid Dynamics, Siberian Federal University, Krasnoyarsk 660041, Russia

E-mail: lifch@hit.edu.cn

Abstract. Rotational Supercavitating Evaporator (RSCE) has been proposed as a new technology for seawater desalination. However, thermodynamic characteristics of rotational supercavitation are still vacant. In this paper, numerical simulations are conducted on the supercavitating flows around a 3D rotating blade of RSCE with different rotational speeds and extraction pressures. Energy effect is taken into consideration in the simulation and thermodynamic characteristics of rotational supercavitation are obtained. Rotational supercavitation has a larger convective heat transfer coefficient than the boiling on a heated wall.

1. Introduction

Water is the source of our lives and the material basis of the human beings' survival. There exists a severe shortage of fresh water around the world, which has led to development of seawater desalination methods. The most widely used technologies are thermal methods and membrane methods, which make up about 95% of worldwide fresh water production. However, the former is limited by the scale formation and heat transfer coefficient, and the later is limited by the membrane fouling and the recovery factor. There is a technological bottleneck in the thermal methods to evaporate water into steam free from scale formation and dramatical increase in heat transfer coefficient. On the surface of supercavity, water evaporates into steam in the same way with boiling of water. However, the heat transfer coefficient of evaporation from the surface of supercavity depends on the heat-flux density and there is no scale formation. So a device named Rotational Supercavitating Evaporator (RSCE) was put forward for desalination based on supercavitation effect. The RSCE model is shown in Figure 1. When the rotational blade rotates at high speed (1000~5000 r/min), supercavitation is generated right behind the blade and the steam is extracted from the inside of supercavity, which would be then condensed to produce fresh water. The RSCE also has a good application foreground in sewage disposal, such as nuclear waste water and polluted water containing heavy metals.

* To whom any correspondence should be addressed.

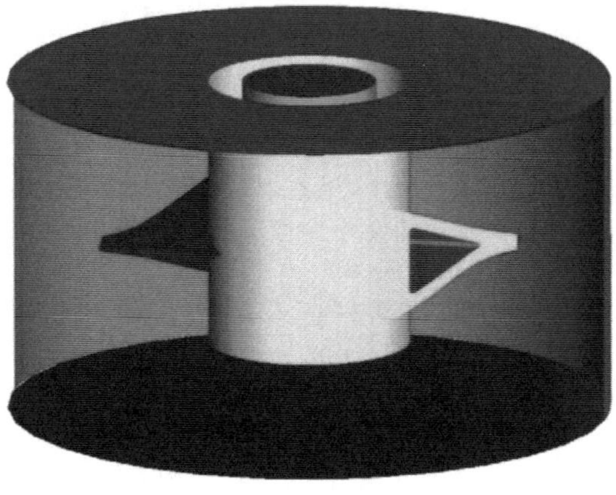

Figure 1. 3D RSCE model

Cavitation can be defined as the breakdown of a liquid medium under very low pressures and applied to cases in which the liquid is either static or in motion. It was first discovered on the propeller in the warship of British Royal Navy in 1897. The study on cavitation primarily focused on the prevention of negative influence of cavitation phenomenon such as noise, vibration and surface erosion. Subsequently the drag reduction effect of supercavitation on an underwater body was found, and it came an upsurge of research on the drag reduction of underwater vehicle based on supercavitation effect. 30 years ago, Machinski [1] introduced the stationary supercavitating cone evaporator, which can be connected to vacuum system for steam extraction for desalination. Recently Likhachev and Li [2-4] put forward RSCE and performed preliminary design and research. Li and Zheng acquired many characteristics on supercavitation of 2D wedge-shaped cavitators and 3D rotational cavitators [3-6]. Even though many characteristics of rotational supercavitation are acquired, the thermal effect on the steam extraction process has never been taken into account. So detailed thermodynamic characteristics of rotational supercavitation, for example, convective heat transfer coefficient, are still vacant.

In order to obtain thermodynamic characteristics of rotational supercavitation, numerical simulations are conducted on 3D RSCE model shown in Figure 1. Energy effect is taken into consideration in numerical simulations to investigate the thermal characteristics of rotational supercavitation. Numerical simulations are performed under different steam extraction pressures to investigate the thermodynamic characteristics of rotational supercavitation with different quantities of steam extraction. As a conclusion, the thermodynamic characteristics of rotational supercavitation are summed up.

2. Numerical method

2.1. Governing equations

Numerical simulations are conducted on the three-dimensional steady supercavitating flows in this paper. In the numerical simulation, supercavitating flow is a multiphase and turbulent flow and mixture model based on the homogeneous equilibrium multiphase flow theory is selected as the multiphase flow model. The mixture of the gas and liquid phases is considered as homogeneous single-phase fluid and the equations in this model are shown below.
Continuity equation of the mixture phase:

$$\nabla\cdot\left(\rho_m\vec{v}_m\right)=0 \tag{1}$$

Momentum equation of the mixture phase:

$$\frac{\partial}{\partial t}(\rho_m\vec{v}_m)+\nabla\cdot(\rho_m\vec{v}_m\vec{v}_m)=-\nabla p+\nabla\cdot\left[\mu_m\left(\nabla\vec{v}_m+\nabla\vec{v}_m^T\right)\right] \tag{2}$$

Transportation equation of the vapour fraction of the gas phase:

$$\nabla\cdot(\alpha_v\rho_v\vec{v}_m)=R_e-R_c \tag{3}$$

where ρ_m is the density of the mixture phase, $\vec{v}_m$ is velocity, α_v is the vapour fraction of the gaseous phase, ρ_v is the density of the gaseous phase, R_e are R_c the production rate and condensation rate of the gaseous phase, respectively. In order to depict R_e and R_c, Schnerr-Sauer model [7] is adopted as the cavitation model.

For $p_v \geq p$,

$$R_e=\frac{\rho_v\rho_l}{\rho_m}\alpha_v(1-\alpha_v)\frac{3}{\Re_B}\sqrt{\frac{2}{3}\frac{(p_v-p)}{\rho_l}} \tag{4}$$

For $p_v \leq p$,

$$R_c=\frac{\rho_v\rho_l}{\rho_m}\alpha_v(1-\alpha_v)\frac{3}{\Re_B}\sqrt{\frac{2}{3}\frac{(p-p_v)}{\rho_l}} \tag{5}$$

where ρ_l is the density of the liquid phase, p_v is the saturation vapor pressure, $\Re_B$ is the radius of the bubble and defined as follows.

$$\Re_B=\left(\frac{\alpha_v}{1-\alpha_v}\frac{3}{4\pi}\frac{1}{n}\right)^{\frac{1}{3}} \tag{6}$$

Herein, n is the number of the microbubbles within the liquid of unit volume and set as 1×10^{13}.
The energy equation for the mixture multiphase model takes the following form:

$$\frac{\partial}{\partial t}\sum_{k=1}^{n}(a_k\rho_kE_k)+\nabla\cdot\sum_{k=1}^{n}\left(a_k\vec{v_k}(\rho_kE_k+p)\right)=\nabla\cdot(k_{eff}\nabla T)+S_E \tag{7}$$

where k_{eff} is the effective conductivity. The first term on the right-hand side represents energy transfer due to conduction. S_E includes any other volumetric heat sources.
where:

$$E_k=h_k-\frac{p}{\rho_k}+\frac{v_k^2}{2} \tag{8}$$

Turbulent flow in the numerical simulation is described by Realizable $k-\varepsilon$ model and Scalable wall function is adopted to solve the flow near the wall.

2.2. Boundary conditions and mesh
The computational domain is shown in Figure 2. It is a cylinder-shape domain with the rotational blade fixed on the top surface. The diameter of the cylinder-shape domain is 200 mm and the height is 100 mm. The maximum diameter of the rotational blade is 120 mm and the diameter of the rotational axle is 60 mm. The diameter of the extraction tunnel is 40 mm.

The annular region on the top surface is set as velocity-inlet and two fluid velocities, 1m/s and 3m/s, are used in the numerical simulation in different cases. The temperature of the water flowing into RSCE from the inlet is set as 379.15K. So, the hot water, flowing into RSCE from the top surface, release the heat through the evaporation on the surface of supercavity and flow out from the bottom surface as cooled water. The bottom surface is set as pressure outlet with the pressure of 134010 Pa, which is the saturated vapor pressure at 381.15 K. In this case, the water entering from the inlet has a very small supercooling degree of 2 K, which contributes to the generation of supercavity. The exit of the extraction tunnel is set as pressure outlet with the same pressure as the extraction pressure of RSCE. In the simulation, three extraction pressure, 84608 Pa、70182 Pa and 47414 Pa, are adopted, respectively. The side surface of the cylinder-shaped domain, the blade and the rotational axle are set as walls.

The thermodynamic properties of water and steam are defined according to international standard for the thermodynamic properties of water and steam IAPWS95 [8] at 379.15 K. Density of water liquid is $954\,\mathrm{kg/m^3}$, viscosity of water liquid is $2.65\times10^{-4}\,\mathrm{Pa\cdot s}$, density of water vapor is $0.728\,\mathrm{kg/m^3}$ and viscosity of water vapor is $1.25\times10^{-5}\,\mathrm{Pa\cdot s}$. The computational domain is meshed by the unstructured grids, which is shown in Figure 3.

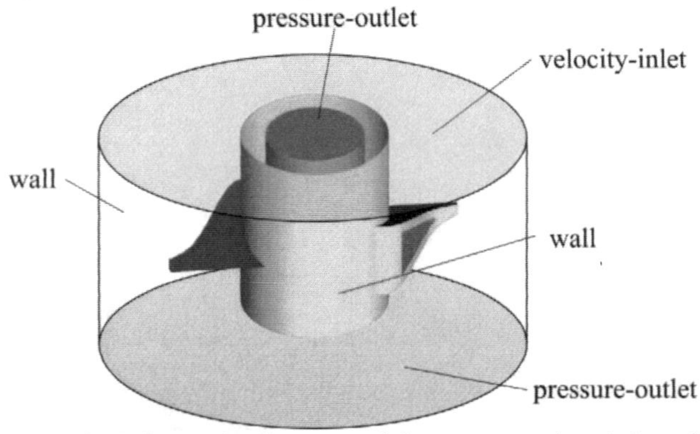

Figure 2. Schematic diagram of the computational domain.

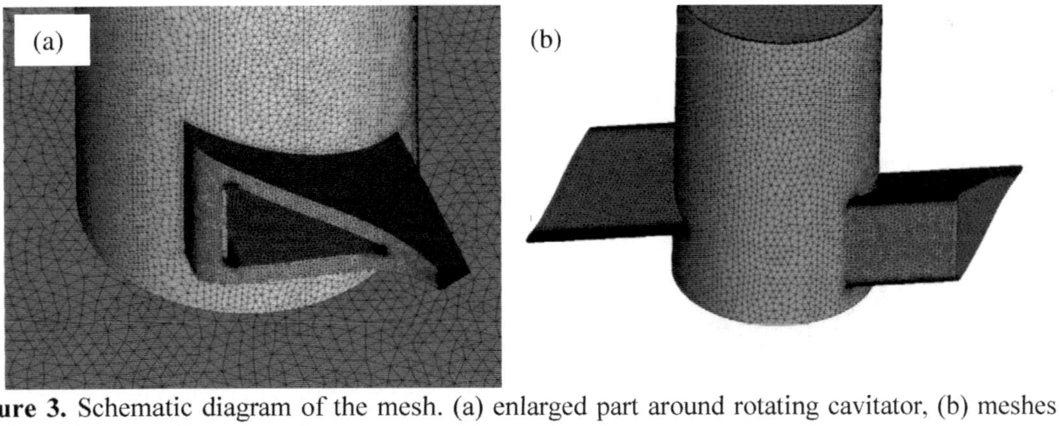

Figure 3. Schematic diagram of the mesh. (a) enlarged part around rotating cavitator, (b) meshes in the extraction tunnel.

3. Results and discussions

In the simulation, saturated vapor pressure varies with temperature. As vapor pressure is one of the key parameters that has a vital influence on the supercavitation, it should be fitted with temperature. The fitting function in the simulation of this paper is as follow:

$$P_v=\begin{cases}519.17+10T & 0\mathrm{K}<T\le301.15\mathrm{K}\\ 7963307-109393T+563.6T^2-1.294T^3+0.00112T^4 & 301.15\mathrm{K}<T\le395.15\mathrm{K}\\ 170734.37+58T & 395.15\mathrm{K}<T\le10000\mathrm{K}\end{cases} \tag{9}$$

When the simulation is convergent, the temperature is within the range $301.15\text{K}<T\leq 395.15\text{K}$, so the fitting function beyond the range is just used to avoid the divergence in the begin of the simulation.

In order to generate supercavity with appropriate size, two rotational speeds of the blade, 2500 r/min and 3000 r/min, are chosen in the simulations. Numerical simulations on RSCE model are conducted with certain rotational speeds and extraction pressures. The hydraulic and thermodynamic characteristics of supercavity formed by blade with different rotational speeds and extraction pressures are obtained.

3.1. hydraulic characteristics

In the numerical simulation, two rotational speeds of the blade, 2500 r/min and 3000 r/min, are chosen and three cases with the extraction pressure of 84608 Pa、70182 Pa and 47414 Pa are simulated for each rotational speed. All these cases with the velocity of 1 m/s at the inlet. The supercavitygenerated by the rotational blade is shown in Figure 4 (for 2500 r/min and 84608 Pa). Contours of vapor volume fraction in each case are shown in Figure 5. The contour is on the rotating plane with the leading edge of the blade on(z=0 plane).

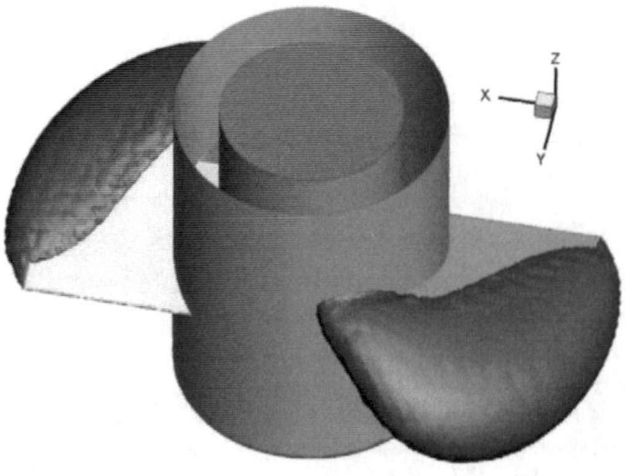

Figure 4. Supercavity generated by the rotating blade (for 2500 r/min and 84608 Pa).

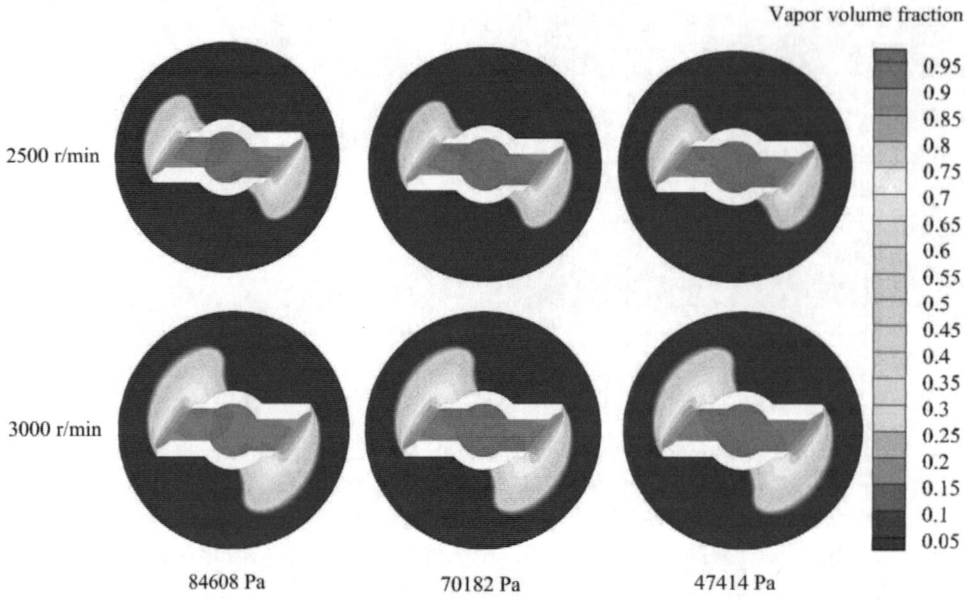

Figure 5. Contours of vapor volume fraction in cases with different rotational speeds and extraction pressures.

As observed, the supercavity is just generated right behind two blades and symmetric about the axle. The blade with a higher rotating speed can generate a larger supercavity, and the shape of the surpercavity is not affected by the extraction pressure at the same rotational speed.

Numerical simulation is also conducted with the inlet velocity of 3m/s, in which the rotational speed is 3000 r/min and the extraction pressure is 47414 Pa. Comparing with the case with the velocity of 1 m/s which is shown in Figure 6, it can be obtained that a higher velocity of the flow that strikes on the side surface of supercavity, more visible the effect on the shape of the supercavity. But both the striking velocities of 1 m/s and 3 m/s induce no prominent change of the supercavity shape, which means the effect of smaller velocity on the side surface is acceptable.

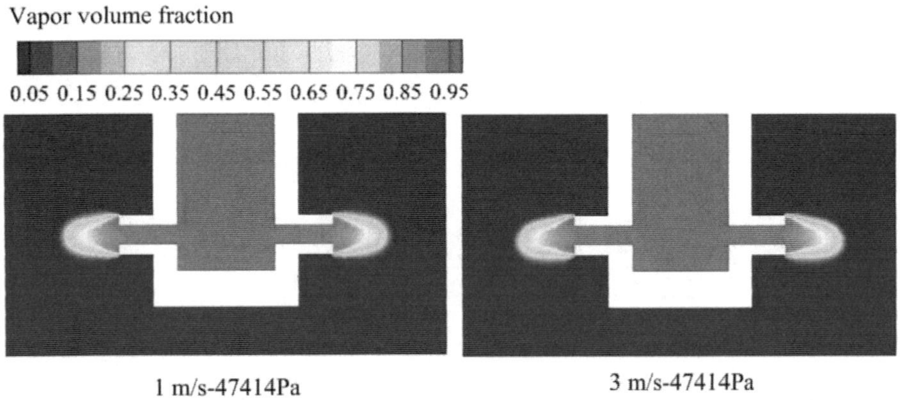

Figure 6. Contours of vapor volume fraction in cases with different inlet velocities (plane y = 0).

3.2. Thermodynamic characteristics

The temperature distribution is shown in Figure 7 (a). As observed, when flowing through the supercavity area, the hot water is subjected to a significant temperature decline. But the low temperature area beneath rotational axle is abnormal, which is caused by the backflow from the bottom surface, as shown in Figure 7 (b).

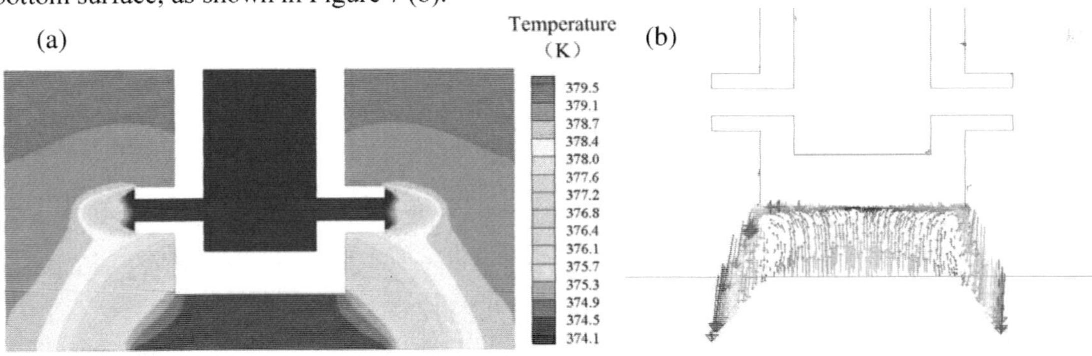

Figure 7. Distribution of temperature (plane y=0). (a) contour of temperature in cases with different inlet velocities (plane y=0). (b) velocity vector in partial area which is coloured by temperature.

As the analysis of the result, convective heat transfer coefficient is calculated as follow:

$$h = \frac{\Phi}{A \cdot \Delta T} \tag{10}$$

where heat flux Φ is calculated from the steam extraction and enthalpy difference between liquid water and vapor. As the area of the supercavity is unobtainable, the area A in the equation is the area of the annular inlet, which is equal to the rotational plane's area of the blade and supercavity. The temperature different ΔT is calculated from the temperature of the inlet and the boiling temperature of water under the extraction pressure. Thermodynamic parameters are shown in Table 1. The convictive heat transfer coefficient is much higher than that of the boiling on a heated wall, which means supercavitation has a great enhancement on heat transfer during the evaporation. As observed, a low extraction pressure can generate a high steam extraction and convective heat transfer coefficient, but a higher rotational speed has no contribution to the steam extraction and convective heat transfer coefficient, even it could lead to a larger supercavity.

Table 1. Thermodynamic parameters of the RSCE

Rotating speed (r/min)	Inlet velocity (m/s)	Extracting pressure (Pa)	Steam extraction (kg/s)	Convective heat transfer coefficient (W/(m²• K))
2500	1	84608	0.00546	39149
	1	70182	0.01199	59059
	1	47414	0.03336	101147
3000	1	84608	0.00413	29598
	1	70182	0.01230	60603
	1	47414	0.03279	99428
	3	47414	0.03391	102840

4. Conclusions

Numerical simulations are conducted on the cavitating flows around the 3D rotating cavitators with different rotational speeds and extraction pressures. In the simulation, energy effect is taken into consideration and thermodynamic characteristics of rotational supercavitation are obtained.

Blade with a higher rotational speed can generate a larger supercavity and extraction pressure affects the shape of the supercavity little. The flow with smaller velocity of that strikes on the side surface dose not affect the supercavity fatally. The convective heat transfer coefficient of supercavitation is much higher than that of the boiling on a heated wall. Extraction pressure has a significant influence on the steam extraction and convective heat transfer coefficient, while rotating speed does not.

Acknowledgments

This work is supported by National Natural Science Foundation of China (51276046). The authors would like to appreciate the valuable discussions with the members of Complex Flow and Heat Transfer Lab.

References

[1] Machinski A S. 1984 Hydrodymanics and thermal transfer characteristics of supercavitating evaporators for water desalination (in Russian). *Dissertation for the Doctoral Degree.* (Moscow: Kiev Order of Lenin Polytechnic Institute) 285

[2] Likhachev D S and Li F C 2013 Large-scale water desalination methods: a review and new perspectives Desalin. Water Treat. 51 2836–49

[3] Likhachev D S and Li F C 2013 Modeling of rotational supercavitating evaporator and the geometrical characteristics of supercavity within Sci. China Phys. Mech. 56 1–14

[4] Likhachev D S, Li F C and Kulagin V A 2014 Sci. China-Technol. Sci. 57(11) 2115–30

[5] Zheng Z Y, Cheng J P, Li F C, et al. 2015 Numerical simulation of the temperature effects on the performance of rotational supercavitating evaporator IOP Conference Series: Materials Science and Engineering. (IOP Publishing) 72(6): 062054

[6] Zheng Z Y, Li F C, Li Q and Kulagin V A 2015 Sci. China-Technol. Sci. 58(6) 1072–83

[7] Schnerr G H and Sauer J 2001 Physical and numerical modeling of unsteady cavitation dynamics 4th International Conference on Multiphase Flow (New Orleans, USA)

[8] Wagner W and Pruß A 2002 The IAPWS formulation 1995 for the thermodynamic properties of ordinary water substance for general and scientific use J. Phys. Chem. Ref. Data 31 387–535

IOP Conf. Series: Materials Science and Engineering **129** (2016) 012045 doi:10.1088/1757-899X/129/1/012045

Micro-grid platform based on NODE.JS architecture, implemented in electrical network instrumentation

M Duque[1], E Cando [1], A Aguinaga, [1] F Llulluna [1], N Jara[2], T Moreno [3]

[1] Department of Mechanical Engineering, Escuela Politécnica Nacional, Quito-Ecuador.
[2] Faculty of Mechanical Engineering, Universidad Politécnica Salesiana, Cuenca-Ecuador.
[3] Department of Mechanical Engineering, Escuela Politécnica del Chimborazo, Riobamba-Ecuador

E-mail: mauricio.duque@epn.edu.ec

Abstract. In this document, I propose a theory about the impact of systems based on micro-grids in non-industrialized countries that have the goal to improve energy exploitation through alternatives methods of a clean and renewable energy generation and the creation of the app to manage the behavior of the micro-grids based on the NodeJS, Django and IOJS technologies. The micro-grids allow the optimal way to manage energy flow by electric injection directly in electric network small urban's cells in a low cost and available way. In difference from conventional systems, micro-grids can communicate between them to carry energy to places that have higher demand in accurate moments. This system does not require energy storage, so, costs are lower than conventional systems like fuel cells, solar panels or else; even though micro-grids are independent systems, they are not isolated. The impact that this analysis will generate, is the improvement of the electrical network without having greater control than an intelligent network (SMART-GRID); this leads to move to a 20% increase in energy use in a specified network; that suggest there are others sources of energy generation; but for today's needs, we need to standardize methods and remain in place to support all future technologies and the best option are the Smart Grids and Micro-Grids.

Introduction

The electrical network was designed a hundred years ago, its goal was to supply electricity to an specific community for basic needs (radio, illumination and cooking), were with, electric need was low; but, now a day necessity on electricity has grown significantly and our society now depends on it, for this reason the networks that were not prepared to asume and satisfy this needs are lack on efficiency due to their unidirectionality, they do not have a feed back system and also have few interaction with the user. Another motivation is the integration of alternative and renewable sources of energy. The Smart-Grid is an autonomous, efficient and optimum system with the capacity of making decisions that will benefit the user and also the business that generates the service; a Smart-Grid allows the integration between multiple, which can interact between them with the goal of optimizing

[1] To whom any correspondence should be addressed.

the energetic use in micro systems like the Smart-Home, Smart-Car, Renewable Energy, Consumer Engagement, Operation Centers, Distribution Intelligence y Plugins and others that together make a Smart-grid system, improving the efficiency of electricity use. Unlike today's trend, that suggest that all devices that are connected to a network are under one system that just verify its function but not the quality the Smart-Grid, provides a whole of independent systems interconnected in one network that have the ability of control some processes but with the efficiency for work under one premise 'Optimization'.

In —'Figure 1', we show the Smart-Grid systems reach word wide in 2010; this shows that the utilization of this system is low for some reasons:

1) High costs of governmental installation.
2) Obsolete electrical networks.
3) Low prices on electricity production.
4) The non-existence of penalty for electrical inefficiency (power stealing leak, foulcault flowing etc).
5) The poor effort of the government for brings in this technology.
6) The low stock of the products that work with this technology.
7) The cost of the equipment for its correct work.
8) Complex urban infrastructure.

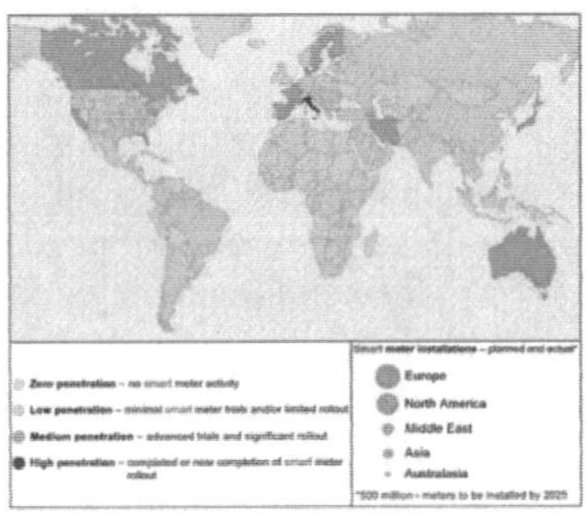

Figure 1. Smart-Grid diffusion in the world wide 2010.

From the above, in this paper we will analyse the points (6), (7) y (8).

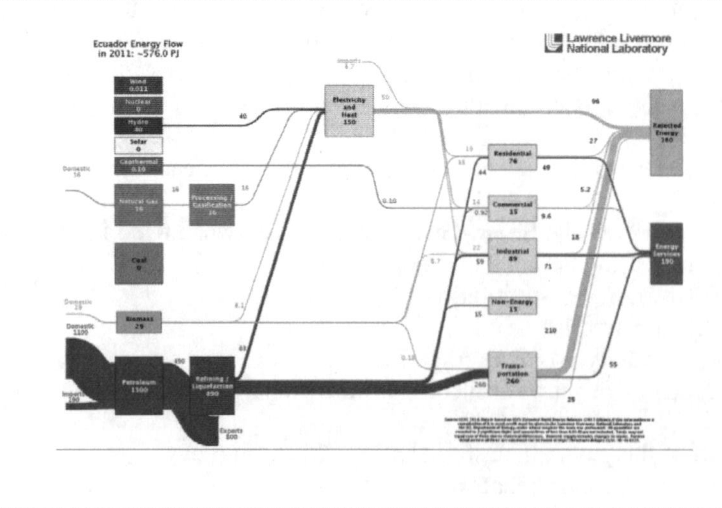

Figure 2. Energy Flow in Ecuador 2011.

The —'Figure 2'shows the electrical flow in Ecuador in 2011. Of the total or energy consumption in Ecuador, the 66.6% is wasted energy and just the 36% of electricity generation is use, el other 64% is wasted energy for many reasons, among these by energy inefficiency, because we could provide other countries of the same size of Ecuador with the amount of energy we waste; for this reason, we see that this country has potential for carrying this Smart-Grid technology mainly for the recent change on the energy matrix, where is expected to generate more electricity for avoid the importation of it, that in 2006 represented the spending of a billion dollar because of the building of hydroelectrics which have the challenge of generate 8740 megawatts that will be working in 2016.

The incorporation of renewable and alternative energy will demand the availability of microgrids. There is a trend to decentralized energy resources that include homeowners connecting to the grid. With this idea of the new energy matrix, Ecuador must consider to have a system that minimize energy losses, which affects directly for sale energy to other countries; in conclusion, through optimize the use of electricity in every home, the capacity for export it will increase.

One way to develop the micro-grid in Ecuador is by creating a self sustaining basis with the purpose of build infrastructure that support systems that are based on in Smart and Micro Grids, that work with modern technology such as NodeJS, IOJS and Django; with this our main goal is to have an place in the word in this area.

Ecuador has initiated a transformation process model, from being an exporter of raw material to become a society that will export products based on the knowledge and the biodiversity of the change of the productive matrix that will provide the product high added value; according to the government: Ecuador is betting to change its production model in a long term, where the change of imports and exports, along with innovation, science, technology and knowledge will allow to give a massive step in to the future.

Even though in Ecuador Smart-Grids systems could be implemented, this kind of technology is available the most of the time for first world countries, which have a bigger economic power than countries that have a poor economy. In the same way, business that owns this kind of technology will create technological dependency due to they are owners of the systems that also impose policies that are beneficial for their industry but not for the industrial development o the country.

When the NodeJS appears in a middle of 2011, change the web world, because this can serve more 64000 connections in a same time, this more than java and PHP, where the users number it's the approximately 4000 concurrent connections, for another hand, NodeJS use the benefits of a v8 engine of the Google Chrome, and plus the asynchronous power, the meaning of this, NodeJS can be change in a real time all variables or independently each ones of them, without need recharge the browser, this let the NodeJS are the really tool for develop our purposes.

Objective

a. Design microgrid architecture and controls.
b. Design a Communications protocol for microgrids.
c. Lay the groundwork for the standardization of control systems for Micro – Grids.

2.1. Specific objective

I. Design a control platform for Micro -Grid systems, with components of low cost, applied to the basic measurement variables, for Internet access.
II. Provide a network tool that can monitor and control signals processes through the analysis of physical variables.
III. Use current technologies such as JavaScript, Python, Django, IOJS and NodeJS for the development of the platform, leveraging the benefits of real-time telemetry and interconnectivity that these have.

3. Methodology

In first place, this proposal includes the development and implements of a web platform of a virtual system for perform the instrumentation and measuring of electrical, climatic and physical variables in order to have a better distribution of the micro-girds in Ecuador. In second hand, I suggest the utilization of low cost devices, like a Raspberry pi, Arduino and others components open hardware that could withstand the idea that I gave above; both proposals would make up a basic design for the development and implementation of the Micro - Grid.

The application will have the following peculiarities:

a) Local and two-way communication with a prototype (hardware) for the control and automation
b) An interface where the user can create an instrumentation panel for monitoring critical variables (Temperature, humidity, electric current and voltage and more).

The control devise must do four basic duties: acquisition, analysis, data reporting and monitoring; this devise has the chance of add more functionality without extra money spending, but it's not all to will doing, the most important, this can be take a decisions on critical moments, connect/disconnect any device in a real time, with intentionally for cause more benefits to electrical network.

From computers use was essential in measuring the analysis of such data in histograms that allows to have control and a feedback of the process, but this was not only the advantage of using virtual instrumentation; now a day we can integrate this technology with different industrial equipment PLCs, inverters, among others, transforming complex manual processes into automated processes, where it is preponderant the use of virtual instrumentation as a principle of decision-making, minimizing the human errors improving quality of peoples live.

Flexibility, low maintenance cost, reusability, customizing each instrument, the fast introduction of new technologies, low cost per channel and function are some of the benefits of the Smart-grid integration. Mainly the greater impact it has on the energy industry was the emergence of the terms "flexible automation" which involves synapses between different elements that are acting on a grid and can have an impact depending on the behavior of different variables.

Next we show the architecture for the platform:

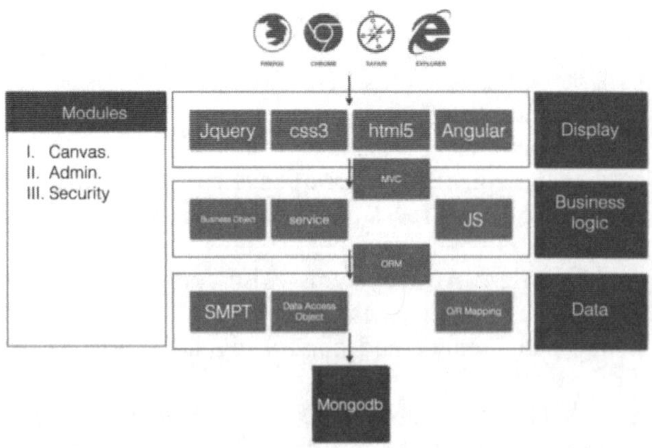

Figure 3. Platform Architecture.

The —'Figure 3', we present the APP; this platform allows: Connect to device, view the variables; personalize the dashboard, determinate behavior the micro-grid shield.
All the app is programmed in NodeJS and Django, are disposed in internet; when the user sign in to the app, he can find the all devices are connected in your net, for example: in one home have three micro-grids (i.e. roof, devices, and smart-car) are all plug in to the micro-grid shield net, each one are controlled independently but the setup is configuring through the app. The user can view in the dashboard variables (volts, currents) for each devices global or independent and in another hand, he can do turn on/off the energy flow.

APP	
OS	Multiplatform
Application Server	NodeJS IOJS Django
Languajes	JavaScript Python C#, C++
Frameworks	Mongoose Express Angular
Data Base	MongoDB Rethink
Dispositives	SmartPhones Tablets PC

Figure 4. Components of APP.

The —'Figure 4', contains all characteristics have the app, if Django and NodeJS are not compatibility but both have a characteristic target in to the APP, for example NodeJS and IOJS handles to all visual in app, while Django to do the signals process; the user never see the difference between three servers.

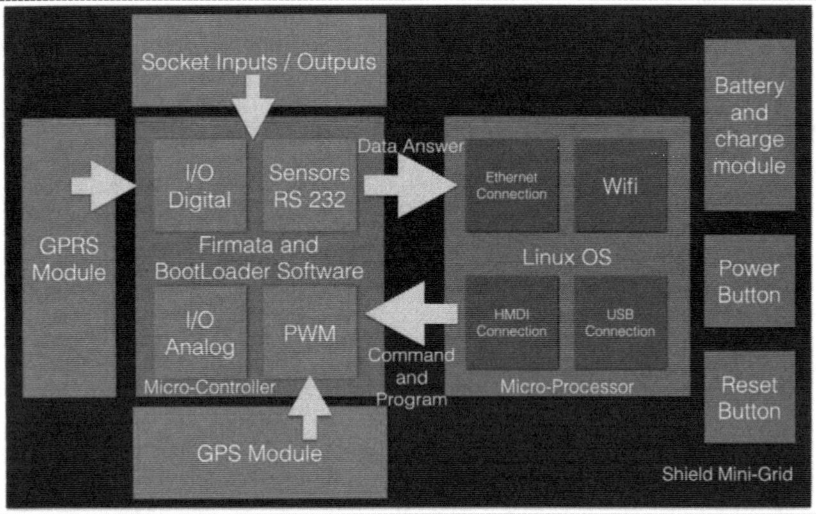

Figure 5. Micro and Mini-grid shield Architecture.

The micro-grid shield allows data acquire signals for the process, for example consumed current, temperature and the lux for the any areas, too it can power on/off any devices and finally and more important are a big capacity for management the switching electric flows, between consumed and produced energy in this way, the micro-grid together conform a mini-grid —'Figure 6', interconnected among all through internet way, and all information are send to the app, where the user can manage the behavior the each ones. The diagram —'Figure 5', represent are components have the shield architecture.

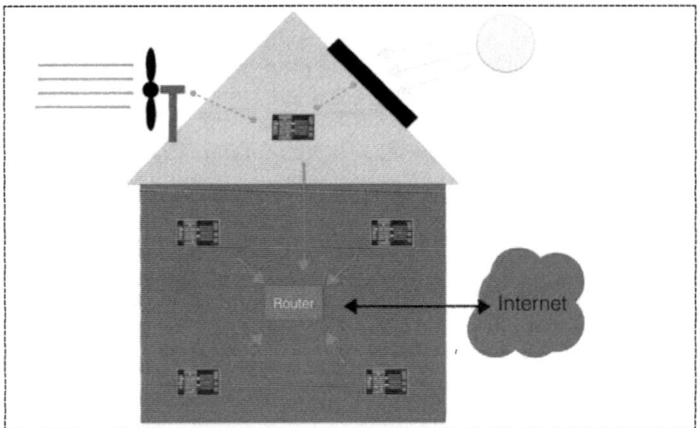

Figure 6. Mini-Grid Architecture.

The Mini-Grid are conformed by many micro-grids set, this behave like a network, where all information send to cloud through Internet for the app, like the picture above —'Figure 6'.

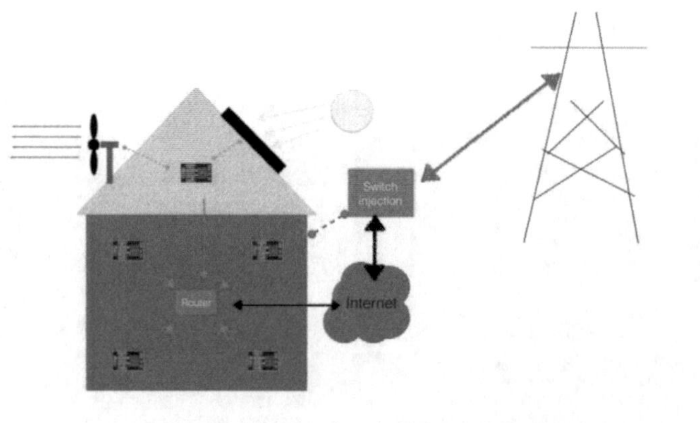

Figure 7. Mini-Grid Architecture connection to the network.

When the Mini-Grid produces more electricity than consumed, the behavior of this are allows the flow energy to Mini-Grid to electrical network.

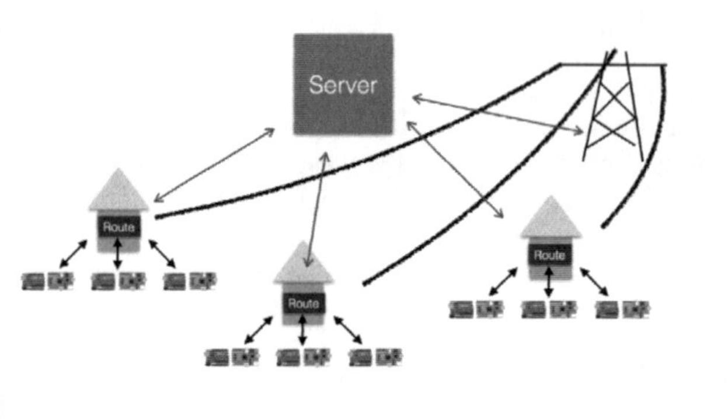

Figure 8. Cell.

Many sets of Mini-Grids are calling Cell, all interconnected Internet way. We hope in the future this system allows take more benefits by increasing breakthrough in Ecuador, this solve the 20% to 30% the not-benefit electricity.

Special Regards

Our special appreciation to **CEDIA (NATIONAL RESEARCH AND EDUCATION NETWORK OF ECUADOR) CEPRA 2015** for the cooperation for the investigation.

Examples taken from published papers:

[1] Academia J, Lorandi Labs 2011 Los Laboratorios Virtuales y Laboratorios Remotos en la Enseñanza de la Ingeniería. Ed.11, **Vol 4**

[2] Acción pedagógica, **Vol 11**, N°1/2002. Rafael Chacón Rugeles, La instrumentación Virtual en la Enseñanza de ka Ingeniería Electrónica, pp. 74-84.
[3] F. J. Quiles, J.I. Benavides, A. Moreno, R. Llamuza, J,C. Garcia, M. A. Ortiz, Unidad Docente Arquitectura de Ordenadores, EU. Politecnica, 14004 Cordoba. Puesto De Instrumentación Virtual Digital.
[4] V Congreso de la Sociedad Cubana de Bioingenieria, Habana 2003, A.Adan Vidal, F. Fernández Pérez Terán, A. Calle Herranz, E. Valdés Zaldivar. Estrategia para una instrumentación virtual de bajo costo con aplicaciones médicas.
[5] G. León, A. Aguinaga, 2011 Simulación de control y monitores vehicular usando tecnología GSM como medio de transmisión. Escuela Politécnica Nacional. ISBN 978-9942-07-048-7.
[6] Aguinaga, C. Ávila, E. Cando; 2014 A Multi-layer based architecture for the development of an open source CAD/CAM integration virtual platform. *3th International Conference on Mechanical Egineering and Mechatronics ICMEM 14*, (Praga República Checa). World Academy of Science Engineering and Technology. ISSN 20103778, 2010376X
[7] Jason Stamp, Sandia National Laboratories, 2014, Microgrid Design Tools (MDT), *2014 Smart Grid R&D Program Peer Review Meeting*.
[8] Michael Stadler, Lawrence Berkeley National Laboratory, 2014, Microgrid design toolset (MDT) development, *2014 Smart Grid R&D Program Peer Review Meeting*. (Oak Ridge), TN 37831-6070
[9] Department of energy, USA 2014, The future on the grid, envolving to meet american's needs. No. GS-10F-0103J, Subtask J3806.0002.
[10] U.S. Department of Energy Office of Electricity Delivery & Energy Reliability 2012, Smart Grid Research & Development, Multi-Year Program Plan (MYPP), pp. 6 -22.

IOP Conf. Series: Materials Science and Engineering **129** (2016) 012046 doi:10.1088/1757-899X/129/1/012046

The heat transfer characteristic of the reactor coolant pump canned motor

X Y Gu[1], R Xu[1], G Tao[2], Y L Yang[3], D Z Wang[1]*

[1]School of mechanical engineering, Shanghai Jiao Tong University, No. 800, Dongchuan Rd., Shanghai, China.

[2] Shanghai Nuclear Engineering Research & Design Institute, No. 29, Hongcao Rd., Shanghai, China.

[3]The UM-SJTU Joint Institute, Shanghai Jiao Tong University, No. 800, Dongchuan Rd., Shanghai, China.

*Email: dzwang@sjtu.edu.cn

Abstract. This paper deals with the heat transfer characteristic of the reactor coolant pump canned motor. The cooling of the canned motor is an important issue for the design of the pump. In order to analyze the heat transfer characteristic of the canned motor, firstly the electromagnetic field of the canned motor is calculated with finite element method, and the magnetic resistance loss is gotten, then the heat distribution of the canned motor is obtained based on the electromagnetic field, finally the flow field and temperature field of the canned motor is calculated with CFD methods. The calculation indicates that the highest temperature and highest temperature rising are both occurred at the end winding.

1. Introduction

The reactor coolant pump is the only long-term continuously running key equipment of the primary coolant system of the nuclear power plant. Recent advanced large-scale pressurized water reactors regularly adopt the canned motor pumps as their reactor coolant pumps. The canned motor pump is driven by a canned motor, in which the coils of the rotor and the stator are both protected by the metal can. As its special structure、 electromagnetic loss distribution、 heat emission condition and work condition, the internal temperature distribution and the heat dissipation is very complicated, so it is necessary to analyze the temperature field of the canned motor..

In fact, many studies have been done on the temperature field analysis of the canned motor [1-6]. Paper [1] analyzed the temperature field of the reactor coolant pump canned motor. Paper [2] proposed a coupled thermal–magnetic analysis of an induction motor (IM) with the primary goal of achieving a rapid and accurate prediction of the IM performance. Paper [3] proposed a simplified thermal model for variable-speed self-cooled induction motors and has been experimentally verified. Paper [4] dealt with the formulations used to predict convection cooling and flow in electric machines. Paper [5] offered a systematic approach to magnetothermal FEA. Paper [6] presented the study of the motor and the converter temperatures at rated and overload working conditions.

Determining the heat transfer characteristics of the walls to the fluid is the key issue for the heat transfer research of the motor. Paper [7-14] have done lots of work on how to determine the heat transfer characteristics of the rotating cylinder wall. But up to now there's still no definite conclusion. More work is needed.

In this paper, firstly the electromagnetic field of the canned motor is calculated with finite element method, and then the heat distribution of the canned motor is obtained based on the electromagnetic field, finally the temperature field of the reactor coolant pump canned motor is analyzed with thermo-hydro-solid coupled method.

2. Calculation Model

The temperature field simulation of the solid domain is based on the three dimensional steady state heat conduction equations which could be expressed by equation (1)

$$\begin{cases} \frac{\partial}{\partial x}\left(k_x \frac{\partial T}{\partial x}\right) + \frac{\partial}{\partial y}\left(k_x \frac{\partial T}{\partial y}\right) + \frac{\partial}{\partial z}\left(k_x \frac{\partial T}{\partial z}\right) = -q \\ T|_{S_1} = T_0,\ k\frac{\partial T}{\partial n}|_{S_2} = q_0 \\ k\frac{\partial T}{\partial n}|_{S_3} = -\alpha(T - T_f) \end{cases} \tag{1}$$

where T is the temperature of the object, k is the thermal conductivity, q is the heat volumetric flow rate, S_1 is a boundary surface with constant temperature while T_0 is the temperature of it, q_0 is the heat flow rate of S_2 and T_f is the temperature of the surrounding fluid.

The temperature field simulation of the fluid domain is based on the equations which are expressed by equation (2)-(4)

$$\frac{\partial u}{\partial x} + \frac{\partial v}{\partial y} + \frac{\partial w}{\partial z} = 0 \tag{2}$$

$$\begin{cases} \frac{\partial(\rho uu)}{\partial x} + \frac{\partial(\rho uv)}{\partial y} + \frac{\partial(\rho uw)}{\partial z} = \frac{\partial}{\partial x}\left(\mu\frac{\partial u}{\partial x}\right) + \frac{\partial}{\partial y}\left(\mu\frac{\partial u}{\partial y}\right) + \frac{\partial}{\partial z}\left(\mu\frac{\partial u}{\partial z}\right) - \frac{\partial p}{\partial x} \\ \frac{\partial(\rho vu)}{\partial x} + \frac{\partial(\rho vv)}{\partial y} + \frac{\partial(\rho vw)}{\partial z} = \frac{\partial}{\partial x}\left(\mu\frac{\partial v}{\partial x}\right) + \frac{\partial}{\partial y}\left(\mu\frac{\partial v}{\partial y}\right) + \frac{\partial}{\partial z}\left(\mu\frac{\partial v}{\partial z}\right) - \frac{\partial p}{\partial y} \\ \frac{\partial(\rho wu)}{\partial x} + \frac{\partial(\rho wv)}{\partial y} + \frac{\partial(\rho ww)}{\partial z} = \frac{\partial}{\partial x}\left(\mu\frac{\partial w}{\partial x}\right) + \frac{\partial}{\partial y}\left(\mu\frac{\partial w}{\partial y}\right) + \frac{\partial}{\partial z}\left(\mu\frac{\partial w}{\partial z}\right) - \frac{\partial p}{\partial z} \end{cases} \tag{3}$$

$$\frac{\partial(\rho uT)}{\partial x} + \frac{\partial(\rho vT)}{\partial y} + \frac{\partial(\rho wT)}{\partial z} = \frac{\partial}{\partial x}\left(\frac{\lambda_l}{c_p}\frac{\partial T}{\partial x}\right) + \frac{\partial}{\partial y}\left(\frac{\lambda_l}{c_p}\frac{\partial T}{\partial y}\right) + \frac{\partial}{\partial z}\left(\frac{\lambda_l}{c_p}\frac{\partial T}{\partial z}\right) \tag{4}$$

where u、 v、 w are the velocity components of x、 y and z direction respectively, ρ is the fluid density, μ is the dynamic viscosity, p is the pressure of the fluid, c_p is the specific heat capacity of the fluid, T is the temperature of the fluid, λ_l is the thermal conductivity of the fluid[15].

3. Electromagnetic Field Calculation

Electromagnetic field calculation is the fundment of heat distribution and temperature field calculation. The modelling of the motor consists of two parts: 1) the general section, which includes the iron core and coils of the stator, the coils and squirrel cage of the rotor; 2) other metal components, which include the shielding of the rotor, finger clamp and the conical ring of the stator. The modelling of the general section is achieved by Maxwell while other metal components is modelled by CAD after which the model is imported to the finite element simulation software Maxwell. The model of the general section and the whole motor are showed in figure 1 and figure 2. The half model is used in the calculation to reduce computational time. Maxwell has applied the self-adaptive grid technique which can refine the grids step by step according to the computational accuracy of setting. But asigning the mesh scale of some special components is also required for faster convergence speed and better solving accuracy. The total simulation time is 1.5s in which there is 0.5s before the motor start to run. The variable time step method is applied in the calculation as the time step was 1ms in the first 0.5s while the time step changed to 0.4ms after that. The 3-D electromagnetic field of the motor is gotten in the calculation. Figure 3 and figure 4 are the magnetic strength distribution of the motor and the magnetic force line distribution of the end-winding respectively. These results are used in the latter calculation.

Figure 1. The 3-D model of the general section.

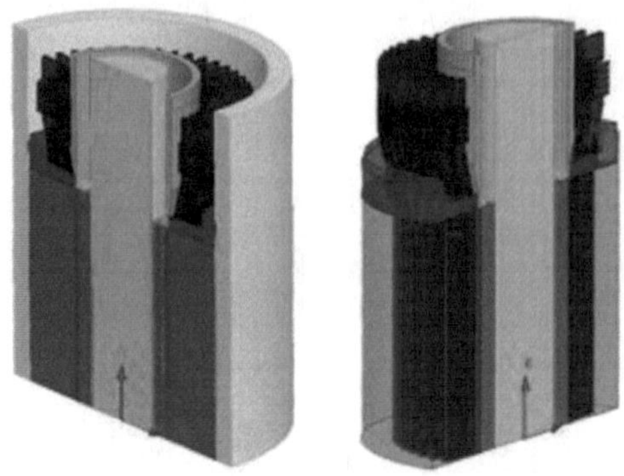

Figure 2. The complete model of the motor.

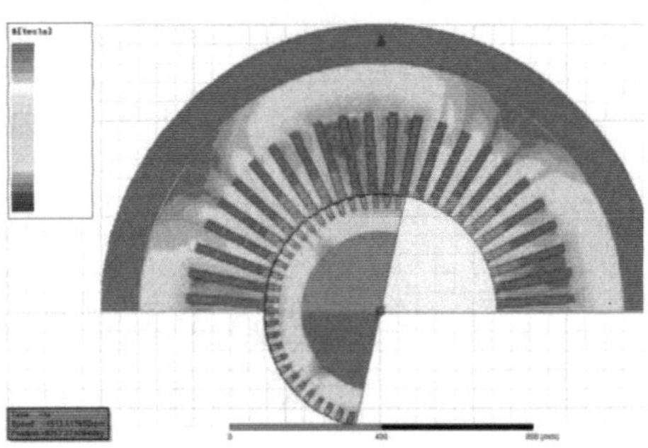

Figure 3. The magnetic strength distribution of the motor.

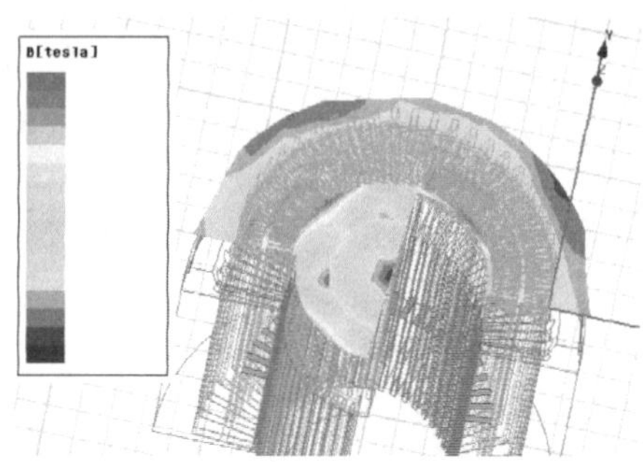

Figure 4. The magnetic force line distribution of the end-winding.

IOP Conf. Series: Materials Science and Engineering **129** (2016) 012046 doi:10.1088/1757-899X/129/1/012046

4. Heat Distribution

As the electromagnetic field of the motor has been gotten in the electromagnetic field calculation, the heat distribution can be calculated, and this is the basis of the temperature field calculation. The power consumption of the motor in different components are converted to heat of these parts, so the heat distribution can be calculated from the power consumption. The power consumption of the motor mainly takes place by three means: iron core loss, eddy current loss and coil loss. Figure 5-8 has given out the results of these losses respectively, which reflect the eddy current loss has the largest contribution to the motor power consumption. Meanwhile, the stator shield loss is the main part of the eddy current loss.

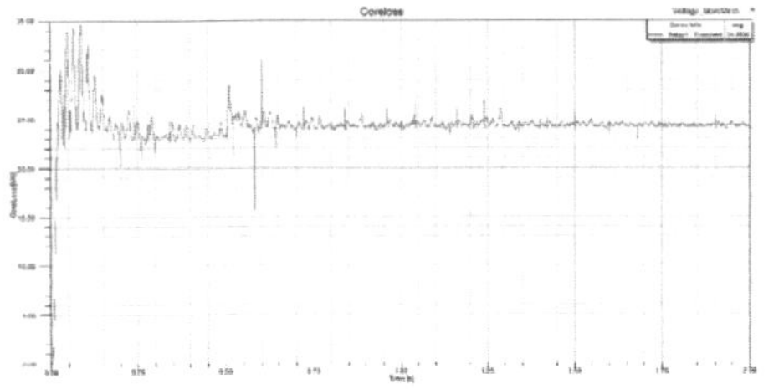

Figure 5. The iron core loss.

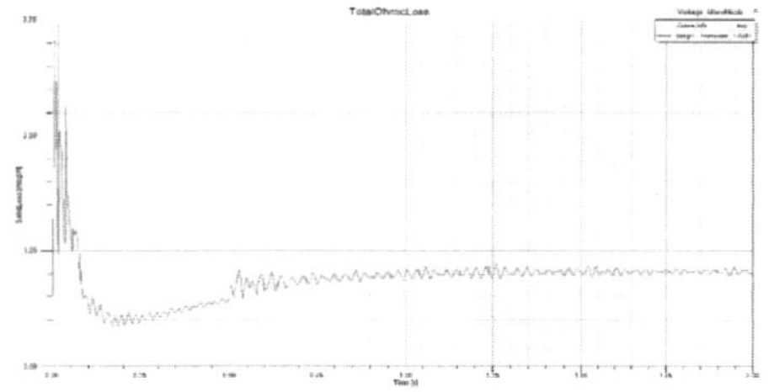

Figure 6. The eddy currnt loss.

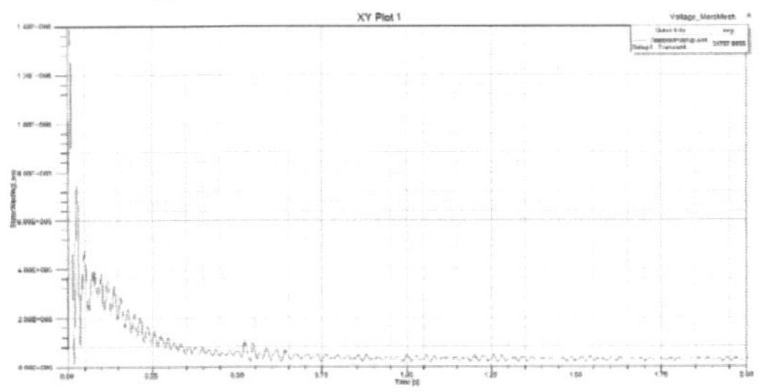

Figure 7. The stator coil loss.

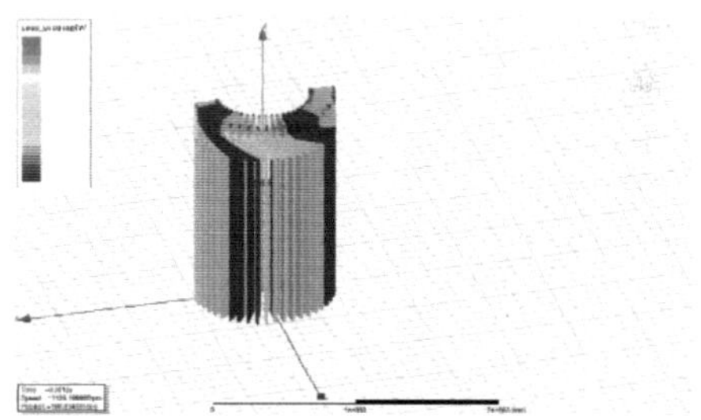

Figure 8. The contour of stator coil loss - axonometric drawing.

5. Temperature Field of The Motor

With the heat distribution of the motor, the temperature field can be analyzed. The calculation has taken the heat convection in the motor gap and the heat conduction in different motor components into consideration. There are some assumptions in the calculation: 1) the flow in the canned motor belongs to the turbulent flow as the Reynolds number of the flow is relatively high (Re>2300), the turbulence model is k-ε model; 2) as the gas flow rate in the motor is far below the sound speed (the Mach number is very low) and the compressibility of water is negligibly small, all of the fluid in the motor are regarded as incompressible; 3) as the calculation is a steady state simulation, the time terms in the calculating equations are ignored.

The calculation is carried by ANSYS FLUENT 14.0. According to the symmetric structure of the motor and the above assumptions, the physical model takes 1/8 part in the circular direction and the whole length in the axial direction as figure 9 shows. Figure 10-12 shows the mesh details of the physical model, the total number of the grid is 6,600,152 while the grid number of the end winding cavity area is 1,605,378, the grid number of the external water jacketed clearance flow area is 172,800 and the grid number of the internal clearance flow area is 2,000,000. The mesh type of the clearance flow is hexahedral and other parts are tetrahedral.

The calculation takes the power consumption of each part as the heat boundary conditions and put them homogeneously on the solid calculation domain. The caculation has also taken the motor's stray-load loss which is considered as 1.5% of the motor's rated power into consideration, this heat source is put on the stator core calculation domain. The boundary conditions of the cooling water outlets are pressure boundary. For the model's heat transfer

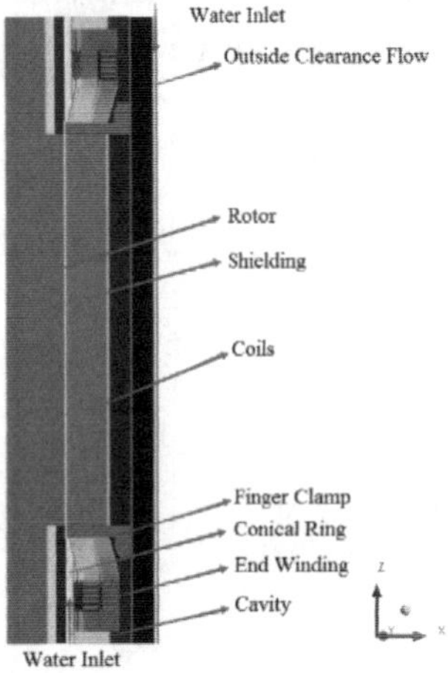

Figure 9. The calculation model.

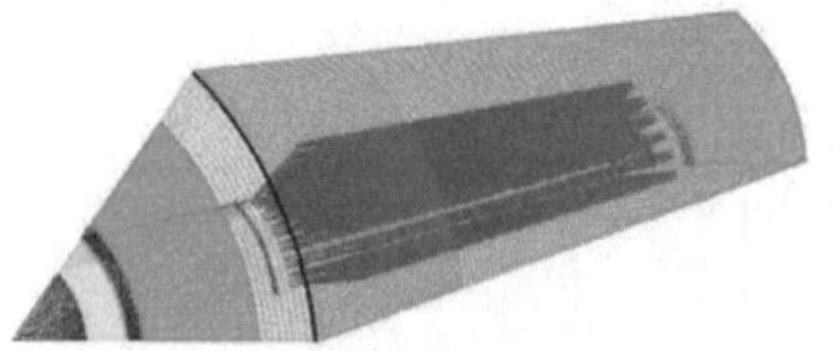

Figure 10. The mesh detail of the model.

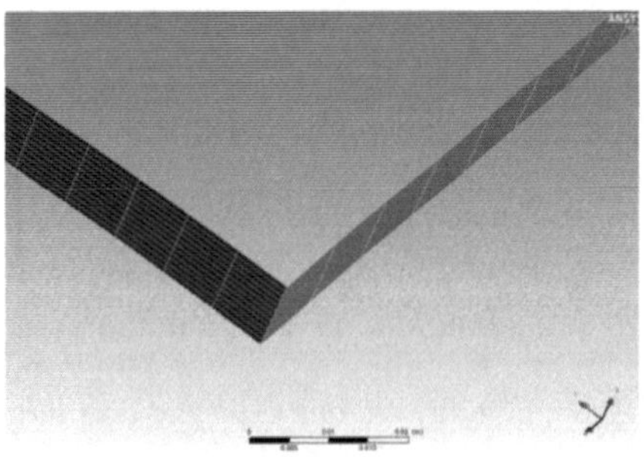

Figure 11. The mesh detail of the external water jacketed clearance flow area.

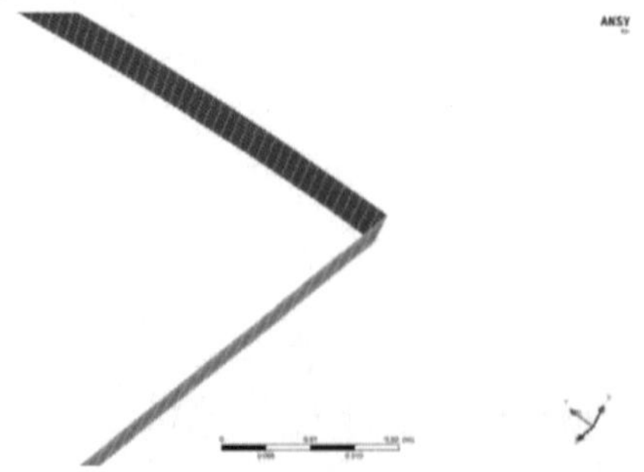

Figure 12. The mesh detail of the internal clearance flow area.

coefficients, the heat release coefficient of the motor's outside surface is 3.2 W/(m2·K) as this coefficient is decided by the characteristic scales and the fluid parameters which can be calculated by specific formulas, the heat transfer coefficients of the solid and fluid interfaces in the motor are aquired by flow-heat coupling method. The pressure coupling method is SIMPLEC which is widely used in incompressible fluid field calculation. The solver is chosen as a double precision solver and the iteration convergence precision is 1e-5.

Figure 13-18 are the calculation results of the temperature field. From the results and the temperature rise distribution characteristic, one can find that the stator iron core, the rotor, the cooling water, the inner and outer jacket and the chassis have a relatively low temperature rising while the highest temperature and highest temperature rising are both occurred at the end winding. The highest temperature rising is as high as 127K.

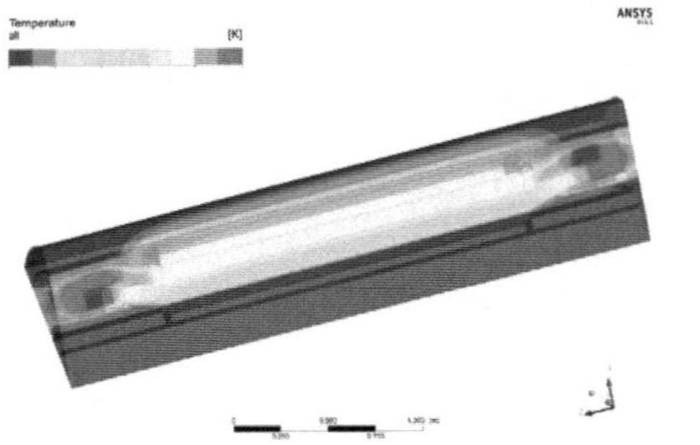

Figure 13. The Temperature field of the motor.

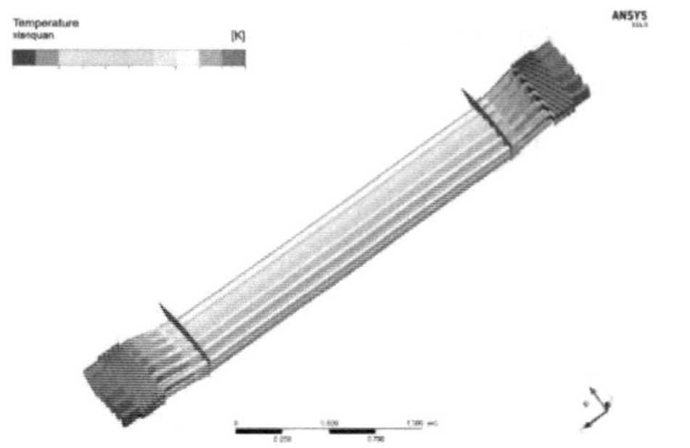

Figure 14. The Temperature field of the stator coil.

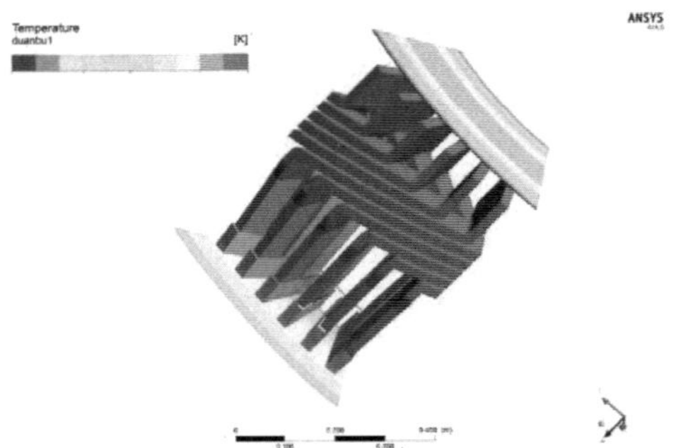

Figure 15. The temperature field of the end winding.

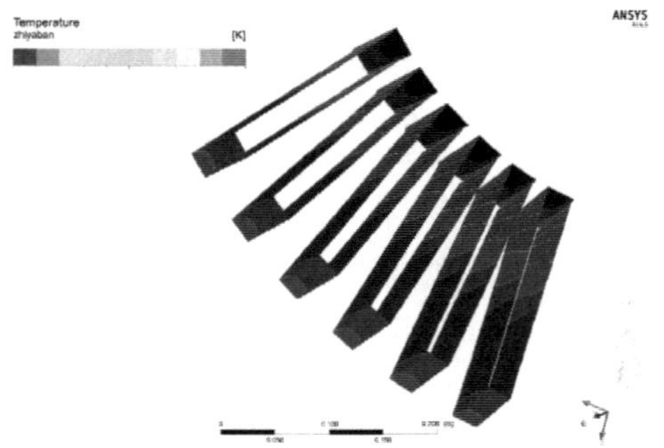

Figure 16. The temperature field of the finger clamp

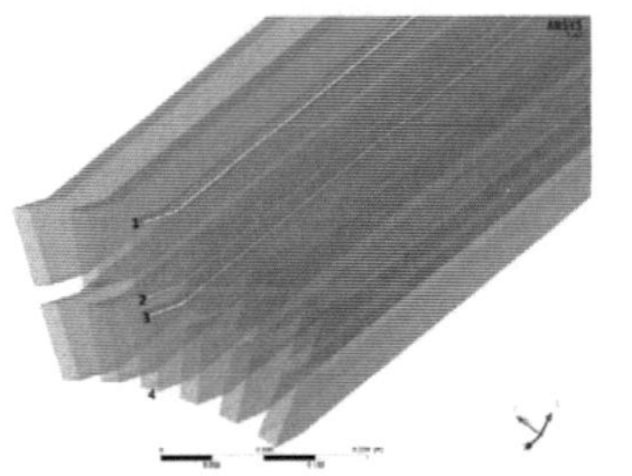

Figure 17. The value location of the stator coil.

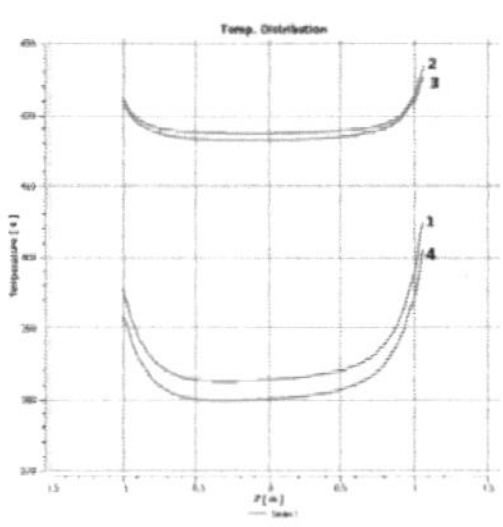

Figure 18. The temperature distribution of the stator coil.

6. Conclusions

In this work a series of computations have been carried to analyze the temperature field of the reactor coolant pump canned motor. The following conclusions could be summarized.

- This paper has achieved the canned motor's numeric analysis of the electromagnetic field, the temperature field and the fluid field, the results are reasonable compared to the practical operation experiences.
- The highest temperature and highest temperature rising are both occurred at the end winding, meanwhile the temperature in the middle are much higher than both ends and it has brought a prodigious temperature gradient, so the heat releasing of the end winding needs significant attention.
- The calculation has not gotten an experiment to validate its computational accuracy, further study is still required.

Acknowledgments
This work is funded by the National Natural Science Foundation of China (No. 51406114 and No. 51576125).

References

[1] Ding S Y, Meng F D and Ge Y Z 2012 Temperature field investigation of canned primary pump motors in nuclear power stations *J Proceedings of the CSEE* **36** 149 (in Chinese)

[2] Luigi A and Nicola B 2008 A coupled thermal–electromagnetic analysis for a rapid and accurate prediction of IM performance *J IEEE Transactions On Industrial Electronics* **55** 3575

[3] Aldo B, Andrea C, Mario L and Michele P 2003 A simplified thermal model for variable-speed self-cooled industrial induction motor *J IEEE Transactions On Industry Applications* **39** 945

[4] David A S and Andrea C 2008 Convection heat transfer and flow calculations suitable for electric machines thermal models *J IEEE Transactions On Industrial Electronics* **55** 3509

[5] Fabrizio M, Vincenzo D C and Yuri C 2008 Design of axial flux PM synchronous machines through 3-D coupled electromagnetic thermal and fluid-dynamical finite-element analysis *J IEEE Transactions On Industrial Electronics* **55** 3591

[6] Alberto T, Francesco P, Stefan E B and Martin D H 2008 Temperatures evaluation in an integrated motor drive for traction applications *J IEEE Transactions On Industrial Electronics* **55** 3619

[7] Lee, Y N., Minkowycz and W J 1989 Heat transfer characteristics of the annulus of two-coaxial cylinders with one cylinder rotating. Int. *J Heat Mass Transfer* **32** 711

[8] Escudier, M.P, Gouldson and I W 1995 Concentric annular flow with centerbody rotation of a newtonian and a shear-thinning liquid. Int. *J Heat Fluid Flow* **16** 15

[9] Nouri, J M, Whitelaw, J H 1994 Flow of newtonian and non-newtonian fluids in a concentric annulus with rotation of the inner cylinder *J Fluid Eng* **116** 821

[10] Bouafia, M , Bertin, Y, Saulnier, J , Ropert, P 1998 Analyse expérimentale des transferts de chaleur en espace annulaire étroit et rainuré avec cylinder intérieur tournant. Int. *J Heat Mass Transfer* **41** 1279

[11] Naser, J A 1997 Prediction of newtonian and non-newtonian flow through concentric annulus with centerbody rotation *In International Conference on CFS in Mineral and Metal Processing and Power Generation* CSIRO

[12] Kuosa, M , Sallinen, P, Larjola, J 2004 Numerical and experimental modelling of gas flow and heat transfer in the air gap of an electric machine *J Therm. Sci.* **13** 264

[13] Giret, A 2009 Transferts Thermiques Convectifs dans le cadre de Machines Tournantes, Ph.D. thesis Université de Poitiers

[14] Chung, S Y, Sung, H J 2005 Large-eddy simulation of turbulent flow in a concentric annulus with rotation of an inner cylinder. Int. *J Heat Fluid Flow* **26** 191

[15] Donatello Annaratone 2009 Engineering heat transfer M ISBN 978-3-642-03931-7

Experiments on the magnetic coupling in a small scale counter rotating marine current turbine

I C Kim[1], N J Lee[1], J Wata[1], B S Hyun[2], Y H Lee[3]

[1]Department of Mechanical Engineering, Graduate School, Korea Maritime and Ocean University, 727 Taejong-ro, Yongdo-gu, Busan, 606-791, Republic of Korea
[2] Division of Naval Architecture and Ocean System Engineeing, Korea Maritime and Ocean University, 727 Taejong-ro, Yongdo-gu, Busan, 606-791, Republic of Korea
[3] Division of Mechanical Engineering, Korea Maritime and Ocean University, 727 Taejong-ro, Yongdo-gu, Busan, 606-791, Republic of Korea

E-mail: lyh@kmou.ac.kr

Abstract. Modern economies are dependent on energy consumption to ensure growth or sustainable development. Renewable energy sources provide a source of energy that can provide energy security and is renewable. Tidal energy is more predictable than other sources or renewable energy like the sun or wind. Horizontal axis marine current turbines are currently the most advanced and commercially feasible option for tidal current convertors. A dual rotor turbine is theoretically able to produce more power than a single rotor turbine at the same fluid velocity. Previous experiments for a counter rotating dual rotor horizontal axis marine current turbine used a mechanical oil seal coupling that caused mechanical losses when water entered through small gaps at the shaft. A new magnetic coupling assembly eliminates the need for a shaft to connect physically with the internal mechanisms and is water tight. This reduces mechanical losses in the system and the effect on the dual rotor performance is presented in this paper.

1. Introduction

Modern economies are dependent on the use of energy in its numerous forms to drive and sustain growth. Whether to provide sufficient heating to its citizens, fuels for transportation or electricity for utilities and vital services, enough energy must be supplied to meet demand. Demand that only increase annually as the population grows. Using energy efficient devices and reducing unnecessary use is one way of reducing demand but more often than not, more fuel is imported to satisfy energy demand. This situation causes dependence on outside sources for supply and concern for energy security. Countries are now considering renewable energy sources, such as the sun, wind and ocean, as alternative and domestic sources of energy. Since these resources are replenished naturally and obtained locally, there are long term benefits of developing infrastructure and technology in for these sources. South Korea is highly dependent on petroleum, coal and liquefied natural gas (LNG) imports to keep up with energy demand. In the government's National Energy Master Plan [1], it has planned to reduce these imports gradually and amongst other things, to have new and renewable energy resources produce 11% of the energy demand by 2030.

The ocean and all the natural process that occur in it are the result of the solar energy that is radiated by the sun. The solar energy drives the temperature gradients in the sea, the winds over the waters and indirectly drives the waves and currents. Energy that is contained in ocean waves and currents can be broadly classified under wave energy and tidal energy respectively. Tidal energy sources have the advantage of being more predictable than other sources of energy such as solar or wind. Tidal energy sources can be further divided into tidal range energy and tidal/marine current energy. Tidal range technologies use the differences in water elevation that are caused by high and low tides to drive hydraulic turbines. Whereas marine current energy convertors use the kinetic energy contained in currents to produce power by turbines or an appropriate mechanism [2]. Among the technologies available in marine current energy, the most advanced is the horizontal axis tidal current turbine.

The horizontal axis tidal current turbine utilizes the same principles as a wind turbine with the same orientation and the advances made in the wind industry have also benefitted the marine current turbine designs. In Betz theorem, the theoretical maximum efficiency that can be produce by a single rotor is 59.3% and by extending this theorem to a dual rotor concept, the maximum that can produced is 64% [3]. Lee et al [4] designs a counter rotating turbine that is based on the Blade Element Momentum Theory (BEMT) and investigates the effect of the distance between the two rotors on the performance of the turbine. This present study utilizes a small scale dual rotor horizontal axis marine current turbine design that gives the highest power coefficient from Lee's study. In addition to this, a new magnetic watertight coupling mechanism is placed at the hub of the counter rotating turbine instead of a mechanical oil seal type coupling system used in the previous study. The new coupling mechanism prevents water from entering the hub and causing mechanical losses. In order to assess the effect the new coupling system had on the performance of the dual rotor turbine, experiments were done one the turbine with the new setup and the results are presented in this paper. Also, the experimental results are also compared to the CFD results obtained in the previous study.

2. Experimental Setup

The experiments were done in a vertically circulating water tank in Korea Maritime and Ocean University. The water tank is 4m in length, 1.8m wide and 1.2m high. The depth of the water used was 0.9m. Before conducting experiments, the water velocity was measured at 15 different equidistant points in plane 1m in front of the turbine. At each point, 5 measurements were taken over a 1 minute duration. The average difference in measures was found to be 2%. The water velocity is controlled by the rotational speed of 2 fans in the tank, which is controlled via control panel. In addition, the water velocity could only be changed from 0.8m/s to 1.4m/s. Over this limit would put the structural integrity of the water tank at risk.

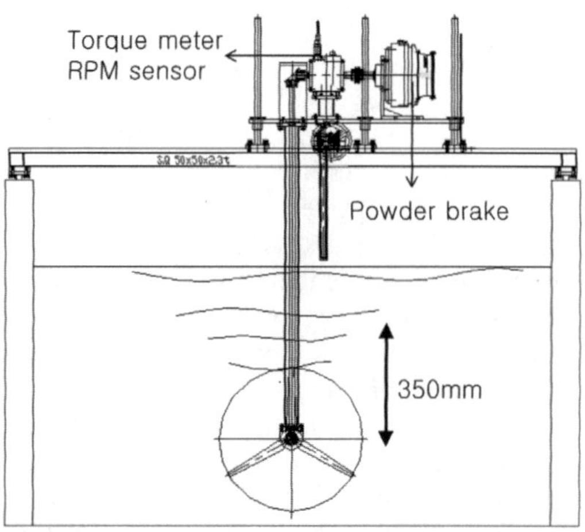

Figure 1. Diagram of experimental setup with experiment equipment highlighted.

The experimental setup is shown the figure 1 below. The torque meter (model SBB) could measure from 0-2 kgfm. An RPM sensor is also contained in the torque meter and could measure up 10,000 RPM. To control the RPM of the turbine, a forced air cooled type powder brake (model PRB-5Y3F) range.

The water velocity is controlled by a fan via control panel and varied from 0.6m/s to 1.3m/s. The torque, RPM and water velocities are measured and then recorded into a data logger. The blockage ratio of the experimental setup was calculated to be about 12% and a blockage correction factor was used to correct the results [5]. The turbine was located 350mm below the free surface.

The counter rotating turbine contains two rotors of 3 blades each. The blades were designed using the BEMT methodology. The BEMT combines the equations of momentum theory and blade element theory equations to predict the forces that act on a 2D airfoil or 3D blade span. By using the BEMT equations and 2D airfoil data, a rotor blade profile can be designed. The airfoil used for the blade profile was NACA-63421 with different chord lengths and twist angles [4]. The diameter of the model turbine was 500mm, and the distance between the two rotors was 250mm. Furthermore, the blade angle on both the front and rear blades was set at 0°.

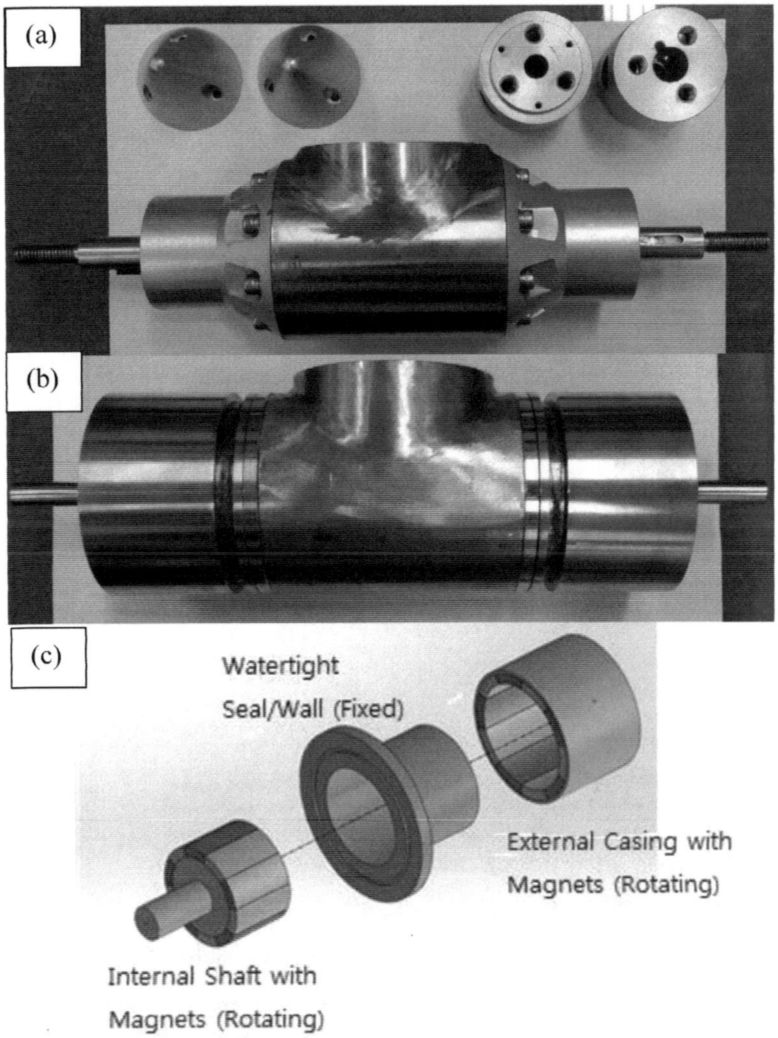

Figure 2. (a) Photo of the Mechanical Oil Seal Assembly (b) Magnetic coupling assembly (c) 3D rendering of the Magnetic coupling.

The mechanical oil seal coupling assembly, magnetic coupling assembly and 3d model of the magnetic coupling is shown in figure 2a-c. Figure 2a shows the mechanical oil seal coupling assembly with the various hub connections and key and shaft design. During earlier experiments, it was observed that water would leak through small openings between the shaft and the hub and cause friction in the internal gearing. Figure 2b shows the new magnetic coupling assembly and figure 2c is the 3D rendering of the magnetic assembly. The external casing of the magnetic coupling is connected to the rotating blades by a shaft seen in the photo of the magnetic coupling assembly. The external casing closely fits and rotates over the fixed watertight seal/wall. This seal prevents any water from entering the internal portion of the hub. The internal shaft also fits a close fit inside the seal and a bearing inside seal ensures the smooth rotation of the internal shaft. As the external casing rotates over the seal, the magnetic attraction between the magnets of the external casing and internal shaft causes the internal shaft to rotate together with the external casing. In the mechanical oil seal design, the shaft has to connect through to internal gearing and a small gap between the shaft and hub casing must be made to accommodate the shaft. The main advantage of the magnetic coupling is that the new method does not require the shaft to physically connect external and internal portions and thus becomes water tight. This will eliminate potential mechanical losses caused by any water leakage through the shaft opening. This could possibly reduce manufacturing costs and potentially reduce maintenance costs due to fewer parts.

3. Results and Discussion

Figure 3 shows the power output obtained from the experiments of the magnetic and mechanical coupling setups. These results are also compared with results obtained from numerical simulations, labelled as CFD, from the previous study. The maximum power output from the previous experiment was 116.6 W and with the magnetic coupling, the maximum power output was 122.84 W. Both the oil seal coupling and magnetic coupling setups show close agreement with the CFD results. However, the magnetic coupling setup shows a higher output than the oil seal setup. From this graph a small difference can be seen and in order to further study the difference, the mechanical losses caused by both setups can be measured. To measure the overall mechanical loss in the two setups, the water tank was fully drained and the powder brake was removed. The torque/RPM meter was rotated at different rotational speeds and the output torque was measured and averaged.

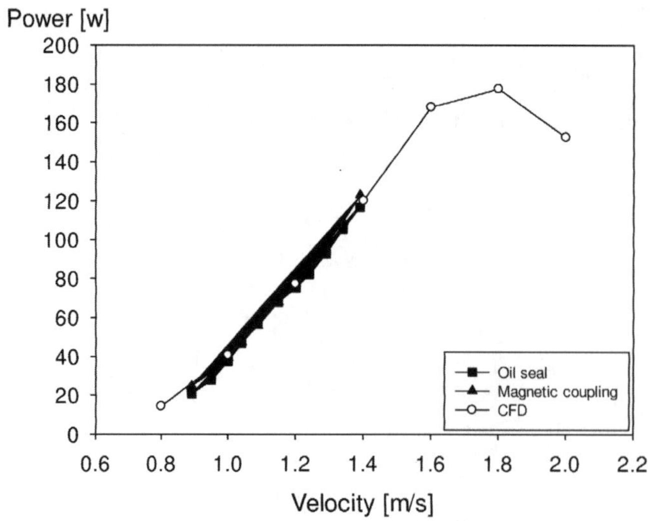

Figure 3. Graph of the power output against water velocity. Comparing the Oil Seal, Magnetic coupling and CFD results.

The difference in overall mechanical loss in torque at different rotational speeds were recorded and presented in figure 4. The figure shows that at all rotational speeds, the averaged mechanical losses that occur in the magnetic coupling setup are lower than the mechanical oil seal setup. The average mechanical loss was 0.035 kgf.m for the oil seal setup and 0.029 kgf.m for the magnetic seal setup. The difference in losses can be also seen by comparing the effectiveness of the new assembly to the previous one.

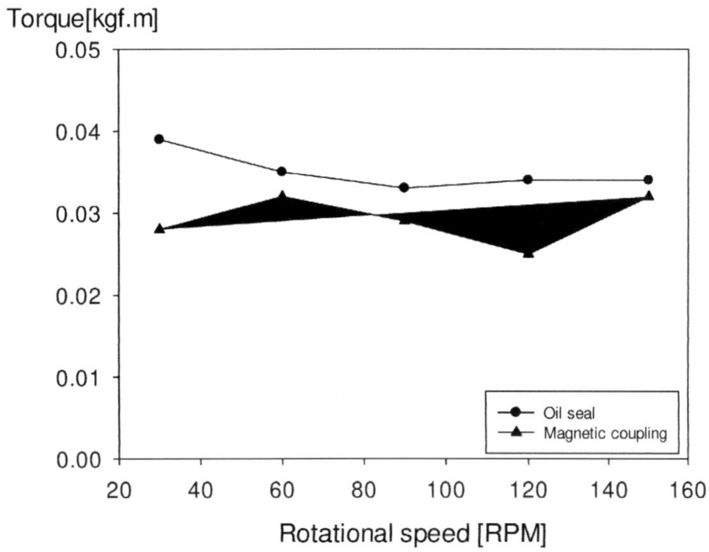

Figure 4: Torque mechanical loss measured for both coupling systems.

In figure 5, the coefficient of power (C_p) calculated for the o ring and magnetic coupling setups are plotted against water velocity. These experimental results are also compared with the numerical result. The coefficient of power is defined as the ratio between the power produced at the turbine and the power available in the water current as shown in equation (1):

$$C_p = \frac{\tau.\frac{2\pi}{60}.n}{0.5\rho A V_\infty{}^3} \tag{1}$$

Where:
τ is the torque measured [N.m]
n is the rotational speed of the turbine [RPM]
ρ is the density of the water [$kg.m^{-3}$]
A is the frontal area of the turbine [m^2]
V_∞ is the free stream velocity of the water [$m.s^{-1}$]

The values calculated from magnetic coupling case show better agreement than the oil seal coupling case. The maximum Cp of the oil seal setup was 0.45 at a water velocity of 1.15 m/s and 0.46 for the magnetic coupling setup at a water velocity of 1.24 m/s. The differences caused by the mechanical losses in the oil seal coupling can be observed to have an effect on how efficient the turbine can produce power. By eliminating potential mechanical losses caused by water leakage, the power produced and the effectiveness of the dual rotor turbine is seen to increase. This small improvement in performance would be more apparent and beneficial for larger turbines designed in the kilo-megawatt range.

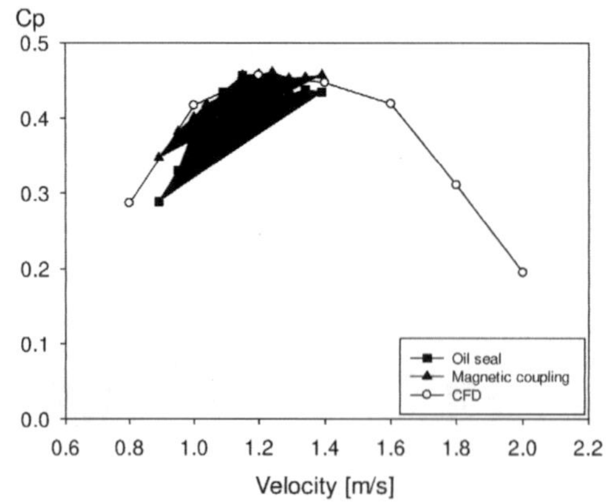

Figure 5: Graph of the CP against water velocity for the Oil seal assembly, Magnetic coupling assembly and CFD results

4. Conclusion

In this study, a magnetic hub coupling design for a dual rotor horizontal axis marine current turbine was tested in small scale experiments. The magnetic design was compared to the mechanical oil seal design of previous studies on the same turbine. In addition, the results were compared with the numerical results of a CFD analysis of the turbine. The best turbine operating parameters according to previous studies such as blade angle and blade gap distance were set at 0° and 250mm, respectively. The magnetic coupling setup showed an increase in power output, lower mechanical losses in torque and higher power coefficient than the oil seal setup. The magnetic coupling case was also in close agreement with the CFD results. The small improvement seen in the small scale experiment could be more apparent and beneficial for larger designs of marine current turbines.

Acknowledgments

This work was supported by the New & Renewable Energy Core Technology Program of the Korea Institute of Energy Technology Evaluation and Planning (KETEP), granted financial resource from the Ministry of Trade, Industry & Energy, Republic of Korea. (20133030000260)

References

[1] Ministry of Trade, Industry and energy 2014 *Korea Energy Master Plan: Outlook & policies to 2035*. South Korea.

[2] Owen A 2006 Tidal current energy: orgins and challenges *Future Energy: Improved, Sustainable and Clean Options for our Planet* ed Letcher T (Oxford: Elsevier) chapter 7 pp 111-128

[3] Newman BG 1983 Actuator-disc theory for vertical-axis wind turbines *J. Wind Eng. Indus. Aerodyn.* **15** 347-355

[4] Lee N J, Kim I C, Kim C G, Hyun B S and Lee Y H 2014 Performance study on a counter-rotating tidal current turbine by CFD and model experimentation *Rene. Ene.* **79** 122-126

[5] Chen T Y and Liou L R 2011 Blockage corrections in wind tunnel tests of small horizontal-axis wind turbines *J Exp. Therm. Fluid Sci.* **35** 565-569

IOP Conf. Series: Materials Science and Engineering **129** (2016) 012048 doi:10.1088/1757-899X/129/1/012048

Numerical simulation of flow past a circular base on PANS methods

J T Liu[1], Y Li[1], Y Gao[1], Q Hu[1] and Y L Wu[2]

[1] Beijing Institute of Control Engineering, Beijing 100090, China
E-Mail: liujintao86@hotmail.com
[2] State Key Laboratory of Hydroscience and Engineering, Tsinghua University, Beijing 100084, China

E-mail:wyl-dhh@mail.tsinghua.edu.cn

Abstract. A nonlinear partially averaged Navier-Stokes (PANS) method based on RNG k-ε turbulence model is evaluated by a moderately high Reynolds number turbulence flow past a circular cylinder. The ratios of unresolved-to total kinetic energy (f_k) and unresolved-to-total disspation (f_ε) for PANS method is sensitive to the simulation result. Simulation results based on different fk are compared with the experimental data. The quantities including mean streamline velocity, mean normal velocity, streamlines and et al. are analyzed. The computational result is reasonable when f_k is less than 0.6. The PANS method can be used in the simulation of high Reynolds number turbulence flow.

1. Introduction

For predictions of turbulence flow with high Reynolds number based on the Unsteady Reynolds Averaged Navier-Stokes (URANS) equations have been well discussed in the literature [1]. Unfortunately, the aforementioned URANS approach couldn't capture the flow with all kinds of scales. Large Eddy Simulation (LES) could improve predictions but is not computationally economical due to the considerable high cost for engineering problems[3]. As an alternative, various hybrid RANS/LES approaches have been proposed. For example, a Partially-Averaged Navier-Stokes (PANS) approach, which changes seamlessly from RANS to the direct numerical solution of the Navier-Stokes equations, was proposed by Girimaji et al.[2].

Lakshmipathy[3] analyzed the flow past a circular cylinder with high Reynolds number via PANS k-ε model. Lakshmipathy[4] proposed a new kind PANS model derived from k-ω turbulence model, which was called k-ω PANS methods. Flow around a marine propeller[5] and a twisted hydrofoil[6] in a non-uniform wake were also investigated using PANS methods. A variable-resolved PANS bridging strategy was applied to the four-equations k-ε-ξ-f turbulence model[7]. Embedded large-eddy simulation using the partially averaged Navier-Stokes model was proposed via Davidson[8].

Computations of flow past a circular cylinder have also been performed to evaluate the numerical as well as physical modeling aspects. Several experimental data sets are available for the flow past a circular cylinder test case in all Reynolds number regimes. Cantwell[9] studied the flow past a circular cylinder at $Re_d=1.4\times10^5$. Breuer[10] was the first to perform LES calculations of flow past a circular cylinder at $Re_d=1.4\times10^5$. It concluded that the dynamic subgrid model gave the best results but was not decisively better than the constant Smogorinsky model for this flow. Travin[11] performed detached eddy simulations (DES) of flow past a circular cylinder at $Re_d=1.4\times10^5$ using a multiblock grid and fifth-order upwind scheme for spatial discretization.

The primary objective of the present study is to validate a nonlinear RNG k-ε PANS mothod[12] in the flow past a circular cylinder at $Re_d=1.4\times10^5$. The PANS results are compared with experimental data at different f_k.

2. Nolinear PANS model

For incompressible flow, the continuity equation and Reynolds averaged Navier-Stokes equations:

$$\frac{\partial U_i}{\partial t}+U_j\frac{\partial U_i}{\partial x_j}+\frac{\partial \tau(V_i,V_j)}{\partial x_j}=-\frac{\partial\langle p\rangle}{\partial x_i}+\nu\frac{\partial^2 U_i}{\partial x_j\partial x_i} \tag{1}$$

$$-\frac{\partial^2\langle p\rangle}{\partial x_i\partial x_i}=-\frac{\partial U_i}{\partial x_j}\cdot\frac{\partial U_j}{\partial x_i}+\frac{\partial \tau(V_i,V_j)}{\partial x_i\partial x_j} \tag{2}$$

Here, V is the instantaneous velocity (subscript i and j indicating components in different directions); $\boldsymbol{U}$ is the partially averaged velocity; t is the time; p is the pressure; $\langle\rangle$ denotes a constant-preserving arbitrary (implicit or explicit) filter commuting with spatial and temporal differentiation; $\mathbf{x}$ denotes the coordinates; ν is the kinematic viscosity. Here, V_i is partitioned into two parts: partially averaged velocity ($\boldsymbol{U}$) unresolved part (u_i).

$$V_i=U_i+u_i \qquad U_i=\langle V_i\rangle \tag{3}$$

where U_i is the resolved velocity field; u_i is the unresolved field. It is used by equation 4 instead of the resolved field.

In Eqs. (1) and (2), the additional non-linear term $\tau(V_i, V_j)$ (i.e. the generalized central second moment) is defined as,

$$\tau(V_i,V_j)=(\langle V_iV_j\rangle-\langle V_i\rangle\langle V_j\rangle) \tag{4}$$

The PANS models are:

$$\frac{\partial(\rho k_u)}{\partial t}+\frac{\partial(\rho\overline{U}_jk_u)}{\partial x_j}=\frac{\partial}{\partial x_j}\left[\alpha_k\left(\mu+\frac{\mu_u}{\sigma_u}\right)\frac{\partial k_u}{\partial x_j}\right]+P_{ku}-\rho\varepsilon_u \tag{5}$$

$$\frac{\partial(\rho\varepsilon_u)}{\partial t}+\frac{\partial(\rho\overline{U}_j\varepsilon_u)}{\partial x_j}=\frac{\partial}{\partial x_j}\left[\alpha_k\left(\mu+\frac{\mu_u}{\sigma_u}\right)\frac{\partial\varepsilon_u}{\partial x_j}\right]+C^*_{\varepsilon1}P_{ku}\frac{\varepsilon_u}{k_u}-C^*_{\varepsilon2}\rho\frac{\varepsilon_u^2}{k_u} \tag{6}$$

Considering the nolinear turbulence flow in the pump-turbine, the shear stress was solved by nolinear turbulence model which was proposed by Ehrhard [12].

$$P_k=-\rho\overline{U_i'U_j'}\frac{\partial\overline{U}_j}{\partial x_j} \tag{7}$$

$$\begin{aligned}\overline{U_i'U_j'}=&\frac{2}{3}k\delta_{ij}-2C_{\mu P}\upsilon^2TS_{ij}+C_1C_{\mu P}\upsilon^2T^2\left(S_{ik}S_{kj}-\frac{1}{3}S_{kl}S_{kl}\delta_{ij}\right)+C_2C_{\mu ke}\upsilon^2T^2\left(\Omega_{ik}S_{kj}-\Omega_{jk}S_{ki}\right)\\&+C_3C_{\mu P}\upsilon^2T^2\left(\Omega_{ik}\Omega_{jk}-\frac{1}{3}\Omega_{lk}\Omega_{lk}\delta_{ij}\right)+C_4C_{\mu ke}\upsilon^2T^3\left(S_{ki}\Omega_{lj}-S_{kj}\Omega_{li}\right)S_{kl}+C_5C_{\mu P}\upsilon^2T^3S_{ij}S_{kl}S_{kl}+C_6C_{\mu P}\upsilon^2T^3S_{ij}\Omega_{kl}\Omega_{kl}\end{aligned} \tag{8}$$

$$C_{\mu P}=\min\left(\frac{1}{0.9S^{1.4}+0.4\Omega^{1.4}+3.5},0.15\right)\quad \Omega_{ij}=\frac{1}{2}\left(\frac{\partial\overline{U}_i}{\partial x_j}+\frac{\partial\overline{U}_j}{\partial x_i}\right)\quad S=\frac{k}{\varepsilon}\sqrt{2S_{ij}S_{ij}}\quad \Omega=\frac{k}{\varepsilon}\sqrt{2\Omega_{ij}\Omega_{ij}}$$

where $C_{\mu ke}=\beta\frac{k^2}{\varepsilon}$, $C_1=-0.2$, $C_2=0.4$, $C_3=2.0-\exp\left(-(S-\Omega)^2\right)$, $C_4=-32.0C_{\mu P}^2$, $C_5=-16.0C_{\mu P}^2$, $C_6=16.0C_{\mu P}^2$, T

is turbulence time scale， v is the turbulence velocity scale.

3. Simulation Setup

Flow past a circular cylinder at $R_{ed}=1.4\times10^5$ was simulated using the nonlinear RNG k-ε PANS model at different resolutions. Values of the f_k were 0.2, 0.6, 0.8 and 1. At this Reynolds number the flow is still sub-critical, i.e., the boundary layers at the cylinder separate laminarly and transition takes place in the free shear layers. In the wake strong vortex shedding is observed.

The computational domain shown in Fig.1 is a square cross section of 30D×30D with the cylinder located at the center. D is the diameter of the circular cylinder at the center of the section. The spanwise lengthen of the domain is 2D with the circular cylinder stretching along the enter domain. The nonlinear RNG k-ε PANS model was performed via the user defined function (UDF) using a commercially available Fluent code.

Enhanced wall boundary function was used at the upper and lower solid wall of the domain. No slip boundary condition was chosen for the calculations. The periodicity of the flow is assumed in the spanwise direction of the cylinder. A constant velocity inlet flow with zero turbulence intensity is imposed. Three dimensional (3D) simulations were performed with segregated and implicit solver. Second order scheme was using for the pressure discretization. SIMPLE was using for the pressure and velocity coupling. Second order upwind was using for discretizing the momentum, turbulence kinetic energy, turbulence dissipation rate. Constant velocity inlet was used at the inlet of the computational domain and outflow was using for the outlet. The time step using for the calculation was set up according to the courant number less than 1. Detailed simulation setups are shown in Table 1. Hexahedral mesh forms the computational domain. Mesh at the vicinity of the cylinder is shown in Fig.2.

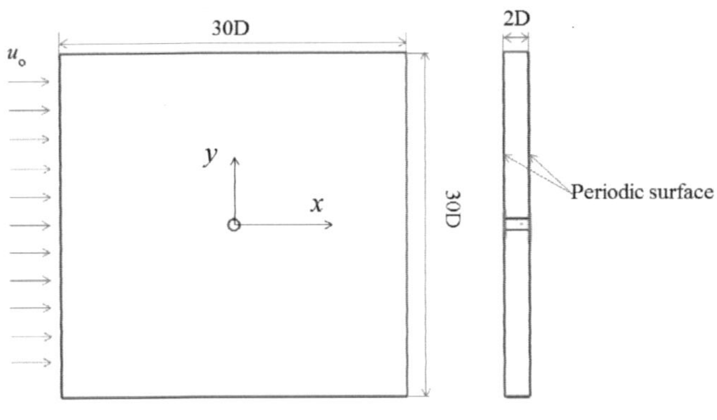

Figure 1. Geometry of computational domain

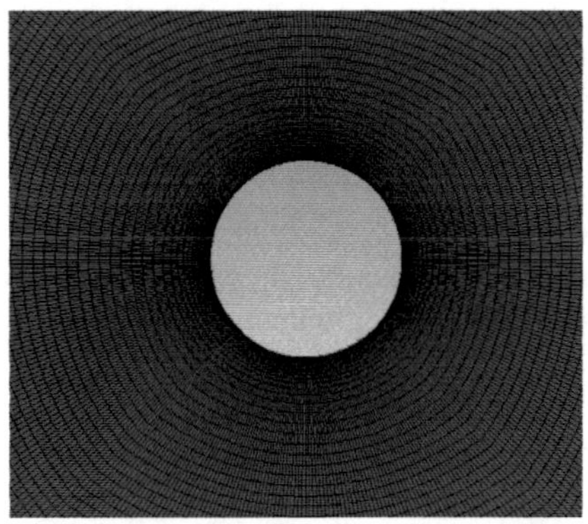

Figure 2. Mesh at the vicinity of the cylinder

4. Results and discussion

4.1. Influence of f_k on velocity statistics

The effects of varying f_k values on the flow statistics are investigated. The experimental results by Cantwell and Coles[22] are plotted alongside for comparison purposes.

The mean streamwise velocity along the wake centerline (y=0) is calculated by the nonlinear PANS model and shown in Fig.3. Normalized streamwise velocity (u/U_0) of various f_k values plotted alongside the experimental result. The nonlinear PANS calculation results with f_k=1.0 and fk=0.8 have large tolerances compared with experimental data. With PANS calculations f_k=0.6 and 0.4, the mean velocity profiles get closer to the experimental results. When f_k=0.4, the calculation result is in good agreement with experimental data. In the wake region(1<x/D<5), the PANS results show monotonic improvement toward experimental data as the f_k value is reduced from 1.0 to 0.4. In the far wake region(x/D>6), the PANS results are all close to the experimental data. This could be attributed to the decrease of the influence of wake flow downstream the cylinder. Results performed by Lakshmipathy and Girimaji [10] using PANS model without nonlinear shear stress are also shown in Fig.3. It can be seen than the nonlinear PANS model is accuracy in the region (1<*x*/*D*<5).

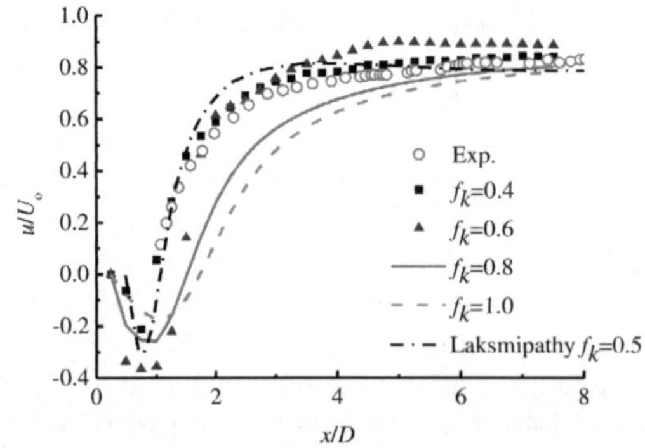

Figure 3. Mean streamwise velocity along the wake centerline for various f_k values

Figure 4 shows the mean normal velocity on *x*/*D*=1.0 line. There has a rotational symmetry of the mean normal velocity on *x*/*D*=1.0 line. The PANS results with f_k=0.6 and 0.4 predict the mean normal velocity at the *x*/*D*=1.0 line with good accuracy. The computations results with f_k=1.0 and 0.8 fail to capture the trend for this velocity statistics at the wake flow. The maximum normal velocity at the *x*/*D*=1.0 line predicted with f_k=0.4 is only a little large than the experimental result. For the mean velocity statistics presented above, a smaller f_k values is better for the nonlinear PANS computations for turbulence flow.

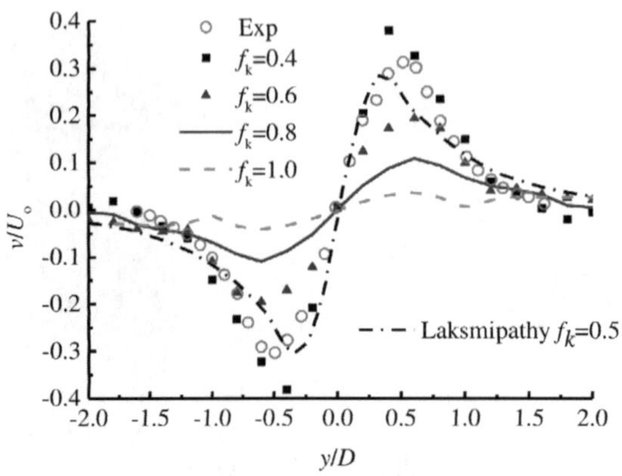

Figure 4. Mean normal velocity on *x*/*D*=1.0 line

Figure 5 shows the mean streamwise velocity at *x*/*D*=1.0 line in the near wake region. The mean streamwise velocity shows a symmetry distribution by *y*/*D*=0. At *x* *D*=1.0, the PANS results with f_k=0.4 are in good agreement with the experimental measurements, which show V-shaped profiles at this location. A V-shaped profile has been indicated by experiments done by Kravchenko and Moin [23], which also shows that simulation results are more accuracy when f_k=0.4. As f_k increases, the maximum mean streamwise velocity increases, while the minimum mean streamwise velocity decreases. When $f_k \geq 0.6$, the mean streamwise velocity profile on *x*/*D*=1.0 line transforms into U-shape. This could be attributed to the fact that the separation in the case of PANS computations is turbulent in nature. The peak velocity for the PANS computations with PANS of f_k=0.4 showing the best agreement with experiments.

Figure 6 shows the mean streamwise velocity at *x*/*D*=3.0 line in the near wake region. The mean streamwise velocity shows a symmetry distribution by *y*/*D*=0. When f_k=0.4, the simulation results of mean streamwise velocity have the same shape with experimental result and it can predict the hump characteristic at 1.0<y/*D*<2.0 and -2.0<*y*/*D*<-1.0. At *x* *D* =3.0, the PANS computations with $f_k \geq 0.6$ over predict the velocity at the center.

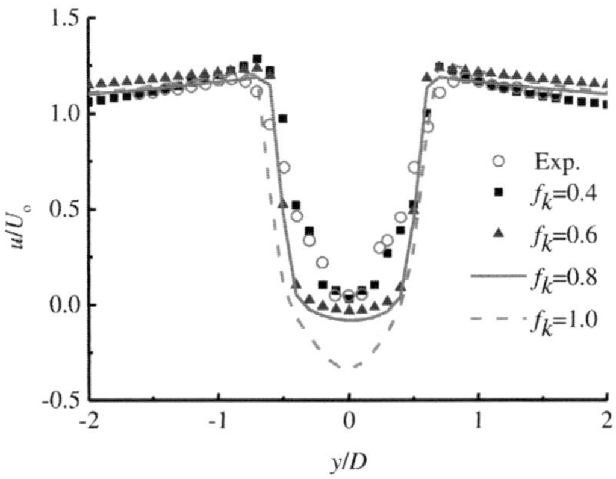

Figure 5. Mean streamwise velocity at *x*/*D*=1.0 line

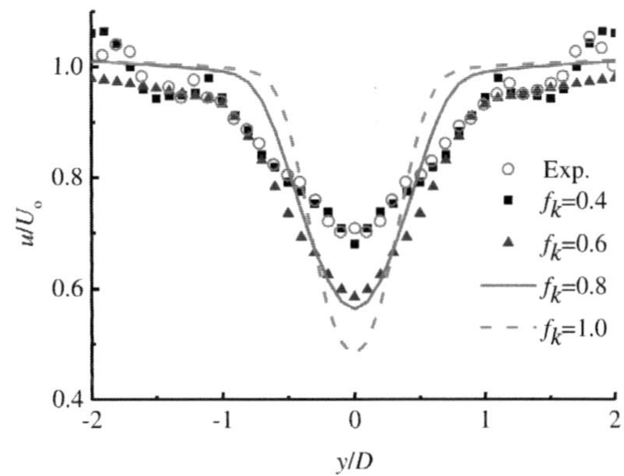

Figure 6. Mean streamwise velocity at *x*/*D*=3.0 line

4.2. pressure coefficient distributions

Figure 7 shows the distribution of pressure coefficient versus θ, the angle measured from the forward stagnation point along the surface of the cylinder. The nonlinear PANS results are compared with experimental data at this high Reynolds number [19]. With the reduce of f_k, the minimum pressure is more and more close to the experimental result. The inflection point of the pressure coefficient line at θ=100° is predicted accurately when $f_k \leq 0.6$. The pressure drop on the cylinder surface increases further as the f_k value is reduced, leading to delayed separation.

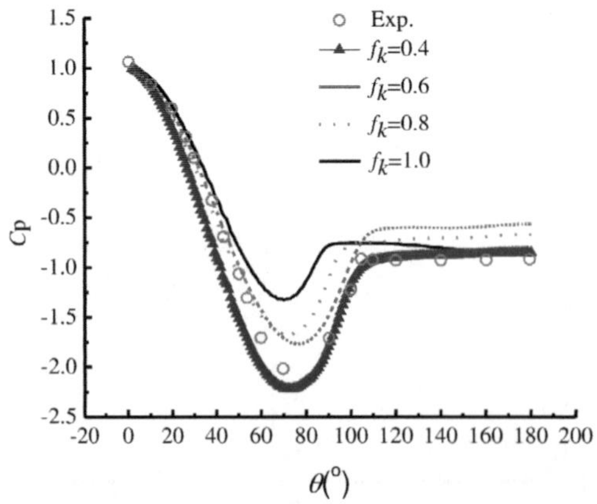

Figure 7. C_p distribution over the cylinder surface

5. Conclusions

The nonlinear PANS model modified from RNG k-ε turbulence model is performed by the commercial CFD package FLUENT using user define function. Turbulence flow past a circular cylinder at Re_D=1.4×10^5 is used to evaluate the accuracy of the nonlinear PANS model. The nonlinear PANS capability of reducing the closure cut-off length-scale by decreasing the f_k value is examined by performing simulations with f_k values of 1.0, 0.8, 0.6, and 0.4. As can be seen from the results presented above, PANS performs very well in computing profiles and flow distributions. Based on the computed results for flow past a circular cylinder, it is reasonable to conclude that the nonlinear PANS bridging method is a theoretically sound and computationally viable variable resolution method for practical flow computations. The computational result is reasonable when f_k is less than 0.6. The PANS method can be used in the simulation of high Reynolds number turbulence flow.

Acknowledgments

The authors would like to thank National Natural Science Foundation of China (No. 51406010) and Open research fund program of State Key Laboratory of Hydroscience and Engineering (Project No. sklhse-2014-16-E-02).

References

[1] Wu Y L, Li S C, Liu S H, Dou H S and Qian Z D, 2013 Vibration induced by hydraulic excitation in Vibration of hydraulic machinery *Springer Germany* 147-233.

[2] S Girimaji, R Srinivasan and E Jeong 2003 PANS turbulence models for seamless transition between RANS and LES: fixed point analysis and preliminary results, *ASME paper FEDSM* 45336.

[3] Lakshmipathy, S and Girimaji, S 2006 Partially-averaged Navier-Stokes method for turbulent flows: k-ω model implementation. *44th AIAA Aerospace Sciences Meeting and Exhibit* (Reno, Nevada) 119.

[4] S Lakshmipathy and S Girimaji 2010 *Journal of Fluids Engineering* **132(12)**:121202-1-9.

[5] Ji B, Luo X W, Wu Y L, Peng X X, Xu H Y 2012. *International Journal of Heat and Mass Transfer* **55 (23-24)**, 6582-6588.

[6] Ji B, Luo X W, Wu Y L, Peng X X, Duan Y L 2013. *International Journal of Multiphase Flow* **51**, 33-43.

[7] B Basara, S Krajnović, S Girimaji, and Z 2011 *Pavlovic. AIAA Journal* **49(12)**:2627-2636.

[8] L Davidson, Peng S H 2013 *AIAA Journal* online first, doi: 10.2514/1.J051864

[9] B Contwell and D Coles 1983 *J. Fluid Mech.* **136(1)**:321-374.

[10] M Breuer 2000 *Int. J. Heat Fluid Flow* **21(5)**:648-654.

[11] A Travin, M Shur, M Strelets, and P. Spalart 2000 *Turbul Combust.* **63(1-4)**:293-313.
[12] Liu J, Zuo Z, Wu Y, Zhuang B and Wang L 2014 *Computers & Fluids* **102**, 32-40.

IOP Conf. Series: Materials Science and Engineering **129** (2016) 012049 doi:10.1088/1757-899X/129/1/012049

Critical cavitation coefficient analysis of a space low specific centrifugal pump with micro gravity

J T Liu[1], Y Li[1], Y Gao[1], Q Hu[1], C Zhou[1] and Y L Wu[2]

[1]Beijing Institute of Control Engineering, Beijing 100090, China
E-Mail: liujintao86@hotmail.com
[2]State Key Laboratory of Hydroscience and Engineering, Tsinghua University, Beijing 100084, China

E-mail:wyl-dhh@mail.tsinghua.edu.cn

Abstract. Centrifugal pump was used in the loop as a baselined unit. The flow rate of the pump was very small, while the head was high. This space pump must work stable for a long time (more than a year), so the performance of the pump attracted public attention. The rotational speed of the impeller was limited for stability, so the pump belonged to low specific centrifugal pump. In this paper, a single-phase centrifugal pump, which was designed for single-phase fluid loops in satellites, was modeled for numerical simulation. The hydraulic region of the pump was discretized by structured mesh. Three dimensional (3-D) flow in the pump was studied by the use of computational fluid dynamics. Partially-Averaged Navier-Stokes (PANS) model based on RNG k-ε turbulence model was developed for the simulation of the unsteady flow. Velocity inlet and pressure outlet was used as the boundary conditions. Interface was used between the impeller and the casing, as well as the impeller and inlet pipe. Performances and pressure fluctuation of the pump were investigated. The dominant frequency of the pressure fluctuation is blade passing frequency at the region close to the tongue of the casing, while it is twice of blade passing frequency at the other region.

1. Introduction

Single-phase mechanically pumped fluid loop (MPFL) is one of active thermal control technology which is used in the field of thermal control in a spacecraft. Precise control of the temperature in the cabin of the spacecraft and key devices can be realized with the help of MPFL. The MPFL has advantages including strong capability of heat exchange, safe and reliable structure and simple fluid mechanism [1]. Since the 1990 s, the MPFL technology has been used in many international spacecraft. The key component of the MPFL is the circulating pump, which should have long life, high reliability and stability, as well as high efficiency. The energy loss of the pump will be change into heat in the spacecraft, so the high efficiency is one of the most important factors for the designation of the pump.

In manned spacecraft, such as skylab space station, "Gemini" spacecraft, "Alliance" spacecraft and "Salute" space station, the MPFL has played an important role in the development of the technology of aerospace [2]. The MPFL also has been used in the field of Mars exploration program

1 .

since the 1990s. Most of pumps in MPFL for NASA space program including Mars Pathfinder in 1997, Mars Exploration Rovers in 2003, Mars Science Laboratory in 2011, were all developed by Jet Propulsion Laboratory (JPL)[3,4].

ESA started to develop their MPFL technology in 2004. The MPFL technology was used in large communications satellite for the demand of heat exchange, and the pump used in MPFL were designed inheriting to the pump developed by ESA in 1980s and 1990s[5]. Rapid development has come true in the past few years with the corporation of Bradford Engineering, Realtechnologie and Netherlands National space lab. The pump developed by Realtechnologie has small diameter with lone life for 5 years [6]. In China, the study of MPFL technology started from 1980s and it has been used in the Shenzhou manned spaceship. Although China had ripe experience of the MPFL technology in manned spacecraft, there hasn't have a report on the MPFL for the heat control used in the satellite [7].

In this paper, a pump used in large satellite for MPFL technology were designed and analyzed. Performances of three kinds of impeller were compared and flows in the impeller were analyzed.

2. Nolinear PANS model

For incompressible flow, V_i is partitioned into resolved and unresolved parts in the instantaneous velocity field, using an arbitrary homogeneous filter.

$$V_i = U_i + u_i \tag{1}$$

where U_i is the resolved velocity field; u_i is the unresolved field. It is used by equation 4 instead of the resolved field.

$$U_i = \langle V_i \rangle \tag{2}$$

The additional non-linear term $\tau(V_i, V_j)$, which is the generalized central second moment, is defined as:

$$\tau\left(V_i, V_j\right) = \left(\langle V_i V_j \rangle - \langle V_i \rangle\langle V_j \rangle\right) \tag{3}$$

For PANS methods,

$$f_k = \frac{k_u}{k} \tag{4}$$

$$\varepsilon_k = \frac{\varepsilon_u}{\varepsilon} \tag{5}$$

Then the PANS model by the modification of RNG k-ε turbulence model are:

$$\frac{\partial(\rho k_u)}{\partial t} + \frac{\partial(\rho \overline{U}_j k_u)}{\partial x_j} = \frac{\partial}{\partial x_j}\left[\alpha_k\left(\mu + \frac{\mu_u}{\sigma_u}\right)\frac{\partial k_u}{\partial x_j}\right] + P_{ku} - \rho\varepsilon_u \tag{6}$$

$$\frac{\partial(\rho \varepsilon_u)}{\partial t} + \frac{\partial(\rho \overline{U}_j \varepsilon_u)}{\partial x_j} = \frac{\partial}{\partial x_j}\left[\alpha_k\left(\mu + \frac{\mu_u}{\sigma_u}\right)\frac{\partial \varepsilon_u}{\partial x_j}\right] + C^*_{\varepsilon 1} P_{ku}\frac{\varepsilon_u}{k_u} - C^*_{\varepsilon 2}\rho\frac{\varepsilon_u^2}{k_u} \tag{7}$$

thus,

$$\mu_u = \rho C_\mu \frac{k_u^2}{\varepsilon_u} \tag{8}$$

$$\sigma_{ku} = \frac{f_k^2}{f_\varepsilon}, \qquad \sigma_{\varepsilon u} = \frac{f_k^2}{f_\varepsilon} \tag{9}$$

$$C^*_{\varepsilon 2} = C^*_{\varepsilon 1} + \frac{f_k}{f_\varepsilon}\left(C_{\varepsilon 2} - C^*_{\varepsilon 1}\right) \tag{10}$$

$$C^*_{\varepsilon 1} = C_{\varepsilon 1} - \frac{\eta(1-\eta/\eta_0)}{1+\beta\eta^3} \tag{11}$$

where $C_\mu = 0.0845$, $\alpha_k = \alpha_\varepsilon = 1.39$, $C_{\varepsilon 1} = 1.42$, $\eta_0 = 4.377$, $\beta = 0.012$, $C_{\varepsilon 2} = 1.68$

Equating the source terms, P_k in RNG k-ε turbulence mode has a relationship with P_{ku} in PANS model shown in equation 12.

$$P_k = \frac{1}{f_k}(P_{ku} - \varepsilon_u) + \frac{\varepsilon_u}{f_\varepsilon} \tag{12}$$

Considering the nolinear turbulence flow in the pump-turbine, the shear stress was solved by nolinear turbulence model which was proposed by Ehrhard [15].

$$P_k = -\rho \overline{U_i' U_j'} \frac{\partial \overline{U_j}}{\partial x_j} \tag{13}$$

$$\overline{U_i' U_j'} = \frac{2}{3}k\delta_{ij} - 2C_{\mu P}\upsilon^2 T S_{ij} + C_1 C_{\mu P}\upsilon^2 T^2\left(S_{ik}S_{kj} - \frac{1}{3}S_{kl}S_{kl}\delta_{ij}\right)$$
$$+C_2 C_{\mu ke}\upsilon^2 T^2\left(\Omega_{ik}S_{kj} - \Omega_{jk}S_{ki}\right) + C_3 C_{\mu P}\upsilon^2 T^2\left(\Omega_{ik}\Omega_{jk} - \frac{1}{3}\Omega_{lk}\Omega_{lk}\delta_{ij}\right)$$
$$+C_4 C_{\mu ke}\upsilon^2 T^3\left(S_{ki}\Omega_{lj} - S_{kj}\Omega_{li}\right)S_{kl} + C_5 C_{\mu P}\upsilon^2 T^3 S_{ij}S_{kl}S_{kl} + C_6 C_{\mu P}\upsilon^2 T^3 S_{ij}\Omega_{kl}\Omega_{kl} \tag{22}$$

$$C_{\mu P} = \min\left(\frac{1}{0.9S^{1.4} + 0.4\Omega^{1.4} + 3.5}, 0.15\right) \tag{23}$$

$$\Omega_{ij} = \frac{1}{2}\left(\frac{\partial \overline{U_i}}{\partial x_j} + \frac{\partial \overline{U_j}}{\partial x_i}\right) \tag{24}$$

$$S = \frac{k}{\varepsilon}\sqrt{2S_{ij}S_{ij}} \tag{25}$$

$$\Omega = \frac{k}{\varepsilon}\sqrt{2\Omega_{ij}\Omega_{ij}} \tag{26}$$

where $C_{\mu ke} = \beta\frac{k^2}{\varepsilon}$, $C_1 = -0.2$, $C_2 = 0.4$, $C_3 = 2.0 - \exp\left(-(S-\Omega)^2\right)$, $C_4 = -32.0C_{\mu P}^2$, $C_5 = -16.0C_{\mu P}^2$, $C_6 = 16.0C_{\mu P}^2$, T is turbulence time scale, ν is the turbulence velocity scale.

3. Rated conditions of the pump

The rated Parameters of the pump are shown in Tab.1. H_r denotes the rated head; n denotes the rotational speed of the impeller; Q_r denotes the rated discharge. n_s denotes the specific speed of the impeller; ρ is the density of the transmission medium. D_o is the outlet diameter of the impeller. D_i is the diameter of the impeller inlet. The structure of the pump is shown in Fig.1.

Table 1. Parameters of the pump

H_r (m)	12
Q_r(L/min)	1
n (r/min)	8000
n_s	18
Transmission medium	ethylene glycol
D_o(mm)	82
D_i(mm)	10
ρ (kg/m^3)	1030
No. of blades	4

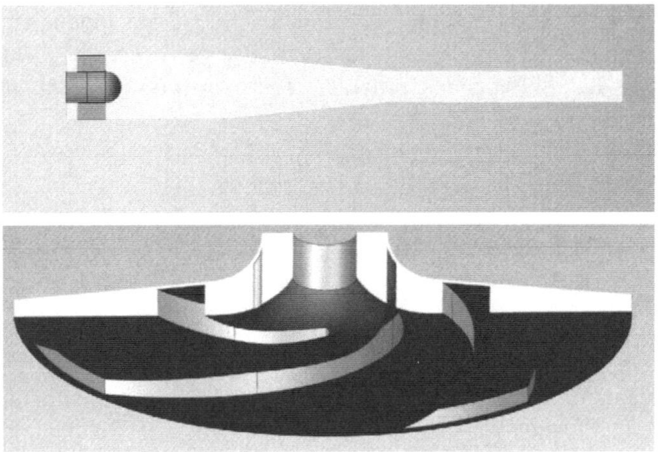

Figure 1. Profile of the pump

4. Simulation conditions

Mesh of the pump was created by hexahedron using ICEM. The total number of mesh of the pump was 5 million.

The runner's hydraulic region was set to rotate at a speed of 8000 rpm. Velocity at pipe inlet was specified. The outlet condition was chosen and the value of static pressure was used. The outlet pressure was specified according to the experimental data. No slip wall boundary condition on the solid walls was specified. The upwind discretization scheme was used to discretize the advection term. The time step was 0.0000208 s. Thus, the converged turbulent flow solutions were obtained by rotating the mesh in the runner region by 1° per time step. The unsteady solutions are formed by the converged solutions at all times. For the unsteady calculation, the pressure and velocity change with time in the whole flow passage.

5. Results and discussions

5.1 Performance analysis

The external characteristic of the pump with different rotational speeds is shown in Fig.2. Hump characteristic can't be seen in the head curve. The head of the pump maintain in a small range as the flow rate changes in the computational range. Constant pressure can be provided by the pump.

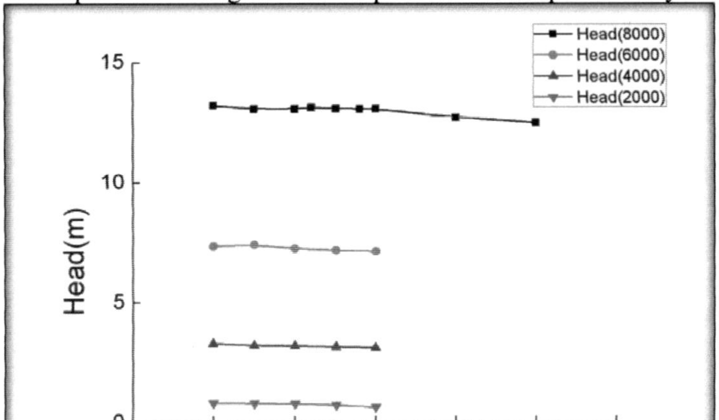

Figure 2. External characteristics of the pump

In order to predict the critical cavitation coefficient, the pressure value of the pressure outlet is reduced and the performance of the pump is calculated. The performance of the pump with different cavitation coefficient is shown in Fig.3. The critical cavitation coefficient of the pump is 0.065.

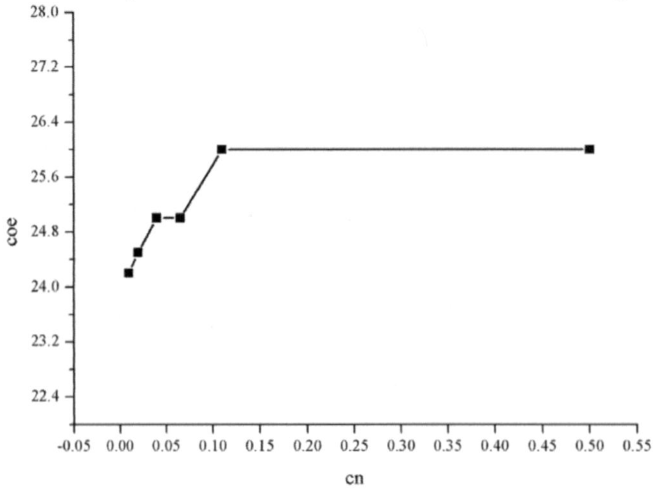

Figure 3. Performance of the pump with different cavitation coefficients

5.2 Cavitating flow in the pump

Figure 4 shows the cavitating flow in the pump with different cavitation coefficients. When the pump operates at critical cavitation coefficient condition, cavitation occurs at the inlet of the blade in the runner and the volume of the cavitating flow is small. The cavitating flow is gradually developed along the suction side of the blade. The cavitating flow almost block the passage of the runner when the cavitation coefficient is 0.01.

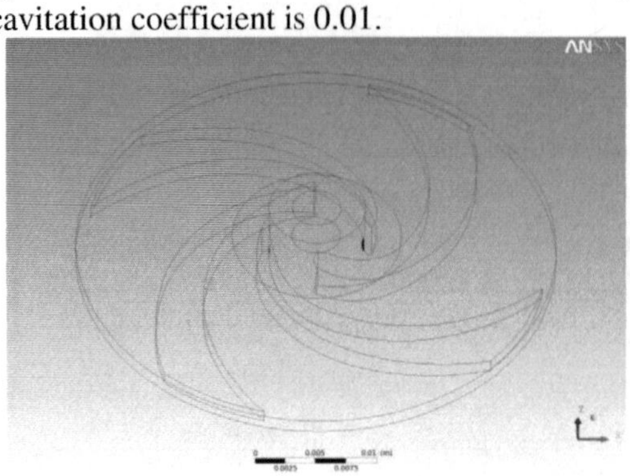

(a) 0.065

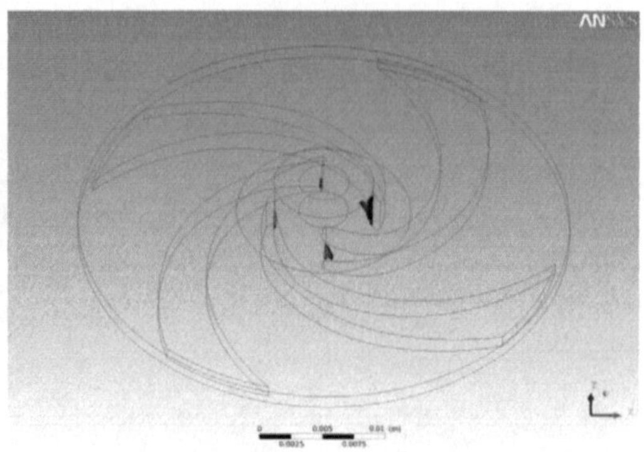

(b) 0.04

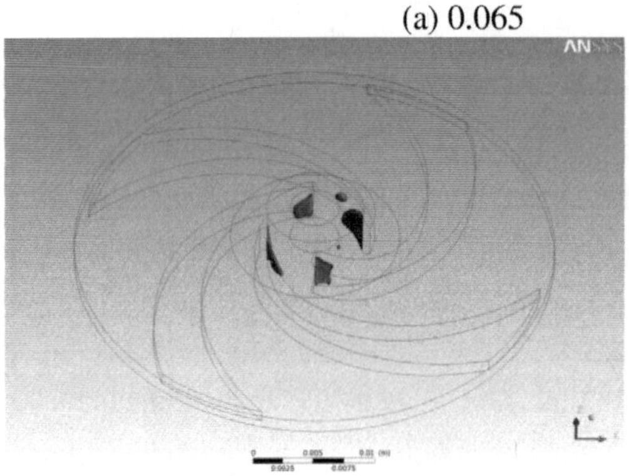

(c) 0.02

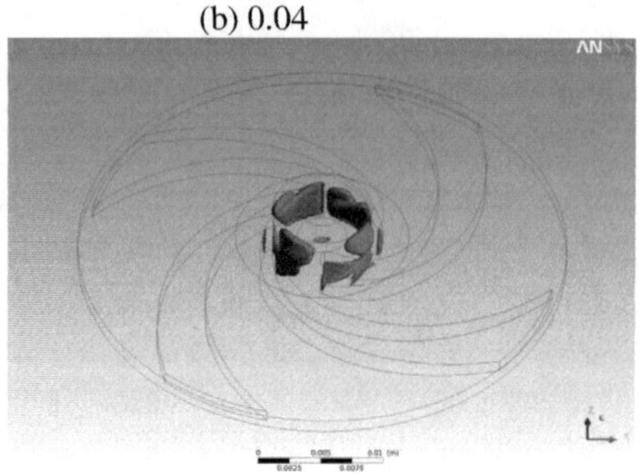

(d) 0.01

Figure 4. Cavitating flow with different cavitation coefficients

5.2 Cavitating flow in the pump

Pressure fluctuation in the casing close to the outlet of the blade is shown in Fig.5. The amplitude of the pressure fluctuation is 20.2%. The dominant frequency of the pressure fluctuation is blade passing frequency at the region close to the tongue of the casing, while it is twice of blade passing frequency at the other region.

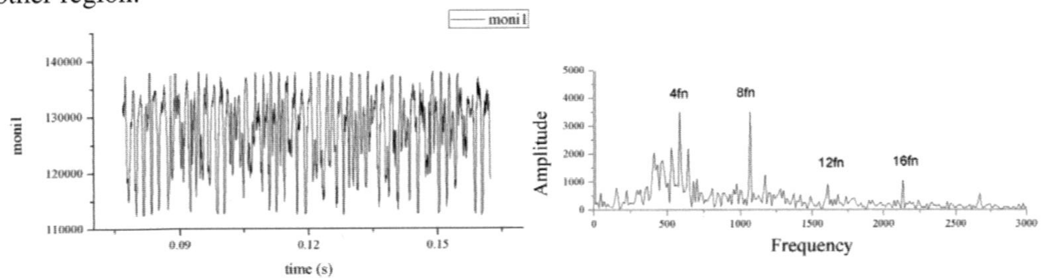

(a) 0.065

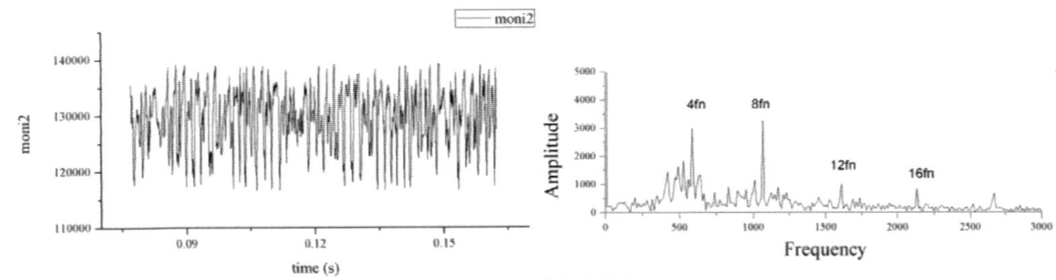

(b) 0.04

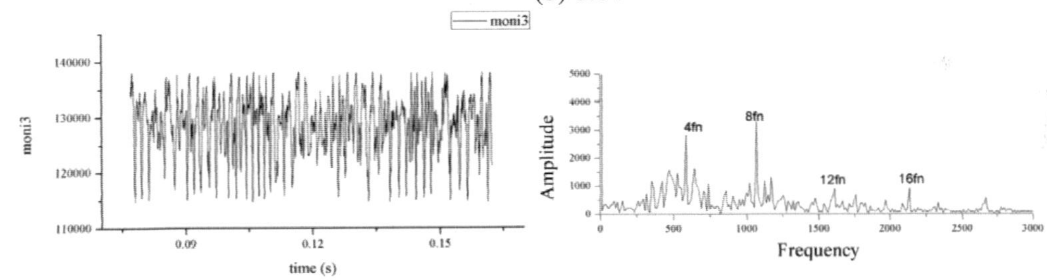

(c) 0.02

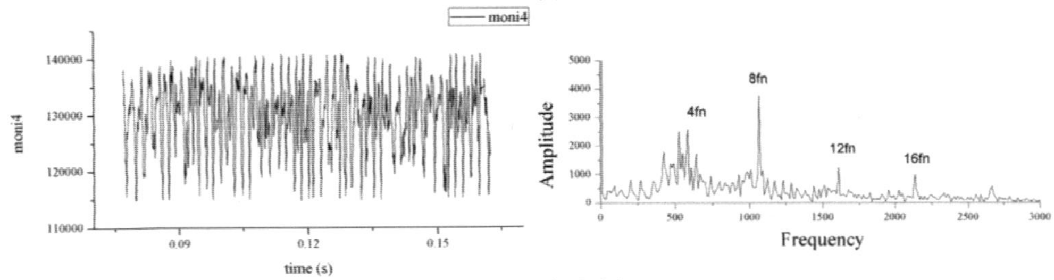

(d) 0.01

Figure 5. Pressure fluctuation in the pump

6. Conclusions

In this paper, a PANS model was used to analyze the instability of a centrifugal pump used in space. The external characteristic of the pump with different rotational speeds is calculated and is could be found that constant pressure can be provided by the pump as the flow rate changes in the computational range. The critical cavitation coefficient of the pump is 0.065. The amplitude of the pressure fluctuation is 20.2%. The dominant frequency of the pressure fluctuation has a relationship with the blade passing frequency.

Acknowledgments
The authors would like to thank National Natural Science Foundation of China (No. 51406010) and Open research fund program of State Key Laboratory of Hydroscience and Engineering (Project No. sklhse-2014-16-E-02).

References

[1] Zhang L, Jiang J and Zhao Q W 2008 *Chinese space science and technology*, 3:65-70.
[2] Gajanana B and Pradeep B 1997 Mars pathfinder active thermal control system: ground and flight performance of a mechanically pumped cooling loop. *AIAA* 200469
[2] Jiang J, Zhang N L, Xu J W 2004, *Chinese space science and technology*, 2:33-37.
[3] Bhandari, P, et al. 2009 Mars science laboratory mechanically pumped fluid loop for thermal control - design, implementation, and testing *SAE Int. J. Aerosp.* **4(1)**: 299- 310.
[4] Jet Propulsion Laboratory 1996 Integrated pump assembly-an active cooling system for mars pathfinder thermal control *SAE Int.*
[5] NASA-Jet Propulsion Laboratory Advanced thermal control technologies for space science missions at JPL
[6] ESA-technology-thermal control: single-phase mechanically-pumped fluid loops [EB/OL] [2007-03-27].http://www.esa.int/TEC/Thermal_control/SEM4CLBE8YE_0.html.
[7] Van Benthem, R C, et al. 2009 Development of a mechanically pumped fluid loop for 3 to 6 kW payload cooling *SAE Int.*

Numerical prediction of pressure fluctuations in a prototype pump turbine base on PANS methods

J T Liu[1], Y Li[1], Y Gao[1], Q Hu[1] and Y L Wu[2]

[1]Beijing Institute of Control Engineering, Beijing 100090, China
E-Mail: liujintao86@hotmail.com
[2]State Key Laboratory of Hydroscience and Engineering, Tsinghua University, Beijing 100084, China

E-mail:wyl-dhh@mail.tsinghua.edu.cn

Abstract. Unsteady flow and pressure fluctuations within a prototypel pump turbine are numerically studied using a nonlinear Partial Averaged Navier Stokes (PANS) model. Pump turbine operating at different conditions with guide vanes opening angle 6° is simulated. Results revealed that the predictions of performance and relative peak-to-peak amplitude by PANS approach agree well with the experimental data. The amplitude of the pressure fluctuation in the vaneless space at turbine mode on a "S" curve increases with the decrease of the flow rate, and it has maximum value when it runs close to runaway line at turbine braking mode. The amplitude of the pressure fluctuation in the vaneless space at turbine braking mode on a "S" curve decreases with the reduce of the flow rate. The above high pressure fluctuations should be avoided during the design of pump turbines especially those operating at high-head condition.

1. Introduction

Pump-turbine played an important role in the optimization of power network. The starting period and stopping period of a pump-turbine ran at a high frequency, leading to a lot of instability problems due to the existence of "S" characteristics. When a pump-turbine ran in the "S" region, pressure fluctuation was very dangerous, even leading to some serious accidents.

Pump-turbines were always running at partial condition with the power grid changing, leading to flow separations and stall phenomena in the pump-turbine. The study of instability of a pump-turbine should overcome the difficulties in the simulation of strong vortex and separation flow. Large-scale coherent structures play a crucial role in the internal instabilities. For predicting the fluctuation of the dominant scale of vortexes, the traditional Reynolds-averaged Navier-Stokes (RANS) method suffers from inherent physical limitations[1], which couldn't capture the flow with all kinds of scales. Large eddy simulation (LES) may not be computationally viable due to the large computational cost. Instead, the mixed computational methods, which combined the desirable aspects of RANS and LES, were used in engineering computation [2]. Speziale [3] proposed a new turbulence model that combines the advantages of RANS method with those of LES.

1

Ji [4] studied the unsteady cavitating turbulence flow around a highly skewed model marine propeller based on the PANS method. Huang [5] used PANS model to investigate the cavitating flow around a hydrofoil. The PANS models mentioned above were all modified from standard *k-ε* turbulence model, while the standard k-ε turbulence model was poor in the simulation of strong swirling flows [6]. Most of the RANS turbulence models solved the shear stress by linear difference scheme and they were isotropic models[7], so they couldn't capture all kinds of vortexes in the pump-turbine well.

In this paper, a new nonlinear PANS turbulence model was proposed. The results based on the new nonlinear PANS turbulence model were compared with experimental results. Then the instability of a pump-turbine was studied by the nonlinear PANS model.

2. Nolinear PANS model

For incompressible flow, the continuity equation and Reynolds averaged Navier-Stokes equations:

$$\frac{\partial U_i}{\partial t}+U_j\frac{\partial U_i}{\partial x_j}+\frac{\partial \tau(V_i,V_j)}{\partial x_j}=-\frac{\partial \langle p\rangle}{\partial x_i}+\nu\frac{\partial^2 U_i}{\partial x_j \partial x_i} \tag{1}$$

$$-\frac{\partial^2\langle p\rangle}{\partial x_i \partial x_i}=-\frac{\partial U_i}{\partial x_j}\cdot\frac{\partial U_j}{\partial x_i}+\frac{\partial \tau(V_i,V_j)}{\partial x_i \partial x_j} \tag{2}$$

V_i is partitioned into resolved and unresolved parts in the instantaneous velocity field, using an arbitrary homogeneous filter.

$$V_i=U_i+u_i \qquad U_i=\langle V_i\rangle \tag{3}$$

where U_i is the resolved velocity field; u_i is the unresolved field. It is used by equation 4 instead of the resolved field.

The additional non-linear term $\tau(V_i, V_j)$, which is the generalized central second moment, is defined as:

$$\tau(V_i,V_j)=\left(\langle V_iV_j\rangle-\langle V_i\rangle\langle V_j\rangle\right) \tag{4}$$

The PANS models are:

$$\frac{\partial(\rho k_u)}{\partial t}+\frac{\partial(\rho\overline{U_j}k_u)}{\partial x_j}=\frac{\partial}{\partial x_j}\left[\alpha_k\left(\mu+\frac{\mu_u}{\sigma_u}\right)\frac{\partial k_u}{\partial x_j}\right]+P_{ku}-\rho\varepsilon_u \tag{5}$$

$$\frac{\partial(\rho\varepsilon_u)}{\partial t}+\frac{\partial(\rho\overline{U_j}\varepsilon_u)}{\partial x_j}=\frac{\partial}{\partial x_j}\left[\alpha_k\left(\mu+\frac{\mu_u}{\sigma_u}\right)\frac{\partial \varepsilon_u}{\partial x_j}\right]+C^*_{\varepsilon1}P_{ku}\frac{\varepsilon_u}{k_u}-C^*_{\varepsilon2}\rho\frac{\varepsilon_u^2}{k_u} \tag{6}$$

Considering the nolinear turbulence flow in the pump-turbine, the shear stress was solved by nolinear turbulence model which was proposed by Ehrhard [6].

$$P_k=-\rho\overline{U_i'U_j'}\frac{\partial\overline{U_j}}{\partial x_j} \tag{7}$$

$$\begin{aligned}\overline{U_i'U_j'}&=\frac{2}{3}k\delta_{ij}-2C_{\mu P}\upsilon^2TS_{ij}+C_1C_{\mu P}\upsilon^2T^2\left(S_{ik}S_{kj}-\frac{1}{3}S_{kl}S_{kl}\delta_{ij}\right)+C_2C_{\mu ke}\upsilon^2T^2\left(\Omega_{ik}S_{kj}-\Omega_{jk}S_{ki}\right)\\&+C_3C_{\mu P}\upsilon^2T^2\left(\Omega_{ik}\Omega_{jk}-\frac{1}{3}\Omega_{lk}\Omega_{lk}\delta_{ij}\right)+C_4C_{\mu ke}\upsilon^2T^3\left(S_{ki}\Omega_{lj}-S_{kj}\Omega_{li}\right)S_{kl}+C_5C_{\mu P}\upsilon^2T^3S_{ij}S_{kl}S_{kl}+C_6C_{\mu P}\upsilon^2T^3S_{ij}\Omega_{kl}\Omega_{kl}\end{aligned} \tag{8}$$

$$C_{\mu P}=\min\left(\frac{1}{0.9S^{1.4}+0.4\Omega^{1.4}+3.5},0.15\right)\quad \Omega_{ij}=\frac{1}{2}\left(\frac{\partial\overline{U_i}}{\partial x_j}+\frac{\partial\overline{U_j}}{\partial x_i}\right)\quad S=\frac{k}{\varepsilon}\sqrt{2S_{ij}S_{ij}}\quad \Omega=\frac{k}{\varepsilon}\sqrt{2\Omega_{ij}\Omega_{ij}}$$

where $C_{\mu ke} = \beta \frac{k^2}{\varepsilon}$, $C_1 = -0.2$, $C_2 = 0.4$, $C_3 = 2.0 - \exp\left(-\left(S-\Omega\right)^2\right)$, $C_4 = -32.0C_{\mu P}^2$, $C_5 = -16.0C_{\mu P}^2$, $C_6 = 16.0C_{\mu P}^2$,T is turbulence time scale, v is the turbulence velocity scale.

3. Pump-turbine geometry

Parameters of the model pump-turbine are shown in Tab.1. D_1 denotes the runner inlet diameter in pump mode; Z_S, Z_G and Z are the numbers of stay vanes, guide vanes and runner blades, respectively; H_d denotes the rated head; n denotes the rotational speed of the runner; Q_d denotes the rated discharge. The pump-turbine's structure is shown in Fig.1.

Table 1. Parameters of the model pump-turbine

H_d (m)	Q_d (m^3/s)	n (rpm)	D_1 (m)	Z	Z_S	Z_G
52.4	0.45	500	0.3	9	20	20

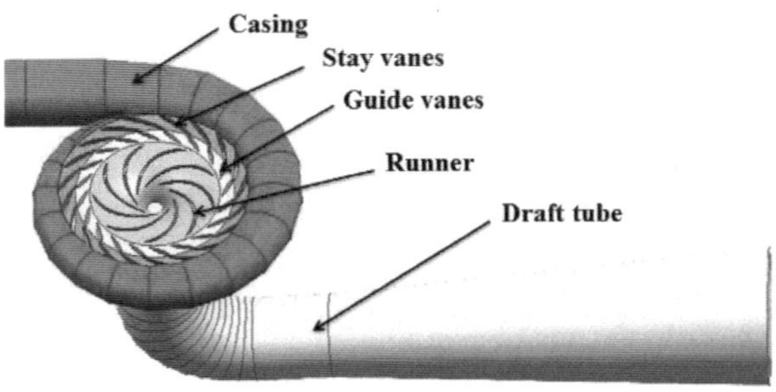

Figure 1. Profile of Pump-turbine

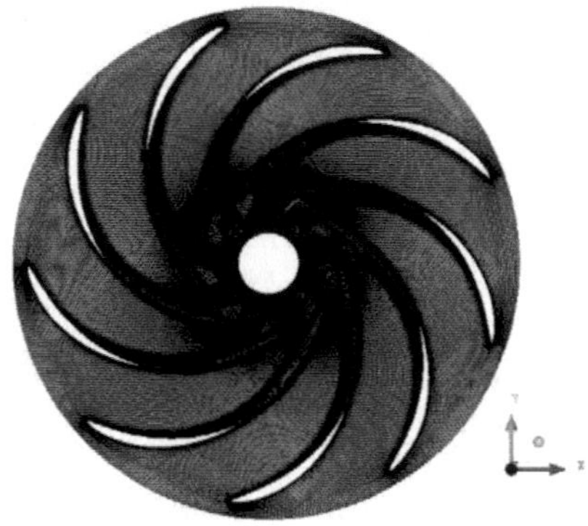

Figure 2. Mesh of the runner

The model's grids, which were composed of an unstructured hexahedron and tetrahedron, were developed using ICEM, which is a commercial software package used for CFD discretization. Hexahedral grids were used for the runner and draft tube, and mixed grids were used for the other components. The mesh of the runner is shown in Fig.2. A mesh with about 9 million cells in total was chosen for the simulations. The number of nodes and elements in each part is shown in Tab.2.

Table 2. Mesh of different hydraulic region

Hydraulic region	Runner	Guide vanes	Stay vanes	Draft tube	Casing	Total
Cells	2,734,700	1,734,700	1,455,324	1,173,888	1,840,506	8,939,118
Nodes	2,842,386	1,904,000	1,585,040	1,196,517	1,952,039	9,479,982

4. Results and discussion

4.1. Flow in the pump-turbine with MGV

Streamline in the runner on blade to blade surface at no-load condition is shown in Fig. 3. No reverse flow can be seen at turbine mode(G1 and G2). Reverse flow appears in the runner at run-away point (G3). The runaway point is the point for starting load during the start-up process, it is very important for the instability of a pump-turbine. The appearance of reverse flow may cause large pressure fluctuation in the pump-turbine. When the pump-turbine runs at turbine braking mode(G4, G5), the stall phenomenon can be seen in each passage of the runner. At zero flow rate point, the stall flow will block all the passages of the runner, and no flow can pass the passage of the runner.

4.2. Pressure fluctuation in the vaneless space

Pressure fluctuations in the vaneless space at different point on S curve are shown in Fig.4. When the pump-turbine runs at turbine mode, the amplitude of the pressure fluctuation in the vaneless space is small, and there has no low frequency. As the flow rate decreases, a low frequency signal can be observed. When the flow rate less than the point G5, the low frequency signal disappears.

The pressure fluctuations at frequency domain are calculated by fast Fourier transform (FFT), which are shown in Fig.5. The amplitude of pressure fluctuation increases as the flow rate decreases when the flow rate large than run-away point(G3). The dominant frequency (DF) of pressure fluctuation at each point on the S curve is the blade passing frequency. The second dominant frequency (SDF) are twice of the blade passing frequency at G1 and G2, while it is rotating frequency at G3 and G6.

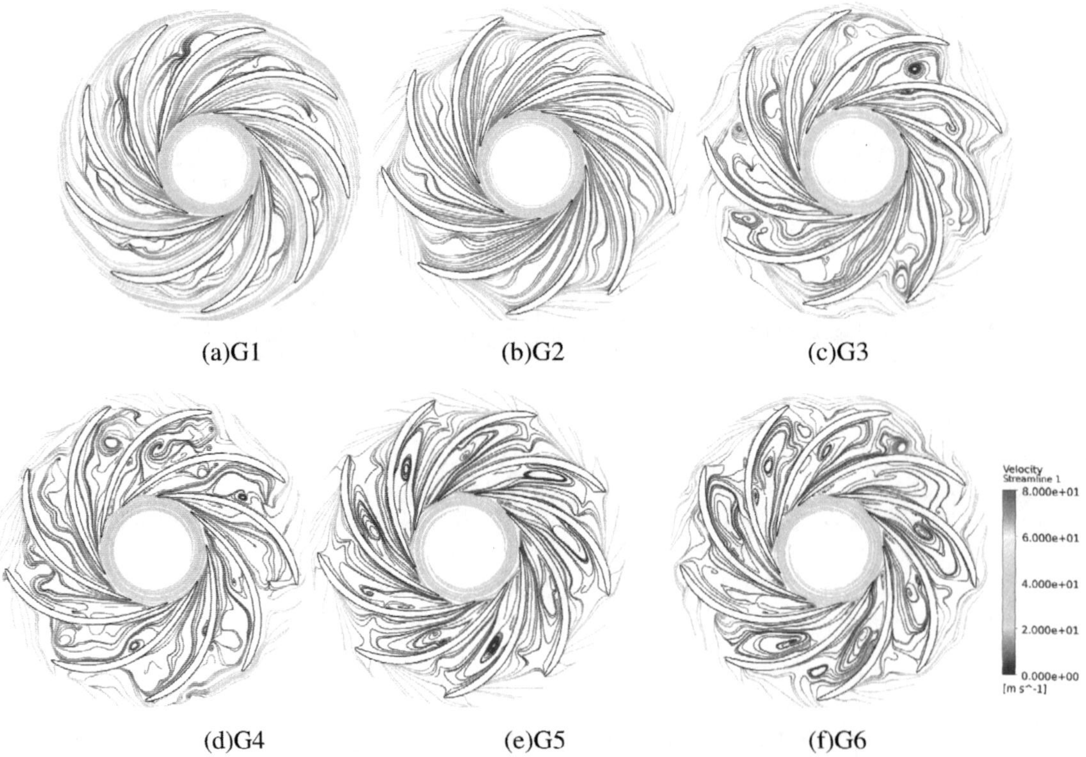

Figure. 3 Streamlines in the runner

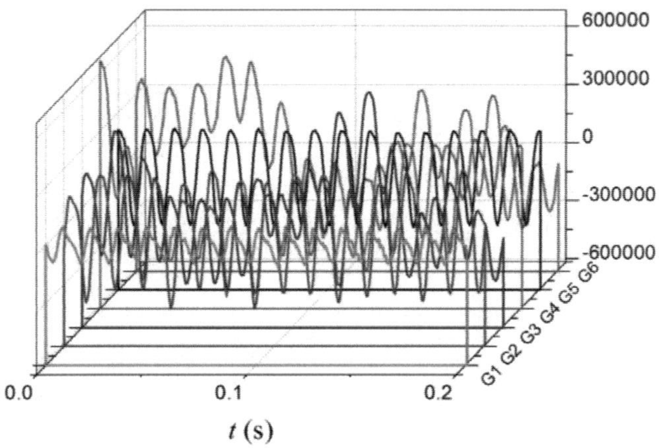

Figure 4. Pressure fluctuation at time domain

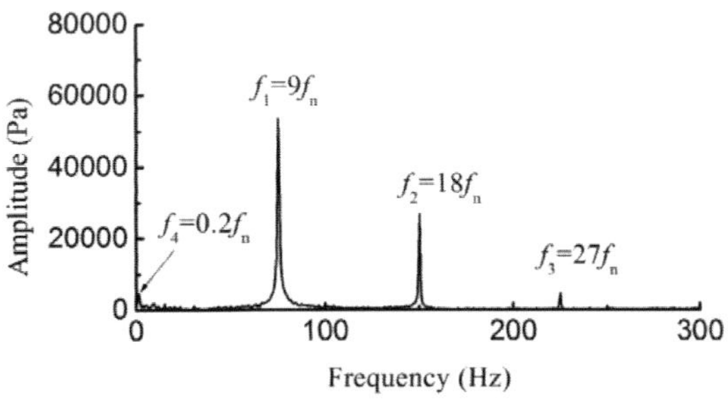

(a)G1

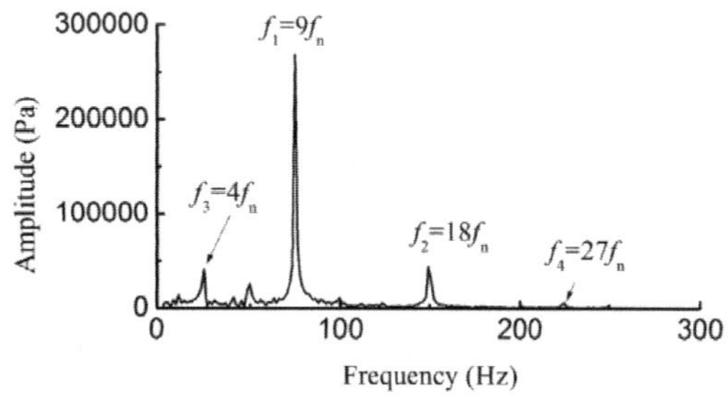

(b)G2

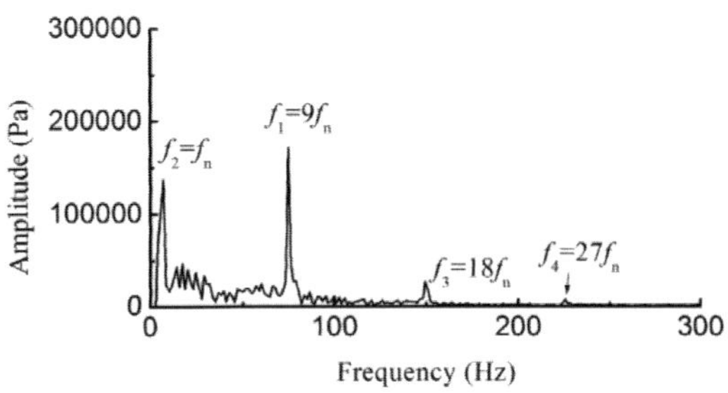

(c)G3

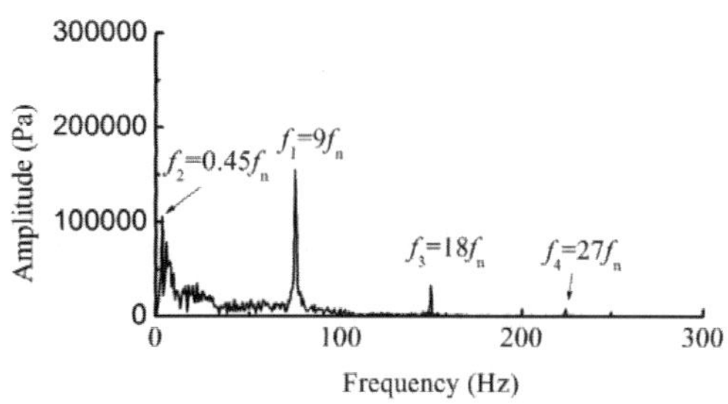

(d)G4

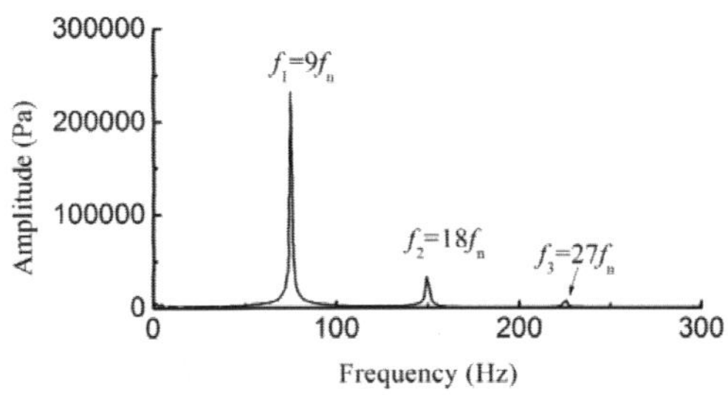

(e)G5

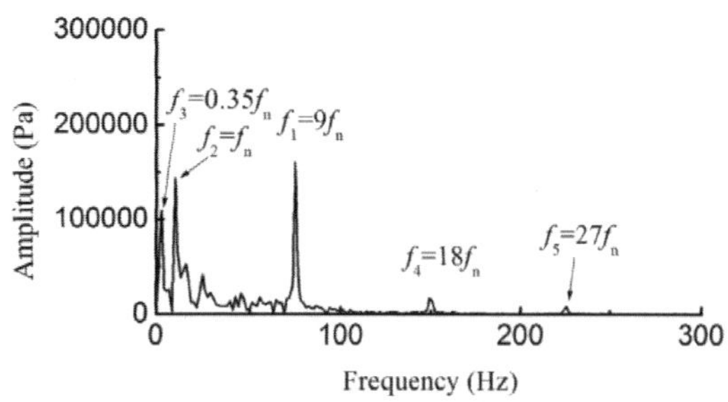

(f)G6

Figure 5. Pressure fluctuation at frequency domain

Table 3. Pressure fluctuation at the vaneless space

		$\Delta H/H$ (%)	DF ($\times f_n$)	SDF ($\times f_n$)
G1		3.6	9	18
G2	Cal.	11.9	9	18
	Exp.	11.3	9	18
G3		15.4	9	1
G4		14.8	9	0.45
G5		8.6	9	18
G6		13.1	9	1

5. Conclusions

In this paper, pressure fluctuations in the vaneless space at each point of the "S" curve of a pump-turbine were investigated. Results revealed that the predictions of performance and relative peak-to-peak amplitude by PANS approach agree well with the experimental data. The amplitude of the pressure fluctuation in the vaneless space at turbine mode on a "S" curve increases with the decrease of the flow rate, and it has maximum value when it runs close to runaway line at turbine braking mode. The amplitude of the pressure fluctuation in the vaneless space at turbine braking mode on a "S" curve decreases with the reduce of the flow rate. The above high pressure fluctuations should be avoided during the design of pump turbines especially those operating at high-head condition.

Acknowledgments

The authors would like to thank National Natural Science Foundation of China (No. 51406010) and Open research fund program of State Key Laboratory of Hydroscience and Engineering (Project No. sklhse-2014-16-E-02).

Reference

[1] Khorrami M R, Singer B, and Berkman M E 2002 *AIAA J.*, 40, **vol.7**, pp.1284–1291.
[2] Batten P, Goldberg U and Chakravarthy, S. 2002 *AIAA* 2002-0427.
[3] Girimaji, S S, Jeong, E, and Srinivasan, R, 2006 *ASME J. Appl. Mech.*, **73**, pp. 422–429.
[4] Ji B, Luo X W, Wu Y L, Peng X X and Xu H Y 2012 *Int J Heat Mass Tran* **vol.55**, pp.6582-6588.
[5] Huang B., Wang G. Y., 2011 *Journal of hydrodynamics* **vol.23(1)** pp.26-33.
[6] Yakhot, V., Orszag, S.A., Thangam, S., Gatski, T.B. & Speziale, C.G. 1992 *Physics of Fluids A*, **4(7)**, pp1510-1520.
[7] Liu J T, Liu S H , Wu Y L et al. 2012 *Computers and Fluids* **vol.68**,pp.105-111.

Simple, stable and reliable modeling of gas properties of organic working fluids in aerodynamic designs of turbomachinery for ORC and VCC

T Kawakubo

Manager, *Turbomachinery and Engine Technology Department,*

IHI Corporation, 1, Shin-nakahara-cho, Isogo-ku, Yokohama 235-8501, Japan

E-mail: tomoki_kawakubo@ihi.co.jp

Abstract. A simple, stable and reliable modeling of the real gas nature of the working fluid is required for the aerodesigns of the turbine in the Organic Rankine Cycle and of the compressor in the Vapor Compression Cycle. Although many modern Computational Fluid Dynamics tools are capable of incorporating real gas models, simulations with such a gas model tend to be more time-consuming than those with a perfect gas model and even can be unstable due to the simulation near the saturation boundary. Thus a perfect gas approximation is still an attractive option to stably and swiftly conduct a design simulation. In this paper, an effective method of the CFD simulation with a perfect gas approximation is discussed. A method of representing the performance of the centrifugal compressor or the radial-inflow turbine by means of each set of non-dimensional performance parameters and translating the fictitious perfect gas result to the actual real gas performance is presented.

1. Introduction

The Vapor Compression Cycle (VCC) or the Organic Rankine Cycle (ORC) makes use of the phase transition of the working fluid, thus the compressor or the turbine in the cycle inevitably operates near the vapor saturation boundary. Due to the high non-ideality (typically expressed by the deviation of the compressibility-factor from unity) of the working fluids near the saturation boundary, the designers of VCC compressors or ORC turbines must take the real gas effects into consideration appropriately for the reliable performance prediction in their designs. Although many current meanline performance prediction tools and commercially available Computational Fluid Dynamics (CFD) tools are capable of incorporating real gas models, simulations with such a gas model tend to be awkward and more time-consuming than those with a perfect gas model.

Figure 1 shows an example of the relative comparisons of accuracy and required computational time of CFD simulations with various degrees of fidelity in the gas modeling [1]. The simulation was conducted on an inviscid convergent-divergent nozzle operating with vapour R245fa which has a supersonic region downstream of the throat terminating with a shock wave. The applied equations of state include a perfect gas (PG) model, van der Waals (VDW) model, Peng-Robinson-Stryjek-Vera (PRSV) model and a real gas (RG) model provided by NIST REFPROP. The Mach number deviation was evaluated at an immediate vicinity of the shock wave on the downstream side. PG is the most attractive in terms of computational time. Its level of Mach number deviation is the highest among all but still acceptably small (less than 2%). VDW is a good compromise between the perfect gas approximation and more sophisticated modelings. PRSV gives highly accurate solution with the half computational time of RG. From this example, it can be stated that by employing RG model we can get a last small improvement in accuracy at the cost of additional high computational burden.

Moreover the CFD simulation with a real gas model can be unstable because it is possible that a part of the internal flow may encounter a mixed-phase condition, especially in off-design conditions, or in the early stage of the CFD simulations when the thermodynamic states can be still far away from the final convergence and may go beyond the saturation boundary erroneously. From these observations, simulations with a perfect gas approximation are still a highly attractive option in order to swiftly and stably conduct design simulations.

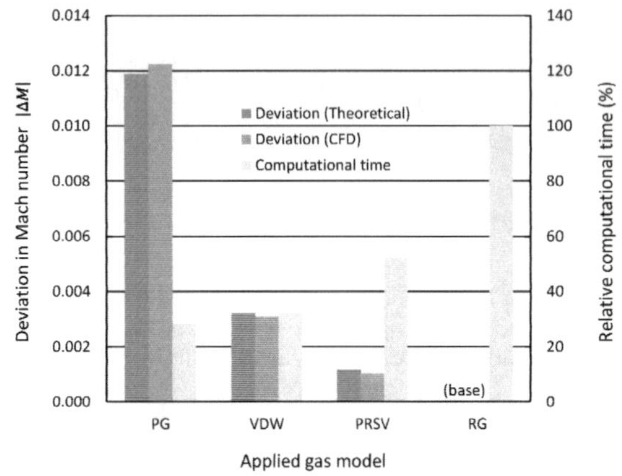

Figure 1. Mach number deviation and computation time with various gas modelings.

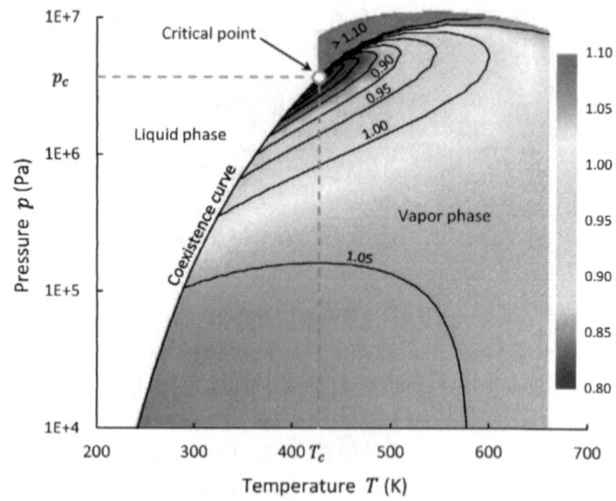

Figure 2. Variation of isentropic exponent of R245fa.

The perfect gas model (more rigorously, the calorically perfect gas model for which both the thermal and the caloric equations of state are "perfect") has only two independent modeling parameters, the gas constant R and the isentropic exponent γ. The isentropic exponent of a general real gas is defined by

$$\gamma = \frac{\rho}{p}\left(\frac{\partial p}{\partial \rho}\right)_s = \frac{\rho a^2}{p} \tag{1}$$

and equal to the specific heat ratio C_p/C_v under the perfect gas assumption.

What we want to do here is to select a set of the gas constant and the isentropic exponent which can best approximate the specific real gas behavior in terms of non-dimensional parameters. According to the similarity theory, the value of the gas constant does not substantially affect the non-dimensional performance and the isentropic exponent is the only gas property which enters into the problem of expressing the non-dimensional performance characteristic of the compressor or the turbine with a perfect gas as its working medium [2].

$$\psi_{is}, \eta_{is} = f\left(\phi, Re, M_{tip}, \gamma\right) \tag{2}$$

Thus, we can select any value as the gas constant. Here we simply define the gas constant by using the actual inlet condition as $R = p_1/(\rho_1 T_1)$ or in the original form $R = R_U/M_W$.

What to do next is to select an appropriate value of the isentropic exponent. Figure 2 shows a variation of the isentropic exponent of R245fa on the p-T space. It varies from place to place in a compressor or a turbine, therefore it is natural to select one value which falls in the range of its variation or which represents an average. The problem is that the discharge condition of compressors or the inlet condition of turbines can be close to the critical point where the value of the isentropic exponent can get smaller than unity. An isentropic exponent less than unity means a negative isobaric specific heat, thus in such a hypothetical fluid the temperature increases while the enthalpy drops.

Another issue associated with the gas modeling is the way the performance parameters should be converted to the actual intended condition when they are acquired in a simulation with a different working fluid than the intended one. Walsh and Fletcher suggest using a set of non-dimensional parameters to convert the performance, such as the tip Mach number, flow coefficient, isentropic load coefficient and efficiency [3]. Roberts and Sjolander [4] took a similar way for a centrifugal compressor but developed a method to correct the effects of the different specific heat ratios.

ASME PTC-10, on the other hand, requests to use the density ratio, flow coefficient, load coefficient and efficiency [5] for the conversion. Similarly, Macchi and Perdichizzi [6] suggest the use of the density ratio to accurately correlate the performance of the axial-flow turbines. Backström [7] described a translation of the compressor characteristic by means of the density ratio.

In the present study, it is intended to provide a way to approximate the real gas nature of an organic working fluid (in this study, R134a for a VCC compressor and R245fa for an ORC turbine) as a perfect gas. Problems in the approximation in the aerodynamic design of the VCC compressor or the ORC turbine were examined. Also discussed is the way to convert the compressor or the turbine performance obtained under a perfect gas approximation to the actual real gas working conditions.

2. Perfect gas approximation

2.1. Isentropic exponent

As mentioned in the introduction, the isentropic exponent is the only parameter to characterize the perfect gas in the non-dimensional term. In this study, the isentropic exponent will take a value close to either the stage inlet value or the stage outlet one in an expectation that such a value can most closely approximate the actual real gas behavior.

2.2. Conversion rule of the performance characteristic between the real gas and the perfect gas

It is common to use a set of non-dimensional parameters to convert the performance data between different operating conditions or between different working fluids [3]. One may use the tip Mach number, flow coefficient, isentropic load coefficient and efficiency (hereafter designated as Type1), or the density ratio in place of the tip Mach number (designated as Type2).

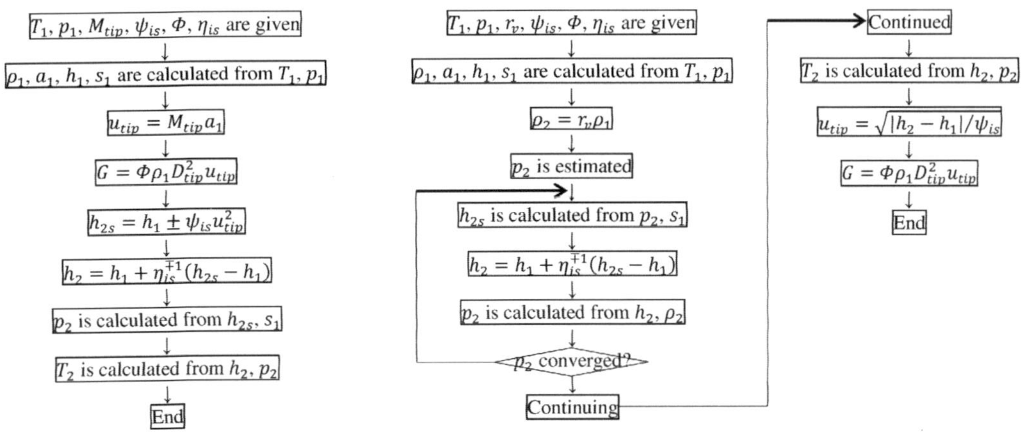

(a) Type1 conversion rule (b) Type2 conversion rule

Figure 3. Conversion rules from PG to RG (upper signs for compressors, lower signs for turbines).

Figure 3 shows how to transform the perfect gas (PG) results into the real gas (RG) condition using these two methods. In both rules, the non-dimensionalization is based on the stage inlet total condition as usual. When Type2 is applied, an iterative procedure is required to get the stage exit thermodynamic condition. Moreover the rotor speed and the mass flow rate in the PG condition which will correspond to the intended values in the RG condition are unknown at the beginning of the simulation. In the design simulations, they are often used as inputs rather than outputs. Therefore Type2 conversion method cannot be readily incorporated into the conventional design procedure.

3. Validation of perfect gas approximation in CFD simulations

3.1. VCC centrifugal compressor stage

First of all, a CFD validation of PG was conducted against a refrigerant centrifugal compressor stage. It comprises a splittered impeller, a vaneless diffuser and a volute, but the volute was not modeled for simplicity. In this study and in all later studies, ANSYS Fluent was used. The convection terms were discretized with a second-order upwind scheme. The fluid was assumed to be viscous and the turbulence closure was modeled by the standard k-ε model with a near-wall treatment. Only one pitch of the impeller passages was modeled with a periodic boundary condition. Approximately 0.5 million hexahedral cells were used. The total pressure, the total temperature and the flow angle were specified at the impeller inlet while the back pressure was specified at the diffuser exit.

The real gas behaviour of the working fluid (R134a) was modeled by a virial equation provided in [8]. In the PG modeling, the physical gas constant and the inlet isobaric specific heat were used. The isentropic exponent was derived from these two values and slightly larger than the inlet value which is greater than unity. As an intermediate modeling, van der Waals model was also examined. As a conversion rule, the Type1 rule was adopted. The non-dimensional parameters were defined based on the blade tip speed and the density and the speed of sound evaluated at the inlet total condition.

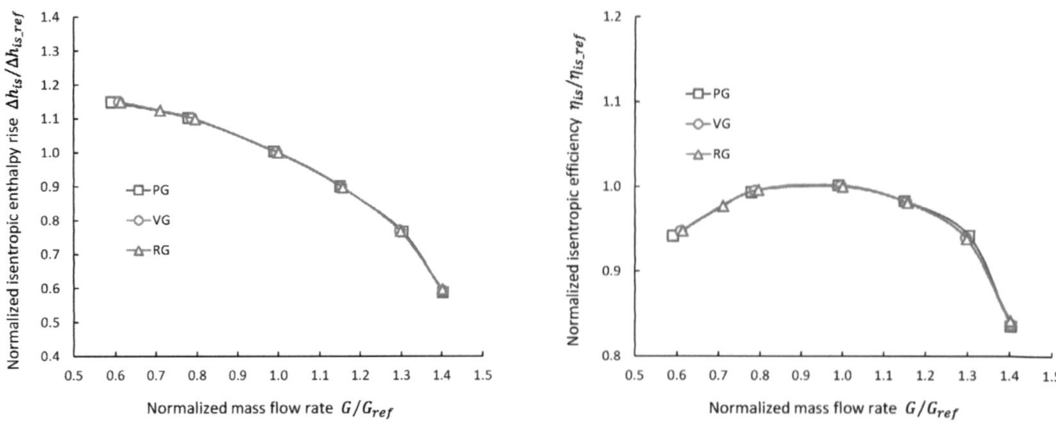

(a) Isentropic enthalpy rise (b) Isentropic efficiency

Figure 4. Overall compressor performance.

Figure 4 shows the comparisons of the overall performance characteristics acquired by three gas models, the perfect gas (PG), van del Waals gas (VG) and the real gas (RG). For PG and VG cases, all these performance parameters were once non-dimensionalized and converted to RG condition by Type1 rule. All the three models predict exactly the same characteristics not only for the mid-range points but also for the highly off-design points. Figure 5 compares the deviation of the blade surface Mach number obtained by PG or VG from the one given by RG. The deviation is the highest at the leading and trailing edges but the level of the deviation is not so high, less than 2% in PG, which is generally acceptable in the usual design simulations. By using VG, the accuracy can be improved over PG considerably and the deviation can be suppressed to the half of PG level.

Thus far, no modeling difficulty has been encountered in the perfect gas approximation, even though the isentropic exponent was selected relatively arbitrarily.

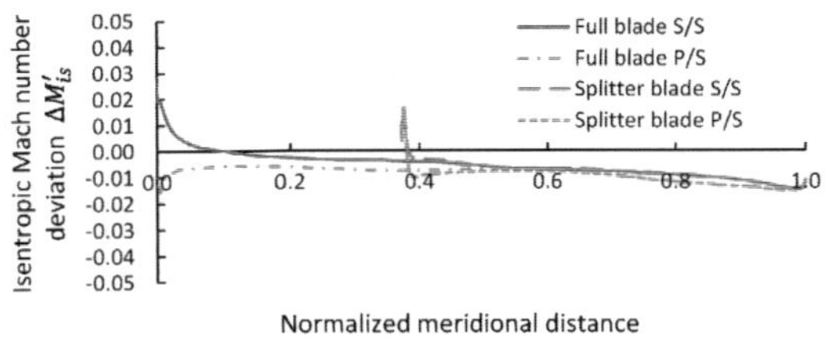

(a) Impeller blade surface Mach number (PG)

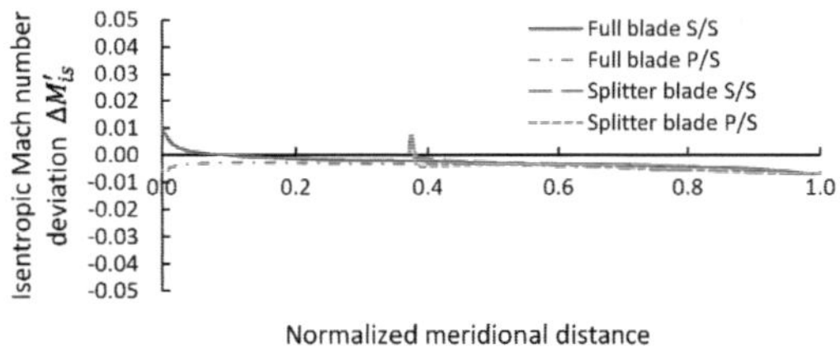

(b) Impeller blade surface Mach number (VG)

Figure 5. Deviations in blade surface isentropic Mach numbers.

3.2. ORC radial-inflow turbine stage

Another CFD validation study about PG was conducted against an ORC radial-inflow turbine stage. In order to select an isentropic exponent and a conversion rule, one-dimensional meanline turbine simulations were first carried out. The meanline solver is based on the conservation laws supplemented with some loss models [9]. It can incorporate either RG or PG. At first the turbine was simulated under a specified design condition with a real gas model of R245fa provided by NIST REFPROP. Then several PG simulations were conducted with an isentropic exponent varying from the actual turbine inlet value $\gamma_1 < 1$ to the actual exit value $\gamma_2 > 1$ with a conversion rule either Type1 or Type2. Hereafter we call the perfect gas model with the inlet value γ_1 PG1, while with exit value γ_2 PG2. The non-dimensionalization was based on the stage inlet total condition.

Table 1 summarizes the meanline results in terms of Mach numbers and flow angles. As far as PG1 is used, two conversion rules do not make any substantial difference and the deviations from the RG results are satisfactorily small. When PG2 is used with Type2, the deviations are still on acceptable level, but when used with Type1, the deviations are much higher than other cases.

Table 1. Deviations of PG solutions from RG one (meanline simulations).

PG type	PG1		PG2	
Conversion rule	Type1	Type2	Type1	Type2
(a) Impeller inlet parameters				
Absolute Mach number	0.02	0.01	0.02	0.00
Relative Mach number	0.01	0.01	0.02	0.00
Relative flow angle (deg)	-0.9	-0.7	-3.7	-1.2
(b) Impeller exit parameters				
Absolute Mach number	0.01	0.01	0.04	0.02
Relative Mach number	0.02	0.01	0.08	0.03
Absolute flow angle (deg)	-0.6	-0.3	-5.5	-1.6

Even though the present meanline method accepts the negative isobaric specific heat as in PG1 with no difficulty, it would turn out later that the CFD simulation can be unstable with such a fictitious specific heat value. So we had to give up PG1. Also we gave up Type2 conversion rule due to the troublesome iterative procedure which would hardly be implemented in the usual CFD design simulations. Though the flow angle deviations are relatively high under the combination of PG2 and Type1, the impeller inlet relative flow is low subsonic so the incidence mismatch may not yield a severe difference in the incidence loss, and the exit absolute flow angle does not strongly affect the impeller component efficiency, therefore we decided to use PG2 with Type1 conversion rule. Any γ value greater than unity but smaller than γ_2 would give better result but not used because the closer the γ value to unity the more unstable the CFD simulation may become.

A comparison of CFD with RG/PG models was conducted next. The turbine stage has a volute, a vaned nozzle and an impeller, but again the volute was not included in the model for simplicity. The rotor-stator interface was connected by a mixing-plane. The realizable k-ε model with the enhanced near-wall treatment was used. The other conditions were similar to those used in the compressor case.

Figure 6 compares the overall turbine performance. For PG case, all these performance parameters were once non-dimensionalized and converted to the RG condition by means of Type1 conversion rule. In these performance comparisons, there can be observed little substantial difference between solutions. Figure 7 shows the deviation of the isentropic Mach number on the impeller blade surface or on the nozzle vane surface. No substantial difference can be observed in the nozzle passage. On the other hand, the deviation in the impeller is as high as 0.1. The level of this deviation is compatible with the one predicted by the meanline analysis and considered non-negligible.

If we simplify the turbine expansion process as isentropic and the working fluid as PG, then the density ratio can be expressed as

$$\rho_{04}/\rho_{01} = \left\{1 - (\gamma - 1)\psi_{is} M_{tip}^2\right\}^{1/(\gamma-1)} \tag{3}$$

This expression suggests that the higher the γ, the lower the exit density. The adopted value of γ in PG2 is the highest one in the actual turbine passage therefore PG2 should predict an excessively low exit density. The deviation in Mach number is mainly caused by this excessive decrease of the density, which results in the higher through-flow velocity.

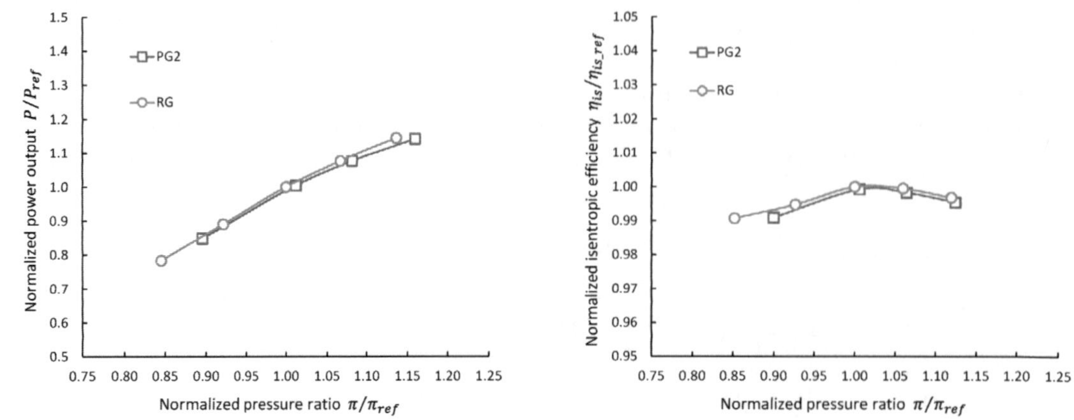

(a) Power output (b) Isentropic efficiency

Figure 6. Overall turbine performance (total-total).

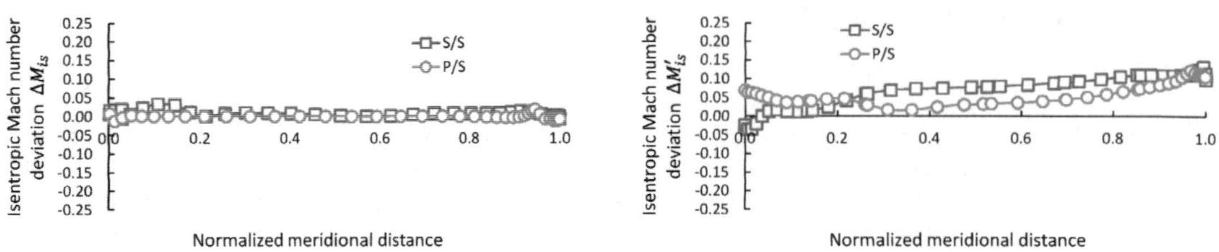

(a) Nozzle vane surface Mach number (b) Impeller blade surface Mach number

Figure 7. Deviations in blade/vane surface isentropic Mach numbers (PG2).

3.3. Improvement of perfect gas approximation for ORC turbine

In an attempt to improve the PG modeling, Type1 conversion rule was modified. Now, as in the compressor case, the non-dimensionalization will be conducted based on the lowest pressure condition (Type3), the isentropic exit condition which is defined by the stage inlet total temperature and total pressure and the stage exit static pressure. These three quantities are the CFD input, so no iterative procedure will be required. The tip Mach number and the flow coefficient are defined as $M_{tip} = u_{tip}/a_{2s}$ and $\Phi = G/\left(\rho_{2s}D_{tip}^2 u_{tip}\right)$, respectively. The gas constant and the isentropic exponent of the revised perfect gas model (PG2S) are defined $R = p_2/(\rho_{2s}T_{2s})$ and $\gamma = \rho_{2s}a_{2s}^2/p_2$, respectively. By defining gas properties and non-dimensional parameters as above, we do not have to translate the rotor speed and the mass flow rate. The inlet condition of PG is defined so that the isentropic loading coefficient and the back pressure will be same as RG. A meanline simulation was again carried out and the maximum Mach number deviations and the flow angle deviations in the nozzle/impeller were 0.02/0.01 and +0.7(deg)/+1.3(deg), respectively. Therefore the expected accuracy of the revised method is better than Type1 and comparable to Type2.

Next the CFD simulation based on the revised method was carried out. Figure 8 compares the performance characteristics. In the previous Figure 6, the pressure ratio and the efficiency were defined by total-total basis, but in order to emphasize the influence of the density on the performance they are now defined by total-static basis. The efficiency discrepancy between RG and PG2 is about 1%, rather a big difference, and is caused by the difference in the discharge velocity and density. By applying PG2S and Type3, the efficiency is predicted quite closely to RG. The output power is not affected whether define in total-total or total-static, and PG2S predicts it as accurately as PG2. Figure 9 displays the Mach number difference between PG2S and RG. The deviation in the nozzle increased only slightly compared with PG2 but still negligible. On the other hand, the deviation in the impeller is greatly suppressed. No deviation can be observed in the leading edge part, implying that the new method can correctly predict the inlet flow vector. The gradual increase of the deviation toward the trailing edge observed in PG2 is also suppressed. From these results, it can be concluded that the modified perfect gas model can effectively substitute the RG simulation for the design purpose.

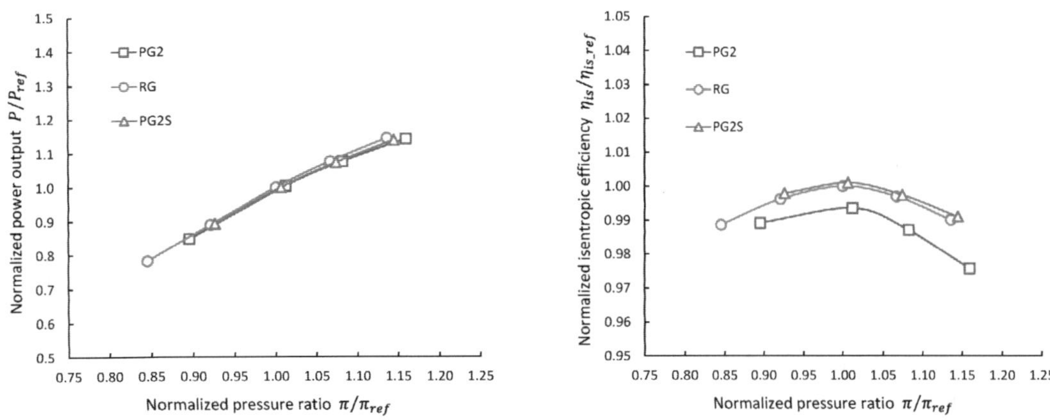

(a) Power output (b) Isentropic efficiency

Figure 8. Overall turbine performance (total-static).

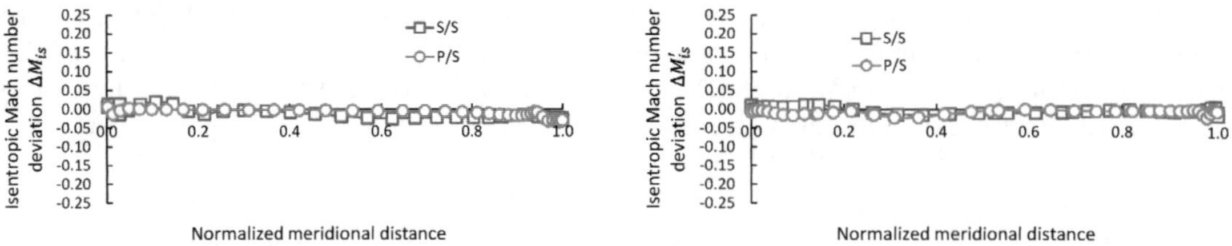

(a) Nozzle vane surface Mach number (b) Impeller blade surface Mach number

Figure 9. Deviations in blade/vane surface isentropic Mach numbers (PG2S).

4. Conclusion

A method to approximate the real gas nature of organic compounds by a perfect gas and the way to convert the CFD results obtained with a perfect gas approximation into the real gas condition were proposed. The validity was examined by comparative CFD simulations. The method proved to be quite reliable in the prediction of the overall performance and very helpful to accelerate and stabilize the design CFD simulations of VCC compressors and ORC turbines.

5. Nomenclature

a	Speed of sound (m/s)	r_v	Density ratio
C_p, C_v	Isobaric/Isochoric specific heat (J/kg/K)	s	Entropy (J/kg/K)
D_{tip}	Impeller tip diameter (m)	T	Temperature (K)
G	Mass flow rate (kg/s)	u_{tip}	Blade tip speed (m/s)
h	Enthalpy (J/kg)	γ	Isentropic exponent
M, M'	Absolute/Relative Mach number	η_{is}	Isentropic efficiency
M_{tip}	Tip Mach number $= u_{tip}/a$	π	Pressure ratio
M_W	Molar mass (kg/kmol)	ρ	Density (kg/m^3)
p	Pressure (Pa)	Φ	Flow coefficient $= G/\left(\rho D_{tip}^2 u_{tip}\right)$
R	Gas constant (J/kg/K)	ψ_{is}	Isentropic load coefficient $= \Delta h_{is}/u_{tip}^2$
R_U	Universal gas constant (J/kmol/K)		

6. References

[1] Kaneko, Y., et.al, 2015, Design method of a radial-inflow turbine for an organic rankine cycle generator using CFD with perfect gas assumption, *Proc. of the 13th Asian Int. Conf. on Fluid Machinery* (Tokyo, Japan) Paper No. AICFM13-198.

[2] Dixon, S. L., 1998, Fluid mechanics, thermodynamics of turbomachinery

[3] Walsh, P. and Fletcher, P., 2004, Gas turbine performance,(Blackwell-Publishing).

[4] Roberts, S. K. and Sjolander, S. A., 2005, Effect of the specific heat ratio on the aerodynamic performance of turbomachinery, *ASME J. of Engg. for Gas Turbines and Power*, **vol.127**, pp.773-780.

[5] ASME, 1997, Performance test code on compressors and exhausters, **ASME**.

[6] Macchi, E. and Perdichizzi, A., 1981, Efficiency prediction for axial-flow turbines operating with nonconventional fluids, *ASME J. of Engg. for Power*, **vol.103**, pp.718-724.

[7] Von Backström, T. W., 2008, The Effect of specific heat ratio on the performance of compressible flow turbo-machines, *Proc. of ASME Turbo Expo 2008*, (Berlin, Germany), Paper No. GT2008-50183.

[8] JAR and JFGA, 1991, Thermophysical properties of environmentally acceptable fluorocarbons – HFC-134a and HCFC-123, *JAR and JFGA*

[9] Rodgers, C., 1987, Mainline performance prediction for radial inflow turbines, VKI Lecture Series 1987-07.

IOP Conf. Series: Materials Science and Engineering **129** (2016) 012052 doi:10.1088/1757-899X/129/1/012052

The research on particle trajectory of solid-liquid two-phase flow and erosion predicting in screw centrifugal pump

Z J Shen, R N Li, W Han, W G Zhao, X H Wang

School of Energy and Power Engineering, Lanzhou University of Technology, Lanzhou 730050, China
Key Laboratory of Fluid machinery and Systems, Gansu Province, Lanzhou 730050, China

E-mail: szj007cool@126.com

Abstract. Use the Discrete Phase Model (DPM) based on Euler-Lagrange method, the internal flow field of screw centrifugal pump was simulated by computational fluid dynamics(CFD) code when transmission medium is solid-liquid two phase flow with large-size particles. The research of liquid phase is under the Euler coordinate system while the solid phase is under the Lagrange coordinate system. The energy change, trajectory characteristic of solid phase particle and its erosion damage rule of solid-phase particle in whole computational domain is analyzed with different density, partical size(d=0.05mm、d=0.2mm、d=2mm) and solid volume fraction(Cv=3%、Cv=5%、Cv=7%).The result shows that within a given diameter range, the low density fine particles trajectory are longer, more collision times with flow passage components, more energy loss and the erosion parts are relatively uniform, but particles which are large-size diameter and high density has a big collision angle with the surface of impeller and volute, even the area of impact and abrasion are quite focus, and easy to be transported. particles will impact with the head of impeller when it enter into impeller domain, the erosion mainly occurs on the work side of impeller.

1. Introduction

The screw centrifugal pump offers unique technological advantages such as a better hydraulic characteristic, solids handling capability, and high efficiency compared with common centrifugal pump [1].Because of the pressure, velocity-slip, temperature and the roughness of passage components are very common in centrifugal pump which transport solid-liquid two-phase flow.

Over view influence factors of erosion, the physical property of particle, hardness, impact velocity and angle have a significant impact on erosion. In order to research the principle of erosion in solid-liquid two phase flow pump, researcher has studied its flow characteristics by numerical simulation [2], and explored the reason of erosion. So far, researchers more likely to use Mixture model [7-8]. There are more researchers use DPM model to simulate the solid-liquid two-phase flow in recent years, but rarely use it to solid-liquid two-phase flow pump[9-10].

In this paper, use the DPM model in Fluent which based on the Euler-Lagrange coordinate, the DPM model can export the single or all of the particle trajectory information in flow field at the corresponding position and time point to more intuitive judgment the law motion of particle, and combine the calculate of erosion rate to comprehensive forecast the erosion law in screw centrifugal pump.

2. Theoretical analysis

2.1 Solid particle motion equation

The trajectory calculate of discrete phase in DPM model is processed under the Lagrangian coordinate system, and the continuous phase is processed under the Eulerian coordinate system. The particle in solid-liquid two-phase flow fields mainly by gravity, additional mass force, passing-flow resistance and the additional force which triggered by pressure gradient, Saffman, Magnus lift force and Basset force and so on. The Thermophoretic Force, Brown force and thermal radiation are not be considered in this paper. For the sake of discussion, we only write down the balance equation in the direction of x.

$$\frac{du_p}{dt} = F_D\left(u-u_p\right) + \frac{g_x(\rho_p-\rho)}{\rho_p} + F_V + F_P + F_x \quad (1)$$

Where $F_D\left(u-u_p\right)$ is the unit quality of draft force

$$F_D = \frac{18\mu}{\rho_p d_p^2}\frac{C_D \mathrm{R_e}}{24} \quad (2)$$

where u is the velocity of liquid phase, u_p is the velocity of solid phase, μ is the kinetic viscosity, ρ is the density of liquid, ρ_p is the density of solid, d_p is the diameter of particle, and the g_x is the acceleration of gravity in the direction of x.

The Reynolds number of particle can be defined as follows:

$$\mathrm{Re} \equiv \frac{\rho d_p\left|u_p-u\right|}{\mu} \quad (3)$$

The draft coefficient C_D can be defined as follows:

$$C_D = a_1 + \frac{a_2}{R_e} + \frac{a_3}{R_e} \quad (4)$$

For the spherical particle, at a given range of Reynolds number, the a_1、a_2、a_3 is constant。

$$F_V = \frac{1}{2}\frac{\rho}{\rho_p}\frac{d}{dt}(u-u_p) \quad (5)$$

$$F_P = \frac{\rho}{\rho_p}u_p\frac{\partial u}{\partial x} \quad (6)$$

Where F_V, F_P is the additional force which triggered by additional mass force and pressure gradient, F_x is the resultant force of the other external force.

Because of the Magnus lift force and the Saffman lift force caused by the self rotation of particle is only effective with submicron size (d=1-10 μm), and in this paper, the diameter of solid particle we simulate will be larger, so the self-rotation effect are ignored [11-13].

2.2 The erosion theory of particle

The erosion rate can be monitored on the flow wall in Fluent by select the Interaction with Continuous Phase option. The erosion rate can be defined as follows:

$$R_{erosion} = \sum_{P=1}^{N_{particles}} \frac{m_p C(d_p) f(a) v^{b(v)}}{A_{face}} \tag{7}$$

Where $C(d_p)$ is the function of solid diameter, a is the particles impact angle with internal face (°), $f(a)$ is the function of impact angle, v is the Relative velocity of particle and internal wall(m/s), $b(v)$ is the relative velocity function of particle, and A_{face} is the unit area of wall.

The unit of erosion is $kg/(m^2/s)$, this equation can be used to forecast the flow passage components erosion caused by particle motion. To judge the distribution of different diameter and volume fraction solid phase in flow channel more directly, and master its erosion position and form on flow passage, the erosion rate on flow passage components corresponds to the diameter and volume fraction were calculated, which can help to find the erosion rules on flow passage components.

3. Computational model and numerical simulation methods

3.1 Basic hypotheses

In this paper, The following assumptions are made about the complex three dimensional solid-liquid two-phase flow : ①liquid phase is incompressible, solid phase is discrete phase, and each of the physical properties is constant. ②solid phase is Spherical, uniform-sized and Regardless of the phase transformation. ③the flow is steady.

3.2 Pump model and parameters

The pump and its impeller under investigation are 150×100LN-32 type screw centrifugal pump, its main parameters are, the quantity Q=165m^3/h, head H=32m, efficiency η=62.5%, shaft power P=23.2kw, impeller speed n=1480r/min, wrapping angle of blade is 781°, which centrifugal part is 112.5°.

The Calculation Region includes suction , impeller, volute and outlet. using ICEM-CFD was to generate unstructured mesh and the grid independence verification was made

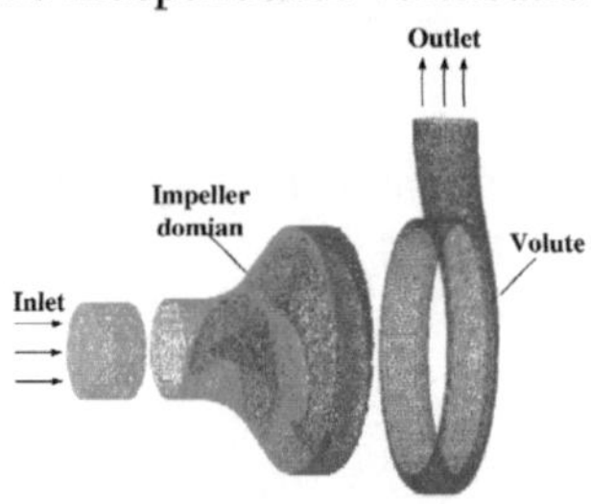

Figure 1. Mesh of computation region

3.3 Numerical simulation methods

Because of the run condition of screw centrifugal pump in this paper is with Large particle size and the solid volume fraction is within 10%, the DPM model based on Euler-Lagrange method is used to research the particle's trajectory and erosion rate. The standard k-epsilon turbulence model is used to simulate the continuous phase, and the track of solid phase use steady method. The iterative algebra equations use under-relaxation factor, and set the convergence precision as 10^{-5}.

3.4 Boundary condition

The model inflow boundary can be assumed to be uniform velocity inlet, and the outlet boundary use free outflow condition. For the discrete phase, both inlet and outlet use escape condition, and the contact with the surface of impeller and volute use reflect condition.

4. Computational results and discussion

By contrasting the classification of particle diameter from The above reference, the particle diameter with 0.05mm,0.2mm and 2mm were selected to analysis the particle trajectory and erosion law in the screw centrifugal pump. picking sand as the solid phase whose density is similar to sediment in yellow river.

4.1 trajectory distribution of particles

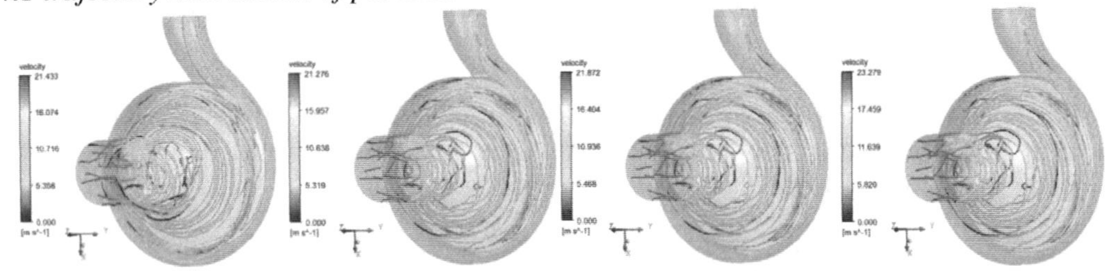

a.*Cv*=3% b.*Cv*=5% c.*Cv*=6% d.*Cv*=7%

Figure 2. Trajectory of solid phase(*d*=0.05mm)

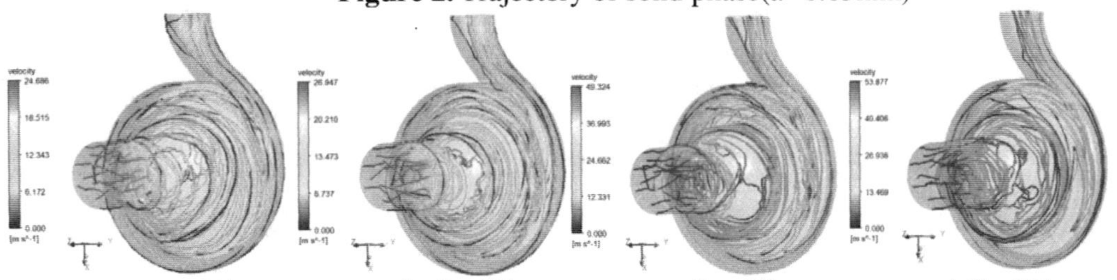

a.*Cv*=3% b.*Cv*=5% c.*Cv*=6% d.*Cv*=7%

Figure 3. Trajectory of solid phase(*d*=0.2mm)

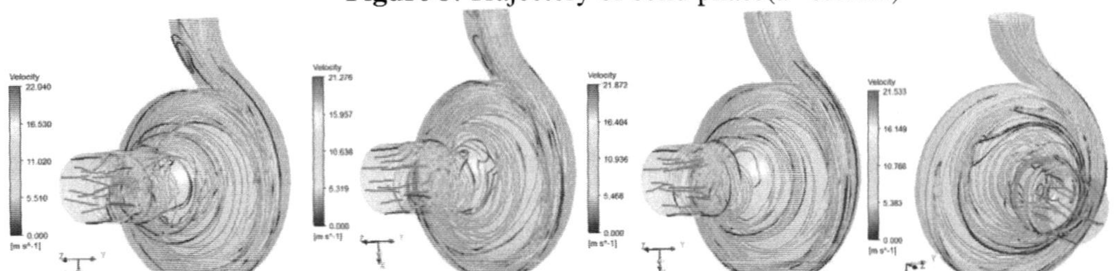

a.*Cv*=3% b.*Cv*=5% c.*Cv*=6% d.*Cv*=7%

Figure 4. Trajectory of solid phase(*d*=2mm)

By observing the trajectory of solid phase with different diameter and volume fraction in the computational domain in the fig2-4, it can be seen as follows:

First of all, because of the small-size particles are completely controlled by water, there are some fluctuated at inlet segment, but the large-size particles are more regular compared with small-size, the incident particle all into the flow area vertical and almost no tangential-velocity component. The velocity of large-size particles are no significant change and no effected by pre-rotation from impeller.

Secondly, with the increasing of particle diameter, flow in screw part of impeller are turbulent, and have a large collision angle with volute. this is because of particles collision with the head of impeller when it coming into impeller zone. the bigger the diameter of particle, they have more inertia, and greater the interference on surrounding fluid. the motion of particle transition from axial to radial led by impeller, and do circular motion after the middle of screw part. it can be seen from the figure that the velocity of solid phase are increasing in impeller, particles get the energy from the rotation of impeller, while in the volute, the velocity has certain decrease, and were thrown out of the pump. The contrast shows that at the same volume fraction, when the diameter of solid d=0.2mm, the maximum speed is faster than other diameter, it shows that this screw centrifugal pump has a excellent transportation performance in this work condition.

Finally, with the increasing of volume fraction, the particle trajectory becoming turbulent, and the flow condition more disorderly under the low volume fraction until the end of screw part. this is because of the increase of solid volume fraction, the greater probability of collision happens, and the hindering effect from water becoming more obvious, so it influenced the liquid flow.

4.2 the energy change of particles in impeller

To quantitative analysis the trajectory of solid phase in the screw centrifugal pump caused by diameter and solid volume fraction, choose the same incidence angle ,velocity and coordinate(x, y, z), all flow from the outlet by way of escape, to observe the velocity distribution in the whole flow passage. use the relative axial length to represent the axial distance of sand in full flow passage, and use 0 represent the inlet surface, also use 1 to represent the base of impeller, and divide the full passage into 4 parts: inlet part, screw part, centrifugal part and the part after the centrifugal part.

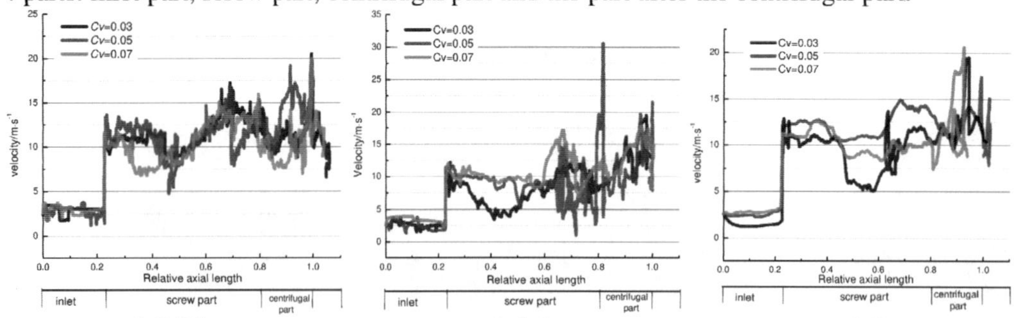

d=0.05mm *d*=0.2mm *d*=2mm

Figure 5. the time-velocity distribution of solid phase

it can be seen from the figure 5, after the solid phase entered into inlet part, the velocity increased faster, and then fluctuated. the sand finished the improvement of energy, and with the energy loss . at the screw part of impeller, On the one hand , the solid phase get energy from impeller, and on the other hand, it collision with the surface of vane, front shroud, and each other, which lead the loss of energy, and all the magnitude and direction of velocity were changed. the velocity of solid decreased in volute is because of the sand move away from the impeller surface and flung into the volute at the centrifugal part, it would do circular motion in volute and collision with volute surface which can lead the energy loss and the abrasion of volute face.

From the velocity distribution of different diameter it can be seen that: first of all, when transport the solid phase which diameter is *d*=0.05mm or *d*=0.2mm, its velocity fluctuated obviously, on the contrary, it's more steady when transport *d*=2mm, it is because of at the same density, the bigger the particle size, the bigger the inertial, and the trajectory less affected by the liquid phase, and keep the original trajectory enter into impeller.

Secondly, the velocity of small size particle increased significantly at the screw part, and the larger the diameter, the more fluctuated of the velocity, and do turn-back return movement in this region. There are about 2 reasons about this result: first, the bigger the particle size is, the hindering effect from water becoming more obvious. Second, the normal velocity of large size particles are relatively high after collision with flow passage surface, in other words, reflect angle becomes large, and the trajectory of solid phase is easy to be changed.

Thirdly, at the same axial length, the velocity of small size particles fluctuated obviously, the large size particles, by contrast, is more regular than small size particles. it is because of the small size particles have a good following with liquid phase, while the velocity of large size particles may slip and not moving with the liquid phase.

4.3 the erosion distribution of flow passage component

4.3.1 the erosion distribution of d=0.05mm

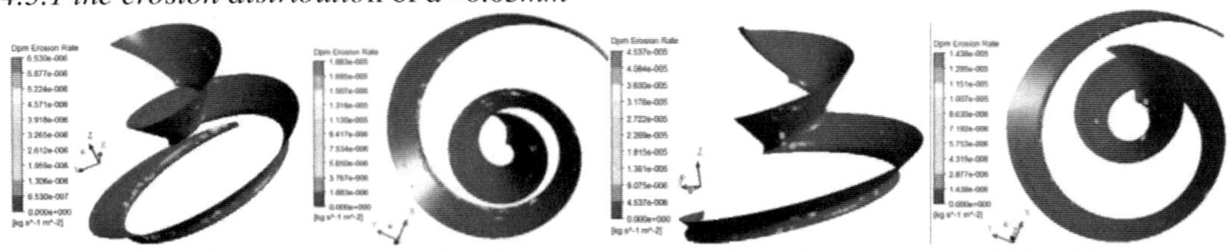

a.*Cv*=3% b.*Cv*=5% c.*Cv*=6% d.*Cv*=7%

Figure 6. Work side of impeller

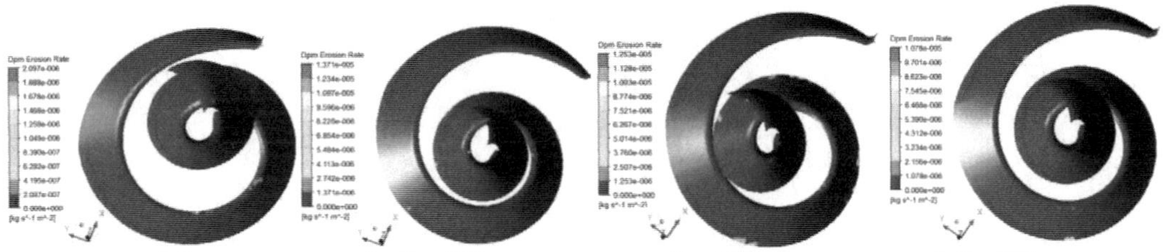

a.*Cv*=3% b.*Cv*=5% c.*Cv*=6% d.*Cv*=7%

Figure 7. Back side of impeller

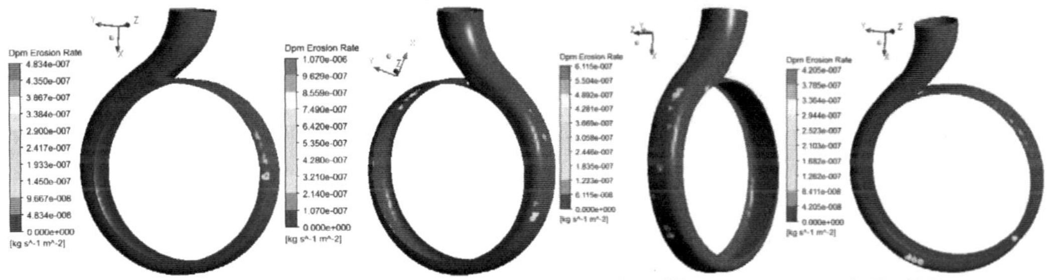

a.*Cv*=3% b.*Cv*=5% c.*Cv*=6% d.*Cv*=7%

Figure 8. Volute

4.3.2 the erosion distribution of d=0.2mm

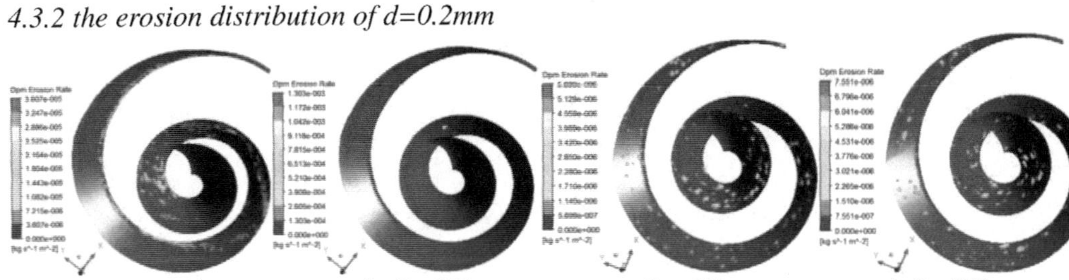

a.*Cv*=3% b.*Cv*=5% c.*Cv*=6% d.*Cv*=7%

Figure 9. Work side of impeller

IOP Conf. Series: Materials Science and Engineering **129** (2016) 012052 doi:10.1088/1757-899X/129/1/012052

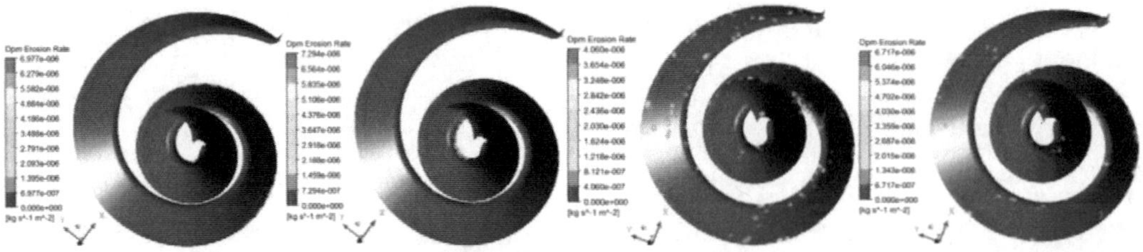

a.Cv=3% b.Cv=5% c.Cv=6% d.Cv=7%

Figure 10. Back side of impeller

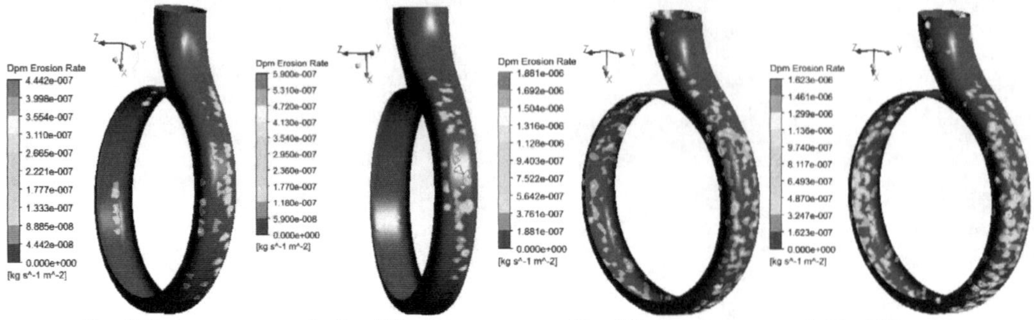

a.Cv=3% b.Cv=5% c.Cv=6% d.Cv=7%

Figure 11. Volute

4.3.3 the erosion distribution of d=2mm

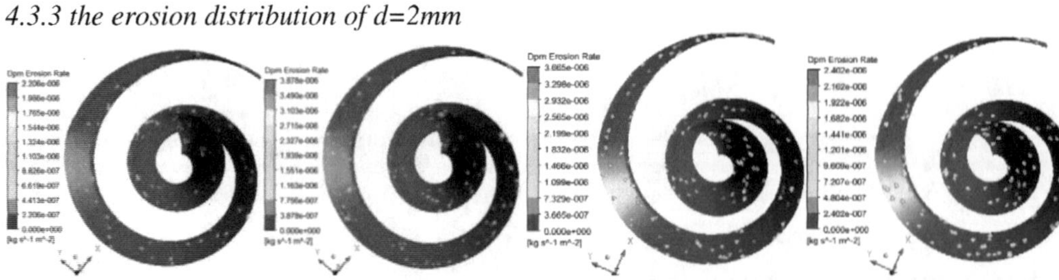

a.Cv=3% b.Cv=5% c.Cv=6% d.Cv=7%

Figure 12. Work side of impeller

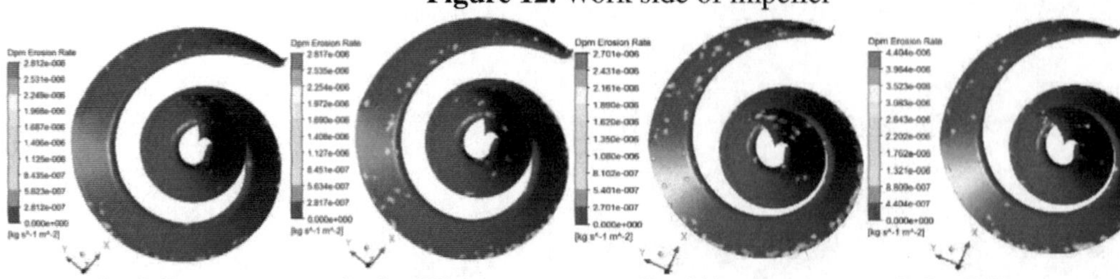

a.Cv=3% b.Cv=5% c.Cv=6% d.Cv=7%

Figure 13. Back side of impeller

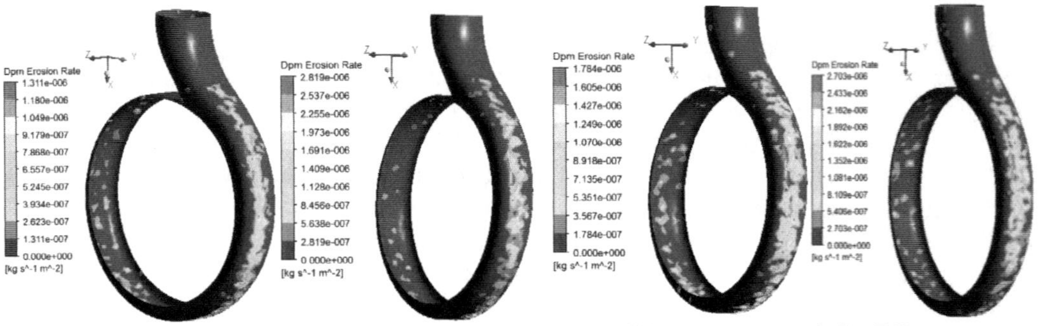

a.*Cv*=3% b.*Cv*=5% c.*Cv*=6% d.*Cv*=7%

Figure 14. Volute

From the distribution of erosion rate with the different diameter and volume fraction on the flow passage components, it can be seen that:

Firstly, the distribution of erosion rate on the impeller surface are mainly concentrated on the work side, while its lower on the back side. the erosion style on the volute surface mainly dominated by abrasive wear, under the centrifugal force, the solid phase out of the outlet of impeller impact and collision with the volute surface, and mainly dominated by erosive wear.

Secondly, as the diameter and the volume fraction increasing, there is a increasing trend of erosion rate while at the work side and back side of impeller. The erosion rate of particles which *d*=0.2mm are higher than other particles at the same diameter and volume fraction. It shows that particles which *d*=0.2mm can highly contacted with impeller and can effectively complete the energy transfer under the rotation of impeller.

Finally, as the size of the particle increase, the range of erosion basically have not change, but the parts of erosion are much more dispersed and distributed. It is because of the particles are discontinuity when the size increased, the impact angle becoming wide with the flow surface and not very frequently. The impact and collision with flow surface mainly hold a dominant position.

5. Conclusion

(1) when transport the large size solid phase, the solid volume fraction has a big influence with the trajectory while with small size particles, it is not very obvious, the impact angle of large size particle are bigger than small size particle under the same volume fraction, impact and friction and easy to lead the flow disorder.

(2) The mainly work part of impeller is the middle of screw part, on one hand, particles get energy under the rotation of impeller, on the other hand, its impact and friction with flow parts and lead the energy loss. there are energy discharge appears near the centrifugal part. solid phase complete the conversion from kinetic energy to static pressure energy and impact with the surface of volute, which can lead the energy loss and the erosion of volute surface.

(3) At the same axial length, the velocity of small size particles are fluctuated and the flow is unsteady while the motion of large size particles are quite uniform. The large size of particles may generate velocity slipping in solid-liquid two phase flow, and not moving together with liquid phase.

(4) solid phase will impact with the head of impeller when it enter into impeller. the effect of erosion mainly appears on work face, with the size of solid phase increasing, the erosion parts are much more dispersed and distributed.

6. Acknowledgments

This project is supported by the National Natural Science Foundation of China (Grant No. 51079066 and 51269001)

7. References

[1] John H. 1999 Screw centrifugal impeller technology and its advantages for sludge *Pumping*

(Toronto: Hayward Gordon)

[2] Li Y, He W Q. 2009 Numerical simulation of solid-Liquid two-phase flow and abrasion characteristics in desulfurization pump *Drainage and Irrigation Machinery* **27(2)**

[3] Vesiliev V. 1970 On evaluation of w ear of centrifugal pump parts in hydro-abrasive mixtures. *1st International* Conference *on Hydraulic Transport of Sol ids in Pipes* 153-169

[4] Tian A M，Tian A J, Xu H Y. 2000 Experimental research on the movement rule of solid particles in centrifugal pump. *Water Conservancy Electric Power Machinery*

[5] Liu J , Xu H Y, Tang S. 2008 Experimental research on the movement rule of solid particles in centrifugal pump. *Journal of* Hydroelectric *Engineering* **27(6)**

[6] Liu J. 2006 The research on solid-liquid abrasive and abrasive rules of hydraulic machines (Bei Jing: Tsing Hua *university*)

[7] Li R N, Wang Q B, Liu C S. 2006 Numerical model for solution of inner flow in screw centrifugal pump. *Journal of Lanzhou University of Technology* **32(1)**：57-60

[8] Li R N，Wang Q B. 2008 Numerical simulation of solid-liquid two-phase flow inthe screw centrifugal pump. *Fluid Machinery* **36(12)**：24-27.

[9] Berlemont A 1990 Particle lagrangian simulation in turbulent flows. *Int J Multiphase Flow*，**16**:19-34.

[10] L A Oliveira 2002 A Lagrangian–Eulerian model of particle dispersion in a turbulent planemixing layer. Fluids 639-693.

[11] Huang S H, Cheng L J. 1996 Analysis of particle Basset force in unsteady flow field. *Mechanics in Engineering* **(3)**:50-52.

[12] Zhao J F, Li W. 1998 An analysis of characteristics of unsteady forces on accelerated articles. *Mechanics in Engineering*，**20(3)**:43-44.

[13] Huang S H, Li W, Cheng L J. 2000 On equation of discrete solid particles motion in arbitrary flow field *and* its properties. *Applied Mathematics and Mechanics*,**21(3)**:45-48.

[14] Fluent Inc.2006 FLUENT User's Guide. Fluent Inc.

IOP Conf. Series: Materials Science and Engineering **129** (2016) 012053 doi:10.1088/1757-899X/129/1/012053

Heat transfer analysis of compound multi-layer insulation for cryogenic tank under different service conditions

H L Zhu, C Yao, Y Li, H L Pan

Beijing Institute of Control Engineering, China Aerospace Science and Technology Corporation, Beijing, 100090, China

E-Mail: hlzhu0001@163.com

Abstract. Future space missions require efficient delivery of large payloads over great distances, necessitating the use of high-energy cryogenic propellant. Therefore, reliable compound multi-layer insulation on cryogenic tank is a crucial part of future space exploration. Compound multi-layer insulation is composed of double-aluminized radiation shielding and separated by a combination of netting and bumper strips, with a foam substrate. Considering conduction, convection, and radiation in heat transfer, the thermal field of multi-layer insulation is analysis by theoretical analysis with different thickness of foam substrate and MLI. Based on the formerly theoretical analysis, the heat flux and apparent thermal conductivity are discussed under the different thickness of foam substrate and MLI. Finally, the optimum design of multi-layer thermal insulation is present in consideration of the thickness and insulation performance of multi-layer insulation.

1. Introduction

Future space programs and missions require efficient delivery of large payloads over great distances. The Cryogenic Fluid Management (CFM) plays a vital role in the development of high-energy upper stage vehicles. Therefore, efficient and reliable insulation materials are a crucial part of future space exploration. The importance of insulation in cryogenics is easily realized by noting that the heat of vaporization of cryogenic liquids, as well as, their specific heats are small, and it takes small amount of heat flux to boil off the cryogenic liquids or to raise the system temperature.

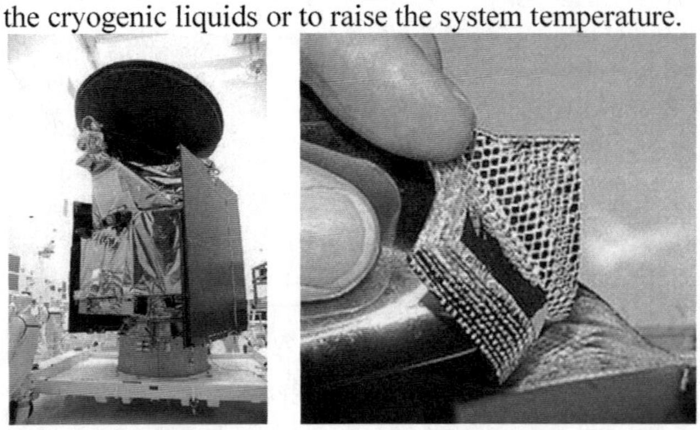

Fig 1 space application of compound multilayer insulation

Multi-layer insulations (MLI) are typically high-vacuum systems made of many radiation shields between the hot and cold boundaries. They normally consist of an assembly of numerous thin plastic films coated on one or both sides by a thin-deposited layer of high reflectance metal, usually aluminium or gold. The multilayer insulation systems can be used in both high-temperature and cryogenic applications, however, the insulation material, the arrangement, and heat transfer characteristics are quite different.

Many groups studied the effects of layer density on MLI throughout the years preceding and following Dr. McIntosh's innovation. The subject of variable density multilayer insulation (VD-MLI) was pioneered by McIntosh in the early 1990's. The theory behind variable density MLI is that since there is an optimal density for MLI as a function of both boundary temperatures, a more optimal solution exists if the MLI system is split into multiple sections or segments. Hyde, Stuckey, Fredrickson, and Spradley have each shown analytically based graphs that suggest the optimal layer density for different systems.

The foam-MLI system performed exceptionally well in the orbit hold testing when compared with historical data and predictions based on constant density MLI concepts. Therefore, the focus in this paper is on analytical modeling of the MLI performance during orbital coast periods.

2. Theory analysis

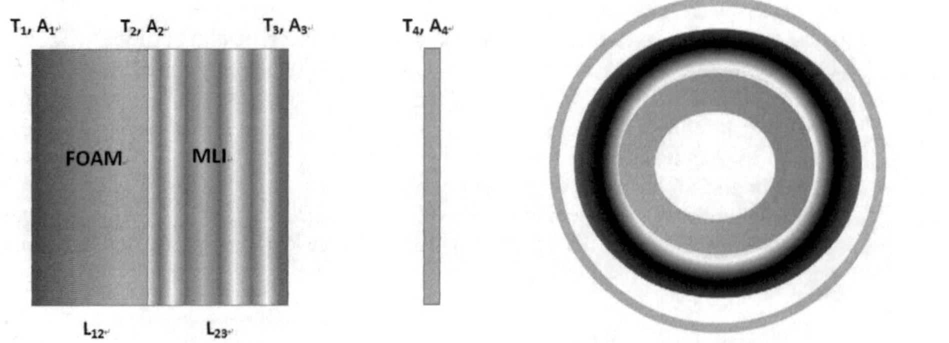

Fig 2 thermal transfer in compound multilayer insulation

To evaluate MLI performance, analytical models is presented. The analytical modeling of the insulation during orbital coast periods is discussed in subsequent sections.

2.1. Foam conduction Q_1

The conduction of the foam obeys Fourier's law

$$Q_1 = K_1 A_{12} \frac{T_2 - T_1}{L_{12}}$$

Where, K_1 is foam conductivity(W/m-K), A_{12} is effective area, $A_{12} = (A_2 - A_1)/\ln(A_2/A_1)$.

2.2. Thermal radiation between MLI Q_2

The thermal of radiation between MLI can represent:

$$Q_2 = K_r A_{23} \frac{T_3 - T_2}{L_{23}}$$

Where, A_{23} is effective area, $A_{23} = (A_3 - A_2)/\ln(A_3/A_2)$, K_r is MLI apparent conductivity

$$K_r = \frac{1}{N}\frac{\sigma(T_1+T_2)(T_1^2+T_2^2)}{\frac{(\alpha+2\gamma)\delta_s}{2n}+\frac{1}{\varepsilon_1}+\frac{1}{\varepsilon_2}-1}$$

Where, n is the numbers of shield, N is the density of MLI, $N = n/\delta_s$, ε_1 and ε_2 are the emissivities of the two faces on the shield, δ_s is the thickness of the spacer, α is the absorptance, γ is the transmittance.

2.3. Thermal transform in vacuum environment Q_3

Thermal transform in vacuum accounts for two modes of heat transfer: gas conduction and thermal radiation in vacuum, $Q_3 = Q_{fm} + Q_r$.

2.3.1. *Gas conduction in vacuum environment.* For the low pressure/vacuum environment, the space between the shields is considered to be in free-molecule regime; therefore, the gas conduction equation is applied,

$$Q_{fm} = KA_3\alpha P(T_4 - T_3)$$

Where, P is gas pressure, Pa,, α is accommodation coefficient,

$$K = \frac{\gamma+1}{\gamma-1}\sqrt{\frac{R}{8\pi MT}}$$

Where, γ is specific heat ratio, R is gas constant, 8.314 kJ/mol-K, M is molecular weight of gas, kg/mol

2.3.2. *Thermal radiation in vacuum environment.* The radiation heat transfer is,

$$Q_r = \sigma\varepsilon_{4-3}A_4\left(T_4^4 - T_3^4\right)\varphi_{34}$$

Where, σ represents Stephan-Boltzmann constant (5.675 E-8 W/m2-K4), φ_{34} is angle factor , ε_{4-3} is comprehensive emissivity

$$\varepsilon_{4-3} = \frac{1}{\frac{1}{\varepsilon_4}+\frac{A_4}{A_3}\left(\frac{1}{\varepsilon_3}-1\right)}$$

3. Result

Based on the conservation laws of energy, the relation of thermal transfer is

$$Q_1 = Q_2 = Q_3$$

Then equation can be rearrange as

$$\begin{cases} K_1A_{12}\dfrac{T_2-T_1}{L_{12}} = K_rA_{23}\dfrac{T_3-T_2}{L_{23}} \\ K_1A_{12}\dfrac{T_2-T_1}{L_{12}} = KA_3\alpha P(T_4-T_3)+\sigma\varepsilon_{4-3}A_4\left(T_4^4-T_3^4\right)\varphi_{34} \end{cases}$$

When the T_1 =90K, T_4=293K and P=0.001Pa, from the formerly theoretical analysis, the heat flux is discussed under the different thickness of foam substrate and MLI.

Table 1 the heat flux under the different thickness of compound MLI.

No.	FOAM+MLI	Thermal leakage/W
1	1cm+5cm	0.0324
2	1.5cm+4.5cm	0.0612
3	2cm+4cm	0.1049
4	2.5cm+3.5cm	0.1736
5	3cm+3cm	0.2867

4. Conclusion

In conclusion, the analytical model can be utilized in multiple applications to predict the performance of the SOFI/ MLI combination. In the higher vacuum, compound multi-layer leak lower thermal for the same thickness when it contain more shields. But the foam insulation not only enabled the elimination of a helium purge system, but also reduced the ground hold heat leak sufficiently to improve the effective as compared with an "MLI" only concept.

Reference

[1] Tein, C L, and Cunnigton, G R 1973 Cryogenic insulation heat transfer, *Advances in Heat Transfer*, **Vol 9,** pp. 349-417.

[2] Hastings, L J and Martin, J J 1998 Experimental testing of a foam/multilayer insulation (FMLI) thermal control system (TCS) for use on a cryogenic upper stage, *Space Technology Applications International Forum, 1st Conference on Orbital Vehicles*,(Albuquerque, NM).

[3] Fredrickson, G O 1973 Investigation of high-performance insulation application problems,final report. Contract NAS8-21400, *McDonnell Douglas Astronautics Co Report* MDCG4722

[4] Keller, C W, Cunnington, G R, and Glassford, A P 1974 Thermal performance of multi-layerInsulations, final report, Contract NAS3-14377, Lockheed Missiles & Space Company.

[5] Corruccini, R J 1959 Gaseous heat conduction at low pressures and temperatures, *Vacuum*, **VIII**, pp. 19-29.

[6] NASA 1967 Thermal insulation systems a survey, **Chapter 4**, NASA Report SP-5027.

[7] E H Hyde 1971 Multilayer insulation thermal protection systems technology, *Research Achievements IV* NASA TM X-64561 **5**-52.

[8] J M Stuckey 1971 Multilayer high performance insulation materials, *Research Achievements IV* NASA TM X-64561 93-98.

Experimental study of liquid-solid two phase flow over a step using PIV

E H Cando[1], X W Luo[1], V H Hidalgo[1], L Zhu[2] and A G Aguinaga[3]

[1] Department of Thermal Engineering, Tsinghua University, Beijing 100084, China
[2]China Institute of Water Resources and Hydropower Research (IWHR), Beijing 100038, China
[3]Department of Mechanical Engineering, Escuela Politécnica Nacional, Quito 17-01-2759, Ecuador

E-mail: luoxw@mail.tsinghua.edu.cn

Abstract. The present investigation focuses on the water-sand flow through a rectangular tunnel with a step using the Particle Image Velocimetry (PIV). Two cameras with appropriate optical filters have been used to capture each phase image separately. The optical filters were selected according to the optical properties of the sand and fluorescent tracers. Through data processing the experimental flow field such as the velocity profiles of sand and water had been obtained. In order to compare with the experiment, the steady state two phase flow fields were simulated using RANS method with k-ω SST turbulence model. It is noted that the numerical results matches the experimental results fairly good. Furthermore, the flow rates obtained from experimental and numerical velocity profiles also have a good match with the measurement by flow meter. The flow analysis shows that the water velocity variation induced by the presence of the step in the water-sand flow is equivalent to those cases with low sand concentration. However, the sand velocity in downstream region is 5% greater than the water velocity when the cross section is reduced in 25%.

Introduction

The dispersed phase flow in which one phase does not constitute a connected continuum is known as liquid-solid two phase flow [1]. It is encountered in various industries including energy generation, foods, pharmaceuticals, chemicals, etc. The flow through the hydraulic systems and its components (such as pump, turbine, pipeline, elbow, etc.) in places where the rivers have high sediment concentration produce the most erosive environments, resulting in high wear and short life to flow components, and leading problems like head and efficiency drop in hydraulic machinery. The wear prediction helps to improve the life of the components but requires the knowledge of the two phase-flows. Numerous experimental and numerical studies have been done; nevertheless the behaviour of the two phases and its interaction in the flow with different geometries has not been fully understood [2]. The flow velocity field, the interaction between two phases and the local particle concentration are considered to determine erosion coefficients [3]. A step is a geometry which is very similar to many structures in hydraulic systems, such as gap between the components of turbines and pumps, sudden

[1] Corresponding author. Tel.:+861062789853, Fax.:+861062785285
E-mail address: luoxw@tsinghua.edu.cn

variation sections in pipelines and geometries inside accessories; therefore the experimental flow velocity fields of continuous and disperse phases in this region allow to understand better the interaction between phases and establish data for numerical analysis.

The advanced and not-intrusive techniques for measuring flow velocity field are recommended because the intrusive measurement methods comprise the flow field by disturbing the flow [4].The most common non-intrusive and advanced technique used for velocity field measurements is particle image velocimetry (PIV) [3], which determine the instantaneous flow velocity field. In a single-phase flow condition, tracer particles are used inside the flow and a light sheet is used to illuminate. The velocity fields are calculated based on the particle images from the scattered light at different time. However, the PIV technique shows complications and challenges for two phases, due to simultaneous measurement, so that, images for each phase should be separated. Furthermore, PIV has one natural limitation which is the presence of phase boundaries in multiphase flows that blocks the optical path to the measurement location, which means that PIV can be used on low volume fractions of solid particles or gas-bubbles.

Due to the growth of industries which need to the transport fluid with solid concentration and the installation of hydroelectric power plant where the rivers have high sand concentration, there has been growing interest in liquid-solid two phase PIV techniques to measure two phase flow using PIV. Numerous solutions are proposed to measure two-phase flow using PIV. Hassan et al. [5] separated the phases using fluorescent particles and record the images with two cameras, both phases can be recorded in conjunction with filters to divide the signal and allow only the wavelength of the fluorescent light to pass. Khalitov and Longmire [3] used an algorithm to discriminated each phase, which is based on the intensity and the diameter of the appearing objects. Towers et al. [6] used a method that implied a laser pulse (λ = 532 nm) corresponding to the first image of PIV couples and laser pulse (λ = 610 nm) corresponding to the second image, it is note that fluorescent particles were used as tracer. A camera recorded single image from the two layers of the field-of view (FoV). Those images allow to separate the temporal information and to discriminate between the fluorescence and non- fluorescence. Rottenkolber et al. [7] used two techniques: First, the masking technique consisted of long-pass filters used in a camera, which reduces the scattering intensity level of the fluorescence emission. After, a binary mask by thresholding the image which preserves grey level was created. Finally some characteristics (size and shape) were deleted in the mask and filled using an average background grey level calculated from the original bitmap [8]. In the second technique, the properties of the correlations peaks were used to differentiate and separate the phases; for this technique is required a sufficient velocity different between the two phase flows which produce a double peak in the correlation plane. This technique is known as correlation peak. These techniques have been widely used for the two-phase flow PIV studies, some of which have been used to liquid-solid flow. Liu et al. [9] used a combination of PIV and image processing method to obtain the velocity information of the liquid-solid two phase flow in a chemical pump; this result show that the presence of the solid particles change the relative velocity of the liquid and the average relative velocity of mixture was larger than the clear water condition. Metha et al. [4] used the PIV in conjunction with refractive index matching to study the particulate flow in the impeller of a centrifugal pump, got the results of flow field at different particle concentration, and concluded that the highest particle velocity was near the suction side of the blade.

The application of the PIV technique to separate phases depends on study case. For liquid-solid two phase flow, all the above techniques could be used when the continuous phase is water and dispersed phase is sand. In this paper, the Hassan et al. [5] technique is used to simultaneous PIV of water-sand flow over a step when the sand concentration is low. Two cameras with appropriate optical filters are used to separate the liquid-solid phases. The first camera takes the images of tracers that represent the behavior of the continuous phase, and the second camera takes the images of second phase. For comparison, numerical simulation was also conducted to analyze the liquid-solid flow over the step.

Method and test rig

The experimental study aims to investigate the differences and similarities between the velocity of sand and water, when liquid-solid two phase flow through a changing section like a step. The close flow pump loop facility and PIV equipment (Figure 1) used in this research is located in China Institute of Water Resources and Hydropower Research. The flow passage components are built in stainless steel. However, the walls of the test section were performed in an optically clear material. The test section is constituted for a rectangle cross section of 80 x 20 mm and the localization of the step is shown in Figure 2. Water at 20° C is the carrier phase and the flow rate is 0.0087 m^3/s. The mixtures of solid and water are prepared in the tank.

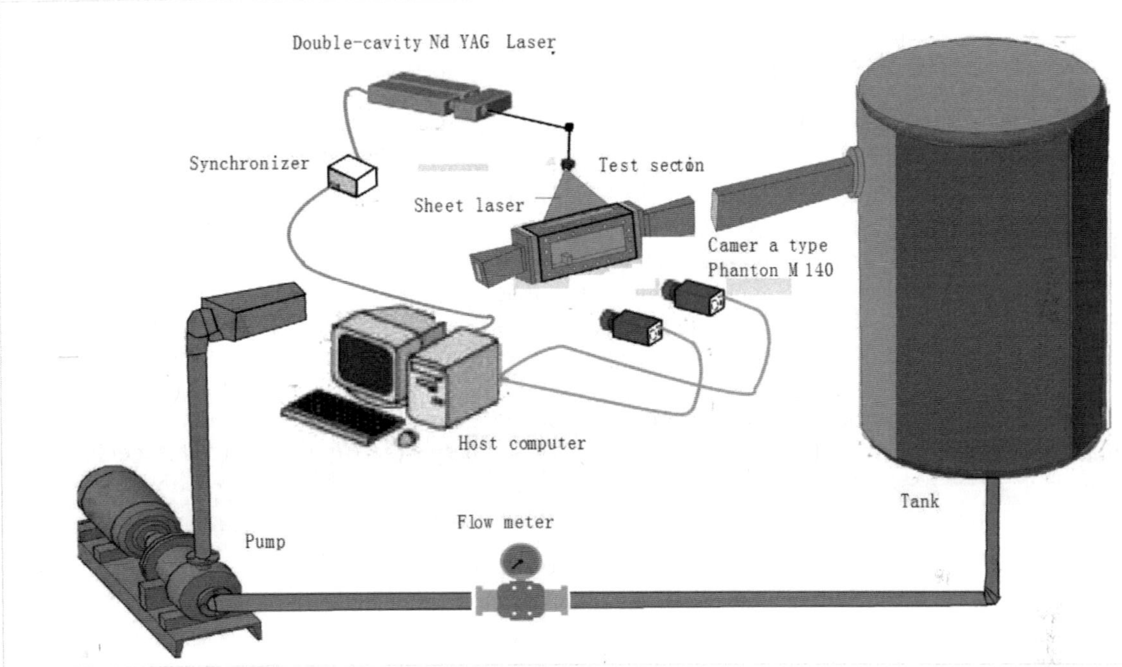

Figure 1. Experimental system.

The solid phase is sand particle with a bulk density of 2650 kg/m^3 with diameter of 20 μm. The volumetric particle concentration is less than 2% and it is assumed sand has a negligible effect on the flow of the carrier phase [1,10]. The tracers are the MMA fluorescence particles, whose diameter is 20 μm, the density is 1050 kg/m^3 and reflective index is 1.49.

1.1. PIV acquisition system

To perform a PIV test, the following equipment (Figure 1) are required: a 120 mJ/pulse Nd:YAG double-cavity laser (532 nm wavelength), laser light sheet optics, two camera Phantom M140, resolution: 2560 x 1600 pixels, a data acquisition system consisting of a PC, two camera filters and a frame grabber card. The laser beam is formed into a 1 mm thick light sheet using a combination of cylindrical and spherical lenses. The cameras focal axis is perpendicular to the plane of the laser light sheet to acquire flow images. Pair of single exposure image frames are required to enable cross correlation data processing. The image pair acquisition and processing is done using the DaVis software license version 8 offered by LaVision .The image pair acquisition is synchronized. The synchronization board and programmable timing unit are applied for image pair acquisition. The Δt used to take each pair of images was 100μs.

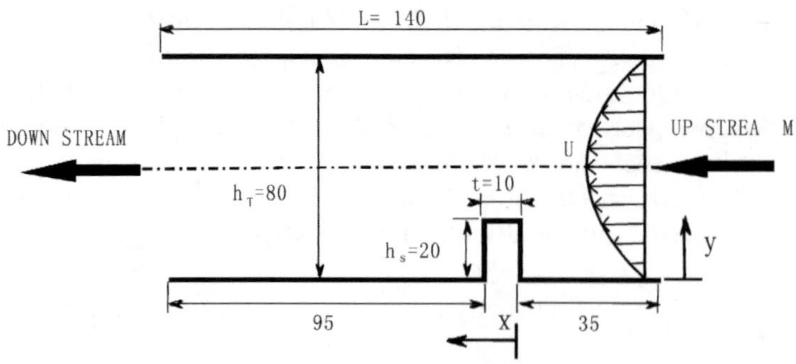

Figure 2. Illustration of test section (dimension in mm).

1.2. Description

The optical properties of the sand and the tracer have been used to separate images of each phase. The middle plane of flow through the test section is illuminated with the laser beam with 532 nm wavelength; this plane has a thickness of 1 mm. The first camera is equipped with an optical filter for LIF 540 nm cut, which takes the measurement of the tracers that represent behaviour of the continuous phase. This filter allows the passage only of light with a wavelength greater than 540 nm. For that reason this camera records only images of the tracers. On the other hand, the second camera has a filter 532 nm which blocks the passage of the light with a wavelength greater than 532 nm, therefore it records the flow of the sand. The geometry configuration of the PIV system is shown in Figure 3. The camera 1 is placed perpendicular to the plane of the object and the camera 2 is not completely perpendicular to the object plane; the position has a deviation of 5 degrees to the perpendicular plane but this deviation was evaluated and it does not affect the results. Finally a Rosemount vortex flow meter with 0.3% accuracy is used in the test rig for comparing the acquired data.

Numerical simulation

The numerical simulation in steady state has been carried out using the commercial software Ansys Fluent 14.0. The turbulence model used is Reynolds Average Navier Stokes (RANS *k*-ω SST) and it is considered that the dispersed phase has a negligible effect on the flow of water because the volumetric particle concentration is less than 2%, according with some authors [1,10].The conservation equations are based on the Eulerian approach for the water and the stochastic-Lagrangian approach for the particles [10-13]. The discrete phase model (DPM) is used to simulate the trajectory and particle velocity. This model can be used for these studies due to the sand low concentration [11].

1.3. Continuous Phase.

The continuity and momentum equations in steady state for continuous phase q (water) are expressed as:

$$\frac{\partial(\rho_q \overline{u_j})}{\partial x_j} = 0\,, \tag{1}$$

$$\frac{\partial(\rho_q \overline{u_j})}{\partial x_j} = -\frac{\partial \bar{P}}{\partial x_i} + \frac{\partial}{\partial x_j}\left(\mu \frac{\partial \overline{u_j}}{\partial x_j} - \rho_q \overline{u_i u_j}\right), \tag{2}$$

whereρ_q,$u_j$are the density and velocity, respectively. The term $-\rho_q \overline{u_i u_j}$ is the Reynolds stress.

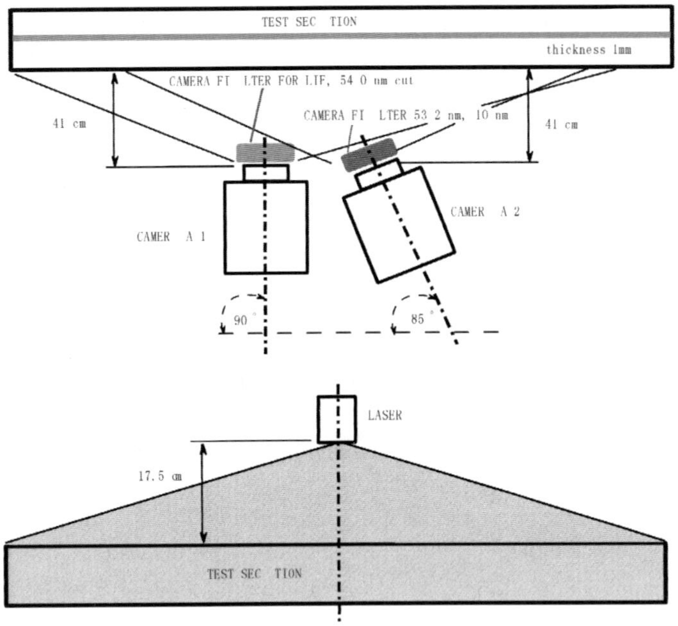

Figure 3. Camera, laser and optical filter positions.

1.4. Dispersed Phase.

DPM model is applied to calculate the trajectory and velocity of each particle injected based on Newton's second law of motion, which is given by Equation (3):

$$\frac{dv_i}{dt} = F_D + F_{pi} + F_{Bi} + F_{mi}, \tag{3}$$

where v_i is the particle velocity andF_D, F_{pi}, F_{Bi}, F_{mi} are the drag, pressure gradient, buoyancy and virtual mass forces, respectively. These forces are indicated from Equation (4) to (7).

$$F_D = \frac{\mu}{\rho_p {D_p}^2} \cdot \frac{18 . C_D . Re}{24} (u_i - v_i), \tag{4}$$

$$F_{pi} = \left(\frac{\rho_f}{\rho_p}\right) v_i \frac{\partial u_i}{\partial x_i}, \tag{5}$$

$$F_{Bi} = \left(1 - \frac{\rho_f}{\rho_p}\right) g, \tag{6}$$

$$F_{mi} = \frac{-1}{2} \frac{\rho_f}{\rho_p} \frac{d}{dt} (u_i - v_i), \tag{7}$$

where$\mu \rho_p$, D_p, u_i, C_D and $Re$are dynamic viscosity, particle density, particle diameter, water velocity, drag coefficient and particle Reynolds number based on the relative velocity between the particle and fluid, respectively.

The drag coefficient $C_D$in Equation (4) is given by Equation (8):

$$C_D = \frac{24}{Re} . (1 + 0.15 . Re^{0.687}), \tag{8}$$

The Equation (5) is the force around the particle in the direction of the pressure gradient. The buoyancy force in Equation (6) is considered when the particle and fluid have significantly different densities and when the inclusion of gravitational effects is desired. The virtual mass force, Equation (7), accounts for the inertia of the fluid surrounding the particle.

Due to the flow characteristic, the following considerations also are applied: the particle-particle interaction and Brownian motion are neglected, the particle shape is considered as a spherical and the physical properties of the solid phase are constant.

Particle distribution was simulated with stochastic trajectory model in conjunction with the discrete random walk mode. The stochastic trajectory considers the fluctuations in the flow field, then the instantaneous value of the fluctuating flow velocity, $u'_i(t)$, due to turbulence is included to predict the dispersion of particles. The instantaneous fluid velocity,$\overline{u_i} + u_i'(t)$, along the particle path is used to predict the turbulent dispersion of particle by the integrating the trajectory equations for each particle. The adequate number of particle injection is tracked repeatedly in order to generate a statistically meaningful sampling; the number of tries for this simulation was 10. Furthermore, the mass flow rate and exchange source terms for each injection are divided equally among multiple stochastic tracks. In the dispersed random walk model the interaction of a particle with a succession of discrete stylize fluid phase turbulent eddies is simulated by:

$$u'_i = \zeta\sqrt{\frac{2k}{3}}, \tag{9}$$

where k is the turbulent kinetic energy and $\zeta$is a normally distributed random number.

1.5. Geometry and solver

The computation domain is shown in Figure 4(a). The software Gambit has been used to generate a structured mesh, of which the mesh around the step is shown in Figure 4(b). The main mesh information is listed in Table 1.The mean calculated Yplus is 12.5.

The boundary conditions for this numerical simulation can be observed in Table 2.

The numerical solution is performed in double precision. The solution methods are: SIMPLE algorithm for pressure-velocity coupling and the second-order up wind spatial discretization have been used for the momentum equation, turbulence kinetic energy and specific dissipation rate. The residual reached of mass and momentum equations is 10^{-4}.

Results and discussions

Tests were conducted using two sand concentrations of 210 g/m^3 and 140 g/m^3. The experiment data processing of 200 pairs of PIV images for a selected area have been conducted. The PIV data were processed with an algorithm generated using Python and the results had been plotted using Tecplot 360. The numerical simulation has been carried out using ANSYS Fluent 14.0 and the data has been processed using ANSYS CFD Post 14.0. In the test, the height and width of a rectangle view are 140 mm and 80 mm, respectively. The selected area covers 60 mm over the step, i.e. 35 mm upstream and 95 mm downstream the step. The measurements were conducted at flow rate of 0.0087 m^3/s. The corresponding Reynolds number (based on the hydraulic diameter of the cross section) was 1.7 x 10^5 and average velocity U_m is calculated by U_m=Q/A, where Q and A are the flow rate and cross section area.

Table 1. Structural meshing.

Description	Quantity
Nodes	314128
Hexahedrons	233478

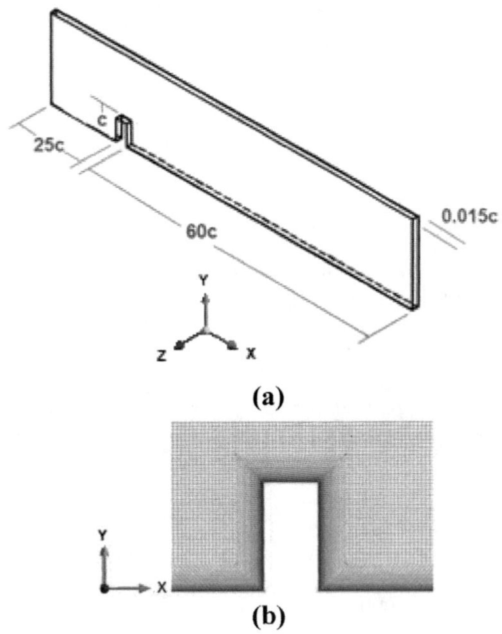

Figure 4. Domain and mesh: a) Computation domain; and b) Meshing around the step.

Table 2. Boundary conditions.

Surface	type	observation
Inlet	Velocity inlet	$U_\infty = 5.438\ m/s$
Outlet	Pressure outlet	
Top wall	Wall	No slip
Bottom wall	Wall	No slip
Side	Symmetry	

1.6. Validation

The numerical results with sand concentration of 210 g/m^3 have been compared with experimental results to validate the PIV technique. Moreover the flow rate calculated from velocity profiles is compare with the measurement of the flow meter. Figure 5 shows the FOV comparison between the experimental and numerical results, the velocity fields of numerical results for water are listed in the right column and the experimental results are listed in the left column. The experimental and numerical velocity profiles of each phase at x/L=0.4 are shown in the Figure 6. Finally, the calculated flow rate of numerical solution and PIV experimental results at x/L=0.4 are compared in the Table 3. The following milestones can be noted:

1) The water behavior has been obtained with good accurate using the fluorescent particles of density 1050 kg/m^3. It is supported because the PIV flow field velocity of water Figure 5(a) is reasonable match with numerical flow field velocity of water in Figure 5(b). Furthermore, the flow rate of the water obtained from experimental and numerical velocity profiles are similar to the flow meter measurement, as show in Table 3.

2) The velocity measurement of sand and water are successfully separated with the technique applied. The comparison between Figures 5(c) and 5(d) has a good match in the region where

the velocity is increased by the chance section. The comparison between experimental and numerical dimensionless profile at x/L=0.4 also shows the excellent match for the increased velocity of the sand due to the presence of step. Additionally, the flow rate obtained of the experimental water velocity profile at x/L=0.4 confirm that this camera takes the measurement only the tracers because the sand has higher velocity that should produce a higher flow rate.

3) The step inside of the geometry induces changes on the velocity profiles for both phases. However, the velocity increases in a different magnitude for sand and water, which is similar to previous studies liquid-solid two-phase flow [4,9].

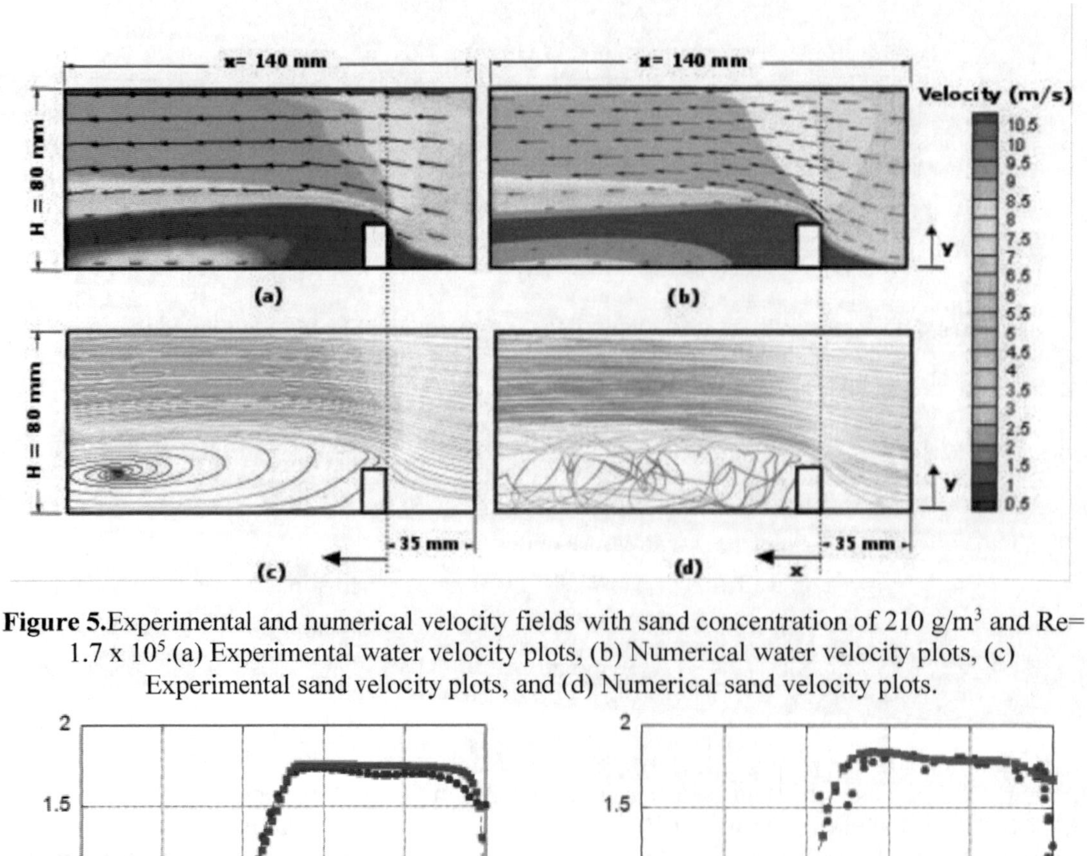

Figure 5. Experimental and numerical velocity fields with sand concentration of 210 g/m^3 and Re= 1.7 x 10^5.(a) Experimental water velocity plots, (b) Numerical water velocity plots, (c) Experimental sand velocity plots, and (d) Numerical sand velocity plots.

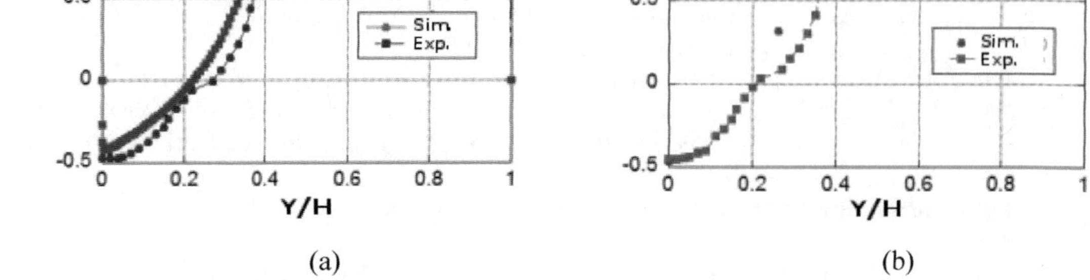

(a) (b)

Figure6. Experimental and numerical dimensionless velocity profiles at X/L=0.4, Re= 1.7 x 10^5. (a) Water and (b) Sand.

Table 3. Experimental and numerical comparison of flow rate.

Description	m^3/s	Abs error
Flow rate measurement	0.00870	–
Numerical profile	0.00864	0.67%
PIV profile	0.00843	3.1%

1.7. General feature

The changes of the velocity profiles obtained from experimental results are plotted in Figure 7 for both phases. Note that the dimensionless velocity, i.e. U/U_m is used to plot the velocity distribution. Both velocities of the sand and the water have similar profiles upstream. However, there is a difference close to the bottom, which can be the buoyancy force effects. In the same way, the velocity profiles change over the step and are higher than other regions due to channel reduction effects, but the magnitude of sand velocity is greater than water. Although the difference between the velocity of sand and water over the step is higher than in any other regions, the sand velocity remains greater than water downstream step.

Since the increased velocity of sand and water is different, it is necessary to study the velocity profiles after the step. Thus the Figure 8 show the experimental velocity profiles obtained by PIV at downstream step distance of x/L=0.4.

In order to evaluate whether a variation in the low concentration influences the velocity variation, the sand concentration of 140 g/m^3 was included in this analysis. The Figure 9 shows a comparison between dimensionless velocity profiles of sand and water with 140 g/m^3 and 210 g/m^3 sand concentration. These figures allow quantifying the increase in the sand velocity for two concentrations.

Based on these outcomes, the following can be described:

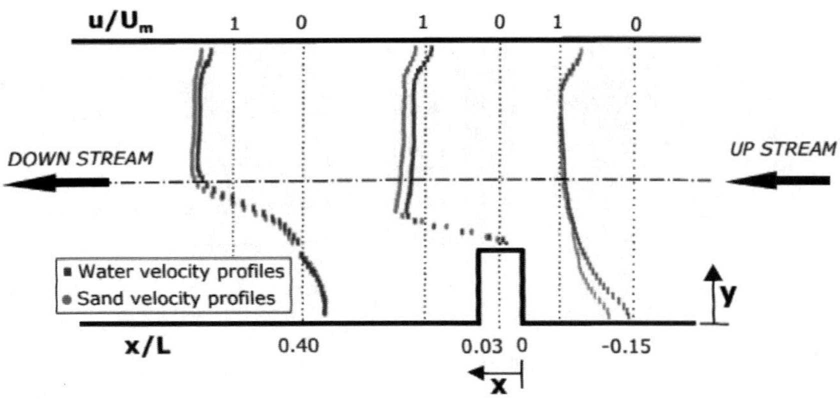

Figure 7. Dimensionless velocity profiles of water and sand, Re= 1.7 x 10^5.

1) The main characteristic of water flow field is the increase of velocity as the water approaches the step combined with large curvature in the streamlines. Furthermore the step downstream zone has a large region of recirculation flow.

2) The velocity profile of the water does not change for different low concentrations of sand, as shown in Figure 9(a)

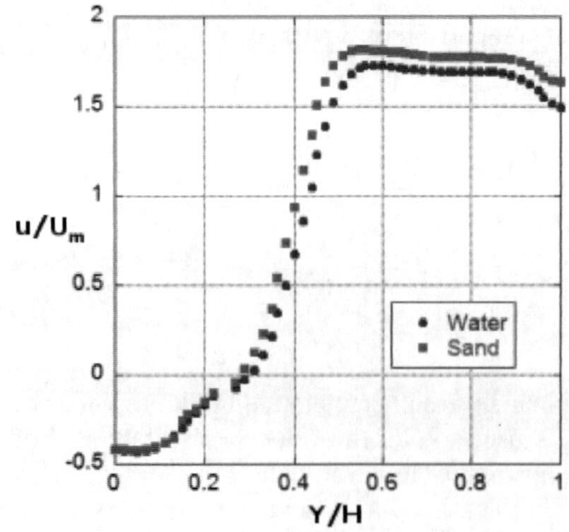

(a)

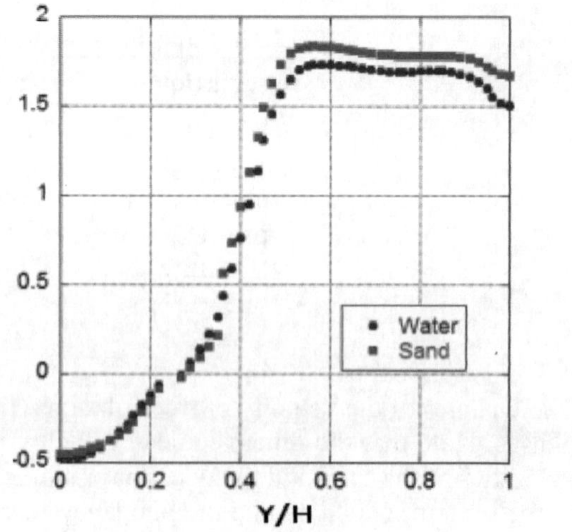

(b)

Figure 8. Dimensionless velocity profiles of sand and water at x/L=0.4and Re=1.7 x 10^5. (a) sand concentration of 140 g/m^3, and (b) sand concentration of 210 g/m^3.

3) In downstream step region, where the velocity increase occurs, the sand velocity increase 5% higher than water velocity, when the cross section is reduced in 25% and the sand velocity does not change for different low concentration.

Conclusions

The Particle Image Velocimetry (PIV) in conjunction with the simultaneous separation phase technique proposed by Hassan et al. [5] was successfully utilized to investigate the water-sand flow. Additionally, the following general conclusions can be obtained:

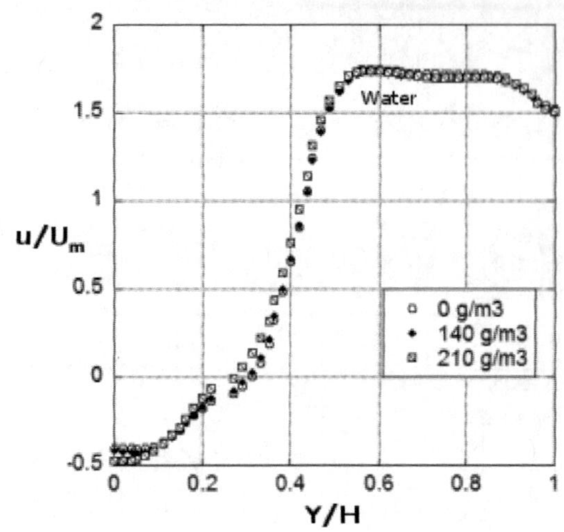

(a)

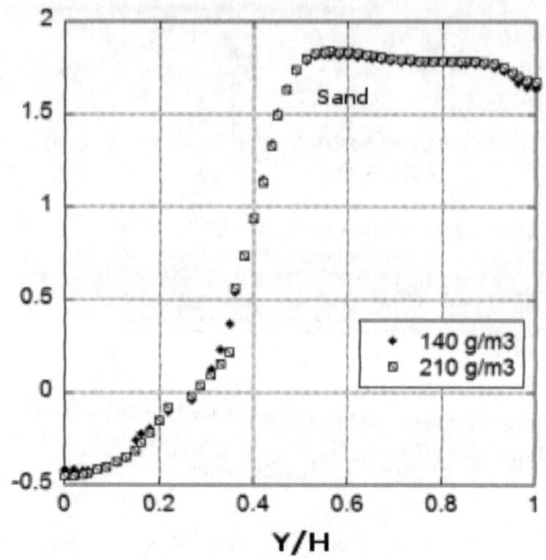

(b)

Figure 9. Dimensionless velocity profiles of water and sand at x/L=0.4, Re=1.7 x 10^5.

1. The particles with high density should not be used as tracers for PIV studies with changes in the cross section.
2. Downstream step, the velocity increase due to a change of section as a step is higher for sand than for water and is not influenced by low sand concentration variation.
3. The sand velocity is 5 % higher than water velocity for a reduction of 25% in the cross section and the water velocity profiles do not change due to variation of low concentrations of sand.

Acknowledgments

This work was financially supported by the Major National Scientific Instrument and Equipment Development Project (Project No. 2011YQ07004901), the National Natural Science Foundation of China (Project Nos. 51306018, 51406010 and 51179091) and State key Laboratory of Hydroscience and Engineering, Tsinghua University (Project Nos.2014-KY-05 and 2015-E-03).

References

[1] Crowe C T, Troutt, T R and Chung J N 1996 Two-phase turbulent flows, *Annu. Rev. Fluid. Mech.* **28**1143
[2] Thapa B S, Dahlhaug O G and Thapa B 2015 Sediment erosion in hydro turbines and its effect on the flow around guide vanes of Francis turbine Renewable and Sustainable Energy Reviews **49** 1100–1113
[3] Khalitov D A and Longmire E K 2002 Simultaneous two-phase PIV by two-parameter phase discrimination Experiments in Fluids **32(2)** 252–268
[4] Mehta M, Kadambi J R, Sastry S, Wernet M, Addie G. and Visintainer R 2004 Study of particulate flow in the impeller of a slurry pump using PIVASME Heat Transfer/Fluids Engineering Summer Conferencevol1(North Carolina:USA) Paper No. HT-FED2004-56684 489-499
[5] Hassan Y A, Blanchat TK, Seeley C.H Jr., and Caanan, RE 1992 Simultaneous velocity measurements of both components of a two-phase flow using particle image velocimetry Int. J. Multiphase Flow **18 (3)** 371-395
[6] Towers D P, Towers C E, Buckberry C H and Reeves M 1999 A colour PIV system employing fluorescent particles for two-phase flow measurements. Measurement Science and Technology **10** 824–830
[7] Rottenkolber G, Gindele J, Raposo J, Dullenkopf K, Hentschel W, Wittig S, Spicher U and Merzkirch W 2002 Spray analysis of a gasoline direct injector by means of two-phase PIVExperiments in Fluids **32** 710–721
[8] Jakobsen M L, Easson W J, Created C A and Glass D H 1996 Particle image velocimetry: simultaneous two-phase flow measurement Meas. Sci. Tech. **7** 1270-1280
[9] Liu D and Yang M 2009 Research of liquid-solid two phase flow in the chemical pump. In Advances in Water Resources and Hydraulic Engineering 2203–2207
[10] Ji S M, Xiao F Q and Tan D P 2010 Analytical method for softness abrasive flow field based on discrete phase model Sci China Tech Sci. **53** 2867-2877
[11] Mostafa A.A. 1987 On the interaction of particles and turbulent fluid flow. Int. J Heat Mass Tran. **30 (12)** 2063–2075
[12] Okita R, Zhang Y. McLaury B S. and Shirazi S A 2012 Experimental and computational investigations to evaluate the effects of fluid viscosity and particle size on erosion damage. Journal of Fluids Engineering **134(6)** 061301
[13] Morsi S A and Alexander A J. 2006 An investigation of particle trajectories in two-phase flow systems Journal of Fluid Mechanics **55(02)** 193

IOP Conf. Series: Materials Science and Engineering **129** (2016) 012055 doi:10.1088/1757-899X/129/1/012055

Effects of shaft supporting structure on performance test of axial flow fan

R Ma[1], S L Liu[2], M X Li[3], S Y Zheng[1]

[1]Institute of Chemical Machinery, Zhejiang University, Hangzhou, China
[2]School of Mechanical and Automotive Engineering, Zhejiang University of Science and Technology, Hangzhou, China
[3]Department of Mechanical Engineering, Zhejiang University, Hangzhou, China

E-mail: zhengshuiying@zju.edu.cn

Abstract. CFD numerical simulation combined with theoretical analysis are used to research and discuss the obstructing effect, caused by the supporting structure of torsion meter and connecting shaft, on the outlet airflow of axial-flow fan in type-C ducted inlet device. The relations between axial flow fan's total pressure efficiency and flow rate are studied when the distance between supporting structure and outlet section is different, which may provide a reference for the proper design of the performance test device.

1. Introduction

All the ventilation fans have characteristic parameters and performance curves that are acquired through performance test when they leave factory, therefore, fan performance test is an indispensable part of design, production and inspection of fans and device of fan performance test is widely used in various factories and some ancillary departments. At present, the domestic standard of fan performance test is GB1236-2000 'Industrial fans–Performance testing using standardized airways', its technical content and writing rules are equal to ISO 5801: 1997. According to the standard, in the performance test conducted in standardized airways, the input power to the fan shaft needs to be measured when the fan is not directly driven by motor, and it can be measured by using reaction dynamometer or torsion meter. By both means, a supporting structure is needed to support the reaction dynamometer or the torsion meter. In addition, the fan and the torsion meter are connected by shaft which needs to be supported by bearing. If a fan performance test is conducted in a type-C device, the supporting structure of shaft and torsion meter is at the free end of fan outlet, which may obstruct the outflow. According to GB1236-2000, of which section 19.3 specify that 'Care should be taken to avoid the presence of any obstruction which might significantly modify the air flow at inlet or outlet. In particular, no wall or other major obstruction should be closer than 2D from the inlet and 5D on the outlet of the airways or the test fan'[1]. So it is necessary to study the influence of the distances between supporting structure and outlet section on fan performance test. This problem may be frequently encountered in the performance test conducted in type-C device, and related materials are not found at present.

In this paper, CFD numerical simulation[2-4] combined with theoretical analysis are used to research and discuss the obstructing effect of the supporting structure, which may provide a theoretical reference for the edit of fan performance test standard.

2. Model

2.1. Geometry and mesh

The device of fan performance test is type-C with ducted inlet and free outlet, which is composed of bellmouth inlet, flow regulating valve, airway, flow straightener, torsion meter, motor, connecting shaft and fan etc, as shown in figure 1.

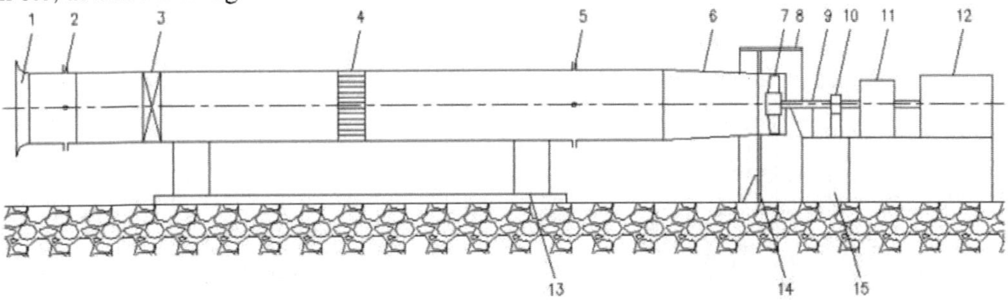

Figure 1. Schematic diagram of performance test device.

1-bellmouth inlet 2-differential pressure tapping 3-flow regulating valve 4-flow straightener 5-static pressure tapping 6-transition duct 7-aixal flow fan 8-protective shield 9-connecting shaft 10-torsion meter 11-high-speed gearbox 12-motor 13-supporting seat of airway 14-vertical plate 15-motor base

The diameter of airway is 700mm. The axial-flow fan has 8 blades. The hub and tip diameter for the rotor is 320mm and 690mm respectively with the hub ratio of 0.46. The shaft diameter is 110mm and the length of fan casing is 200mm. The width of the supporting structure is 50mm and that of the motor base is 200mm.

The computational model is established according to the structure shown in figure 1. It is properly simplified that pressure tapping, flow regulating valve and flow straightener are neglected[5]. Flow rate can be directly specified in CFD commercial software. The computational domain is divided into three parts: inlet region, compeller region and outlet region. The inlet region is the airflow field between static pressure tapping and fan inlet, which is three times the length of airway diameter. The outlet region is the airflow field surrounded by protective shield, and the size is 570mm×1280mm× 1940mm.

The supporting structure of connecting shaft and motor would inevitably affect the pressure of the airflow at outlet, and further affect the testing efficiency results. So it is necessary to reasonably design the shaft supporting structure and find an appropriate distance between it and fan outlet. Building computational models when this distance is different, and structure diagram along with 3-D model of the computational domain are shown in figure 2. The value of L in figure 2(a) is the distance between shaft supporting structure and fan outlet. To compare with the situation that the fan outlet is not obstructed, the computational model without any supporting structure is established, as shown in figure 2(b).

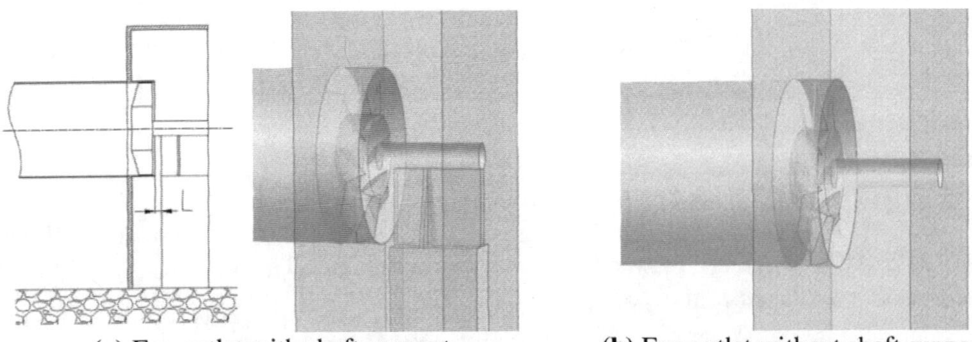

(a) Fan outlet with shaft support **(b)** Fan outlet without shaft support

Figure 2. 3-D model of computational domain.

The modeling process is completed in Turbo module of GAMBIT[6]. Contour lines of the 5 airfoil sections are generated first, and surfaces of the rotor blade are constructed through these contour lines. With these surfaces, we can further stitch out body of the blade and carry out some Boolean operations to create the final computational fluid domain. Impeller model is shown in figure 3.

Structured hexahedral grids are applied in the inlet region with size of 30mm and unstructured tetrahedral grids are applied in the outlet region with size of 20mm. Due to the complexity of blade geometry, tetrahedral meshes are used around the blade with size of 10mm, in terms of the balance of calculation accuracy and computation time. EquiSize Skew of the elements is controlled within 0.9 and the total number of grids is about 2×10^6. The mesh of computational domain is presented in figure 4.

Figure 3. Geometric model of impeller.

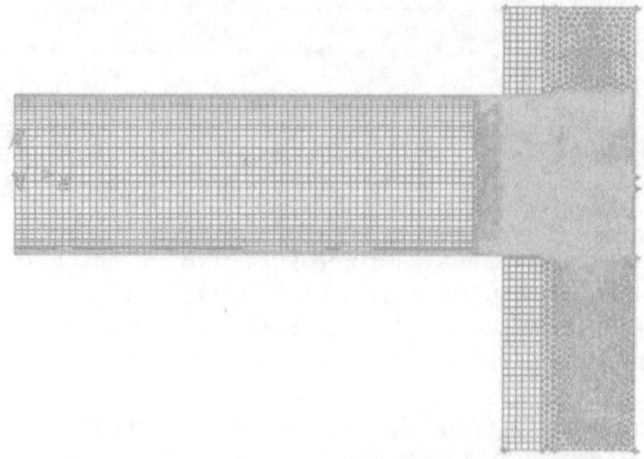

Figure 4. Mesh of computational domain.

2.2. Parameter setting

Following basic assumptions are made: the flow field is taken as three-dimensional steady turbulent flow and the heat conduction problem is neglected, as well as the compressibility of the air. Standard $k-\varepsilon$ turbulence model is enabled. The boundary conditions near the solid walls are analyzed using the method of standard wall function. The coupling equations of pressure and velocities are solved with the SIMPLE algorithm[7]. The numerical simulation of this model was analyzed through first-order upwind difference scheme. Convergence criterion for the equations is 10^{-4}.

The inlet flow uses mass flow rate boundary condition and the velocity direction is normal to boundary. The outlet flow uses pressure boundary condition and the static pressure equals to zero. No-slip boundary condition and multiple reference frame model (MRF) are employed in the simulation[8].

5 operating condition points specified at rational speed of 1750 r/min are selected as shown in table 1.

Table 1. Different operating condition points.

	Point 1	Point 2	Point 3	Point 4	Point 5
Flow rate (m^3/h)	14000	16000	18000	20000	22000

3. Results

3.1. Theoretical calculation of total pressure and total pressure efficiency

According to GB1236-2000, the fan total pressure is defined as the sum of static pressure difference and axial dynamic pressure difference between inlet and outlet of the fan. The rotary dynamic pressure isn't taken account of in the calculation of total pressure[9]. The calculation formula is

$$p_t = p_{st2} - p_{st1} + \frac{1}{2}\rho\left[\left(\frac{q_v}{A_2}\right)^2 - \left(\frac{q_v}{A_1}\right)^2\right] \tag{1}$$

The calculation formula of fan power N is

$$N = T \cdot n \times \frac{2\pi}{60} \quad (2)$$

The calculation formula of total pressure efficiency η_t is

$$\eta_t = \frac{q_v \cdot P_t}{N} \quad (3)$$

In the formula, p_{st2} and p_{st1} is the static pressure of outlet flow and inlet flow of the fan respectively. As fan outlet is directly connected with atmosphere, the outlet static pressure equals to zero. ρ is air density and q_v is volume flow rate. A_2 and A_1 is outlet area and inlet area of the fan respectively. According to the stand, 'the fan outlet area shall be taken as the gross area at the outlet flange or the outlet opening in the casing without deduction for motors, fairings or other obstructions', so A_2 is the difference value of cross-sectional area between airway and shaft and equals to 0.375m^2, A_1 is the cross-sectional area of airway and equals to 0.386m^2.

3.2. Effects of shaft supporting structure on inlet and outlet pressure

Based on the structure shown in figure 2, the flow field in the test device is calculated by using FLUENT. Changing the distance L and then analysing the pressure distribution to obtain a proper distance at which the obstruction has negligible influence on the outlet pressure of axial-flow fan. The value of L is set as 50mm, 70mm and 100mm. For the purpose of comparison, the pressure distribution is also calculated without any supporting structure.

Figure 5 is the contours of static pressure in vertical section and axial cross section of the outlet with flow rate of 22000m^3/h in the airway of 700mm in diameter.

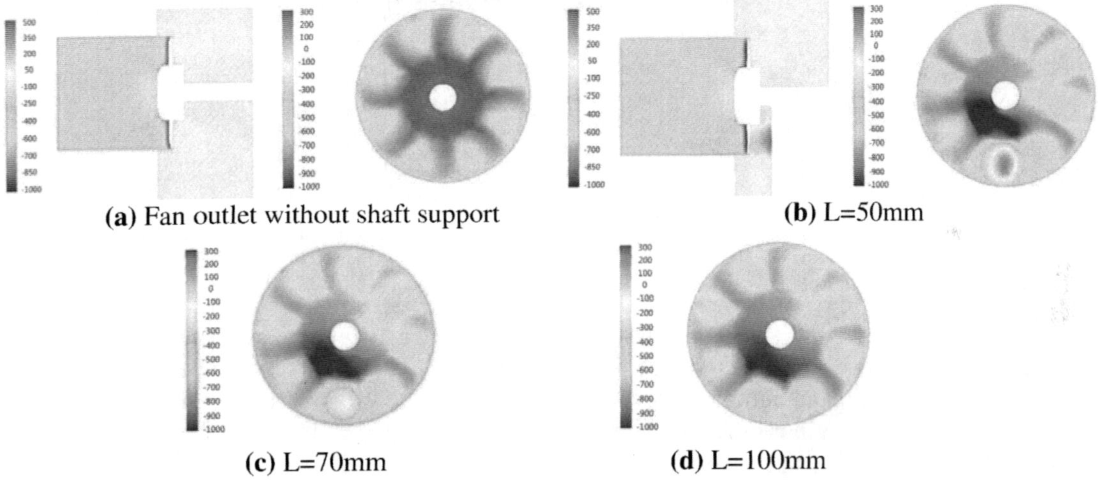

(a) Fan outlet without shaft support **(b)** L=50mm

(c) L=70mm **(d)** L=100mm

Figure 5. Contours of static pressure.

It can be seen from figure 5 that, when L=50mm, the dynamic pressure of airflow transforms into static pressure as obstructed by the supporting structure, which makes the static pressure at the bottom area of outlet section increases obviously. Meanwhile, performance property of the impeller remains the same, thus making the inlet static pressure increase correspondingly. However, the outlet static pressure equals to zero according to the standard, so the increased value of outlet static pressure isn't taken account of in the calculation of total pressure. According to formula (1), p_{st2} remains unchanged and p_{st1} increases, making the test value of total pressure below the actual value. As L increases, the outlet static pressure reduces gradually. When L=100mm, we can see from figure 4(a) in comparison with 4(d) that the outlet static pressure still rises with the influence of supporting structure, but the value is not very large. Therefore, it can be seen that the outlet obstruction doesn't influence the measurement of total pressure when L is greater than 100mm in this device.

However, the shaft rigidity may not meet requirements when the supporting structure is too far from the fan outlet, so the shaft rigidity and the influence of obstruction are considered together to improve the design of supporting structure. As shown in figure 6, the supporting structure is far from the outlet with a wedge-shaped axial stiffened plate to guarantee the rigidity of shaft. The distance between stiffened plate and fan outlet equals to 50mm, which remains consistent with the original design. The distance between motor base and fan outlet equals to 280mm, which is obtained through trial and error.

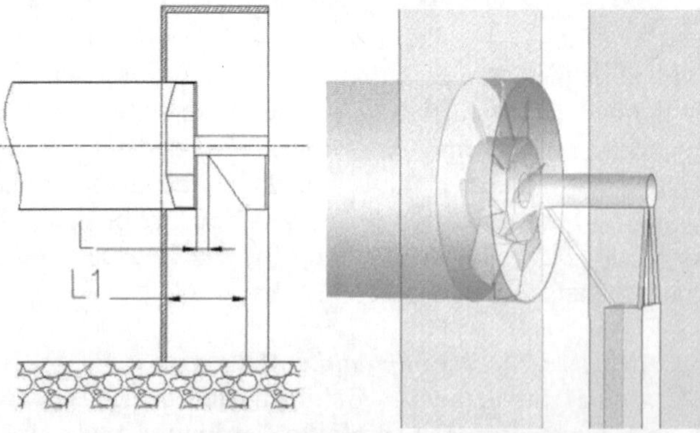

Figure 6. 3-D model of computational domain after improvement of supporting structure.

Similarly, figure 7 is the contours of static pressure in vertical section and axial cross section of the outlet with flow rate of 22000m^3/h. Compare figure 5(a) and 5(d) with figure 7, we can see that improved structure has little influence on outlet pressure when the distance is 280mm.

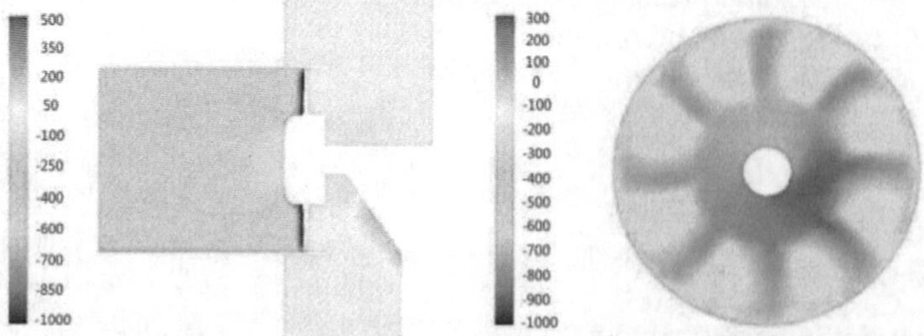

Figure 7. Contours of static pressure after improvement of supporting structure.

3.3. Effects of shaft supporting structure on total pressure efficiency

Figure 8 is the comparison chart of total pressure efficiency versus flow rate at different distances of supporting structure shown in figure 2. As L increases, total pressure efficiency of the fan is getting closer to the value when the fan outlet is not obstructed, and with the increase of flow rate, the influence of obstruction on total pressure efficiency becomes greater. When L=100mm, total pressure efficiency is very close to the value of no obstruction, and a larger distance may lead to an even better coincidence. To meet the requirement of rigidity, the extension length of the shaft shouldn't be too big, and the comparison of total pressure efficiency with different supporting structures is shown in figure 9. We can see that the fan total pressure efficiency with the wedge-shaped supporting structure is very close to the value without any supporting structure, and its obstruction effect can be neglected approximately. Through simulation, it's shown that the improved supporting structure can ensure the shaft rigidity and have little influence on the pressure distribution at the fan outlet at the same time.

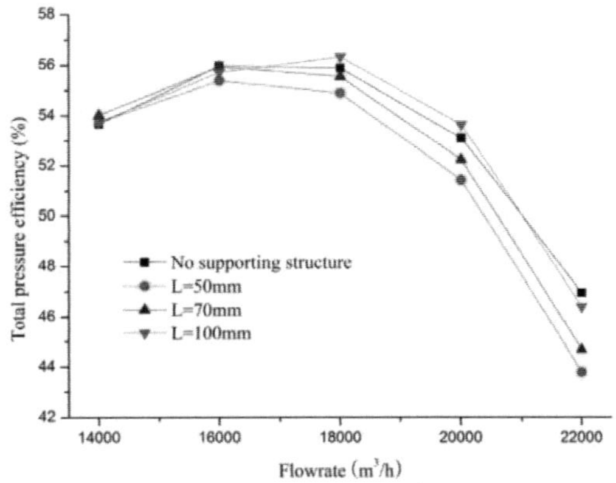

Figure 8. η-Q curves at different distances between supporting structure and fan outlet.

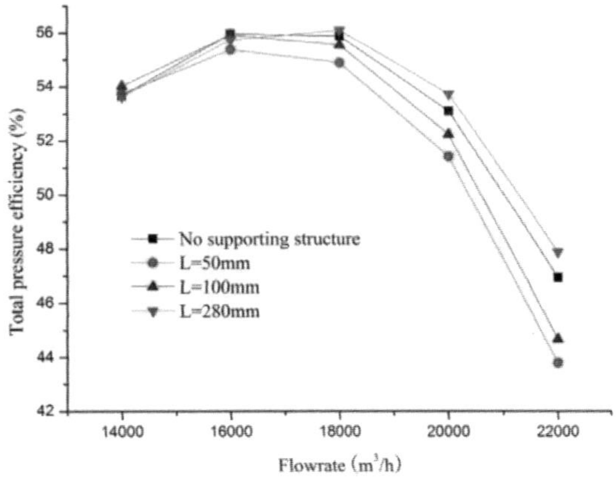

Figure 9. η-Q curves after improvement of supporting structure.

4. Conclusions

- For the performance test of an axial-flow fan conducted in a type-C ducted inlet device, if the shaft supporting structure is too close to the fan outlet, the test values of total pressure and total pressure efficiency decrease.
- Under the premise of ensuring the rigidity of connecting shaft, we can properly increase the distance between supporting structure and fan outlet, which may decrease the influence of obstruction on test results.

5. Acknowledgment

This work was supported by the National Natural Science Foundation of China (grant number 512754 52).

6. References

[1] GB/T1236-2000 2001 *Industrial fans - Performance testing using standardized airways.* (Beijing: Standards Press of China).
[2] Li J F, Lv J F. 2006 Simulation of air flow in a blower fluid machinery, **34(4)**: 11-13.
[3] Chang Z Z, Luo H. 2009 Application of CFD software in the design of fan. *Compressor,*

Blower and Fan Technology, **(2)**, 60-64.

[4] Liu S, Wu S F, Zhang L X. 2009 Discussion of 3-D modeling for axial-flow fan and its internal flow field based on CFD. *Compressor, Blower and Fan Technology*, **(2)**, 16-19.

[5] Shangguan W B, Wu M, Wang Y Y. 2010 Calculation method of aerodynamic performances of engine cooling fans. *Automotive Engineering*, **32(9)**:800-802.

[6] Liu Y, Zheng S Y, Shen H T. 2013 Effects of airfoil thickness on performances of axial-flow fan. *Compressor, Blower and Fan Technology*, **(2)**, 15-18.

[7] Ashcroft G, Nurnberger D.2009 A computational investigation of broadband noise generation in a low-speed axial fan. *Proceedings of the 15th AIAA/CEAS Aeroacoustics Conference*, AIAA-2009-3219, 2009.

[8] Zhu Z Q. 1998 Computational fluid dynamics. (Beijing: Beihang University Press).

[9] Wu Y L, Chen Q G, Liu S H. 2011 Fans and Compressors. (Beijing: Tsinghua University Press).

Flow field distribution of liquid film of water lubricated bearing-rotor coupling systems

Q L Hu [1], J N Hu [1], X Y Ye [1], D S Zhang [1] and J B Zheng [1]

[1]National Research Center of Pumps, Jiangsu University, Zhenjiang, Jiangsu 212013, China

E-mail: hu_jingning@qq.com, 473535919@qq.com, 1074282764@qq.com, 108044392@qq.com, 250286355@qq.com

The National Natural Science Foundation of China（51276083)
National Science and Technology Support Program(2013BAB08B02)

Abstract. According to the desalination high-pressure pump water lubricated bearing-rotor coupling systems flow field distribution of liquid film in the starting transient process and its power transmission mechanism can lay the foundation of further exploring and judging lubrication state at the boot process. By using the computational fluid dynamics Fluent secondary development platform and calling the relevant DEFINE macro function to achieve the translation and rotation movement of the journal, we will use the dynamic grid technique to realize the automatic calculation and grid update of water lubricated bearings 3d unsteady liquid film flow field, and finally we will dispose the results of numerical simulation and get the pressure. When the eccentricity is large, film thickness was negatively correlated with the pressure, and positive with the velocity. Differential pressure was negatively correlated with velocity. When the eccentricity is small, film thickness is no significant relationship with differential pressure and velocity. Differential pressure has little difference with velocity.

1. Introduction

Reverse osmosis technology is a desalination technology which is high-efficiency and easy to use on a large scale. It is in broad application prospects especially in a time of scarce water resources [1]. High-pressure pump is the key equipment of reverse osmosis desalination [2], its typical structure is a new type of axial suction multistage sectional pumps, with the supporting of a rotor bearing. In the high-pressure multi-stage pumps, water-lubricated bearings at both ends of the rotor system simplify the structure of the pumps, which also improve the efficiency and service life [3-6]. Water lubricated bearing is the main elastic support components, whose performance will directly affect the formation of dynamic pressure liquid film, lubrication and the stability of the entire rotor system in the startup process. Due to the low viscosity of water (1/100 ~ 1/20 of oil viscosity), and the low film bearing capacity, the formation of hydrodynamic lubrication is difficult. As well as the pump unit startup, stop and the unit running speed changes, the water lubricated bearings is easy in boundary lubrication or dry friction state. It leads to some extreme phenomena such as unpredictable instability and growth of vortex, which will seriously affect the dynamic stability of the pump bearing-rotor system [7]. In some desalination systems [8-12], the high-pressure pump startup and stop frequently. Meanwhile, because the rotor speed, film strength and the moment of the bearing - rotor system suffered a transient state in every startup process. Which results water lubricated bearing journal position in a transient state. In the startup state, the above various transient factors will directly affect the dynamic distribution of water lubricated bearing liquid film flow field, which further cause water lubrication complicated. The power transmission between water-lubricated bearings and rotor system is not only related to the distribution of film flow field and lubrication, but also closely related to dynamics of rotor system. Eventually a dynamic coupling system is formed [13].

It is CFD was used in this paper. With using FLUENT secondary development platform for programming journal of the shaft of the translational and rotational motion, combined with real-time updates of dynamic mesh grid. Finally FLUENT post-processing techniques were applied to deal with the startup transient numerical result. Then the transient distribution of three-dimensional film (velocity field, pressure field, etc.) of water-lubricated bearings flow field was obtained. It concluded the dynamic distribution of the film flow field of water lubricated bearing of high-pressure desalination pump in start transients.

2. Geometric modeling and meshing of three-dimensional film flow field [14]

This study will apply a type of high-pressure pumps as a sample to establish a physical model of FEM for water desalination lubricated bearings - rotor coupling system, which consists of six-stage impellers, balance wheel, balance disc and the front sliding bearings (B1), the back sliding bearing (B2), the shaft. The back balance wheel is applied as the slide shaft and rotor. The physical assembly model of the coupling system was shown in Figure 1.

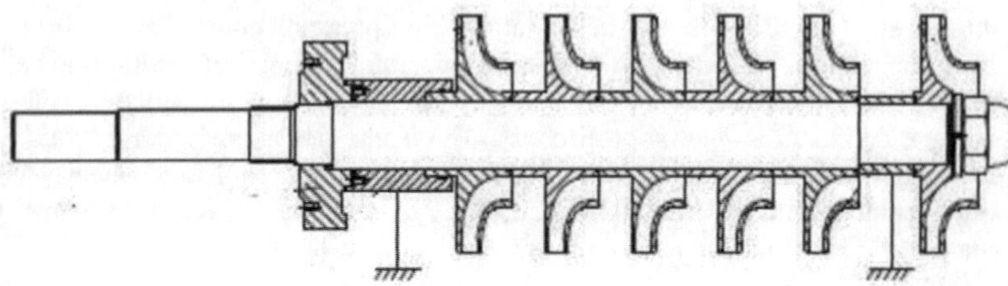

Figure 1. The desalination high-pressure pump's water lubricated bearing-rotor coupling systems

This paper will apply the water lubricated bearing B1 as a sample, of which the main parameters is: bearing width B is 60 mm, bearing diameter D is 90 mm, journal diameter D is 89.7 mm, the radius clearance c is 0.15 mm, the rotate speed is 2980 r/min.

The geometrical modeling will use Geometry of ANSYS ICEM, which establishes a 3D flow field. As shown in Figure 2(a). Compared with the circumferential and axial dimensions, the film thickness is too small. Blocking of ANSYS ICEM is chosen. As shown in Figure 2(b).

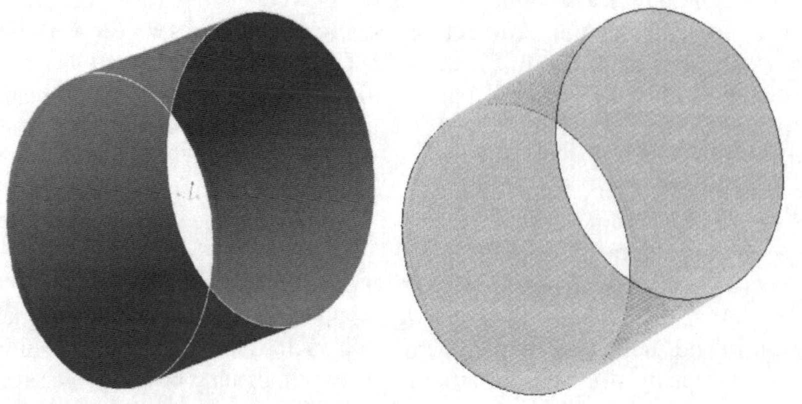

(a) (b)

Figure 2. The geometric modeling and mesh of three-dimensional film flow field

3. UDF control of rigid body motion

The rigid body motion included the translation and the variable speed rotation. This paper will write the UDF through two DEFINE macro provided by the FLUENT. Inc.

3.1 Using DEFINE_ CG_ MOTION macro to realize translation at journal

DEFINE_ CG_ MOTION macro, which mainly used for rigid body motion, is one of the ways to realize the control of dynamic boundary motion. It uses the above theoretical calculation data of discrete axis trace during writing the UDF, and the interval between each discrete point is 0.05s, the total time is 7.85s. Parameters are set individually in each time step to advance the movement of journal gradually. It can be determined as follows: 'omega' is set as 0; 'dtime' is set as 0.05s; the total time is 7.85s; 'vel' in the z axis is set as 0, namely vel [2] = 0, velocity component in the direction of x and y vel [0] and vel [1] can be determined by the ratio of interpolation and the time step of discrete axis trace coordinate(x, y), it assumed that the velocity component in x and y direction is near to a constant in each micro time step. The UDF will conduct translational motion of journal along the known axis after the above the specific set of parameters in each time step.

3.2 Using the DEFINE_PROFILE macro to realize the variable speed rotation at journal

DEFINE_PROFILE macros can define boundary conditions and variables, such as velocity, pressure, temperature. In this study, the main idea is to make the speed of each unit center of the surface on the border of cylinder along their tangent direction of the circumference and contra-rotating, setting different rotating angular velocity within each time step iteration to implement change angular motion in the startup transient phase. Motion moving boundary journal schematic diagram was shown in Figure 3.

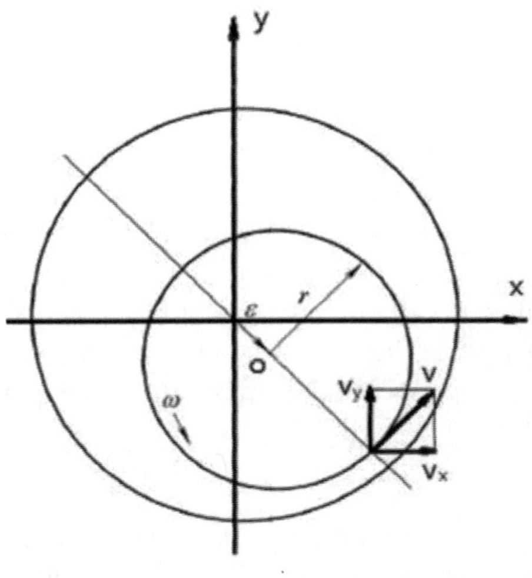

Figure 3. Journal moving boundary motion schematic

It assumes that the coordinates of journal on the moving boundary is center, centrey, center coordinates of either mess plane on moving boundary are x, y, either mess center and the journal on moving boundary center is angle, r, w, time, v_x, v_y. x velocity of tangential linear velocities of rotation in moving boundary are written by DEFINE_PROFILE macros, so as y velocity components. Water, of which viscosity is small, is the working media of water lubricated bearings. There are 3D incompressible turbulence characteristics in water lubricated bearing clearance flow. In this situation,

SST turbulence model, which takes the transmission of turbulent shear stress into account, will be chosen. Defined inlet pressure as 0.6MPa (absolute pressure), outlet pressure as 0.1 MPa and the outer wall as the stationary wall. Besides, inner wall is defined as the velocity boundary condition, compiled UDF, with x, y velocity component of tangential linear velocities of rotation in moving boundary, is selected in the drop-down list.The UDF calculation program chart of x and y velocity component is shown in Figure 4.

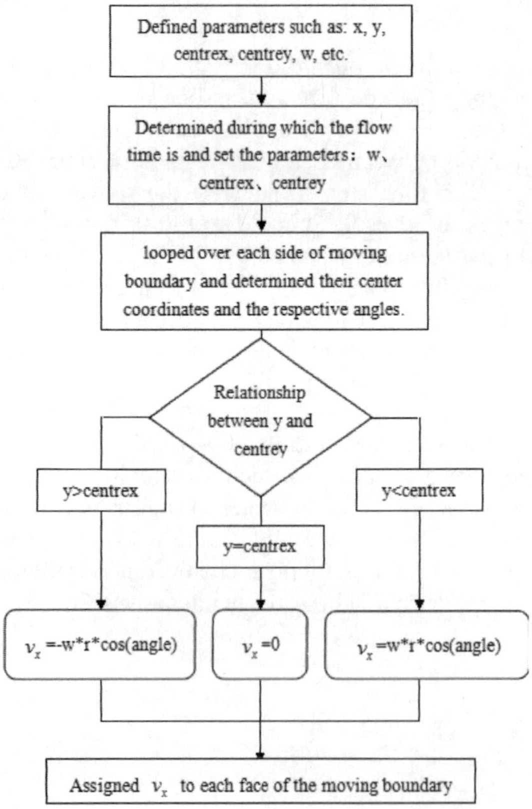

(a) x velocity

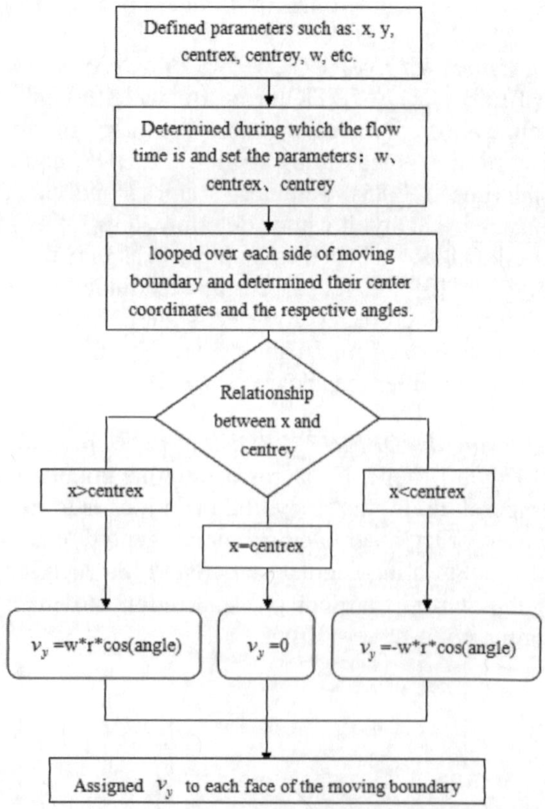

(b) y velocity

Figure 4. x and y velocity component of tangential linear velocities of rotation in moving boundary

Eventually, compiled UDF, with x, y velocity component of tangential linear velocities of rotation in moving boundary, will be loaded in the FLUENT. In addition, internal face should be set as boundary condition, two compiled UDF as x, y velocity, then transient change of angular velocity movement can be realized.

4.Results and discussion

After automatically calculating, a series of 0.05s intervals samples and data can be gotten, then import these into FLUENT for post processing to get the rules of hydrostatic pressure, absolute pressure and velocity distributions in whole startup transient transition process.

4.1 The rules of static pressure distributions in startup transient transition process.

After post-processing, the static pressure distribution of the three-dimensional film flow field of water lubricated bearings in each time of the startup can be gotten. Parts major changes characteristics, which reflect static pressure distribution of the three-dimensional film flow field in the startup, will be listed. Figure 5. shows the rules.

The static pressure distribution rules of three-dimensional flow field in startup phase are summarized below.

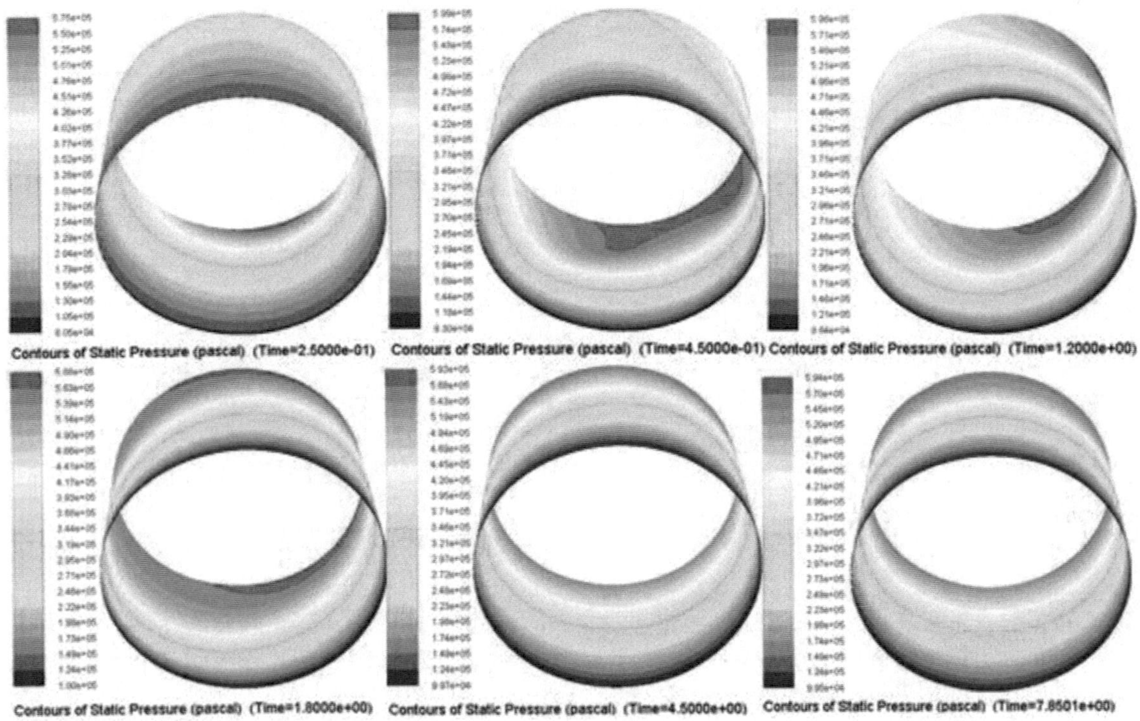

Figure 5. The rules of hydrostatic pressure distributions in startup transient process

4.1.1 The static pressure of the inlet and the outlet. In startup, The static pressure of water lubricated bearing film both in the inlet and the outlet, are slightly lower than stop, and the static pressure were fluctuations during stop.

4.1.2 Variation of axial static pressure distribution. From the axial view, static pressure is gradually decreasing from import to export. When the eccentricity is large, which range from 0 to 1.5s, the minimum thickness of the film at the inlet and outlet differential pressure reaches a maximum value. With the increasing of film thickness on both sides, the differences between inlet and outlet pressure are gradually smaller; when the eccentricity is small, the differences will be significantly reduced.

4.1.3 Variation of circumference static pressure distribution. When the eccentricity is large, In the front of the inner wall, static pressure circumferentially distributed roughly parabolic, and reached its maximum in the minimum thickness, which increases with the film thickness on both sides of the static pressure decreases; In the end part of the inner wall, the static pressure remains constant. When the eccentricity is small, the static pressure nearly remains the same.

4.2 The rules of absolute pressure distributions in startup transient transition process.
Because the inlet and the outlet pressure boundary conditions are absolute pressure, variation of absolute pressure distribution is the same as static pressure.

4.3 The rules of velocity distributions in whole startup transient transition process.

4.3.1 Radial velocity distribution of the inlet and the outlet. To illustrate the main features of the radial velocity distribution changes, the radial velocity distribution in 0.4s and 2s is being taken. As shown in Figure 6.

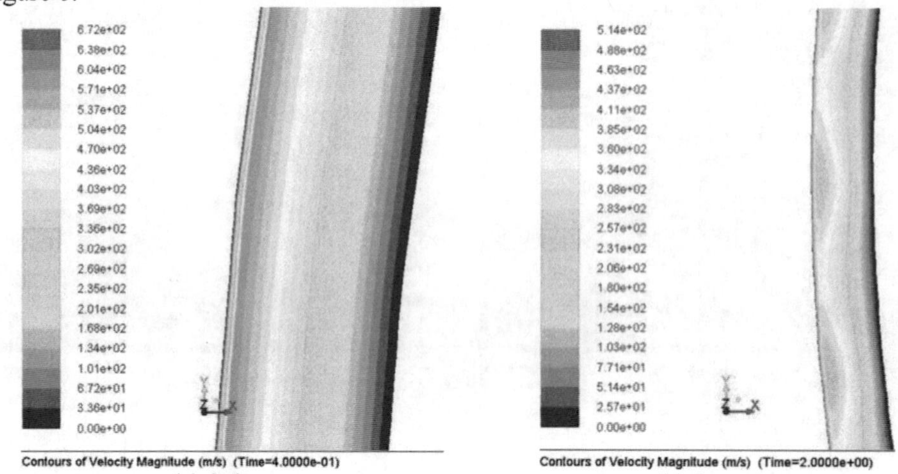

(a) 0.4s (b) 2s

Figure 6. The main features of the radial velocity distribution in startup phase

4.3.1.1 Radial velocity distribution of inlet

As shown in Figure 6(a), the speed radial from the inner wall to the outer wall increases first and then decreasing, the increasing rate is less than the decreasing rate. Due to the viscosity of water is small, the impact of the rotation around its journal circumferential velocity of the fluid is small, so the velocity distribution above may be differential pressure between the inlet and the outlet causing a greater axial flow, which is somewhat similar to the two-dimensional parabolic Poiseuille flow distribution [15] speed.

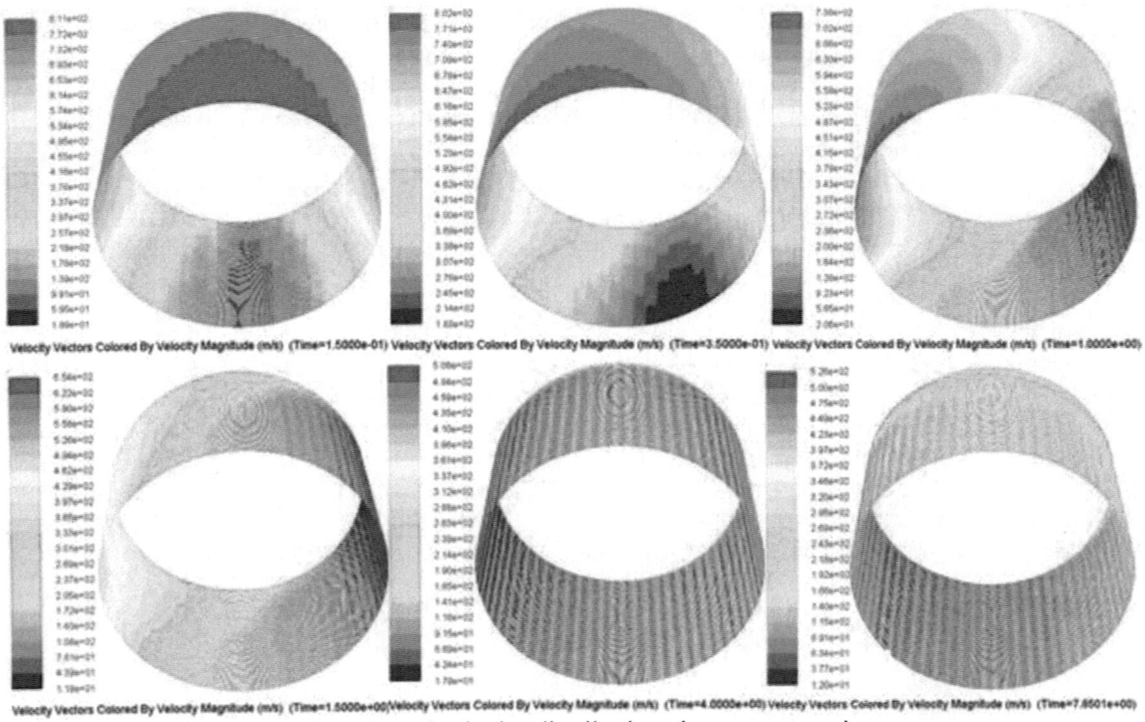

Figure 7. The rules of velocity distributions in startup transient process

4.3.1.2 Radial velocity distribution of the outlet

In the beginning of the startup, the radial velocity distribution of the outlet is similar to the inlet, but the speed is smaller than the inlet. then, the outer wall will appear high-speed area alternately, as shown in Figure 6(b).

To summarize the variation of axial and circumferential velocity distribution, the main distribution will be shown in Figure 7.

4.3.2 The rules of circumferential velocity distribution

When the eccentricity is large: The velocity distribution of circumferential velocity is substantially "concave" parabolic, and reached a minimum at the minimum thickness of the film. However, if not, it presents as a wave with little amplitude.

5. Conclusion

3D dynamic distribution of the film flow field of water lubricated bearing of high-pressure desalination pump in start transients is as follows:

1) When the eccentricity is large, film thickness was negatively correlated with the pressure, and positive with the velocity. Differential pressure was negatively correlated with velocity. When the eccentricity is small, film thickness is no significant relationship with differential pressure and velocity. Differential pressure has little difference with velocity.

2) It will dispose the results of numerical simulation and get the pressure and velocity distribution of 3d unsteady liquid film flow field in the startup phase, it provides certain numerical reference for revealing the development of water lubrication status and power transmission mechanism in the boot process.

6. Reference

[1] Hu J N, Xiao X P, Zhou S G, et al. 2009 Optimal design of high pressure pump for 10000tons reverse osmosis desalination system. *Journal of Drainage and Irrigation Machinery Engin*, **27(1)**: 25-29.

[2] Ye X Y,Xu G L, Hu J N. 2008 Optimization selection of reverse osmosis desalination high-pressure pump. *Technology of Water Treatment*, 79-81+91.

[3] Wen S Z, Huang P. 2008 Principles of Tribology. (Beijing:Tsinghua University Press).

[4] Wen S Z . 2007 Progress of research on nanotribology. *Journal of Mechanical Engineering*, **10(43)**:1—8.(in Chinese).

[5] Ye X Y,Yang S S,Xu J Q et al. 2008 Optimi-zation design for hydraulic end of high-pressure pump.*Drainage and Irrigation Machinery*, **26(6)** :31-35. (in Chinese)

[6] Baumgarten S.2005 Hydraulic comparison of high pressurepumps for R O desalination system.*World Pumps*, **(463)** :52-55.

[7] Jiang X P, Shi W D, Li W, et al. 2012 Dynamic stability analysis on water lubricated bearing rotor system of high pressure multistage pump. *Transactions of the Chinese Society of Agricultural Engineering (Transactions of the CSAE)*,**28(22)**:45-50.(in Chinese with English abstract)

[8] Fan X,Zhang W.2010 Design of reverse osmosis system in coal-fired power plant.*Technology of Water Treatment*,**36(6)**:90-94.(in Chinese with English abstract)

[9] Al-Kharabsheh S. 2006 An innovative reverse osmosis desalination system using hydrostatic pressure. *Desalination*, **196 (1)**: 210—214.

[10] Liang C H, Jiang H, Xing H H. 2007 Development and application of reverse-osmosis seawater desalination technology. *Journal of Naval Aeronautical Engineering Institute*, **22(4)**;494—496.(in Chinese with English abstract)

[11] Zhang P, Hu J T, Li W, et al. 2013 Research progress of brackish water desalination by reverse osmosis. *Journal of Water Resource and Protection*, **5(3)**: 304—309.

[12] Voros N G ,Maroulis Z B,Marinos-Kouris D. 1997 Short-cut structural design of reverse osmosis desalination plants. *Journal of Membrane Science*, **127(1)**: 47—68.

[13] Hu J N, Xue Y, Zhang D S, et al. 2015 Research on transient axis trajectory of high-pressure multistage centrifugal pump on starting. *Transactions of the Chinese Society of Agricultural Engineering (Transactions of the CSAE)*, **31(Supp.1)**: 61—70. (in Chinese with English abstract)

[14] Ji B B, Chen J P. 2012 ANSYS ICEM CFD Detailed examples of meshing technology. (Bei Jing:China Water&Power Press).

[15] Anna K, Stella P, Georgios C. G, 2011 Incompressible Poiseuille flows of Newtonian liquids with a pressure-dependent viscosity, *Journal of Non-Newtonian Fluid Mechanics*,**Vol.166**, Issues 7–8, ISSN 0377-0257.

IOP Conf. Series: Materials Science and Engineering **129** (2016) 012057 doi:10.1088/1757-899X/129/1/012057

Development of a model counter-rotating type horizontal-axis tidal turbine

B Huang*, **K Yoshida, T Kanemoto**

Faculty of Engineering, Kyushu Institute of Technology

Sensui 1-1, Tobata, Kitakyushu 804-8550, Japan

huang.bin404@mail.kyutech.jp

Abstract. In the past decade, the tidal energies have caused worldwide concern as it can provide regular and predictable renewable energy resource for power generation. The majority of technologies for exploiting the tidal stream energy are based on the concept of the horizontal axis tidal turbine (HATT). A unique counter-rotating type HATT was proposed in the present work. The original blade profiles were designed according to the developed blade element momentum theory (BEMT). CFD simulations and experimental tests were adopted to the performance of the model counter-rotating type HATT. The experimental data provides an evidence of validation of the CFD model. Further optimization of the blade profiles was also carried out based on the CFD results.

1. Introduction

Ocean energy resources have attracted great attention to get sustainable society with accompanying hydro, wind and solar resources, and then many kinds of the power unit have been provided for the tidal stream [1]-[4].

The unique tidal stream power unit called "Counter-Rotating Type Tidal Stream Power Unit" has been proposed in the previous studies [5, 6]. As shown in Fig.1, tandem propellers counter-drive double rotational armatures respectively, in keeping the rotational torque counter-balance between both propellers/armatures. The counter-rotating type tidal stream power unit has fruitful advantages as follows: (1)The tandem propellers may be easy to suppress the cavitation rather than the single propeller, because the individual rotational speed can be reduced by about half while the armature diameter and the induced voltage are the same as those of the traditional power unit with single propeller. (2)The individual rotational speed can be designed freely in response to the tidal circumstances. (3)The rotational moment hardly acts on the casing and/or the pile because the rotational torque between both propellers/armatures is counter-balanced in the unit. That is, the unit can be mounted easily on the mono-pile. (4)The flow has scarcely the swirling component and the disturbed wave downstream of the unit, because the angular momentum change, namely the rotational torque, through the front propeller is almost the same as that through the rear propeller. That is, the flow discharges in the axial direction while the tidal stream has no swirling component.

In the present work, the front and rear propellers of counter-rotating type tidal stream power unit were designed according to BEMT. And then CFD simulations and experimental tests were adopted to the performance of the model counter-rotating type HATT.

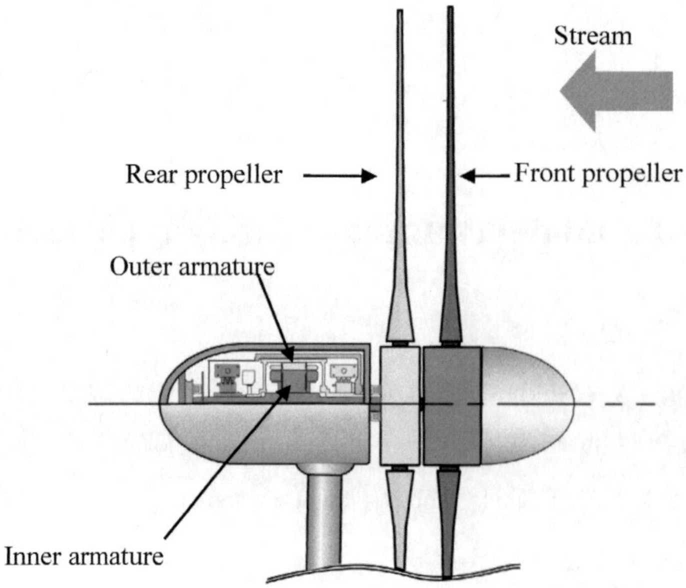

Fig.1 Concept of counter-rotating type tidal stream power unit (upstream type)

2. Turbine design

A model counter-rotating type HATT with the diameter of 500 mm was designed as a trade-off between maximizing the Reynolds number and not incurring excessive tunnel blockage correction [7, 8]. The blade profiles were composed of the blade element KIT001 with three different relative thicknesses as shown in Fig. 2. The blade profiles of the model counter-rotating type HATT designed by BEMT are illustrated in Fig.3.

The diameter of the front propeller is d_F = 500mm and the rear propeller is d_R = 420 mm, namely the diameter ratio [$=d_R/d_F$] is 0.84. The number of blades for upstream propeller and downstream propeller is Z_F = 3 and Z_R = 5, respectively. The axial distance between the front and the rear propellers is set to 80mm, and the diameter of the hub is set to 90 mm.

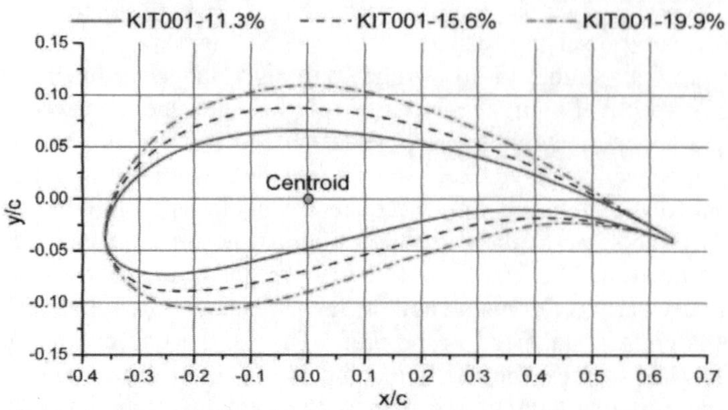

Fig.2 Blade element KIT001 with three different relative thicknesses

IOP Conf. Series: Materials Science and Engineering **129** (2016) 012057 doi:10.1088/1757-899X/129/1/012057

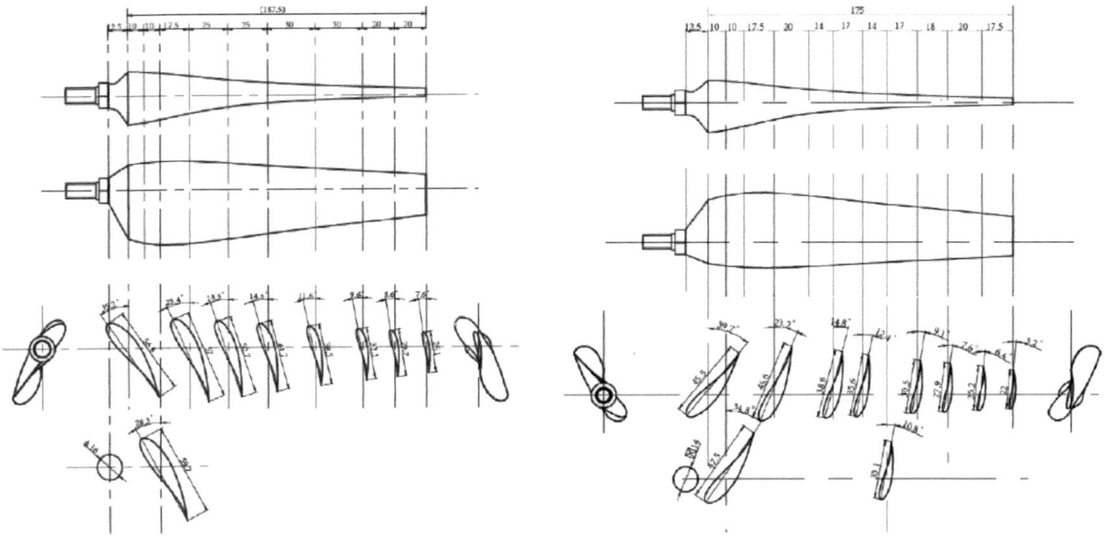

(a) Front blade (b) Rear blade

Fig.3 Blade profiles of model turbine

3. Performance obtained by CFD and experimental tests

In order to obtain the hydrodynamic performance of the model counter-rotating type HATT designed in the previous section, CFD simulations and experimental tests were carried out in this section. The detailed introduction of CFD simulations can refer to the published works [5, 6]. And at this stage the experimental tests were conducted in the wind tunnel in Kyushu Institute of Technology as shown in Fig.4.

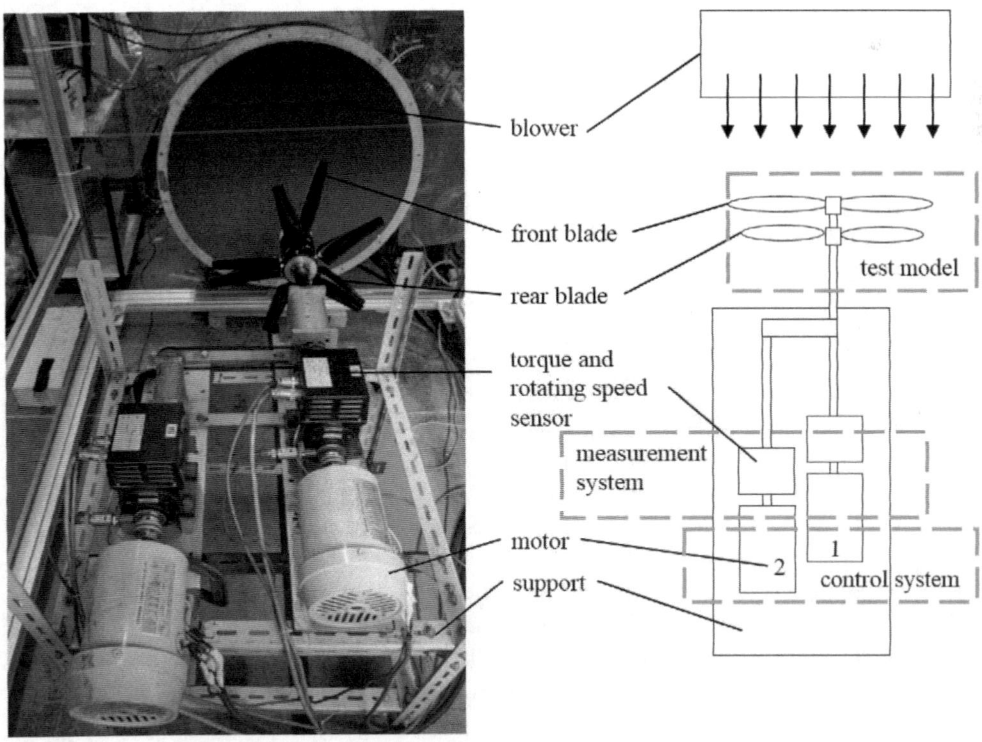

Fig.4 Schematic diagram of test rig in a wind tunnel

Fig.5 shows the performance of the designed model counter-rotating type HATT in terms of power coefficient. It can be seen that the trend of the power coefficient shows good agreement over a range of tip speed ratios. However, the CFD over predicts the power coefficient in the high tip speed ratio region. It should be noted that in the wind tunnel tests, there is no blockage effect and lower Reynolds number than that in the water tunnel. These differences lead to the deviation between CFD predictions and experimental results.

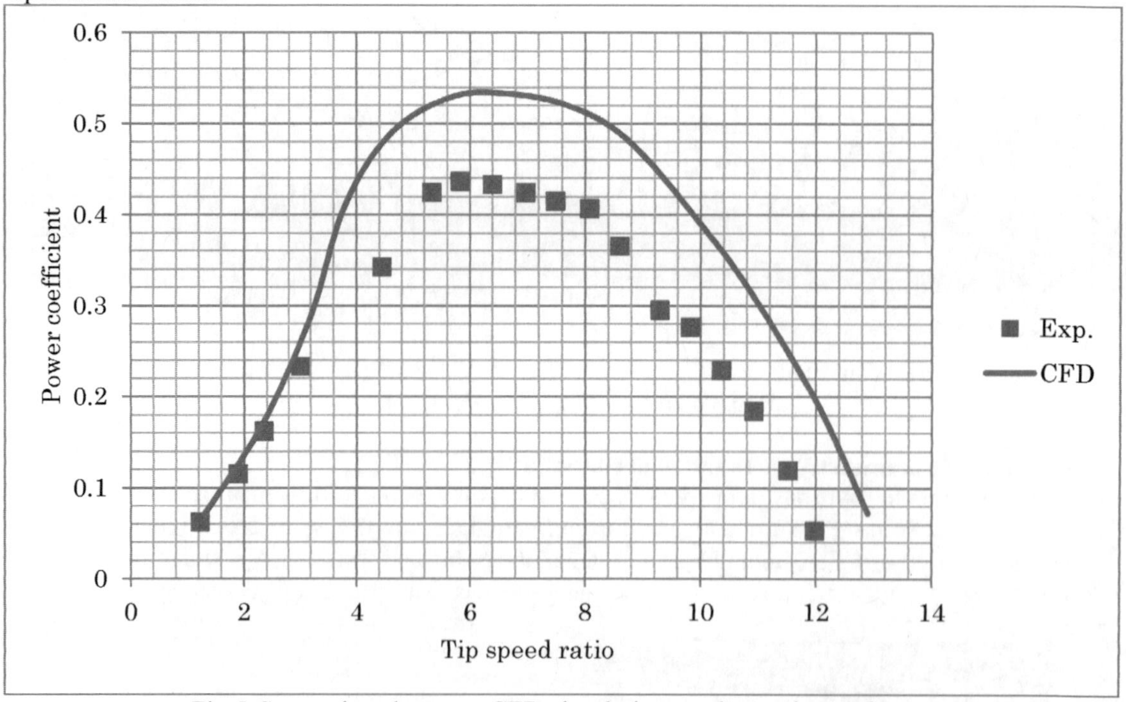

Fig.5 Comparison between CFD simulations and experimental tests

4. Effects of Reynolds number

Consideration was tried about the way to presume the output coefficient of the actual machine size from model test outcome. When converting from model test to actual performance, it is necessary to know the difference from the similarity law (size effect) due to the fact that the Reynolds number is different. The efficient conversion formula to predict efficient difference between a model and an actual machine has been announced from the old days. Among these the next Moody 1/5th power formula obtained empirically to a water turbine has been also used much to a pump as well as a water turbine about the total efficiency.

$$\eta = 1 - (1 - \eta')\,(D/D')^{-1/5} \qquad (4\text{-}1)$$

Here η: The total efficiency and the D: representative size and the symbol ' indicates a model.

Formula (4-1) can be transformed into following equation.

$$(1-\eta)/(1-\eta') = (Re/Re')^{-1/5} \qquad (4\text{-}2)$$

Here -1/5 of the index corresponds to a relation between friction coefficient λ on a flat smooth board and Reynolds number Re ($\lambda \propto$ Re -1/5).

Based on the above, we examined the effect of Reynolds number based on the flow analysis around the two-dimensional blade profile. For a two-dimensional blade profile, we designed the camber blade profile (Basic thickness form: NACA65-010, Mean line: NACA Mean Line 65, Chord length: 50mm, Trailing edge radius: 0.5mm) to show in figure 4-1. With analysis code, ANSYS 14.5 was used .

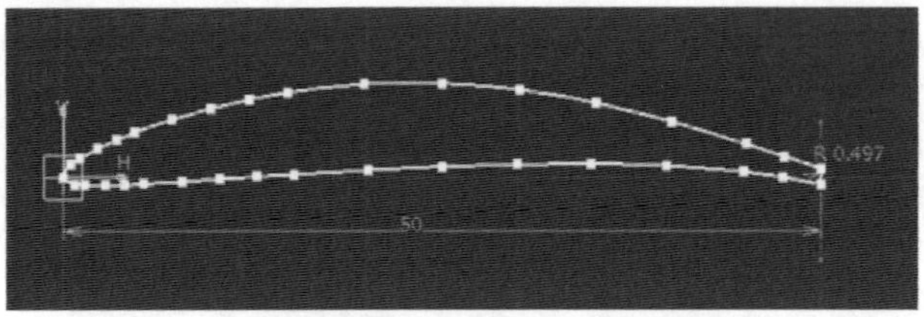

Fig.6 Blade profile used for analysis

The analysis condition did the attack angle (α=0~12 deg) and Reynolds number (Re=10^4~10^7). Re=10^4 ~ 10^5 was equivalent to the former model test range [9], and Re=10^6 ~ 10^7 assumed a future actual power unit (Propeller diameter is on the order of 10m).

An analysis result of lift coefficient Cl, drag coefficient Cd, and pressure coefficient Cp=ΔP/ (0.5 ρ v^2) is indicated in Fig.7.

As shown in Fig.7(a), Cl of Re=10^4 is the small value extremely compared with other Re . According to the analysis result of the streamline, in the case of Re = 10^4, separation is seen near trailing edge in an analysis range of α.But this separated region greatly decreases if it becomes more than Re=10^5. Further, when it'll be the actual power unit size level of Re=10^6~10^7, the curves are overlap each other, and influence of Reynolds number doesn't show.

As shown in Fig.7(b) and (c), Cd and Cp of Re=10^4 is the big value extremely compared with other Re. Cp also shows the same trend as Cd, the decrease amount becomes smaller as the Reynolds number increases. It has a tendency to approach saturation.

Here, we consider the relation between above-mentioned CFD result and Moody 1/5th power formula. The left-hand side of (1-η)/(1-η') in equation (4-2) represents a loss ratio of actual and model. So this parameter is considered to be physically associated with pressure coefficient ratio Cp/Cp' of actual and model. We show a rearranging result in Fig.7(d). In addition, Cp/Cp' varies a little according to an attack angle, but the tendency is almost the same.

By the two dimensional blade profile which simplified that effects of the centrifugal force are not considered at all, the CFD result is similar of Moody 1/5th power formula as shown in Fig.7 (d). From this ,efficiency improvement by the scale effect of the future actual power unit (Re = 10^6~10^7) is expected.

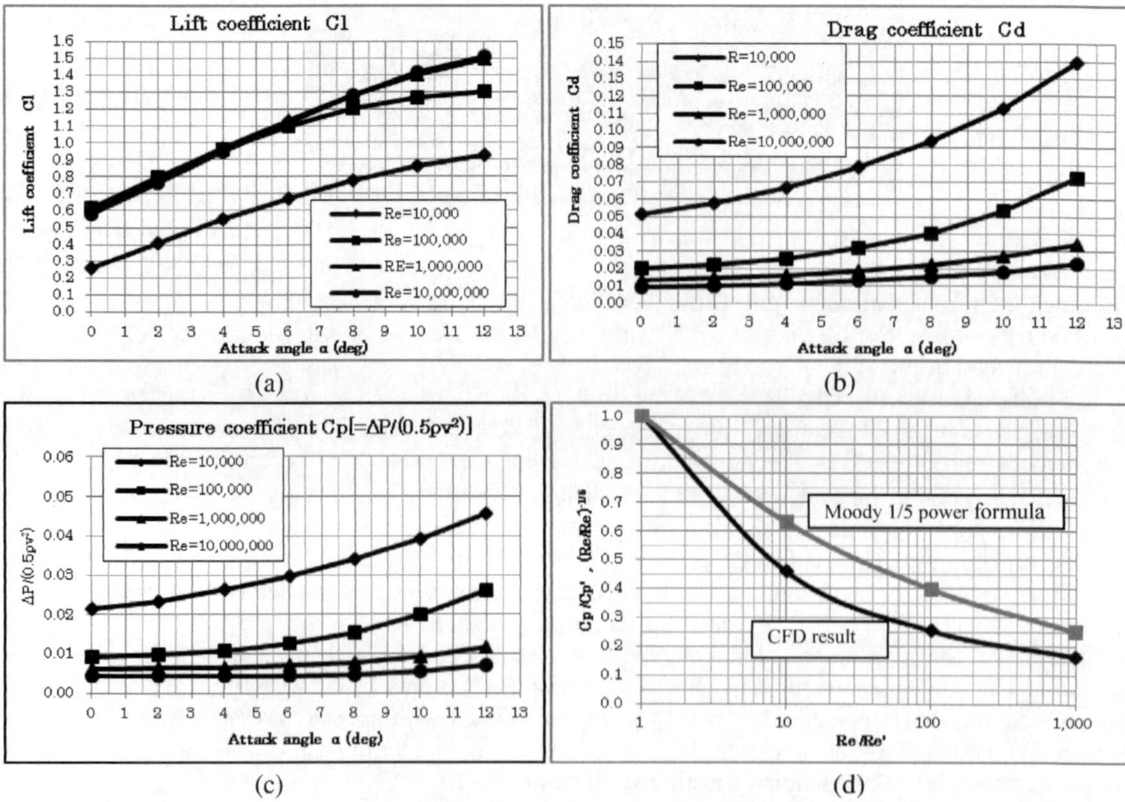

Fig.7 Results of CFD analysis

5. Conclusions

In the present study, a unique counter-rotating type HATT was proposed and the blade profiles were designed based on the BEMT. Both of CFD simulations and experimental tests in the wind tunnel have been carried out to verify our design. Effects of Reynolds number have also been conducted on a 2D hydrofoil, which provides a reference for the efficiency conversion from model to prototype turbine.

References

[1] Fraenkel, P.L., 2007 Marine current turbines: pioneering the development of marine kinetic energy converters, *Journal of Power and Energy*, **Vol. 221**, No. 2, pp. 159-169.

[2] Bahaj, A.S., Molland, A.F., Chaplin, J.R. and Batten, W.M.J., 2007 Power and thrust measurements of marine current turbines under various hydrodynamic flow conditions in a cavitation tunnel and a towing tank, *Renewable Energy*, **32**, pp. 407-426.

[3] Clarke, J., Connor, G., Grant, A., Johnstone, C., Ordonez-Sanchez, S., 2008 Contra-rotating marine current turbines: performance in field trials and power train developments, *Proceedings of 10th World Renewable Energy Congress*, (Glasgow, Scotland, UK).

[4] Hiraki, K., Wakita, R., Kanemoto, T., 2012 Demonstrative power generation by twin-runner darrieus turbine in Kanmon strait, *Proceedings of the 22nd International Offshore and Polar Engineering Conference*, (Rhodes, Greece), pp.725-729.

[5] Huang, B., Kanemoto, T., 2015 Multi-objective numerical optimization of the front blade pitch angle distribution in a counter-rotating type horizontal-axis tidal turbine, *Renewable Energy*, **81**, pp.837-844.

[6] Huang, B., Kanemoto, T., 2015 Performance and internal flow of a counter-rotating type tidal stream turbine, *Journal of Thermal Science*, **24(5)**, pp.1-7.

[7] Bahaj, A.S., Molland, A.F., Chaplin, J.R., Batten, W.M.J., 2007 Power and thrust measurements of marine current turbines under various hydrodynamic flow conditions in a cavitation tunnel and a towing tank, *Renewable Energy*, **32(3)**, pp.407-426.

[8] Bahaj, A.S., Batten, W.M.J., McCannb, G., 2007 Experimental verifications of numerical predictions for the hydrodynamic performance of horizontal axis marine current turbines, *Renewable Energy*, **32(15)**, pp.2479-2490.

[9] Usui, Y., Kanemoto, T.,Takaki, k. and Hiraki,K., 2014 Counter-rotating type tidal stream power unit mounted on a mono-pile, *Journal of Energy and Power Engineering*, pp.1748-1755.

IOP Conf. Series: Materials Science and Engineering **129** (2016) 012058 doi:10.1088/1757-899X/129/1/012058

Study on the influence of back blade shape on the wear characteristics of centrifugal slurry pump

X Cai, S P Zhou and S Li

Key Laboratory of Pressure Systems and Safety of Ministry of Education, East China University of Science and Technology
NO.130 Meilong Road, Shanghai, 200237, China

E-mail: shpzhou@ecust.edu.cn

Abstract. CFX particle inhomogeneous model was introduced for the mechanism analysis of a centrifugal slurry pump which is equipped with back blades on impeller shrouds. Combining with the total efficiency correction, the simulation showed good prediction accuracy of external characteristics results compared with the experimental values. Vorticity and Q-Criterion were chosen as the variables to illustrate the abrasion morphology and wear mechanism by contrasting simulation result with worn impeller in engineering. The analysis showed that the large vorticity intensity areas are distributed at the edge of impeller shroud and intensively behind the back blades. Moreover, the vorticity scattered on suction surface of back blade shows the largest intensity. The contour of Q-Criterion demonstrated that the swirl scale in front cavity is obviously larger than that in back cavity. The distribution of vorticity on both front and back shrouds can reasonably explain the impeller wear characteristics. Finally, the forward curved back blade proved to be excellence performance in vorticity distribution.

1. Introduction

Centrifugal slurry pump is one of the most important fluid machines, which has been widely used in industries such as mine, coal and power [1-2]. Generally, the shroud of centrifugal slurry pump impeller is equipped with the back blades that rotate synchronously with the impeller, and the back blades can prevent the medium of solid and liquid being leaked through the cavity clearance in pump body for the high pressure around the outlet of impeller. Moreover, the back blades also adjust the axial force by increasing the movement speed of mediums in the clearances. However, in the engineering application, the impeller blades and the back blades trend to be worn even invalid due to the influence of hard solid particles with complex multiple flows. Therefore, it is of great importance to investigate the effect of back blade shape on the wear characteristics of centrifugal slurry pump. In the past few years, it has been studied by simulation and experiment research. Wang [3] investigated the wear characteristics of centrifugal slurry pump by analyzing the solid volume fraction, solid phase slip velocity and the motion laws of different diameters of solid particles. Wu [4] adopted the discrete phase model to study the effect of rotation speed, the diameters of solid particles and the parameters of impeller on the features of the particle erosion behavior. Tao [5] used the multiphase model to simulate the abrasion conditions of the impeller in a semi-open centrifugal pump, and the simulation results were also confirmed by the rapid wear test.

All the representative researches mentioned above utilized the conventional variables to characterize the abrasion on impeller blades and passages, but there are not many articles that discuss the back blade on shrouds. Therefore, in this paper, numerical simulation for the slurry pump with the radial back blade is conducted by ANSYS CFX 14.0. The medium is the mixture of water and solid particles, whose volume fraction is 10%. This study analyzes the wear characteristics on impeller back blades and the improvement project is also put forward to ameliorate the abrasion.

2. Physical model

2.1 Slurry pump model and the main parameters
The main parameters of slurry pump model are presented in table 1.

Table 1. Design parameters of slurry pump.

Parameters	Q_d	H_d	η	Shaft power P_e	Rotate speed n	Specific speed
	500 m³/h	72 m	65.2%	150.2 kW	1450 r/min	81.5
	Outlet diameter D_2	Outlet width b_2	Number of impeller blade	Number of back blades on front shroud	Number of back blades on back shroud	Back blade width
	515 mm	28 mm	5	16	8	12 mm

The structure diagram of centrifugal slurry pump channel is shown in figure 1, the clearance distance between the top of back blade and the pump body is 2mm. The impeller inlet channel is sealed up by angular contact seal, and labyrinth seal is designed in the place closed to the impeller inlet to prevent the mediums from being leaked out. The back blade of the original model is radial type ($\beta_2' = 90^o$, and β_2' represents the outlet angle of back blade) , as shown in figure 2.

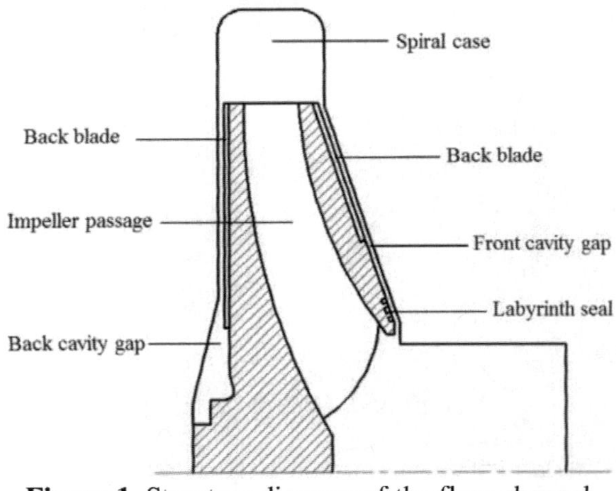

Figure 1. Structure diagram of the flow channel.

2.2 Calculation model
All the pump components are modeled by Solidworks, which are based on the 2D engineering drawings. The 3D models include: inlet section, inlet liner, impeller, volute and shroud liner. T the whole domain is built by the process of assembling and Boolean operation. To set the boundary conditions explicitly, the whole domain is divided into four parts: inlet, cavity, impeller and volute. All the full flow channels of the slurry pump are shown in figure 3.

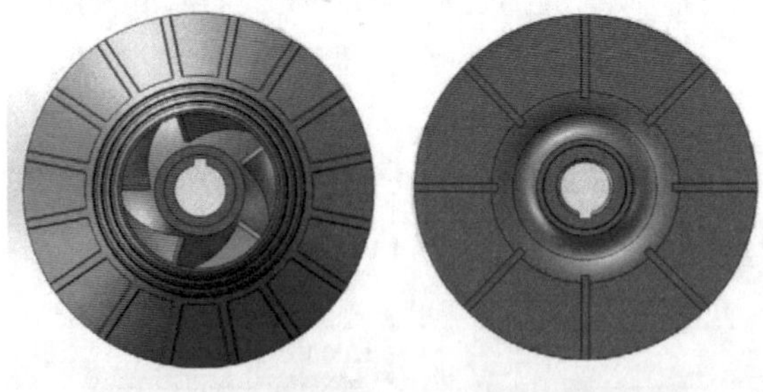

Front shroud Back shroud
Figure 2. The prototype impeller model.

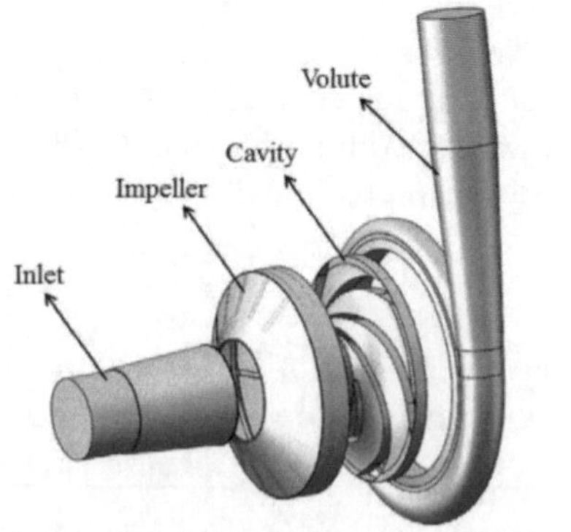

Figure 3. The whole machine model.

2.3 Grid division

As the twisted spatial structure of computational domains are rather complex, the unstructured tetrahedron grids for all the models are generated by ICEM, then the verification for grid independence is also carried out. When the total number of the grids is more than 2812520, the performance parameters are stabilized, and the relative errors of head and efficiency are respectively 0.68% and 0.78% when the total number is 3655673. Therefore, conclusion can be drawn that the quality of grid and the precision of simulation can satisfy the requirements when the total number of divisor grids is about 2.8 millions. The grid models are shown in figure 4.

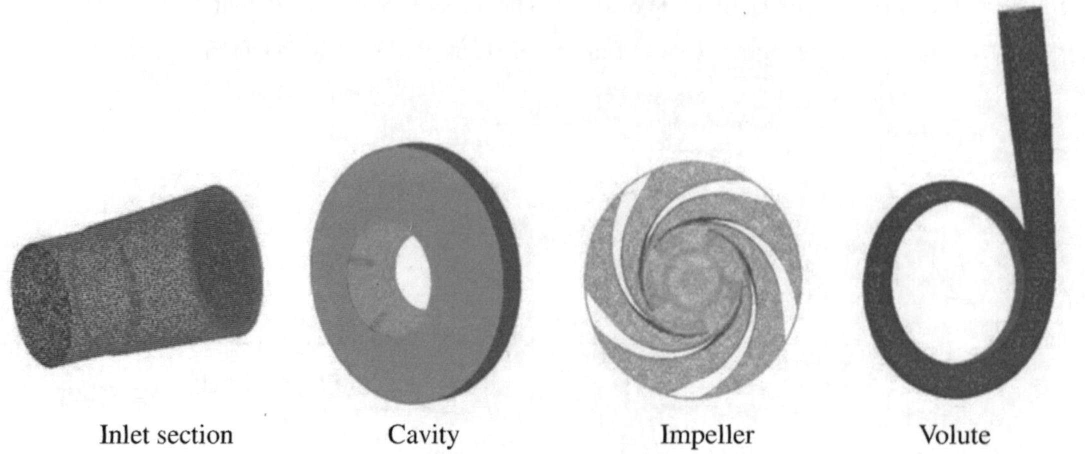

Inlet section Cavity Impeller Volute
Figure 4. Grid meshing.

3. Numerical simulation method

The volume fraction of solid particle in the mixture is 10% and another phase is water. The solid particles are set as the spherical pulverized coal which assumed to be spherical, whose diameter d=0.2 mm, density ρ=1350 kg/m^3. Considering the complex flow conditions such as streamline curvature and whirlpool in the pump, the continuous phase adopts the RNG k-ε turbulence model, while the discrete phase adopts the zero equation model. Particle model is defined in phase transport model, and drag coefficient adopts the Schiller Naumann model which is suitable for the spherical particles. The interphase turbulent diffusion is Favre Averaged Drag model and turbulent diffusion coefficient is 0.9.

Velocity inlet is used as the boundary condition for the inlet. The Opening Pressure and Direction is chosen as the boundary condition for the volute outlet, where the static pressure is given by the theoretical head. The gradient of solid volume fraction on the outlet is 0. The interfaces between the static and motive areas are set by "Transient Rotor Stator" model in transient calculation.

The initial value of transient calculation comes from the result of steady calculation. The physical timescale $\Delta t=1/n$ = 0.000689655s (n represents the rotate speed of impeller), and impeller rotates for 3 circles totally. High resolution is chosen for the transient solution model, which ensures the accurate calculate process. Monitor points set on the inlet and outlet surfaces are used to judge the level of computation convergence. The max residual of transient calculation is 0.00001.

4. Simulation results and analysis

4.1 Model validation

Test stand was set up in certain fluid machinery Co., Ltd. in Shanghai to perform the pure water performance experiment for the prototype model. Test standard is referred to the standard of centrifugal pump, francis pump, axial pump and vortex pump (GB3216-1989). Simulation and experimental results are shown in figure 5 and figure 6.

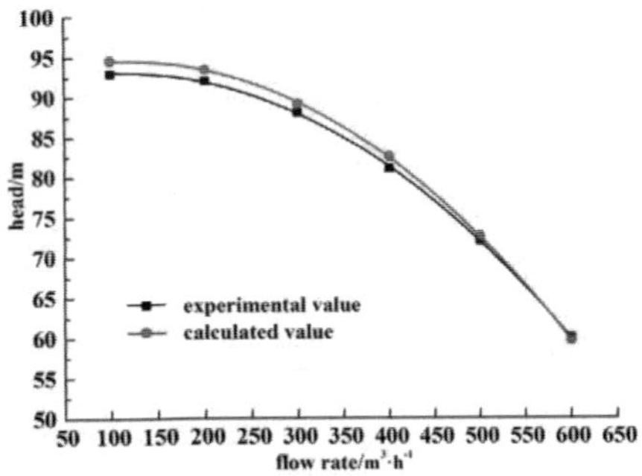

Figure 5. H-Q curve.

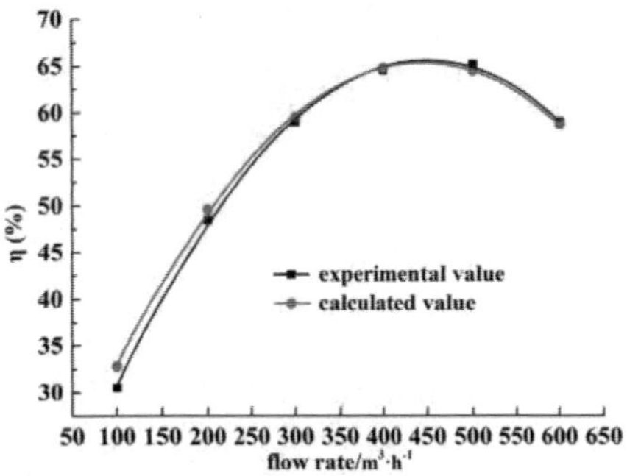

Figure 6. η-Q curve.

The formula of efficiency:

$$\eta = \frac{1}{\frac{1}{\eta_v \eta_h} + \frac{\Delta P_d}{P_e} + 0.03} \qquad (1)$$

Where, ΔP_d represents the disk friction power loss, Kw. The calculation method of ΔP_d can be referred by related references [6]. $\eta_v \eta_h$ and P_e can be calculated by simulation.

The trend of simulated head value changes consistent with the experimental value and the average relative error is 1.28%. The total efficiency is corrected by formula (3-1), which takes the hydraulic efficiency, volumetric efficiency and mechanical efficiency into consideration. The average relative error of corrected efficiency is 1.93%. The available comparison above demonstrates that physical model and numerical simulation method applied in the research are reliable.

4.2 Wear analysis of impeller back blades

CFD-Post is utilized to analyze the calculation results. Vorticity distribution contour of solid phase is illustrated to analyze the wear situation on shroud and back blade. Vorticity is one of the most important parameters to describe the feature of eddy, which can directly characterize the swirl strength in the flow space. Solid particle is a kind of high hardness, strong abrasive medium. When the solid particles flow fast over the surface of flow components, especially for the impeller shrouds and back blades, they will lead to severe erosion and abrasion. According to the vorticity contour showed in figure 7, it can be observed that most of the vorticity intensity emerge around the edge of shroud and they are comparatively concentrated behind the back blades, which are closed to the suction surface of back blades. Although the vorticity intensity are slightly different in their distribution, location and magnitude, obvious laws are still visible on the whole. The impeller dismantled from centrifugal pump unit in overhaul is shown in figure 8. From overview of the wear appearances on shroud, we find that long spiral-fluted and cured eyebrow grooves are consistently observed on each back blade passage, and they just look like the hooks along the peripheral direction approach to outlet edge, but carved into the shroud. Along the radical direction, the back blades are worn severely with the increase of impeller radius and the height of back blade becomes flat gradually, so we can deduce that the particle moment velocity is raised with the increase of radial distance, and wear extent is positive correlated with the particle velocity. It is obvious that the worn areas display bright metallic luster, which is resulted from the cutting action of the high speed particle on impeller shroud.

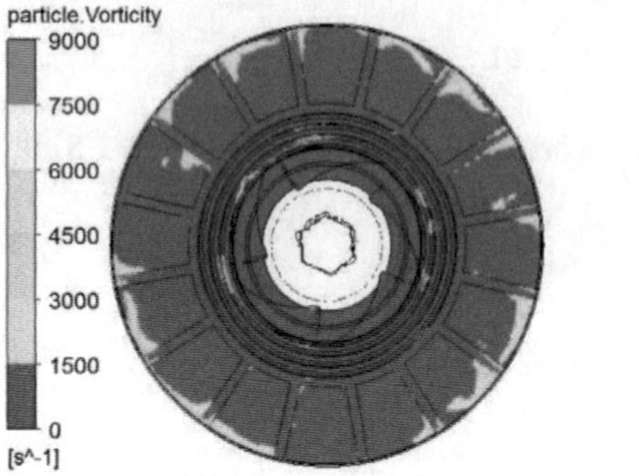

Figure 7. Vorticity distribution on front shroud.

Figure 8. Wear morphology on front shroud.

It can also be observed from figure 9 that there are some inconsistent areas of strong vorticity intensity on the suction surface of back blades (small range of yellow area), while continuous vorticity intensity are unlikely to be emerged on the pressure surface of back blades. The severe worn back blade in engineering is showed in figure 10. The outlet circumference edge is skived to be flake. The actual worn morphology in different back blades passages seems different and there are two or three grooves on most of the back blades. Furthermore, Q-Criterion is also defined as a variable to analyze the instantaneous vortex structure in three dimensional spaces, as is showed in figure 11. The closed envelopes in front cavity gap are obvious larger than that in back cavity, which indicates that the vortex scale is larger in the front cavity gap. Based on the worn morphology combined with simulation results, it can be deduced that the abrasion happened around the shrouds and back blades are caused by the strong vortex flow with high hardness solid particle medium.

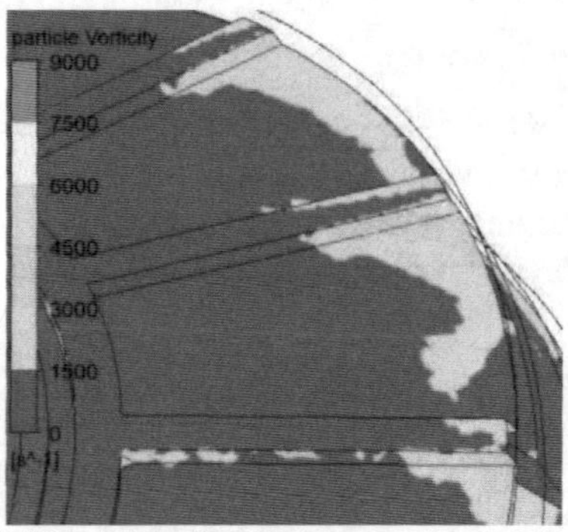

Figure 9. Vorticity distribution on back blade (front shroud).

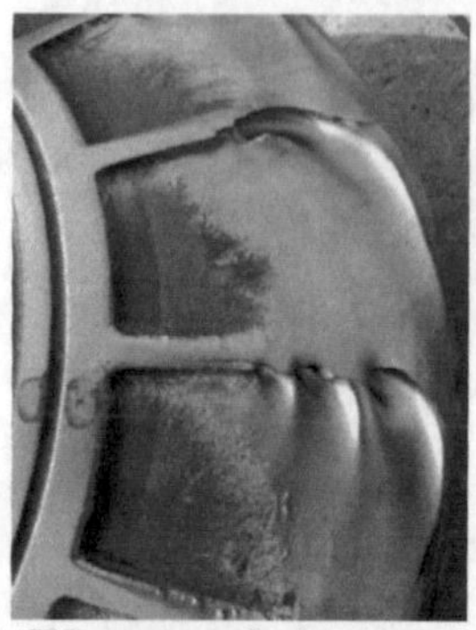

Figure 10. Wear morphology on back blade (front shroud).

From figure 12 to 15, it is obvious that the back shroud and its back blades are seriously worn than that of the front shroud. The number of blade on back shroud is less than that of the front shroud, and this means that heavier load must be burdened for each back blade on the back shroud, which will increases the probability for the solid particle colliding to back blade surfaces. The actual abrasion circumstances show that the back blade passages have been linked into the impeller passages. However, any impeller structure out of work must be rejected in centrifugal pump. Differ from the bright metallic luster displayed on the front shroud, corrugated or scaly morphology are the main surface features on the back shroud. This comes from the reason that the swirl scale on back shroud is smaller and more dispersed. In general, the abrasion characteristics on back shroud are similar to the front shroud. The analysis above can reflect and illustrate the impeller abrasion. Moreover, the structure integrity of impeller will also greatly affect the imbalance and vibration of the rotor. Therefore, structure optimization and anti-wear performance are urgently studied in the next.

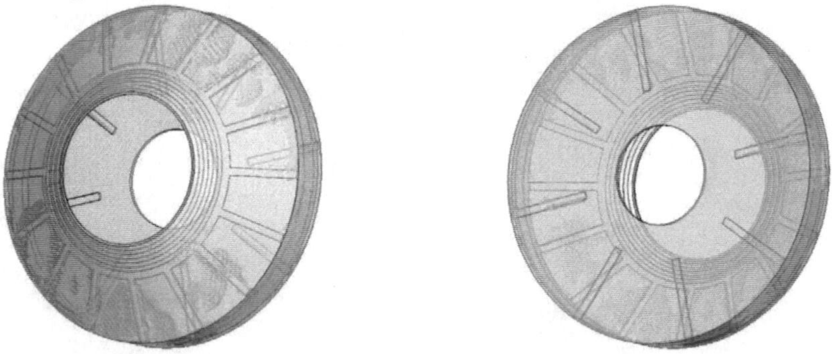

Front cavity gap Back cavity gap

Figure11. Contour of Q-Criterion.

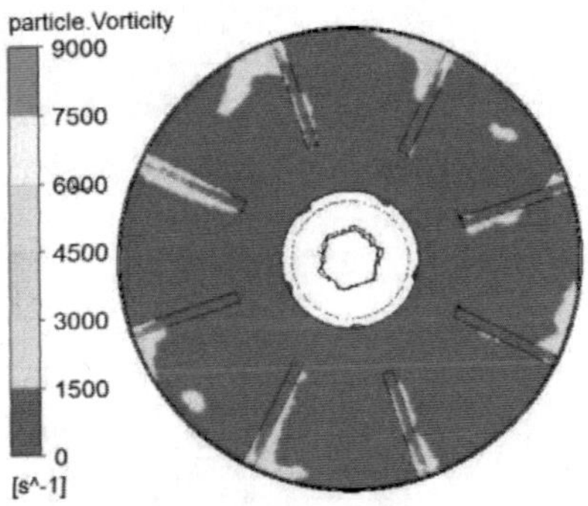

Figure 12. Vorticity distribution on back shroud.

Figure 13. Wear morphology on back shroud.

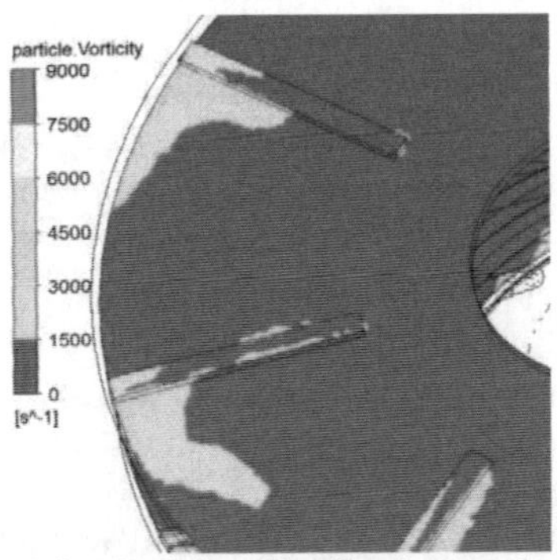

Figure 14. Vorticity distribution on back blade (back shroud).

Figure15. Wear morphology on back blade (back shroud).

4.3 The optimization of back blade shape

The analysis result shows that the back sides of back blades suffer the most severe abrasion. In order to reduce the interference effect of back blade on the vortex, forward curved back blade is optimized to compare with the radial blade. Number, width and length along radial direction of all the back blades on front and back shroud remain unchanged. Then the same grid meshing and numerical boundary conditions are consistent with the original model.

As we can see in figure 16 and 17, the number of back blade passage seized by large vorticity intensity has been decreased in quantity. There nearly no continue distributed vorticity intensity around the edge of shroud, and distribution range of vorticity is also markedly shrunk on the suction surface of back blade. The figure 18 and figure 19 show that the value of particle vorticity on suction surface (such as range of the yellow area) also has been reduced. The integral improvement of vorticity intensity is acceptable compared with the situation before. In the meanwhile, the influence of back blade optimized on the performance should be considered to expound the feasibility of the optimization. For this reason, $0.2Q_d$, $0.4Q_d$, $0.6Q_d$, $0.8Q_d$, $1.0Q_d$ and $1.2Q_d$ flow rate are calculated for the new model, and then the performances, like head and efficiency curves, are contrasted with the original model, as showed in figure 20.

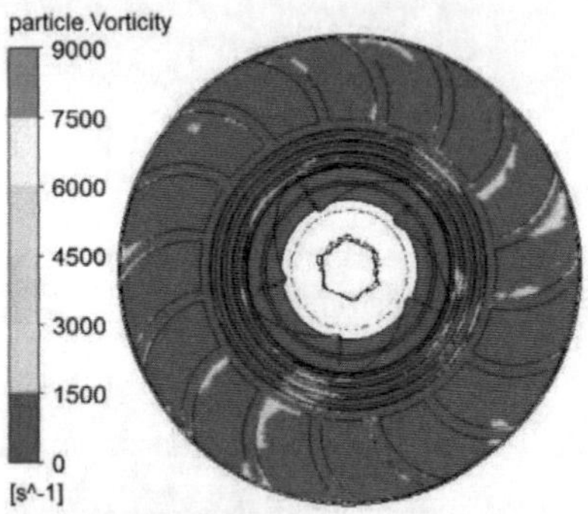

Figure 16. Vorticity distribution on front shroud.

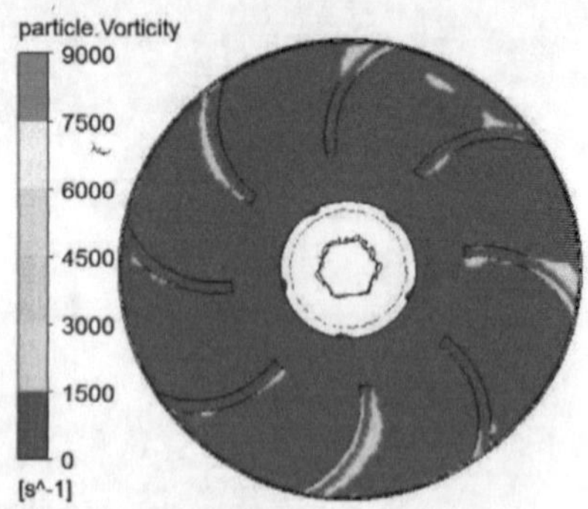

Figure 17. Vorticity distribution on back shroud.

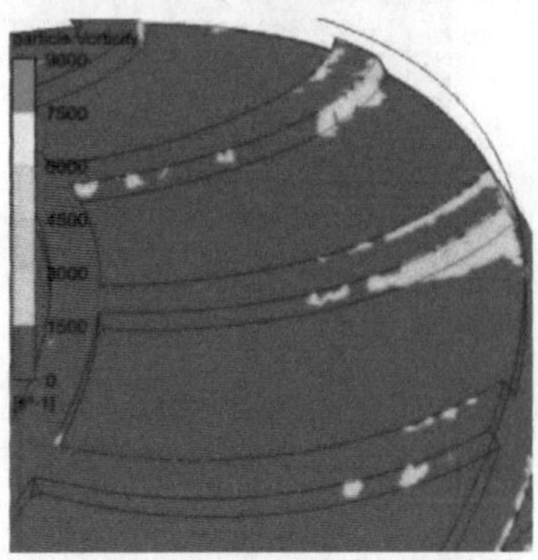

Figure 18. Vorticity distribution on back blade (front shroud)

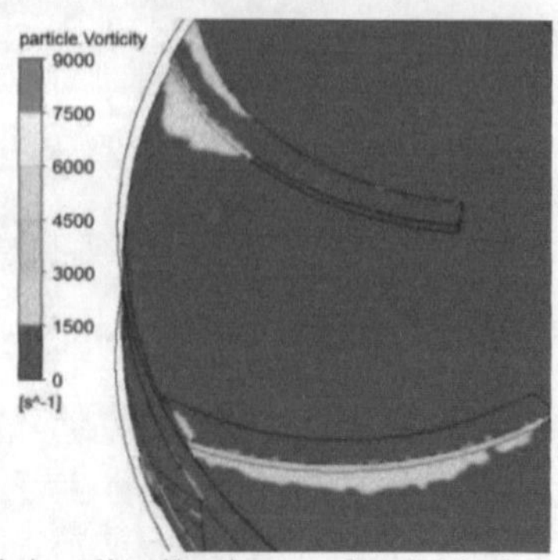

Figure 19. Vorticity distribution on back blade (back shroud)

The average values of head and efficiency in six different operation conditions are taken into account. The H-Q curve and η-Q curve of two models both show the similar variation law. Compared with the head of radial back blade model, the forward curved back blade model decreases by 3.25%. In the same way, the efficiency of forward curved back blade model decreases by 0.86% compared with the radial back blade model. So the effect of change of the model on head and efficiency are limited, and the forward curved back blade can effectively improve the distribution of vorticity, which is directly related to the back blade abrasion. Therefore, sacrificing few amounts of head and efficiency to prolong service life of impeller is a feasible way in engineering.

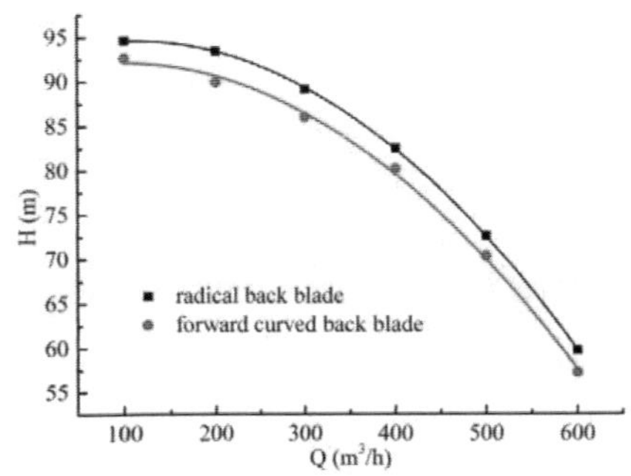

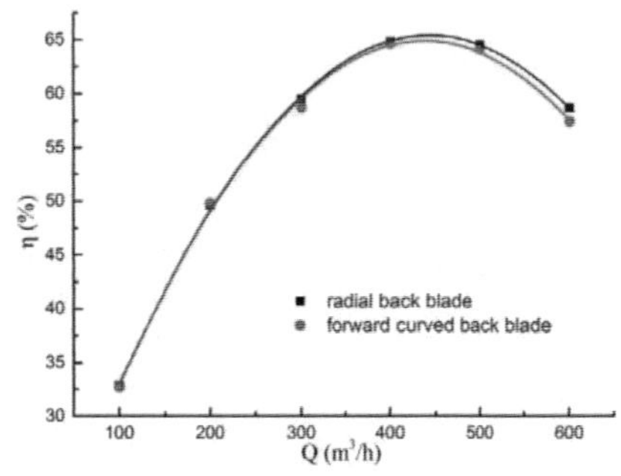

H-Q curve η-Q curve

Figure 20. External characteristics verification.

5. Conclusion

(1) The transient numerical simulation used in the research can get consistent results with experimental values. In the calculated flow rate, the average relative errors of head and efficiency between calculation and simulation are 1.28% and 1.93%. This comparison indicates the accuracy of calculation model and applicability of efficiency correction.

(2) The vorticity distribution contours and envelopes of swirl scale based on Q-Criterion can reasonably explain the wear mechanism. It can be speculated from the abandoned impeller that long spiral-fluted and cured eyebrow grooves are caused by strong vortex flow behind the suction surface of back blade. The simulation results of vorticity distribution have a good agreement with the severe abrasion morphology on worn impeller.

(3) According to the shape of worn grooves on shroud, forward back blade model is designed to verify the effectiveness on the improvement of vorticity distribution. The optimization result shows that the range of vorticity intensity on shroud has been shrunk largely. In the meanwhile, the influences brought by geometry change on external performances of centrifugal pump have also been considered. It is worthy of sacrificing few amount of head and efficiency to improve the abrasion on shrouds and back blades.

Acknowledgments

The authors are grateful for the support of National Key Technology Program of China (Grant No. 201 3BAF01B01).

References

[1] Wang H S 2006 *Coal Preparation Techology* **(z1)** 60-62

[2] Shaha S R, Jain S V, and Patel P N 2013 *Chemical, Civil and Mechanical Engineering Tracks of 3rd Nirma University International Conference on Engineering* **51** 715-720

[3] Wang J Q, Jiang W M, and Kong F Y 2013 *Transactions of the Chinese Society for Agricultural Machinery* **44**(11) 53-60

[4] Wu B, Yan H Z, Xu H L, and Tan Y Q 2012 *Journal of Central South University* **43**(1) 124-129

[5] Tao Y, Yuan S Q, Zhang J F, Zhang F, and Tao J P 2014 *Transactions of the Chinese Society of Agricultural Engineering* **21** 63-69

[6] Liu H L, Tan M G, Yuan S Q 2007 *Transactions of the Chinese Society of Agricultural Engineering* **22**(12) 107-109

Effect of attack angle on flow characteristic of centrifugal fan

Y Wu[1], H S Dou[*1], Y K Wei[1], X P Chen[1], Y N Chen[2], W B Cao[2]

[1] Key Laboratory of Fluid Transmission Technology of Zhejiang Province, Zhejiang Sci-Tech University, Hangzhou, 310018, China
[2] Zhejiang Yilida Ventilator Co. Ltd., Taizhou, 318056, China

*E-mail: huashudou@yahoo.com

Abstract. In this paper, numerical simulation is performed for the performance and internal flow of a centrifugal fan with different operating conditions using steady three-dimensional incompressible Navier-Stokes equations coupled with the RNG k-e turbulent model. The performance curves, the contours of static pressure, total pressure, radial velocity, relative streamlines and turbulence intensity at different attack angles are obtained. The distributions of static pressure and velocity on suction surface and pressure surface in the same impeller channel are compared for various attack angles. The research shows that the efficiency of the centrifugal fan is the highest when the attack angle is 8 degree. The main reason is that the vortex flow in the impeller is reduced, and the jet-wake pattern is weakened at the impeller outlet. The pressure difference between pressure side and suction side is smooth and the amplitude of the total pressure fluctuation is low along the circumferential direction. These phenomena may cause the loss reduced for the attack angle of about 8 degree.

1. Introduction

Centrifugal fan is widely used in various industries of the national economy, and it is one of the main energy consuming equipment in the industrial and agricultural production. Thus, it is very important to enhance its performance [1]. According to statistics, the power consumption of various types of fans and pumps in China accounted for more than a third of the total power consumption. From the viewpoint of saving energy and reducing noise pollution, it is very important to design efficient centrifugal fans for saving energy [2]. Numerical simulation using computational fluid dynamics has gradually become an important method to understand the internal flow of fluid machinery. Only in the past few decades, the research of internal flow in the turbomachinery has evolved from the simplified model to the actual flow model, and from inviscid model to viscous model [3]. Liu et al [6] make statistical analysis of attack angle of the highest efficient point on the existing fifty kinds of centrifugal fans, reach a conclusion that the airflow attack angle of the highest efficient point in the actual fan is negative ten degree to seventy degree. The airflow attack angle of the highest efficient point for the high efficient backward airfoil blade fan and backward plate blade fan is generally not more than positive six degree or negative six degree [7]. This is in good agreement with the design values recommended in the literatures [8-9]. Dou [10] simultaneously considered the effects of blade blocking and axial vortex, put forward the method of determining the geometrical angle of blades at impeller entrance. He gained that the efficiency of the centrifugal compressor is the highest when the airflow attack angle at the impeller entrance is six to eight degree, and gave an example for the calculation of attack angle. The optimal attack angle was about six degree obtained by the experiment of literature [10]. It has been proved that the optimal blade attack angle is about seven degree when considering the effect of the prewhirl angle by the experiment of literature [10]. The practical experience shows that the airflow attack angle at the highest efficient point of the fan is not necessarily zero degree due to the complexity of the actual air flow. Therefore, it is of great practical significance to design a highly efficient fan by the experiment and simulation.

In the paper, the flow characteristics of the fluid in the fan are studied using three-dimensional numerical simulation. The parameters of the whole flow field are obtained by using the finite volume method and SIMPLE algorithm to calculate Reynolds average Navier-Stokes equation. Then, the efficiency of a fan is calculated at various attack angles. And relevant analysis is carried out to clarify the origin of increasing efficiency at non-zero attack angle.

2. Governing equation

In this paper, the internal fluid flow characteristics of the 4-79-C type centrifugal fan used in Zhejiang Yilida Ventilator Incorporated Corporation are studied. For this type of fan, the specific speed is 67.4, the pressure coefficient is 0.91, and backward plate blade is adopted for the impeller vane. The governing equations are three-dimensional Reynolds average N-S equation and RNG $k-\varepsilon$ turbulence modle.

$$\begin{cases} \dfrac{\partial \overline{u_i}}{\partial x_i} = 0 \\ \rho u_j \dfrac{\partial \overline{u_i}}{\partial x_j} = \rho F_i - \dfrac{\partial \overline{p}}{\partial x_i} + \mu \dfrac{\partial^2 \overline{u_i}}{\partial x_j \partial x_j} - \rho \dfrac{\partial}{\partial x_j}(\overline{u_i u_j}) \end{cases} \tag{1}$$

The Realizable-$k-\varepsilon$ model is used for the turbulence model, which was proposed by Shih in 1995, which was considered as the coefficient C_μ of the calculation formula of the turbulent eddy viscosity coefficient μ_t was not constant [4-5], but it was related to the strain rate, in comparison with the standard $k-\varepsilon$ model, the ε equation had changed greatly, the production term in the k equation no longer contained the generated term P_k; the constraint equations of the turbulent kinetic energy k and the turbulent dissipation rate are as follows:

$$\frac{\partial(\rho k)}{\partial t} + \frac{\partial(\rho k u_j)}{\partial x_j} = \frac{\partial}{\partial x_j}\left[\left(\mu + \frac{\mu_t}{\sigma_k}\right)\frac{\partial k}{\partial x_j}\right] + \rho(P_k - \varepsilon) \tag{2}$$

$$\frac{\partial(\rho\varepsilon)}{\partial t}+\frac{\partial(\rho\varepsilon u_j)}{\partial x_j}=\frac{\partial}{\partial x_j}\left[\left(\mu+\frac{\mu_t}{\sigma_\varepsilon}\right)\frac{\partial\varepsilon}{\partial x_j}\right]+\rho C_1 E\varepsilon-\rho C_2\frac{\varepsilon^2}{k+\sqrt{v\varepsilon}} \quad (3)$$

where: $C_1=\max\left(0.43,\frac{\eta}{\eta+5}\right)$; $\eta=\left(2E_{ij}\cdot E_{ij}\right)^{1/2}\frac{k}{\varepsilon}$; $E_{ij}=\frac{1}{2}\left(\frac{\partial u_i}{\partial x_j}+\frac{\partial u_j}{\partial x_i}\right)$

constant values in the model: $\sigma_k=1.0$; $\sigma_k=1.2$; $C_2=1.9$.

3. Numerical method

In this paper, using Navier – Stokes equation and RNG $k-\varepsilon$ turbulence model, the governing equation is discretized in space by finite volume method and two order upwind scheme. SIMPLE algorithm is used to solve the discrete equation, the convergence criterion is that the relative residual error of each variable is decreased by three orders of magnitude and the pressure difference is stable.

The inlet condition is velocity and outlet condition is outflow for the fan, the static and dynamic regions are set respectively by the multiple reference frame model(MRF), the wall condition is no slip boundary.

4. Verification of numerical method

4.1. Comparison of experiment and simulation

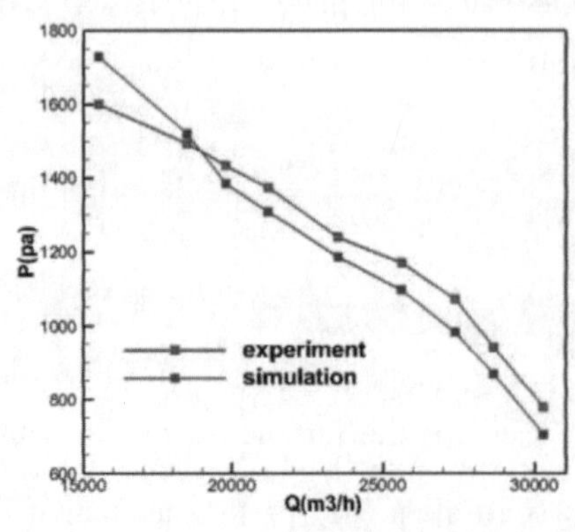

Fig 1. Curve of total pressure and flow rate.

The curve of total pressure and flow rate is shown in Figure 1. In this picture, the horizontal coordinate represents the flow rate, the vertical coordinate represents the total pressure. The red line is the experimental value, the blue line is the simulated value. As can be seen from Figure 1, the simulated value and experimental value of total pressure have same trend with flow rate. It is that the total pressure decreases with the increase of the flow rate, and they are in good agreement at different conditions. Under the condition of small flow rate, the difference between the simulated value and experimental value is relatively large. The flow rate at design condition is small, and the difference between the simulated value and the experimental value is relatively small near design condition. The cause of above phenomenon may be that the non-steady composition is small under the condition of large flow rate and design flow rate, it has the little effect on the simulation of the flow field; under the condition of small flow rate, the non-steady composition is large, the effect and error are also large, but the error of simulation is within allowable error range, the result of simulation is correct.

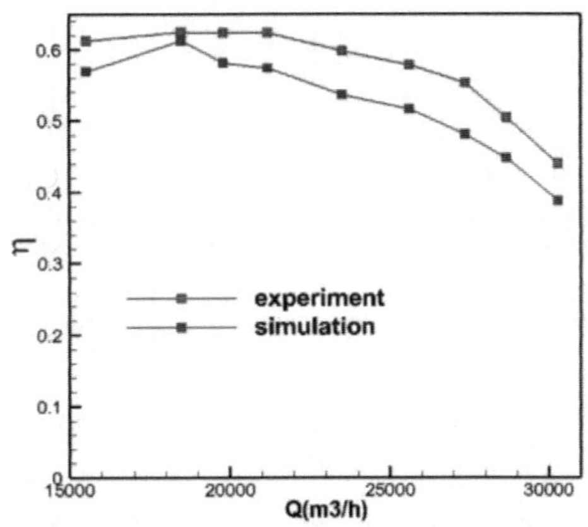

Fig 2. Curve of efficiency and flow rate.

The curve of efficiency and flow rate is shown in Figure 2. In the picture, the horizontal coordinate represents the flow rate, the vertical coordinate represents the efficiency. The red line is the experimental value, the blue line is the simulated value. As can be seen from Figure 2, the experimental efficiency and the simulated efficiency are in good agreement. The error is within allowable error range, the error may be caused by the simplified simulation model, the simulated efficiency is consistent with the experimental efficiency. Generally the numerical simulation method used in this paper is effective.

4.2. Verification of grid independence

The grid number is 2170000, 3620000 and 4900000, and the total pressure under design condition is calculated respectively, the result is shown in Figure 3.

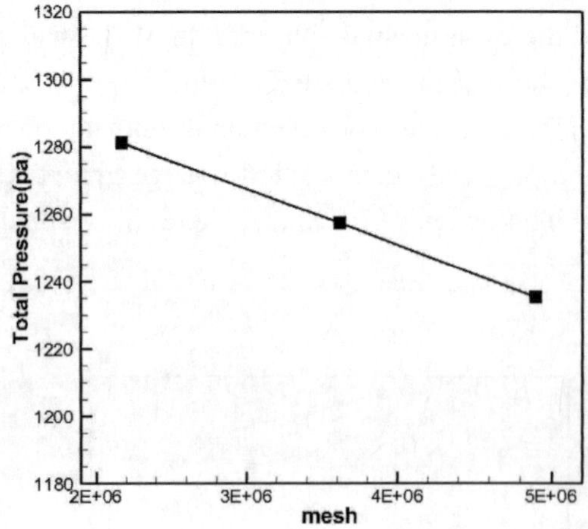

Fig 3. Total pressure under different mesh numbers.

The total pressure values under different grid numbers are shown in Figure 3, the relative errors between the simulated values and the experimental values are within three percent. It shows that the grid number has no effect on the calculation result, grid independence is proved.

5. Results and analysis of calculation

5.1. Example of calculation method

The geometrical parameters of the impeller are as follows: the impeller outlet diameter is $D_2 = 700mm$, the impeller inlet diameter is $D_1 = 488mm$, the impeller blade inlet width is $b_1 = 253.61mm$, the blade thickness is $S = 7mm$, the blade number is $Z = 16$. The rated rotation rate of impeller is $n = 1600r/m$, the blade inlet volume flow rate at design point is $V_1 = 6m3/s$. Assuming that there is no airflow prewhirl at the blade inlet, that is $C_{1u} = 0$, the calculation results are as follows:

The circumferential velocity at blade inlet:

$$\mathrm{U}_1 = \frac{\pi D_1 n}{60} = \frac{\pi \times 0.488 \times 1600}{60} = 40.88m/s$$

The airflow radial velocity at blade inlet:

$$C_{1r} = \frac{V_1}{\pi D_1 b_1} = \frac{6}{\pi \times 0.488 \times 0.25361} = 15.43m/s$$

The airflow angle at blade inlet:

$$tg\beta_1 = \frac{C_{1r}}{U_1} = \frac{15.43}{40.88} = 0.37745$$

$$\beta_1 = 20.68^\circ$$

The blocking factor at blade inlet:

$$\tau_1 = 1 - \frac{ZS}{\pi D_1 \sin\beta_{1A}} = 1 - \frac{16 \times 0.007}{\pi \times 0.488 \times \sin 25^\circ} = 0.827$$

Assuming that $\beta_{1A} = 25^\circ$ in the above formula, if the difference between the back calculation result and the assumption is large, the calculation needs to be repeated until they are consistent.

The airflow radial velocity at blade inlet only considering blade blocking:

$$C_{1r}' = \frac{C_{1r}}{\tau_1} = \frac{15.43}{0.827} = 18.66 m/s$$

The airflow angle at blade inlet only considering blade blocking:

$$tg\beta_1' = \frac{C_{1r}'}{U_1} = \frac{18.66}{40.88} = 0.45646$$

$$\beta_1' = 24.53^\circ$$

The airflow angle at blade inlet simultaneously considering the effect of blade blocking and axial vortex:

$$tg\beta_1'' = \frac{tg\beta_1}{\tau_1(1 - \frac{\pi}{Z}\sin\beta_{1A})} = \frac{0.37745}{0.827 \times (1 - \frac{\pi}{16} \times \sin 25^\circ)} = 0.4977$$

$$\beta_1'' = 26.46^\circ$$

$$\lambda = \beta_1'' - \beta_1 = 26.46^\circ - 20.68^\circ = 5.78^\circ$$

According to the reference [10], it is reasonable that the attack angle i is zero to two degrees smaller than λ. So we take $i = 4.78^\circ$

$$\beta_{1A} = \beta_1 + i = 20.68^\circ + 4.78^\circ = 25.46^\circ$$

The relative error between them is within two percent, which indicates that the hypothesis is reasonable.

5.2. Curve of efficiency and attack angle

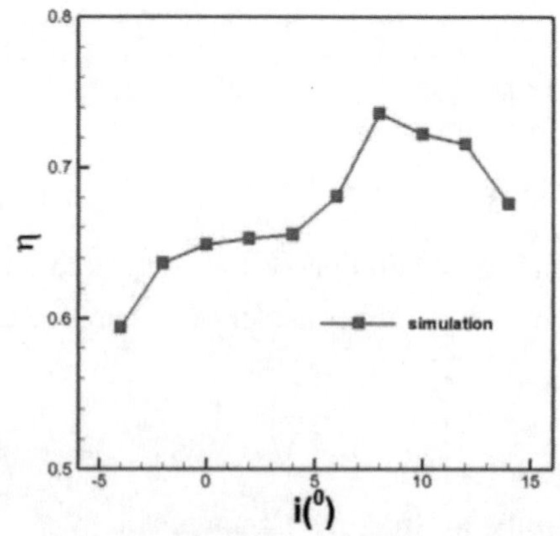

Fig 4. Broken line of efficiency and attack angle.

The broken line of efficiency and attack angle is shown in Figure 4. As can be seen from Figure 4, the efficiency is not the highest when the attack angle is zero degree under design condition. The efficiency is the highest when the attack angle is eight degree and not zero degree. The losses in the centrifugal fan are mainly composed of friction loss in the boundary layer of the impeller, loss caused by separation of boundary layer on the blades, loss from secondary flow, loss caused by the jet-wake pattern resulted from the nonuniform distribution of radial impeller outlet velocity [13]. In theory, the relative velocity angle β_1 of the airflow into the blades is in agreement with that the geometrical angle β_{1A} of blades when the attack angle is zero degree. The airflow enters the blade passage and doesn't impact into it, it makes the shock loss reduce to zero. But the practical experience shows that the airflow attack angle at the highest efficient point of the fan is not necessarily zero degree due to the complexity of the actual airflow. The larger the attack angle, the smaller the flow rate. In practical calculation, the efficiency of the fan is the highest when the flow rate is less than the design flow rate, which is consistent with the conclusions of relevant experiments in the literatures.

5.3. Total pressure distributions of different attack angles

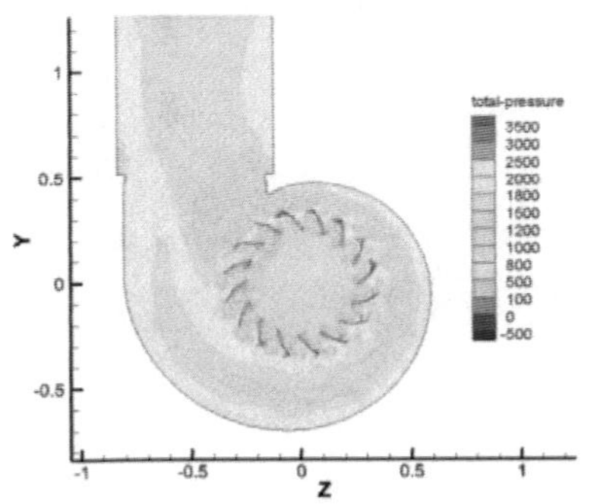

$(a) i = -4^{\circ}$

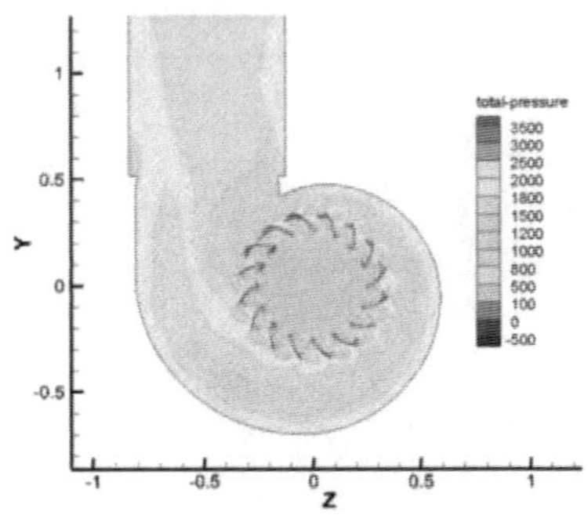

$(b) i = 0^{\circ}$

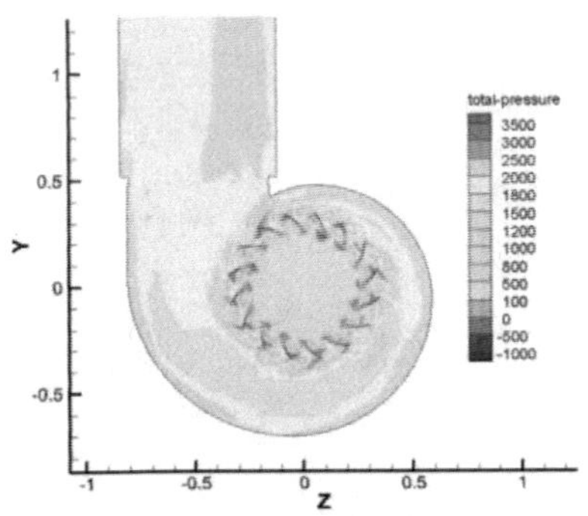

$(c) i = 8^{\circ}$

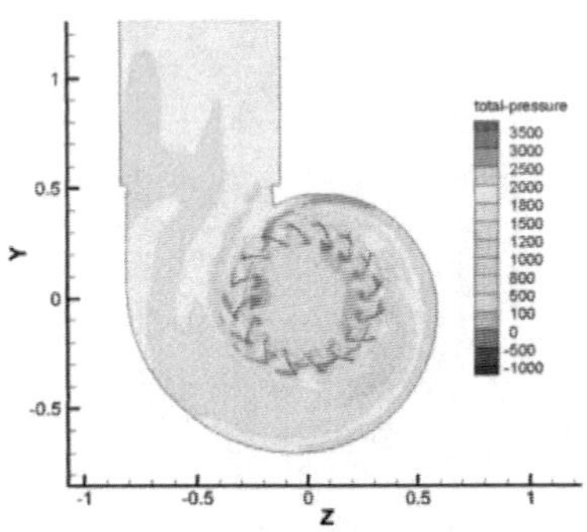

$(d) i = 14^{\circ}$

Fig 5. Total pressure charts of different attack angles.

The total pressure distribution at the same cross section is shown in Figure 5. As can be seen from Figure 5, the average total pressure of the fan cross section is the highest when the attack angle is eight degree, it indicates that the total energy is the largest. It takes the second place when the attack angle is fourteen degree, the average total pressure is lower when the attack angle is negative four degree and zero degree. Combined with the chart of efficiency and attack angle, the various energy losses of the airflow in the centrifugal fan are the least when the attack angle is eight degree, the various energy losses of the airflow are the largest when the attack angle is negative four degree. In the picture, it also can be seen that the total pressure is the lowest at the inner edge of the impeller in the centrifugal fan. And it increases gradually from the inside to the outside, the total pressure is the largest at the outer edge of the impeller. As far as the whole impeller channel is concerned, the total pressure is higher near the outlet side, the total pressure on the pressure side is the largest near the volute tongue, it is especially obvious when the attack angle is eight degree and fourteen degree. The total pressure in the volute wall is higher, and the total pressure changes of different attack angles in the volute are not uniform.

5.4. Static pressure distributions of different attack angles

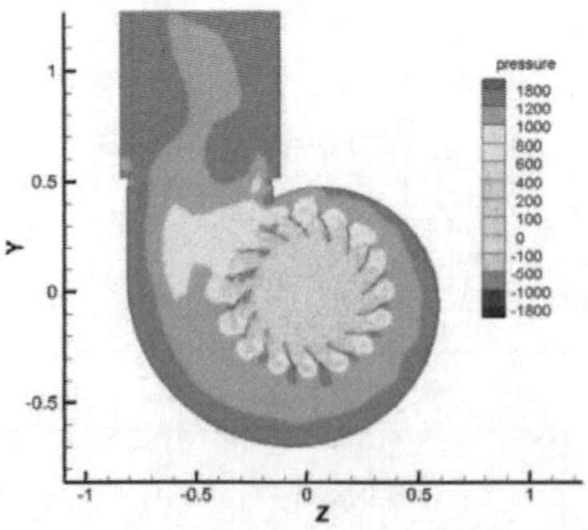

$(a) i = -4^{\circ}$

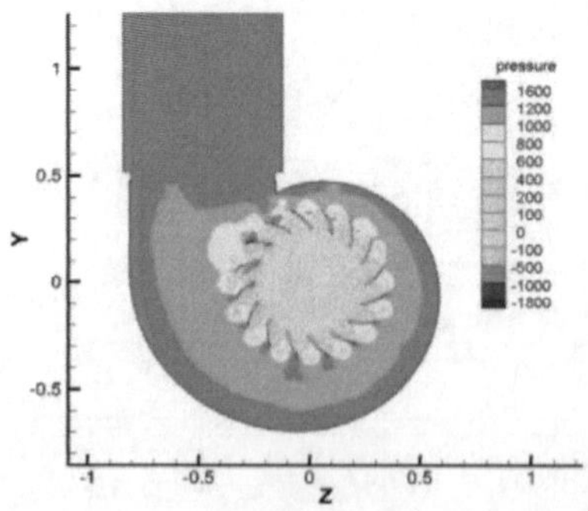

$(b) i = 0^{\circ}$

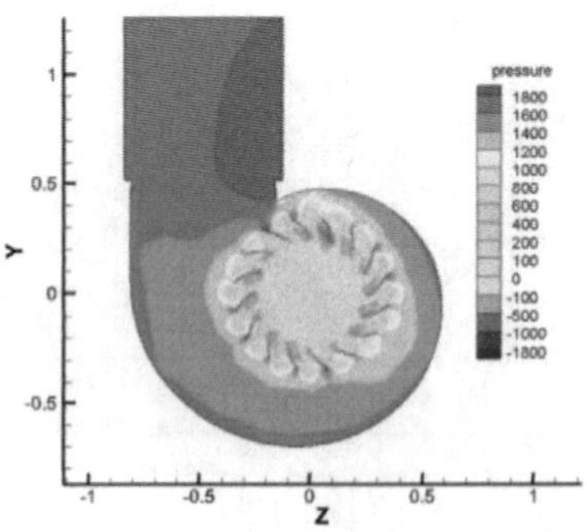

$(c) i = 8^{\circ}$

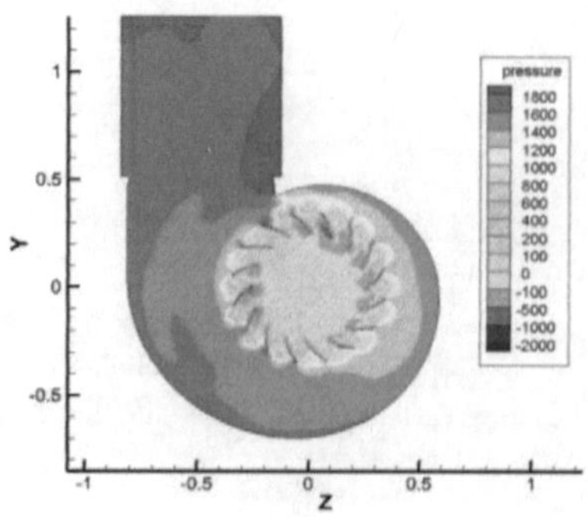

$(d) i = 14^{\circ}$

Fig 6. Static pressure charts of different attack angles.

The static pressure distribution at the same cross section is shown in Figure 6. As can be seen from Figure 6, the static pressure is the largest in the extended section of the volute outlet, the volute wall and the volute tongue, the static pressure is the least near the impeller inlet. The static pressure is lower at the inner edge of the impeller, and it is higher at the outer edge of the impeller. The low pressure zone in the impeller channel is obvious when the attack angle is eight degree, and it extends from the impeller inlet to the middle part of the impeller channel. In the single impeller channel, the static pressure of the pressure surface is higher than that of the suction surface, which provides the condition for the formation of the secondary flow in the impeller channel. The hydrostatic pressure increases gradually from the center to the wall in the volute, the extended section of the volute outlet has an effect of diffusion, the hydrostatic pressure is the highest near the volute outlet.

5.5. Relative streamline charts of the impeller at different attack angles

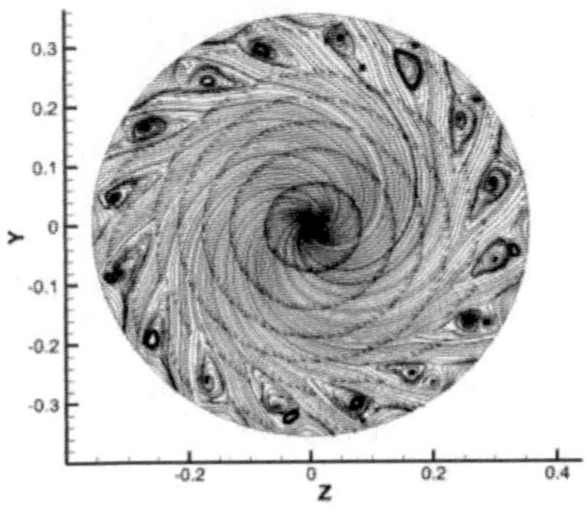

$(a)i = -4^{\circ}$

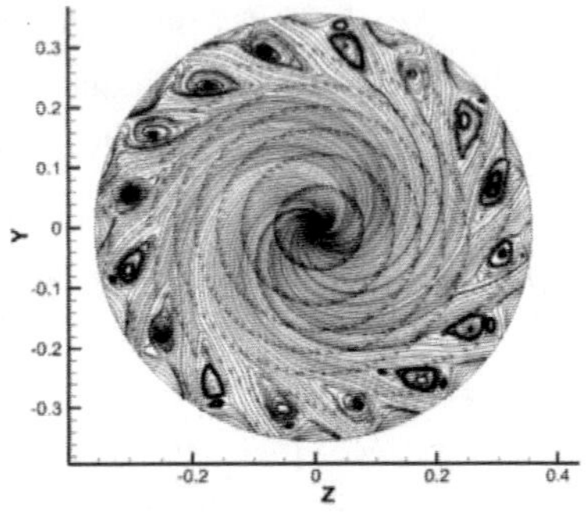

$(b)i = 0^{\circ}$

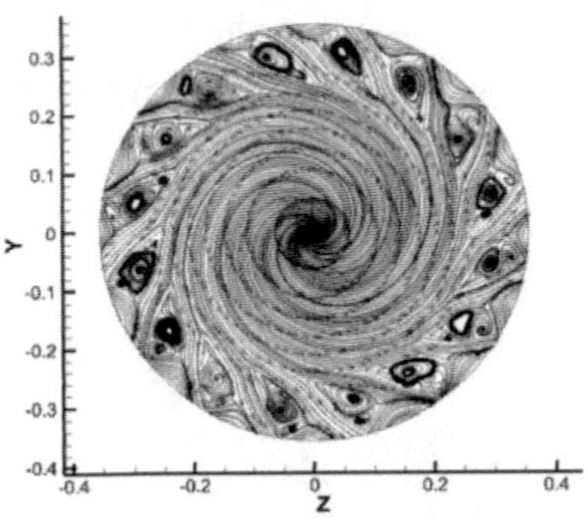

$(c)i = 8^{\circ}$

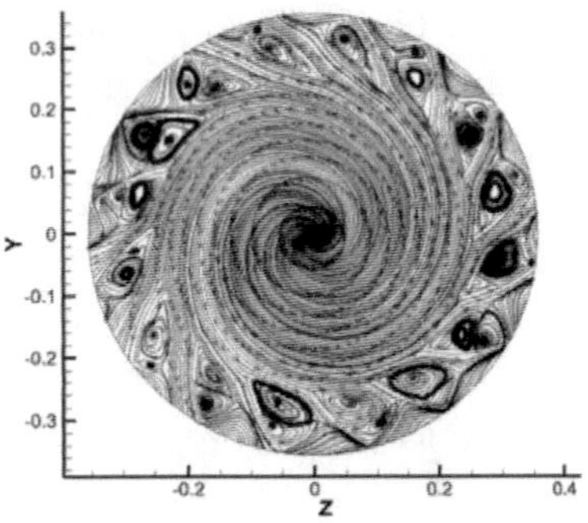

$(d)i = 14^{\circ}$

Fig 7. Relative streamline charts of different attack angles.

The relative streamline distribution at the same cross section is shown in Figure 7. From the figure 7, it can be seen that the flow in the impeller channel is not stable at different attack angles, and there is an obvious phenomenon of vortex and back flow. The flow in some impeller channels is not stable, which appears in the impeller channel outlet and the middle position of the impeller channel, the vortexes increase from the outlet to the inlet in some impeller channels, the flow is very unstable. Compared with the other three kinds of attack angles, when the attack angle is eight degree, the vortexes near the suction side are less in some impeller channels, the energy losses are smaller and the flow is smooth. At this time, the streamlines in the impeller channel are very uneven, the streamlines are not entirely along the curvature of the blade.

In fact, in the centrifugal impeller, the pressure of the blade pressure side is high, the velocity is low; the pressure of the blade suction side is low, the velocity is high, this pressure gradient causes the fluid micelle to be subjected to a force that points to the suction side. In the main flow region, the pressure gradient is balanced with the inertia force which is formed by the circular motion of the fluid micelle and the curvature of the blade. The velocity of the fluid micelle which is in the boundary layer of the blade and the wheel cap surface and the blade and the wheel plate surface is much smaller than that in the main flow area. The inertia force formed can not be balanced with the above pressure gradient. Therefore, the flow is generated from the pressure surface to the suction surface in the boundary layer, the flow is perpendicular to the mainstream, it is the secondary flow [14]. The secondary flow makes the low velocity fluid micelles near the pressure surface sucked in the boundary layer of the wheel cap and that of the wheel plate, the low velocity fluid micelles are used to supplement the micelles with large kinetic energy, it makes the boundary layer of the suction surface thicker and leads to the separation of the boundary layer, this will form some vortex zones and cause some losses. The secondary flow causes the boundary layer thickness and separation of the suction surface, it inevitably affects the mainstream flow and reduces the relative velocity near the suction surface, the jet-wake is formed at the outlet of the impeller [15].

In the impeller channel, the relative streamlines of the fluid are close to the pressure surface. In the suction side, the fluid forms some vortexes, the fluid breaks away from the surface of the blade, which makes the flow separation at the end of the suction side and the wake loss. In some impeller channels, the vortexes severely block the flow channel when they are more intense in the impeller channel, the fluid flow along the radial outflow is seriously squeezed, the fluid produces relatively strong jet-wake at the impeller outlet, it causes the energy loss and affects the efficiency of the fan.

5.6. Turbulence intensity analysis of different attack angles

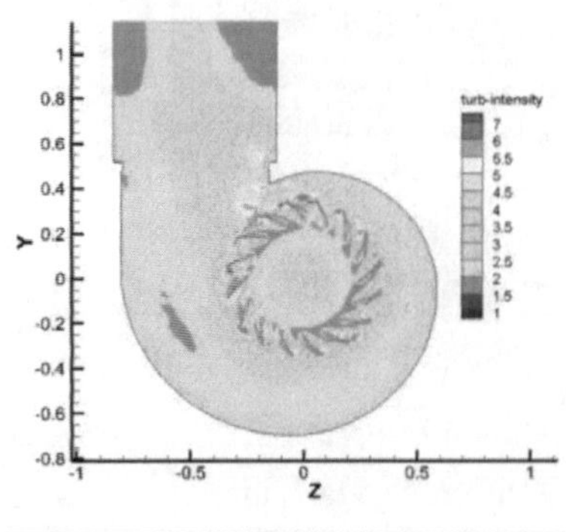

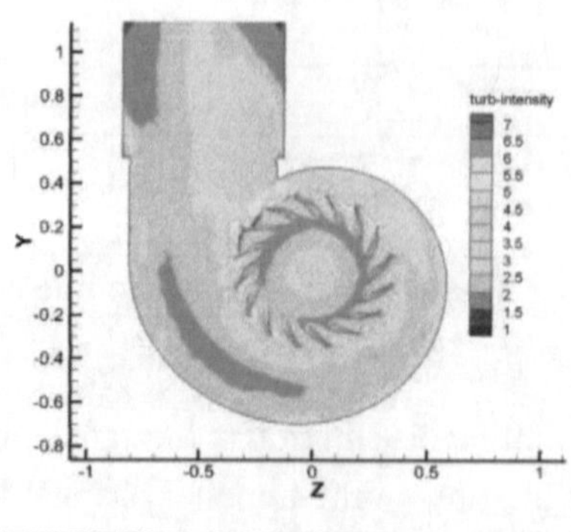

$(a) i = -4^{\circ}$

$(b) i = 0^{\circ}$

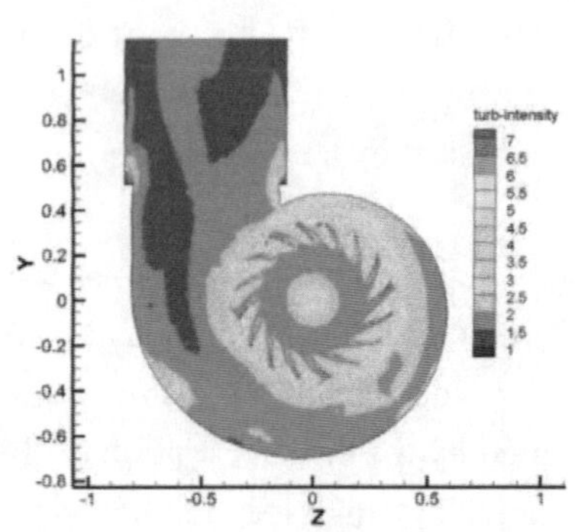

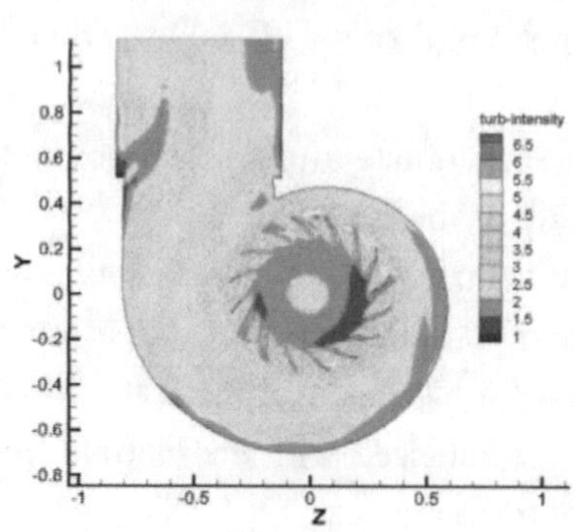

$(c)i = 8^{\circ}$	$(d)i = 14^{\circ}$

Fig 8. Turbulence intensity charts of different attack angles.

The turbulence intensity distribution at the same cross section is shown in Figure 8. The turbulence intensity is the degree that the wind velocity changes with time and space, which reflects the relative intensity of the fluctuating wind velocity, it is the most important feature of the atmospheric turbulence characteristics. The turbulence intensity is equal to the ratio of the mean square root of the turbulence fluctuating velocity and the average value of the turbulence fluctuating velocity [16]. As can be seen clearly from Figure 8, the turbulence intensity of the impeller is higher than that of the volute in the centrifugal fan, the turbulence intensity of the volute has a decreasing trend from the inside to the outside. In the impeller flow passage, the turbulence intensity of the suction surface is high, and the turbulence intensity of the pressure surface is low. The turbulence intensity of the volute is high, and the turbulence intensity at the outer edge of the suction surface of the impeller outlet is high near the volute tongue, it is particularly obvious when the attack angle is negative four and zero degree. When the attack angle is eight degree, the average turbulence intensity of the fan cross section is lower, and it is high at the other three angles.

5.7. Static pressure distribution of suction surface and pressure surface at different attack angles

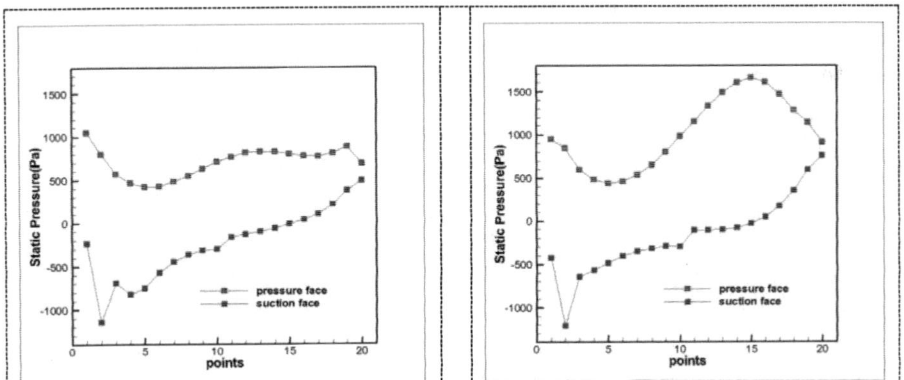

$(a)i = -4^{\circ}$	$(b)i = 0^{\circ}$

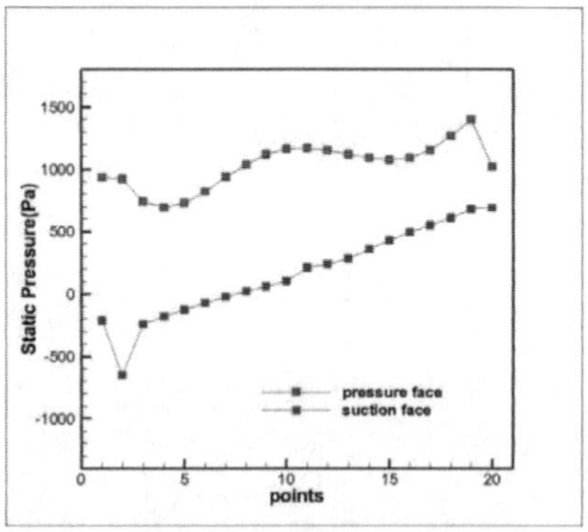

$(c) i = 8^{\circ}$

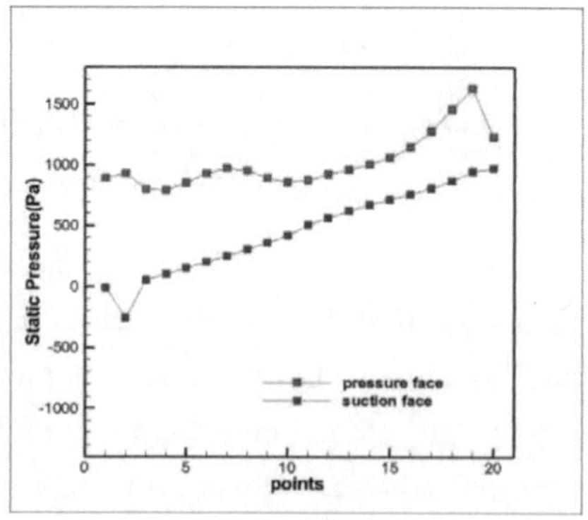

$(d) i = 14^{\circ}$

Fig 9. Static pressure charts of suction surface and pressure surface at different attack angles.

The pressure distribution on pressure and suction sides in a selected blade channel is shown in Figure 9. As can be seen from Figure 9, the horizontal coordinate represents the distance from the blade leading edge with interval on suction surface and pressure surface in the same impeller flow passage. The vertical coordinate indicates the static pressure, the red line is the static pressure of pressure surface, the blue line is the static pressure of suction surface. The distribution trend of the static pressure of the suction surface and the pressure surface in the impeller flow passage is generally consistent at different attack angles. The static pressure of the pressure surface is higher than that of the suction surface. Under the design condition, from the 15th point, the static pressure on the pressure side drops sharply, the static pressure on the suction side increases rapidly, this reduces the pressure difference between them. This causes the decrease of the work that the blades exert on the fluid. The comprehensive comparison of four pictures shows that the average static pressure difference between the pressure surface and suction surface is large and uniform. This causes the increase of the effective work that the blades exert on the fluid and the efficiency of the fan is also improved.

5.8. Pressure difference distribution between pressure side and suction side at different attack angles

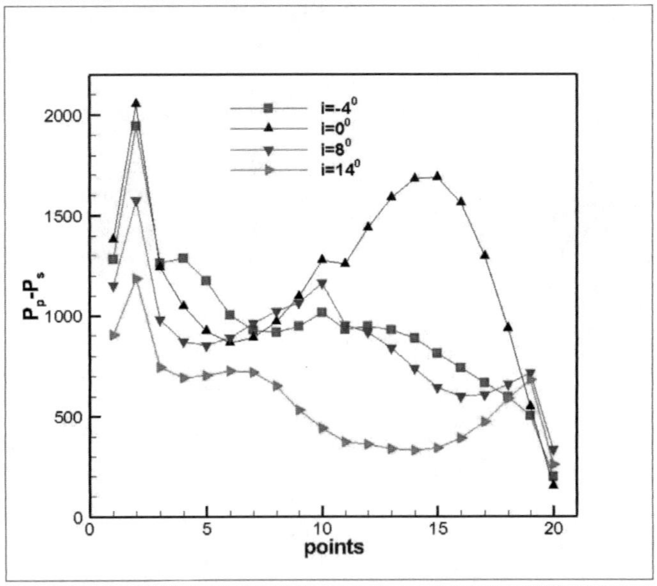

Fig 10. Pressure difference chart between pressure side and suction side at different attack angles.

The pressure difference in the selected blade channel is shown in Figure 10, which is adapted from Figure 9. As can be seen from Figure 10, the horizontal coordinate represents the distance from the blade leading edge with interval on suction surface and pressure surface in the same impeller flow passage. The vertical coordinate indicates the pressure difference between the pressure side and the suction side. In fact, the pressure surface is the working surface and the suction surface is the non-working surface in the impeller flow passage. The pressure is high and the velocity is low in the pressure side and the pressure is low and the velocity is high in suction side. The fluid is subjected to the force which direction is from the pressure side to the suction side, and the force exerts work on the fluid. It increases the velocity of the fluid. The pressure distribution at the pressure side and the suction side should be more uniform, and the pressure difference between them can not be too large. If the pressure difference is too large, the fluid will flow from the pressure side to the suction side. It will produce the back flow and vortex in the suction side, causing the energy loss and affecting the efficiency of the centrifugal fan, that we can see from Figure 11. When the attack angle is eight degree, the pressure difference between pressure side and suction side is smooth. While the attack angle is zero degree, the pressure difference in the second half is too large, which increases the load of the blades.

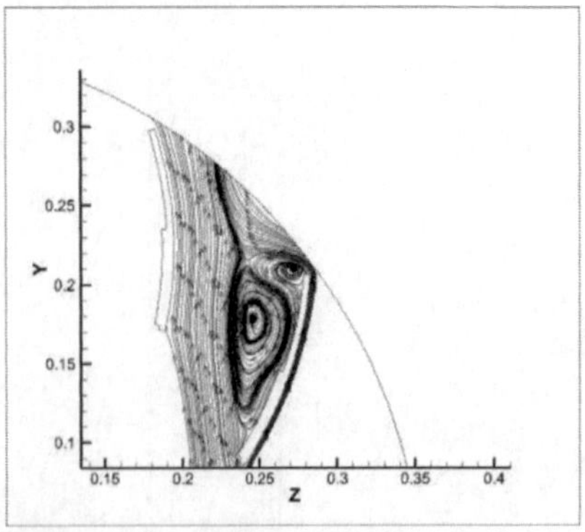

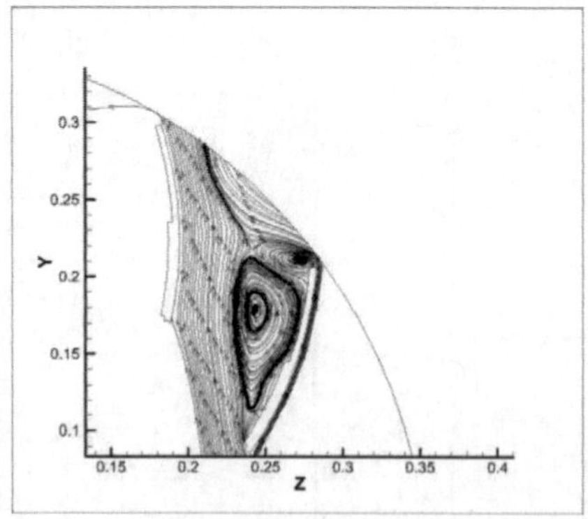

$(a) i = -4^{\circ}$ $(b) i = 0^{\circ}$

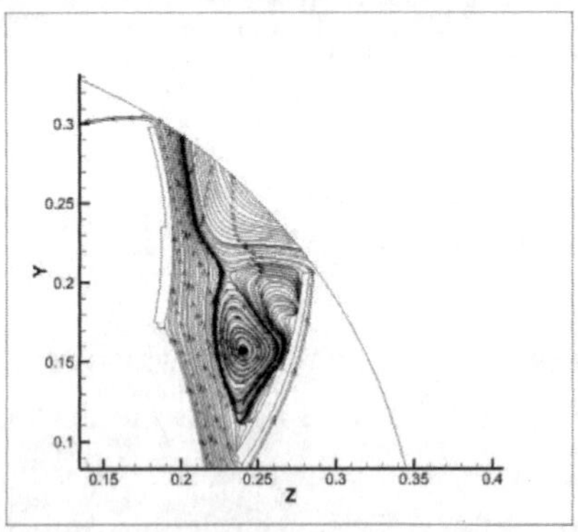

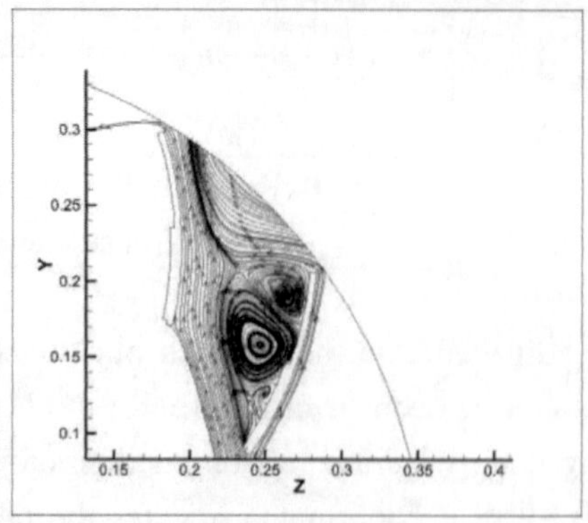

$(c) i = 8^{\circ}$ $(d) i = 14^{\circ}$

Fig 11. Relative streamlines charts of suction surface and pressure surface in the same impeller channel at different attack angles.

5.9. Velocity distribution of suction surface and pressure surface at different attack angles

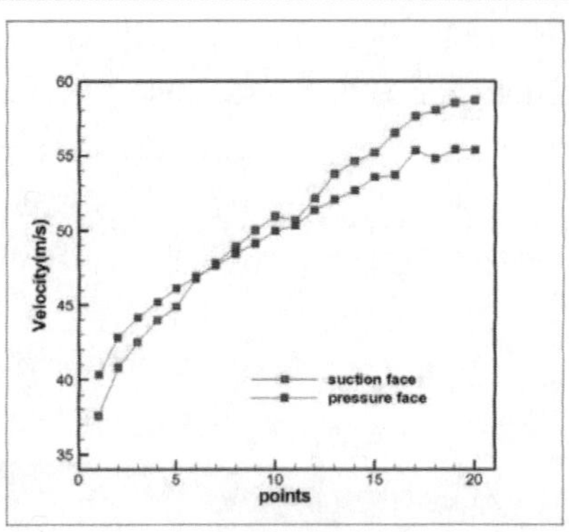

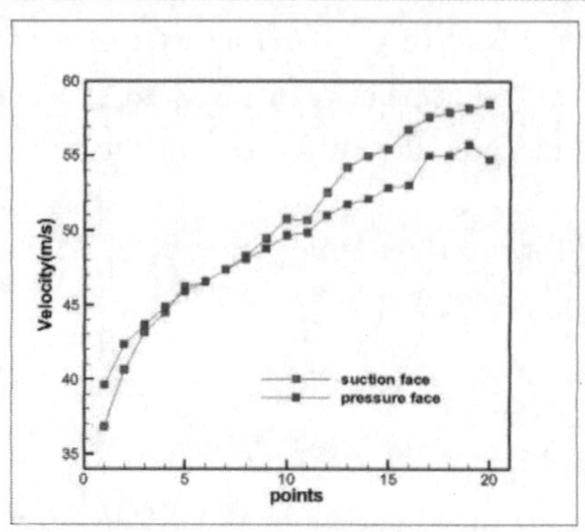

$(a) i = -4^{\circ}$	$(b) i = 0^{\circ}$

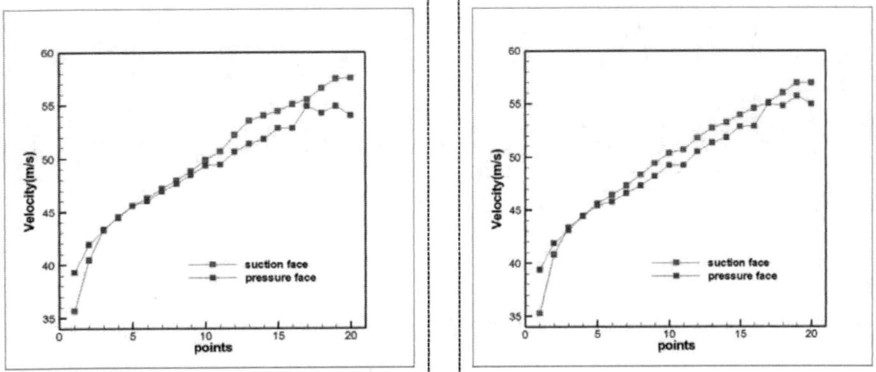

$(c) i = 8^{\circ}$	$(d) i = 14^{\circ}$

Fig 12. Velocity charts of suction surface and pressure surface at different attack angles.

The velocity distribution on pressure and suction sides in a selected blade channel is shown in Figure 12. As can be seen from Figure 12, the horizontal coordinate represents the distance from the blade leading edge with interval on suction surface and pressure surface in the same impeller flow passage. The vertical coordinate represents the velocity, the red line is the velocity of suction surface, the blue line is the velocity of pressure surface. The velocity distribution of suction surface and pressure surface in the same impeller flow passage is similar at different attack angles. The velocity of the pressure surface is higher than that of the suction surface near the inlet of the impeller, the velocity of the suction side is higher than that of the pressure side in the back section. When the fluid enters the impeller channel, the flow is relatively uniform, the velocity of the suction surface is close to that of the pressure surface. But the velocity of the pressure side is almost equal at the outlet of the impeller, the velocity of the suction side increases significantly at the inlet of the impeller. The velocity of the blade root is very small, it is because the flow is hindered when the fluid flows through the blade, it leads to a sudden drop in the velocity. While this position is close to the suction surface, so the velocity distribution of the starting part of the suction surface is affected. When the attack angle is eight degree, the average velocity difference between working face and non-working face for the blade is small. Which is particularly evident in the first section, it shows that the flow loss is smaller and the flow field distribution of the impeller outlet is more uniform. While the attack angle is zero degree, the velocity difference between suction side and pressure side in the second half is larger, the flow loss is larger and this will affect the efficiency of the fan.

5.10. Radial velocity distribution of the impeller channel outlet along the circumferential direction at different attack angles

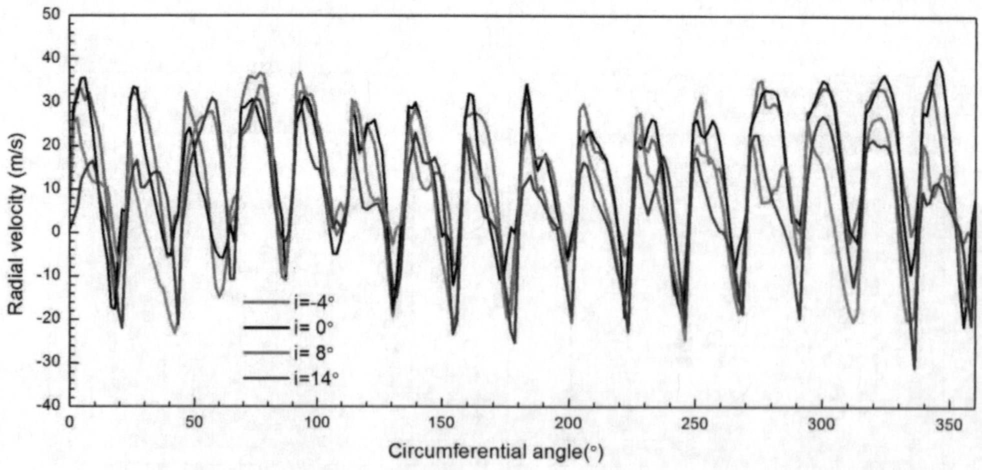

Fig 13. Chart of radial velocity and circumferential angle at different attack angles.

The radial velocity distribution of the impeller channel outlet along the circumferential direction is shown in Figure 13. As can be seen from Figure 13, when the attack angle is zero degree, the amplitude of the radial velocity fluctuation is large along the circumferential direction, this phenomenon can cause the jet-wake loss increased.

5.11. Static pressure distribution of the impeller channel outlet along the circumferential direction at different attack angles

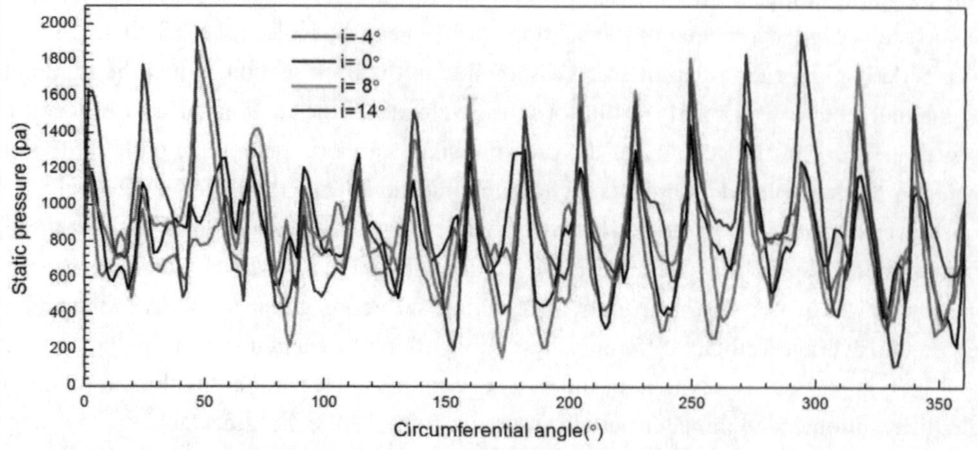

Fig 14. Chart of static pressure and circumferential angle at different attack angles.

The static pressure distribution of the impeller channel outlet along the circumferential direction is shown in Figure 14. As can be seen from Figure 14, when the attack angle is fourteen degree, the amplitude of the static pressure fluctuation is large along the circumferential direction, it may cause the instability in the flow and the loss increased.

IOP Conf. Series: Materials Science and Engineering **129** (2016) 012059 doi:10.1088/1757-899X/129/1/012059

5.12. Total pressure distribution of the impeller channel outlet along the cirumferential direction at different attack angles

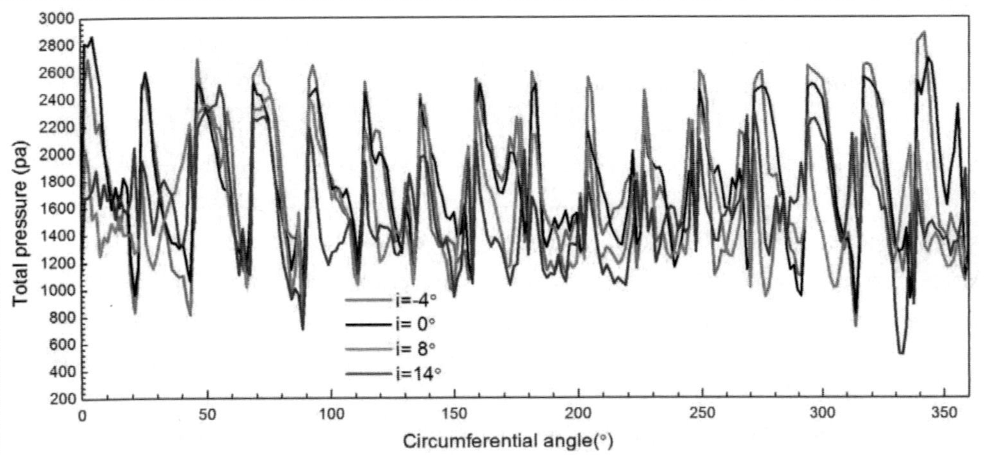

Fig 15. Chart of total pressure and circumferential angle at different attack angles.

The total pressure distribution of the impeller channel outlet along the circumferential direction is shown in Figure 15. As can be seen from Figure 15, when the attack angle is eight degree, the amplitude of the total pressure fluctuation is low along the circumferential direction. This phenomenon may cause the loss reduced.

5.13. Velocity distribution of the impeller channel outlet along the circumferential direction at different attack angles

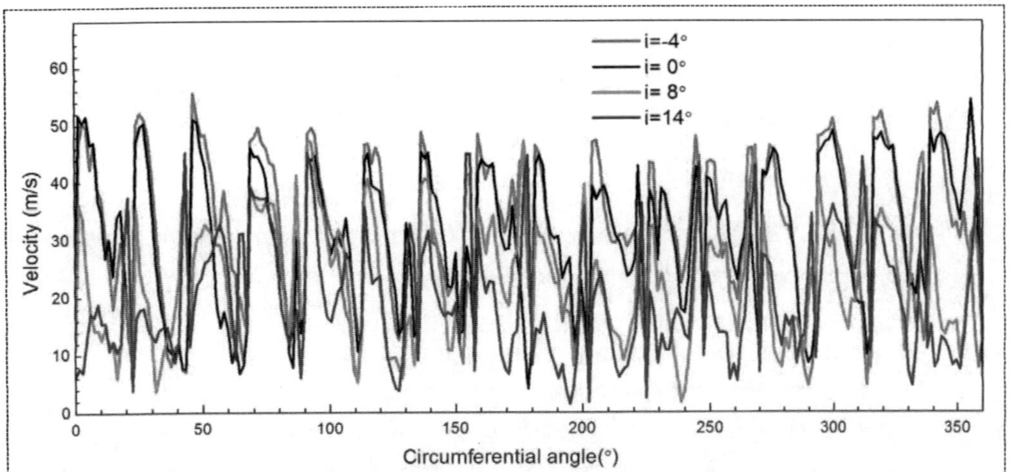

Fig 16. Chart of velocity and circumferential angle at different attack angles.

The velocity distribution of the impeller channel outlet along the circumferential direction is shown in Figure 16. As can be seen from Figure 16, when the attack angle is eight degree, the amplitude of the velocity fluctuation is low along the circumferential direction, which indicates the velocity is stable and the loss is less.

5. Conclusions

(1) In this paper, the efficiency of the centrifugal fan is calculated at various attack angles. The efficiency is the highest when the attack angle is about eight degree. The numerical simulation of the fan model shows that the maximum efficiency point of the centrifugal fan is not at the zero attack angle condition, while the efficiency is the highest when the attack angle is eight degree.

(2) The average total pressure is the highest when the attack angle is six to eight degree, which indicates that the total energy is highest, it has a direct impact on the efficiency of the fan.

(3) When the attack angle changes, the flow has obvious vortex and reflux in the impeller channel. The reflux occurs in the outlet and intermediate position in some impeller channels, the vortex gradually increases from the outlet to the inlet in partial impeller channels. Some separated vortexes form near the suction side of the impeller channel. The jet-wake forms in the outlet of the impeller channel. The vortex loss, the separation loss of boundary layer and the jet-wake loss in the impeller will affect the efficiency of the centrifugal fan.

(4) When the attack angle is in the range of six to eight degree, the flow is smooth and uniform in some impeller channels, the vortexes are reduced and weaker in the suction side.

(5) When the attack angle is eight degree, the pressure difference between pressure side and suction side is smooth. While the attack angle is zero degree, the pressure difference in the second half is too large, which increases the load of the blades. If the pressure difference is too large, the fluid will flow from the pressure side to the suction side, it will produce the back flow and vortex in the suction side, cause the energy loss and affect the efficiency of the centrifugal fan.

(6) When the attack angle is eight degree, the amplitude of the total pressure fluctuation is low along the circumferential direction. This phenomenon may cause the loss reduced.

Acknowledgments
This work is supported by Zhejiang Province Key Science and Technology Innovation Team Project (2013TD18) and the National Natural Science Foundation of China (51579224).

References

[1] Li X H, He H W, Gong W Q and Huang S J, 2002, Numerical simulation of steady flow for the whole machine in centrifugal fan, *Journal of engineering thermal physics*, **23**(4):453-456 (in Chinese).

[2] Zhao L F, 2005, Flow field simulation optimization design of centrifugal fan, *PhD Thesis of Dalian Jiaotong University* (in Chinese).

[3] Peng F, 2004, Three dimensional numerical calculation and research of centrifugal fan, *PhD Thesis of Wuhan University of Technology* (in Chinese).

[4] Shih T H, Liou W W, Shabbir A, Yang Z and Zhu J, 1995, A new k-ε Eddy-viscosity model for high Reynolds number turbulent flows, *Computer & Fluids*, **24**(3): 227-238.

[5] Bhope D V, Padole P M, Bhope D V and Padole P M, 2004, Experimental and theoretical analysis of stresses, noise and flow in centrifugal fan impeller, *Mechanism & Machine*

Theory, **39**(12): 1257-1271.

[6] Liu G Z, Wang X J and Gao Y, 2004, Function curve of the optimum attack angle of centrifugal fan, *Fan technology*, (3): 1-3 (in Chinese).

[7] Pan D L, 2009, Discussion on the optimum attack angle of centrifugal fan, *Fan technology*, (1): 13-15 (in Chinese).

[8] Li Q Y, 1981, Fan, *Mechanical industry press*, 20-25 (in Chinese).

[9] Shenyang Institute of blower and Northeast Engineering Institute, 1984, Centrifugal fan, *Mechanical industry press*, 7-11 (in Chinese).

[10] Dou H S, 1986, The method of determining the geometrical angle of blades at impeller entrance of centrifugal compressors, *Fan technology*, (4): 6-10 (in Chinese).

[11] Gu F, Engeda A, Cave M and Liberti J L D, 2001, A Numerical Investigation on the Volute/Diffuser Interaction Due to the Axial Distortion at the Impeller Exit, *Journal of Fluids Engineering*, **123**(3):475-483.

[12] Jeon W H and Lee D J, 2003, A numerical study on the flow and sound fields of centrifugal impeller located near a wedge,*Journal of Sound & Vibration*, **266**(4):785-804.

[13] Wu C, Zhao J, Sai Q Y, Li M, Zhou D S and Hu S G, 2001, Study on the influence of impeller structure on the flow characteristic and performance of centrifugal fan, *Mechanical research and application*, (6): 34-36 (in Chinese).

[14] Cheng X D, 2006, Centrifugal fan, *Beijing: Chemical Industry Press*, 4-8 (in Chinese).

[15] Wu Y L, Chen Q G and Liu S H, 2011, Fan and compressor, *Tsinghua University press*, 20-24 (in Chinese).

[16] Yu J M, Li S and Hao J Z, 2011, Numerical analysis of the flow characteristic of centrifugal fan, *Fluid machinery*, **39**(9): 34-37 (in Chinese).

IOP Conf. Series: Materials Science and Engineering **129** (2016) 012060 doi:10.1088/1757-899X/129/1/012060

Numerical simulation and analysis of the flow in a two-staged axial fan

J Q Xu [1], H S Dou*[1], H X Jia[1], X P Chen[1], Y K Wei[1] and M W Dong[2]

[1] Faculty of Mechanical Engineering and Automation, Zhejiang Sci-Tech University, Hangzhou 310018, China

[2] Zhejiang Shuangyang Fan CO. Ltd. Shangyu, 312300, China

*Email: huashudou@yahoo.com

Abstract. In this paper, numerical simulation was performed for the internal three-dimensional turbulent flow field in the two-stage axial fan using steady three-dimensional in-compressible Navier-Stokes equations coupled with the Realizable turbulent model. The numerical simulation results of the steady analysis were combined with the flow characteristics of two-staged axial fan, the influence of the mutual effect between the blade and the vane on the flow of the two inter-stages was analyzed emphatically. This paper studied how the flow field distribution in inter-stage is influenced by the wake interaction and potential flow interaction of mutual effect in the impeller-vane inter-stage and the vane-impeller inter-stage. The results showed that: Relatively, wake interaction has an advantage over potential flow interaction in the impeller-vane inter-stage; potential flow interaction has an advantage over wake interaction in the vane-impeller inter-stage. In other words, distribution of flow field in the two inter-stages is determined by the rotating component.

1. Introduction

In recent years, the rotor-stator interaction phenomenon of the axial fan has attracted a lot of researchers. The aerodynamic performance and the strength of structure such as the efficiency, operating range ,the running stability and etc. And all the performance of an axial fan is influenced by the relative motion between the blade and guide blade . The rotor-stator interaction in fluid machine mainly contains the wake interaction in the upstream row , channel and gap and the in-viscid potential flow interaction in the blade and vane [1]. Studying the influence of the different aspects of the rotor-stator interaction can be helpful to design of axial fan and can be improve aerodynamic performance of the fan .

Many scholars have done a lot of research. In the field of experiment .Wang et al [2] used dynamic pressure sensor to measure the upstream vane rotor blade wake effects on unsteady surface pressure spectral characteristics. The measurement results showed that the trailing edge separation flow generated by the pressure fluctuation spectrum contains the upstream blade wake passing frequency and its harmonics component . Wu et al [3] used particle image velocity on static wake and dynamic blades mutual interference flow field with experimental measurements it is concluded that with the effect of the stator wake, the fluctuations of the inlet flow angle on roter-blade become larger, resulting in the intake airflow uneven .The passage [4] used the hot-film to do the research on the flow field of rotor-stator interference and the pressure fluctuation of the axial turbo-machinery .The study of the experiment showed that the influence of the stator wake on the downstream of the rotor blade under two different situations of static and dynamic axial clearance . The passage [5] studied vane wake's effects on the blade row between the quasi-orthogonal plane flow structure, under the condition of the vanes and the rotor axial distance changes. In the field of numerical simulation, Mao et al [6] studied the influence of wake and potential effects on the dynamic and static blade surface aerodynamic load. The results showed that upstream dynamic blade wake had great impact on the generation and transportation of the blade boundary layer flow losses. Qi et al [7] studied the influence of the rotor-stator interaction on the aerodynamic performance of the movement of axial turbine stages. Yang et al [8] analysised superposition effect of the potential flow and wake interference of the flow compressor . As for the analysis of the flow the fluid mechanical inter-stages. Xiao and Wang [9-10] from Huazhong University of Science and Technology have done the numerical simulation on the mutual interference between rotating axial flow fan stages, the conclusion that in the counter rotating axial flow fan, wake interference of the front stage impeller is stronger than that of the end stage impeller. Yang et al [11] from Jiangsu University had done the large eddy simulation on the axial flow band with the guide vane.

The former mainly studied the rotor-stator interaction from the following aspects: inter-stage spacing, number of the impeller blades and vanes blades, and the relative location of static and dynamic blades. The majority focused on single-stage pump or fan and mainly discussed of the rotor-stator interaction influence in the flow channel. But the numerical simulation about the rotor-stator interaction of two-stage axial fan with vanes in the middle is few.

In this paper, numerical simulation is performed for the internal three-dimensional turbulent flow field in the two-stage axial fan using steady three-dimensional in-compressible Navier-Stokes equations coupled with the Realizable turbulent model, the influence of the mutual effect between the blade and the vane on the flow of the two stages is analyzed emphatically. This paper studied how the flow field distribution in stage is influenced by the wake interaction and potential flow interaction of mutual effect in the impeller-vane stage and the vane-impeller stage. It also provide reliable information to the further research of wake and potential flow interaction in impeller and vane.

2. Governing equation and numerical method

2.1. Geometric model

The two-stage axial fan with vane is produced by a Zhe Jiang Company. The main structural features show that its impeller design considered the pressure loss changing along the radial direction . The value of the chord length of the blade, adjustable mounting angle and curvature of the center line of the plate-shaped blades are changing according to certain rules. Its characteristics are that the maximum efficiency of point is consistent with the design point. The two-stage axial fan axial fan can not only show the large flow characteristics, but also improve the pressure head of the axial fan through gas working to impellers.

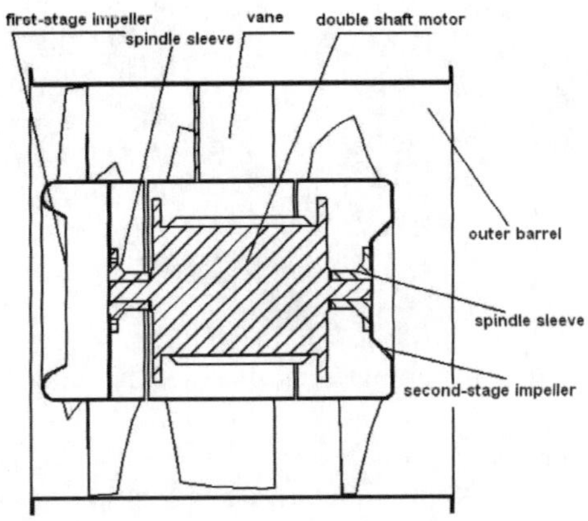

Figure 1. Two-staged Axial Fan structure diagram

The fan mainly contains impeller, guide vane motor, and the outer cylinder. The specific circumstances of the structure are shown in Figure 1. There are six blades in the first stage, nine blades in the guide vane, and five blades in the second-stage. Specific parameters are shown in Table 1.

Table 1. Main parameters of axial fan.

parameters	values
Hub ratio	0. 54
Blade number of first-stage fan	6
Blade number of second-stage fan	5
Blade numbers of vane	9
Rotational speed（r/min）	960
Volume rate of flow（m^3/h）	68000
total pressure （pa）	710
Total pressure efficiency (%)	52

2.2. Computational domain mesh

Numerical Simulation used mesh, as shown in Figure 2. For clearly show the axial fan mesh, here were omitted inlet fluid computational domain and outlet fluid computational domain. Each computational domain was generated the appropriate distribution of mesh nodes and appropriate size of mesh separately, for mesh quality control, improve accuracy in numerical simulation. Entire two-staged axial fan was divided into five computational domain, including inlet, first-stage impeller, vane impeller, second-stage impeller and outlet computational domain. Mesh number of each part were 0.19 millions, 4.5 million, 1.5millions, 3.6million and 0.25 million, giving a total of 10.04 million. Inlet domain, vane domain, outlet domain were static. First-stage impeller domain and second-stage impeller domain were rotating.

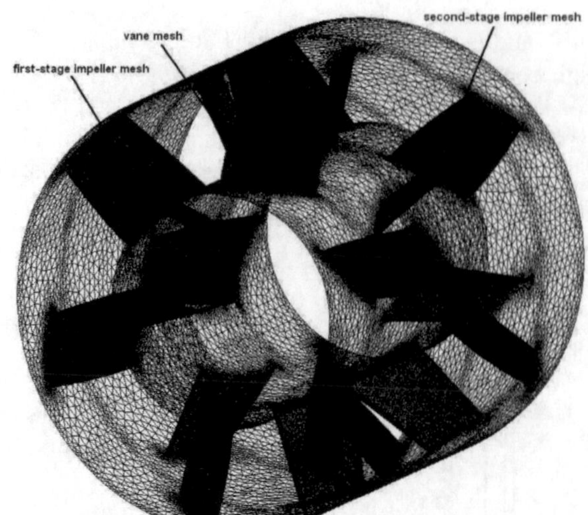

Figure 2. Computational mesh of the axial fan

2.3. Computational method and boundary conditions

The inlet boundary conditions was mass flow inlet on design condition, the outlet boundary conditions was an atmospheric pressure. Turbulent kinetic energy and turbulent kinetic energy dissipation rate was estimated by using empirical formula in inlet and outlet. In this model, governing equation was dispersed using finite volume method in the space, Steady computing used Second order upwind difference, The method of solving equations used SIMPLE algorithm. Computational fluid is 25℃ of air. The outlet boundary conditions was no slip boundary condition [12], Near-wall region used the standard wall function. Using this turbulence model could get better numerical simulation results for the problem of involving rotating, adverse pressure gradient, boundary-layer separation, reflux, and secondary flow.

2.4. Governing equation

Governing equation is three-dimensional in-compressible Navier－Stokes equations:

$$\begin{cases} \dfrac{\partial \overline{u_i}}{\partial x_i} = 0 \\ \rho u_j \dfrac{\partial \overline{u_i}}{\partial x_j} = \rho F_i - \dfrac{\partial \overline{p}}{\partial x_i} + \mu \dfrac{\partial^2 \overline{u_i}}{\partial x_j \partial x_j} - \rho \dfrac{\partial}{\partial x_j} (\overline{u_i u_j}) \end{cases} \tag{1}$$

Turbulence model was Realizable k-ε model. This model was proposed by Shih et al [13] in 1995. Constraint equation is as follows:

$$\frac{\partial(\rho k)}{\partial t} + \frac{\partial(\rho k u_j)}{\partial x_j} = \frac{\partial}{\partial x_j}\left[\left(\mu + \frac{\mu_t}{\sigma_k}\right)\frac{\partial k}{\partial x_j}\right] + \rho(P_k - \varepsilon) \tag{2}$$

$$\frac{\partial(\rho \varepsilon)}{\partial t} + \frac{\partial(\rho \varepsilon u_j)}{\partial x_j} = \frac{\partial}{\partial x_j}\left[\left(\mu + \frac{\mu_t}{\sigma_\varepsilon}\right)\frac{\partial \varepsilon}{\partial x_j}\right] + \rho C_1 E \varepsilon - \rho C_2 \frac{\varepsilon^2}{k + \sqrt{v\varepsilon}} \tag{3}$$

Among:

$$C_1 = \max\left(0.43, \frac{\eta}{\eta + 5}\right)$$

$$\eta = \left(2E_{ij} \cdot E_{ij}\right)^{1/2} \frac{k}{\varepsilon}$$

$$E_{ij} = \frac{1}{2}\left(\frac{\partial u_i}{\partial x_j} + \frac{\partial u_j}{\partial x_i}\right)$$

Constant in the equation: $\sigma_k = 1.0$, $\sigma_\varepsilon = 1.2$, $C_2 = 1.9$

2.5. Mesh independent verification

Computational mesh was independent verification on design condition of the two-stage axial fan. Three sets of mesh were designed; the mesh numbers were 8.06 million, 10.04 million and 12.03 million. Total pressure is shown in Figure 3 at the different number of mesh. From the figure, the total pressure is preferably Convergence on design condition. The total variation range of total pressure was about 3%. It meets mesh-independent; 10.04 million number of computational Mesh was selected in this paper.

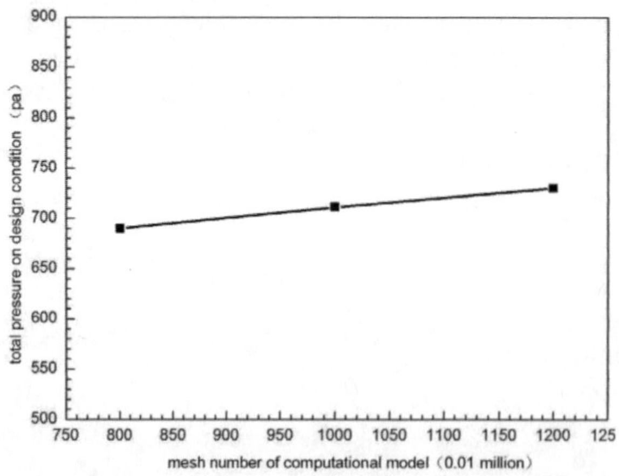

Figure 3. Mesh independent verification

3. Calculation results and discussions

The three dimensional flow in a two-staged axial fan has its inherent characteristic of non constant. Since a two-staged axial fan is composed of two impellers and a middle guide vane, there are two stages: the impeller-vane inter-stage and the vane-impeller inter-stage. Because of the rotor-stator interaction, the flow of the two-staged axial fan is very complex. This paper will focus on the influence of flow field in inter-stage by the wake and the potential flow.

To analyze how the distribution of flow field in inter-stage is influenced by the wake interaction and the potential flow interaction of mutual effect in the impeller-vane inter-stage and the vane-impeller inter-stage, axial distance in two inter-stages is divided into 4 equal parts, gotA1、A2、A3、A4 and B1、B2、B3、B4 sections, as shown in Figure 4.

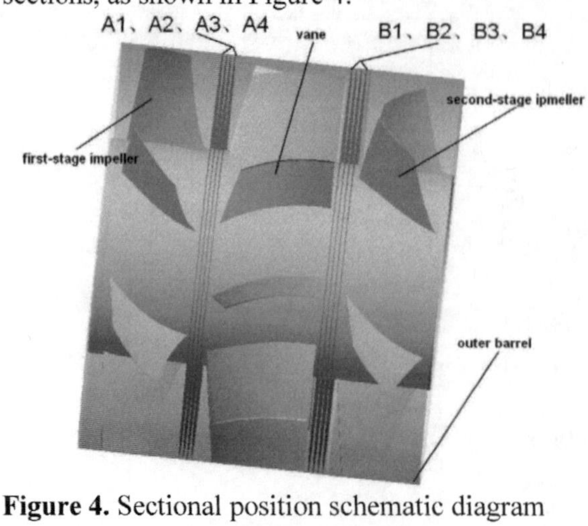

Figure 4. Sectional position schematic diagram

3.1. Analysis of the pressure, velocity and vorticity distribution of the axis in the two inter-stage

In order to study the fluid flow of the impeller-vane inter-stage and the vane-impeller inter-stage, this section analyzes and studies the total pressure, static pressure, velocity and vorticity of the fluid flowing through two stages.

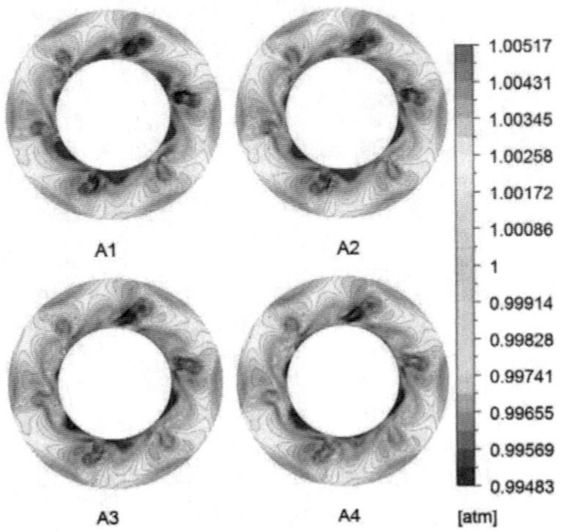

Figure 5. Total pressure distribution in the impeller-vane inter-stage.

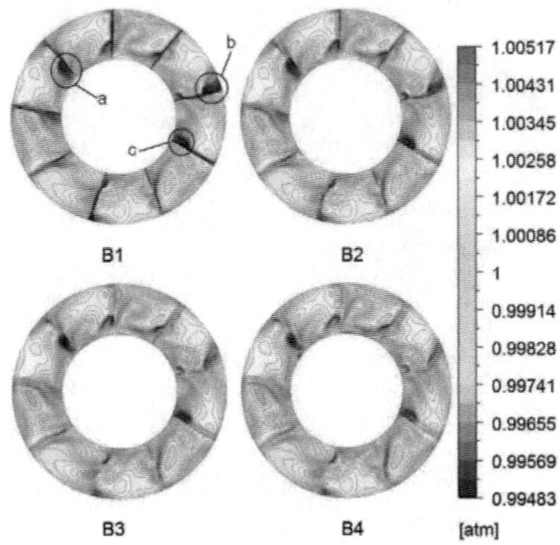

Figure 6. Total pressure distribution in the vane-impeller inter-stage.

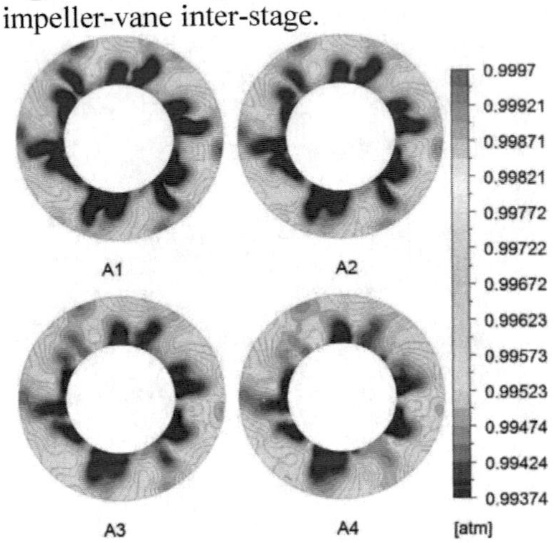

Figure 7. Static pressure distribution in the impeller-vane inter-stage.

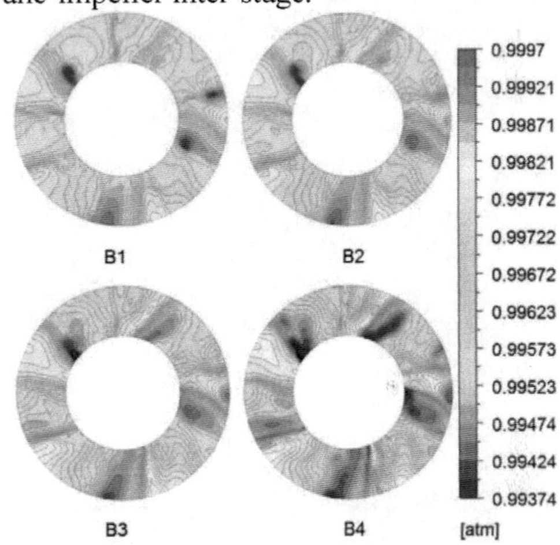

Figure 8. Static pressure distribution in the vane-impeller inter-stage.

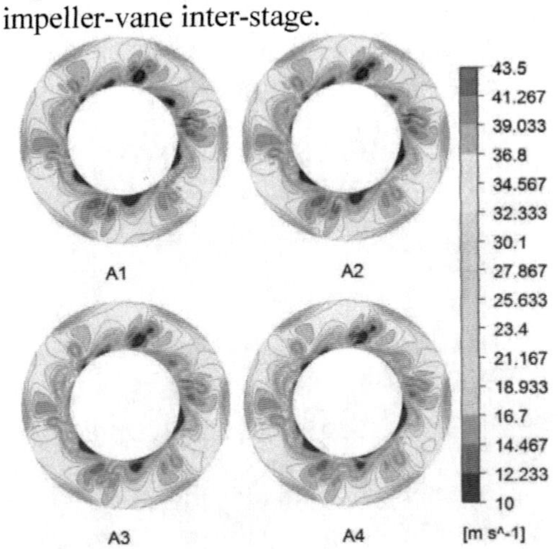

Figure 9. Velocity distribution in the impeller- vane inter-stage.

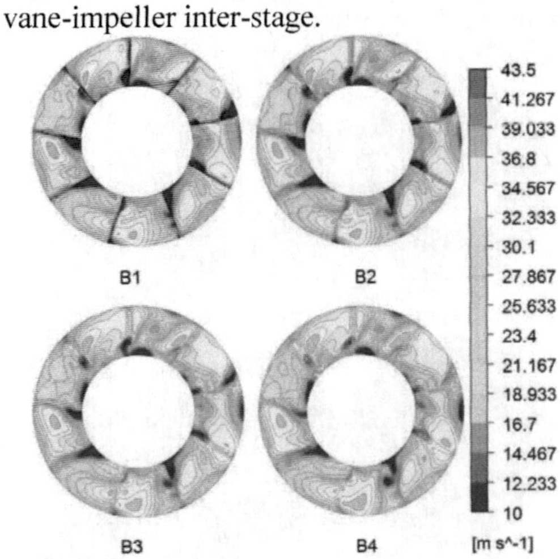

Figure 10. Velocity distribution in the vane-impeller inter-stage.

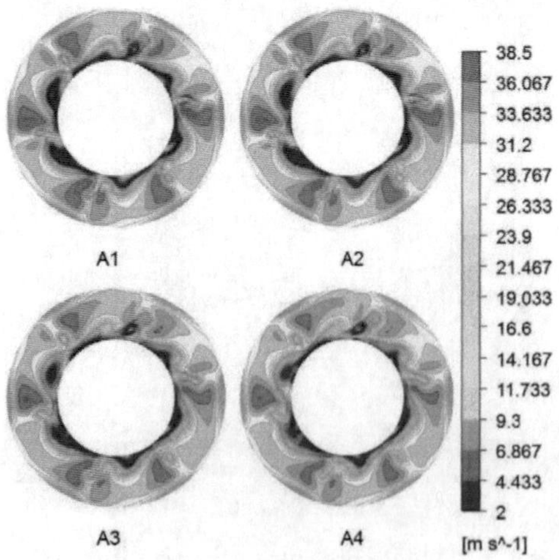

Figure 11. Axial velocity distribution in the impeller-vane inter-stage.

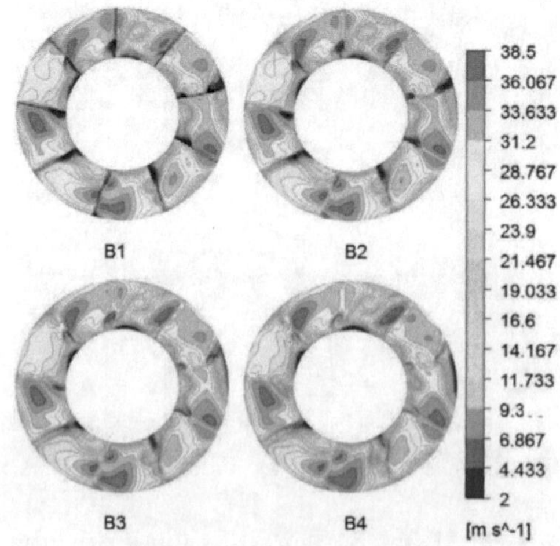

Figure 12. Axial velocity distribution in the vane-impeller inter-stage.

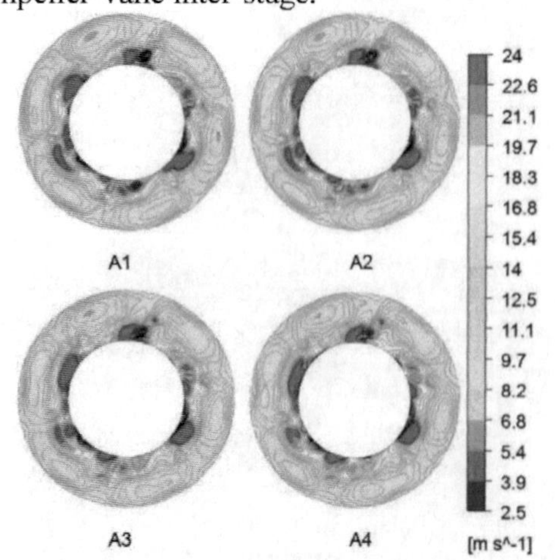

Figure 12. Circumferential velocity distribution in the impeller-vane inter-stage.

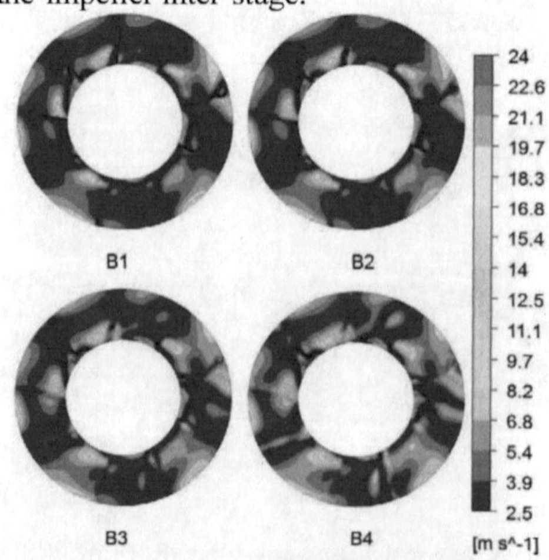

Figure 14. Circumferential velocity distribution in the vane-impeller inter-stage.

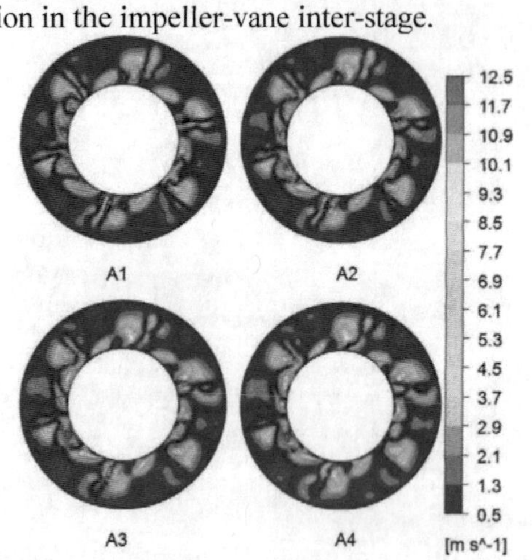

Figure 15. Radial velocity distribution in the

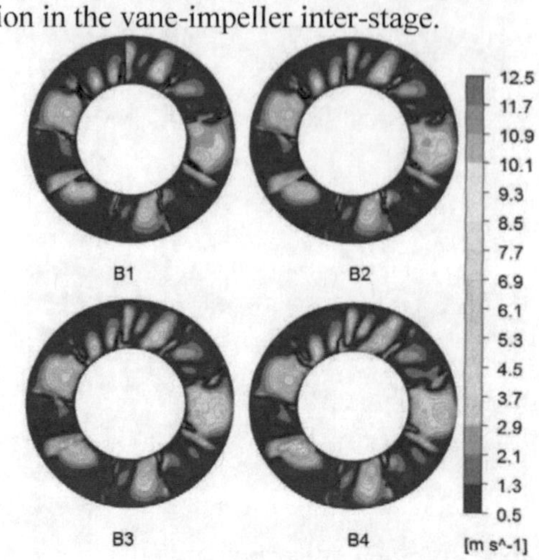

Figure 16. Radial velocity distribution in the

IOP Conf. Series: Materials Science and Engineering **129** (2016) 012060 doi:10.1088/1757-899X/129/1/012060

impeller-vane inter-stage.

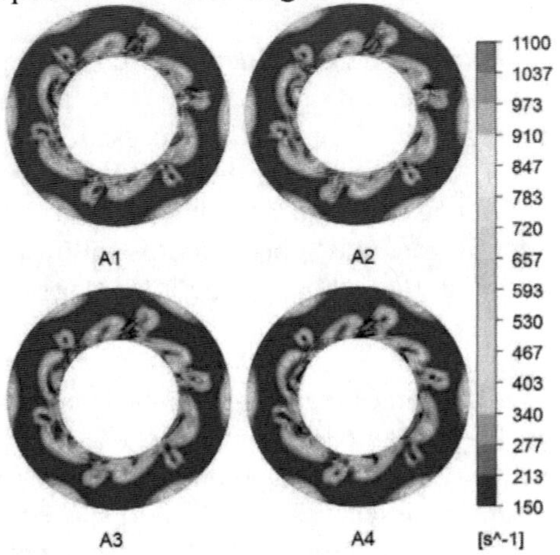

Figure 17. Vorticity distribution in the impeller-vane inter-stage.

vane-impeller inter-stage.

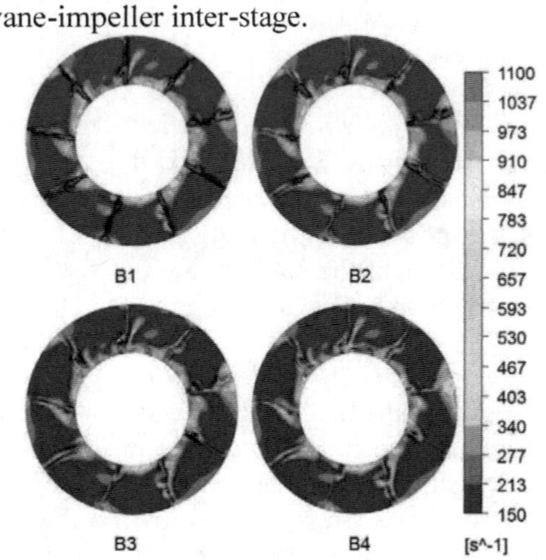

Figure 18. Vorticity distribution in the vane-impeller inter-stage.

3.1.1. Distribution of flow field in the impeller-vane inter-stage

From Figure 5, the total pressure distribution cloud chart of the section of the impeller-vane inter-stage can be clearly seen, and the total pressure distribution is symmetric along the circumferential direction, showing the first stage impeller structure characteristics (the number of the first stage impeller blade is 6). There are 6 non mainstream areas and 6 main flow areas. This trend has continued into the leading edge of the inlet. This shows that the characteristics of the first stage impeller are the main influence of the unsteady influence of the impeller-vane inter-stage.

From Figure 7, the static pressure distribution cloud chart of the section of the impeller-vane inter-stage can be seen, and the static pressure distribution is symmetric along the circumferential direction, showing the first stage impeller structure characteristics. There are 6 wake areas and 6 main flow areas. The structural characteristics of the first stage impeller of the A2 to A3 and A4 section are obviously weakened. This shows that the static pressure of the fluid in the A1 and A2 section of the first stage impeller has a major impact on the flows; the A3 and A4 section are influenced by the vane potential flow. From Figure 7 it can be seen that the static pressure in the axial direction is not obvious, and the reverse pressure gradient is very small. This also shows that the influence of the first stage impeller on the flow of the impeller is stronger than that of the vanes.

From Figure 9 the velocity distribution cloud chart of impeller-vane inter-stage and Figure 11 the axial velocity distribution cloud chart of impeller-vane inter-stage which can be clearly seen, the distribution of velocity and axial velocity is the same as that in Figure 5. The structural characteristics of the first stage impeller are presented in each section. There are 6 main areas and 6 non mainstream areas. Figure 13 the section of impeller-vane inter-stage is the velocity chart, which can be clearly seen, 6 main flow areas and 6 non mainstream areas are demonstrated by the section images. The structural characteristics of the first stage impeller are the main position, and the position of the circumferential velocity is at the root of the leaf. From Fig. 15 the radial velocity of section contours of impeller-vane inter-stage which can be clearly seen, there are the 12 main flow areas and 12 non mainstream areas strong or weak showed in the four sections. The number of the first stage impeller blade is 6, which is just 2 times. This shows that the characteristics of the first stage impeller in the radial velocity field of the impeller-vane inter-stage are the main effects. Even in A3 and A4 sections, the structural characteristics of the first stage impeller are obvious, and the radial velocity is disturbed by the vane.

From Figure 17, it is not difficult to find the characteristics of the first stage impeller 6 blades in each section. The vorticity distribution in each section is symmetric, and the influence of the vane is not affected. It also shows that the first stage impeller has a major influence on the distribution of the vorticity field.

From the above analysis it can be concluded that the total pressure, velocity, circumferential velocity, radial velocity and vorticity in all of the profiles show a strong first level of structural characteristics of the impeller, and the static pressure in the A1 and A2 section also presents a very strong first stage impeller structure characteristics, only the A3, A4 section are influenced by the vane, that is, the influence of the first stage impeller of the impeller-vane inter-stage is the main.

The analysis of the section cloud of the impeller-vane inter-stage can be concluded as follows: the wake interaction has an advantage over potential flow interaction in the impeller-vane inter-stage.

3.1.2. Distribution of flow field in the vane-impeller inter-stage

From Figure 6, the total pressure distribution cloud chart of the section of the vane-impeller inter-stage can be clearly seen, and the total pressure distribution on the B1 and B2 sections show a strong unsteady nature. This is due to the vane-impeller inter-stage flow field in the vane outlet flow not only by the vane wake effect, but also by the effect of the first stage impeller wake. A, B, and C are three distinct asymmetric non mainstream regions from the total pressure contours of B1 section. This is because the mixed results of the first impeller wake which is cut by the vane and the vane wake. The total pressure distribution in B1 and B2 section is symmetric along the circumferential direction, showing vane structure characteristics (The number of vane blade is 9). There are 9 wake areas and 6 main flow areas. While the B3 and B4 sections can also be clearly seen in the structural characteristics of the vanes, the structure features of the second stage impeller has already begun to appear on the section by the influence of the second stage (the number of the second stage impeller blade is 5). There are 5 high pressure zones and 5 low pressure zones. Even the structural features of the impeller on the upstream B1 section are clearly visible.

From Figure 8, the static pressure distribution cloud chart of the section of the vane-impeller inter-stage can be clearly seen. Except of the first stage impeller and vane wake superposition area, the vane blade wake has little influence to the static pressure of the vane-impeller inter-stage. The static pressure distribution along the circumferential direction is symmetrical, and the structure of the second stage impeller is presented. There are 5 high pressure zones and 5 low pressure zones. This shows that the characteristics of the second stage impeller are the main influence of the unsteady influence of the impeller.

From Figure 10 and Figure 12, the velocity distribution cloud chart and axial velocity distribution cloud chart of the section of the vane-impeller inter-stage can be clearly seen. The distribution of velocity and axial velocity is the same as that in Figure 6. Only the B3 and B4 section show a stronger structure of the second stage impeller, Even in the B1 and B2 section, there are 5 regions has the maximum velocity and the axial flow velocity. The structural characteristics of the second stage impeller are presented. From Figure 14 the section circumferential velocity distribution cloud chart of the vane-impeller inter-stage we can see that the structural characteristics of the vane are not. And a clear structure of the second stage impeller can be seen in this section, and this feature closed to the second stage impeller is also more obvious. From Figure 16, the radial velocity distribution cloud chart of the section of the vane-impeller inter-stage can be clearly seen; radial velocity field in each section shows a strong structural characteristic of the second stage impeller. Only in the radial B1 and B2 section, the structure of the vane can be seen not clearly.

From Figure 18, the vorticity distribution cloud chart of the section of the vane-impeller inter-stage can be clearly seen; sections from B1 to B4 have been showing the characteristics of leaf structure. But in the B3 and B4, the structure of the vane is obviously weakened due to the interference of the second stage impeller potential flow. This shows that in the area near the second level of the impeller, the amount of the vortex field is affected by the second stage impeller.

From the above analysis it can be concluded that Static pressure, circumferential velocity and radial velocity in all of the profiles show a strong second level of structural characteristics of the impeller, and the total pressure, velocity, vorticity, in the A1 and A2 section also presents a very strong vane structure characteristics,. But because of the impact of the second stage impeller, the structural characteristics of the second stage impeller are presented in the A3, A4 section. In relative terms, the influence of second - stage impeller in the inner flow field of the vane-impeller inter-stage is stronger than that of the vanes.

The analysis of the section cloud of the vane-impeller inter-stage can be concluded as follows: the potential flow interaction has an advantage over the wake interaction in the impeller-vane inter-stage. The reason leading to this phenomenon is that the second stage impeller acts on the gas, increases the gas pressure, and produces high adverse pressure gradient.

3.2. Quantitative analysis of distribution of flow field in the inter-stages

To study more clearly on how the distribution of flow field in inter-stage is influenced by the wake interaction and the potential flow interaction. Parameters are extracted in the two inter-stages at the half of Radius, plotted and analyzed. As shown in Figure 19 of position of curve diagram. Method to get Curves is same method to get sections as shown in Figure 4.

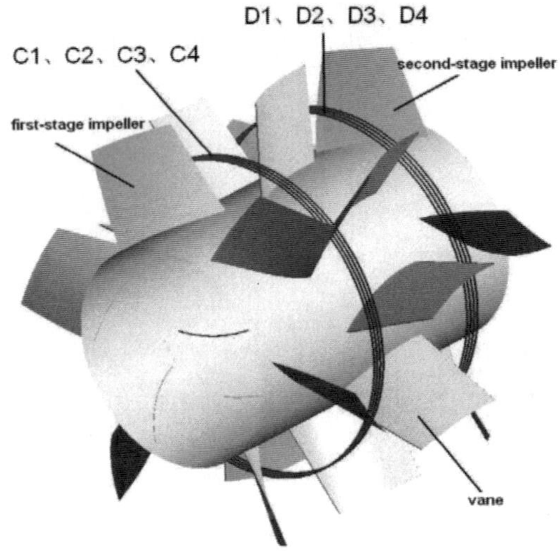

Figure 19. Position of curve schematic diagram

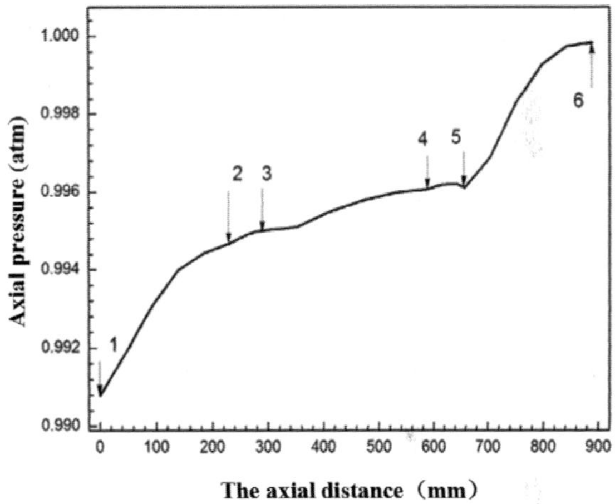

Figure 20. Pressure changes along the axial direction

From the Figure 20 (1-2 is the first-stage impeller, 2-3 is the impeller-vane interstage3-4 is vane, 4-5 is the vane-impeller inter-stage, 5-6 is the second-stage impeller), It is obvious that pressure gradient changes gently near the impeller-vane inter-stage, and pressure gradient changes sharply near the vane-impeller inter-stage. That is the reason why the wake interaction produced dominant effect in the impeller-vane inter-stage, the potential flow interaction produced dominant effect in the vane-impeller inter-stage. This is consistent with the qualitative analysis of the foregoing.

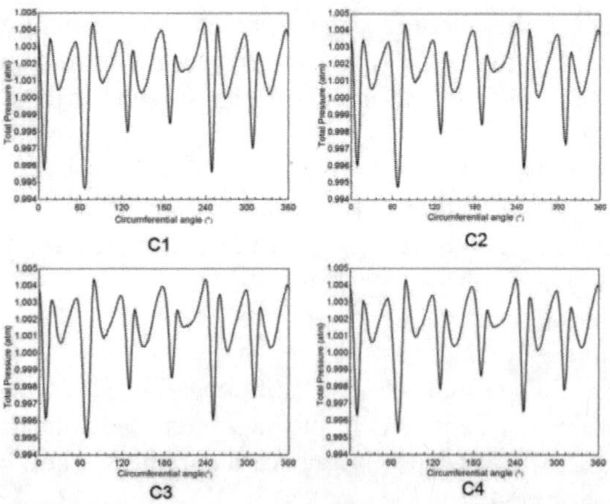

Figure 21. Total pressure distribution along the axial direction in the impeller-vane inter-stage.

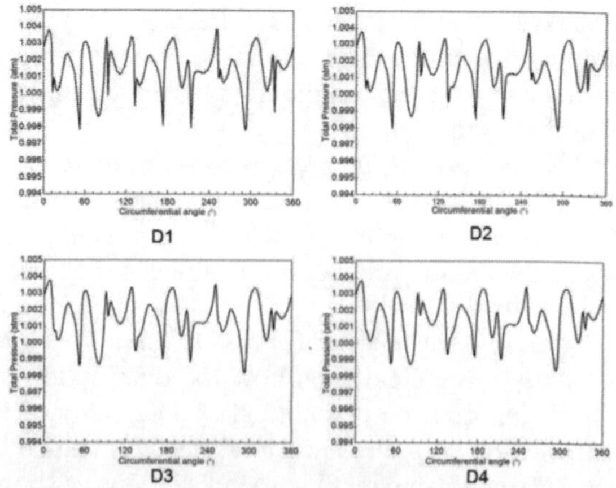

Figure 22. Total pressure distribution along the axial direction in the vane-impeller inter-stage.

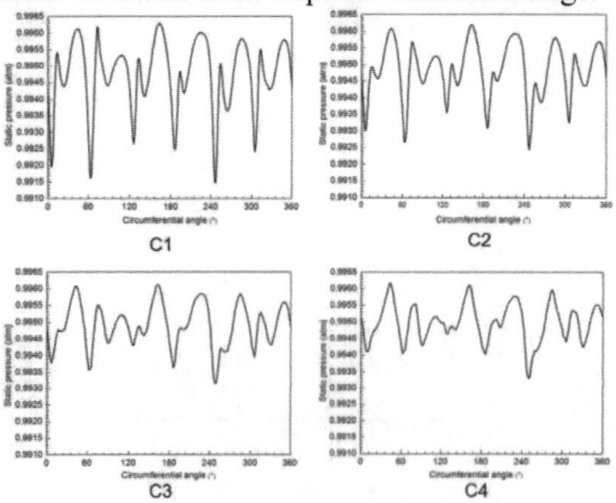

Figure 23. Static pressure distribution along the axial direction in the impeller-vane inter-stage.

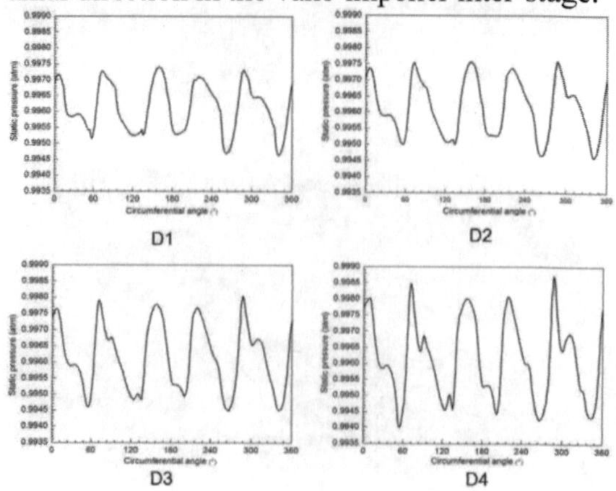

Figure 24. Static pressure distribution along the axial direction in the vane-impeller inter-stage.

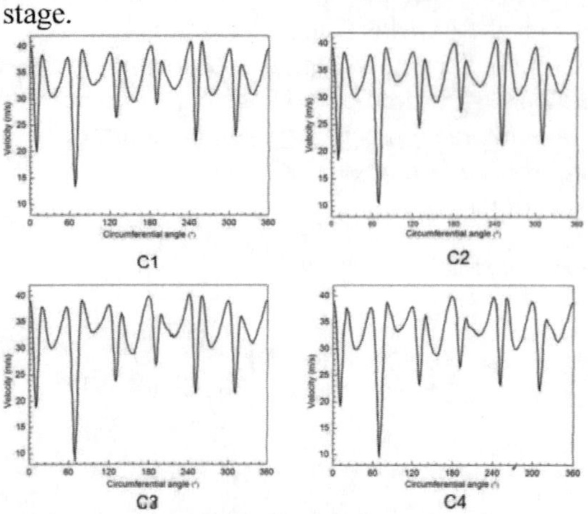

Figure 25. Velocity distribution along the axial direction in the impeller-vane inter-stage.

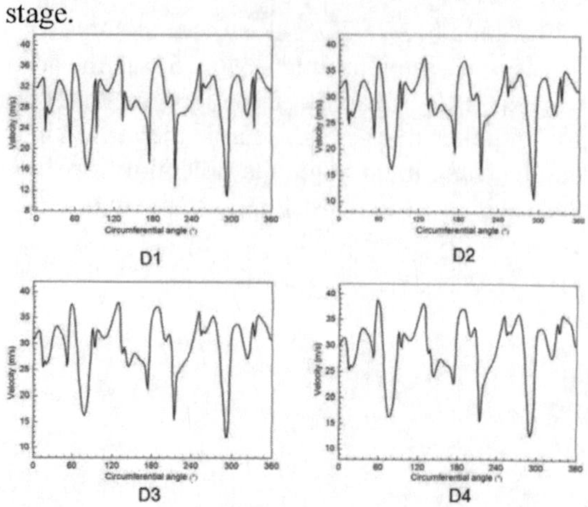

Figure 25. Velocity distribution along the axial direction in the vane-impeller inter-stage.

From the Figure 21, it is obvious that total pressure distribution along the circumferential direction show regularity of 12 peaks and 12valleys, and has 6 main peaks and 6 main valleys, Presents structural feature of first-stage impeller. From the Figure 22, It shows that total pressure distribution along the circumferential direction is very complicated, But It's obvious that in each section, there are 10 main peaks and 10 main valleys, the number is just about double in blade number of second-stage impeller, so it shows structural feature of second-stage impeller.

From the Figure 23, it is obvious that Static pressure distribution along the circumferential direction, show regularity of 12 peaks and 12valleys, and have 6 main peaks and 6 main valleys, Presents structural feature of first-stage impeller. From the Figure 24, it is obvious that in each curve, there are 5 peaks and 5 valleys, shows structural feature of second-stage impeller.

From the Figure 25, It is obvious that velocity distribution along the circumferential direction show regularity of 12 peaks and 12valleys, Presents structural feature of first-stage impeller. From the Figure 26, It shows that total velocity distribution along the circumferential direction is very complicated, But It's obvious that in each section, there are 5 main peaks and 5 main valleys, the number is just same in blade number of second-stage impeller, so it shows structural feature of second-stage impeller.

In summary, total pressure distribution, Static pressure distribution and velocity distribution along the circumferential direction is consistent with the qualitative analysis of the foregoing. Relatively, wake interaction has an advantage over potential flow interaction in the impeller-vane inter-stage; Potential flow interaction has an advantage over wake interaction in the vane-impeller inter-stage.

4. Conclusions

In this paper, Numerical simulation is performed for the internal three-dimensional turbulent flow field in the two-stage axial fan. To analyze how the distribution of flow field in inter-stage is influenced by the wake interaction and the potential flow interaction in the impeller-vane inter-stage and the vane-impeller inter-stage. This paper has the following conclusions:

(1) The wake interaction has an advantage over potential flow interaction in the impeller-vane inter-stage.

(2) The potential flow interaction has an advantage over wake interaction in the vane-impeller inter-stage.

(3) Distribution of flow field in the two inter-stages is determined by the rotating component.

Acknowledgments

This work is supported by Zhejiang Province Key Science and Technology Innovation Team Project (2013TD18) and the National Natural Science Foundation of China (51579224).

References

[1] Yang J, Qiao W Y, Shi P J, Wang P W.2012 Primitive staged stator solidity of axial flow turbine to the rotor-stator interaction effects , *Journal of Aerospace Power*, **27**(8):1832-1840 (in Chinese).

[2] Wang Y f, Hu J, Luo B N, Li C P. 2006 Effects of the Up-stream Blade Wakes on the Spectrum of Rotor Blade Unsteady Surface Pressure, *Journal of Aerospace Power*, **21**(4):693-699 (in Chinese).

[3] Wu Y D, Zhu X C, OuY H, Tian J, Du C H. 2009 Experimental study of flow field of dynamic and static interference, DaLian: The Conference of Pneumatic thermodynamic heat engine of Chinese Society of Engineering Thermophysics (in Chinese).

[4] Png R, Joselyn H D, Hardin L W, Wagner J H. 1982 Turbine rotor-stator interaction, *Journal of Engineering for Gas Turbines & Power*, **204**(4):729-742.

[5] Gaetani P, Persico G, Osnaghi C. 2012 Effects of Axial Gap on the Vane-Rotor Interaction in a Low Aspect Ratio Turbine Stage, *Journal of Propulsion & Power,* **26**(2):325-334.

[6] Mao M M, Zhu Y P, Wang Z Q. 2007 Numerical research on influence of rotor-stator interactions in transonic compressor , *Journal of Aerospace Power*, **22**(9):1468-1474 (in Chinese).

IOP Conf. Series: Materials Science and Engineering **129** (2016) 012060 doi:10.1088/1757-899X/129/1/012060

[7] Qi M X, Feng Z P, Kang Shun, HIRSCH Charles. 2003 Influences of Rotor-stator Interaction on the Performance of Axial-flow Turbine Stages, *Journal of Engineering Thermophysics,* **24**(1):39-42 (in Chinese).

[8] Yang C, Zhang H Z, Zhao B, Ma Z C Lao D Z. 2015 Analysis of the Interaction Superposition Effect Between Potential Flow and Wake in Axial Compressor Blade Rows, *Journal of Engineering Thermophysics,* **36**(2):269-273 (in Chinese).

[9] Xiao P and Wang J. 2006 Numerical Study of Interference of Interstage Flows in a Counter rotating Cascade, *Journal of Engineering for Thermal Energy and Power,* **21**(3):249-254 (in Chinese).

[10] Xiao P. 2005 Steady Numerical Simulation of three-dimensional for the Mutual interference between multi-stage cascades, WuHan： *Huazhong University of Science and Technology* (in Chinese).

[11] Yang Y. 2013 LES-Based Simulation of Unsteady Flow Within the Axial Gap between Impeller and Diffuser of Axial-flow Pump, JiangSu： *Jiangsu University* (in Chinese).

[12] Qian Z D, Yang J D, Huai W X. 2007 Numerical simulation And Analysis Of Pressure Pulsation In Francis Hydraulic Turbine With Air Admission, *Journal of Hydrodynamics*, **19**(7):467-472.

[13] Shih T H, Liou W W, Shabbir A, Yang Z, Zhu J. 1995 A New $k-\varepsilon$ Eddy Viscosity Model For High Reynolds Number Turbulent Flows, *Comput Fluids,* **24**(3):227-238.

IOP Conf. Series: Materials Science and Engineering **129** (2016) 012061 doi:10.1088/1757-899X/129/1/012061

Experimental study on gas-liquid flow characteristics of submerged air jets

S J Qin, J T Liu, T C Miao and D Z Wu[1]

Institute of Process Equipment, College of Chemical and Biological Engineering, Zhejiang University
Zheda Road 38#, 310027 Hangzhou, Zhejiang, China

E-mail:qinshijie@zju.edu.cn

Abstract. The gas-liquid flow structure and interfacial behavior of submerged air jets were investigated experimentally using high speed digital video camera and image processing techniques. The jet pressure ratio varied from 1.8 to 4.8 in the experiment. And results from different jet nozzles were processed and compared. Statistical characteristics of the jet diameters along the axial distance were obtained and analyzed. Time series analysis was implemented to study the interface unsteadiness by calculating the gas-liquid interface deviation. The results showed that the jet diameters increase first linearly then nonlinearly and its growth rate decreases along the axial distance. The reason for the divergence between the result of this experiment and those done by other researchers was analyzed. Comparing the results of different pressure ratios and nozzle diameters, we found that larger jet pressure ratios have larger jet diameters and nozzle diameters nearly have no bearing on the distribution of dimensionless jet diameters. The interface unsteadiness in low and high pressure ratios exhibited totally distinct properties. And a minimum unsteady value was found along the axis of the air jets.

1. Introduction

Air jets submerged in liquids are quite common in many industrial applications. In chemical and metallurgical industry, such as aeration, fluidization, liquid metal stirring and gas-metal reactions, gas injections have been widely used for enhancing reaction efficiency, gas-liquid mixing, heat and mass transfer and etc. For underwater equipment applications, such as the conceptual new generation missile launching platform for submarines proposed by Yagla[1], the submerged gaseous region structure and motion are crucial for this unique design. Hence, understanding the flow structure and interface motion of submerged gas jets is relatively important[2].

Numerous previous studies[3-5] have been trying to gain a better understanding of the gas-liquid flow structure and motion at low and moderate injection pressure. Up to now, the characteristics of flow structure and interfacial behavior of submerged gas jets at low or moderate jet pressure have been widely researched and well understood. However, our understanding to the properties of submerged gas jets at high jet pressure is still obscure, because the flow structure and gas-liquid interface motion

[1] Corresponding author. Tel.: +86 571 87952406.
E-mail address: wudazhuan@zju.edu.cn (D. Wu).

are quite complicated owing to the extreme turbulence and vortex. Until recent decades, increasing investigations have been conducted at high pressure submerged gas jets.

Mori et al. (1982)[6] experimentally studied the bubbling/jetting transition point for nitrogen injected into a mercury bath. And the transonic flow was identified as the transition point to distinguish bubbling and jetting flow. McNallan and King (1982)[7] investigated the processes of several gas injected into several liquids. They also found that the transition point is near the sonic point. Then, using the measurements of pressure probe, γ -ray absorption and laser-Doppler anemometer, Loth and Faeth (1989)[8, 9] studied the structure and mixing properties of round gas jets at various under-expanded jetting conditions. Mean properties such as static pressure distributions, void-fraction distributions and entrainment rates were compiled and analyzed experimentally. Prediction models based on experimentally results were established and compared. Interestingly, flow widths based on void fractions exhibited unusually large flow widths, compared to typical single-phase jets. And they argued that the main reason was the multiphase mixing thick layer around. Besides, they also discovered that using intrusive probe to detect the properties of submerged gas jets may cause inaccuracy.

Recently, with the development of photographic and image processing techniques, method called flow visualization was applied by the researchers like Christopher[10] and Harby[11] to investigated the jet penetration distance, pinch-off events and the unsteadiness of the gas-liquid jet interface. Compared to the results of probe measurements, these non-invade measurement results can be more precise. Christopher considered that RT and KH instability mechanisms are both essential for driving the instability and mixing along the jet interface. Harby found a linear relationship between the dimensionless jetting half-width and axial distance. A function was delivered to predict this relationship using different coefficient for different nozzles. And the Froude number in their view has little effect on this relationship for a constant nozzle.

In this paper, compressed air with different pressure ratios injected from a vertical nozzle in stagnant water were investigated using high speed digital video camera and image processing techniques. Statistical characteristics of jet diameters along the axis of the nozzle were quantified. The influence of nozzle diameters and pressure ratios were analyzed. And the interface unsteadiness based on the calculated boundary of each frame was compared between low and high pressure jets.

2. Experimental method

2.1. Experimental apparatus

An experimental setup was applied to study the characteristics of two-phase flow from an underwater air jet. Figure 1 shows the schematic of the experimental Apparatus. Experiment was carried out in a water-filled tank (3000mm×1500mm×2000mm) with transparent windows on the sidewalls, which allows for visualization and optical supplement. The air jet was generated from a vertically upward nozzle installed on the bottom of the tank. The nozzle was located near the center of the tank and the exit plane of nozzle was kept 450 mm below the liquid surface, in order to ensure that the sidewalls had minimal effects on the flow and the camera can directly record the flow through the window. Two round nozzles made from stainless steel with inner diameters of 6 and 8.7mm were used in this investigation. All the nozzles had a constant length of 90mm and a uniform thickness of 1.5mm. The compressed air was delivered from an air line. Prior to each test, air was compressed into a gas tank (0.8m^3) to a high pressure (0.7Mpa), which could be regarded as a stable gas source during the test. Air volume flow rate was monitored by a pressure regulator valve and metered by a target flowmeter (SBL-BQ50CCZTPX). And the pressure and temperature of compressed air were recorded by transducers synchronously. Experimental parameters consisted stagnation pressure (P_S) inside the injector, hydrostatic pressure (P_H) in the nozzle exit plane, air temperature (T_0), pressure ratio (R_P) of stagnation pressure to hydrostatic pressure and mass flow (Q_M) are showed in Table 1. Here the hydrostatic pressure (P_H) was defined by equation (1). Pa and h refer to the barometric pressure and the constant depth of water.

$$P_H = P_a + \rho g h = 1.01 \times 10^5 + 1000 \times 9.8 \times 0.215 Pa = 0.105 MPa \quad (1)$$

A high-speed digital video camera (NAC Memrecam HX-3) was employed in this experiment. Flow images were taken by this camera at the frame rate of 1000 fps in conjunction with a Nikon AF Nikkor 60mm f 1.8d camera lens. And the exposure time was set automatic mode to obtain distinct images. Two 80w halogen lamps were put on the same side of the camera for illuminating the jet, while the surrounding stagnant water produced a dark background. Figure 2 (a) shows a flow image taken by high-speed digital camera. It is clear that the gas flow and the stagnant water can be distinguished significantly. All the images had a resolution of 768×960 pixels, and the pixels of per millimeter could be calculated from the nozzle's pixels on the image. Over 10 seconds of the gas-liquid flow were recorded under a steady flow rate, among which 2 seconds with 2000 images were taken to ensure a typical analysis.

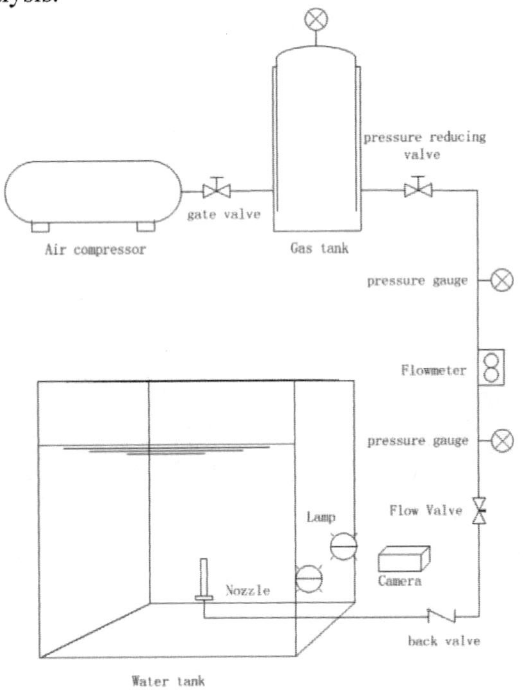

Figure 1. Schematic diagram of experimental set-up.

Table 1. Summary of experimental parameters.

D / mm	T_0 / K	P_S / MPa	P_H / MPa	R_P	Q_M / g/s	ρ / kg/m^3
6	289	0.185	0.105	1.8	147.4	2.24
	289	0.215	0.105	2.0	212.2	2.61
	289	0.295	0.105	2.8	368.0	3.58
	289	0.365	0.105	3.5	552.7	4.43
	289	0.425	0.105	4.0	760.7	5.16
	289	0.485	0.105	4.6	983.7	5.89
	289	0.505	0.105	4.8	1130.3	6.13
8.7	288	0.190	0.105	1.8	360.6	2.61
	288	0.215	0.105	2.0	545.8	3.22
	288	0.305	0.105	2.9	882.0	4.07
	288	0.370	0.105	3.5	1243.0	4.80
	288	0.425	0.105	4.0	1732.7	5.65

288	0.480	0.105	4.6	2292.1	6.50
288	0.505	0.105	4.8	2590.3	6.74

2.2. Image processing method

As the main purpose of this work was to study the characteristics of the gas-liquid flow, images taken by camera were processed in MATLAB using the image processing module, just as the method mentioned by other researchers[12, 13] to obtain the interface of gas and liquid. Figure 2 shows the principal steps to detect the jet interface. Firstly, the image was changed into a gray image and then converted into a binary image using a threshold of pixel intensity. An appropriate threshold was selected to obtain a precise jet (white) and the background (black) area. Then, a median filter and "imfill" function were applied to the images, aiming to eliminate the noise caused by non-uniform illumination and separated gas bubbles. After that, two morphological functions called dilate and erode were used to smooth the boundary of the air jet and background. Finally, edge detection function detected the boundary and stored the coordinates in an array for further analysis. As saw in the final image of figure 2, the detected boundary (red line) agrees quite well with the original image. Namely, the interface of gas and liquid from a submerged air jet is precisely extracted.

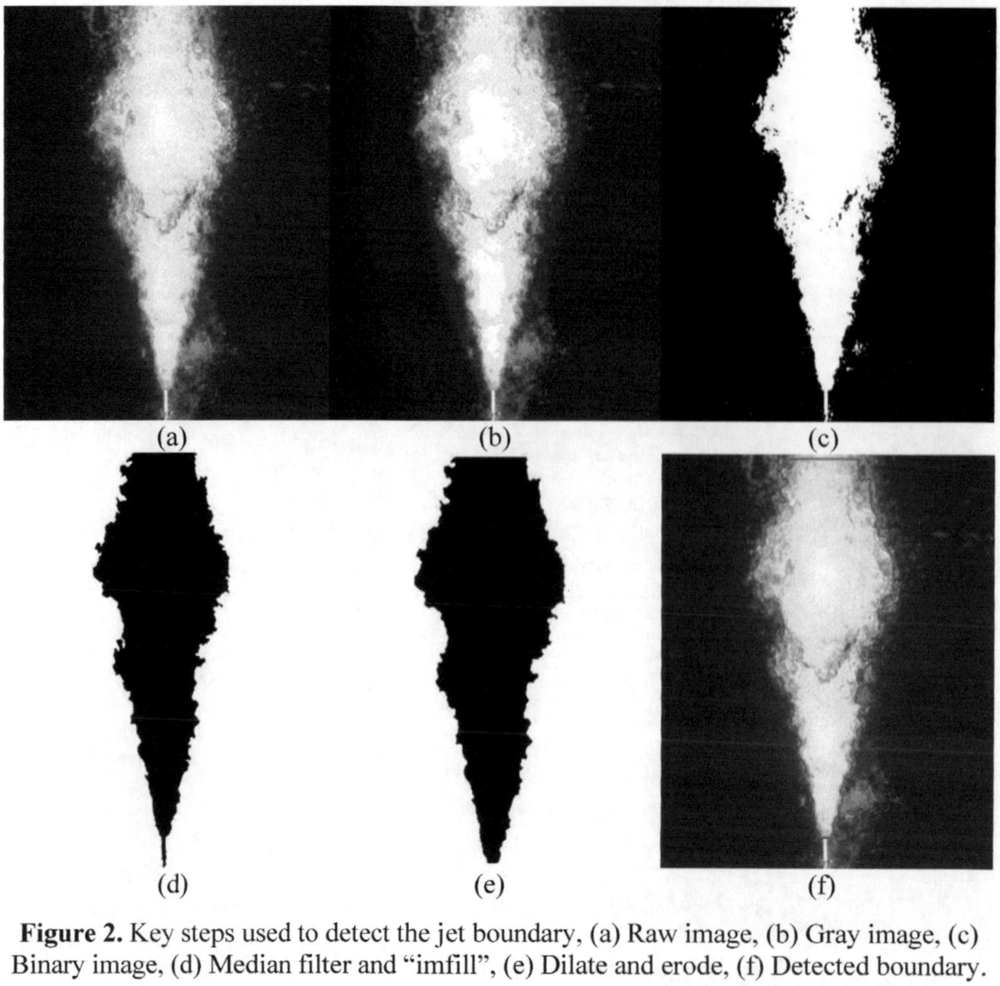

Figure 2. Key steps used to detect the jet boundary, (a) Raw image, (b) Gray image, (c) Binary image, (d) Median filter and "imfill", (e) Dilate and erode, (f) Detected boundary.

3. Result and discussion

3.1. Jet diameter

Jet diameters along the axis with pressure ratio from R_P=1.8 to R_P=4.8 are shown in figure 3. Results of nozzle diameter 6 and 8.7mm are showed, respectively. Axial distance (Y/Lq), or can be referred to as the downstream distance to the nozzle, plots against jet diameters (D/Lq) defined in this paper using the 80% contour of added images. Axial distance and jet diameters are both normalized by Lq, so that results of different nozzle diameters can be compared. Here, Lq is a geometric length scale associated with nozzle structure which for a round jet is the square root of the nozzle area[14]. Resulting from the fact that image sizes taken in experiments with different diameters are alike, the normalized axis distances in nozzle diameter d=6 and 8.7mm are normally unlike. The smaller nozzle diameter has a wider normalized axial distance. It can be obviously seen in figure 3 that normalized axial distance ranges of d=6 and 8.7mm are 0~30 and 0~52 respectively.

It is known that gas-liquid interface of submerged air jets is extremely unsteady, so instantaneous jet diameter detected from single images are meaningless[15, 16]. Therefore, statistical jet diameters are calculated by summing all the images in 2s. In the binary images, region occupied by gas is given a value of 0 and all other positions occupied by water are given a value of 1. So in the binary image, we can see that the gaseous region is black and the surrounding water is white. Before all the images are summed, the value of each pixel must be reversed. Next, values of all the pixels over time are summed and averaged, which produces an added image with the value of each pixel between 0 and 1. Thus, a gray image is generated and a color bar is added for a clearly presentation, just like the example showed in figure 4 (a) and (b). The added image generates an image with a spatially varying intensity and the spatial distribution can be distinctly identified with different colors. According to the statistical results above, statistical characteristics of jet diameter under different pressure ratios and different nozzle diameters can be analyzed.

As can be seen from figure 3, the relation that jet diameters increase with increasing axial distance is consistent for all cases. However, the growth rate of jet diameter decreases along the axis distance. Near the nozzle exit, jet diameter grows fast and the relationship of jet diameter and axis distance is almost linear. On the downstream, it increases slowly and even begins to decrease at some positions (d=6mm, Y/Lq>47). The relationship of jet diameter and axial distance described above is contradictory to the work of Tross[4] who obtained the flow widths along the axis by detecting the time-averaged void fraction using electrical conductivity probe. This difference mainly because of the fact that jet diameters reported here and the flow widths defined by Tross are only nominally similar, but they are fundamentally different. The flow widths depending on time-averaged void fraction are the radius of the gas-containing region without the bubble cloud around it. However, jet diameters calculated from added images in this paper contain both the gas-containing region and the bubble cloud around it. In the near nozzle region, the discharged flow is predominantly governed by the jet momentum force. Jet flow in this region is a continuous gas with fast jet speed and high internal pressure and its boundary layer thickness which mainly consists of bubbles and drops is relatively thin. Therefore, in this region jet diameters calculated from flow images and the flow width detected by probes are normally alike. With the development of air jet, it disintegrates at the position far from the nozzle exit, where the jet breaks up into a column of rising bubbles. This region is named for the plume regime region by other researchers[17, 18]. In this region, bubbles and drops dominated by gravitational or buoyancy effects form a thick boundary layer. This dense layer can be exhibited on flow images, but can't be detected by the probe. The difference that we contain the boundary layer in jet diameters while Tross didn't, leads to a divergent conclusion.

The second observation is that the higher pressure ratio has a wider jet diameter, particularly in the downstream region. And low pressure ratio air jets (R_P<3) are more susceptible to pressure ratio than high pressure ratio air jets (R_P >3). As can be seen in figure 3, curves of low pressure ratio air jets generally have a wide distribution, and jet diameters at the same position obviously increase with increasing pressure ratio. However, curves of high pressure ratio air jets are relatively close. Therefore, we can presume that the jet diameters increase with pressure ratio nonlinearly. Initially, it grows fast, while across the pressure ratio R_P=3, it slows down. Both in the experiment of 6 and 8.7mm, jet diameters of R_P=3 do not obey the regularity above. And the reason for this discrepancy is not yet clear.

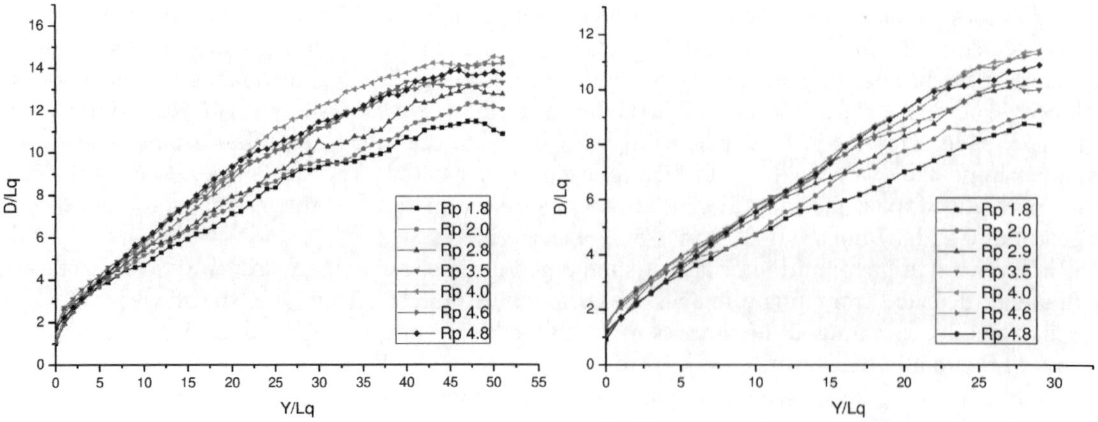

(a) (b)

Figure 3. Jet diameters along the axis under different pressure ratio, (a) d=6mm, (b) d=8.7mm.

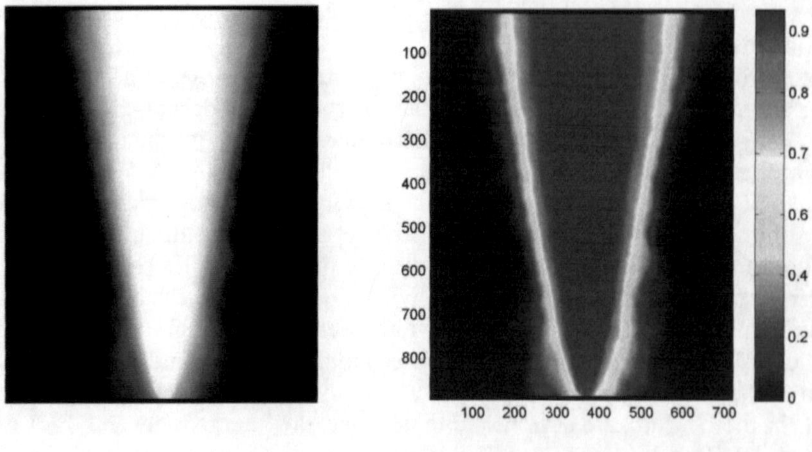

(a) (b)

Figure 4. Spatial distribution of an added image in 2s (d=8.7mm, R_P=4.8), (a) gray distribution image, (b) RGB distribution image with color bar.

To investigate the influence of nozzle diameters on the relation of jet diameter and axial distance, normalized jet diameter versus axial distance of nozzle diameters d=6 and 8.7mm are shown together. Seven graphs are exhibited in figure 5 (a-g) with pressure ratio varying from 1.8 to 4.8. It can be observed in the graphs in figure 5 that nozzle diameters do not have impact on the relation of dimensionless jet diameter and axial distance. Both the curves of d=6 and 8.7mm under certain pressure ratio in the graph nearly grow in a similar path. First, the dimensionless jet diameter (D/Lq) increases linearly with the dimensionless axial distance (Y/Lq). Then, from certain position downstream, its growth becomes nonlinearity and the growth rate decreases. It is possible to suppose that a definite growth function may exist between the dimensionless jet diameter and axial distance

IOP Conf. Series: Materials Science and Engineering **129** (2016) 012061 doi:10.1088/1757-899X/129/1/012061

even though under different nozzle diameters. Further work is required to explore this growth function. The results reported above corroborate the findings of the previous work done by Harby et al[11]. A linear relationship of normalized jet half widths and centerline path is also obtained from horizontal gas jets, but what they didn't report was the nonlinear relationship downstream. A probable explanation for this might be that the downstream flow which chiefly controlled by the buoyancy force and characterized by the production of bubbles, rising vertically, which deviate from the horizontal centerline. Therefore, only the rectilinear relationship demonstrated. As to vertical gas jets, the bubbles controlled by the buoyancy force still rise and expand along the axial direction only the growth rate of jet diameter becomes moderate.

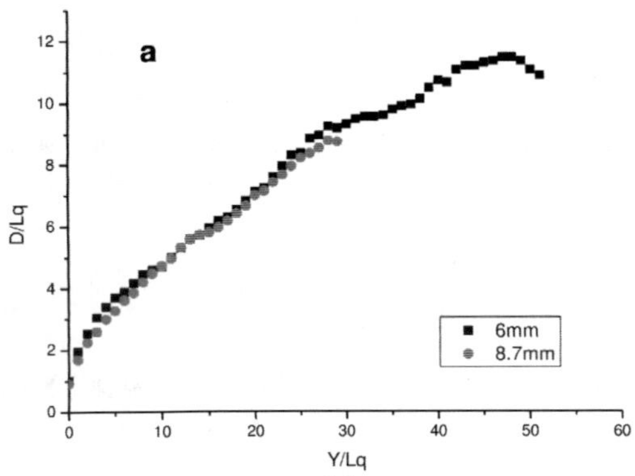

(a)

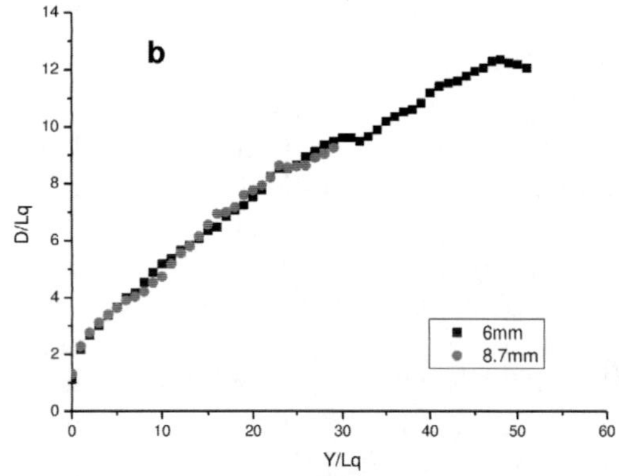

(b)

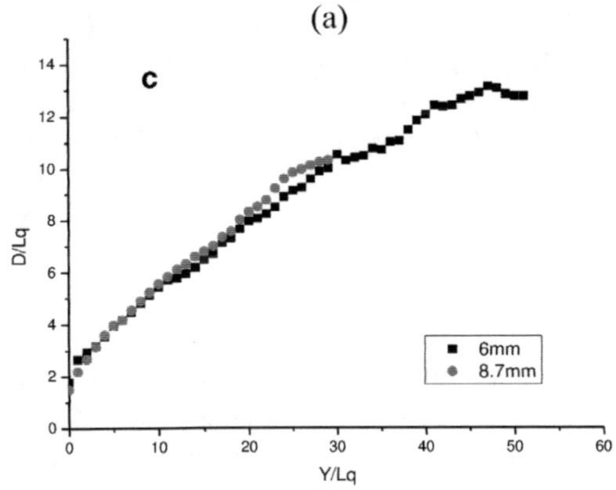

(c)

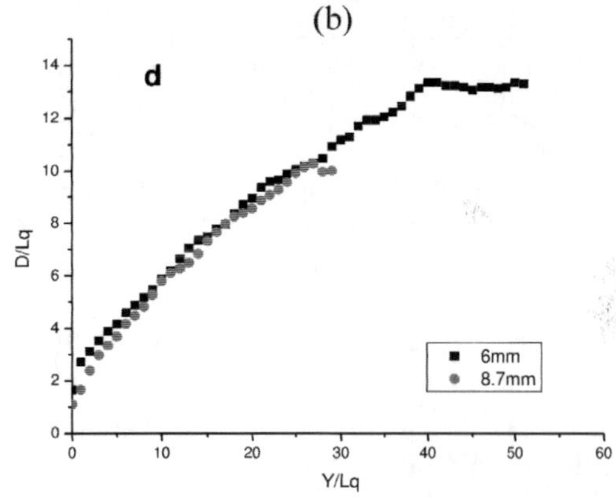

(d)

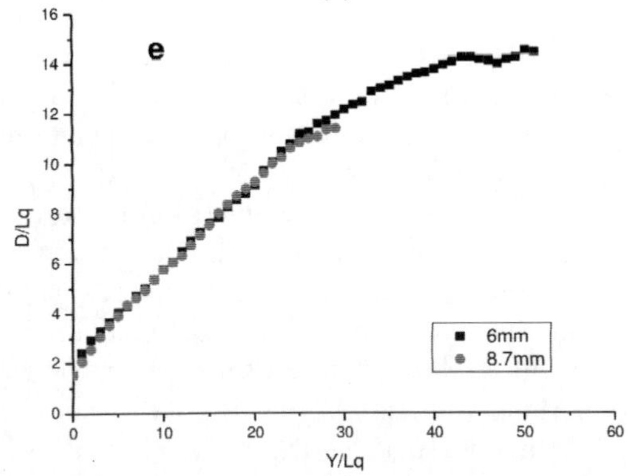

(e)

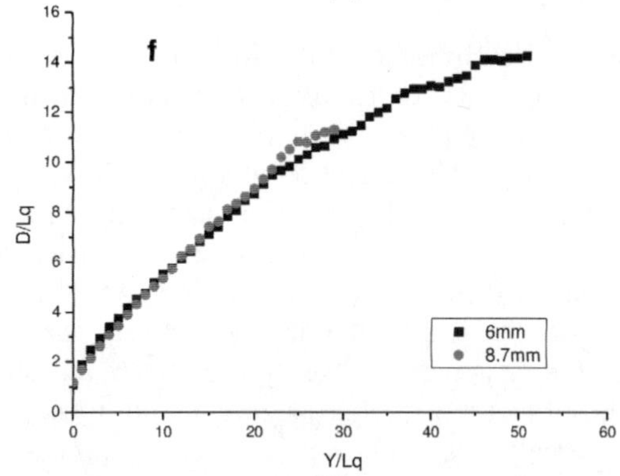

(f)

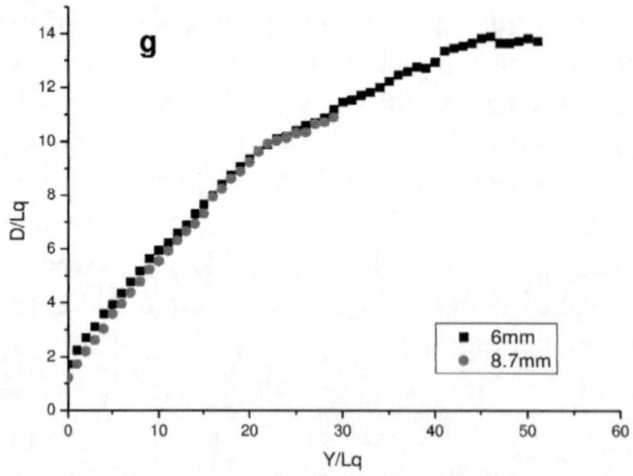

(g)

Figure 5. Jet diameters along the axis of nozzle diameters d=6 and 8.7mm at certain pressure ratio. (a) R_P=1.8, (b) R_P =2.0, (c) R_P=2.8, (d) R_P=3.5, (e) R_P =4.0, (f) R_P =4.6, (g) R_P =4.8

3.2. Characteristics of interface unsteady

The flow structure and the process of a submerged air jets are essentially unsteady and turbulent. The unsteadiness of the submerged air jets is accompanied by appreciable pressure fluctuations in the flow passage and beyond the jet interface. By detecting the pressure fluctuations with pressure probes, prior studies have confirmed the innate unsteadiness of gas-liquid jet interface which is not normally found in typical single-phase jets[8].

It has been established in section 2.2 that the gas-liquid interface of each frame is detected and positioned for 2 seconds. Hence, a time series analysis of the right interface unsteadiness is displayed in figure 6 with nozzle diameter d=8.7mm. As showed in figure 6 and figure 7, normalized deviations at different position along the axis are illustrated for R_P= 1.8, 2.0, 4.6 and 4.8. The normalized deviation is calculated by equation (2). In this equation, Xi is the jet boundary position, X is the mean jet boundary position and D refers to the mean jet diameter. The deviation of relative deviation (RD) which is commonly used is divided by the mean value. But the deviation of normalized deviation (ND) defined in this paper is divided by mean jet diameter not the right mean boundary. It is explained by the fact that that the deviations between different pressure ratio can be compared by using the normalized deviation (ND). Thus, all the deviations at a certain position under a certain pressure ratio are normalized by the mean jet diameters. The normalized deviation $\delta_i(y)$ denotes unsteadiness, or fluctuations of the jet boundary for the chosen axial positions. Eight typical positions are chosen for analysis under a certain pressure ratio, from Y/Lq=3.5 to 28.1.

$$\delta_i(y)=(X_i-X)/D \qquad (2)$$

Time series interface deviations of low pressure ratio (RP= 1.8 and 2.0) and high pressure ratio (RP= 4.6 and 4.8) vertical air jets are shown in figure 6 and figure 7. The results exhibited here are the deviations of right boundary. As can be observed in figure 6 (a) and (b), normalized deviations are relatively large. The deviation gradually increases along the axis from the nozzle exit to downstream, while it decreases as the pressure ratio being larger. It means that the near nozzle region is more stable than the downstream in low pressure ratio air jets and larger pressure ratio yields steadier jets. And this phenomenon is more clearly in figure 8 (a). The mean absolute deviation (MAD) is plotted against the positions. It is because that the momentum force is gradually weakened by mixing drops which have eroded the flow pattern. Losing the power force of momentum, the interface of the jet can be readily influenced by the waves propagating from upstream or surface. In addition, positions where deviations are extremely large indicate the so called pinch-off phenomenon[10,19], such as the certain positions at 271ms, 698ms, 986ms and 1630ms in figure 6 (a). It is evident in the graphs that higher pressure ratio have less pinch-off in low pressure ratio jets. And high pressure ratio jets, no pinch-off is observed.

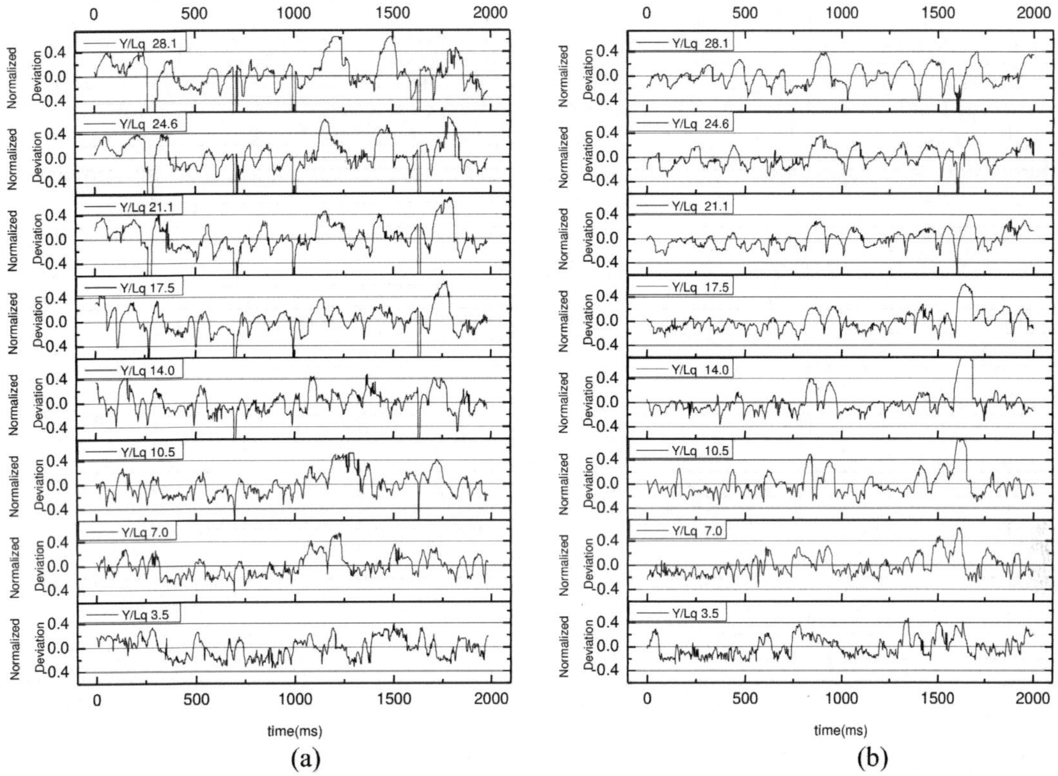

Figure 6. Time series normalized deviations along certain axis positions of low pressure ratio jet, (a) R_P=1.8, (b) R_P=2.0.

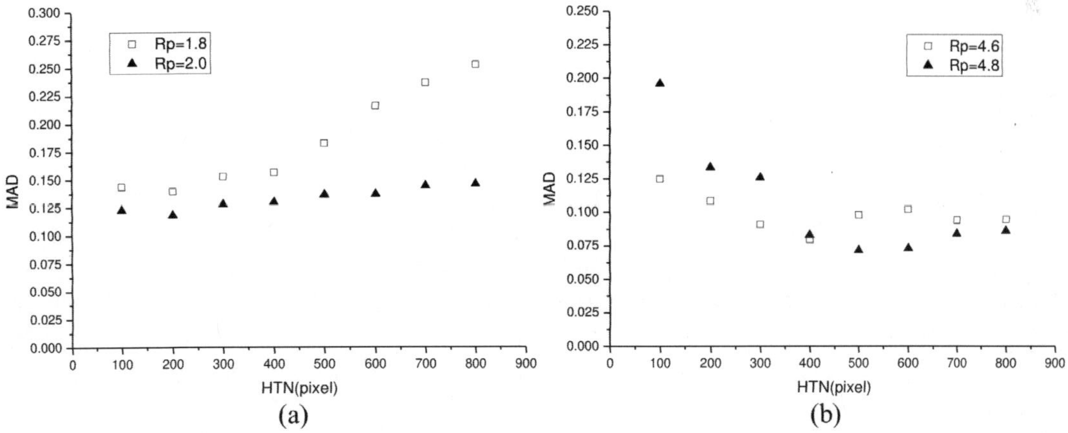

Figure 7. The mean absolute deviation along axis positions, (a) low pressure ratio jet, (b) high pressure ratio jets.

Dislike the features exhibited in low pressure ratio jet, the maximum deviation of high pressure ratio jet shown in figure 7 (a) and (b) is close to the nozzle, and then it decreases along the axis and later increases downstream. Between Y/Lq=14.0 and 17.5, the deviation reaches to a minimum value. And the position of this minimum value is somewhat related to the pressure ratio. When R_P=4.6 and

4.8, as can be seen from the graph in figure 8 (b), the minimum position is Y/Lq=14.0 and 17.5 pixels respectively. As to the reason why the maximum deviation of a supersonic jet is near the nozzle may be the existence of the shock cell in under-expanded jets[9, 20]. When the gas jet pressure is high near the nozzle, pressure equalization is achieved through a complex external expansion region containing repeating shock cells, which have both compression and expansion waves[20]. The compression and expansion waves cause huge unsteadiness of the boundary. Thus, the deviation near the nozzle is incredibly large. However, when the jet pressure near the nozzle exit is low, pressure disturbances can propagate upstream of the passage to equilibrate the exit pressure with the lower ambient pressure. So deviations of low pressure ratio jets are naturally smaller than the high pressure ratio jets. Now, we can explain the reason why position of minimum deviation appears between Y/Lq=14.0 and 17.5, namely, in the intermediate region of the flow. It is due to the combined effect of two factors: the existence of the shock cell near nozzle and the gradually diminished momentum force along the stream wise.

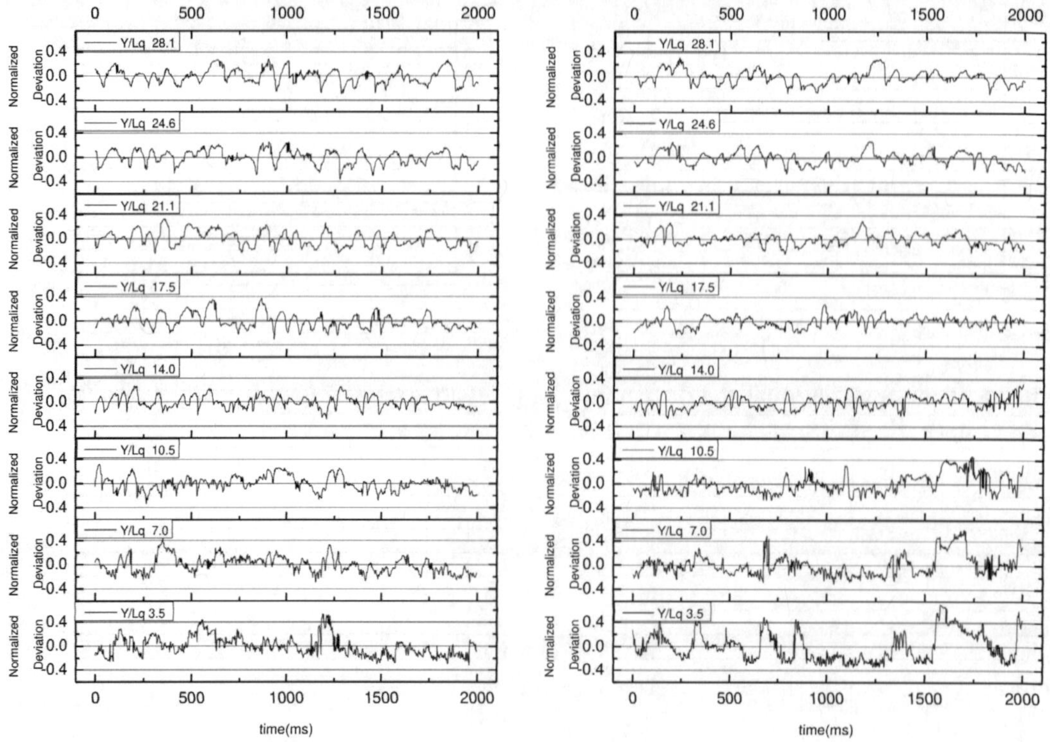

Figure 8. Time series normalized deviations along certain axis positions of high pressure ratio jet, (a) R_P =4.6, (b) R_P =4.8.

4. conclusion

An experimental study on submerged air jets was performed to investigate jet diameters along axis and unsteadiness of gas-liquid interface. The tests included low pressure air jets with pressure ratio, ranging from R_P=1.8 to R_P=3, and high pressure air jets with pressure ratio, ranging from R_P=3.5 to R_P=4.8. And two different diameter nozzles, d=6 and 8.7mm, were used and compared. Flow images were recorded by high speed digital video camera and processed using image processing method. The main conclusions of this work are:

1. Jet diameter increases with increasing axial distance, but its growth rate decreases along the axis distance. Near the nozzle exit, jet diameter grows fast and the relationship of jet diameter and axis distance is almost linear. On the downstream, it increases slowly and even starts to decrease at some positions. The relationship of jet diameter and axis distance detected by flow visualization in this paper is different from the result obtained by Tross using probes. The difference that we contain the boundary layer in jet diameters while Tross didn't, leads to this divergence.

2. A larger pressure ratio air jet generally has a wider jet diameter. Jet diameters of low pressure ratio air jets (R_P <3) are more sensitive to pressure ratio than those in high pressure ratio air jets (R_P>3).

3. From the result of this experiment, we can see that relations of dimensionless jet diameter and axial distance of different nozzles under certain pressure ratio nearly grow in a similar path. And a growth function of certain pressure ratio was assumed existence without the influence of nozzle diameters.

4. The unsteadiness of gas-liquid interface is exhibited by calculated deviation. In low pressure ratio air jets, the minimum deviation is close to the nozzle exit. And the deviation which means the interface unsteadiness gradually increases along the axis from the nozzle exit to downstream. Pinch-off phenomenon reported by other researchers also can be clearly seen in the time series interface. The minimum deviation in a high pressure ratio air jet is in the middle region of the jet. And this position is a bit concerned with pressure ratio. This mainly because of the combined effect of these two factors: the existence of the shock cell near nozzle and the gradually diminished momentum force along the stream wise. The difference of minimum position between low and high pressure ratio air jets mainly due to the complex interaction of shock cell, momentum force and buoyancy force.

5. The shock cell and pinch-off mentioned in some literatures is illustrated in the time series interface deviation, which has extremely large deviation in the near nozzle region in high pressure ratio air jets and downstream in the low pressure air jets, respectively.

5. References

[1] Yagla J J, Busic J and Koski S 2008 Launcher Dynamics Environment of a Water Piercing Missile Launcher *24th Int Symp on Ballistics (New Orleans, Louisiana)*

[2] Cieslinski J T and Mosdorf R 2005 Gas bubble dynamics-experiment and fractal analysis *Int J Heat Mass Transf.* **48** 1808-1818

[3] Dimotakis P E, Miake Lye R C and Papantoniou D A 1983 Structure and dynamics of round turbulent jets *Phys Fluids.* **26** 3185-3192

[4] Tross S R 1974 Characteristics of a turbulent, two-phase, submerged, free jet *Diss. Pennsylvania State University*

[5] Michael Epsteina, Hans K Fauskea, Shigenobu Kubob, Toshio Nakamurac and Kazuya Koya-mad 2001 Liquid entrainment by an expanding core disruptive accident bubble-a Kelvin/Helmholtz phenomenon *Nucl Eng Des.* **210** 53-77

[6] MORI, Kazumi, Yasuhisa OZAWA and Masamichi SANO 1982 characterization of gas-jet behavior at a submerged prifice in liquid-metal *T Iron Steel I Jan.* **22** 377-384

[7] McNallan M J, and T B King 1982 Fluid dynamics of vertical submerged gas jets in liquid metal processing systems *Metall Trans .B* **13** 165-173

[8] Loth E and Faeth G M 1989 Structure of underexpanded round air jets submerged in water *Int J Multiphas Flow.* **15** 589-603

[9] Loth E and Faeth G M 1990 Structure of plane underexpanded air jets into water *AIChE J.* **36** 818-826

[10] Weiland, Chris and Pavlos P Vlachos 2013 Round gas jets submerged in water *Int J Multiphas Flow.* **48** 46-57

[11] Harby K, S Chiva and J L Munoz-Cobo 2014 An experimental investigation on the characteristics of submerged horizontal gas jets in liquid ambient *Exp Therm Fluid Sci.* **53** 26-39

IOP Conf. Series: Materials Science and Engineering **129** (2016) 012061 doi:10.1088/1757-899X/129/1/012061

[12] Drew B, Charonko J and Vlachos P 2011 Liquid Entrainment by Round Turbulent Gas Jets Submerged in Water *ASME-JSME-KSME Joint* Fluids *Engineering Conf*

[13] Weiland C J 2010 Characteristics of the high speed gas-liquid interface *Diss. Virginia Polytechnic Institute and State University*

[14] Hugo B Fischer，John E List，C Robert Koh，Jorg Imberger and Norman H 1979 *Mixing in Inland and Coastal Waters*(New York: Academic Press)

[15] Chen K and Richter H J 1997 Instability analysis of the transition from bubbling to jetting in a gas injected into a liquid *Int J Multiphas Flow.* **23** 699-712

[16] Crow S C and Champagne F H 1971 Orderly structure in jet turbulence *J Fluid Mech.* **48** 547-591

[17] Brevik I, Kristiansen 2002 The flow in and around air-bubble plumes *Int J Multiphas Flow.* **28** 617-634

[18] Neto I E L, Zhu D Z and Rajaratnam N 2008 Bubbly jets in stagnant water *Int J Multiphas Flow.* **34** 1130-1141

[19] OZAWA Y and MORI K 1986 Effect of physical-properties of gas and liquid on bubbling jetting phenomena in gas injection into liquid *T Iron* Steel *I Jan.* **26** 291-297

[20] Dai Z, Wang B, Qi L and Shi H 2006 Experimental study on hydrodynamic behaviors of high-speed gas jets in still water *Acta Mech Sin.* **22** 443-448

IOP Conf. Series: Materials Science and Engineering **129** (2016) 012062 doi:10.1088/1757-899X/129/1/012062

Performance analysis on solid-liquid mixed flow in a centrifugal pump

C Ning, Y Wang

National Research Center of Pumps, Jiangsu University
Zhenjiang, Jiangsu, China

E-mail:dream0830@163.com

Abstract. In order to study the solid-liquid mixed flow hydraulic characteristics of centrifugal pump, the Pro/E software was used for three-dimensional modeling of centrifugal pump chamber. By using the computational fluid dynamics software CFX, the numerical simulation calculation of solid-liquid two-phase flow within whole flow passage of centrifugal pump was conducted. Aim at different particle diameters, the Reynolds-averaged N-S equations with the RNG k-Ɛ turbulence model and SIMPLEC algorithm were used to simulate the two-phase flow respectively on the condition of different volume fraction. The influence of internal flow characteristic on pump performance was analyzed. On the conditions of different particle diameter and different volume fraction, the turbulence kinetic energy and particle concentration are analyzed. It can be found that the erosion velocity ratio on the flow channel wall increases along with the increasing of the volume fraction

1. Introduction

Solid-liquid two-phase transmission centrifugal pump has a very wide range of applications in the national economy departments, but the key technical problems exist in this kind of pump. One is due to the existence of solid material lead to low efficiency. The second is due to the erosion problem of poor reliability. Always restricts the solid-liquid two-phase transportation development and application of centrifugal pump. [1-2] Due to the different parameters of size, density and the concentration of the solid, combined with complex geometry and impeller pump flow channel in high-speed rotating factors, lead to the solid-liquid two-phase flow inside of the pump is very complex. [3-4]

This article mainly research to solid-liquid mixed transportation flow inside the channel-type pump. Using two phase flow model a numerical simulation of the flow inside the pump, analysis of solid phase volume fraction and the influence of particle diameter to the internal flow characteristics of channel-type centrifugal pump.

Commonly used methods for solid-liquid two-phase flow simulation are the Euler-Euler method and Euler-Lagrange method. [5-6]

For accurate and effective calculation analysis, the establishment of the model in the setup of each phase and solve, this paper puts forward the following hypothesis: [7-9]

- The continuous phase and discrete phase is incompressible fluid;
- Continuous phase for the same fluid medium, the discrete phase for spherical particles with the same diameter and the same density;
- The coupling between the continuous phase and discrete phase method for two-way coupling;

IOP Conf. Series: Materials Science and Engineering **129** (2016) 012062 doi:10.1088/1757-899X/129/1/012062

- Ignore the collision between particles and broken, there is no phase transition phenomenon;
- .Don't consider the effect of temperature change flow field.

2.Calculation parameters and meshing

Table 1. Main parameters of the channel-type centrifugal pump

Parameters	Value	Unit
Q	8	m3/h
H	25	m
n	2989	r/min
D1	35	mm
D2	150	mm
Q	8	m3/h
H	25	m

The basic parameters are shown below: flow rate of 8m3/h, head of 25m, speed of 2898r/min, the impeller diameter of 150mm, outlet width of 15mm and the discharge diameter of 50mm.The computational domain uses tetrahedral unstructured grids, totally 572,879 elements.

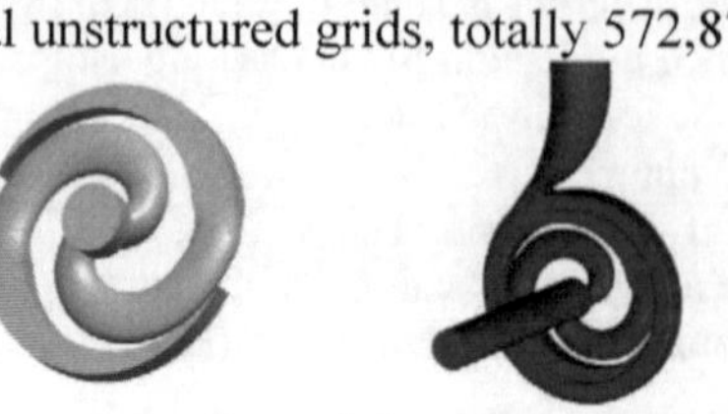

Figure 1. The geometric and the whole flow passage grid

3.Boundary conditions

Mass conservation equation, momentum conservation equation, three-dimensional incompressible pressure Reynolds N-S equations and RNG turbulence model equations have been chosen. Considering the commonly used two-phase flow model, Euler-Euler model is used to study the flow field. Each phase must satisfy the law of momentum and mass conservation. Conservation equation can be obtained by local instantaneous conserved overall average or mixed law of each phase. The velocity inlet, outflow and no slip boundary conditions are used. The given speed of initial value is v=2.3m/s.

4.Results and analysis

4.1.Numerical Simulation of Water Single-Phase Flow

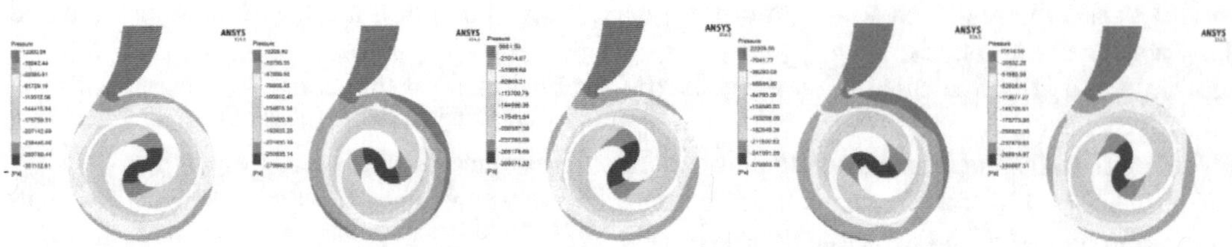

t_1=0s t_2=0.0046s t_3=0.0096s t_4=0.0138s t_5=0.0184s

Figure2. The pressure distributions on the middle section

The pressure distributions in different moment when the impeller around volute tongue a cycle on the middle section are shown in the figure2. It shows that: the moment of t1-t2, the blade passage just pass the volute tongue, the pressure around the volute tongue decline sharply, the pressure decline fastest at the moment of t3.

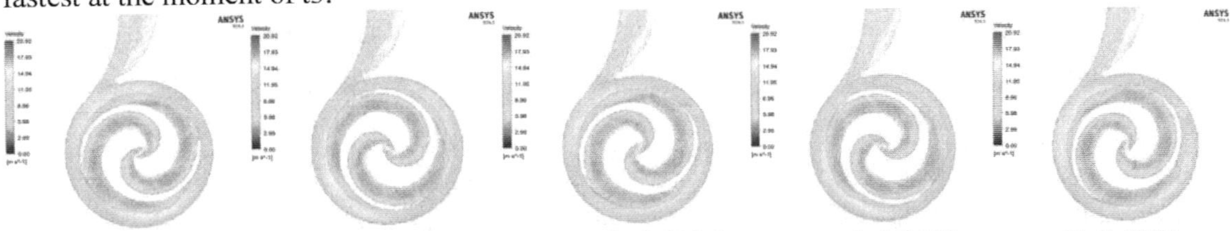

t1=0s t2=0.0046s t3=0.0096s t4=0.0138s t5=0.0184

Figure3. The velocity distribution on the middle section

Figure3 shows the velocity distribution in different moment when the impeller around volute tongue a cycle on the middle section. It shows that the velocity distribution is almost uniformity in the impeller and volute. The velocity in impeller is lower than volute, but the velocity at volute tongue fluctuate all the time.

4.2 .Numerical Simulation Solid-Liquid Two-Phase Flow

Table2. The program of numerical simulation

Number	d(mm)	Solid phase volume fraction Cv(%)
1	0.05	5,10,30
2	0.10	5,10,30
3	0.30	5,10,30

4.2.1.Influence of Solid Size

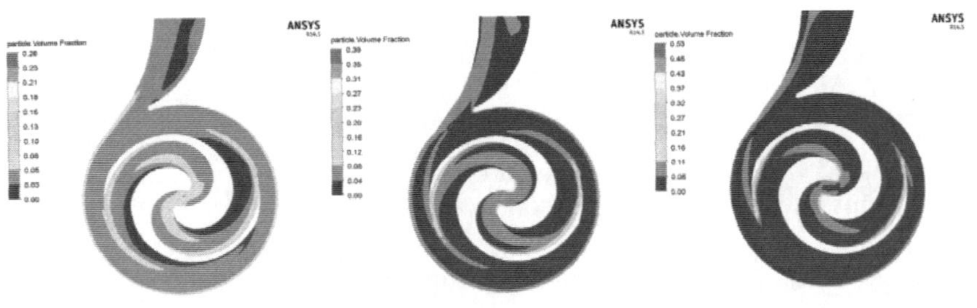

d=0.05mm d=0.10mm d=0.30mm

Figure4. The particle volume fraction distribution on the middle section

Figure4 shows the particle volume fraction distribution on the middle section under the influence of different particle sizes when the concentration of solid is 5%.With the increase of particle diameter, the volume distribution of solid phase obvious changes. Under different particle size, particle concentration on the working surface is greater than the back surface of the impeller, thus the erosion on working face of blade is relatively higher than the back face.

4.2.2 Influence of the concentration of Solid

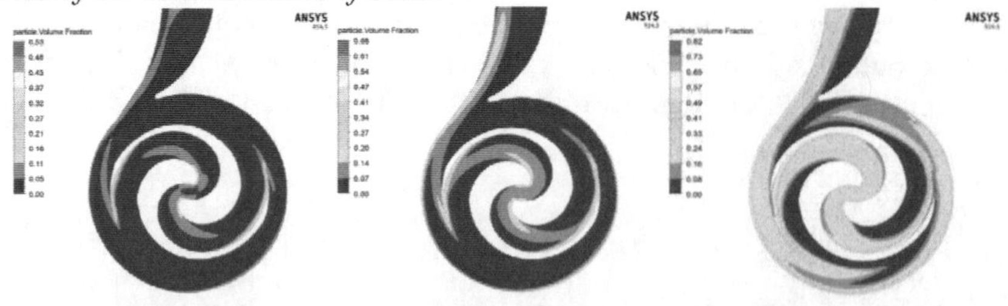

Cv=0.05 **Cv=0.1** **Cv=0.3**

Figure5. The particle volume fraction distribution on the middle section

Figure5 shows the particle volume fraction distribution on the middle section under the influence of different particle concentration when the particle size is set as 0.30mm.With the increase of particle concentration, the particle concentration of the inlet of the impeller is higher, due to centrifugal force, particles in the flow field deflect towards the working face of the impeller.

4.2.3. Turbulence Kinetic Energy of Flow

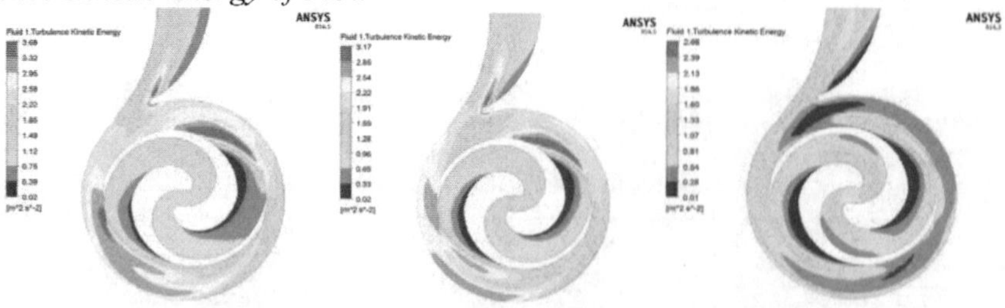

d=0.05mm **d=0.10mm** **d=0.30mm**

Figure6. The turbulence kinetic energy on the middle section

Figure6 shows the turbulence kinetic energy on the middle section under the influence of different particle sizes when the particle concentration is 5%. With the increase of particle diameter, the value of turbulence kinetic energy decline

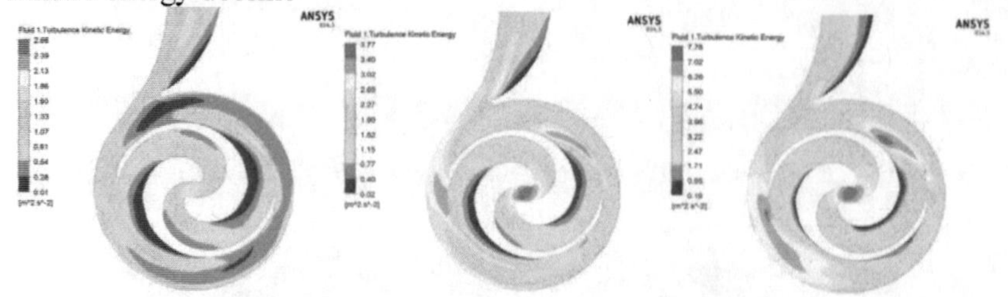

Cv=0.05 **Cv=0.1** **Cv=0.3**

Figure7. The turbulence kinetic energy on the middle section

Figure7 shows the turbulence kinetic energy on the middle section under the influence of different particle concentration when the particle size is 0.30mm. With the increase of particle concentration, the value of turbulence kinetic energy increase.

4.3 Unsteady Numerical Simulation of Solid-Liquid Two-Phase Flow

The particle size is set as 0.05mm and the concentration of solid is 5%.

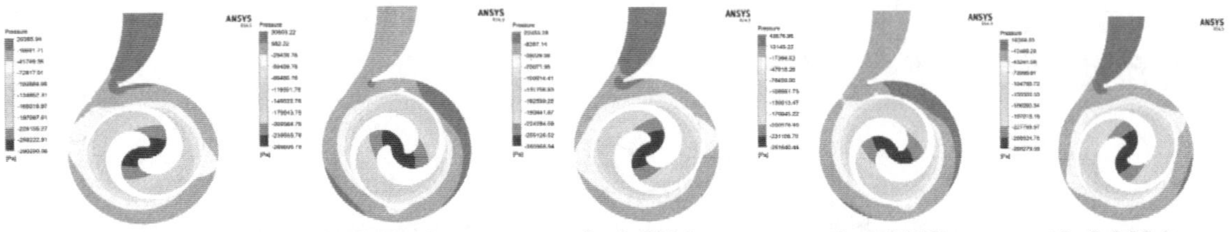

t1=0s **t2=0.0046s** **t3=0.0096s** **t4=0.0138s** **t5=0.0184s**

Figure8. The pressure distribution on the middle section

Figure8 shows the pressure distributions in different moment when the impeller around volute tongue a cycle on the middle section. We can see that the distribution of static pressure in the impeller and volute flow is relatively uniform at the moment of t1, and compared with the single phase unsteady calculation, the value of the pressure at the outlet is higher; At the moment of t1-t2, the impeller outlet is on the volute tongue, an increased pressure gradient near the volute tongue, but static pressure values near the volute tongue of this moment is lower than the static pressure value of unsteady single-phase calculate at the same time. At the moment of t2-t3, the pressure near the volute tongue began to recover, but the uniformity of pressure distribution inside the volute flow obvious deterioration.

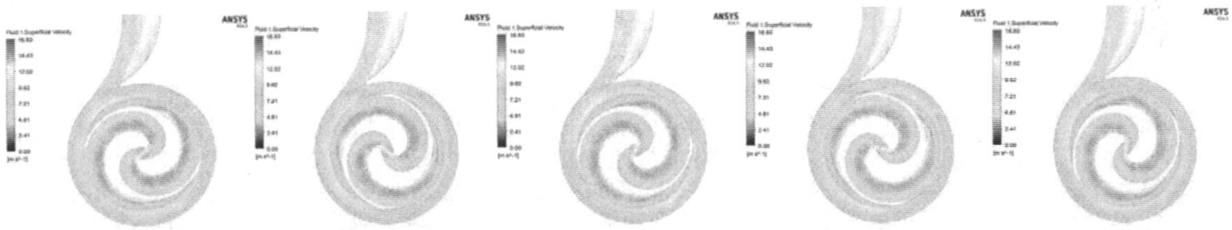

t1=0s **t2=0.0046s** **t3=0.0096s** **t4=0.0138s** **t5=0.0184s**

Figure9. The velocity distribution on the middle section

Figure9 shows the velocity distribution in different moment when the impeller around volute tongue a cycle on the middle section. At the moment of t1-t2, the impeller outlet is on the volute tongue. Due to the hinder of the volute tongue, the flow where the impeller outlet meets the volute is very complicated, the velocity gradient is larger. At the moment of t2, the velocity near the volute tongue is low. At the moment of t2-t3, the area of low velocity increase fast until the moment of t4.

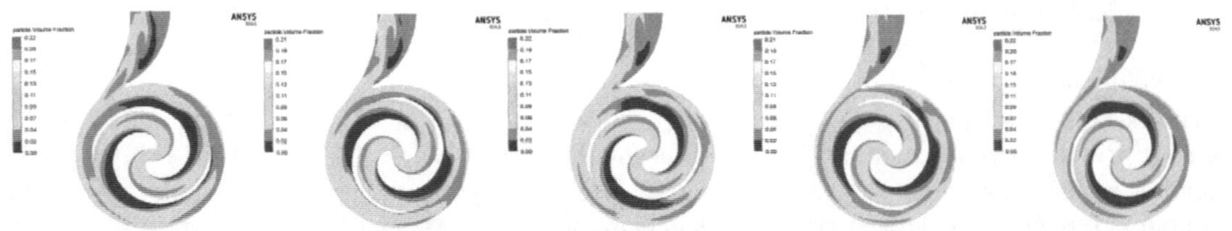

t1=0s **t2=0.0046s** **t3=0.0096s** **t4=0.0138s** **t5=0.0184s**

Figure10. The particle volume fraction distribution

Figure10 shows the particle volume fraction distribution in different moment on the middle section when the particle sizes is 0.05mm and the concentration of solid is 5%. At the moment of t1, the solids in the flow passage are relatively few. At the moment of t1-t2, Particles in outward movement at the same time under the influence of centrifugal force and inertia to favor the impeller of the working face and impeller hub side which cause they are serious damaged. At the moment of t2-t3, due to the mutual influence of gravity and centrifugal force, solid more concentrated on the wall of the volute which is in the larger impeller radius. At the moment of t3-t4, a large number of solid particles are still on the wall out of the volute.

4.4 Comparison of Predicted Results

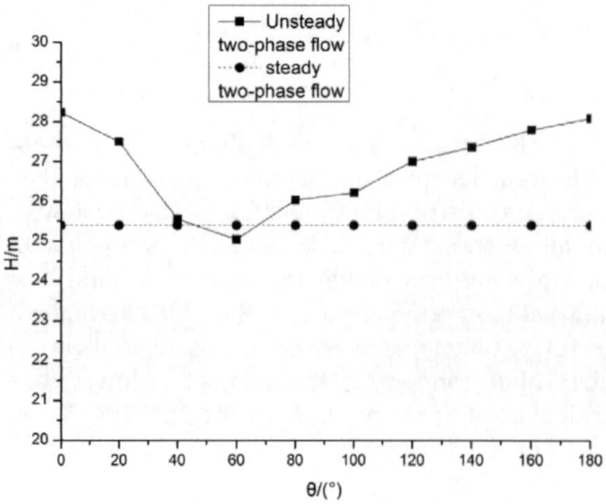

Figure11. The total head of unsteady two-phase flow and steady two-phase flow

Figure11 shows the total head of unsteady two-phase flow and steady two-phase flow. The predicted value of unsteady two-phase flow is higher than it of steady two-phase flow. Under the two phase medium, the total head of steady flow is 25.40m and the total head of unsteady flow is 26.89m.

5. Conclusions

(1)Under different particle size, particle concentration on the working surface is greater than the back surface of the impeller, thus the erosion on working face of blade is relatively higher than the back face.

(2)With the increase of particle concentration, the particle concentration of the inlet of the impeller is higher. Due to centrifugal force, particles in the flow field deflect towards the working face of the impeller.

(3)With the increase of particle diameter, the value of turbulence kinetic energy decline. But with the increase of particle concentration, the value of turbulence kinetic energy increase.

(4)Solid phase mixed have a certain influence on the static pressure fluctuation, velocity distribution and volume fraction distribution near the volute tongue.

Acknowledgement

The authors gratefully acknowledge the financial support of Ordinary University Graduate Student Scientific Research Innovation Projects of Jiangsu Province (SJLX_0465) and PAPD (A Project Funded by the Priority Academic Program Development of Jiangsu Higher Education Institutions).Conflict of interests: The authors declare no conflicts of interest. The support of Ordinary University Graduate Student Scientific Research Innovation Projects of Jiangsu Province(SJLX_0465) and PAPD have no conflicts of interest.

References

[1] Li Y, The 2014 Research on numerical simulation and abrasion property of solid-liquid two-phase-flow centrifugal pump.Zhejiang Sci-tech University.

[2] Song X G, Park J H, Kim S G, et al. 2013 Performance comparison and erosion prediction of jet pumps by using a numerical method. Mathematical *and Computer Modeling*, **57(1)**: 245-253.

[3] Yuan S Q, Zhang P F, Zhang J F.2004 Numerical simulation of 3-D dense solid- liquid two-phase turbulent flow in a non-clogging mud pump. Chinese Journal of Mechanical Engineering, **17(4)**:623-627

[4] Liu H L, Lu B B, Tan M G, Wang Y and Wang K 2009 Drainage and Irrigation Machinery **05** 297–301
[5] Cui Q L, Li Y and Jin S Q 2012 Journal of Zhejiang Sci-Tech University **29** 543–8
[6] Liu J R, Xu Y G, Wang D M and Su Q Q 2010 Transactions of the Chinese Society of Agricultural Engineering **41** 86–90
[7] Liu J H and Zhu M Y 2011 Applied Mechanics and Materials 44-47 345–8
[8] Wei L, Shi W D, Jiang X P, Chen B and Wu Y L 2012 Procedia Engineering **31** 170–5
[9] Zhao B J, Yuan S Q, Liu H L, Huang F Z and Tan M G 2008 Transactions of the Chinese Society of Agricultural Engineering **24** 7–12

IOP Conf. Series: Materials Science and Engineering **129** (2016) 012063 doi:10.1088/1757-899X/129/1/012063

The application of dichotomy in equilibrium position of journal bearing

G K Wu [1, 2], **X Q Luo** [2]

[1] Institute of Chemical Machinery, Zhejiang University, Hangzhou, 310027, China
[2] Institute of Water Resources and Hydro-electric Engineering, Xi'an University of Technology, Xi'an, 710048, China

E-mail: wuguangkuan@163.com

Abstract: The fluid lubricant force in the journal bearing is an important factor for the stability and dynamic characteristics of rotating machine. In order to obtain the dynamic coefficients of journal bearing, the equilibrium position must be known for further calculation. In this paper, the Reynolds equation is solved by finite difference method and the dichotomy is applied to acquire the equilibrium position of journal bearing by means of double loop. The effects of length, radius and clearance of journal bearing on the equilibrium position are also researched. The calculated results show that the dichotomy is an effective method for the equilibrium position of journal bearing and the geometry parameters play an important effect on the equilibrium position.

1. Introduction

It is well known that the oil-film force of journal bearing plays an important role in the dynamical characteristics of modern rotating machines [1-5]. The lubricant film force is usually simplified to 8 dynamical coefficients acting on the rotating shaft. In order to obtain the dynamical coefficients, many researchers have done a lot of works in the previous literature. Klit and Lund [6] calculated the dynamic coefficients of a journal bearing by a variational approach. Zheng and Hasebe [7] calculated the dynamic characteristics and coefficients of a journal bearing using free boundary theory. They found that the numerical results worked satisfactorily compared with finite difference method. Rao et al. [8] developed the analytical expressions for accurate evaluation and predicted the post whirl orbits by computing the nonlinear journal trajectory of a cylindrical journal bearing. In order to simplify the calculating process of dynamical coefficients, Muszynska and Bently [9, 10] proposed a nonlinear fluid dynamics model to predict the dynamical coefficients on the basis of lots of experiments, but the model only could be applied on condition that the fluid film was continuous theoretically. In fact, for the finite length journal bearing, the common method for solving dynamic coefficients included finite difference method, partial derivative method and finite element method [11-13]. In addition, as one of the important process to obtain the dynamic coefficients, the solving accuracy of equilibrium position determined the accuracy of dynamic coefficients and the vibration characteristics of rotor system.

In this paper, the dichotomy is adopted to obtain the equilibrium position of journal bearing by means of double loop. The target of inner loop is to acquire the attitude angle and the outer loop is to calculate the eccentricity. The Reynolds equation was solved by finite difference method in order to describe the pressure distribution and fluid film force of journal bearing. The effects of geometric parameters on the equilibrium position are also researched. The calculating results show that dichotomy is an effective method for the equilibrium position of journal bearing.

2. The solving of equilibrium position

2.1. The model of journal bearing

The following fig.1 shows the geometry model of a journal bearing. The lubricant region is divided into convergent wedge and divergent wedge on the basis of the change trend. The journal rotates in the bearing at the speed of Ω_j and is in equilibrium state under interaction of external load and oil film pressure.

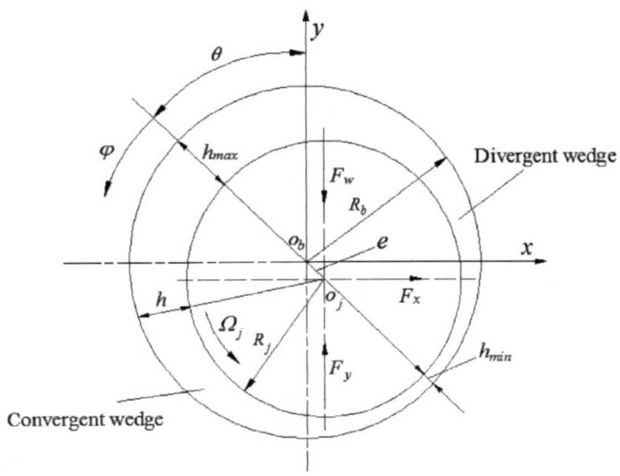

Figure 1. Geometry model of a journal bearing.

The generalized Reynolds equation of lubricant medium under turbulent condition in cylindrical coordinates for cylindrical journal bearing can be written as [14]

$$\frac{1}{R_b^2}\frac{\partial}{\partial\varphi}\left(\frac{h^3}{12\mu}\frac{\partial p}{\partial\varphi}\right)+\frac{\partial}{\partial z}\left(\frac{h^3}{12\mu}\frac{\partial p}{\partial z}\right)=\frac{\Omega_j}{2}\frac{\partial h}{\partial\varphi}+\frac{\partial h}{\partial t} \tag{1}$$

where R_b is the radius of bearing, R_j is the radius of journal, φ is the angular direction, θ is the attitude angle, h is the thickness of lubricant medium, e is the eccentric distance, μ is the viscosity of lubricant medium, p is the pressure of the lubricant film, Ω_j is the rotating angular speed, F_W is the external vertical load, F_x and F_y are x- and y- components of lubricant film force, respectively, the subscript 'max' and 'min' denote the maximal and minimum thickness, respectively.

Introducing the following dimensionless parameters for further calculations

$$\bar{z}=\frac{2z}{L_b},\ \bar{c}=\frac{c}{R_b},\ \bar{h}=1+\varepsilon\cos\varphi,\ \bar{p}=\frac{\bar{c}^{-2}}{6\mu\Omega_j}p,\ \lambda=\frac{L_b}{2R_b},\ \bar{t}=\Omega_j t \tag{2}$$

where L_b is the length of bearing, c is the clearance between bearing and journal, z is the axes in axial direction, ε is the eccentricity.

Considering $d\theta/d\bar{t} = d\varepsilon/d\bar{t} = 0$ for the stable motion, Equation (1) can be further simplified as

$$\frac{\partial}{\partial \varphi}\left(\bar{h}^{3}\frac{\partial \bar{p}}{\partial \varphi}\right)+\frac{\partial}{\partial \bar{z}}\left(\frac{\bar{h}^{3}}{\lambda^{2}}\frac{\partial \bar{p}}{\partial \bar{z}}\right)=-\varepsilon\sin\varphi \tag{3}$$

Equation (3) can be solved by FDM, once the pressure distribution of the lubricant film is obtained, the final lubricant force can be calculated by compound Simpson quadrature formula.

2.2. The static equilibrium position

The static equilibrium position is one of the important steps for dynamic coefficients. In fact, the essence of searching static equilibrium position is the eccentricity and attitude angle.

The dichotomy has been used in the computational process of static equilibrium position [15], but it was only applied to determine the eccentricity and ignored the attitude angle. In order to solve the problem, the double loop is used. In the paper, once the calculated eccentricity and attitude angle satisfy the following convergence criterions, the journal bearing is considered in the equilibrium state. The convergence criterions can be given as

$$f(\theta)=\left|\frac{F_x}{F_y}\right|<10^{-2} \tag{4.a}$$

$$g(e)=\left|F_w-F_y\right|<1 \tag{4.b}$$

The iterative process of static equilibrium position is shown in Fig. 2.

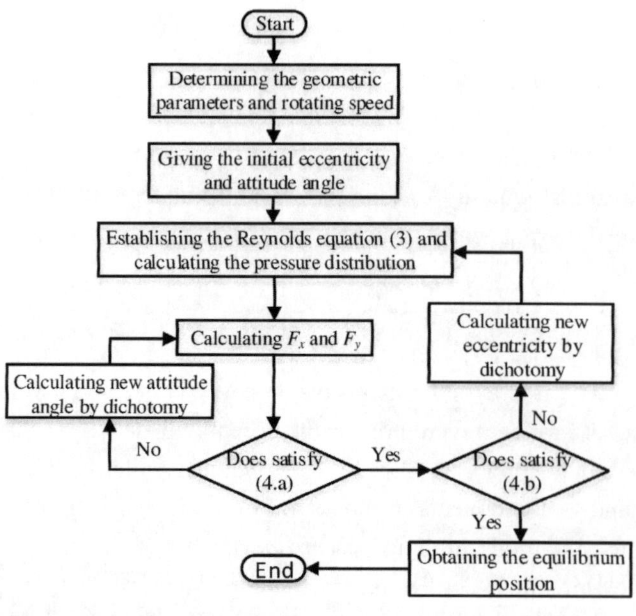

Figure 2. Iterative process of static equilibrium position.

3. Numerical results and discussions

In order to present dichotomy's effectiveness and research the effect of geometric parameters on the equilibrium position, the radius of bearing R_b, length of bearing L_b and clearance c are selected in the following calculation. The pressure distribution under different equilibrium position corresponding with the previous geometric parameters are also presented.

Assign the following parameters

$$R_b = 0.3m,\ L_b = 0.3m,\ F_w = 10000kg,\ \Omega_j = 3000r/min,\ c = 2mm$$

3.1. Influence of the radius of bearing

Figure 3(a) presents the influence of radius of bearing on the equilibrium position. The radius of bearing is selected from 0.1 m to 0.5 m and the other parameters remain the same. It can be seen that the eccentricity increases with the decrease of the radius, but on the contrary, the attitude angle increases as the radius increases. In addition, it also should be note that the changing trend becomes more obvious with larger radius values (from 0.4 m to 0.5 m) compared with smaller radius values (from 0.1 m to 0.2 m).

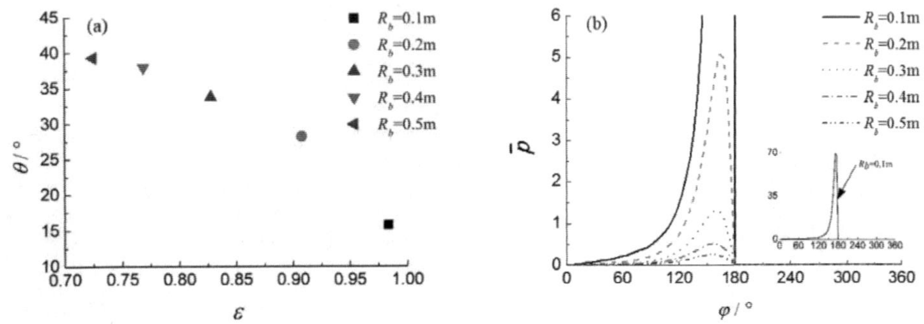

Figure 3. Influence of R_b on the equilibrium position and corresponding pressure distribution.

Figure 3(b) shows the relations of radius and pressure distribution at different equilibrium positions. From the figure, the maximal value of pressure distribution with $R_b = 0.1$ m is considerably larger than other radii and the changing trend of maximal value is the same as equilibrium position in Fig. 3(a).

3.2. Influence of the length of bearing

The influence of length of bearing on the equilibrium position is shown in Fig. 4(a). The length of bearing is employed as the variable parameter, ranging from 0.1 m to 0.5 m. The changing trend is similar to that of radius. However, the difference between adjacent equilibrium positions is more balanced compared with Fig. 3(a).

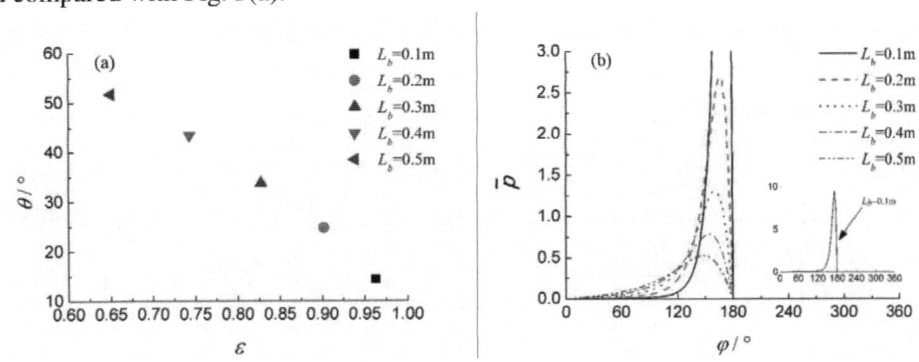

Figure 4. Influence of L_b on the equilibrium position and corresponding pressure distribution.

Fig. 4(b) presents the relations of lengths and pressure distribution at different equilibrium positions. The maximal values of pressure distribution increase as the lengths decrease. It is need to note that when the φ is less than 120°, the pressure distribution values with larger length are greater than these of smaller length.

3.3. Influence of the clearance

Figure 5(a) presents the influence of clearance on the equilibrium position. The clearance between bearing and journal is selected from 1 mm to 3 mm and other parameters also keep the same. On the contrary to the influence of radius and length, the eccentricity increases with the increase of the clearance and the attitude angle decreases as the clearance increases. Moreover, the difference between adjacent equilibrium positions with smaller clearance values (from 1 mm to 1.5 mm) is greater than the larger clearance values (from 2.5 mm to 3 mm).

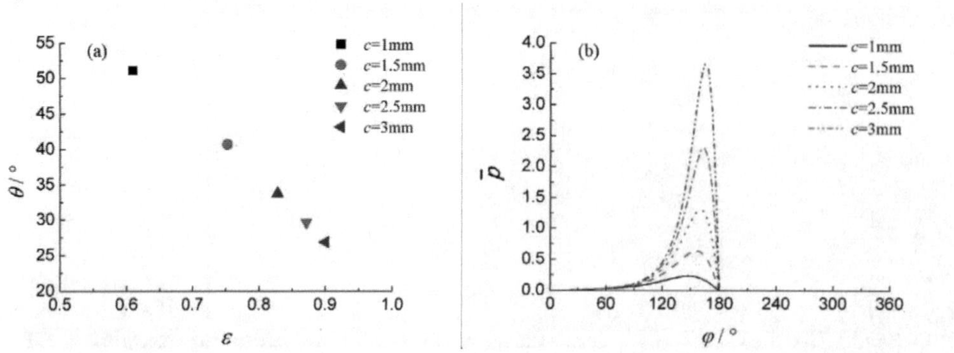

Figure 5. Influence of c on the equilibrium position and corresponding pressure distribution.

The pressure distribution at different equilibrium positions with the change of clearance is shown in Fig. 5(b). From the figure, it can be seen that the pressure values and the maximal values increase as the clearance increases, but the difference of adjacent maximal values is more average compared with that of radius and length.

Conclusions

In the paper, in order to further calculation of dynamic coefficients of journal bearing, the dichotomy with double loop is applied to acquire the equilibrium position of journal bearing. The Reynolds equation is solved by FDM and the final lubricant force is calculated by compound Simpson quadrature formula. The calculated results show that the dichotomy is an effective method for the equilibrium position of journal bearing. The changing trend of equilibrium position with different radius and length is different with that of clearance and the difference values of maximal pressure with different clearances are more average.

Acknowledgement

The financial support by National Natural Science Foundation of China (Grant No. 51479167, No. 51339005) is gratefully acknowledged.

References

[1] Wang C C 2006 *Acta. Mech.* **183** 41
[2] De Castro H F, Cavalca K L and Nordmann R 2008 *J. Sound. Vib.* **317** 273
[3] Zhao S, Xu H, Meng G and Zhu J 2005 *Tribol. Int.* **38** 749
[4] Chang-Jian C W and Chen C K 2007 *Chaos. Soliton. Fract.* **34** 1160
[5] Zhu C, Robb D A and Ewins D J 2003 *J. Sound. Vib.* **265** 469
[6] Klit P and Lund J W 1986 *J. Tribol.* **108** 421
[7] Zheng T and Hasebe N 2000 *J. Tribol.* **122** 616
[8] Rao T, Biswas S, Hirani H and Athre K 2000 *Tribol. T.* **43** 109
[9] Muszynska A 1988 *J. Vib. Acoust.* **110** 129
[10] Muszynska A and Bently D E 1990 *J. Sound. Vib.* **143** 103
[11] Chang S H, Chan C W and Jeng Y R 2015 *Tribol. Int.* **90** 157
[12] Yang L H, Wang W M, Zhao S Q, Sun Y H and Yu L 2014 *Appl. Math. Model.* **38** 5239
[13] Avramov K V and Borysiuk O 2013 *Mech. Mach. Theory.* **70** 474
[14] Wang J L, Cao D Q and Huang W H 2010 *P. I. Mech. Eng. J-J. Eng.* **224** 595
[15] Ye X, Shen H, Hu J, Zhu J and Zhang J 2013 *T. Chin. Soc. Agr. Mach.* **44** 74 (in Chinese)

IOP Conf. Series: Materials Science and Engineering **129** (2016) 012064 doi:10.1088/1757-899X/129/1/012064

Stability and bifurcation analysis of rotor-bearing-seal system

G Y Ying[1] S L Liu[2] R Ma[3] S Y Zheng[3]

[1] Electric Power Scientific Research Institute of Zhejiang Province, State Grid Corporation of China, Hangzhou, China
[2] School of Mechanical and Automotive Engineering, Zhejiang University of Science and Technology, Hangzhou, China.
[3] Institute of Chemical Machinery, Zhejiang University, Hangzhou, China

E-mail: liushul21@aliyun.com
Zhengshuiying@zju.edu.cn

Abstract. Labyrinth seals were extensively used in turbine units, and the seal fluid forces may induce self-excited vibrations of rotor under certain conditions. It has become the main factor to instability of rotor system. In this paper Muszynska seal fluid force model is used to investigate the stability of the rotor system. Nonlinear equations are numerically solved by Newmark integration method. The effect of different seal clearances and differential pressures on system stability is studied. The calculation results show that the dominant vibration component leading to instability changes with different seal clearance. With the differential pressure increased, the unstable speed is reduced. Then the bifurcation behavior of the system with and without seal force is calculated. Results show that the rotor vibration becomes severe and complicated, and the bifurcation behavior of the system has been changed when seal force is considered.

1.Introduction

Labyrinth seals were extensively used in turbine units. The purpose of a seal in turbo-machinery is to prevent the leaking of gas from a high--pressure region to a low-pressure region and to maintain the pressure difference between components rotating at high speed. It's well known that the seal fluid forces may induce self-excited vibrations of rotor under certain conditions. With the pressure increasing of medium in turbine units the fluid excitation problems become increasingly prominent. It has become the main factor to instability of rotor system, even induce the destabilization. A lot of literatures have focused on seal fluid force models and its influence on rotor system, which generally propose that seal forces are a product of asymmetric pressure in seal caused by rotor deflection and the seal forces are linearized under the hypothesis of small disturbances. The force model is expressed by eight-parameters [1-3]. In small disturbance conditions the eight-parameter theory can be used to describe motion characteristics of the rotor system, but there is a lot limitation under large disturbance circumstance. It can not reflect the nonlinear characteristics of sealing force. The distributed pressure and fluid characteristics can be better acquired by simulating the flow inside labyrinth seal with the computational fluid dynamic (CFD) analysis [4-5], by the method it is crucial how to couple the gas excitation force into the rotor system. Muszynska seal fluid force model [6-7], not only inherits the characteristics of the

liushul21@aliyun.com

traditional eight parameters in certain degree, but also reflect some nonlinear characteristic of Alford force. So Muszynska seal fluid force model is mostly used to study the stability of the rotor system.

The Muszynska model of seal fluid forces is applied in this paper to investigate stability of rotor. There are two types of model parameters in the Muszynska model. Some parameters are related to seal structure and media parameters, others are experience parameters which are related to loss and obtained from the experiment or the experience. This paper will focus on the nonlinear bifurcation characteristics under the seal fluid forces, and the effect of Muszynska model parameters on the stability of rotor system is analyzed.

2 Seal Force Muszynska Model

According to Muszynska model, the nonlinear character of seal fluid force may be expressed as follow[6~8] in fixed reference coordinates X, Y.

$$\begin{Bmatrix} F_x \\ F_y \end{Bmatrix} = -\begin{bmatrix} K_s - m_f\tau^2\omega^2 & \tau\omega d \\ -\tau\omega D & K_s - m_f\tau^2\omega^2 \end{bmatrix}\begin{Bmatrix} x \\ y \end{Bmatrix} - \begin{bmatrix} D & 2\tau\omega m_f \\ -2\tau\omega m_f & D \end{bmatrix}\begin{Bmatrix} \dot{x} \\ \dot{y} \end{Bmatrix} - \begin{bmatrix} m_f & 0 \\ 0 & m_f \end{bmatrix}\begin{Bmatrix} \ddot{x} \\ \ddot{y} \end{Bmatrix} \quad (1)$$

In Eq(1), K_s is the stiffness of rotor disturbed by fluid; D is the damping; m_f is the inertial effect. τ, K and D are the nonlinear function of disturbing displacements, x and y, respectively.

$K_s = K_0(1-e^2)^{-n}$ $D = D_0(1-e^2)^{-n}$ $\tau = \tau_0(1-e)^b$

$e = \sqrt{x^2+y^2}/c_s$, e is the relative eccentricity displacement of rotor, c_s is the seal clearance.

$$K_0 = \mu_3\mu_0; \quad D_0 = \mu_3\mu_1; \quad m_f = \mu_3\mu_2 T$$

$$\mu_0 = \frac{2\sigma^2}{1+\xi+2\sigma}E(1-m_0); \quad \mu_1 = \frac{2\sigma^2}{1+\xi+2\sigma}\left(\frac{E}{\sigma}+\frac{B}{2}\left(\frac{1}{6}+E\right)\right);$$

$$\mu_2 = \frac{\sigma}{1+\xi+2\sigma}\left(\frac{1}{6}+E\right); \quad \mu_3 = \frac{\pi R\Delta P}{\lambda}$$

$$T = \frac{l}{v_a}; \quad \sigma = \frac{\lambda l}{c};$$

$$E = \frac{1+\xi}{1+\xi+2\sigma}; \quad B = 2 - \frac{(\frac{R_v}{R_a})^2 - m_0}{(\frac{R_v}{R_a})^2+1}$$

$$\lambda = n_0 R_a^{m_0}\left[1+\left(\frac{R_v}{R_a}\right)^2\right]^{\frac{1+m_0}{2}}; \quad R_v = \frac{Rc\omega}{\upsilon}; \quad R_a = \frac{2v_a c}{\upsilon}$$

$n=0.5\sim3$, $0<b<1$, $\tau_0<0.5$

τ_0 is the average circumferential speed ratio; ξ is the inlet port loss coefficient of gas flow; n, τ_0 , n_0, and m_0 are the experience parameter and determined by experiment and the seal structure; ΔP is the axial differential pressure; R is the radius of the seal; λ is the factor of friction; l is the seal length; va is the axial flow velocity in seal clearance; Ra is the axial flow Reynolds numbers; Rv is the circumferential flow Reynolds numbers; ω is the angular velocity of rotor, ν is the fluid viscosity coefficient.

3. Rotor-Bearing-Seal System

The numerical calculations were performed in the model, as illustrated in Figure.1 .The rotor is supported by two sliding bearings. The seal rig is designed in the left of the rotor. All the disks along the axle are in 170mm diameter, 25mm width. The diameter of the axle is 44mm, and the journal diameter is 32 mm. The whole rotor weights 37.02 kg. The sliding bearings are cylindrical bearings. And the short bearing theory is applied to calculate the oil forces.

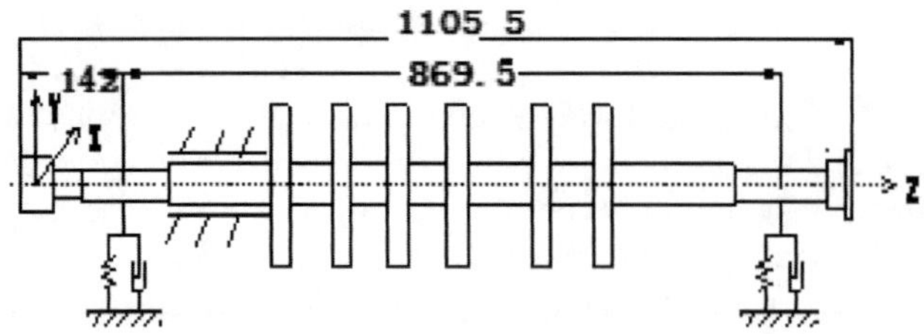

Figure.1 Rotor-bearing-seal model

The bearing parameters: width L = R, clearance c= 0.004×R, and unbalance located on the second disks. The seal construct parameters: seal diameter R=44mm, length l=130mm. Muszynska experience parameters by Child experiment are showed in Tab1 [9]. The axial flow velocity in seal clearance va is 5m/s; The fluid viscosity coefficient ν is 1.48e-5; The Differential pressure is 0.65MPa.

Table 1 Muszynska experience parameters

m_0	n_0	τ_0	n	b	ξ
0.24	0.079	0.28	2.11	0.3	0.093

4. Dynamic Equations of Rotor-Bearing-Seal System

The dynamic equations of rotor-bearing-seal system were built up by d'Alemdert principle, and a continuous shaft was treated as infinite thin disks that may be considered as rigid bodies. For simplification, the dynamic equations of system were treated by Rize way. So it is easy to deal with the integral and derivation of the dynamic equations for the three orders polynomial being selected as the location function. Because of the overlap of the three-order polynomial in the axial direction, not only the displacement and velocity but also the bending moment and transverse force are continuous. Then the nonlinear oil film forces based on steady short bearing model, unbalanced excitation. The dynamic equations of the rotor-bearing system derived in inertia coordinate system and could be written as:

$$\boldsymbol{M\ddot{x}} + \boldsymbol{C\dot{x}} + \boldsymbol{Kx} = \boldsymbol{B}_1\boldsymbol{F}_1 + \boldsymbol{B}_2\boldsymbol{F}_2 + \boldsymbol{B}_3\boldsymbol{F}_3 + \boldsymbol{B}_4\boldsymbol{F}_4 \quad (2)$$

Where $\boldsymbol{M}$ is a mass matrix; $\boldsymbol{C}$ and $\boldsymbol{K}$ are the damping matrix and stiffness matrix respectively; $\boldsymbol{F}_1$ is the unbalanced force; $\boldsymbol{F}_2$ is the gravity, F*3* is the oil force; $\boldsymbol{F}_4 = [F_{gx}, F_{gy}]$ is the gas force; B_i are the location matrix of different forces.

Considering the change of the shaft diameter and the existence of additional mass, the shaft is divided into 47 portions. That is to say there are 47 position functions. Negligible axial vibrations, there are approximate 100 degrees of freedom in the system. For simplification purposes, it is necessary to decrease the degrees of freedom by modal transformation on the basis of accuracy assurance. Such system equations are local nonlinear and have high order. The equations are solved by Newmark integration method.

5. Results and discussion

5.1. Effect of Differential Pressure on Stability

The unstable speeds changing with differential pressure are shown in Figure.2 where the diameter clearance equals to 1.0 mm and eccentric distance located on the second disk is 50μm located. With the differential pressure increased, the unstable speed is reduced. When the differential pressure is small the unstable speed reduces very small with differential pressure changing, but when the differential pressure to a certain value the unstable speed drops accelerately. The unstable speed changing with differential pressure shows obvious nonlinear characteristics. It indicates that the differential pressure is the main factor to affect the gas flow forces.

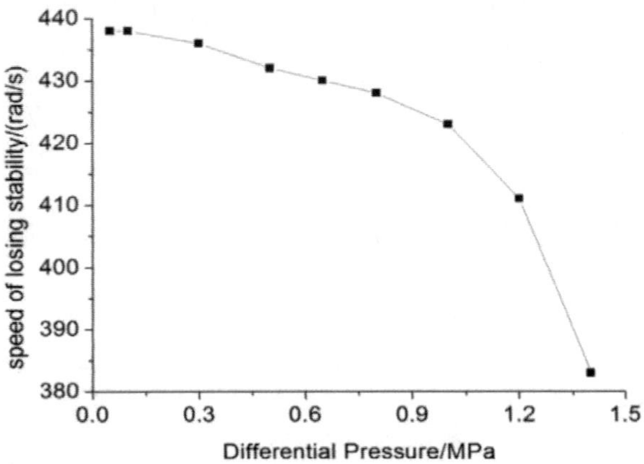

Figure.2 Effect of differential pressure on speed of losing stability

5.2. Effect of seal clearance on Stability

The effect of different seal clearances on system stability are shown in Figure.3 where the differential pressure equals to 0.65MPa and eccentric distance located on second disk is 50μm. When the seal clearance is 0.6mm, the threshold speed of stability is 429rad/s (Figure. 3(a)). It indicates that the trajectory of journal is already too large and amplitude of the half frequency vibration is much larger than that of the rotating frequency vibration. The half-frequency vibration plays the major role which is also known as oil whip. Changing the seal clearance value until to 0.18mm, the half-frequency vibration is still dominant (Figure. 3(b)). When the seal clearance is as small as 0.17mm, the threshold speed of stability is 421rad/s, but the vibration becomes complicated, i.e. there are two kinds of frequency vibrations (Figure. 3(b)). When the seal clearance continuously reduces to 0.15mm, the threshold speed of stability drops to 290rad/s. The working-frequency vibration now dominates. The vibration characteristics are quite different.

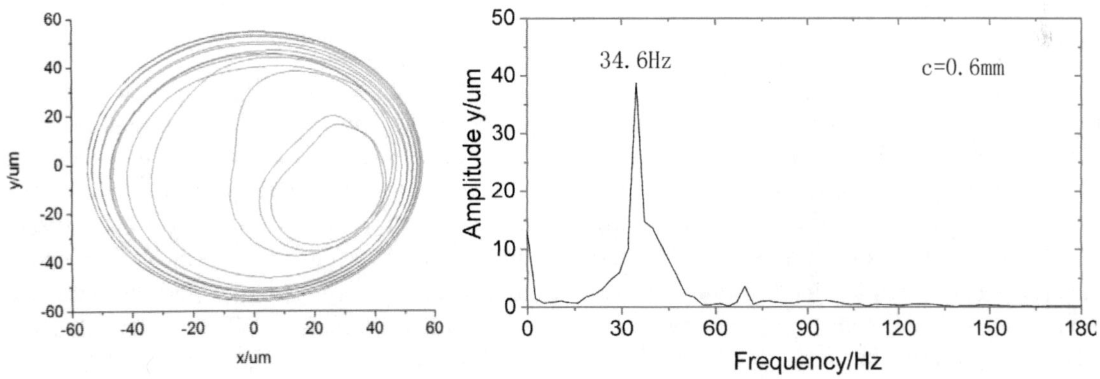

(a)The trajectory and spectrum(c=0.6m, rotating speed=429rad/s)

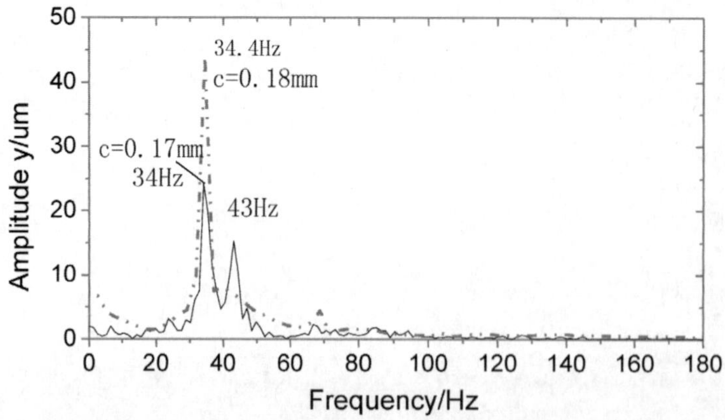

(b) The spectrum(c=0.17m rotating speed=421rad/s and c=0.18m rotating speed=426rad/s)

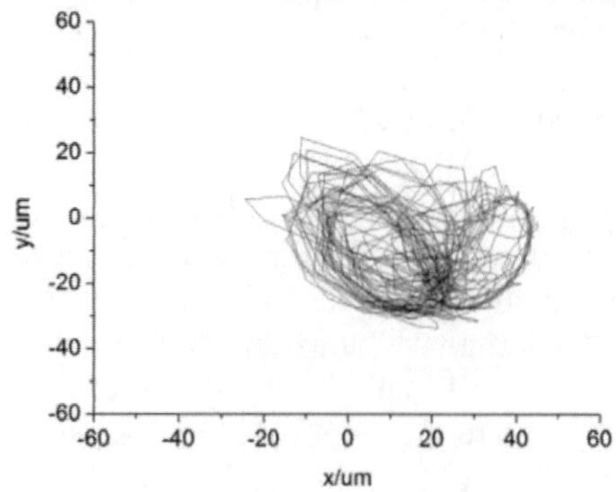

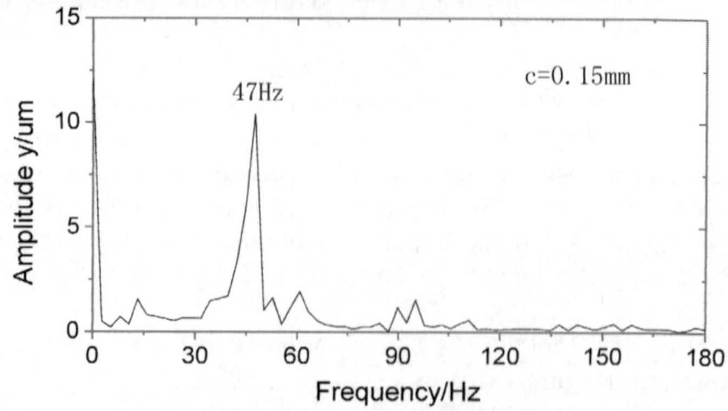

(c)The trajectory and spectrum(c=0.15m, rotating speed=290rad/s)

Figure.3 The trajectory and spectrum under different seal clearance

5.3. Nonlinear bifurcation characteristic with and without the seal force

In order to illuminate the relation between rotation speed and nonlinear dynamic characteristic of rotor with and without the seal force, and the bifurcation map of the right journal displacement in the vertical direction y is plotted(Figure.4 where the diameter clearance equals to 1.0mm, eccentric distance located on the second disk is to 80μm). Figure.4(a) reveals that the rotor first experiences periodic vibration, then comes into quasi-periodic vibration and finally lost stability when there are no seal force on the rotor system. At the same condition if the seal force is considered, Figure.4(a) shows that the rotor come from quasi-periodic vibration into periodic vibration, then returns to quasi-periodic vibration again eventually losing stability. The unstable speed with the seal force is lower. The calculations show that the rotor vibration becomes severe and complicated when there is the seal force acting on the rotor system, and often becomes quasi-periodic.

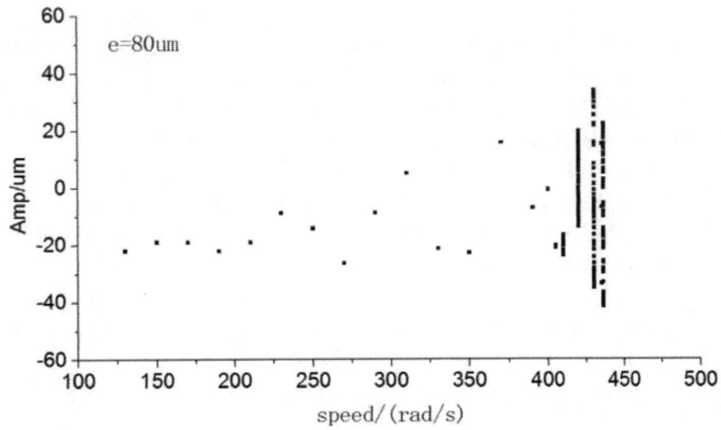

(a) Bifurcation map of the vertical direction displacement without seal force

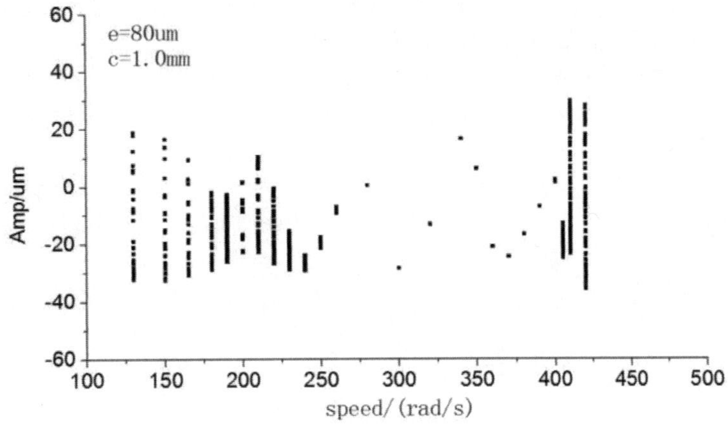

(b) Bifurcation map of the vertical direction displacement after seal force

Figure.4 Bifurcations map of the right bearing displacement in vertical direction y varying with speed

6. Conclusions

(1) The calculations show that the rotor vibration becomes severe and complicated when there is the seal force acting on the rotor system, and often become quasi-periodic. The bifurcation behavior of the system has been changed when seal force is considered.

(2)With the differential pressure increased, the unstable speed is reduced. The differential pressure is the main factor to affect the gas flow forces.

(3) The seal clearance has much influence on vibration characteristics. When the seal clearance is large the half-frequency vibration plays the major role in inducing to instability, but when the seal clearances reduces to a certain value, the characteristic of instability is dominated by the working frequency vibration.

Acknowledgments

Authors would like to acknowledge Science and Technology Project Funding for Electric Power of Zhejiang Province, State Grid Corporation of China (Grant No.5211DS14005B) and Scientific Research Foundation for the Returned Overseas Chinese Scholars, Ministry of Education of China, and The National Natural Science Foundation of China (grant number 51275452)

References

[1] He L D, Gao J J, D Y,Yuan X 2003 Study on flow-induced vieration for a three dimensional roter-seal system. *Journal of mechanical engineering*. **39** No.3,p. 100-104.

[2] Childs D W 1983 Dynamic analysis of turbulent annular seals based on hirs lubrication equation[J] .ASME *Journal of Lubrication technology*. **105(3)**.p.429-436.

[3] Then Qinggen 1996 A Two—Control Volume M odel and Dynamic Characteristics Research for Labyrinth Seal. *Journal of Vibration Engineering*. **9(1)**. p24-50.

[4] Muszynska A A 1986 A whirl and whip rotor/bearing stability problem[J] . *Journal of sound and Vibration*. **110(3)** p.443-463.

[5] Bai C Q, Xu Q Y, Zhang X L.2006 Effect of seal and internal damping on dynamic stability of liquid hydrogen turbopump rotor system in rocket engine. *Journal of mechanical engineering*. **42**（**3**）p.150-155.

[6] Muszynska 1986 A Whirl and Whip in Rotor/Bearing Stability Problems，*Journal of Sound and Vibtation*，**110(3)**：443-462.

[7] Muszynska A and Bently D E 1990 Frequency-swept Rotating Input Perturbation Techniques and Identification of The Fluid Force Models in Rotor/Bearing/Seal Systems and Fluid Handling Machines，*Journal of Sound and Vibration*，143(1)：103-124。

[8] Muszynska 1986 A Model Testing of Rotor/Bearing Systems，*The International Journal of Analytical and Experimental Model Analysis*, **1(3)**：15 -34

[9] Ye J H, Liu Z S and Zheng G H 2008 Nonlinear dynamic characteristics model of labyrinth seal based on Muszynska mode1. *Journal of Harbin Institute of Technology* **159(3)**, 351-355

IOP Conf. Series: Materials Science and Engineering **129** (2016) 012065 doi:10.1088/1757-899X/129/1/012065

Theoretical study on a water muffler

T Du[1], Y W Chen[2], T C Miao[1], D Z Wu[1,3]

[1]Institute of Process Equipment, Zhejiang University, Hangzhou, 310027, China
[2]Nuclear and Radiation Safety Center, Beijing, 100082, China

E-mail: wudazhuan@zju.edu.cn

Abstract. Theoretical computation on a previously studied water muffler is carried out in this article. Structure of the water muffler is composed of two main parts, namely, the Kevlar-reinforced rubber tube and the inner-noise-reduction structure. Rubber wall of the rubber tube is assumed to function as rigid wall lined with sound absorption material and is described by a complex radial wave number. Comparison among the results obtained from theoretical computation, FEM (finite element method) simulation and experiment of the rubber tube and that of the water muffler has been made. The theoretical results show a good accordance in general tendency with the FEM simulated and the measured results. After that, parametric study on the diameter of the inner structure and that of the rubber tube is conducted. Results show that the diameter of the left inner structure has the most significant effect on the SPL of the water muffler due to its location and its effect on the diameter ratio D2/D1.

1. Introduction

Mufflers or silencers are widely used as passive noise reduction measures and has been studied thoroughly. The major research contents include the design and optimization method of muffler structure, computational method on the acoustic performance of mufflers.

Selamet [1] presented a quasi-one-dimensional approach to analyse three-pass mufflers with perforated elements using numerical decoupling. The approach was further developed to include mufflers with ducts extended into the end cavities. Dokumaci [2] applied the distributed parameter method to the analysis of mufflers employing multiple co-axial pipes. Wu [3] proposed a direct mixed-body boundary element method to obtain the acoustic performance of silencers lined with bulk-reacting sound absorbing materials. Selamet [4] investigated the acoustic behavior of a circular dual-chamber muffler in detail by a two-dimensional (2-D) axisymmetric analytical approach based on the mode-matching technique for the concentric configurations, the finite element method and experimental work. A number of effects were studied and some of these effects were shown to modify the acoustic behavior drastically, suggesting potential means to improve the acoustic performance. Denia [5] investigated the acoustic behavior of perforated dissipative circular mufflers with empty extended inlet/outlet in detail by means of a two-dimensional (2D) axisymmetric analytical approach that matches the acoustic pressure and velocity across the geometrical discontinuities, and the finite element method (FEM). Panigrahi [6] suggested a one-dimensional (1-D) scheme based on an algorithm that used user-friendly visual volume elements along with the theory of transfer matrix

[3] To whom any correspondence should be addressed.

based muffler analysis. Xu [7] established acoustic characteristics of a dual Helmholtz resonator which consists of a pair of cylindrical neck and cavity connected in series in terms of a lumped-parameter theory. Mimani [8] analysed the acoustical behavior of an elliptical chamber muffler having an end-inlet and side-outlet port with a 3-D impedance matrix approach. Vijayasree [9] proposed a transfer matrix based approach namely the Integrated Transfer Matrix (ITM) method to analyse a general commercial muffler that consists of both perforated pipes and baffles.

Most studies on mufflers/silencers focus on mufflers with gas medium. However, water mufflers have been generally ignored because the early application fields of mufflers have largely concentrated on the internal combustion engine and relevant components in the automotive industry. Noise from gas-medium-conveying equipment is larger than that of liquid-medium-conveying equipment. However, given that the acoustic performance of liquid-conveying system is gaining increasing attention, methods should be developed to effectively control the noise generated by such systems.

Water muffler presented in previous studies [10-14] are functional only in the problems with single frequencies or narrow frequency bands, and appear with assembly difficulties due to the oversized structure against the limited installation spaces. Moreover the steel tubes in the major parts of muffler are of high rigidity and more helpful to the vibration propagation. Based on these considerations, a water muffler has been proposed by the authors and studied both numerically and experimentally in a previous article [15].

Theoretical computation on the previously studied water muffler is carried out in this article. Structure of the water muffler is composed of two main parts, namely, the Kevlar-reinforced rubber tube and the inner-noise-reduction structure. Rubber wall of the rubber tube is assumed to function as rigid wall lined with sound absorption material and is described by a complex radius wave number. Comparison among the results obtained from theoretical computation, FEM simulation and experiment of the rubber tube and that of the water muffler has been made. The theoretical results show a good accordance in general tendency with the FEM simulated and the measured results which validates the efficiency of the theoretical computation. After that, parametric study on the diameter of the inner structure and that of the rubber tube is conducted. Conclusion has been made and discussed in later sections.

2. Basic equations

Waves in a circular duct with stationary medium are governed by [16]:

$$\left[\frac{\partial^2}{\partial t^2}-a_0^2\left(\frac{\partial^2}{\partial r^2}+\frac{1}{r}\frac{\partial}{\partial r}+\frac{1}{r^2}\frac{\partial^2}{\partial \theta^2}+\frac{\partial^2}{\partial z^2}\right)\right]p=0 \quad (1)$$

Where r is the radius of the duct, a_0 is the sound speed, z is the axial direction of the tube, θ is the circular direction of the tube and p is the sound pressure.

Upon making use of the method of separation of variables, and writing time dependence as $e^{j\omega t}$ and θ dependence as $e^{j\omega\theta}$, the general solution to equation (1) is given by:

$$p(r,\theta,z,t)=\sum_{m=0}^{\infty}\sum_{n=0}^{\infty}J_m(k_{r,m,n}r)e^{jm\theta}e^{j\omega t}\left\{C_{1,m,n}e^{-jk_{z,m,n}z}+C_{2,m,n}e^{+jk_{z,m,n}z}\right\} \quad (2)$$

Where $k_{r,m,n}$ and $k_{z,m,n}$ is the transmission wave number for the (m, n) mode in radius direction and axial direction respectively.

Substituting in the equation of dynamical equilibrium for the axial direction:

$$\rho_0\frac{\partial u_z}{\partial t}+\frac{\partial p}{\partial r}=0 \quad (3)$$

Yields the acoustic particle velocity:

$$u(r,\theta,z,t)=\sum_{m=0}^{\infty}\sum_{n=0}^{\infty}J_m(k_{r,m,n}r)e^{jm\theta}e^{j\omega t}\times\frac{k_{z,m,n}}{k_0}\frac{1}{\rho_0 a_0}\left\{C_1e^{-jk_{z,m,n}z}+C_2e^{+jk_{z,m,n}z}\right\} \quad (4)$$

According to the cut off frequency relation

$$f<1.84a_0/(\pi D) \quad (5)$$

one can assume that only plane waves could propagate in the main ducts under the considered frequency band (20 Hz-5000 Hz with a_0=1500 m/s and D_{max}=0.125 m). The sound pressure and acoustic particle velocity of the (0, 0) mode is given by:

$$p(r,\theta,z,t)=e^{j\omega t}\left\{C_1e^{-jk_0z}+C_2e^{+jk_0z}\right\} \tag{6}$$

$$u(r,\theta,z,t)=\frac{e^{j\omega t}}{\rho_0 a_0}\left\{C_1e^{-jk_0z}+C_2e^{+jk_0z}\right\} \tag{7}$$

For conical tube with diameter being proportional to z (distance from the hypothetical apex), the sound pressure and acoustic particle velocity of the (0, 0) mode is denoted by:

$$p(z,t)=\left\{C_1e^{-jk_0z}+C_2e^{+jk_0z}\right\}e^{j\omega t}/z \tag{8}$$

$$u(z,t)=e^{j\omega t}\times\frac{1}{\omega\rho_0 z}\left\{(-jk_0-\frac{1}{z})C_1e^{-jk_0z}+(jk_0-\frac{1}{z})C_2e^{+jk_0z}\right\} \tag{9}$$

By assuming that the rubber wall functions as rigid wall lined with sound absorption material, the difference with duct of rigid wall is that $k_{r,m,n}$ is now determined from the boundary condition that the wall ($r=r_0$) has a finite impedance Z_w while the rigid wall have infinite impedance.
The momentum equation along the radial direction

$$\rho_0\frac{\partial u_r}{\partial t}+\frac{\partial p}{\partial r}=0 \tag{10}$$

$$u_r=-\frac{\partial p/\partial r}{j\omega\rho_0} \tag{11}$$

Therefore

$$Z_w\equiv\left(\frac{p}{u_r}\right)_{r=r_0}=\frac{-j\omega\rho_0 p}{\partial p/\partial r}=\frac{-j\omega\rho_0 J_m(k_{r,m,n}r_0)}{k_{r,m,n}J_m^{'}(k_{r,m,n}r_0)} \tag{12}$$

Where

$$J_m^{'}(k_{r,m,n}r_0)=\left[\frac{dJ_m(k_{r,m,n}r)}{d(k_{r,m,n}r)}\right]_{r=r_0} \tag{13}$$

Thus $k_{r,m,n}$, n=0,1,2… are the infinite roots of the transcendental eigenequation:

$$\frac{J_m(k_r r_0)}{(k_r r_0)J_m^{'}(k_r r_0)}=j\frac{Z_w}{\rho_0 a_0}\frac{1}{k_0 r_0} \tag{14}$$

For the acoustic particle velocity:

$$u(r,\theta,z,t)=\sum_{m=0}^{\infty}\sum_{n=0}^{\infty}J_m(k_{r,m,n}r)e^{jm\theta}e^{j\omega t}\times\frac{k_{z,m,n}}{k_0}\frac{1}{\rho_0 a_0}\left\{C_1e^{-jk_{z,m,n}z}+C_2e^{+jk_{z,m,n}z}\right\} \tag{15}$$

It can be seen that ducts lined with acoustic absorptive material (that is, with complex wall impedance) would result in complex value of $k_{r,m,n}$ and hence k_z. The imaginary component of k_z would introduce attenuation in the axial direction, and that is the basic principle of dissipative mufflers. For the (0, 0) mode, the sound pressure and acoustic particle velocity are given by:

$$p(r,\theta,z,t)=J_0(k_r r)e^{j\omega t}\left\{C_1e^{-jk_z z}+C_2e^{+jk_z z}\right\} \tag{16}$$

$$u(r,\theta,z,t)=J_0(k_{r0}r)e^{j\omega t}\times\frac{k_{z0}}{k_0}\frac{1}{\rho_0 a_0}\left\{C_1e^{-jk_{z0}z}+C_2e^{+jk_{z0}z}\right\} \tag{17}$$

3. Models and computation

3.1. Muffler structure

The water muffler (shown in figure 1) is composed of two main parts, namely, the Kevlar-reinforced rubber tube (simplified as the rubber tube in later discussions) and the inner-noise-reduction structure (simplified as the inner structure in later discussions). Principle of the noise reduction of the rubber tube is based on the dissipative muffler with compliant walls. As the rubber tube is a Kevlar-reinforced tube, the coupling effect between the wall of the rubber tube and the flow is assumed to be weak and that can be neglected in the analysis. Then the rubber tube is functioned like a duct lined with acoustic absorptive material. The rubber wall is assumed to be locally reacting which means the impedance Z_w is independent of z. The principle of the inner structure is the combination of conical tube and the extended-tube resonators, which are typical reactive muffler structures. These structures will reflect a substantial amount of the incident power back to the source by creating a mismatch of characteristic impedances [16]. The inner structure is symmetrically mounted on both sides of the muffler.

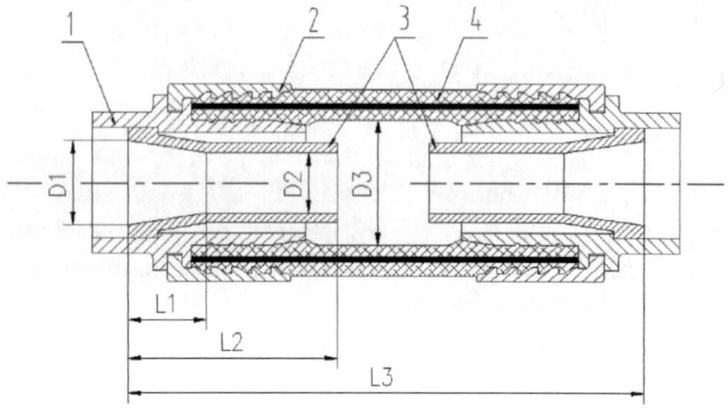

Figure 1. Structure of the water muffler.

3.2. Computational model

Computational model for the rubber tube and the water muffler is shown in figure 2 in which the small annular cavity in the muffler is omitted. The inner flow passage is divided into several elements according to the variation of the cross area and material (numbered in figure2). These elements are connected through the boundary conditions at the discontinuities. Two extended pipes are connected to these models to simulate the inlet and outlet pipe, the same as that modelled in the FEM simulation.

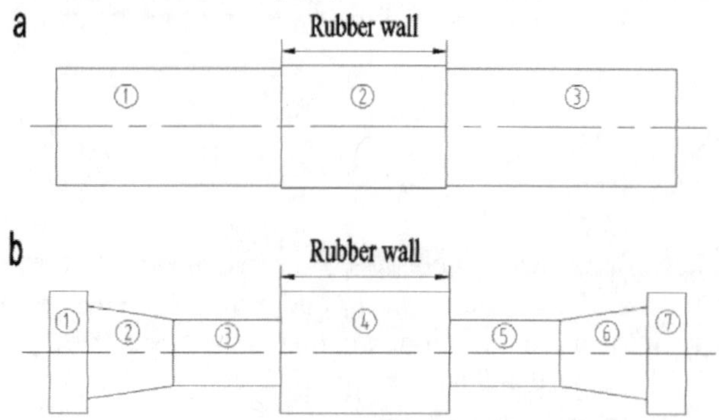

Figure 2. Simplified inner flow passage of: (a) rubber tube, (b) water muffler.

3.3. Boundary conditions

At the discontinuities between two elements, the continuity of acoustic pressure and acoustic mass velocity should be fulfilled, that is:

$$p_1 = p_2$$
$$v_1 = v_2$$

At the inlet, the sound source is set as the test results of the reference tube. At the outlet, outlet impedance obtained from the FEM simulation is applied.

4. Results and discussion

4.1. Verification of the theoretical computation

Comparisons among the results obtained from theoretical computation, FEM simulation and the experiment of the rubber tube and that of the water muffler are shown in figure3 and figure 4 respectively. It can be seen from these figures that the theoretical results show a good accordance in general tendency with the FEM simulated and the measured results which validate the efficiency of the proposed computation method.

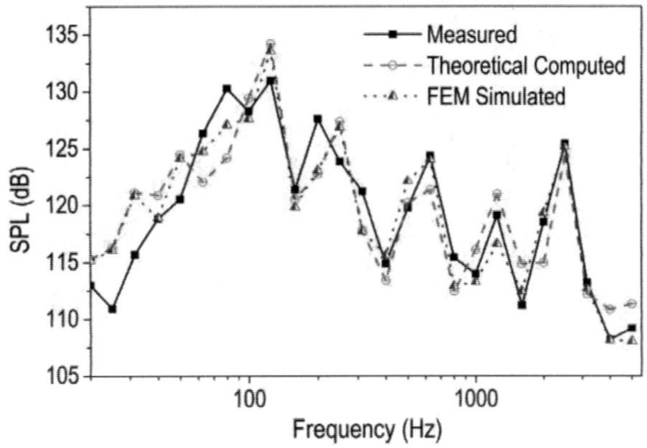

Figure 3. Results of the rubber tube.

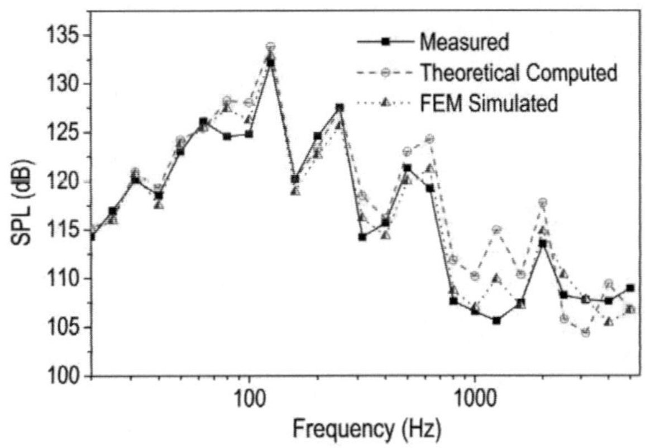

Figure 4. Results of the water muffler.

4.2. Parametric study

The length and diameter of the rubber tube and the inner structure will significantly affect the acoustic performance of the water muffler. Because the length of the tube is normally determined due to the limited installation spaces, more attention is given to the diameter. Parametric study on the diameter of the inner structure (D2) and that of the rubber tube (D3) is conducted in this section. Comparisons on acoustic performance of the water muffler with symmetrically installed inner structures are shown in subfigure 5a and subfigure 5b. From these subfigures one can tell that the SPL increases with the increase of the diameter of the inner structure while shows little difference with the increase of the diameter of the rubber tube. Further comparisons on acoustic performance of the water muffler with asymmetrically installed inner structures in which the left inner structure has different radius with the right one (with $D2_L$ and $D2_R$ denotes the D2 of the left inner structure and that of the right one respectively) are shown in subfigure 5c and subfigure 5d. It can be seen from these subfigures that the SPL increases with the increase of the diameter of the left inner structure while shows little difference with the increase of the diameter of the right one. An overall view on the four subfigures in figure 5 shows that the radius of the left inner structure has the most significant effect on the SPL. This is because it is located near the sound source and the increase of D2 also results in the increase of the diameter ratio D2/D1 which will affect the sound reflection at the conic tube of the inner structure.

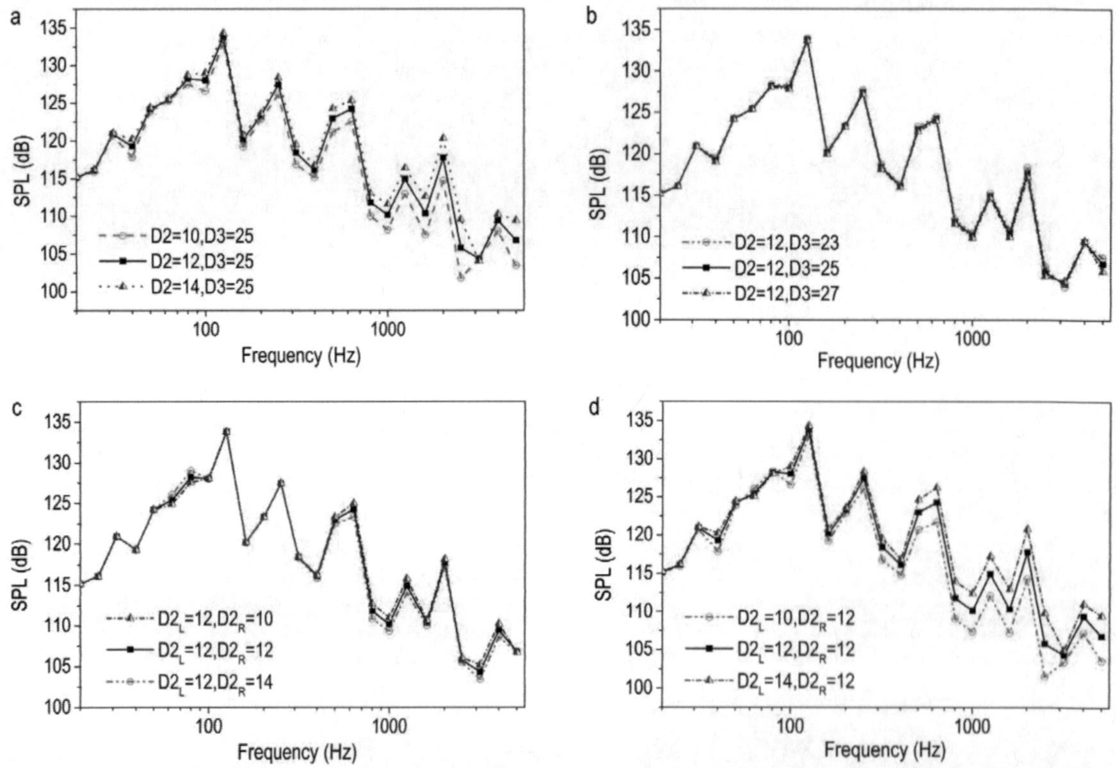

Figure 5. Parametric study on acoustic performance of the water muffler:(a) Symmetric with inner structure in different radiuses; (b) Symmetric with rubber tube in different radiuses; (c) Asymmetric with the right inner structure in different radiuses; (d) Asymmetric with the left inner structure in different radiuses.

5. Conclusion

Theoretical computation on a previously studied water muffler is carried out in this article. Structure of the water muffler is composed of two main parts, namely, the Kevlar-reinforced rubber tube and the inner-noise-reduction structure. Rubber wall of the rubber tube is assumed to function as rigid wall lined with sound absorption material and is described by a complex radius wave number. Comparison among the results obtained from theoretical computation, FEM simulation and experiment of the rubber tube and that of the water muffler has been made. The theoretical results show a good accordance in general tendency with the FEM simulated and the measured results which confirm the efficiency of the computation method.

Parametric study on the diameter of the inner structure and that of the rubber tube is conducted. Results show that the radius of the left inner structure has the most significant effect on the SPL of the water muffler due to its location and effect on the diameter ratio D2/D1.

Acknowledgments

The authors would like to thank for the support of National Nature Science Foundation of China (No. 51276213).

6. References

[1] Selamet A, Easwaran V and Falkowski A G 1999 Three-pass mufflers with uniform perforations *J. Acoust. Soc. Am.* **105** 1548
[2] Dokumaci E 2001 Sound transmission in mufflers with multiple perforated co-axial pipes *J. Sound. Vib.* **247** 379
[3] Wu T W, Cheng C Y R and Zhang P. 2002 A direct mixed-body boundary element method for packed silencers *J. Acoust. Soc. Am.* **111** 2566
[4] Selamet A, Denia F D and Besa A J 2003 Acoustic behavior of circular dual-chamber mufflers *J. Sound. Vib.* **265** 967
[5] Denia F D, Selamet A, Fuenmayor F J, et al. 2007 Acoustic attenuation performance of perforated dissipative mufflers with empty inlet/outlet extensions. *J. Sound. Vib.* **302** 1000
[6] Panigrahi S N and Munjal M L 2007 A generalized scheme for analysis of multifarious commercially used mufflers APAC **68** 660
[7] Xu M B, Selamet A and Kim H 2010 Dual Helmholtz resonator. *APAC* **71** 822
[8] Mimani A and Munjal M L 2012 3-D acoustic analysis of elliptical chamber mufflers having an end-inlet and a side-outlet: An impedance matrix approach *WAVE MOTION* **49** 271
[9] Vijayasree N K and Munjal M L. 2012 On an Integrated Transfer Matrix method for multiply connected mufflers *J. Sound. Vib.* **331** 1926
[10] Strunk R D 1991 Silencer for hydraulic piston pump pressure pulsations *SAE Technical Paper*
[11] Dodson J M, Dowling D R and Grosh K 1998 Experimental investigation of quarter wavelength silencers in large-scale hydraulic systems *Noise Control Engr. J.* **46** 15
[12] Kojima E and Ichiyanagi T 1998 Development research of new types of multiple volume resonators *Bath Workshop on Power Transmission and Motion Control* 193
[13] Mikota J 2001 A novel, compact pulsation compensator to reduce pressure pulsations in hydraulic systems *World Scientific* **45** 69
[14] Earnhart N E and Cunefare K A 2012 Compact Helmholtz Resonators for Hydraulic Systems International Journal of Fluid Power **13** 41
[15] Du T, Li S Y, Liu J T and Wu D Z 2015 Acoustic performance of a water muffler *Noise Control Engr. J.* **63** 239
[16] M L Munjal 1987 Acoustics of Ducts and Mufflers (New York :Wiley New York) p 65

Numerical study on the effect of a lobed nozzle on the flow characteristics of submerged exhaust

T C Miao, T Du, D Z Wu[1], L Q Wang

Institute of Process Equipment, College of Chemical and Biological Engineering, Zhejiang University
Zheda Road 38 #, 310027 Hangzhou, Zhejiang, China

E-mail: jjjjjyes@163.com

Abstract. In order to investigate the effecting mechanism of nozzle structure on the flow characteristics of submerged exhaust, the processes of air exhausted from a lobed nozzle and a round nozzle into water have been numerically simulated using realizable $k-\varepsilon$ model under the framework of the volume of fluid (VOF) model. Both the flow structure and the upstream pressure fluctuations are taken into consideration. The calculated results are in good agreement with the experimental results, showing that gas exhausted from the lobed nozzle would flow along the axial direction easier. Flow structure of the gas exhausted from the lobed nozzle is more continuous and smoother. The pressure fluctuations in the upstream pipeline would also be reduced when gas exhausted from the lobed nozzle. The resulting analysis indicates that the lobed structure could deflect water flow into the gas jet. The induced water would be mixed into the gas jet in form of small droplets, making the jet more continuous. As a result, the mixed jet flow would be less obstructed by the surrounding water, and the upstream pressure fluctuation would be reduced. The work in this paper partly explained the effecting mechanism of nozzle structure on the flow characteristics of submerged exhaust. The results are useful in the designing of exhaust nozzles.

1. Introduction

Submerged exhaust can be found in many industry fields, such as metallurgy, chemical engineering, environmental engineering, etc. It can also be used for underwater propulsion. All these applications are based on the understanding of the submerged gas jet, especially the flow characteristics of it [1-6]. Submerged exhaust would produce unsteady turbulent flow which is not well understood. The big differences in physical properties between gas and liquid phases can result in complicated flow structures, which make relative studies remain challenging [7].

[1] Dazhuan Wu, Email: wudazhuan@zju.edu.cn

Recent studies show that nozzle structures would efficiently effect on the flow structures of submerged exhaust. Linck have found that conical projecting nozzle and corrugated converging diverging nozzle would have significant damping effects on submerged gas flow [8]. Arghode studied the characteristics of submerged exhaust using nozzles with circular, square, triangular and elliptical exit cross-sections. He found that the two phase flow structure and the pressure fluctuation in the upstream pipeline would be effected by these different nozzle structures [9]. Similar studies were also done by Hao who found that both the nozzle size and structure would effect on the gas behavior near the nozzle exit [10, 11]. All these studies indicated that it would be possible to control the flow structure and its associated characteristics by designing different nozzle structures. However, how the nozzle structure would effect on the flow structure is still not clear, relative designs are lack of guidance.

This paper aims to study the effecting mechanism of nozzle structure on the flow characteristics of submerged exhaust using numerical methods. The processes of air exhausted from a lobed nozzle and a round nozzle into water have been numerically simulated using realizable k- ε model under the framework of the volume of fluid (VOF) model. Both the flow structure and the upstream pressure fluctuations are considered. How the nozzle structure would effect on the flow characteristics of submerged exhaust is analyzed.

2. Calculation model

2.1. Numerical method

In order to calculate the gas-liquid two phase flow, 3D N-S equations are used. The control equations are as follows:

$$\frac{\partial}{\partial t}(\rho_m) + \nabla(\rho_m \vec{v}_m) = 0 \quad (1)$$

$$\frac{\partial}{\partial t}(\rho_m \vec{v}_m) + \nabla(\rho_m \vec{v}_m \vec{v}_m) = -\nabla p + \nabla[\mu_m(\nabla \vec{v}_m + \vec{v}_m^T)] + \rho_m g + \vec{F} + \nabla\left(\sum_{k=1}^{n} \alpha_k \rho_k \vec{v}_{dr,k} \vec{v}_{dr,k}\right) \quad (2)$$

$$\frac{\partial(\rho T)}{\partial t} + \frac{\partial(\vec{v}_k \rho T)}{\partial x_k} = \frac{\partial}{\partial x_k}\left(\frac{k}{c_p}\frac{\partial T}{\partial x_k}\right) + S_T \quad (3)$$

Where $\vec{v}_m$ is the average velocity, ρ_m is the mixed density, ρ_k is the density of phase k, α_k is the volume fraction of phase k, $\vec{v}_m = \frac{\sum_{k=1}^{n} \alpha_k \rho_k \vec{v}_k}{\rho_m}$, $\rho_m = \sum_{k=1}^{n} \alpha_k \rho_k$, $\sum_{k=1}^{n} \alpha_k = 1$, $\vec{F}$ is volume force, μ_m is mixed viscosity coefficient, $\mu_m = \sum_{k=1}^{n} \alpha_k \mu_k$, $\vec{v}_{dr,k}$ is the drift velocity of phase k, c_p is specific heat, T is temperature, k is heat transfer coefficient, S_T is the viscous dissipation term.

In order to simulate this situation effectively, a simplified physical model should be used. The following basic assumptions are put forward:1) The influence of the wall is ignored; 2)As the inlet gas velocity is 79.6m/s, the corresponding Maher number is 0.23<0.3, hence the compressibility of the gas medium can be ignored; 3) The mass and heat transfer between phases are ignored.

The turbulence model used in this paper is realizable $k-\varepsilon$ model. The multiphase model used here is VOF (Volume of Fluid) model which can capture the phase interface effectively. The gas medium is air and the liquid medium is water.

2.2. Calculation model parameter and boundary conditions

A schematic of the calculation model is shown in figure.1. The computation region is a cylinder with height of 1000 mm and radius of 600 mm. It was filled with 660 mm depth of water. The direction of gravity is along the axil of the cylinder towards the exhaust nozzle. The exhaust nozzle is placed in the bottom of the cylinder vertically. The diameter of the nozzle inlet is 20mm. Detailed parameters of the lobed nozzle are shown in Fig.1 (b). The trailing edge of the lobed nozzle is formed by six lobe structures. Lobe width and height are 4 mm and 10 mm respectively. The outer and inner penetration angles are both 15°. The mesh generation result of this model is shown in figure.2. The computation region contains 887946 structured grids. We use velocity inlet and pressure outlet as the boundary conditions, the others are set as wall.

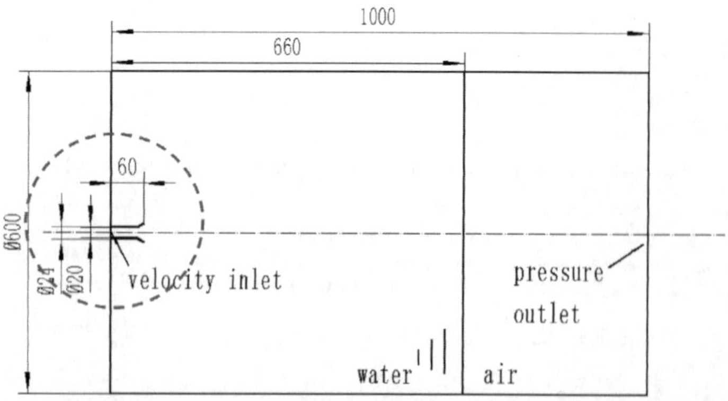

(a)

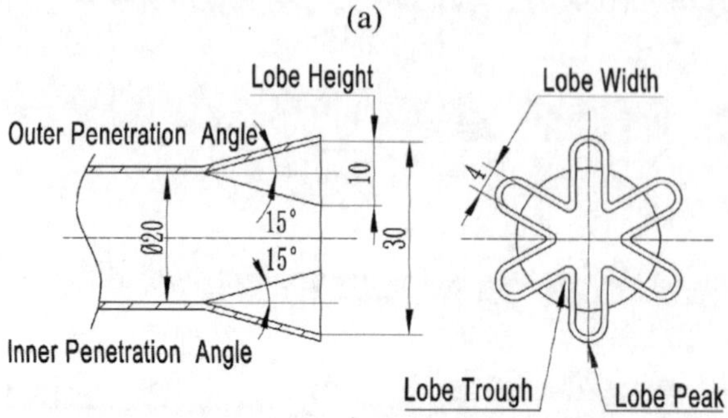

(b)

Figure.1 Schematic of the computation model

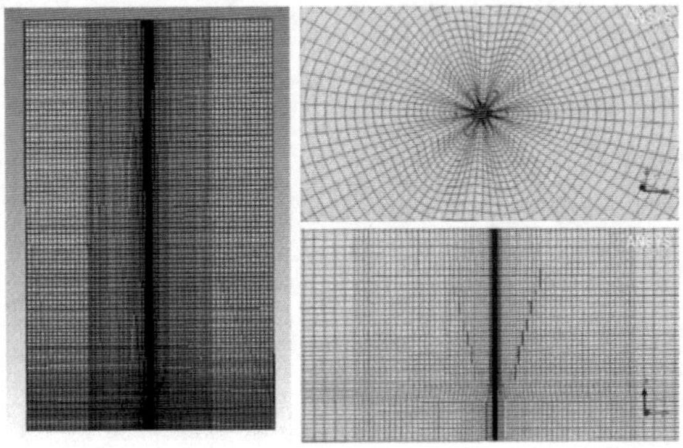

Figure.2 Schematic of the mesh generation result

3. Results and analysis

Firstly, the calculating results are compared with the experiment results. As shown in Fig.3, the calculated flow structures near the nozzle exit are in good agreement with the experiment results (DN20 nozzle, the gas flow rate is 90 m^3/h). This means the numerical methods we used in this paper is suitable to calculate the submerged exhaust process.

Fig.3. Comparison between the calculating results and the experiment results

Then we calculated the flow structures of gas exhausted from the round nozzle and the lobed nozzle respectively. As we can see, an obvious stem like structure can be observed when bubble detach from the round nozzle (Fig.4 (a) and (b)). In general, this structure would occur when the axial flow is obstructed. However, similar structure cannot be observed when bubble detach from the lobed nozzle, which means the axial flow via the lobed nozzle is more continuous. Bubbles detached from the round nozzle have smooth shapes due to the regular cross section shape of the nozzle exit. However, bubbles detached from the lobed nozzle have unstable surfaces which are easily break into small bubbles. Additionally, many small droplets can be seen in the bubbles generated from the lobed nozzle. This is because the lobe trough can guide the surrounding liquid flow towards the gas jet, hence the bubble bottom would be broken up and small droplets would be induced into the gas flow. The small bubbles and droplets exist in the flow field of submerged exhaust through the lobed nozzle would promote the momentum exchange between gas and liquid. As a result, gas exhausted from the lobed nozzle would be less obstructed in the axial direction. This could be proofed by Fig.5. As we can see, gas exhausted from the lobed nozzle (Fig.5 (a)) has thinner jet width and would flow a longer distance along the axial direction within 0.33 s than that from the round nozzle (Fig.5 (b)).

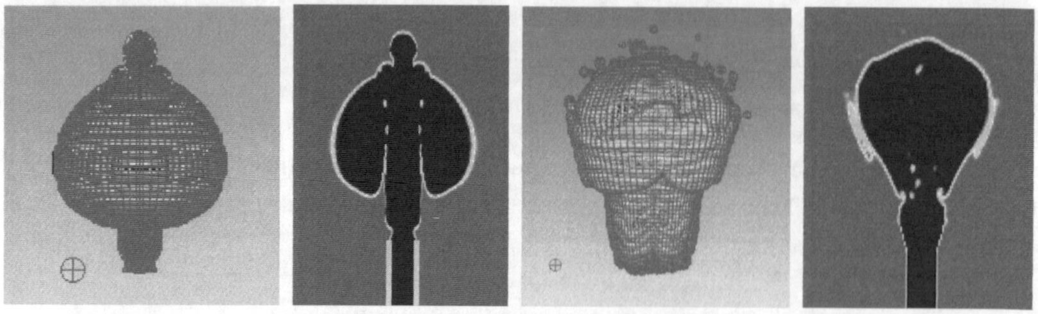

(a) (b) (c) (d)

Fig.4. The calculated flow structures at t=0.15 s, (a), (b) round nozzle, (c), (d) lobed nozzle.

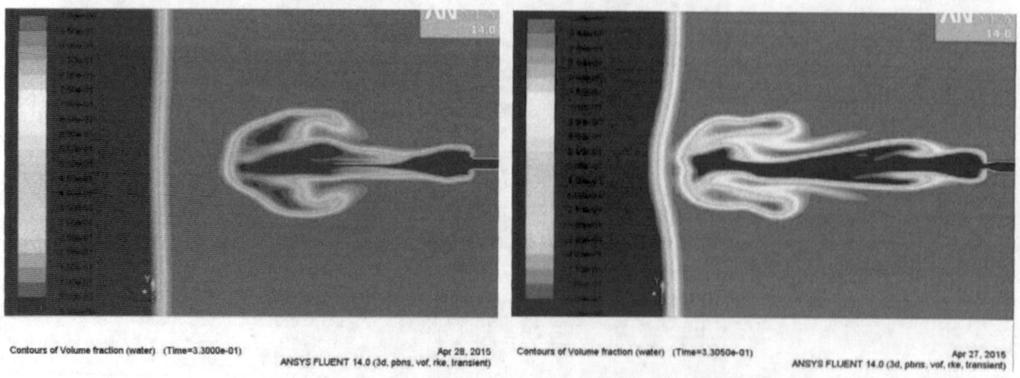

(a) (b)

Fig.5. The calculated flow structures at t=0.33 s, (a) round nozzle, (b) lobed nozzle.

The pressure fluctuations in the exhaust nozzle are measured at point of (0, 0, 30). The measurement results are shown in Fig.6. It can be seen that the pressure fluctuation of gas exhausted from the round nozzle has larger amplitude and higher frequency (Fig.6 (a)). In order to investigate the reason of this phenomenon, the flow structures corresponding to the pressure peaks are also shown in Fig.6. As we can see, the pressure peaks would occur when the gas flow is obstructed in the axial direction. Since the gas exhausted from the lobed nozzle is less obstructed, the corresponding pressure fluctuations are weaker.

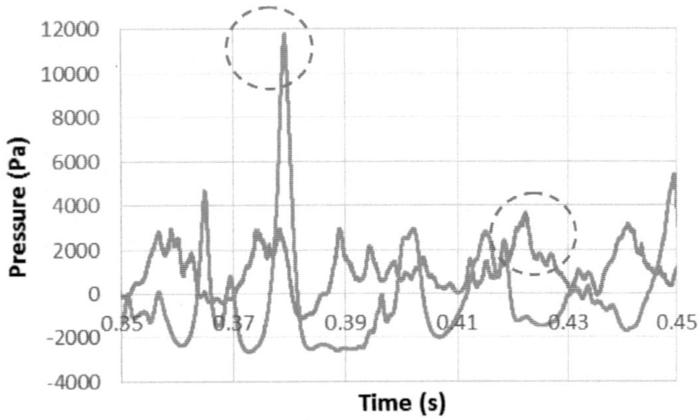

(a)

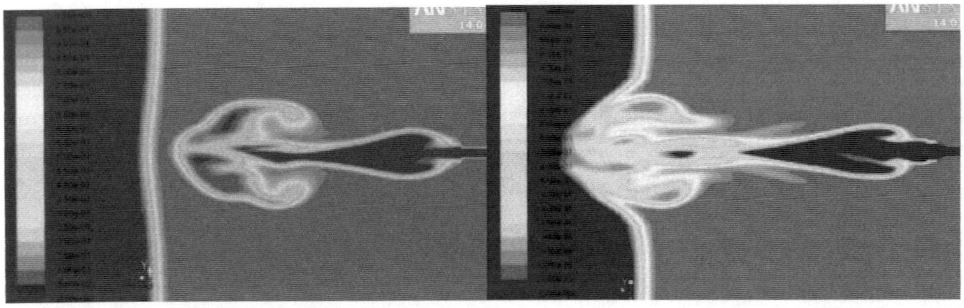

(b) (c)

Fig.6. The upstream pressure fluctuations and the corresponding flow structures, (a) pressure fluctuations, (b) flow structure corresponding to peak A, (c) flow structure corresponding to peak B

4. Conclusions

In this paper, the submerged exhaust through a round nozzle and a lobed nozzle are numerically calculated using realizable $k-\varepsilon$ model and VOF model. The calculating results coincide well with the experiment results, which means the calculating methods we used here are suitable and reliable. Based on that, the flow structures and pressure fluctuations are analyzed. The main conclusions are as follows:

1) The lobed nozzle would induce small bubbles and droplets into the flow field. This is helpful to make the gas flow more continuous and less obstructed in the axial direction.

2) The pressure peaks in the upstream pipeline are caused by the obstructed gas flow. Since gas exhausted from the lobed nozzle is less obstructed, the pressure fluctuations of it have lower amplitude and frequency.

References

[1] Rue D 2004 *Am. Ceram. Soc. Bull.* **83** 18-20.

[2] Pioro L S, Osnach A M, Pioro I L 2003 *Ind. Eng. Chem. Res.* **42** 6697-6704.

[3] Ozawa, Y., Mori, K. 1986 *Trans. Iron Steel Inst. Jpn.* **26** 291-297.

[4] M. Jamialahmadi, M.R. Zehtaban, H. Muller-Steinhagen, et al. 2001 *Chem. Eng. Res. Des.* **79** 523-532.

[5] Sanada, T., Watanabe, M., Fukano, T., Kariyasaki A. 2005 *Chem. Eng. Sci.* **60** 4886-4900.

[6] S. S. Gulawani, S. S. Deshpande, J. B. Joshi, et al. 2007 *Ind. Eng. Chem. Res.* **46** 3188-3218.

[7] Tang J N, Tseng C C, Wang N F, Shyy W 2011 Flow structures of gaseous jets injected into water for underwater propulsion *AIAA* **pp** 185.

[8] Linck, Martin B., Ashwani K. Gupta, and Kenneth H. 2009 *J. Propul. Power* **25.2** 522-532.

[9] Arghode, V. K., Gupta, A. K. 2012 *Appl. Energy.* **89** 246-253.

[10] Hao Z R, Wang L Q, Wu D Z, et al. 2011 *J. Eng. Thermophys.* **32** 949-952.

[11] Hao Z R, Zhou Z H 2012 *App. Mech. Mater.* **204** 4620-4623.

IOP Conf. Series: Materials Science and Engineering **129** (2016) 012067 doi:10.1088/1757-899X/129/1/012067

Design method of water jet pump towards high cavitation performances

L L Cao[1], B X Che, L J Hu, D Z Wu

Institute of Process Equipment, College of Chemical and Biological Engineering, Zhejiang University
Zheda Road 38 #, 310027 Hangzhou, Zhejiang, China

E-mail: caolinlin@zju.edu.cn

Abstract. As one of the crucial components for power supply, the propulsion system is of great significance to the advance speed, noise performances, stabilities and other associated critical performances of underwater vehicles. Developing towards much higher advance speed, the underwater vehicles make more critical demands on the performances of the propulsion system. Basically, the increased advance speed requires the significantly raised rotation speed of the propulsion system, which would result in the deteriorated cavitation performances and consequently limit the thrust and efficiency of the whole system. Compared with the traditional propeller, the water jet pump offers more favourite cavitation, propulsion efficiency and other associated performances. The present research focuses on the cavitation performances of the waterjet pump blade profile in expectation of enlarging its advantages in high-speed vehicle propulsion. Based on the specifications of a certain underwater vehicle, the design method of the waterjet blade with high cavitation performances was investigated in terms of numerical simulation.

Introduction

Developing towards much higher advance speed, the underwater vehicles make more critical demands on the performances of the propulsion system. Basically for the traditional propeller system, the increased advance speed requires the significantly raised rotation speed, which would result in the deteriorated cavitation performances and consequently limit the thrust as well as the efficiency of whole system. Instead of directly increasing the rotation speed, the alternative structures of the propulsion system give more beneficial choices. Distinguished from the traditional propellers which make use of the blade lift to drive, the waterjet pump takes advantages of the momentum difference between the inflow and outflow of the pump to thrust vehicles forward. Hence, under the same design specifications, the water jet pump appears with obviously reduced rotation speed as well as more favourite cavitation and vibration performances.

While further increase the vehicle advance speed, the inherent superiorities in cavitation performances of waterjet pump are also insufficient. Commonly, to relieve the cavitation problem, the underwater vehicles would try to increase their diving depth as well as the substantial pressure during high speed advancing. Plenty of researches have been done to improve the inherent cavitation performances of the waterjet pump. Since the cavitation performances of rotor blade are highly

[1] The correspondence author: Cao Linlin

associated with its pressure distribution, it is essential to make the pressure field optimized but with the blade loading unchanged. The optimization of the inlet condition to raise the inflow static pressure by adjusting the geometries of pump is also effective to avoid or delay the cavitation inception. Thurston and Amsler [4] proposed that it is quite beneficial to narrow the scope between the subcavitating and supercavitating status due to the high instabilities here, and they have made a detailed presentation about the cavitation control methods by adjusting the duct geometries as well as optimizing the inflow static pressure in the waterjet pump. Additionally, the local vortices, the abrupt flow acceleration and other types of flow structures which result in sharp static pressure drop and associated losses also contribute to the cavitation problem significantly. The axial and mixed flow pumps are usually characterized by the tip leakage flows, which are always accompanied with sharp pressure drop as well as cavitation inception inside the vortex cores. Wu et al. [5] simulated the flow vortices structure in the waterjet pump, and carried out the researches on the mechanism and the cavitation inception in the tip leakage vortex. Generally, the loading situation along the span and at the blade tip region must be carefully considered to improve the pump cavitation performances while keeping the work capacities there.

In the present paper, the cavitation performances of the waterjet pump blade profile in high-speed vehicle propulsion are concentrated. Refer to the pump performances under high-speed advance, the blade geometries and associated loading situations were studied to improve the pressure distributions as well as the cavitation performances. Based on the specifications of a certain underwater vehicle, the waterjet blade design method with the consideration of cavitation performances was investigated in terms of numerical simulation.

Key Parameters and Performance Simulation Method of the Waterjet pump

In the present paper, all of the model waterjet pumps were designed and studied based on a certain underwater vehicle with the specifications described in table 1, which could be simplified as an axisymmetric cylindrical body.

Table 1. Major performances specifications of the underwater vehicle

Advance speed [Kn]	Diameter [m]	Length [m]	Thrust [KN]
50	0.324	6	5

Refers to the distribution properties of the wake flow at the vehicle trail [6,7], the inflow profile of the waterjet pump described as equation (1) was specified to simulate the inlet velocity. Herein the axial velocity at the pump inlet V_m [m/s] was assumed to vary linearly against the pump inlet radius D, and the circumferential and radial components are assumed to be negligible.

$$V_m = 73.4D + 7 \tag{1}$$

To offer sufficient power supply to the vehicle while keep the effect to its hydrodynamics properties minimized, the dimension and mechanical structure of propeller must be carefully determined [6]. Since the present paper concentrate on the internal performances of waterjet as the propulsion device, the specifications of the waterjet pump were majorly decided based on the vehicle required thrust and the pump models in other related researches [5,7]. As listed in Table 2, the rotor outer diameter was specified as 67% of that of vehicle, which was supposed to keep the hydrodynamic influence on a relatively low level.

Table 2. Major specifications of the model waterjet pumps

Head rise [m]	20	Blade tip clearance [mm]	0.5

Rotation speed [min^{-1}]	3000	Inlet area [m^2]	0.0278
Flow rate [m^3/s]	0.5477	Outlet area [m^2]	0.0196
Thrust [KN]	5	Rotor hub diameter [m]	0.22
Power [KW]	138	Rotor tip diameter [m]	0.116
Rotor blade number	7	Inflow axial velocity [m/s]	19.70
Stator blade number	11	Outflow axial velocity [m/s]	27.94

In the present study, all of the simulation works were conducted with the commercial Computational Fluid Dynamics (CFD) code ANSYS CFX, and the modelling and computational grid generations were done by ANSYS Turbogrid. The entire annuluses model is illustrated in figure 1 to better visualize the waterjet pump concept. For the preliminary evaluation of the waterjet pump internal and external performances, only the steady-state computations with one rotor-stator passage were carried out. As the most prominent two-equation models proficient in the flow separation predictions, the *k-ω* based SST (Shear Stress Transfer) Model was employed in all of our simulation works.

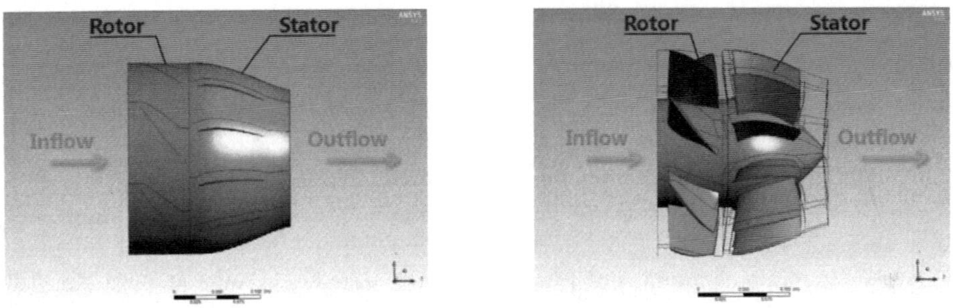

(a) Entire annuluses model with pump casing

(b) Entire annuluses model without pump casing

Figure 1. Computational model of waterjet pump

Blade Loading Distribution Design Method Based on Cavitation Performances

To evaluate the cavitation performances quantitatively, the definition of Required Net Positive Suction Head (NPSHr) is given in equation (1), in which V_L and W_L are specified as the absolute and the relative velocities at the blade leading edge point 'L', and λ_1 and λ_2 as the loss coefficients decided by the inlet condition and the blade profile respectively.

$$\mathrm{NPSH}_r = \lambda_1 \frac{{C_L}^2}{2g} + \lambda_2 \frac{{W_L}^2}{2g} \tag{2}$$

In the traditional axial flow pump designed with the loading profile of 'free vortex', the cavitation inception is ordinarily found at the leading edge around the tip profile due to the low static pressure as well as the rather high peripheral speed and relative velocity there.

Regardless of the inlet condition, the cavitation performances of rotor blade are highly associated with its pressure distribution, it is essential to make the pressure field optimized but with the blade loading unchanged. Generally, the aerofoil with heavy loading is usually fore-loaded with lowest pressure point and sharp pressure gradient at the Leading Edge (LE) on the Suction Surface (SS), which could offer high work capacity but quite unfavourable to the cavitation performances. In the present paper, the blade loading situation with the compromise between hydraulic and cavitation performances was studied. Based on the related research results [8-10], the most decisive parameters for the blade loading situation could be concluded as: 1) camber line function, 2) max camber height location x_f/l, 3) incidence angle i.

1.1. Camber line function

In the present study, the quadratic parabola equation (2) was applied to specify the camber geometry, in which a, b, and c respectively denote any arbitrary constants, the chord ratio x ranges from 0 to 1, and θ represents the local gradient angle of the camber curve (as shown in figure 2) with the condition of $tan\,\theta = dy/dx$.

$$\theta(x) = ax^2 + bx + c \tag{3}$$

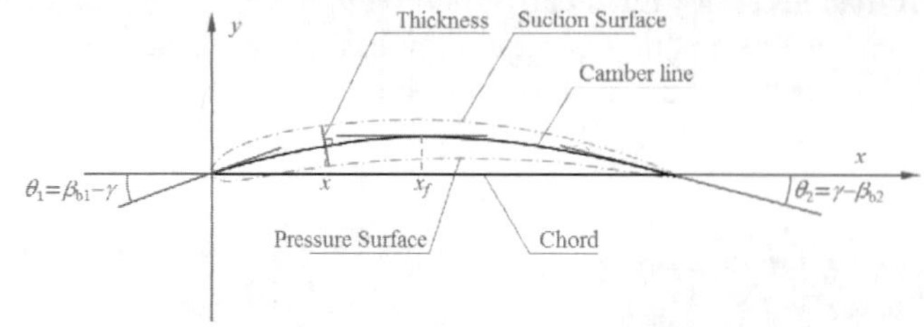

Figure 2. Schematic camber line geometry

1.2. Other factors related to camber geometry

Moreover, the max camber height location x_f/l and the incidence angle i are among the important decisive factors to the blade loading and pressure distribution along the PS and SS, consequently influential to the cavitation performances. The large work capacity is usually accompanied with large attack α and incidence angle i, and the location of x_f/l close to the LE, which ordinarily results in the fore-loading situation. In that case, the SS is characterized by a region with rather low pressure to offer considerable pressure difference as well as large lift force, in which the cavitation inception usually occurs. In the related researches [8, 9], Zangeneh, M. et al. proposed the optimum pressure distribution model for the impellers in terms of secondary flow suppression, and indicated that it is beneficial to respectively increase the blade aft part loading at the hub section and decrease that at the tip area.

Generally, the blade pressure distribution should be optimized with the min pressure raised for the cavitation consideration but the blade loading maintained for the hydraulic performance consideration. The combination of x_f/l, i and other parameters should be carefully specified to get more flat blade loading curve at the SS. In the present research, the several combinations were evaluated for the blade loading situation, hydraulic performances and cavitation performances.

$$\delta = m_c(\beta_{b1} - \beta_{b2})\sigma^{-n_c} \tag{4}$$

$$\delta = \Delta\beta + 13.05\left(\frac{x_f}{l} - 0.4\right) \tag{5}$$

The design specifications of three of these models are given in table 3. For the first two models G1, the classic deviation angle model of equation (4) was adopted, where m_c=0.26, n_c=0.5 as in the axial flow compressors and σ represents the cascade solidity; while for the G2 and G3, to improve the work capacity as well as the flow turning under the aft-loading situation, an alternative empirical formula equation (5) was applied, which has been validated in the previous researches [11].

Table 3. Design specifications of three of rotor models

	Section	1(Hub)	2	3	4	5(Tip)
G1	Incidence [°]	2.00	1.00	1.00	1.00	0.00
	x_f/l	0.40	0.40	0.40	0.40	0.40
	Deviation angle [°]	3.60	3.71	2.54	2.10	1.96
	Attack angle [°]	10.824	7.996	5.368	3.804	2.856
G2	Incidence [°]	2.00	1.00	1.00	1.00	0.00
	x_f/l	0.40	0.45	0.50	0.55	0.60
	Deviation angle [°]	5.00	5.15	4.91	4.95	6.03
	Attack angle [°]	11.898	6.424	3.622	2.079	0.429
G3	Loading coefficient L_c	0.915	1.007	1.060	1.115	1.129
	Incidence [°]	2.00	1.00	1.00	1.00	0.00
	x_f/l	0.500	0.525	0.550	0.575	0.600
	Deviation angle [°]	10	7.10	6.09	5.54	6.03
	Attack angle [°]	7.158	4.114	2.502	1.559	0.429

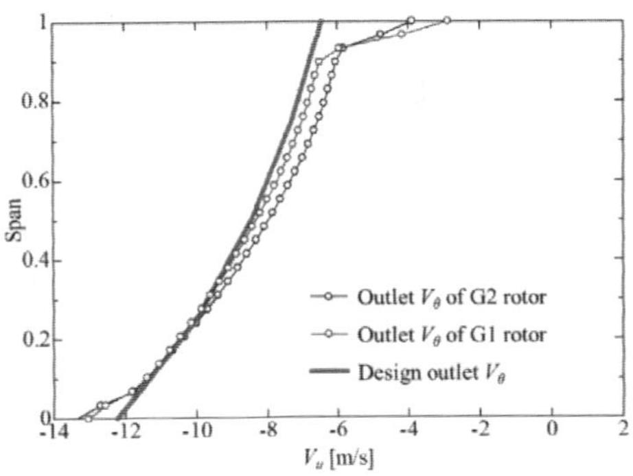

(a) Outlet flow swirl V_θ of G1 and G2 rotors

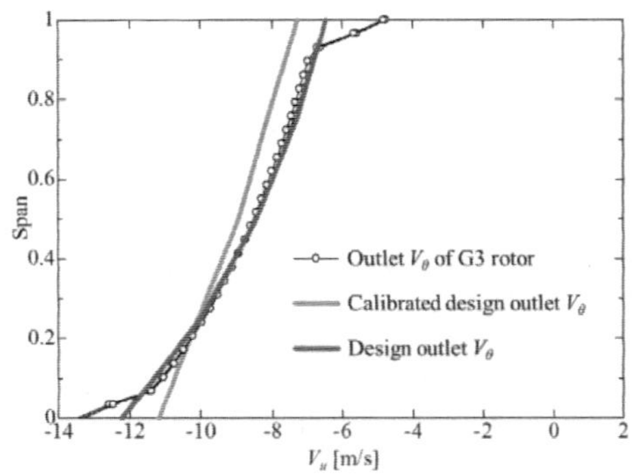

(b) Outlet flow swirl V_θ of G3

Figure 3. Schematic camber line geometry

The third model type G3 was designed based on the simulation results of the first two models. From figure 3(a), it can be seen that expect for the defect due to the tip leakage flow, the distributions of outflow swirl V_θ of G1 and G2 rotors are deviated from the design expectation denoted in green. Compared to the G1 blade designed with x_f/l close to LE (40% of chord length) along the whole span, the G2 rotor shows the more pronounced defect of V_θ with x_f/l moving backward (from 40% at hub to 60% at tip), which indicates the insufficient work capacity of the G2 profile section under the aft-loading situation. To obtain the flow turning as well as the hydraulic head satisfying the design specification, the loading coefficient L_c was introduced to calibrate the distribution of design head along the blade span. In equation (6), subscript d denotes under design condition, and G2 denotes the simulation results of G2 rotor.

$$L_c = V_{\theta,\mathrm{d}}/V_{\theta,\mathrm{G2}} \tag{6}$$

Combining the calibration methods above, the G3 rotor was obtained as described in table 3. It is clear that compared with those profiles designed with equal x_f/l, G3 showed obviously increased attack angles as well as improved work capacity. In this case, the G3 blade profile was supposed to reach a compromise between the cavitation performances and hydraulic performances.
The thickness evolution law of NACA 4* series [12] as expressed by Eq. (3-16) was adopted,

$$\frac{y_t(x)}{l} = 5\frac{t}{l}\left\{a_0\sqrt{\frac{x}{l}} + a_1\frac{x}{l} + a_2\left(\frac{x}{l}\right)^2 + a_3\left(\frac{x}{l}\right)^3 + a_4\left(\frac{x}{l}\right)^4\right\} \tag{7}$$

where $y_t(x)/l$ represents the relative thickness at any arbitrary position of x/l, and $a_0 \sim a_4$ are given by the NACA4* profile data. With the calculated thickness perpendicularly growing to both sides at the corresponding camber point, the outlines of the pressure and suction surfaces are generated as shown in figure 2.

Hydraulic and Cavitation Performances of Rotor Models
With the steady CFD simulation methods mentioned in Section 2, the preliminary performances of three types of rotors at the design flow rate $Q = 0.5477$ [m^3/s] as presented in table 4. The properties of cavitation inception of these rotors were studied with the parameter of NPSH_r (Required Net Positive Suction Head), which was calculated under their respective design conditions by the following equation (8),

$$\mathrm{NPSH}_r = \frac{P_{t,1} - P_{min}}{\rho g} \tag{8}$$

where $P_{t,1}$ represents the total pressure at the rotor inlet, and P_{min} the minimum pressure obtained around the blades of each rotor. And the positions with rather low pressure caused by the local vortices and losses were ignored in the evaluation of NPSH_r, such as the PS-tip corner where the great pressure drop is induced by the tip leakage flow.

Table 4. External Performances of Three Types of Rotors

	H [m]	η	*T* [KN]	NPSHr [m]
G1	18.714	90.43%	4.05	37.88
G2	20.080	92.54%	4.27	37.89
G3	20.304	91.70%	4.5	20.74

From the simulation results represented in table 4, it can be seen that even the profile of G1 rotor was designed as fore-loaded, it can hardly meet the design specification due to the insufficient deviation angles at the blade trail. With the deviation angles renewed, the latter two rotors G2 and G3 show well improved head rises, which are highly close to the design specifications H = 20 [m]. Compared with the G2 rotor, the G3 rotor with max camber height location moved backward represents the well increased P_{min} but can still offer comparable work capacity, which indicates that the G3 blade has the more flat loading curve and do work on the fluid more evenly from LE to TE.

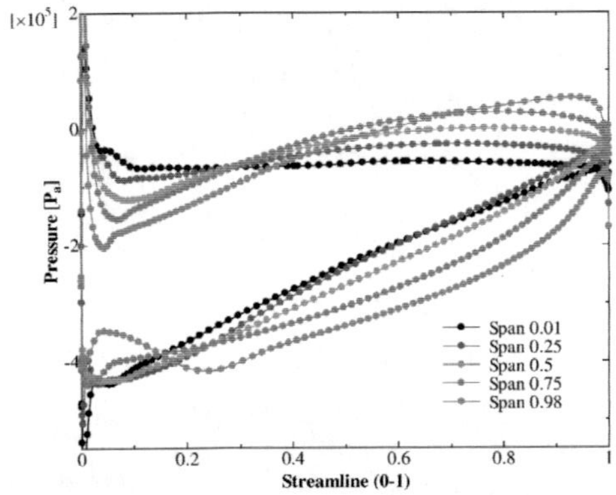

(a) Loading curves of G2 rotor blade

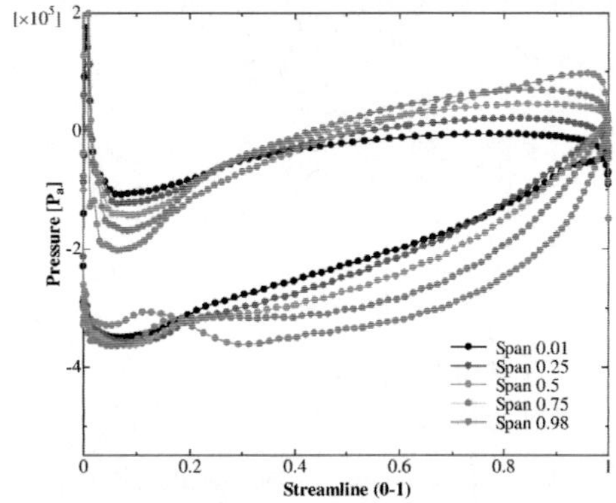

(b) Loading curves of G3 rotor blade

Figure 4. Loading curves of G2 and G3 rotor blades

The loading curves of the G2 and G3 rotor blades at various span locations are given in figure 4, from which the above assumption could be validated. In the area near to hub, the G2 blade shows rather steep pressure growing gradient on SS with a region of lowest pressure at LE as shown in figure 4 (a), which implies that the loading of these sections is concentrated at LE. This region usually corresponds to the location where cavitation inception occurs. With the max camber height location moving backward and the calibration of the head distribution in spanwise, the G3 blade shows the aft-loading situation and relatively flat pressure growing curves on SS through the whole span. Hence, the G3 rotor was believed to have well delayed cavitation inception compared with the G2 rotor under the equal ambient pressure.

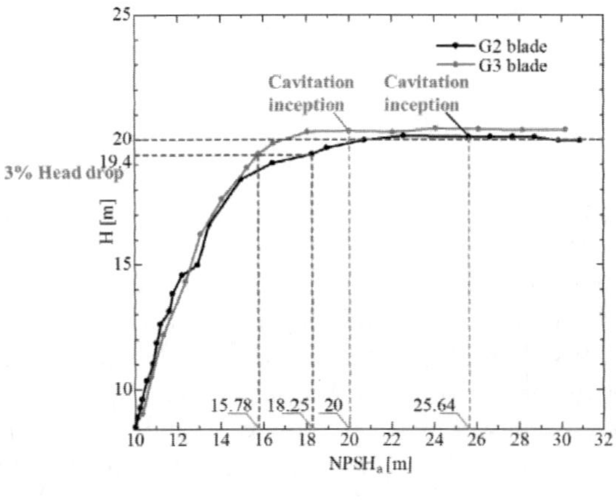

Figure 5. Drop curves of G2 and G3 rotor blades

The drop curves of two rotors are presented in figure 5, which were calculated in steady simulations with cavitation model of Rayleigh Plesset. It can be seen that the G3 rotor blade appears with superior cavitation performances compared to the G2 rotor blade.

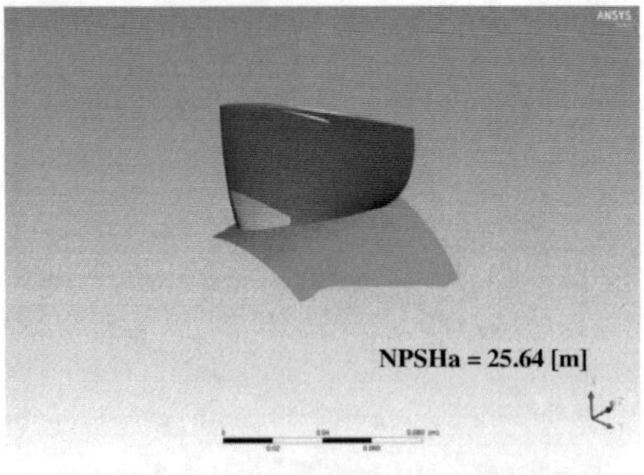

(a) Cavitation inception of G2 rotor blade

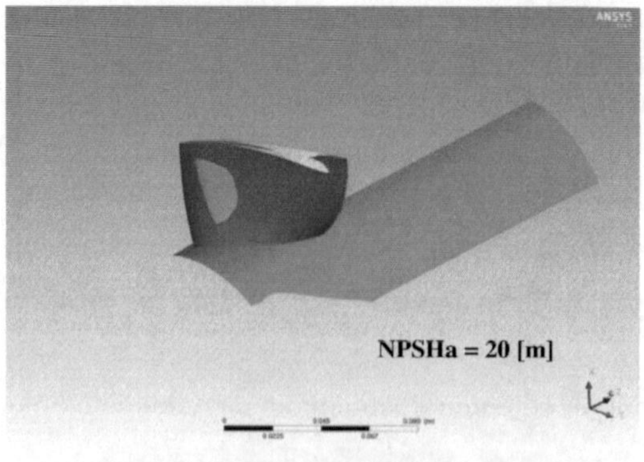

(b) Cavitation inception of G3 rotor blade

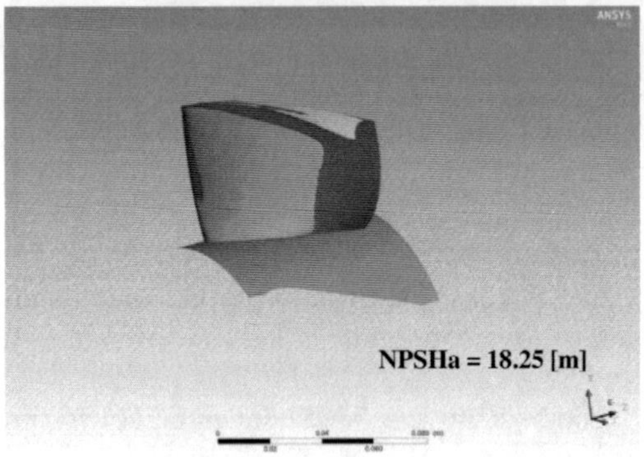

(c) 3% head drop situation of G2 rotor blade

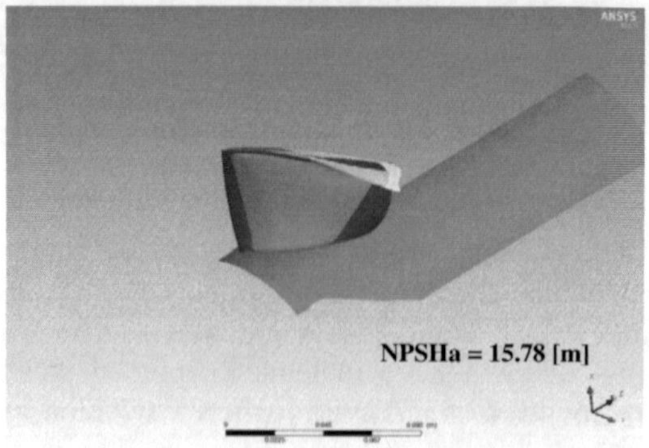

(d) 3% head drop situation of G3 rotor blade

Figure 6. Cavitation visualization of G2 and G3 rotor blades with vapour volume fraction as 10%

In figure 6, the cavitation range under various ambient pressure is presented with the vapour volume fraction of 10% as shown in grey. Regardless of the cavitation occurred in tip leakage flow, due to the different consideration in the design of blade loading, the two rotors differ from each other obviously in the locations of cavitation inception. As shown in figure 5, the G2 rotor was found to have cavitation at LE of the hub section, where the blade loading was designed to be rather large. The G3 rotor with the relatively flat pressure distribution curve at SS shows cavitation inception in the mid part of blade, and with the blade loading distribution optimized the NPSHa at the cavitation inception moment ($NPSH_{a,G3}$ = 20 [m]) is found to be well reduced from that in G2 rotor ($NPSH_{a,G2}$ = 25.64 [m]), indicating the much lower possibility of cavitation under same working condition.

With the ambient pressure further reduced, both two rotors suffer from the hydraulic deterioration as the steep drop of head rise and efficiency. In this paper, the 3% head drop point was determined as the critical cavitation point. As shown in figure 6 (c) and (d), the suction surfaces of both rotor blades are covered with a massive of vapour cloud, and the blade loading is significantly decreased. Even in this case, the G3 rotor blade still shows relatively superior ability to maintain the head rise at a high level, while the G2 rotor suffers from rather early head rise reduction as shown in figure 5.

Conclusion

In the present paper, the cavitation performances of the waterjet pump blade profile in high-speed vehicle propulsion are concentrated. Refer to the pump performances under high-speed advance, the blade geometries and associated loading situations were studied to improve the pressure distributions as well as the cavitation performances.

1) With the loading point moving backward, the flow turning in the blade passages as well as the work capacity of the blades are obviously weakened, and the deviation angles should be increased to maintain the hydraulic performances of the rotor.

2) The appropriate combination of the design incidence, camber geometry and loading distribution in spanwise could obtain the optimized rotor cavitation performances while maintaining the work capacity, efficiency and other hydraulic performances.
3) With the blade loading optimization in the design stage, the G3 rotor blade appears with superior cavitation performances compared to the G2 rotor blade. And in the severe cavitation situation, the G3 rotor blade still shows relatively superior ability to maintain the head rise at a high level.

Reference

[1] Chen B Y and Reed A M 1990 A design method and an application for contrarotating propellers, DTIC-90/003, David Taylor Research Centre, Bethesda, U. S. A.

[2] Van Gunsteren L A 1971 Application of momentum theory in counterrotating propeller design, *International Shipbuilding Progress*, **206(18)**, pp. 359-372.

[3] Caster E B and LaFone T A 1975 A computer program for the preliminary design of contrarotating propellers, No. SPD-596-01, David W. Taylor Naval Ship Research and Development Centre.

[4] Thurston S and Amsler R C 1966 Review of marine propellers and ducted propeller propulsive devices, *Journal of Aircraft*, **3(3)**, pp.255-261.

[5] Wu H, Tan D, Miorini R L and Katz J 2011 Three-dimensional flow structures and associated turbulence in the tip region of a waterjet pump rotor blade, *Experiments in Fluids*, **51(6)**, pp.1721-1737.

[6] Xiao J P and Jin H 2001 Power effect research on torpedo model with counter-rotation pusher propeller, *Experiments and Measurements in Fluid Mechanics*, **15(3)**, pp. 48-53.

[7] Thad J M, Seth D S and Alan J B 2008 Design of the ONR AxWJ-2 axial flow water jet pump, Hydromechanics Department Report, NSWCCD-50-TR-2008/066.

[8] Zangeneh M 1991 A compressible three-dimensional design method for radial and mixed flow turbomachinery blades. *International Journal for Numerical Methods in Fluids*, **13(5)**, pp. 599-624.

[9] Zangeneh M, Goto A and Harada H 1998 On the design criteria for suppression of secondary flows in centrifugal and mixed flow impellers. ASME, *Journal of Turbomachinery*, **120(4)**, pp. 723-735.

[10] Nishiyama T 1998 *Aerofoil Flow Dynamics* [in Japanese]. Nikkan Kogyo Shimbun.

[11] Cao L L, Watanabe S, Momosaki S, Imanishi T and Furukawa A 2013 Low speed design of rear rotor in contra-rotating axial flow pump. *International Journal of Fluid Machinery and Systems*, **6(2)**, pp. 105-112.

[12] Nishiyama T 1998 *Aerofoil Flow Dynamics* [in Japanese]. Nikkan Kogyo Shimbun.

IOP Conf. Series: Materials Science and Engineering **129** (2016) 012068 doi:10.1088/1757-899X/129/1/012068

Fundamental studies on cavitation melt processing

I Tzanakis[1], M Hodnett[2], G S B Lebon[3], D G Eskin[1], K Pericleous[3]
1) Brunel Centre for Advanced Solidification Technology (BCAST), Brunel University, Uxbridge, Middlesex, UB8 3PH, UK
2) Acoustics and Ionising Radiation Division, National Physical Laboratory, Hampton Road, Teddington, Middlesex TW11 0LW, UK
3) Centre for Numerical Modelling and Process Analysis, University of Greenwich, London, SE10 9LS, UK

E-mail: itzanakis@brunel.ac.uk

Abstract. The application of ultrasound to industrial casting processes has attracted research interest during the last 50 years. However, the transfer and scale-up of this advanced and promising technology to industry has been hindered by difficulties in treating large volumes of liquid metal due to the lack of understanding of certain fundamentals. In the current study experimental results on ultrasonic processing in deionised water and in liquid aluminium (Al) are reported. Cavitation activity was determined in both liquid environments and acoustic pressures were successfully measured using an advanced high-temperature cavitometer sensor. Results showed that highest cavitation intensity in the liquid bulk is achieved at lower amplitudes of the sonotrode tip than the maximum available, suggesting nonlinearity in energy transfer to the liquid, while the location of the sonotrode is seen to substantially affect cavitation activity within the liquid. Estimation of real-time acoustic pressures distributed inside a crucible with liquid Al was performed for the first time.

1. Introduction

In recent years, there has been an increased interest in fundamental and applied investigations on metal solidification with the use of ultrasound. Ultrasonic treatment of metallic alloys is driven by cavitation and bubble dynamics and it has been proven to be effective and promising for degassing and refining grain structure. However, its industrial application remains rather limited. Commercial-scale ultrasonic melt treatment is hindered by the lack of fundamental knowledge and practical models needed to optimize the ultrasonic treatment conditions, particularly those concerning, (i) the interaction between ultrasound-induced cavitation and the processed volume as well as (ii) the characteristics of cavitation zone in molten metals, which currently limits this advanced processing to laboratory scale. The development of such knowledge is required for a major technological breakthrough.

Pioneering research work using vibrations in liquid metals can date back to 1878, when Chernov [1] was trying to improve the quality of cast metal by elastic oscillations. Since that time, a significant amount of relevant research [2–5] has been conducted showing that the use of ultrasound in liquid melt alloys can refine material's microstructure by the formation of nondendritic structures by enhanced nucleation, improved wetting, dispersing and distributing solid or immiscible phases, and reducing porosity by degassing. As a result of these effects the downstream properties of metallic alloys and their products are significantly improved [3]. Cast components with refined grain structure have many advantages including significant improvement of product quality and mechanical properties, structural integrity and reduced grain size. At the same time, the mechanisms of ultrasound-induced grain refinement are still under scrutiny [3, 6]. Further experimental work is required in order to clearly understand the fundamental mechanisms of sono-crystallization in liquid metals.

In the current study, two different approaches were used. The first one is related to the characterisation of cavitation activity within a particular water vessel; while the second one investigates cavitation activity within molten Al. Ultrasonic excitation for both cases is achieved by utilising an ultrasonic transducer and a horn (sonotrode) submerged into the melt. Measurements at various locations across the water vessel revealed noticeable spatial variations in the cavitation activity. The local cavitation phenomena in the vessel were explained based on the spectral characteristics of acoustic emission recorded by a high-temperature cavitometer sensor calibrated as an acoustic pressure sensor under free-field conditions, at specific frequencies at the National Physical Laboratory [7].

2. Methodology

In this study two different experimental setups for water and Al were deployed, as shown in figure 1. In the case when water was treated by sonication a titanium sonotrode with 40-mm tip diameter was driven by a 1-kW piezoelectric transducer which oscillates at a nominal fundamental frequency of 20 kHz producing maximum peak-to-peak amplitude of 17 μm (Hielscher/Germany) at the tip. The tip of the sonotrode was submerged 20 mm below the water surface. A 3 litre cylindrical vessel filled with 2 litres of deionised water was used. The reason for choosing this geometry is that the geometrical features of this water vessel are very close to the crucible's geometry where Al melt is treated (table 1). Water temperature was adjusted at 22±1 °C. Experiments were performed at various distances across the central axis of the sonotrode, with the sonotrode placed at two different positions i) in the centre of the vessel and ii) near to the vessel's wall in order to investigate the effect of positioning on the cavitation intensity. The relationship between cavitation activity and depth change of the sonotrode is also examined. All experiments were carried out with sonication amplitude adjusted at 50% (8.5 μm) and 100% (17 μm).

Table 1: Geometrical characteristics of the experimental vessels a) water beaker b) Al crucible

Fluid Phase	Diameter (cm)	Height (cm)	Treated Volume (cm^3)
Water	16	15	2000
Aluminum	15	21	1925

Pure Al was selected as a material for study because it has been extensively studied and widely used in metallurgical, automotive and aerospace industry as an alloy base. Additionally, Al and water liquid phases have very close kinematic viscosities while their Newtonian behaviour is similar [8] making water a frequently used candidate in physical modelling of liquid Al. The charge of 5.2 kg of Al was melted in a clay–graphite crucible coated with boron nitride (BN) from inside, with the size and geometry being close to the corresponding water vessel. The molten Al charge heated to 760 °C. The ultrasonic equipment consisted of a 5-kW generator and a 5-kW water-cooled magnetostrictive transducer (Reltec/Russia). The sonotrode consisted of a Ti concentrator and Nb tip driven by the transducer which oscillates at a nominal fundamental frequency of 17.5 kHz. The peak-to-peak amplitude was measured by a contactless vibrometer (BSUIR/Belorussia) in air and then was divided by a factor of 1.5 [3]. The sonication was performed by dipping the Nb tip (sonotrode) from the top of the melt to a depth of approximately 5 mm. The tip was preheated and the melt temperature was controlled during the process. There was no controlled atmosphere. The input power of the generator was adjusted at 3.5 kW with the corresponding amplitude at the Nb tip at 43 μm. The melt temperature was continuously monitored by a K-type thermocouple.

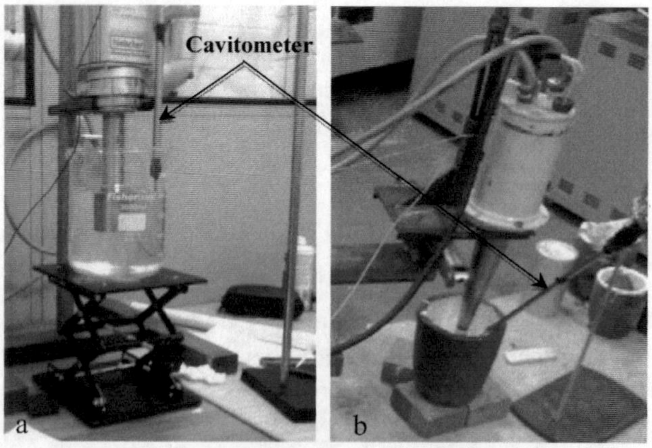

Figure 1: The principle diagram of the experimental test rigs for a) Deionised Water b) Molten Al.

The intensity of cavitation and the extent of the cavitation region were directly measured using an advanced high-temperature cavitometer which is stated by the manufacturer to measure equally well in water and in liquid Al. The cavitometer used in this study is primarily designed for immersion into molten metals (made at the Belorussian State University of Radioelectronics, Dr N.V. Dezkunov) [7]. It consists of a tungsten probe with a diameter of 4 mm and length of 500 mm connected to a piezoelectric receiver mounted within a metallic enclosure. The receiving area of the tungsten waveguide at the tip of the cavitometer probe clearly forms a surface which is potentially sensitive to cavitation signals over a wide area. The receiver is then connected to analogue electronics that process the electrical signals in response to acoustical excitation. Various settings and outputs are available for optimising the detection regime. The signal acquisition and processing was carried out on a dedicated external digital oscilloscope device "Picoscope" (Pico Technology/UK) that allowed real-time signal monitoring of the cavitometer sensor's data and ultrasonic parameters. Measurements were set to acquire 200 snapshots of the acoustic spectrum at 1 ms period of time. Each experiment was repeated to assess reproducibility of results.

3. Experimental Results and Discussion

Figure 2 shows a typical acoustic spectrum for water and liquid Al as received by the cavitometer. The acoustic source in water and in Al working at 20 and 17.5 kHz frequency, respectively, produces broadband signals well into the MHz region, and so examining acoustic signals up to MHz frequencies might be appropriate for analysing cavitation activity, however this is not within the scope of this paper. The range of signals obtained by the cavitometer for both cases are marginally different although the general shape of the spectrum plot is similar reinforcing further the opinion that water and Al can share similar behaviour under ultrasonic treatment. Specifically, the prominent fundamental frequency component is clearly shown in both graphs with further contributions from sub- and ultra-harmonic frequencies. The determinant is that harmonics in molten Al are not prominent implying that cavitation activity has not been fully developed yet (unlikely, judging from the amplitude of vibrations), or a more mild regime has been established with mainly stable cavitation bubbles present (likely) or acoustic signals received are much lower than expected due to the attenuation effects from the liquid (very likely). Additionally, another reason is that in molten Al there are not as many nuclei or pre-existed bubbles as in water, making the inception and development of powerful cavitation more difficult [3]. However, at higher frequencies, specifically in the range of 200–250 kHz prominent peaks are shown, suggesting nonlinear activity from numerous cavitation bubbles. This leads us to a preliminary conclusion that in the studied melt the activity of bubbles with resonant sizes of 10–15 μm (according to Minnaert equation [9]) prevails in the cavitation regime. This outcome is in agreement with our recent in-situ study of cavitation in Al melts, performed in the Diamond Light Source Synchrotron (National UK facility) jointly with University of Manchester, showing after the analysis of more than 30,000 bubbles at a 30-kHz driving frequency, the majority of cavitation bubbles can be also found in that particular range [10]. For both graphs in figure 2, the broadband signal is generated by the collapsing bubbles of a wide range of sizes with their shock emissions and liquid jets to contribute further. The shape of the spectrum with this hump in the area of 160-170 kHz is attributable to the variation in the sensitivity response of the cavitometer.

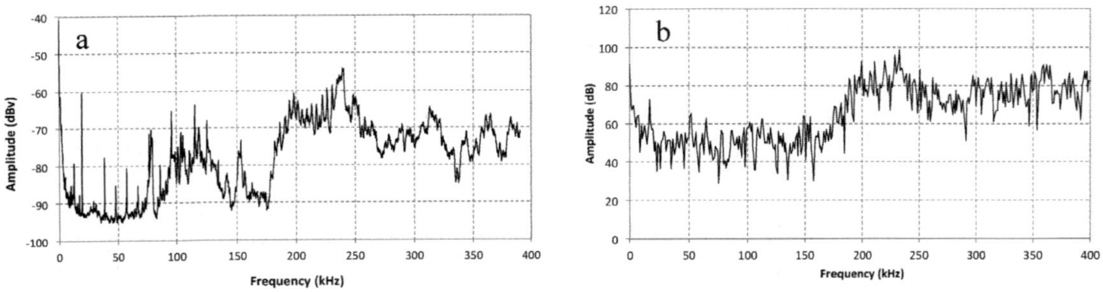

Figure 2: Typical example of acoustic spectrum generated by a) 20-kHz piezoelectric ultrasonic transducer in water and b) 17.5-kHz magnetostrictive ultrasonic transducer measured with the cavitometer tip positioned about 3–4 cm below the sonotrode tip surface

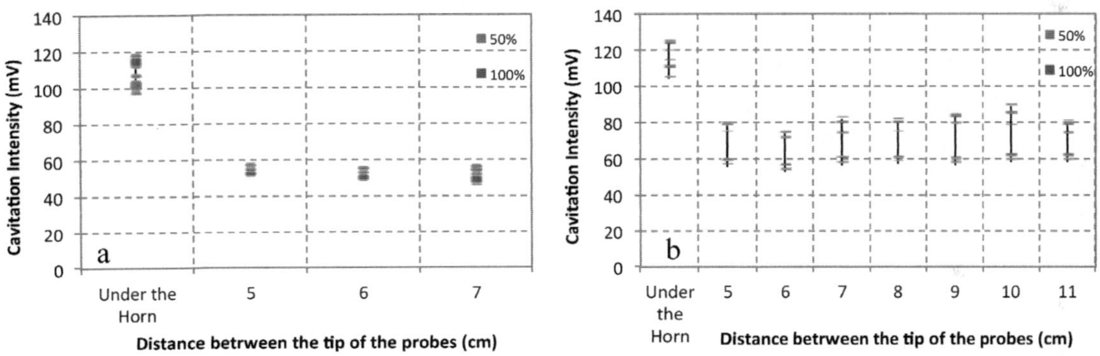

Figure 3: Cavitation activity measured in deionised water for two positions of the sonotrode a) near to the side wall b) in the centre of the cylindrical vessel, as a function of distance between the centre tip of the cavitometer and the sonotrode respectively under mechanical amplitudes of 50% and 100%.

Figure 3 shows the spatial variation data of the cavitation activity for different positions of the sonotrode within the water vessel: in the centre and near to the wall. Each point is represented by the minimum, maximum and average cavitation intensity value measured with the cavitometer at a particular distance from the tip of the sonotrode. It can be clearly seen that less power is more effective in producing higher cavitation activity. In the case where sonotrode was placed in the centre of the cylindrical vessel, 50% power was producing slightly higher cavitation activity levels than 100%. However, when sonotrode was moved near to the side wall the 50% output produces about 30% more cavitation activity as compared to 100%. These amplitude differences regarding the power output and the positioning of the sonotrode are attributable to two things; a) to the shielding and scattering effects; where the higher output level, i.e. 100%, produces an intense region of cavitation very close to the sonotrode tip surface, which prevents the further propagation of ultrasound into the liquid, thus attenuating the signal received by the cavitometer sensor and b) to the fact that activity can be higher near to the wall as bubbles can collapse easier due to the solid interface and also cavitation nuclei or many microscopic air bubbles are more available due to the surface roughness of the glass wall. Results are in a good agreement with the work of Hodnett et al. [11]. Additionally, when cavitometer was placed below the sonotrode, intensity was measured to be similar to that at 100%.

Collected data in liquid Al was numerically filtered into two groups, at 1 kHz higher and 1 kHz lower the fundamental frequency of 17.5 kHz, ensuring that all the prominent peak amplitudes will be captured. Then using the calibration data for the cavitometer obtained in the National Physical Laboratory in water at room temperature, under free-field conditions at the frequencies of interest (assuming that sensitivity does not change at temperatures up to 800 °C, as indicated by the manufacturer) these amplitudes were converted into acoustic pressures. The environment in which the sensor is calibrated is clearly significantly different from that in which it used (high temperatures, non free-field) and so computing the pressures according to the measured voltages received at 17.5 kHz is done with caution. The computed peak negative acoustic pressures were revealed across different areas of the crucible giving physical meaning to the likelihood of cavitation activity in this extreme environment (figure 4a). Acoustic pressures gradually increase as cavitometer approaches the sonotrode. This is in a good agreement with the experiments in water as intensity was also observed to be higher under the sonotrode (figure 3). Interestingly, it can be seen that the existing cavitation regime generates pressures which are relatively small compared to previous investigations that showed that the cavitation threshold is close to 0.5 MPa [6, 12]. A possible reason is that for multi bubble structures the signal level can be significantly reduced due to scattering and shielding effects occurring in the bubbly cloud along with the fact that molten Al is very dense with high surface tension.

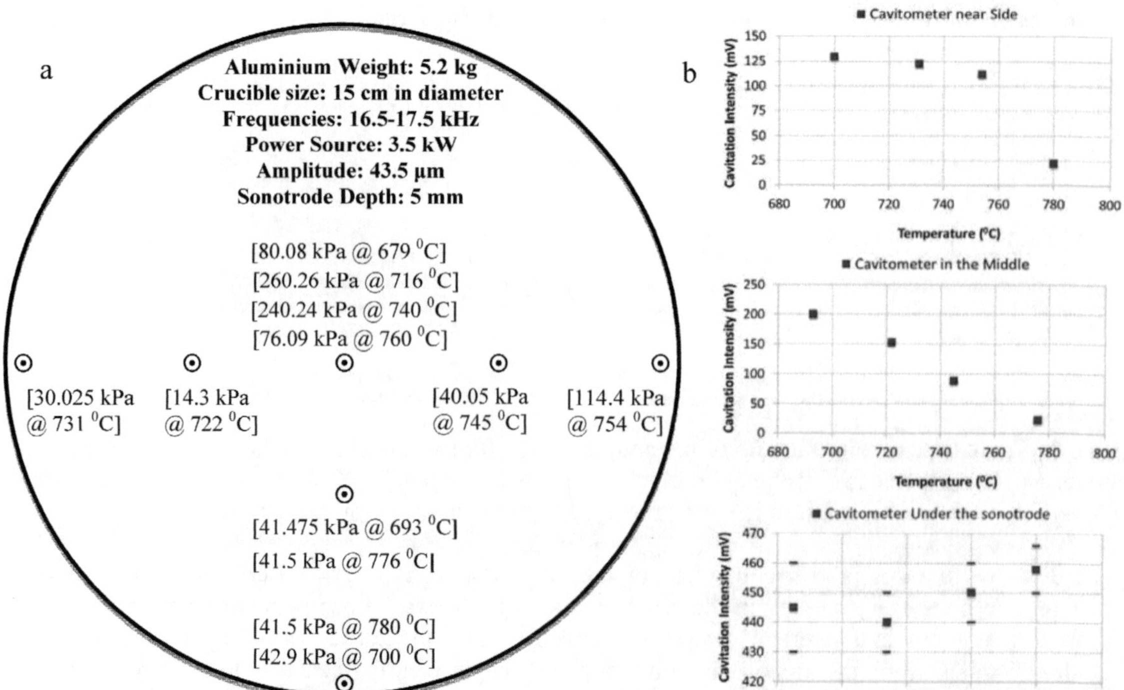

Figure 4: a) Top view of the crucible showing locations of the acoustic measurements and the peak negative acoustic pressures at these locations b) Cavitation activity versus temperature measured at various positions of the cavitometer within the melt as depicted in figure 4a.

Also another interesting feature is that cavitation activity significantly increases with the drop of temperature as shown in figure 4b. This clearly appears in the areas away from the main cavitation zone. Below the sonotrode the intense region of cavitation maintains high levels of cavitation activity regardless of temperature variation.

4. Conclusions

Investigation of cavitation activity in liquid metals is a very complicated task due to high temperatures, opaqueness and chemical activity of the melts however with the use of appropriate equipment, as shown in this study, these obstacles can be overcome and useful data can be drawn. The main conclusions are:

1) Molten Al has presented acoustic spectrum graph comparable to water. Areas with higher cavitation activity were revealed showing that the majority of the cavitation bubbles in Al melts can be found in the range of 10-15 μm.
2) Lower power outputs i.e. 50%, seems to be the best setting for producing the higher cavitation activity. Advantages, in most of the cases, are related with the generation of the maximum cavitation intensity utilizing low power/energy. This in turns can potentially have an impact on environmental savings and on economic benefits.
3) To the best of our knowledge this is the first time when characterisation of real-time acoustic pressures inside liquid Al volume was performed. Additionally, cavitation intensity significantly increases with the drop of temperature except in the area below the sonotrode where intense cavitation is maintained regardless temperature variations.

Acknowledgments: This work is performed within the Ultramelt Project supported by the EPSRC Grants EP/K005804/1 and EP/K00588X/1.

References:

1. D. K. Chernov "Nauka o Metallakh" Moscow: Metallurgizdat, 1950.
2. G.I. Eskin. "Ultrasonic Processing of Molten Aluminium" Moscow: Metallurgiya, 1965.
3. G. I. Eskin "Ultrasonic Treatment of Light Alloys Melts", Amsterdam, Gordon and Breach Science Publishers, 1998.
4. J. Campbell "Effects of vibration during solidification" Intern. Met. Rev. 26 (1981) 71-108.
5. O. V. Abramov "Ultrasound in Liquid and Solid Metals" Boca Raton: CRC Press, 1994.
6. T.V. Atamanenko, D. G. Eskin, L. Zhang, L. Katgerman "Criteria of Grain Refinement Induced by Ultrasonic Melt Treatment of Aluminum Alloys Containing Zr and Ti" Metall. Mater. Trans. A, 41A (2010) 2056–2066.
7. I. Tzanakis, M. Hodnett, D.G. Eskin, B. Lebon "Advanced calibration of a high-temperature cavitometer" the 14th meeting of the European Society of Sonochemistry (ESS 2014), Avignon, France, 2-6 June 2014.
8. D. Xu, W.K. Jones, Jr., J.W. Evans "The Use of PIV in the Physical Modeling of Flow in EM or DC Casting of Aluminum: Part I. Development of the Physical Model" Metall. Mater. Trans. B, 29B (1998) 1281-1288.
9. M. Minnaert "On musical air bubbles and the sound of running water."Phil. Mag., 16 (1933) 235–248.
10. W. Wu, I. Tzanakis, P. Srirangam, S. Terzi, W. Mirihanage, D. G. Eskin, P. Lee "In situ Synchrotron Radiography of Ultrasound Cavitation in a molten Al–10Cu alloy", TMS Annual Meeting, Orlando, USA, 15-19 March 2015 (submitted).
11. M. Hodnett, R. Chow, B. Zeqiri "High-frequency acoustic emissions generated by a 20 kHz sonochemical horn processor detected using a novel broadband acoustic sensor: a preliminary study" Ultrason. Sonochem. 11 (2004) 441–454.
12. T.J. Mason. Advances in Sonochemistry, Volume 4, JAI Press, 1996.